Computational Methods in Engineering

To
IIT Madras
for providing unlimited opportunities

Computational Methods in Engineering

S.P.Venkateshan
Professor Emeritus
Mechanical Engineering
Indian Institute of Technology Madras

Prasanna Swaminathan
Postdoctoral researcher
EM2C Laboratory
Ecole Centrale Paris, France

June 2013

Academic Press is an imprint of Elsevier
The Boulevard, Langford Lane, Kidlington, Oxford OX5 1GB, UK
225 Wyman Street, Waltham, MA 02451, USA

First edition 2014

British Library Cataloguing in Publication Data
A catalogue record for this book is available from the British Library

Library of Congress Cataloging-in-Publication Data
A catalog record for this book is availabe from the Library of Congress

ISBN–13: 978-0-12-416702-5

For information on all Academic Press publications
visit our web site at store.elsevier.com

Working together to grow
libraries in developing countries

www.elsevier.com | www.bookaid.org | www.sabre.org

ELSEVIER   BOOK AID
            International   Sabre Foundation

Preface

Researchers and engineers use commercial computing software in their day to day professional activities. User friendly commercial software is a *black box* for many of them with input being given by clicking appropriate boxes and expecting great results to come out. Generally it is helpful if the user of such software has a background in numerical methods so that he/she will avoid some pitfalls.

The emphasis of the book is the application of numerical methods for the solution of equation(s) - linear, non-linear (algebraic), differential equations (ordinary and partial) - that occur in most engineering disciplines. There is a common theme running through all these topics and hence the book has been arranged in a particular way, the chapters being grouped into four Modules. The authors recommend that the reader go through Modules I and II serially and then take up Modules III and IV in any order. The reader is encouraged to do the worked examples on his/her own using any accessible computer resource so as to get a feel for the numerical methods discussed in the text. The reader may also go beyond the book and perform grid sensitivity, parametric study, stability analysis using the worked examples given in the book. The reader may also try different algorithms and study their relative merits and demerits. Spreadsheet programs such as **EXCEL** (part of Microsoft Office suite) operating under Windows or free software **LibreOfficeCalc** working on Linux platform, may be used for simple problems. More complex problems may be approached better by **MATLAB** about which an introduction has been provided in the book. Several MATLAB programs have also been given in the book. The reader may use these and also develop his/her own MATLAB codes.

This book is definitely not for the mathematician but targets the users of mathematics, specifically from engineering disciplines. The emphasis is not on mathematical rigor (even though it is very important) but utility in the context of engineering (applied mathematics). Important principles of computational methods are brought out via worked examples with obvious connection to real life engineering applications.

The junior author (PS) has assisted the senior author, as a teaching assistant for a few years while pursuing his doctoral research. He brings a student's point of view to the book. His contribution has been substantial. He has written and debugged all the MATLAB codes presented in the book. He has also contributed material relating to interpolation in more than one dimension, finite element and finite volume methods, conjugate gradient methods. He has also contributed by working out many example problems using MATLAB that are included in the book. Format of the book and creation of class file called "bookSpvPras.cls" is his other important contribution.

It may be difficult to cover the entire book in a one semester course. Some advanced topics - such as interpolation in two and three dimensions, multiple integral, FEM,FVM, collocation method, solution of stiff equations, parts of PDE - may be dropped in a first level course.

Exercise problems of varying levels of difficulty are given at the end of each module. The reader will benefit most from the book if he/she puts in effort to solve them.

Chennai   S.P.VENKATESHAN
3 June 2013   PRASANNA SWAMINATHAN

Acknowledgements

The writing of a book requires the help and support of many people. Ambiance provided by Indian Institute of Technology Madras (IIT Madras) has made the writing of the book a pleasant activity. Feedback has come in plenty from students who have taken my (SPV) course in Computational Methods in Engineering through the years. The authors acknowledge the comments from research scholars of the Heat Transfer and Thermal Power Laboratory of IIT Madras, on earlier versions of some chapters of the book, which helped in improving the text significantly. Comments from reviewers of one chapter of the book has led to many desirable changes in the book. The senior author owes a debt of gratitude to excellent teachers (specifically Late Prof. Y.G. Krishnaswamy), who created an abiding interest in mathematics in him, when he was an undergraduate student in Bangalore University in the 1960's.

The authors particularly thank Ane Books, New Delhi for bringing out this text in an expeditious manner. This book is co-published by Elsevier, Inc., as an Academic Press title, based on additional market feedback provided by the Publisher.

Chennai
3 June 2013

S.P.VENKATESHAN
PRASANNA SWAMINATHAN

Contents

Module II   Interpolation, differentiation and integration

Module III Ordinary differential equations

Module IV Partial differential equations

Chapter 1

Preliminaries

*This chapter is a stand alone that deals with numerical preliminaries. Specifically this chapter talks about precision in numerical calculations made using a digital computer and its effect on numerical calculations. This is followed by a brief introduction to MATLAB programming. The rest of the book is arranged in **four** modules as follows:*

- *Module I: System of equations and eigenvalues*
- *Module II: Interpolation, differentiation and integration*
- *Module III: Ordinary differential equations*
- *Module IV: Partial differential equations*

1.1   Introduction

We discuss in brief some general issues which will bear on the material presented in the rest of the book. Computational methods require the use of a computing machine such as a calculator, if the calculations are not too many, or a digital computer, if calculations are many and possibly repetitive and hence dull. While studying the behavior of physical systems we may want to study the effect of many parameters that describe or model the physical system. This may require simulation involving the same set of calculations to be repeated, but with different parameter values, each time. The calculations should have adequate precision to show the smallest effect the parameters may have on the result(s).

We also introduce MATLAB so that the reader will appreciate many short programs written in MATLAB that are found throughout the book. We do not recommend a particular program or programming language to be used by the reader. This is left to the choice of the reader. Interested reader may use a spreadsheet program for many simulations and avoid writing a computer program. The authors have, in fact, used spreadsheet to work out many of the worked examples presented in the book. When necessary MATLAB programs were written. Specially useful are MATLAB programs when large number of iterations are involved. MATLAB also provides easy access to graphics without leaving the MATLAB environment.

1.1.1   Floating point numbers in binary form

Calculators and computers work with binary representation of numbers. In binary system, the numbers are represented by digits 0 and 1. Hence, binary system is also referred to as base-2 number system. The commonly used decimal representation is base-10 system. A number p in a system using base b is represented by p_b and is given by

$$p_b = \sum_i^N d_i b^i = d_0 b^0 + d_1 b^1 + \cdots d_i b^i \cdots + d_N b^N \tag{1.1}$$

where d_i is the ith digit and $N = 1$ is the number of digits. In case the number involves a fraction i can take negative values. One can easily convert from decimal to binary and binary to decimal representation. For example

$$10101_2 \quad = \quad 10101_2 = 1 \times 2^4 + 0 \times 2^3 + 1 \times 2^2 + 0 \times 2^1 + 1 \times 2^0 = 21_{10} = 21$$
$$11.01_2 \quad = \quad 11.01 = 1 \times 2^1 + 1 \times 2^0 + 0 \times 2^2 - 1 + \times 2^{-2} = 3.25_{10} = 3.25$$

The decimal number is familiar to us and hence the subscript 10 is normally not indicated. Similarly binary form is common and hence the subscript 2 would not normally be indicated. We commonly use the scientific notation to represent a number i.e $21 = 2.1 \times 10^1$ (1 is the exponent) and $3.25 = 3.25 \times 10^0$ (0 is the exponent). The computer also uses a similar representation in binary notation.

Sometimes we use a number system based on base 8 (octal) and base 16 (hexadecimal) also. In the case of the former the numerals are 1 to 7 while in the case of the latter we have 1 to 9 followed by A to F representing decimal numbers 10 through 15. For example $3407_8 = 3 \times 8^3 + 4 \times 8^2 + 0 \times 8^1 + 7 \times 8^0 = 1799$ and $3F_{16} = 3 \times 16^1 + 15 \times 16^0 = 63$.

Precision with which a number is represented in binary is determined by the number of bits used. IEEE654[1] defines the format for numbers as well as the precision.

A floating point number in single precision is represented by 32 bits as indicated below:

1	8	23	
s	e	f	
(*sign*)	(*biased exponent*)	(*fraction*)	(1.2)
	msb *lsb*	*msb* *lsb*	

msb – most significant bit, lsb – least significant bit

The above format has three fields - a 23-bit fraction, f; an 8-bit biased exponent, e; and a 1-bit sign, s. These fields are stored in one 32-bit word, as shown in 1.2. Bits 0:22 contain the 23-bit fraction f, with bit 0 being the lsb of the fraction and bit 22 being the msb; bits 23:30 contain the 8-bit biased exponent, e, the bias being 127, with bit 23 being the lsb of the biased exponent and bit 30 being the msb; and the highest-order bit 31 contains the sign bit, s. The number is then given by $(-1)^s 2^{e-127}(1.f)$ where f is the fraction represented by the bits from 0 to 22 (total number of bits is 23).

Consider, as an example, a floating point number in single precision given by

1	10000001	010 1000 0000 0000 0000 0001

Here we have $s = 1$ and hence $(-1)^s = (-1)^1 = -$. Hence the number is a negative number. The exponent is given by $e = 2^7 + 2^0 = 129$. Hence the biased exponent is $e - 127 = 129 - 127 = 2$. f is given by $2^{-2} + 2^{-4} + 2^{-23} = 0.3125001$. Hence $1.f = 1 + 0.3125001 = 1.3125001$. Finally we have the number as $-2^2 \times 1.312500119 = -5.2500005$.

A floating point number in double precision is represented by 64 bits as indicated below:

1	11	52	
s	e	f	
(*sign*)	(*biased exponent*)	(*fraction*)	(1.3)
	msb *lsb*	*msb* *lsb*	

msb – most significant bit, lsb – least significant bit

The above format has three fields - a 52-bit fraction, f; an 11-bit biased exponent, e; and a 1-bit sign, s. These fields are stored in one 64-bit word, as shown in 1.3. Bits 0:51 contain the 52-bit fraction f, with bit 0 being the lsb of the fraction and bit 51 being the msb; bits

---

[1]IEEE standard for Binary Floating-Point Arithmetic, ANSI/IEEE Std 754-1985, Published by the Institute of Electrical and Electronics Engineers, N.Y., U.S.A.

52:62 contain the 11-bit biased exponent, e, the bias being 1023, with bit 52 being the lsb of the biased exponent and bit 62 being the msb; and the highest-order bit 63 contains the sign bit, s. The number is then given by $(-1)^s 2^{e-1023}(1.f)$. The portion $(1.f)$ represents the fraction in the form of 1 followed by the decimal point and the fraction to the right of the decimal point.

Apart from the above extended formats are also included in IEEE654. Table below summarizes the precision possible with these formats.

IEEE format	Machine precision	Number decimal digit	Maximum $e-bias$	Minimum $e-bias$
Single	$2^{-23} \sim 1.2 \times 10^{-7}$	7	128	-127
Double	$2^{-52} \sim 2 \times 10^{-16}$	16	1024	-1023
Double Extended	$2^{-64} \sim 1.1 \times 10^{-19}$	19	16383	-16382

Underflow or overflow is decided by the largest and the smallest values of the exponent given in the table. Precision of the number is limited by the values shown in the table. In the case of single the precision is limited by $2^{-23} \sim 1.2 \times 10^{-7}$. One may possibly assume that six digit accuracy is realizable with single precision arithmetic. Similarly double precision guarantees precision up to 16 digits after decimals if machine is allowed to round the numbers as calculations proceed.

It is thus clear that digital computers represent a number (a variable) by a single or double representation in digital form. The continuous variable is thus represented by a digital variable that is **discrete**, in the sense that two neighboring values of the function have at least a difference of one lsb, with a difference of one precision unit. Since precision is finite the digital representation of data is inherently an approximate or discrete representation of the variable.

1.1.2   Rounding errors and loss of precision

To demonstrate the effect of rounding we take a simple example. Assume that it is desired to calculate $y = e^{\sqrt{2}}$. Calculator gives $\sqrt{2} = 1.414213562$ with nine digits after decimals. Presumably the number is uncertain in the 10th place after decimals. We then calculate the exponential of this number using the calculator to get $y = e^{1.414213562} = 4.113250379$. In case the number were to be rounded while determining the square root to six digits, we would have $\sqrt{2} = 1.414214$. Exponential of this would then be given by the calculator as 4.113252179 which differs from the more accurate calculation by 0.000002 when rounded to six digits after decimals. We may imagine a chain of such calculations that involve rounding after each calculation to see that the error will surely propagate towards more significant digits i.e. from right to left.

A spreadsheet program that uses a better precision (double precision arithmetic) yields the result $y = e^{\sqrt{2}} = 4.1132503787829300$ which on rounding to nine digits would yield 4.113250379 in agreement with that obtained using the calculator. The error between the calculator value and the spreadsheet value is $-2.1707258213155 \times 10^{-10}$ or roughly digit 2 in the 10th place after decimals.

Example 1.1

It is desired to calculate $f(x) = \dfrac{\cosh(1-x)}{\cosh(1)}$ *at* $x = 0.5$. *Study the effect of rounding. This function is required in calculating a second function* z *given by* $z = \dfrac{1}{1+y^{2.5}}$. *How does the rounding of* y *affect the value of* z?

Solution :

We make use of a spreadsheet program to perform the calculations. A built in function will round the calculated number to user specified precision. The function is ROUND(*number*, *integer*). For example ROUND(*number*, 3) will round the number to 3 digits after decimals. Rounding follows the normal practice of rounding up if the digit is ≥ 5 in the specified (fourth) place.

$x =$	0.5	Round to 3 digits	Error due to rounding	Round to 7 digits	Error due to rounding
$\cosh(1) =$	1.543080635	1.543	-8.1×10^{-5}	1.5430806	-3.5×10^{-8}
$\cosh(0.5) =$	1.127625965	1.128	3.7×10^{-4}	1.1276260	3.5×10^{-8}
$f(x = 0.5) =$	0.730762826	0.731	2.4×10^{-4}	0.7307628	-2.6×10^{-8}

Now we shall see how the calculation of z is affected by the rounding in the calculation of y. The calculations are again shown as a table below.

		Round to 3 digits	Error due to rounding	Round to 7 digits	Error due to rounding
$f(x) =$	0.730762826	0.686	-5.8E-004	0.6865773	5.4E-008
$z =$	0.686577246	0.686	-5.8E-004	0.6865773	5.4E-008

At once we see that the errors have increased while calculating z. Hence error due to rounding *accumulates* and increases as the number of arithmetic manipulations increase.

1.1.3   Effect of rounding on numerical computation

We have seen that rounding errors propagate and may make results of numerical calculations unreliable. Rounding errors are like perturbations or errors that affect all calculations that follow. In case of differential equations rounding errors may grow and throw the calculation method out of gear by making the solution diverge. The numerical scheme may become unstable. Such a behavior imposes restrictions as we shall see later.

Sometimes we use an iterative method to calculate the roots of a set of equations. Iterations are expected to stop when some stopping criterion is satisfied. For example, if iterations stop when change in a quantity is less than 10^{-6}, it is necessary that the numerical scheme and the precision in the calculation makes it possible to achieve this condition. If rounding of numbers takes place at every iteration we may never achieve this condition and hence the solution may never converge!

We conclude that we will be able to use adequate precision in the calculations such that the above type of situation does not arise.

1.1.4   Taylor series and truncation

 Very often numerical methods make use of Taylor[2] series representation of functions. A function is then numerically evaluated by summing terms in the Taylor series - an infinite series - truncating it to a manageable number of terms. Sometimes only a small number of terms such as 2 or 3 may be made use of to represent the function. This is referred to as truncation. The accompanying error, the difference between the Taylor series calculated value and the exact value (if known) is referred to as truncation error. In practice the truncation error may be estimated by calculating the leading order term that was dropped. An example will clarify these issues.

Example 1.2

Taylor expansion around $x = 0$ of the function $y = e^{-x}$ is written down as $y(x) = 1 - x + \dfrac{x^2}{2} - \dfrac{x^3}{6} +$

$- \cdots + (-1)^i \dfrac{x^i}{i!} \cdots$. Evaluate the function at $x = 0.1$ and study the consequences of truncation if arithmetic is performed with 9 digit precision.

Solution :

To 9 significant digits (for example, a typical calculator[3] will display this precision) the function may be calculated as $e^{-0.1} = 0.904837418$. This value will be used for comparison purposes in discussing truncation and rounding errors. First we use nine digit precision to obtain the following results:

No. of terms	$y = -0.1$	Truncation Error
2	0.900000000	-0.004837418
3	0.905000000	0.000162582
4	0.904833333	-0.000004085
5	0.904837500	0.000000082
6	0.904837417	-0.000000001
7	0.904837418	0.000000000
Exact	0.904837418	0.000000000

 It is apparent that the truncation error i.e. $y_{Truncated\ Taylor} - y_{Exact}$ differs by less than 1 digit in the ninth place after decimals for number of terms in Taylor expansion greater than 6. Hence if 9 digit precision is what is available there is no point in including more than 6 terms in the Taylor series. Hence we may *truncate* the Taylor series to the first 6 terms. It is also seen that the truncation error is as much as 5 digits in the fourth place if Taylor series is truncated to just 2 terms.

 We shall now consider the effect of rounding. Let us assume that we perform arithmetic that has five digit precision. The state of affairs will change as given below.

---

[2]Brook Taylor, 1685-1731, English mathematician

[3]Texas Instruments TI-30X calculator has option to display functions with number of digits ranging from 0 to 9 after decimals

No. of terms	$y = -0.1$	Truncation Error
2	0.90000	-0.00484
3	0.90500	0.00016
4	0.90483	-0.00001
5	0.90483	-0.00001
6	0.90483	-0.00001
7	0.90483	-0.00001
Exact	0.904837418	0.00000

It is clear this time that the rounding has made its effect felt! The 5th term in the Taylor series has no effect on the result. This simply means that truncation errors are resolved only when rounding errors are not severe.

1.1.5   Effect of digital calculations on iteration

We have mentioned earlier that many numerical schemes involve iteration i.e. repeated use of a formula till the result converges according to a stopping criterion. The stopping criterion usually involves observation of change in a quantity from iteration to iteration and stopping when the difference is smaller than a prescribed value. It is clear that rounding will have an effect on such calculations. We take a simple example to demonstrate this.

Example 1.3

Perform iteration given by the formula $x_{i+1} = \dfrac{x_i}{2} + \dfrac{1}{x_i}$ *starting with* $x_0 = 1$. *Discuss the effect of rounding and specified stopping criterion on the result.*

Solution :

We first use arithmetic with 9 digit (decimal) precision. The starting value is specified as $x_0 = 1.000000000$ to obtain $x_1 = 1.500000000$ after one iteration. The difference these two values is $x_1 - x_0 = 0.500000000$. We continue the iterations as shown in the table below.

x_i	Difference
1.000000000	
1.500000000	0.500000000
1.416666667	-0.083333333
1.414215686	-0.002450981
1.414213562	-0.000002124
1.414213562	0.000000000

In this case iterations perforce stop after $i = 5$ since the difference between two consecutive iterations has reached machine precision.

If we round the calculations to 5 digits during the calculations we get the following tabulation.

x_i	Change	Difference with respect to 9 digit final value
1.00000		-4.1E-001
1.50000	0.50000	8.6E-2
1.41667	-0.08333	2.5E-3
1.41422	-0.00245	6.4E-6
1.41421	-0.00001	-3.6E-6
1.41421	0.00000	-3.6E-6

We are forced to stop iterations with $i = 5$ since rounding has now limited the precision. If we were to calculate using double precision we would get a converged value for 1.4142135623731000.[4]

1.2   Mathematical and computational modeling

Mathematical models are used extensively in science and engineering. A model may be composed of simple or complex operations which approximates an application. A mathematical model could be a set of linear equations or algebraic equations or differential equations. The model is constructed based on practical observations. For example, Joseph Fourier surmised that the heat flux q is proportional to temperature gradient and hence proposed that (in one space dimension)

$$q = -k\frac{dT}{dx}$$

where k is known as the thermal conductivity of the medium and T is the temperature that varies with one independent space variable x. The above equation is known as the Fourier law of conduction. Other examples are Newton's laws of motion, seepage flow across porous media (Darcy law), Faraday's law of electromagnetic induction, Maxwell equations in electromagnetism etc. Mathematical models may also be constructed by correlating cause and effect, based on experimental observations. For example, the population growth rate of a country or the ups and downs of financial markets.

Mathematical models may be of any of the types given below.

1. **Linear or nonlinear**: A model is said to be linear if cause and effect are linearly related. Otherwise the model is nonlinear.
2. **Static or dynamic**: A model in which the dependent variable is a function of time is known as dynamic. Otherwise the system is static.

---

[4]This is an approximation of $\sqrt{2}$. Value of $\sqrt{2}$ to one million digits is available at the NASA web site http://apod.nasa.gov/htmltest/gifcity/sqrt2.1mil

3. **Lumped or distributed**: A model which describes the average quantity in a spatial domain is said to be lumped. If the model describes the quantity at every location in the spatial domain, we have a distributed model. The pressure drop characteristics of a pump is a lumped representation. On the other hand, modeling the fluid flow through the pump represents a distributed system. A lumped system is often used in the simulation of complex systems.

Invariably one has to use computational tools to solve the mathematical models. Computational modeling approximates the mathematical model suitably. As the computational methods are discrete in nature, the *continuous* mathematical models are approximated by *discrete* computational representations.

Basically we have divided the book into four modules:

- Module I: System of equations and eigenvalues
- Module II: Interpolation, differentiation and integration
- Module III: Ordinary differential equations
- Module IV: Partial differential equations

1.3 A brief introduction to MATLAB

A program is a set of instructions to be executed by the computer written in a language that the machine can understand. Popular high-level programming languages are C, C++ and Fortran. These programming languages form an interface between the user and the machine.

MATLAB is a popular software package for performing numerical computations. It provides interactive user interface with several built in functions as well as programming capabilities. The libraries in MATLAB cover a wide range of areas including linear algebra, computation of eigenvalues, interpolation, data fit, signal analysis, optimization, solving ODEs and PDEs, numerical quadrature and many more computational algorithms. MATLAB has incorporated several popular libraries such as LAPACK. It also provides graphical and plotting tools.

We have adopted MATLAB to demonstrate the computational methods discussed in this textbook. This is because of its relatively simple and easy programming capabilities. However, we expect the reader to have some basic experience in high-level programming such as C, Fortran or MATLAB. We will briefly introduce MATLAB to the readers.

1.3.1 Programming in MATLAB

When MATLAB is launched, we get a command window. The command window forms the main interface for the user. Command window can be used to execute simple commands or call functions. For example

```
>> u= sqrt(2)            Command

u =                      Output

      1.4142
```

u is the name of the variable and sqrt is an inbuilt MATLAB function to calculate square root. The names of the variables should always start with an alphabet. In MATLAB, by default, variable names are case sensitive (**a** and **A** are different). Unlike C or Fortran, we need not declare the data type of variable (integer, real, double , character, complex). MATLAB automatically assigns the appropriate type of variable. MATLAB also offers inbuilt commands to convert from one data type to another.

MATLAB offers a range of inbuilt MATLAB functions. Some of the common functions have been listed below

abs   sin   cos   tan   exp   log   acos   asin   atan   cosh   sinh

Scripts

When a large number of commands have to be used, which may require frequent editing, one can write the set of commands as a file called *M-files* (extension **.m**). MATLAB provides an editor of its own which can be used to write these programs. Such programs are also known as scripts. An example of a script file named "quadraticroots.m" has been provided below to determine the roots of a quadratic equation.

```
a   = 1;    % constants of quadratic equation
b   = 2;    % a command ending with semi-colon ; would not
c   = 1;    % print the output in the command window
root1 = (-b+sqrt(b^2-4*a*c))/(2*a);   % root 1
root2 = (-b-sqrt(b^2-4*a*c))/(2*a);   % root 2
```

The above program can be executed from the command prompt as follows

```
>> quadraticroots
```

or by clicking on the execute button provided in the MATLAB editor window. The character % is used for commenting. Anything written after % will not be executed by MATLAB. To execute the program successfully, one has to ensure the desired file is in the current directory. The current directory window provided by MATLAB lists all the files in the current directory. All the variables that have been generated since opening MATLAB are listed in the workspace window. In addition to this MATLAB has a command history window which lists all the commands that have been executed using the Command window. There are several more features offered by the MATLAB interface and the reader is encouraged to explore them.

Functions

In some applications, we require a piece of code has to be executed more than once. Then the same piece of code can be written as a "function" and saved as a *M-file* (the name of the function is same as the M-file).

Program : A sample program

```
1 function [root1,root2] = quadraticroots(a,b,c)
2 % Input   : a,b,c
3 % Output  : root1,root2
4 root1 = (-b+sqrt(b^2-4*a*c))/(2*a);   % root 1
5 root2 = (-b-sqrt(b^2-4*a*c))/(2*a);   % root 2
```

The user defined function, just like inbuilt MATLAB functions, can be called either from command window or another function. Calling quadraticroots from command window

```
[root1,root2] = quadraticroots(1,2,1);
```

Again one has to ensure that the function file is present in the current directory. To make a function available universally to all directories, the user can set the path of the folder containing the function.

MATLAB offers two other ways of creating customized user defined functions. These functions can be written in command prompt or can be part of a M-file. For example

```
% name of function is ''a'' with input variable x
>> a = @(x) x.^3;
>> a(4)
ans =
    64
% the dot preceding ^ is an array operation
>> a([4 5])      % operating on an array
ans =
    64    125
>> b = @(x,y) x.^3 + y.^2  % multiple input variables
>> b(4,5)
ans =
    89
```

inline operator in MATLAB also serves the same purpose

```
>> a = inline('x.^3','x');
>> b = inline('x.^3+y.^2','x','y')
```

1.3.2 Array and matrices

Array and matrices form the basis of computational methods and MATLAB offers an extensive range of tools to carry out array operations. An array is a one dimensional set of data where as matrix is a multidimensional data set. MATLAB treats every variable as an array or matrix. Then a scalar is an array of dimension 1×1. Arrays can be declared in the following ways

```
>> a = [1 1 1]                % creates a row vector of size 1×3
a = 1   1   1
>> a = [1; 1; 1]              % creates a column vector of size 3×1
a =
     1
     1
     1
```

While dealing with arrays, one has to make sure that the dimensions of array are properly defined. This becomes important for performing array operations such as addition, multiplication or division. A column vector can be converted into a row vector and vice versa using transpose

```
>> a = [1 1 1]; % creates a row vector of size 1×3
>> b = a'       % transpose of vector a
b =
     1
     1
     1
```

The following command

```
>> a = 0:0.2:1
```

produces an array with 6 equi-spaced elements between 0 and 1

```
a = 0   0.2   0.4   0.6   0.8   1
```

Similar to arrays, matrices can be declared as follows

```
>> a = [1 2 3 4; 5 6 7 8];    %  a matrix of size 2×4
```

One can access or declare individual or group of elements of a matrix in the following way

```
a(1,2) = 2  % element at first row and second column
%   first index refers to row and second to column
a(2,2:4) = 6   7   8    % elements in second row and
%   columns between 2 and 4
a(2,:) = 5 6 7 8  % elements in second row
a(:,2) = 2
          6    %  elements in second column
```

Matrices can also be generated using built in functions such as

```
a = zeros(4);   % creates a matrix of size 4×4 filled with zeros
a = ones(3,2);  % creates a matrix of size 3×2 filled with ones
a = eye(4);     % creates an identity matrix of size 4×4
```

MATLAB offers the following scalar arithmetic operations

operator	operation	function
+	addition	adds two matrices/arrays of same size
−	subtraction	subtracts two matrices/arrays of same size
*	multiplication	$A * B$ provided number of columns of A is equal to number of rows of B
/	right division	A/B is equivalent to AB^{-1}
\	left division	$A \setminus B$ is equivalent to $A^{-1}B$
^	power	$A\text{^}2$ is equivalent to AA

The above operations are valid even when A or B is a scalar as indicated below

```
>> a = [1 2 3 4; 5 6 7 8]    %  a matrix of size 2×4
>> b = 2*a + 1
b =               % output
     3      5      7      9
    11     13     15     17
```

The operator left division \ can be used to solve a system of linear equation of the form
Ax = b

```
x = A\b;
```

More details on this operation is given in Chapter 2.

Array operations

Sometimes, it is required to perform element by element operations between two matrices. A good example is a dot product between two vectors. Such operations can be performed by just adding a *dot* before the corresponding arithmetic operator. For example, a dot product between two vectors can be achieved by

```
>> a = [1 1 1];     b = [0.2 1 0.5];
>> c = a.*b
c = 0.2   1   0.5
```

Similarly

```
>> d = a./b
d = 5   1   2
>> e = b.^2
e = 0.04   1   0.25
```

For the operations to be valid, the dimensions of the two arrays/ matrices should be equal. The above operations are still valid if any one of the arrays is a scalar.

1.3.3 Loops and conditional operations

Conditional statements and loop statements are similar to other programming languages. There are two ways of defining a loop

```
for i = starting point:increment:last point
    % some executable statements
end
```

and

```
while (condition for loop)
    % some executable statements
end
```

break command can be used within a loop to exit the loop. The conditional statements can be defined as

```
if  condition
    % some executable statements
elseif  condition
    % some executable statements
else
    % some executable statements
end
```

Relational and logical operators are useful for such operations. They are as follows

Relational operators	
<	less than
<=	less than or equal to
>	greater than
>=	greater than or equal to
==	equal to
~=	not equal to
Logical operators	
&	logical **and**
|	logical **or**
~	logical **not**

The relation and logical operators can also be used outside the conditional and loop statements. Please refer to MATLAB reference for more details.

1.3.4 Graphics

MATLAB can be used for plotting and graphical purpose. plot command can be used to produce a simple two dimensional plot. An example follows

```
x = 0:0.1:1; y = sin(x);z = cos(x);
plot(x,y);        % plots x,y
hold on           % command to overlap plots
plot(x,z,'r');    % plots x,z as a red line
```

The plots can also be customized by defining labels, line width, line color, fonts etc. The customization can be done either using the command prompt or the figure window itself. One can also make two dimensional plots such as contour plot as well as three dimensional plots such as surface plots.

The reader may study several MATLAB codes presented at various places in this book. We have provided suitable comments in these codes so that the reader will be able to learn writing MATLAB scripts and functions by studying these.

Suggested reading

There are several books, references and on line material available on MATLAB programming. The following references are suitable for getting started with MATLAB.

1. *MATLAB help* - resource available with MATLAB
2. **Rudra Pratap**, *Getting started with MATLAB*. Oxford university press, 2010
3. More books available in the following link
   http://www.mathworks.in/support/books/

Module I

System of equations and eigenvalues

Calculation of roots of equations, both linear and nonlinear, are at the heart of all computational methods used in solving algebraic as well as differential equations. Essentially all equations- algebraic, ordinary differential equations(ODE), partial differential equations(PDE) - are converted to yield a set of linear equations which are solved by the various methods given in Chapter 2 of this module. Methods such as Newton Raphson (introduced in Chapter 4) essentially converts nonlinear equation(s) to the linear form and facilitates the solution using methods applicable to a set of linear equations. These are also useful in optimization and system simulation.

Eigenvalues form the basic quantities which characterize many different types of differential equations. Particularly periodic systems are governed by parameters referred to as eigenvalues, which are associated with matrices. The evaluation of eigenvalues is of considerable importance and hence is dealt with in detail in Chapter 3 of this module.

In the case ODE as well as PDE the governing equations are converted to a set of linear or nonlinear equations which are essentially solved using the methods dealt in this module. Therefore this module forms the most important part of the book.

Chapter 2

Solution of linear equations

Linear equations are encountered in all branches of engineering and hence have received much attention. Geometric applications include the determination of the equation of a plane that passes through three non-collinear points, determination of the point of intersection of three non-parallel planes. Engineering applications are to be found in diverse areas such as analysis of electrical networks, conduction of heat in solids, solution of partial differential equations by finite difference and finite element methods. When the number of equations are small solution may be obtained by elementary methods. For example, two or three equations may be solved easily by the use of Cramer's rule[a]. When the number of equations become larger Cramer's rule is cumbersome and one of several alternate methods may be used. The present chapter considers some elementary cases amenable to traditional methods followed by more complex applications that require advanced techniques.

---

[a]after Gabriel Cramer (French, 1704-1752, a Swiss mathematician

2.1   Analytical methods of solving a set of linear equations

In this section we consider analytical methods of solving a set of linear equations in real three dimensional space $\mathbf{R}^{(3)}$.[1] We discuss Cramer's rule and calculation of inverse of the coefficient matrix.

2.1.1   Cramer's rule

Cramer's rule gives solution for a system of linear equations having unique solutions, through calculation of determinants of square matrices. Consider a system of linear equations having 3 unknowns.

$$\begin{pmatrix} a_{1,1} & a_{1,2} & a_{1,3} \\ a_{2,1} & a_{2,2} & a_{2,3} \\ a_{3,1} & a_{3,2} & a_{3,3} \end{pmatrix} \begin{pmatrix} x_1 \\ x_2 \\ x_3 \end{pmatrix} = \begin{pmatrix} b_1 \\ b_2 \\ b_3 \end{pmatrix} \tag{2.1}$$

The root may now be obtained by the use of Cramer's rule as

$$x_1 = \frac{\begin{vmatrix} b_1 & a_{1,2} & a_{1,3} \\ b_2 & a_{2,2} & a_{2,3} \\ b_3 & a_{3,2} & a_{3,3} \end{vmatrix}}{\begin{vmatrix} a_{1,1} & a_{1,2} & a_{1,3} \\ a_{2,1} & a_{2,2} & a_{2,3} \\ a_{3,1} & a_{3,2} & a_{3,3} \end{vmatrix}}, x_2 = \frac{\begin{vmatrix} a_{1,1} & b_1 & a_{1,3} \\ a_{2,1} & b_2 & a_{2,3} \\ a_{3,1} & b_3 & a_{3,3} \end{vmatrix}}{\begin{vmatrix} a_{1,1} & a_{1,2} & a_{1,3} \\ a_{2,1} & a_{2,2} & a_{2,3} \\ a_{3,1} & a_{3,2} & a_{3,3} \end{vmatrix}}, x_3 = \frac{\begin{vmatrix} a_{1,1} & a_{1,2} & b_1 \\ a_{2,1} & a_{2,2} & b_2 \\ a_{3,1} & a_{3,2} & b_3 \end{vmatrix}}{\begin{vmatrix} a_{1,1} & a_{1,2} & a_{1,3} \\ a_{2,1} & a_{2,2} & a_{2,3} \\ a_{3,1} & a_{3,2} & a_{3,3} \end{vmatrix}} \tag{2.2}$$

Obviously, the solution exists only if the denominator in Equation 2.2 is non zero. A matrix whose determinant is non zero is referred to as a non-singular matrix. Such a matrix also has an inverse $\mathbf{A}^{-1}$ such that $\mathbf{A}\mathbf{A}^{-1} = \mathbf{I}_n$ where $\mathbf{I}_n$ is an identity matrix of rank n. All the diagonal elements of an identity matrix are 1 while all its non-diagonal elements are 0.

$$\mathbf{I}_n = \begin{pmatrix} 1 & 0 & 0 & \cdots & 0 \\ 0 & 1 & 0 & \cdots & 0 \\ 0 & 0 & 1 & \cdots & 0 \\ \vdots & \vdots & \vdots & \ddots & \vdots \\ 0 & 0 & 0 & \cdots & 1 \end{pmatrix} \tag{2.3}$$

Rank of a matrix is defined as the number of independent equations in the given set of equations. It is also the dimension of the largest non-zero determinant obtained by removing equal number of rows and columns from the given matrix. If the rank of the matrix is equal to the dimension of the square matrix (i.e. the coefficient matrix is non-singular), a unique solution exists. If the rank of the coefficient matrix is less than the dimension of the coefficient matrix (determinant of the coefficient matrix is zero), there exist an infinite number of solutions satisfying the set of linear equations.

---

[1]We assume that the reader has had an exposure to elementary matrix algebra and vector analysis

Plane passing through non-collinear points

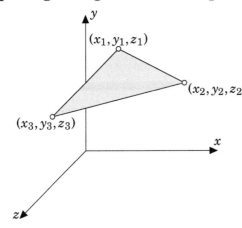

Figure 2.1: *Plane passing through three non-collinear points*

Consider three non-collinear points $A(x_1, y_1, z_1)$, $B(x_2, y_2, z_2)$ and $C(x_3, y_3, z_3)$ lying on a plane whose equation is sought in the form $a'x + b'y + c'z = d$. When d is a non-zero constant we may divide the equation of the plane through by d to get the standard form $ax + by + cz = 1$. Since the given three points lie on the plane we should have

$$
\begin{aligned}
x_1 a + y_1 b + z_1 c &= 1 \\
x_2 a + y_2 b + z_2 c &= 1 \\
x_3 a + y_3 b + z_3 c &= 1
\end{aligned}
\tag{2.4}
$$

The above three equations may be written in matrix form

$$
\underbrace{\begin{pmatrix} x_1 & y_1 & z_1 \\ x_2 & y_2 & z_2 \\ x_3 & y_3 & z_3 \end{pmatrix}}_{\text{Coefficient matrix}} \underbrace{\begin{pmatrix} a \\ b \\ c \end{pmatrix}}_{\text{Vector of unknowns}} = \begin{pmatrix} 1 \\ 1 \\ 1 \end{pmatrix}
\tag{2.5}
$$

We use Cramer's rule to write down the solution as

$$
a = \frac{\begin{vmatrix} 1 & y_1 & z_1 \\ 1 & y_2 & z_2 \\ 1 & y_3 & z_3 \end{vmatrix}}{\begin{vmatrix} x_1 & y_1 & z_1 \\ x_2 & y_2 & z_2 \\ x_3 & y_3 & z_3 \end{vmatrix}}, \quad b = \frac{\begin{vmatrix} x_1 & 1 & z_1 \\ x_2 & 1 & z_2 \\ x_3 & 1 & z_3 \end{vmatrix}}{\begin{vmatrix} x_1 & y_1 & z_1 \\ x_2 & y_2 & z_2 \\ x_3 & y_3 & z_3 \end{vmatrix}}, \quad c = \frac{\begin{vmatrix} x_1 & y_1 & 1 \\ x_2 & y_2 & 1 \\ x_3 & y_3 & 1 \end{vmatrix}}{\begin{vmatrix} x_1 & y_1 & z_1 \\ x_2 & y_2 & z_2 \\ x_3 & y_3 & z_3 \end{vmatrix}}
\tag{2.6}
$$

Example 2.1

Obtain the equation of a plane that passes through the three points $A(2,1,3)$, $B(4,6,9)$ and $C(2,4,6)$

Solution :

We use the procedure given above to solve for the constants a, b and c. Cramer's rule requires the evaluation of the determinants indicated in Equation 2.6. The determinant of the coefficient matrix is obtained as

$$\begin{vmatrix} 2 & 1 & 3 \\ 4 & 6 & 9 \\ 2 & 4 & 6 \end{vmatrix} = 2(6 \times 6 - 4 \times 9) - 1(4 \times 6 - 2 \times 9) + 3(4 \times 4 - 2 \times 6) = 6$$

The other three determinants required to complete the solution are obtained below:

$$\begin{vmatrix} 1 & 1 & 3 \\ 1 & 6 & 9 \\ 1 & 4 & 6 \end{vmatrix} = 1(6 \times 6 - 4 \times 9) - 1(1 \times 6 - 1 \times 9) + 3(1 \times 4 - 1 \times 6) = -3$$

$$\begin{vmatrix} 2 & 1 & 3 \\ 4 & 1 & 9 \\ 2 & 1 & 6 \end{vmatrix} = 2(1 \times 6 - 1 \times 9) - 1(4 \times 6 - 2 \times 9) + 3(4 \times 1 - 2 \times 1) = -6$$

$$\begin{vmatrix} 2 & 1 & 1 \\ 4 & 6 & 1 \\ 2 & 4 & 1 \end{vmatrix} = 2(6 \times 1 - 4 \times 1) - 1(4 \times 1 - 2 \times 1) + 1(4 \times 4 - 2 \times 6) = 6$$

The constants are then obtained as $a = \dfrac{-3}{6} = -0.5$, $b = \dfrac{-6}{6} = -1$ and $c = \dfrac{6}{6} = 1$. The equation of the plane is thus given by $-0.5x - y + z = 1$ or $-x - 2y + 2z = 2$.

Intersection of three non-parallel planes

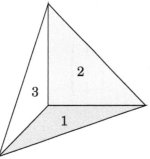

Figure 2.2: *Intersection of three non-parallel planes*

Three non-parallel planes will intersect at a point. The three planes are specified as follows.

$$\begin{aligned} a_{1,1}x_1 + a_{1,2}x_2 + a_{1,3}x_3 &= b_1 \\ a_{2,1}x_1 + a_{2,2}x_2 + a_{2,3}x_3 &= b_2 \\ a_{3,1}x_1 + a_{3,2}x_2 + a_{3,3}x_3 &= b_3 \end{aligned} \tag{2.7}$$

If the coefficient matrix is non-singular, i.e. the rank of the coefficient matrix is 3 the three planes will intersect at a point. This condition requires that the column vectors of the coefficient matrix be linearly independent. We have to solve for the root x_1, x_2, x_3 such that the following matrix equation is satisfied.

$$\begin{pmatrix} a_{1,1} & a_{1,2} & a_{1,3} \\ a_{2,1} & a_{2,2} & a_{2,3} \\ a_{3,1} & a_{3,2} & a_{3,3} \end{pmatrix} \begin{pmatrix} x_1 \\ x_2 \\ x_3 \end{pmatrix} = \begin{pmatrix} b_1 \\ b_2 \\ b_3 \end{pmatrix} \tag{2.8}$$

Basically we have to solve three simultaneous linear equations. Hence the problem is no different from the previous case where we were finding a plane passing through three non-collinear points in $\mathbf{R}^{(3)}$. The point of intersection of the three non parallel planes may now be obtained by the use of Cramer's rule as

$$x_1 = \frac{\begin{vmatrix} b_1 & a_{1,2} & a_{1,3} \\ b_2 & a_{2,2} & a_{2,3} \\ b_3 & a_{3,2} & a_{3,3} \end{vmatrix}}{\begin{vmatrix} a_{1,1} & a_{1,2} & a_{1,3} \\ a_{2,1} & a_{2,2} & a_{2,3} \\ a_{3,1} & a_{3,2} & a_{3,3} \end{vmatrix}}, x_2 = \frac{\begin{vmatrix} a_{1,1} & b_1 & a_{1,3} \\ a_{2,1} & b_2 & a_{2,3} \\ a_{3,1} & b_3 & a_{3,3} \end{vmatrix}}{\begin{vmatrix} a_{1,1} & a_{1,2} & a_{1,3} \\ a_{2,1} & a_{2,2} & a_{2,3} \\ a_{3,1} & a_{3,2} & a_{3,3} \end{vmatrix}}, x_3 = \frac{\begin{vmatrix} a_{1,1} & a_{1,2} & b_1 \\ a_{2,1} & a_{2,2} & b_2 \\ a_{3,1} & a_{3,2} & b_3 \end{vmatrix}}{\begin{vmatrix} a_{1,1} & a_{1,2} & a_{1,3} \\ a_{2,1} & a_{2,2} & a_{2,3} \\ a_{3,1} & a_{3,2} & a_{3,3} \end{vmatrix}} \tag{2.9}$$

2.1.2 Inverse of a square matrix

Solution of Equation 2.5 or 2.1 may also be accomplished by using the inverse of the coefficient matrix. Consider the general case of n linear equations in n unknowns. The coefficient matrix $\mathbf{A}$ is an $n \times n$ square matrix[2] that is assumed to be non-singular. This requires that the determinant of the coefficient matrix be non-zero and hence the matrix rank be n. The inverse of the coefficient matrix is defined as an $n \times n$ matrix $\mathbf{A}^{-1}$ such that $\mathbf{AA}^{-1} = \mathbf{I_n}$ where $\mathbf{I_n}$ is the $n \times n$ identity matrix.

Inverse of a matrix may be found by the following method. We take a 3×3 matrix as an example. The given matrix is written down as

$$\mathbf{A} = \begin{pmatrix} a_{1,1} & a_{1,2} & a_{1,3} \\ a_{2,1} & a_{2,2} & a_{2,3} \\ a_{3,1} & a_{3,2} & a_{3,3} \end{pmatrix} \tag{2.10}$$

We form the matrix of the minors where the minor is the determinant of 2×2 matrix obtained by deleting the elements in the ith row and j^{th} column. For example, minor $M_{2,1}$ is given by

$$\begin{pmatrix} a_{1,1} & \boxed{a_{1,2}\quad a_{1,3}} \\ \boxed{a_{2,1}} & a_{2,2}\quad a_{2,3} \\ a_{3,1} & \boxed{a_{3,2}\quad a_{3,3}} \end{pmatrix} \implies M_{2,1} = \begin{vmatrix} a_{1,2} & a_{1,3} \\ a_{3,2} & a_{3,3} \end{vmatrix}$$

---

[2]Matrices (upper case) and vectors (lower case) are indicated by bold roman symbols

The minor is then multiplied by -1^{i+j}. Thus the above minor element will be multiplied by $-1^{2+1} = -1^3 = -1$. We then form the transpose of the matrix of minors to get the adjoint matrix Adj $\mathbf{A}$. Transpose operation requires interchange of the rows by columns. For example, in the present case the adjoint will be given by

$$\text{Adj}\,\mathbf{A} = \begin{pmatrix} M_{1,1} & -M_{2,1} & M_{3,1} \\ -M_{1,2} & M_{2,2} & -M_{3,2} \\ M_{1,3} & -M_{2,3} & M_{3,3} \end{pmatrix} \tag{2.11}$$

Finally the inverse is obtained by dividing each element of the adjoint matrix by the determinant of $\mathbf{A}$ represented as $|A|$. Thus we get

$$\mathbf{A}^{-1} = \frac{1}{|A|} \begin{pmatrix} M_{1,1} & -M_{2,1} & M_{3,1} \\ -M_{1,2} & M_{2,2} & -M_{3,2} \\ M_{1,3} & -M_{2,3} & M_{3,3} \end{pmatrix} \tag{2.12}$$

The solution to the given equation $\mathbf{Ax} = \mathbf{b}$ may then be written down as $\mathbf{x} = \mathbf{A}^{-1}\mathbf{b}$.

Example 2.2

Equations of three non-parallel planes are given as

$$\begin{aligned} x_1 + 3x_2 + 0.5x_3 &= 3 \\ 2x_1 + 5x_2 + 3x_3 &= 5 \\ 0.4x_1 + x_2 - 0.5x_3 &= 2.3 \end{aligned}$$

Determine the intersection point by matrix inversion.

Solution :

The coefficient matrix is seen to be

$$\mathbf{A} = \begin{pmatrix} 1 & 3 & 0.5 \\ 2 & 5 & 3 \\ 0.4 & 1 & -0.5 \end{pmatrix}$$

Step 1 The minors are all calculated to get the matrix of the minors as

$$\mathbf{M} = \begin{pmatrix} -5.5 & -2.2 & 0 \\ -2 & -0.7 & -0.2 \\ 6.5 & 2 & -1 \end{pmatrix}$$

Step 2 We multiply each term by -1^{i+j} and transpose the matrix to get the adjoint matrix.

$$\text{Adj}\,\mathbf{A} = \begin{pmatrix} -5.5 & 2 & 6.5 \\ 2.2 & -0.7 & -2 \\ 0 & 0.2 & -1 \end{pmatrix}$$

Step 3 The determinant of the coefficient matrix $|A|$ is easily evaluated as 1.1.

Step 4 Dividing term by term the Adj **A** by this value we then get the inverse matrix given by

$$\mathbf{A}^{-1} = \begin{pmatrix} -5 & 1.8182 & 5.9091 \\ 2 & -0.6364 & -1.8182 \\ 0 & 0.1818 & -0.9091 \end{pmatrix}$$

In the above we have truncated all the elements to 4 digits after the decimal point.

Step 5 The solution is then obtained as

$$\begin{pmatrix} x_1 \\ x_2 \\ x_3 \end{pmatrix} = \begin{pmatrix} -5 & 1.8182 & 5.9091 \\ 2 & -0.6364 & -1.8182 \\ 0 & 0.1818 & -0.9091 \end{pmatrix} \begin{pmatrix} 3 \\ 5 \\ 2.3 \end{pmatrix} = \begin{pmatrix} 7.6818 \\ -1.3636 \\ -1.1818 \end{pmatrix}$$

Both Cramer's rule as well as finding the inverse by the method given above require too many arithmetic operations (addition, subtraction, multiplication and division).

Cost of calculations of Cramer's rule:

1. Cost of calculating determinant of rank $n \sim O(n!)$
2. Number of determinants to be calculated for Cramer's rule $= n + 1$
3. Total number of operations $\sim O(n + 1)!$

Cost of calculations of inverse:

1. Cost of calculating minor determinants $\sim O((n-1)!)$
2. Number of minors to be calculated n^2
3. Number of operations to calculate the Adjoint matrix $\sim O(n^2(n-1)!)$
4. Number of operations to calculate determinant of matrix $\sim O(n!)$
5. Total number of operations required to calculate the inverse $\sim O(n+1)!$

n	No. of operations
2	6
3	24
5	720
8	362880

It would be worth while looking for alternate methods that require less computational effort.

2.2 Preliminaries

Before we proceed onto the solution of matrix equations using numerical methods, we shall look into some preliminaries concerning matrix operations and properties which will aid us in understanding the numerical procedures better.

2.2.1 Row operations

Basically elementary row operations are at the heart of most numerical methods. Consider any one of the equations in a set of linear equations such as $a_{i,1}x_1 + a_{i,2}x_2 + \cdots + a_{i,j}x_j + \cdots + a_{i,n}x_n = b_i$. This equation may be written as

$$\underbrace{\left[\begin{array}{cccc} a_{i,1} & a_{i,2} & \cdots & a_{i,n} \end{array} \right]}_{1 \times n} \underbrace{\left(\begin{array}{c} x_1 \\ x_2 \\ \cdots \\ x_n \end{array} \right)}_{n \times 1} = \underbrace{b_i}_{1 \times 1} \tag{2.13}$$

- If we multiply or divide the coefficients and the right hand term by a non-zero number the equation remains unchanged. **Multiplication or division** of a row with a non-zero number as above is an elementary row operation.
- The solution of a system of linear equations is independent of the order in which the equations are arranged. We may therefore **interchange any two rows** without affecting the solution, giving thus another elementary row operation.
- We may **add or subtract two rows** after multiplying each by a chosen number. This constitutes another elementary row operation.

Elementary row operations may also be explained as follows. Consider, as an example, a 3×3 matrix given by

$$\mathbf{A} = \left(\begin{array}{ccc} a_{1,1} & a_{1,2} & a_{1,3} \\ a_{2,1} & a_{2,2} & a_{2,3} \\ a_{3,1} & a_{3,2} & a_{3,3} \end{array} \right) \tag{2.14}$$

Consider also the identity matrix $\mathbf{I}_3$ given by

$$\mathbf{I}_3 = \left(\begin{array}{ccc} 1 & 0 & 0 \\ 0 & 1 & 0 \\ 0 & 0 & 1 \end{array} \right) \tag{2.15}$$

To swap two rows, say, rows 1 and 3

$$\left(\begin{array}{ccc} a_{1,1} & a_{1,2} & a_{1,3} \\ a_{2,1} & a_{2,2} & a_{2,3} \\ a_{3,1} & a_{3,2} & a_{3,3} \end{array} \right) \quad \left(\begin{array}{ccc} a_{3,1} & a_{3,2} & a_{3,3} \\ a_{2,1} & a_{2,2} & a_{2,3} \\ a_{1,1} & a_{1,2} & a_{1,3} \end{array} \right)$$

we pre-multiply the given matrix by the permutation matrix $\mathbf{P}_{1,3}$ obtained by swapping the corresponding rows of $\mathbf{I}_3$, i.e. by using the matrix

$$\mathbf{P}_{1,3} = \left(\begin{array}{ccc} 0 & 0 & 1 \\ 0 & 1 & 0 \\ 1 & 0 & 0 \end{array} \right) \tag{2.16}$$

We may verify that the following holds.

$$\mathbf{P}_{1,3}\,\mathbf{A} = \begin{pmatrix} 0 & 0 & 1 \\ 0 & 1 & 0 \\ 1 & 0 & 0 \end{pmatrix} \begin{pmatrix} a_{1,1} & a_{1,2} & a_{1,3} \\ a_{2,1} & a_{2,2} & a_{2,3} \\ a_{3,1} & a_{3,2} & a_{3,3} \end{pmatrix} = \begin{pmatrix} a_{3,1} & a_{3,2} & a_{3,3} \\ a_{2,1} & a_{2,2} & a_{2,3} \\ a_{1,1} & a_{1,2} & a_{1,3} \end{pmatrix} \tag{2.17}$$

If we intend to replace a row by an elementary row operation involving elements in two rows we may use a suitable matrix to pre-multiply $\mathbf{A}$. Suppose we want to replace the second row by subtracting 2 times element in the second row with 3 times the element in the first row, the matrix required is

$$\mathbf{E} = \begin{pmatrix} 1 & 0 & 0 \\ -3 & 2 & 0 \\ 0 & 0 & 1 \end{pmatrix} \tag{2.18}$$

Thus we will have

$$\mathbf{E}\,\mathbf{A} = \begin{pmatrix} 1 & 0 & 0 \\ -3 & 2 & 0 \\ 0 & 0 & 1 \end{pmatrix} \begin{pmatrix} a_{1,1} & a_{1,2} & a_{1,3} \\ a_{2,1} & a_{2,2} & a_{2,3} \\ a_{3,1} & a_{3,2} & a_{3,3} \end{pmatrix}$$

$$= \begin{pmatrix} a_{1,1} & a_{1,2} & a_{1,3} \\ -3a_{1,1}+2a_{2,1} & -3a_{1,2}+2a_{2,2} & -3a_{1,3}+2a_{2,3} \\ a_{3,1} & a_{3,2} & a_{3,3} \end{pmatrix} \tag{2.19}$$

Elementary row operations are useful in transforming the coefficient matrix to a desirable form that will help in obtaining the solution. For example, the coefficient matrix may be brought to upper triangle form (or row echelon form)[3] by elementary row operations. In the upper triangle form all the elements along the diagonal and above it are non-zero while all the elements below the diagonal elements are zero. The coefficient matrix may be brought to diagonal form (or reduced echelon form) by elementary row operations. In this case only diagonal elements are non-zero with zero off diagonal elements.

Example 2.3
Show the steps involved in transforming the following matrix to upper triangle form.

$$\mathbf{A} = \begin{pmatrix} 5 & 4 & 3 \\ 3 & 2 & 2 \\ 2 & 1 & -2 \end{pmatrix}$$

Solution :
The goal is to make the elements below the leading diagonal zero.

---

[3]Matrix in row echelon form, in general, may look like the following:

$$\begin{pmatrix} * & * & * & * & * \\ 0 & 0 & * & * & * \\ 0 & 0 & 0 & * & * \\ 0 & 0 & 0 & 0 & * \end{pmatrix}$$

where $*$ is a non zero entry

Step 1 If we subtract 3/5 of elements in the first row from the elements in the second row we will obtain a zero in the position 2,1. This is equivalent to pre-multiplication of $\mathbf{A}$ by the matrix

$$\mathbf{E_1} = \begin{pmatrix} 1 & 0 & 0 \\ -\dfrac{3}{5} & 1 & 0 \\ 0 & 0 & 1 \end{pmatrix}$$

We may easily verify that

$$\mathbf{E_1 A} = \begin{pmatrix} 1 & 0 & 0 \\ -\dfrac{3}{5} & 1 & 0 \\ 0 & 0 & 1 \end{pmatrix} \begin{pmatrix} 5 & 4 & 3 \\ 3 & 2 & 2 \\ 2 & 1 & -2 \end{pmatrix} = \begin{pmatrix} 5 & 4 & 3 \\ 0 & -\dfrac{2}{5} & \dfrac{1}{5} \\ 2 & 1 & -2 \end{pmatrix}$$

Step 2 Similarly we may make the element in position 3,1 of $\mathbf{A}$ zero by pre-multiplication by the matrix

$$\mathbf{E_2} = \begin{pmatrix} 1 & 0 & 0 \\ 0 & 1 & 0 \\ -\dfrac{2}{5} & 0 & 1 \end{pmatrix}$$

Thus we have

$$\mathbf{E_2 E_1 A} = \begin{pmatrix} 1 & 0 & 0 \\ 0 & 1 & 0 \\ -\dfrac{2}{5} & 0 & 1 \end{pmatrix} \begin{pmatrix} 5 & 4 & 3 \\ 0 & -\dfrac{2}{5} & \dfrac{1}{5} \\ 2 & 1 & -2 \end{pmatrix} = \begin{pmatrix} 5 & 4 & 3 \\ 0 & -\dfrac{2}{5} & \dfrac{1}{5} \\ 0 & -\dfrac{3}{5} & -3\dfrac{1}{5} \end{pmatrix}$$

Step 3 It is easily seen that the final step that puts a zero in the position 3,2 is obtained by the following elementary row operation.

$$\mathbf{E_3 E_2 E_1 A} = \begin{pmatrix} 1 & 0 & 0 \\ 0 & 1 & 0 \\ 0 & -\dfrac{3}{2} & 1 \end{pmatrix} \begin{pmatrix} 5 & 4 & 3 \\ 0 & -\dfrac{2}{5} & \dfrac{1}{5} \\ 0 & -\dfrac{3}{5} & -3\dfrac{1}{5} \end{pmatrix} = \begin{pmatrix} 5 & 4 & 3 \\ 0 & -\dfrac{2}{5} & \dfrac{1}{5} \\ 0 & 0 & -3\dfrac{1}{2} \end{pmatrix} = \mathbf{U}$$

The matrix $\mathbf{A}$ has now been brought to the upper triangle form $\mathbf{U}$.

Pre-multiplication of $\mathbf{A}$ by $\mathbf{E_3 E_2 E_1}$ converts it to upper triangle form $\mathbf{U}$.

2.2.2   Some useful results

Consider a diagonal matrix, such as, for example

$$\mathbf{D} = \begin{pmatrix} d_{1,1} & 0 & 0 \\ 0 & d_{2,2} & 0 \\ 0 & 0 & d_{3,3} \end{pmatrix}$$

One may easily verify that the inverse of this matrix is given by

$$\mathbf{D}^{-1} = \begin{pmatrix} d_{1,1}^{-1} & 0 & 0 \\ 0 & d_{2,2}^{-1} & 0 \\ 0 & 0 & d_{3,3}^{-1} \end{pmatrix}$$

Consider now a typical $\mathbf{E}$ matrix such as

$$\mathbf{E} = \begin{pmatrix} 1 & 0 & 0 \\ 0 & 1 & 0 \\ e_{3,1} & 0 & 1 \end{pmatrix}$$

It is easily verified that the inverse of this matrix is given by

$$\mathbf{E}^{-1} = \begin{pmatrix} 1 & 0 & 0 \\ 0 & 1 & 0 \\ -e_{3,1} & 0 & 1 \end{pmatrix}$$

2.2.3 Condition number of a matrix

The matrix equation $\mathbf{Ax} = \mathbf{b}$ may be interpreted as an operation that transforms a column vector $\mathbf{x}$ to a column vector $\mathbf{b}$ that involves a change in its magnitude and direction. For quantifying the transformation we introduce the norm of a vector and norm of a matrix. The former is defined as the absolute value of the largest element in the vector. Thus the norm of a vector is defined as

$$\underbrace{\|\mathbf{x}\|}_{\text{Norm}} = \max_{k=1}^{n} |x_k| \tag{2.20}$$

The vector norm satisfies the following properties.

$$\begin{aligned} \|\mathbf{x}\| &> 0 \text{ for } x_k \neq 0 \text{ for all } k \\ \|\mathbf{x}\| &= 0 \text{ for } x_k = 0 \text{ for all } k \\ \|\alpha\mathbf{x}\| &= |\alpha|\|\mathbf{x}\|; \ \alpha \text{ a scalar} \\ \|\mathbf{x}+\mathbf{y}\| &\leq \|\mathbf{x}\| + \|\mathbf{y}\| \end{aligned} \tag{2.21}$$

In the case of a matrix the norm is defined as given below.

$$\|\mathbf{A}\| = \max_{i=1}^{n} \sum_{j=1}^{n} |a_{ij}| \tag{2.22}$$

In addition to satisfying all the properties of the vector norm, matrix norm satisfies also the condition

$$\|\mathbf{Ax}\| \leq \|\mathbf{A}\|\|\mathbf{x}\| \tag{2.23}$$

Using the concept of norm defined above we may look at the linear transformation alluded to above. Consider a perturbation in the solution vector because of a small change in the right hand side (i.e. vector **b**) such that

$$\mathbf{A}(\mathbf{x} + \delta\mathbf{x}) = (\mathbf{b} + \delta\mathbf{b}) \tag{2.24}$$

Since $\mathbf{A}\mathbf{x} = \mathbf{b}$ we have

$$\mathbf{A}\delta\mathbf{x} = \delta\mathbf{b} \tag{2.25}$$

This may be recast, assuming **A** to be non-singular, in the form

$$\delta\mathbf{x} = \mathbf{A}^{-1}\delta\mathbf{b} \tag{2.26}$$

Using the property 2.23 we then have

$$\|\delta\mathbf{x}\| \leq \|\mathbf{A}^{-1}\| \|\delta\mathbf{b}\| \tag{2.27}$$

Noting that $\|\mathbf{b}\| \leq \|\mathbf{A}\| \|\mathbf{x}\|$ we may then obtain

$$\frac{\|\delta\mathbf{x}\|}{\|\mathbf{x}\|} \leq \|\mathbf{A}\| \|\mathbf{A}^{-1}\| \frac{\|\delta\mathbf{b}\|}{\|\mathbf{b}\|} \tag{2.28}$$

The fractional change in **x** is seen to be $\leq \kappa(\mathbf{A}) = \|\mathbf{A}\| \|\mathbf{A}^{-1}\|$ times the fractional change in **b**. $\kappa(\mathbf{A})$ is known as the condition number of the matrix **A**. In general $\kappa(\mathbf{A}) \geq 1$. If the condition number is large the relative change in the solution vector is a magnified version of the change in the vector **b**. A matrix whose condition number is very large compared to unity is referred to as an ill-conditioned matrix.

   The reader may note that the above arguments leading to the condition number may also be obtained when we are looking for the fractional change in the solution vector when there is a perturbation of elements of the coefficient matrix. We can show that the following holds:

$$\frac{\|\delta\mathbf{x}\|}{\|\mathbf{x} + \delta\mathbf{x}\|} \leq \kappa(\mathbf{A}) \frac{\|\delta\mathbf{A}\|}{\|\mathbf{A}\|} \tag{2.29}$$

In numerical calculations inevitably rounding of numbers takes place. Rounding itself may be looked upon as leading to a perturbation of the elements in the coefficient matrix or the right hand vector. This will immediately affect the solution vector through a magnification factor equal to the condition number. Solution to an ill conditioned system is thus prone to errors due to rounding. Partial pivoting or pivoting may be used to mitigate this effect to some extent.

Example 2.4

Determine the condition number of the following matrix and comment on it.

$$\mathbf{A} = \begin{pmatrix} 0.63 & 0.68 & 0.42 \\ 0.65 & 0.72 & 0.44 \\ 2.2 & 0.44 & 0.32 \end{pmatrix}$$

Solution :

In order to calculate the condition number of the matrix we first obtain the inverse by the analytical method discussed above. The inverse thus obtained is given by

$$\mathbf{A}^{-1} = \begin{pmatrix} -7.1097 & 6.3369 & 0.6182 \\ -146.8315 & 139.5672 & 0.8114 \\ 250.7728 & -235.4714 & -2.2411 \end{pmatrix}$$

The row sums required for calculating the norm are given by $1.73, 1.81, 2.96$ for matrix $\mathbf{A}$ while those for $\mathbf{A}^{-1}$ are given by $14.0649, 287.2102, 488.4853$. The maximum values are identified as 2.96 and 488.4853. The condition number for matrix $\mathbf{A}$ is then given by

$$\kappa(\mathbf{A}) = 2.96 \times 488.4853 \approx 1446$$

Since the condition number is very large compared to unity $\mathbf{A}$ is an ill conditioned matrix.

Intersection of two lines

Consider two lines $\sin\theta_1\, x - \cos\theta_1\, y = c_1$ and $\sin\theta_2\, x - \cos\theta_2\, y = c_2$ in $\mathbf{R}^{(2)}$. Equation of lines are essentially in slope intercept form if it is noted that θ is the angle made by the line with the x axis. The point of intersection of these two lines has to be determined. Expressing in matrix form

$$\begin{pmatrix} \sin\theta_1 & -\cos\theta_1 \\ \sin\theta_2 & -\cos\theta_2 \end{pmatrix} \begin{pmatrix} x \\ y \end{pmatrix} = \begin{pmatrix} c_1 \\ c_2 \end{pmatrix}$$

the inverse of coefficient matrix is given by

$$\mathbf{A}^{-1} = \frac{1}{(\sin\theta_2\, \cos\theta_1 - \cos\theta_2\, \sin\theta_1)} \begin{pmatrix} -\cos\theta_2 & -\sin\theta_2 \\ \cos\theta_1 & \sin\theta_1 \end{pmatrix}$$

The condition number of the matrix $\mathbf{A}$ is mainly determined by $\sin\theta_2\, \cos\theta_1 - \cos\theta_2\, \sin\theta_1$ or $\sin(\theta_2 - \theta_1)$ in the denominator. (Both $\|\mathbf{A}\|$ and $\|Adj\,\mathbf{A}\|$ are of order unity).

- For $\theta_2 - \theta_1 \approx 0$, $\sin(\theta_2 - \theta_1) \approx 0$ and therefore condition number is very large. The point of intersection corresponds to the location where both the equations are satisfied.

  When computed using a digital computer, the precision (i.e. number of digits after decimal point) is limited and therefore this would limit the accuracy. When the angle between two lines is very small, the error in the coefficient matrix can change the point of intersection by a large value. This is represented by the ellipse around the point of intersection in Figure 2.3(a). Although, the system of equations has a unique solution, due to round off, there would be large number of points satisfying the equations within the specified tolerance and this means the matrix is ill conditioned.

- In the special case, $\theta_2 - \theta_1 = 0$, the two lines become parallel and no unique solution exists. The condition number becomes ∞.

- As the angle between the two lines become larger, the condition number decreases. The region where the points satisfy the equations within the precision/tolerance also decreases as shown in Figures 2.3 (b) and (c)

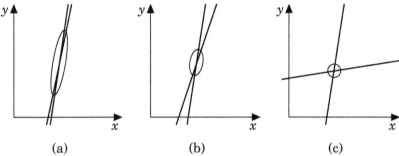

$$\text{(a)} \qquad\qquad \text{(b)} \qquad\qquad \text{(c)}$$

Figure 2.3: *Effect of condition number on the uncertainty in the location of the intersection point between two lines*

2.2.4 Pivoting

If the coefficient matrix **A** is ill conditioned, pivoting would be required to get accurate solution. In general it is good if the absolute magnitude of the pivot element is $\geq$ to the sum of absolute values of all other elements in the row, leading to diagonal dominance. The idea of pivoting is to improve the magnitudes of the diagonal elements by performing elementary row operations.

- If any diagonal element of the coefficient matrix $a_{i,i}$ (also called pivot element) is zero we interchange rows such that each diagonal element is non-zero.
- We also make sure that the pivot element in a particular row is larger than any other element in the same column but in rows **below** it by choosing to swap rows that satisfy this. Note that interchanging rows is an elementary row operation. This is called as **partial pivoting**
- The coefficient matrix can be further improved by performing **Complete pivoting** where both rows and columns are interchanged. When we interchange columns the solution vector should also be rearranged accordingly. For most applications **partial pivoting** would be sufficient to achieve good accuracy.
- Alternatively, we may apply **scaled partial pivoting**. The procedure is similar to partial pivoting but here the pivot element is chosen such that the pivot element has the largest magnitude relative to other entries in its row.

2.2.5 Triangular matrices

A triangular matrix is a special type of square matrix where all the values above or below the diagonal are zero.

$$L = \begin{pmatrix} l_{1,1} & & & \mathbf{0} \\ l_{2,1} & l_{2,2} & & \\ \cdots & \cdots & \ddots & \\ l_{n,1} & l_{n,2} & \cdots & l_{n,n} \end{pmatrix}$$
$$U = \begin{pmatrix} u_{1,1} & u_{1,2} & \cdots & u_{1,n} \\ & u_{2,2} & \cdots & u_{2,n} \\ & \mathbf{0} & \ddots & \cdots \\ & & & u_{n,n} \end{pmatrix}$$

$\mathbf{L}$ is called a lower triangular matrix and $\mathbf{U}$ is called an upper triangular matrix. Matrix equations of above form can be easily solved using backward substitution or forward substitution.

Example 2.5

Solve the following sets of equations

$$\begin{array}{rcrcrcl} x_1 & & & & & = & 2 \\ 2x_1 & + & 3x_2 & & & = & -2 \\ 4x_1 & + & 2x_2 & + & 5x_3 & = & 10 \end{array}$$

Solution :

The above set of equations can be written with the coefficient matrix in **Lower triangular form L** as

$$\begin{pmatrix} 1 & 0 & 0 \\ 2 & 3 & 0 \\ 4 & 2 & 5 \end{pmatrix} \begin{pmatrix} x_1 \\ x_2 \\ x_3 \end{pmatrix} = \begin{pmatrix} 2 \\ -2 \\ 10 \end{pmatrix}$$

Step 1 The first row of L contains coefficient pertaining to x_1, hence we get $x_1 = 2$.

Step 2 The second row contains coefficients related to x_1 and x_2 of which as x_1 is already known from previous step, we can determine x_2 as

$$x_2 = \frac{-2 - 2x_1}{3} = -2$$

Step 3 In the third row, only x_3 is unknown and can hence be determined as

$$x_3 = \frac{10 - 4x_1 - 2x_2}{5} = 1.2$$

The above set of operations is referred to as **Forward substitution** i.e. determining the unknowns starting from the first row and marching towards the last row.

The system of equations can also be represented with the coefficient matrix in **Upper triangular form** U. For a system in Upper triangular form, one has to perform the same steps as that for Lower triangular matrix, except that the unknowns are determined at the bottom first and we march back towards first row. This operation is called as **Backward substitution**.

Consider a Upper Triangular matrix $\mathbf{U}$ of dimension n, $\mathbf{Ux = b}$

$$
\begin{pmatrix}
u_{1,1} & u_{1,2} & \cdots & u_{1,n} \\
0 & u_{2,2} & \cdots & u_{2,n} \\
\vdots & \vdots & \ddots & \vdots \\
0 & 0 & \cdots & u_{i,n} \\
\vdots & \vdots & \ddots & \vdots \\
0 & 0 & \cdots & u_{n,n}
\end{pmatrix}
\begin{pmatrix}
x_1 \\ x_2 \\ \vdots \\ x_i \\ \vdots \\ x_n
\end{pmatrix}
=
\begin{pmatrix}
b_1 \\ b_2 \\ \vdots \\ b_i \\ \vdots \\ b_n
\end{pmatrix}
\tag{2.30}
$$

Step 1 The last row contains only one coefficient related to x_n. Therefore

$$
x_n = \frac{b_n}{u_{n,n}}
$$

Step 2 Moving to $n-1^{th}$ row, there are two coefficients related to x_n (already known) and x_{n-1} (unknown).

$$
u_{n-1,n-1}x_{n-1} + u_{n-1,n}x_n = b_{n-1}
$$

Step 3 Solving for unknown x_{n-1} we have

$$
x_{n-1} = \frac{b_{n-1} - u_{n-1,n}x_n}{u_{n-1,n-1}}
$$

Step 4 Similarly for x_{n-2} we have

$$
x_{n-2} = \frac{b_{n-2} - u_{n-2,n-1}x_{n-1} - u_{n-2,n}x_n}{u_{n-2,n-2}}
$$

Generalizing for i^{th} row

$$
x_i = \frac{1}{u_{i,i}}\left(b_i - \sum_{j=i+1}^{n} u_{i,j}x_j\right)
$$

A MATLAB function to solve a Upper Triangular matrix $\mathbf{U}$ of dimension n, $\mathbf{Ux = b}$ has been provided.

Program 2.1: *Backward Substitution*

```
1  function X = backwardsubstitution(U,B)
2  % Input     U = upper triangular matrix
3  %           B = right hand vector (force vector)
4  % Output    X = solution vector (output vector)
5  n = size(U,1);                    % n = rank of matrix U
6  for i=n:-1:1                       % loop to move from last row
7                                     % towards first row
8      sum = 0;                       % initialize sum
9      for j = i+1:n
10         sum = sum + U(i,j)*X(j)/U(i,i);
11     end
12     X(i) = B(i)/U(i,i)-sum;        % calculate solution vector
13 end
14 X = X';
15 %end of function backsubstitution
```

Use Program 2.1 and solve Example 2.5.

```
A = [5 2 4;0 3 2; 0 0 1];
B = [10;-2;-2];
x = backwardsubstitution(A,B)
```

The output is

```
x = [3.3333;     0.6667;      -2.0000]
```

Note: Rearranging the rows and columns the same matrix can be represented either in upper triangular or lower triangular form.

Number of operations for performing back/ forward substitution:
Number of operations performed for i^{th} row :
Multiplication '×' = $(i-1)$ operations
Addition '+' = $(i-2)$ operations
Subtraction '−' = 1 operation
Division '/' = 1 operation
Total no of operations for i^{th} row : $2i-1$
Total no of operations for substitution = $\sum_{i=1}^{n}(2i-1) = 2\dfrac{n(n+1)}{2} - n = n^2$

For a Lower Triangular matrix **L** of dimension n, $\mathbf{L}.\mathbf{x} = \mathbf{b}$

$$
\begin{pmatrix}
l_{1,1} & 0 & \cdots & 0 \\
l_{2,1} & l_{2,2} & \cdots & 0 \\
\cdots & \cdots & \ddots & \cdots \\
l_{i,1} & l_{i,2} & \cdots & 0 \\
\cdots & \cdots & \ddots & \cdots \\
l_{n,1} & l_{n,2} & \cdots & l_{n,n}
\end{pmatrix}
\begin{pmatrix}
x_1 \\ x_2 \\ \vdots \\ x_i \\ \vdots \\ x_n
\end{pmatrix}
=
\begin{pmatrix}
b_1 \\ b_2 \\ \vdots \\ b_i \\ \vdots \\ b_n
\end{pmatrix}
\tag{2.31}
$$

the procedure for **Forward Substitution** can be summed up as

$$
x_i = \frac{1}{l_{i,i}}\left(b_i - \sum_{j=1}^{i-1} l_{i,j}x_j\right)
$$

Philosophy of solving the linear system of equations using methods such as Gauss elimination, LU decomposition and QR decomposition (all to be presented later)
When we have a system of equations of the form $\mathbf{Ux} = \mathbf{b}$ (Back substitution) or $\mathbf{Lx} = \mathbf{b}$ (Forward substitution), the unknown quantities are estimated by substituting the known quantities already determined and marching along one direction. If the system of equations is converted to a Lower triangular form or an Upper triangular form using simple row operations already discussed, we can determine the solution of the system. This would be much simpler compared to the analytical methods already discussed such as Cramers rule and the determination of inverse matrix using minors. The idea behind numerical methods such as Gauss elimination and LU decomposition is to convert the original matrix into a triangular matrix and solve for the variables. Steps involved can be summarized as

> 1. Perform pivoting
> 2. Reduce coefficient matrix to row echelon form
> 3. Backward substitution or forward substitution

2.3 Gauss elimination method

Gauss elimination method[4,5] performs elementary row operations on the augmented matrix where the $n \times n$ coefficient matrix is augmented by a column containing the right hand vector. Thus we consider $n \times n + 1$ augmented matrix given by

$$
\left(
\begin{array}{ccccc|c}
a_{1,1} & a_{1,2} & \cdots & a_{1,n} & & b_1 \\
a_{2,1} & a_{2,2} & \cdots & a_{2,n} & & b_2 \\
a_{3,1} & a_{3,2} & \cdots & a_{3,n} & & b_3 \\
\cdots & \cdots & \cdots & \cdots & & \cdots \\
a_{i,1} & a_{i,2} & \cdots & a_{i,n} & & b_i \\
\cdots & \cdots & \cdots & \cdots & & \cdots \\
a_{n-1,1} & a_{n-1,2} & \cdots & a_{n-1,n} & & b_{n-1} \\
a_{n,1} & a_{n,2} & \cdots & a_{n,n} & & b_n
\end{array}
\right)
\tag{2.32}
$$

Elementary row operations are performed so as to put the augmented matrix in the row echelon form. Thus we expect to have $a'_{i,j} = 0$ for $j < i$, $a'_{i,i} \neq 0$ and in general, $a'_{i,j} \neq 0$ for $j \geq i$. The process of putting the augmented matrix in to upper triangle form has been shown to be equivalent to the pre-multiplication of the coefficient matrix by appropriate $\mathbf{E}'$s. In case partial pivoting is used the process will yield $\mathbf{PEA} = \mathbf{U}$ where $\mathbf{P}$ is the permutation matrix that keeps track of all the row swaps that have been made during the process of bringing the coefficient matrix to upper triangle form. Note that Gauss elimination with/without partial pivoting yields $\mathbf{Ux} = \mathbf{b}'$. Once the augmented matrix is reduced to row echelon form, the solution to the system of equations is determined using Back substitution.

Example 2.6
Solve the system of equations given below using Gauss Elimination method

$$
\begin{pmatrix}
5 & 4 & 3 \\
3 & 2 & 2 \\
2 & 1 & -2
\end{pmatrix}
\begin{pmatrix}
x_1 \\
x_2 \\
x_3
\end{pmatrix}
=
\begin{pmatrix}
50 \\
25 \\
30
\end{pmatrix}
$$

Solution :
The coefficient matrix is the same as in **Example 2.3**.

---

[4]after Johann Carl Friedrich Gauss, 1777-1855, a German mathematician and physical scientist

[5]The historical development of Gauss elimination has been given by J. Grcar, "Mathematicians of Gaussian Elimination" Notices of the AMS, 2011, 58, 782-792

Step 1 Performing row operations we obtain the coefficient matrix in the Upper triangular form as

$$\mathbf{U} = \begin{pmatrix} 5 & 4 & 3 \\ 0 & -\dfrac{2}{5} & \dfrac{1}{5} \\ 0 & 0 & -3\dfrac{1}{2} \end{pmatrix}$$

Step 2 The row operation is also performed on the right hand column vector, which is obtained by multiplying column vector by the chain of $\mathbf{E}$ matrices.

$$\mathbf{E_3 E_2 E_1 b} = \mathbf{b'} = \begin{pmatrix} 1 & 0 & 0 \\ 0 & 1 & 0 \\ 0 & -\dfrac{3}{2} & 1 \end{pmatrix} \begin{pmatrix} 1 & 0 & 0 \\ 0 & 1 & 0 \\ -\dfrac{2}{5} & 0 & 1 \end{pmatrix} \begin{pmatrix} 1 & 0 & 0 \\ -\dfrac{3}{5} & 1 & 0 \\ 0 & 0 & 1 \end{pmatrix} \begin{pmatrix} 50 \\ 25 \\ 30 \end{pmatrix} = \begin{pmatrix} 50 \\ -5 \\ 17\dfrac{1}{2} \end{pmatrix}$$

The modified set of equations are

$$\begin{pmatrix} 5 & 4 & 3 \\ 0 & -\dfrac{2}{5} & \dfrac{1}{5} \\ 0 & 0 & -3\dfrac{1}{2} \end{pmatrix} \begin{pmatrix} x_1 \\ x_2 \\ x_3 \end{pmatrix} = \begin{pmatrix} 50 \\ -5 \\ 17\dfrac{1}{2} \end{pmatrix}$$

Step 3 Performing back-substitution we obtain the solution as

$$\mathbf{x}^T = \begin{pmatrix} 5 & 10 & -5 \end{pmatrix}$$

Algorithm for Gauss elimination without pivoting

Consider the system of equations, written in augmented matrix form (Equation 2.32) We have to reduce the augmented matrix to an upper triangular matrix $\mathbf{U}$.

- Let us start with the first column. All the elements below $a_{1,1}$ have to be reduced to zero. Here $a_{1,1}$ is the **Pivot element**. This is to be done by performing elementary row operation by subtracting first row, multiplied by a factor, from other rows. Operating on the second row of the augmented matrix.

$$R_2' = R_2 - \frac{a_{2,1}}{a_{1,1}} R_1$$

- Similarly, for j_{th} row, the multiplication factor is $\dfrac{a_{j,1}}{a_{1,1}}$ and the row operation would be

$$R_j' = R_j - \frac{a_{j,1}}{a_{1,1}} R_1$$

Now, all the elements below $a_{1,1}$ are zero. Similar procedure has to be followed for other columns as well. This procedure is called as **Forward elimination**.

For i^{th} column the pivot element would be $a_{i,i}$. The row operations can be generalized as

$$R'_j = R_j - \frac{a_{j,i}}{a_{i,i}} R_i : 1 \le i < n \text{ and } i < j \le n \tag{2.33}$$

Program 2.2: *Gauss elimination (without pivoting)*

```
1  function X = gausselimination(A,B)
2  %      Input : A: coeffieient matrix
3  %              B: right hand vector
4  %      Output: X: output vector
5  n = size(A,1);   % rank of matrix A
6  Aug = [A B];     % augmented matrix
7  for i=1:n-1      % loop to perform forward elimination
8      for j=i+1:n
9          r = Aug(j,i)/Aug(i,i);           % factor
10         Aug(j,:) = Aug(j,:) - r*Aug(i,:); % elimination
11     end
12 end
13 %call function backwardsubstitution to calculate X
14 X = backwardsubstitution(Aug(1:n,1:n),Aug(1:n,n+1));
```

The above program has been applied to Example 2.6.

```
A =  [5 4   3;
      3 2   2;
      2 1  -2;];
B =  [50; 25; 30];
x = gausselimination(A,B);
```

The output of the program is

```
x = [5;  10;  -5;]
```

Number of operations involved in Gauss Elimination method

Consider i^{th} column. All elements below $a_{i,i}$ have to become zero.

Number of rows undergoing transformation : $n - i$

Number of operations for each row:

Division '/' : 1 computation

Multiplication $\times$: $n - i + 1$ computations

addition '+' : $n - i + 1$ computations

Total number of computations for each row: $2n + 1 - 2i$

Total number of computations for i^{th} column: $(n - i)(2(n - i + 1) + 1)$

Total number of computations for elimination:

$$\sum_{i=1}^{n-1} (n-i)(2(n-i+1)+1) = \frac{2n^3}{3} + \frac{n^2}{2} - \frac{7n}{6}$$

Number of operations for back substitution: n^2

Total number of operations for Gauss elimination: $\frac{2n^3}{3} + \frac{3n^2}{2} - \frac{7n}{6}$.

For large values of n, the total number of computations for performing Gauss elimination would be $\sim \frac{2n^3}{3}$.

In the following table we compare the operation counts of Gauss elimination with the analytical methods discussed previously.

n	No. of operations	
	Gauss Elimination	Analytical methods
3	28	24
10	805	39916800
25	11325	4.0329×10^{26}
50	87025	1.5511×10^{66}
100	681550	9.4259×10^{159}
1000	6.6817×10^8	Too large!!!

We see that the analytical methods are completely ruled out even for a matrix of size 25×25.

Example 2.7

Solve the following set of equations by Gauss elimination method.

$$
\begin{aligned}
x_1 - 4x_2 + 4x_3 + 7x_4 &= 4 \\
2x_2 - x_3 &= 5 \\
2x_1 + x_2 + x_3 + 4x_4 &= 2 \\
2x_1 - 3x_2 + 2x_3 - 5x_4 &= 9
\end{aligned}
$$

Solution :

The augmented matrix is written down as

$$
\begin{array}{c}
R_1 \to \\
R_2 \to \\
R_3 \to \\
R_4 \to
\end{array}
\left(
\begin{array}{cccc|c}
1 & -4 & 4 & 7 & 4 \\
0 & 2 & -1 & 0 & 5 \\
\mathbf{2} & 1 & 1 & 4 & 2 \\
\mathbf{2} & -3 & 2 & -5 & 9
\end{array}
\right)
$$

where the rows of the augmented matrix are identified as $R_1 - R_4$.

Step 1 We use partial pivoting[6], and hence rearrange the augmented matrix by swapping rows 3 and 1 to get

$$
\begin{array}{c}
R_1' = R_3 \to \\
R_2' = R_2 \to \\
R_3' = R_1 \to \\
R_4' = R_4 \to
\end{array}
\left(
\begin{array}{cccc|c}
\mathbf{2} & 1 & 1 & 4 & 2 \\
0 & 2 & -1 & 0 & 5 \\
1 & -4 & 4 & 7 & 4 \\
2 & -3 & 2 & -5 & 9
\end{array}
\right)
$$

(There is another possibility in this step. We could have, instead, swapped the fourth and first rows to put a 2 in the pivot position of the first row.) This is equivalent to premultiplication of the augmented matrix by the permutation matrix

$$
\mathbf{P} =
\begin{pmatrix}
0 & 0 & 1 & 0 \\
0 & 1 & 0 & 0 \\
1 & 0 & 0 & 0 \\
0 & 0 & 0 & 1
\end{pmatrix}
$$

---

[6]The reader is encouraged to redo the steps without employing partial pivoting.

We have managed to make the diagonal element in the first row the largest i.e. 2 (shown in **bold face**).

Step 2 We now perform elementary row operations on this rearranged augmented matrix. We make column elements in rows 2 - 4 zero by suitable elementary row operations as given below.

$$
\begin{array}{l}
R_1' \to \\
R_2' \to \\
R_3'' = 2R_3' - R_1' \to \\
R_4'' = R_4' - R_1' \to
\end{array}
\left(
\begin{array}{cccc|c}
2 & 1 & 1 & 4 & 2 \\
0 & 2 & -1 & 0 & 5 \\
0 & \mathbf{-9} & 7 & 10 & 6 \\
0 & -4 & 1 & -9 & 7
\end{array}
\right)
$$

The above step is equivalent to pre-multiplication by the following matrix.

$$
\mathbf{E_1} =
\left(
\begin{array}{cccc}
1 & 0 & 0 & 0 \\
0 & 1 & 0 & 0 \\
-1 & 0 & 2 & 0 \\
-1 & 0 & 0 & 1
\end{array}
\right)
$$

Step 3 We move on to second column of the augmented matrix. Another partial pivoting step is required now. We note that -9 is the largest element in the second column and we swap rows such that this element is the diagonal element in row 2.

$$
\begin{array}{l}
R_1' \to \\
R_2'' = R_3'' \to \\
R_3''' = R_2' \to \\
R_4'' \to
\end{array}
\left(
\begin{array}{cccc|c}
2 & 1 & 1 & 4 & 2 \\
0 & \mathbf{-9} & 7 & 10 & 6 \\
0 & 2 & -1 & 0 & 5 \\
0 & -4 & 1 & -9 & 7
\end{array}
\right)
$$

At this stage the permutation matrix is given by

$$
\mathbf{P} =
\left(
\begin{array}{cccc}
0 & 0 & 1 & 0 \\
1 & 0 & 0 & 0 \\
0 & 1 & 0 & 0 \\
0 & 0 & 0 & 1
\end{array}
\right)
$$

Step 4 Now we make second column elements in the third and fourth rows zero by suitable elementary row operations.

$$
\begin{array}{l}
R_1' \to \\
R_2'' \to \\
R_3^{iv} = -9R_3''' - 2R_2'' \to \\
R_4''' = -9R_4' - (-4)R_2'' \to
\end{array}
\left(
\begin{array}{cccc|c}
2 & 1 & 1 & 4 & 2 \\
0 & -9 & 7 & 10 & 6 \\
0 & 0 & -5 & -20 & -57 \\
0 & 0 & \mathbf{19} & 121 & -39
\end{array}
\right)
$$

The above step is equivalent to pre-multiplication by the following matrix.

$$
\mathbf{E_2} =
\left(
\begin{array}{cccc}
1 & 0 & 0 & 0 \\
0 & 1 & 0 & 0 \\
0 & -2 & -9 & 0 \\
0 & 4 & 0 & -9
\end{array}
\right)
$$

Step 5 Moving to the third column, partial pivoting requires another row swap as below. The largest element 19 is now made to take the position 3,3.

$$
\begin{array}{l}
R_1' \to \\
R_2'' \to \\
R_3^v = R_4''' \to \\
R_4^{iv} = R_3^{iv} \to
\end{array}
\left(
\begin{array}{cccc|c}
2 & 1 & 1 & 4 & 2 \\
0 & -9 & 7 & 10 & 6 \\
0 & 0 & \mathbf{19} & 121 & -39 \\
0 & 0 & -5 & -20 & -57
\end{array}
\right)
$$

At this stage the permutation matrix is given by

$$\mathbf{P} = \begin{pmatrix} 0 & 0 & 1 & 0 \\ 1 & 0 & 0 & 0 \\ 0 & 0 & 0 & 1 \\ 0 & 1 & 0 & 0 \end{pmatrix}$$

Step 6 Now we make third column element in the fourth row zero by suitable elementary row operation.

$$\begin{array}{c} R'_1 \rightarrow \\ R''_2 \rightarrow \\ R^v_3 \rightarrow \\ R^v_4 = 19R^{iv}_4 - (-5)R^v_3 \rightarrow \end{array} \left(\begin{array}{cccc|c} 2 & 1 & 1 & 4 & 2 \\ 0 & -4 & 1 & -9 & 7 \\ 0 & 0 & 2 & 18 & -34 \\ 0 & 0 & 0 & 225 & -1278 \end{array} \right)$$

The above step is equivalent to pre-multiplication by the following matrix.

$$\mathbf{E_3} = \begin{pmatrix} 1 & 0 & 0 & 0 \\ 0 & 1 & 0 & 0 \\ 0 & 0 & 1 & 0 \\ 0 & 0 & 5 & 19 \end{pmatrix}$$

Step 7 The augmented matrix has been brought to the row echelon form. The set of given equations is written in the equivalent form

$$\begin{pmatrix} 2 & 1 & 1 & 4 \\ 0 & -4 & 1 & -9 \\ 0 & 0 & 2 & 18 \\ 0 & 0 & 0 & 225 \end{pmatrix} \begin{pmatrix} x_1 \\ x_2 \\ x_3 \\ x_4 \end{pmatrix} = \begin{pmatrix} 2 \\ 7 \\ -34 \\ -1278 \end{pmatrix}$$

Back substitution will now yield the solution vector. From the last row of the augmented matrix in row echelon form, we get

$$x_4 = \frac{-1278}{225} = -5.68$$

From the third row of the augmented matrix in row echelon form, we get

$$x_3 = \frac{-34 - 18 \times (-5.68)}{2} = 34.12$$

From the second row of the augmented matrix in row echelon form, we get

$$x_2 = \frac{7 - 34.12 - (-9) \times (-5.68)}{-4} = 19.56$$

From the first row of the augmented matrix in row echelon form, we get

$$x_1 = \frac{2 - 19.56 - 34.12 - 4 \times (-5.68)}{2} = -14.48$$

Thus the solution vector is given by

$$\mathbf{x} = \begin{pmatrix} -14.48 \\ 19.56 \\ 34.12 \\ -5.68 \end{pmatrix}$$

Example 2.8

Solve the following system of equations by Gauss elimination method.

$$\begin{bmatrix} 0.63 & 0.68 & 0.42 \\ 0.65 & 0.72 & 0.44 \\ 2.2 & 0.44 & 0.32 \end{bmatrix} \begin{bmatrix} x_1 \\ x_2 \\ x_3 \end{bmatrix} = \begin{bmatrix} 2 \\ 1.6 \\ 3.2 \end{bmatrix}$$

Use partial pivoting and obtain the solution vector by finally rounding to three digits. Compare with the solution set obtained by no pivoting but with rounding to three digits during the Gauss elimination process.

Solution :

The coefficient matrix in this problem is the one that was shown to be an ill conditioned matrix in Example 2.4.

Case 1: Solution obtained by the use of partial pivoting and subsequent rounding

Step 1

$$\begin{array}{c} R_1 \to \\ R_2 \to \\ R_3 \to \end{array} \left[\begin{array}{ccc|c} 0.63 & 0.68 & 0.42 & 2 \\ 0.65 & 0.72 & 0.44 & 1.6 \\ \mathbf{2.2} & 0.44 & 0.32 & 3.2 \end{array} \right]$$

Noting that element 3,1 is the largest one in the first column, we swap the first row with the third row to get

$$\begin{array}{c} R_1' = R_3 \to \\ R_2' = R_2 \to \\ R_3' = R_1 \to \end{array} \left[\begin{array}{ccc|c} \mathbf{2.2} & 0.44 & 0.32 & 3.2 \\ 0.65 & 0.72 & 0.44 & 1.6 \\ 0.63 & 0.68 & 0.42 & 2 \end{array} \right]$$

Step 2 We now perform elementary row operations to obtain zero elements in 2,1 and 3,1.

$$\begin{array}{c} R_1' \to \\ R_2'' = 2.2 \times R_2' - 0.65 \times R_1' \to \\ R_3'' = 2.2 \times R_3' - 0.63 \times R_1' \to \end{array} \left[\begin{array}{ccc|c} 2.2 & 0.44 & 0.32 & 3.2 \\ 0 & 1.2188 & 0.7224 & 2.384 \\ 0 & \mathbf{1.298} & 0.76 & 1.44 \end{array} \right]$$

Step 3 Note that element 3,2 is now the largest element in the second column. We swap the second and third rows to introduce partial pivoting now.

$$\begin{array}{c} R_1' \to \\ R_2''' = R_3'' \to \\ R_3''' = R_2'' \to \end{array} \left[\begin{array}{ccc|c} 2.2 & 0.44 & 0.32 & 3.2 \\ 0 & \mathbf{1.298} & 0.76 & 1.44 \\ 0 & 1.2188 & 0.7224 & 2.384 \end{array} \right]$$

Step 4 Now we perform a final elementary row operation to bring the augmented matrix to row echelon form.

$$\begin{array}{c} R_1' \to \\ R_2''' \to \\ R_3^{iv} = 1.298 \times R_3''' - 1.219 \times R_2''' \to \end{array} \left[\begin{array}{ccc|c} 2.2 & 0.44 & 0.32 & 3.2 \\ 0 & \mathbf{1.298} & 0.76 & 1.44 \\ 0 & 0 & 0.0113872 & 1.33936 \end{array} \right]$$

Step 5 The solution vector is obtained by back substitution and rounded to three digits to get the following:

$$x_1 = -2.102 \qquad x_2 = -67.759 \qquad x_3 = 117.62$$

Case 2: Solution with no pivoting and rounding

The Gauss elimination process uses elementary row operations with no rearrangement of the three equations. Rounding to three digits is employed at every stage. The augmented matrix is given by

$$
\begin{array}{c}
R_1 \rightarrow \\
R_2 \rightarrow \\
R_3 \rightarrow
\end{array}
\left(
\begin{array}{ccc|c}
0.63 & 0.68 & 0.42 & 2 \\
0.65 & 0.72 & 0.44 & 1.6 \\
2.2 & 0.44 & 0.32 & 3.2
\end{array}
\right)
$$

Step 1 Elementary row operations are used now to get zeros at 2,1 and 3,1.

$$
\begin{array}{c}
R_1 \rightarrow \\
R_2' = 0.63 \times R_2 - 0.65 \times R_1 \rightarrow \\
R_3' = 0.63 \times R_3 - 2.2 \times R_1 \rightarrow
\end{array}
\left(
\begin{array}{ccc|c}
0.63 & 0.68 & 0.42 & 2 \\
0 & 0.012 & 0.004 & -0.292 \\
0 & -1.219 & -0.722 & -2.384
\end{array}
\right)
$$

Step 2 An elementary row operation is used now to get zero at 3,2 and the matrix is reduced to row echelon form.

$$
\begin{array}{c}
R_1 \rightarrow \\
R_2' \rightarrow \\
R_3'' = 0.012 \times R_3' - (-1.219) \times R_2' \rightarrow
\end{array}
\left(
\begin{array}{ccc|c}
0.63 & 0.68 & 0.42 & 2 \\
0 & 0.012 & 0.004 & -0.292 \\
0 & 0 & -0.004 & -0.385
\end{array}
\right)
$$

Step 3 The solution vector is obtained by back substitution and rounded to three digits to get the following:

$$x_1 = -0.098 \qquad x_2 = -56.417 \qquad x_3 = 96.25$$

The solution is grossly in error! We may also calculate the residual (difference between the left hand side and right hand side of an equation) to get the following:

Equation No.	Residual pivoting - final rounding	Residual no pivoting - rounding
1	0.000	0.000
2	0.000	0.066
3	0.000	2.561

The above example demonstrates how partial pivoting is desirable while dealing with an ill conditioned coefficient matrix.

2.4 Gauss Jordan method of determining the inverse matrix

We have presented earlier analytical method involving minors for inverting a non-singular square matrix. The method involves too many steps and is certainly not used in practice, especially when the matrix is very large. An alternate and computationally more economical method is the Gauss Jordan method that will be presented now. We write the given matrix $\mathbf{A}$ and augment it by writing $\mathbf{I_n}$ along side to get a $n \times 2n$ matrix.

$$
\left(
\begin{array}{ccccc|ccccc}
a_{1,1} & a_{1,2} & \cdots & a_{1,i} & \cdots & a_{1,n} & 1 & 0 & \cdots & 0 & \cdots & 0 \\
a_{2,1} & a_{2,2} & \cdots & a_{2,i} & \cdots & a_{2,n} & 0 & 1 & \cdots & 0 & \cdots & 0 \\
\vdots & \vdots & \ddots & \vdots & \ddots & \vdots & \vdots & \vdots & \ddots & \vdots & \ddots & \vdots \\
a_{i,1} & a_{i,2} & \cdots & a_{i,i} & \cdots & a_{i,n} & 0 & 0 & \cdots & 1 & \cdots & 0 \\
\vdots & \vdots & \ddots & \vdots & \ddots & \vdots & \vdots & \vdots & \ddots & \vdots & \ddots & \vdots \\
a_{n,1} & a_{n,2} & \cdots & a_{n,i} & \cdots & a_{n,n} & 0 & 0 & \cdots & 0 & \cdots & 1
\end{array}
\right)
\tag{2.34}
$$

The process of Gauss elimination is to be applied to the augmented matrix such that the first half of the augmented matrix is reduced to an identity matrix. Eliminating all the elements below $a_{1,1}$, we obtain

$$
\left(
\begin{array}{ccccc|ccccc}
1 & a_{1,2} & \cdots & a_{1,i} & \cdots & a_{1,n} & a_{1,1}^{1} & 0 & \cdots & 0 & \cdots & 0 \\
0 & a_{2,2} & \cdots & a_{2i} & \cdots & a_{2,n} & a_{2,1}^{1} & 1 & \cdots & 0 & \cdots & 0 \\
\vdots & \vdots & \ddots & \vdots & \ddots & \vdots & \vdots & \vdots & \ddots & \vdots & \ddots & \vdots \\
0 & a_{i,2} & \cdots & a_{i,i} & \cdots & a_{i,n} & a_{i,1}^{1} & 0 & \cdots & 1 & \cdots & 0 \\
\vdots & \vdots & \ddots & \vdots & \ddots & \vdots & \vdots & \vdots & \ddots & \vdots & \ddots & \vdots \\
0 & a_{n,2} & \cdots & a_{n,i} & \cdots & a_{n,n} & a_{n,1}^{1} & 0 & \cdots & 0 & \cdots & 1
\end{array}
\right)
\tag{2.35}
$$

We perform elementary row operations and eventually transform the matrix $\mathbf{A}$ to $\mathbf{I_n}$. Thus we will have

$$
\mathbf{E_k E_{k-1} \ldots E_2 E_1 A = I_n}
\tag{2.36}
$$

The resultant augmented matrix (Equation 2.34) would become

$$
\left(
\begin{array}{ccccc|ccccc}
1 & 0 & \cdots & 0 & \cdots & 0 & a_{1,1}^{1} & a_{1,2}^{1} & \cdots & a_{1,i}^{1} & \cdots & a_{1,n}^{1} \\
0 & 1 & \cdots & 0 & \cdots & 0 & a_{2,1}^{1} & a_{2,2}^{1} & \cdots & a_{2,i}^{1} & \cdots & a_{2,n}^{1} \\
\vdots & \vdots & \ddots & \vdots & \ddots & \vdots & \vdots & \vdots & \ddots & \vdots & \ddots & \vdots \\
0 & 0 & \cdots & 1 & \cdots & 0 & a_{i,1}^{1} & a_{i,2}^{1} & \cdots & a_{i,i}^{1} & \cdots & a_{i,n}^{1} \\
\vdots & \vdots & \ddots & \vdots & \ddots & \vdots & \vdots & \vdots & \ddots & \vdots & \ddots & \vdots \\
0 & 0 & \cdots & 0 & \cdots & 1 & a_{n,1}^{1} & a_{n,2}^{1} & \cdots & a_{n,i}^{1} & \cdots & a_{n,n}^{1}
\end{array}
\right)
\tag{2.37}
$$

Since $\mathbf{A^{-1}A = I_n}$ we conclude that

$$
\mathbf{A^{-1} = E_k E_{k-1} \ldots E_2 E_1}
\tag{2.38}
$$

Thus elementary row operations transform the identity matrix to the inverse of matrix $\mathbf{A}$.

The solution for the system of linear equations is found as $\mathbf{x = A^{-1} b}$. We use the matrix of Example 2.2 to demonstrate this procedure.

Example 2.9

Invert the coefficient matrix of Example 2.2 by the Gauss Jordan method.

Solution :

Step 1 We form the augmented matrix as

$$
\begin{array}{c}
R_1 \rightarrow \\
R_2 \rightarrow \\
R_3 \rightarrow
\end{array}
\left(
\begin{array}{ccc|ccc}
1 & 3 & 0.5 & 1 & 0 & 0 \\
2 & 5 & 3 & 0 & 1 & 0 \\
0.4 & 1 & -0.5 & 0 & 0 & 1
\end{array}
\right)
$$

Rows of the augmented matrix are indicated as R_1-R_3. Elementary row operations consist in multiplication or division of elements in any row by a number and subtraction or addition with corresponding elements in another row. In the present case we would like to make element 2,1 zero. This may be accomplished by subtracting elements in the first row after multiplication by 2 from the corresponding elements in the second row. Thus the second row will become

$$
R_2' = (R_2 - 2 \times R_1) \rightarrow \quad [0 \quad -1 \quad 2 \mid -2 \quad 1 \quad 0]
$$

Similarly we make element 3,1 zero by multiplying each element of first row by 0.4 and subtracting from the corresponding element in the third row. Thus the third row becomes

$$
R_3' = (R_3 - 0.4 \times R_1) \rightarrow \quad [0 \quad -0.2 \quad -0.7 \mid -0.4 \quad 0 \quad 1]
$$

At this stage the augmented matrix appears as

$$
\begin{array}{c}
R_1' = R_1 \rightarrow \\
R_2' \rightarrow \\
R_3' \rightarrow
\end{array}
\left(
\begin{array}{ccc|ccc}
1 & 3 & 0.5 & 1 & 0 & 0 \\
0 & -1 & 2 & -2 & 1 & 0 \\
0 & -0.2 & -0.7 & -0.4 & 0 & 1
\end{array}
\right)
$$

The elementary matrix that accomplishes this is given by

$$
\mathbf{E_1} = \left(
\begin{array}{ccc}
1 & 0 & 0 \\
-2 & 1 & 0 \\
-0.4 & 0 & 1
\end{array}
\right)
$$

Step 2 The next operation is intended to make the element 2,2 equal to unity. This is accomplished by dividing each term of the second row in the above matrix by -1. The augmented matrix appears now as shown below.

$$
\begin{array}{c}
R_1' \rightarrow \\
R_2'' = R_2' \div (-1) \rightarrow \\
R_3' \rightarrow
\end{array}
\left(
\begin{array}{ccc|ccc}
1 & 3 & 0.5 & 1 & 0 & 0 \\
0 & 1 & -2 & 2 & -1 & 0 \\
0 & -0.2 & -0.7 & -0.4 & 0 & 1
\end{array}
\right)
$$

The elementary matrix that accomplishes this is given by

$$
\mathbf{E_2} = \left(
\begin{array}{ccc}
1 & 0 & 0 \\
0 & -1 & 0 \\
0 & 0 & 1
\end{array}
\right)
$$

Step 3 We can make the element 3,2 equal to zero by multiplying each term in row 2 by -0.2 and subtracting from the corresponding term in row 3. Similarly element 1,2 may be made equal to zero by multiplying each term in the second row by 3 and subtracting from the corresponding term in the first row. At this stage we have

$$
\begin{array}{l}
R_1'' = R_1' - 3 \times R_2'' \rightarrow \\
R_2'' \rightarrow \\
R_3'' = R_3' - 0.2 \times R_2'' \rightarrow
\end{array}
\left(
\begin{array}{ccc|ccc}
1 & 0 & 6.5 & -5 & 3 & 0 \\
0 & 1 & -2 & 2 & -1 & 0 \\
0 & 0 & -1.1 & 0 & -0.2 & 1
\end{array}
\right)
$$

The elementary matrix that accomplishes this is given by

$$
\mathbf{E_3} =
\begin{pmatrix}
1 & -3 & 0 \\
0 & 1 & 0 \\
0 & 0.2 & 1
\end{pmatrix}
$$

Step 4 We now divide each term in the third row by -1.1 to get

$$
\begin{array}{l}
R_1'' \rightarrow \\
R_2'' \rightarrow \\
R_3''' = R_3'' \div (-1.1) \rightarrow
\end{array}
\left(
\begin{array}{ccc|ccc}
1 & 0 & 6.5 & -5 & 3 & 0 \\
0 & 1 & -2 & 2 & -1 & 0 \\
0 & 0 & 1 & 0 & 0.1818 & -0.9091
\end{array}
\right)
$$

The elementary matrix that accomplishes this is given by

$$
\mathbf{E_4} =
\begin{pmatrix}
1 & 0 & 0 \\
0 & 1 & 0 \\
0 & 0 & -0.9091
\end{pmatrix}
$$

Step 5 Now we make the term 1,3 take on the value of zero by multiplying the third row by 6.5 and subtracting from the corresponding elements in the first row. Also we make the term 2,3 take on the value of zero by multiplying the third row by -2 and subtracting from the corresponding elements in the second row.

$$
\begin{array}{l}
R_1''' = R_1'' - R_3''' \times 6.5 \rightarrow \\
R_2''' = R_2'' - (-2) \times R_3''' \rightarrow \\
R_3''' \rightarrow
\end{array}
\left(
\begin{array}{ccc|ccc}
1 & 0 & 0 & -5 & 1.8182 & 5.9091 \\
0 & 1 & 0 & 2 & -0.6364 & -1.8182 \\
0 & 0 & 1 & 0 & 0.1818 & -0.9091
\end{array}
\right)
$$

The elementary matrix that accomplishes this is given by

$$
\mathbf{E_5} =
\begin{pmatrix}
1 & 0 & -6.5 \\
0 & 1 & 2 \\
0 & 0 & 1
\end{pmatrix}
$$

The coefficient matrix has thus been reduced to $\mathbf{I_3}$. The inverse of the matrix $\mathbf{A}$ is then given by

$$
\mathbf{A}^{-1} =
\begin{pmatrix}
-5 & 1.8182 & 5.9091 \\
2 & -0.6364 & -1.8182 \\
0 & 0.1818 & -0.9091
\end{pmatrix}
$$

This agrees with the inverse found earlier in Example 2.2. The reader may easily verify that $\mathbf{E_5 E_4 E_3 E_2 E_1} = \mathbf{A}^{-1}$.

MATLAB function to determine the inverse of a matrix

```
Ainv = inv(A);
```

Solving Example 2.4 using MATLAB

```
A = [0.63 0.68 0.42; 0.65 0.72 0.44;2.2 0.44 0.32];
B = [2; 1.6; 3.2];
x = inv(A)*B
```

The output of the program is

```
x =  [    -2.1020;          -67.7589;          117.6198]
```

Number of operations involved in Gauss Jordan method for determining inverse of a matrix

Consider i^{th} column. All elements except $a_{i,i}$ have to become zero and $a_{i,i}$ has to become 1.

Number of rows undergoing transformation : $n-1$

Number of operations for each row:

Division '/' : 1 computations

Multiplication × : n computations

addition '+' : n computations

Total number of computations for each row: $2n+1$

Number of operations to normalize the pivot element: $n-1$ divisions

Total number of computations for i^{th} column: $(n-1)(2n+1)+n-1$

Total number of computations for elimination:

$$\sum_{i=1}^{n}(n-1)(2n+1)+(n-1)=2n^3-2n$$

Cost of multiplying the inverse matrix with right hand vector: $(2n-1)^2=4n^2-4n+1$ Total number of computations for determining the solution: $2n^3+4n^2-6n+1$

For large values of n, the total number of computations for performing Gauss Jordan method for inverse would be $\sim O(2n^3)$

The method of inverse is slower than Gauss elimination method

Alternatively, MATLAB defines a operator "\" which can directly solve system of linear equations

```
x = A\B;
x =  [    -2.1020;          -67.7589;          117.6198]
```

The above operation is much faster than inverse operation. MATLAB automatically chooses an appropriate algorithm based on the type of coefficient matrix **A**. The reader is encouraged to read through the MATLAB help menu for more information on the algorithms used by MATLAB.

2.5 LU decomposition or LU factorization

Gauss elimination requires that all the elementary row operations that are performed operate both on the coefficient matrix and the right hand side vector. The solution obtained is specific to the right hand side vector that is used. However it will be advantageous if one can obtain the solution for *any* right hand vector. This is possible if the coefficient matrix is decomposed into $\mathbf{A} = \mathbf{LU}$ form. There are three ways of doing this:

1. $\mathbf{L}$ is unit lower triangular (diagonal elements are 1) and $\mathbf{U}$ is upper triangular - the procedure is called Doolittle LU decomposition
2. $\mathbf{L}$ is lower triangular and $\mathbf{U}$ is unit upper triangular (diagonal elements are 1) - the procedure is called Crout LU decomposition
3. In the case of Cholesky decomposition (applicable under some conditions) we choose $l_{i,i} = u_{i,i}$.

2.5.1 Doolittle decomposition

In Example 2.3 we have seen how a matrix may be transformed to upper triangle form by elementary row operations. We thus had $\mathbf{E_3 E_2 E_1 \cdot A} = \mathbf{U}$. Taking inverse, we have $\mathbf{A} = \mathbf{E_1^{-1} E_2^{-1} E_3^{-1} U}$. Using the numerics in that example, we have

$$\mathbf{E_1^{-1} E_2^{-1} E_3^{-1}} = \begin{pmatrix} 1 & 0 & 0 \\ \dfrac{3}{5} & 1 & 0 \\ 0 & 0 & 1 \end{pmatrix} \begin{pmatrix} 1 & 0 & 0 \\ 0 & 1 & 0 \\ \dfrac{2}{5} & 0 & 1 \end{pmatrix} \begin{pmatrix} 1 & 0 & 0 \\ 0 & 1 & 0 \\ 0 & \dfrac{3}{2} & 1 \end{pmatrix}$$

$$= \begin{pmatrix} 1 & 0 & 0 \\ \dfrac{3}{5} & 1 & 0 \\ \dfrac{2}{5} & \dfrac{3}{2} & 1 \end{pmatrix} = \mathbf{L}$$

The above shows that the product of the three elementary matrices gives a single unit lower diagonal matrix $\mathbf{L}$. Thus we have managed to write the square matrix $\mathbf{A}$ as the product of a unit lower triangle matrix $\mathbf{L}$ and an upper triangle matrix $\mathbf{U}$ i.e. $\mathbf{A} = \mathbf{LU}$, basically constituting the Doolittle decomposition.

> Doolittle decomposition is no different from Gauss elimination. Doolittle decomposition is advantageous when the same coefficient matrix is used for several systems of linear equations.

We may deduce the elements of $\mathbf{L}$ and $\mathbf{U}$ as follows. Consider the general case of an $n \times n$ coefficient matrix $\mathbf{A}$. We then write it as

$$\mathbf{A} = \begin{pmatrix} a_{1,1} & a_{1,2} & \cdots & a_{1,n} \\ a_{2,1} & a_{2,2} & \cdots & a_{2,n} \\ \cdots & \cdots & \ddots & \cdots \\ a_{i,1} & a_{i,2} & \cdots & a_{i,n} \\ \cdots & \cdots & \ddots & \cdots \\ a_{n,1} & a_{n,2} & \cdots & a_{n,n} \end{pmatrix}$$

$$
= \begin{pmatrix}
1 & 0 & \cdots & 0 \\
l_{2,1} & 1 & \cdots & 0 \\
\cdots & \cdots & \ddots & \cdots \\
l_{i,1} & l_{i,2} & \cdots & 0 \\
\cdots & \cdots & \ddots & \cdots \\
l_{n,1} & l_{n,2} & \cdots & 1
\end{pmatrix}
\begin{pmatrix}
u_{1,1} & u_{1,2} & \cdots & u_{1,n} \\
0 & u_{2,2} & \cdots & u_{2,n} \\
\cdots & \cdots & \ddots & \cdots \\
0 & 0 & \cdots & u_{i,n} \\
\cdots & \cdots & \ddots & \cdots \\
0 & 0 & \cdots & u_{nn}
\end{pmatrix}
$$

$$
= \begin{pmatrix}
u_{1,1} & u_{1,2} & \cdots & u_{1,n} \\
l_{2,1}u_{1,1} & l_{2,1}u_{1,2}+u_{2,2} & \cdots & l_{2,1}u_{1,n}+u_{2,n} \\
\cdots & \cdots & \ddots & \cdots \\
l_{i,1}u_{1,1} & l_{i,1}u_{1,2}+l_{i,2}u_{2,2} & \cdots & \sum_{p=1}^{i-1} l_{i,p}u_{p,n}+u_{i,n} \\
\cdots & \cdots & \ddots & \cdots \\
l_{n,1}u_{1,1} & l_{n,1}u_{1,2}+l_{n,2}u_{2,2} & \cdots & \sum_{p=1}^{n-1} l_{n,p}u_{p,n}+u_{n,n}
\end{pmatrix}
\quad (2.39)
$$

Equate term by term the elements in **A** and the product of matrices **L** and **U** to determine the two triangular matrices.

Step 1 Comparing first row of matrix **A** and **LU**, elements of first row of **U** can be determined

$$u_{1,j} = a_{1,j},\ 1 \le j \le n$$

Step 2 Comparing first column of matrix **A** and **LU**, $u_{1,1}$ is already known and hence the elements in the first column of L ($l_{i,1}$) can be determined

$$l_{i,1} = \frac{a_{i,1}}{u_{1,1}},\ 2 \le i \le n$$

Step 3 Elements of first row of **U** and first column of **L** are known. Comparing elements of second row of matrix **A** and **LU**, elements of second row of U can be determined.

$$u_{2,j} = a_{2,j} - l_{2,1}u_{1,j},\ 2 \le j \le n$$

Step 4 Now, second column of matrix **L** can be determined by comparing second column of matrix **A** and **LU**

$$l_{i,2} = \frac{a_{i,2} - l_{i,1}u_{1,2}}{u_{2,2}},\ 3 \le i \le n$$

Step 5 The other elements of the matrix **L** and **U** may similarly be determined.

Generalized formula to determine the two matrices can be written as

$$u_{i,j} = a_{i,j} - \sum_{p=1}^{i-1} l_{i,p}u_{p,j},\ i \le j$$

$$l_{ij} = \frac{a_{i,j} - \sum_{p=1}^{j-1} l_{i,p} u_{p,j}}{u_{j,j}}, i > j \tag{2.40}$$

These expressions are useful in writing a code.

In case pivoting is employed, LU decomposition will lead to $\mathbf{LUx} = \mathbf{Pb}$ as mentioned earlier. In either case the final form of equation to be solved is

$$\mathbf{LUx} = \mathbf{b} \tag{2.41}$$

This equation is solved in two steps. The first step involves forward substitution during which we solve

$$\mathbf{Ly} = \mathbf{b} \tag{2.42}$$

for the vector $\mathbf{y}$ where $\mathbf{y} = \mathbf{Ux}$.
In the second back substitution step we solve

$$\mathbf{Ux} = \mathbf{y} \tag{2.43}$$

for the vector $\mathbf{x}$, thus obtaining the desired solution.

The general algorithmic structure of Doolittle decomposition is as follows

- Start with pivot element $a_{1,1}$, check if pivoting is necessary
- Determine all the elements of first row of $\mathbf{U}$
- Determine all the elements of first column of $\mathbf{L}$
- Move on to the next pivot element $a_{2,2}$ and check if pivoting is necessary. Determine the elements of second row of $\mathbf{U}$ and second column of $\mathbf{L}$.
- Continue the process for other pivot elements $a_{i,i}$ for $1 \leq i \leq n$ and determine the matrices $\mathbf{L}$ and $\mathbf{U}$
- Using forward substitution, determine $\mathbf{y}$
- Using backward substitution, determine $\mathbf{x}$

Example 2.10

Solve the following set of equations by LU (Doolittle) decomposition. Show the chain of elementary matrices that will bring the coefficient matrix to LU form. Do not use pivoting.

$$
\begin{aligned}
5x_1 + 4x_2 + 3x_3 &= 12 \\
3x_1 + 2x_2 + 2x_3 &= 7 \\
2x_1 + x_2 - 2x_3 &= 1
\end{aligned}
$$

Solution :

The coefficient matrix is given by

$$\mathbf{A} = \begin{pmatrix} 5 & 4 & 3 \\ 3 & 2 & 2 \\ 2 & 1 & -2 \end{pmatrix}$$

Step 1 Determine matrices **L** and **U**

Three elementary matrices are required to bring the coefficient matrix to LU form. They may easily be shown to be given by

$$\mathbf{E_1} = \begin{pmatrix} 1 & 0 & 0 \\ -0.6 & 1 & 0 \\ 0 & 0 & 1 \end{pmatrix} \quad \mathbf{E_2} = \begin{pmatrix} 1 & 0 & 0 \\ 0 & 1 & 0 \\ -0.4 & 0 & 1 \end{pmatrix} \quad \mathbf{E_3} = \begin{pmatrix} 1 & 0 & 0 \\ 0 & 1 & 0 \\ 0 & -1.5 & 1 \end{pmatrix}$$

The Lower unit triangle matrix is then given by

$$\mathbf{L} = \mathbf{E_1^{-1}E_2^{-1}E_3^{-1}} = \begin{pmatrix} 1 & 0 & 0 \\ 0.6 & 1 & 0 \\ 0.4 & 1.5 & 1 \end{pmatrix}$$

The reader may then show that the Upper triangle matrix is

$$\mathbf{U} = \begin{pmatrix} 5 & 4 & 3 \\ 0 & -0.4 & 0.2 \\ 0 & 0 & -3.5 \end{pmatrix}$$

Step 2 Forward substitution: We have

$$\begin{pmatrix} 1 & 0 & 0 \\ 0.6 & 1 & 0 \\ 0.4 & 1.5 & 1 \end{pmatrix} \begin{pmatrix} y_1 \\ y_2 \\ y_3 \end{pmatrix} = \begin{pmatrix} 12 \\ 7 \\ 1 \end{pmatrix}$$

From the first equation $y_1 = 12$. From the second equation, forward substitution gives

$$y_2 = 7 - 0.6 \times 12 = -0.2$$

Using the last equation, forward substitution yields

$$y_3 = 1 - 0.4 \times 12 - 1.5 \times (-0.2) = -3.5$$

We thus have

$$\begin{pmatrix} 5 & 4 & 3 \\ 0 & -0.4 & 0.2 \\ 0 & 0 & -3.5 \end{pmatrix} \begin{pmatrix} x_1 \\ x_2 \\ x_3 \end{pmatrix} = \begin{pmatrix} 12 \\ -0.2 \\ -3.5 \end{pmatrix}$$

Step 3 Backward substitution: From the last equation we get

$$x_3 = \frac{-3.5}{-3.5} = 1$$

From the second equation we then get

$$x_2 = \frac{-0.2 - 0.2 \times 1}{-0.4} = 1$$

Lastly, the first equation gives

$$x_1 = \frac{12 - 4 \times 1 - 3 \times 1}{5} = 1$$

Thus the solution vector is

$$\mathbf{x}^T = \begin{pmatrix} 1 & 1 & 1 \end{pmatrix}$$

Cost of solution of equations using LU factorization
Consider i^{th} row/column (pivot element = $a_{i,i}$).
Number of operations for determining i^{th} row of U
Additions '+' = $i - 1$ computations
Multiplications '×' = $i - 1$ computations
Total number of elements in i^{th} row of **U** = $n - i + 1$
Total number of operations for i^{th} row of **U** = $(n - i + 1)(2i - 2)$
Number of operations for determining i^{th} column of L
Additions '+' = $i - 1$ computations
Multiplications '×' = $i - 1$ computations
Divisions '/' = 1 computation
Total number of elements in i^{th} row of **L** = $n - i$
Total number of operations for i^{th} row of **L** = $(n - i)(2i - 1)$
Total number of operations to determine **L** and **U** =

$$\sum_{i=1}^{n}(2i-1)(n-i)+(2i-2)(n-i+1) = \frac{2}{3}n^3 + \frac{1}{2}n^2 - \frac{1}{6}n$$

Number of operations for forward and backward substitution = $2n^2$

Total number of operations for LU decomposition method = $\frac{2}{3}n^3 + \frac{5}{2}n^2 - \frac{1}{6}n$

For large values of n, the number of computations involved in LU decomposition method is $\sim \frac{2}{3}n^3$

Example 2.11

Perform Doolittle decomposition with partial pivoting on the following set of equations and obtain the solution vector.

$$\begin{aligned} 2x_1 + x_2 - 2x_3 &= 1 \\ 3x_1 + 2x_2 + 2x_3 &= 7 \\ 5x_1 + 4x_2 + 3x_3 &= 12 \end{aligned}$$

Solution :

Step 1 Determination of matrices **L** and **U** with partial pivoting.
We swap rows and 1 and 3 by using a permutation matrix

$$\mathbf{P} = \begin{pmatrix} 0 & 0 & 1 \\ 0 & 1 & 0 \\ 1 & 0 & 0 \end{pmatrix}$$

The coefficient matrix is then written as

$$\begin{pmatrix} 5 & 4 & 3 \\ 3 & 2 & 2 \\ 2 & 1 & -2 \end{pmatrix}$$

We use the following $\mathbf{E}_1$ to convert the first column elements in the rows 2 and 3 to zero.

$$\mathbf{E}_1 = \begin{pmatrix} 1 & 0 & 0 \\ -\dfrac{3}{5} & 1 & 0 \\ -\dfrac{2}{5} & 0 & 1 \end{pmatrix}$$

The coefficient matrix is then given by

$$\begin{pmatrix} 5 & 4 & 3 \\ 0 & -\dfrac{2}{5} & \dfrac{1}{5} \\ 0 & -\dfrac{3}{5} & -3\dfrac{1}{5} \end{pmatrix}$$

Pivoting is now accomplished by swapping rows 2 and 3. The permutation matrix becomes

$$\mathbf{P} = \begin{pmatrix} 0 & 0 & 1 \\ 1 & 0 & 0 \\ 0 & 1 & 0 \end{pmatrix}$$

and the coefficient matrix becomes

$$\begin{pmatrix} 5 & 4 & 3 \\ 0 & -\dfrac{3}{5} & -3\dfrac{1}{5} \\ 0 & -\dfrac{2}{5} & \dfrac{1}{5} \end{pmatrix}$$

Also we then have

$$\mathbf{E_1} = \begin{pmatrix} 1 & 0 & 0 \\ -\dfrac{2}{5} & 1 & 0 \\ -\dfrac{3}{5} & 0 & 1 \end{pmatrix}$$

We now use the following $\mathbf{E_2}$ to convert the second column element in row 3 to zero.

$$\mathbf{E_2} = \begin{pmatrix} 1 & 0 & 0 \\ 0 & 1 & 0 \\ 0 & -\dfrac{2}{3} & 1 \end{pmatrix}$$

The coefficient matrix then becomes

$$\begin{pmatrix} 5 & 4 & 3 \\ 0 & -\dfrac{3}{5} & -3\dfrac{1}{5} \\ 0 & 0 & 2\dfrac{1}{3} \end{pmatrix}$$

This is identified as $\mathbf{U}$. Also we then have

$$\mathbf{L} = \mathbf{E_1}^{-1}\mathbf{E_2}^{-1} = \begin{pmatrix} 1 & 0 & 0 \\ \dfrac{2}{5} & 1 & 0 \\ \dfrac{3}{5} & 0 & 1 \end{pmatrix} \begin{pmatrix} 1 & 0 & 0 \\ 0 & 1 & 0 \\ 0 & \dfrac{2}{3} & 1 \end{pmatrix} = \begin{pmatrix} 1 & 0 & 0 \\ \dfrac{2}{5} & 1 & 0 \\ \dfrac{3}{5} & \dfrac{2}{3} & 1 \end{pmatrix}$$

We finally have

$$\mathbf{LUx} = \mathbf{Pb}$$

$$
\mathbf{LUx} = \begin{pmatrix} 1 & 0 & 0 \\ \dfrac{2}{5} & 1 & 0 \\ \dfrac{3}{5} & \dfrac{2}{3} & 1 \end{pmatrix} \begin{pmatrix} 5 & 4 & 3 \\ 0 & -\dfrac{3}{5} & -3\dfrac{1}{5} \\ 0 & 0 & 2\dfrac{1}{3} \end{pmatrix} \begin{pmatrix} x_1 \\ x_2 \\ x_3 \end{pmatrix} = \begin{pmatrix} 0 & 0 & 1 \\ 1 & 0 & 0 \\ 0 & 1 & 0 \end{pmatrix} \begin{pmatrix} 1 \\ 7 \\ 12 \end{pmatrix} = \begin{pmatrix} 12 \\ 1 \\ 7 \end{pmatrix}
$$

Step 2 The solution vector may now be obtained by forward substitution using **L** and backward substitution using **U** as

$$
\mathbf{x}^T = \begin{pmatrix} 1 & 1 & 1 \end{pmatrix}
$$

MATLAB function for performing LU decomposition is

```
[L,U,P] = lu(A)
% L = lower triangular matrix
% U = upper triangular matrix
% P = permutation matrix
```

Applying the function to the above example

```
A = [2 1 -2; 3 2 2; 5 4 3];
B = [1; 7 ; 12];
[L,U,P] = lu(A)
y = L\P*B;
x = U\y;
```

The output of the program is

```
L =
    1.0000         0         0
    0.4000    1.0000         0
    0.6000    0.6667    1.0000
U =
    5.0000    4.0000    3.0000
         0   -0.6000   -3.2000
         0         0    2.3333
P =
    0    0    1
    1    0    0
    0    1    0
y = [    12.0000;         -3.8000;      2.3333]
x = [     1.0000;          1.0000;      1.0000]
```

Example 2.12

Solve the following set of equations by LU (Doolittle) decomposition using Expressions 2.40

$$
\begin{aligned}
4x_1 + x_2 + x_3 + x_4 &= 15.9 \\
x_1 + 4x_2 + x_3 + x_4 &= 17.7 \\
x_1 + x_2 + 4x_3 + x_4 &= 13.2
\end{aligned}
$$

$$x_1 + x_2 + x_3 + 4x_4 \quad = \quad 9.9$$

Solution :

The coefficient matrix is written down as

$a_{1,1} = 4$	$a_{1,2} = 1$	$a_{1,3} = 1$	$a_{1,4} = 1$
$a_{2,1} = 1$	$a_{2,2} = 4$	$a_{2,3} = 1$	$a_{2,4} = 1$
$a_{3,1} = 1$	$a_{3,2} = 1$	$a_{3,3} = 4$	$a_{3,4} = 1$
$a_{4,1} = 1$	$a_{4,2} = 1$	$a_{4,3} = 1$	$a_{4,4} = 4$

Using expressions based on Equation 2.40 the elements in the LU matrix may be written down as under.

$u_{1,1} = a_{1,1}$	$u_{1,2} = a_{1,2}$	$u_{1,3} = a_{1,3}$	$u_{1,4} = a_{1,4}$
$l_{2,1} = \dfrac{a_{2,1}}{u_{1,1}}$	$u_{2,2} = a_{2,2} - l_{2,1}u_{1,2}$	$u_{2,3} = a_{2,3} - l_{2,1}u_{1,3}$	$u_{2,4} = a_{2,4} - l_{2,1}u_{1,4}$
$l_{3,1} = \dfrac{a_{3,1}}{u_{1,1}}$	$l_{3,2} = \dfrac{a_{3,2} - l_{3,1}u_{1,2}}{u_{2,2}}$	$u_{3,3} = a_{3,3} - l_{3,1}u_{1,3}$ $-l_{3,2}u_{23}$	$u_{3,4} = a_{3,4} - l_{3,1}u_{1,4}$ $-l_{3,2}u_{2,4}$
$l_{4,1} = \dfrac{a_{4,1}}{u_{1,1}}$	$l_{4,2} = \dfrac{a_{4,2} - l_{4,1}u_{1,2}}{u_{2,2}}$	$l_{4,3} = \dfrac{1}{u_{3,3}}\big(a_{4,3} - l_{4,1}u_{1,3}$ $-l_{4,2}u_{2,3}\big)$	$u_{4,4} = a_{4,4} - l_{4,1}u_{1,4}$ $-l_{4,2}u_{2,4} - l_{4,3}u_{3,4}$

Note that $l_{1,1} = l_{2,2} = l_{3,3} = l_{4,4} = 1$. Calculations are performed row-wise, from left to right. The resulting tabulation is given below.

$u_{1,1} = 4$	$u_{1,2} = 1$	$u_{1,3} = 1$	$u_{1,4} = 1$
$l_{2,1} = \dfrac{1}{4}$ $= 0.25$	$u_{2,2} = 4 - 0.25 \times 1$ $= 3.75$	$u_{2,3} = 1 - 0.25 \times 1$ $= 0.75$	$u_{2,4} = 1 - 0.25 \times 1$ $= 0.75$
$l_{3,1} = \dfrac{1}{4} =$ 0.25	$l_{3,2} =$ $\dfrac{1 - 0.25 \times 1}{3.75} = 0.2$	$u_{3,3} = 4 - 0.25 \times 1 -$ $0.2 \times 0.75 = 3.6$	$u_{3,4} = 1 - 0.25 \times 1 -$ $0.2 \times 0.75 = 0.6$
$l_{4,1} = \dfrac{1}{4}$ $= 0.25$	$l_{4,2} =$ $\dfrac{1 - 0.25 \times 1}{3.75}$ $= 0.2$	$l_{4,3} =$ $\dfrac{1 - 0.25 \times 1 - 0.15}{3.6}$ $= 0.166667$	$u_{4,4} = 4 - 0.25 \times 1 -$ $0.2 \times 0.75 - 0.166667 \times$ 0.6 $= 3.5$

Thus we have

$$\mathbf{L} = \begin{pmatrix} 1 & 0 & 0 & 0 \\ 0.25 & 1 & 0 & 0 \\ 0.25 & 0.2 & 1 & 0 \\ 0.25 & 0.2 & 0.166667 & 1 \end{pmatrix} \text{ and } \mathbf{U} = \begin{pmatrix} 4 & 1 & 1 & 1 \\ 0 & 3.75 & 0.75 & 0.75 \\ 0 & 0 & 3.6 & 0.6 \\ 0 & 0 & 0 & 3.5 \end{pmatrix}$$

The solution vector may now be obtained by forward substitution using $\mathbf{L}$ and backward substitution using $\mathbf{U}$ as

$$\mathbf{x}^T = \begin{pmatrix} 2.6 & 3.2 & 1.7 & 0.6 \end{pmatrix}$$

2.5.2 Crout decomposition

Crout decomposition is an alternate method of LU decomposition where $\mathbf{L}$ is a lower triangle matrix and $\mathbf{U}$ is a unit upper triangle matrix.

$$\mathbf{A} = \begin{pmatrix} a_{1,1} & a_{1,2} & \cdots & a_{1,n} \\ a_{2,1} & a_{2,2} & \cdots & a_{2,n} \\ \cdots & \cdots & \ddots & \cdots \\ a_{i,1} & a_{i,2} & \cdots & a_{i,n} \\ \cdots & \cdots & \ddots & \cdots \\ a_{n,1} & a_{n,2} & \cdots & a_{n,n} \end{pmatrix}$$

$$= \begin{pmatrix} l_{1,1} & 0 & \cdots & 0 \\ l_{2,1} & l_{2,2} & \cdots & 0 \\ \cdots & \cdots & \ddots & \cdots \\ l_{i,1} & l_{i,2} & \cdots & 0 \\ \cdots & \cdots & \ddots & \cdots \\ l_{n,1} & l_{n,2} & \cdots & l_{n,n} \end{pmatrix} \begin{pmatrix} 1 & u_{1,2} & \cdots & u_{1,n} \\ 0 & 1 & \cdots & u_{2,n} \\ \cdots & \cdots & \ddots & \cdots \\ 0 & 0 & \cdots & u_{i,n} \\ \cdots & \cdots & \ddots & \cdots \\ 0 & 0 & \cdots & 1 \end{pmatrix}$$

$$= \begin{pmatrix} l_{1,1} & l_{1,1}u_{1,2} & \cdots & l_{1,1}u_{1,n} \\ l_{2,1} & l_{2,1}u_{1,2}+l_{2,2} & \cdots & l_{2,1}u_{1,n}+l_{2,2}u_{2,n} \\ \cdots & \cdots & \ddots & \cdots \\ l_{i,1} & l_{i,1}u_{1,2}+l_{i,2} & \cdots & \sum_{p=1}^{i-1} l_{i,p}u_{p,n}+l_{i,i}u_{i,n} \\ \cdots & \cdots & \ddots & \cdots \\ l_{n,1} & l_{n,1}u_{1,2}+l_{n,2} & \cdots & \sum_{p=1}^{n-1} l_{n,p}u_{p,n}+l_{n,n} \end{pmatrix} \qquad (2.44)$$

The elements of $\mathbf{L}$ and $\mathbf{U}$ are obtained by the following relations.

$$l_{i,j} = a_{i,j} - \sum_{p=1}^{j-1} l_{i,p}u_{p,j}, \, i \geq j$$

$$u_{i,j} = \frac{a_{i,j} - \sum_{p=1}^{i-1} l_{i,p}u_{p,j}}{a_{i,i}}, \, i < j \qquad (2.45)$$

It may easily be verified that the $\mathbf{L}_{Crout} = \mathbf{U}_{Doolittle}^{T}$ and $\mathbf{U}_{Crout} = \mathbf{L}_{Doolittle}^{T}$. The reader is encouraged to solve Example 2.12 by using Crout decomposition.

2.5.3 Cholesky decomposition

When the coefficient matrix is positive definite and symmetric it is possible to use Cholesky decomposition. A positive definite symmetric matrix is a matrix that satisfies the condition $\mathbf{x}^T \mathbf{A} \mathbf{x} > 0$ and the elements of $\mathbf{L}$ and $\mathbf{U}$ are such that $l_{i,j} = u_{j,i}$ i.e. $\mathbf{U} = \mathbf{L}^T$ and hence $\mathbf{A} = \mathbf{L}\mathbf{L}^T$. A symmetric matrix is positive definite if all its diagonal elements are positive and the diagonal element is larger than the sum of all the other elements in that row (diagonal dominant).

Consider as an example, a 4×4 symmetric positive definite matrix

$$\mathbf{A} = \begin{pmatrix} a_{1,1} & a_{2,1} & a_{3,1} & a_{4,1} \\ a_{2,1} & a_{2,2} & a_{3,2} & a_{4,2} \\ a_{3,1} & a_{3,2} & a_{3,3} & a_{4,3} \\ a_{4,1} & a_{4,2} & a_{4,3} & a_{4,4} \end{pmatrix} \tag{2.46}$$

This will be written as a product of $\mathbf{L}$ and $\mathbf{U}$ given by

$$\mathbf{A} = \begin{pmatrix} l_{1,1} & 0 & 0 & 0 \\ l_{2,1} & l_{2,2} & 0 & 0 \\ l_{3,1} & l_{3,2} & l_{3,3} & 0 \\ l_{4,1} & l_{4,2} & l_{4,3} & l_{4,4} \end{pmatrix} \begin{pmatrix} l_{1,1} & l_{2,1} & l_{3,1} & l_{4,1} \\ 0 & l_{2,2} & l_{3,2} & l_{4,2} \\ 0 & 0 & l_{3,3} & l_{4,3} \\ 0 & 0 & 0 & l_{4,4} \end{pmatrix} \tag{2.47}$$

Performing the indicated multiplication we then have $\mathbf{A} =$

$$\begin{pmatrix} l_{1,1}^2 & l_{1,1}l_{2,1} & l_{1,1}l_{3,1} & l_{1,1}l_{4,1} \\ l_{1,1}l_{2,1} & l_{2,1}^2 + l_{2,2}^2 & l_{3,1}l_{2,1} + l_{3,2}l_{2,2} & l_{4,1}l_{2,1} + l_{4,2}l_{2,2} \\ l_{1,1}l_{3,1} & l_{3,1}l_{2,1} + l_{3,2}l_{2,2} & l_{3,1}^2 + l_{3,2}^2 + l_{3,3}^2 & l_{4,1}l_{3,1} + l_{4,2}l_{3,2} + l_{4,3}l_{3,3} \\ l_{1,1}l_{4,1} & l_{4,1}l_{2,1} + l_{4,2}l_{2,2} & l_{4,1}l_{3,1} + l_{4,2}l_{3,2} + l_{4,3}l_{3,3} & l_{4,1}^2 + l_{4,2}^2 + l_{4,3}^2 + l_{4,4}^2 \end{pmatrix} \tag{2.48}$$

Using Equations 2.46 and 2.48, equating element by element and rearranging we get the following:

$$l_{1,1} = \sqrt{a_{1,1}} \qquad l_{2,1} = \frac{a_{2,1}}{l_{1,1}} \qquad l_{3,1} = \frac{a_{3,1}}{l_{11}} \qquad l_{4,1} = \frac{a_{4,1}}{l_{1,1}}$$

$$l_{2,2} = \sqrt{a_{2,2} - l_{2,1}^2} \qquad l_{3,2} = \frac{a_{3,2} - l_{3,1}l_{2,1}}{l_{2,2}} \qquad l_{3,3} = \sqrt{a_{3,3} - l_{3,1}^2 - l_{3,2}^2}$$

$$l_{4,2} = \frac{a_{4,2} - l_{4,1}l_{2,1}}{l_{2,2}} \qquad l_{4,3} = \frac{a_{4,3} - l_{4,1}l_{3,1} - l_{4,2}l_{3,2}}{l_{3,3}}$$

$$l_{4,4} = \sqrt{a_{4,4} - l_{4,1}^2 - l_{4,2}^2 - l_{4,3}^2} \tag{2.49}$$

Note that the entries in Equation 2.48 are in the order in which calculations are performed (row-wise from left to right). We may generalize and write the following expressions for the case of a positive definite symmetric $n \times n$ matrix ($1 \le i \le n$, $1 \le j \le n$).

$$l_{1,1} = \sqrt{a_{1,1}}$$

$$l_{i,1} = \frac{a_{i,1}}{l_{1,1}}$$

$$l_{i,j} = \frac{a_{i,j} - \sum_{p=1}^{j-1} l_{i,p}l_{j,p}}{l_{j,j}}, \, 1 < j < i$$

$$l_{i,i} = \sqrt{a_{i,i} - \sum_{p=1}^{i-1} l_{i,p}^2} \ , i \neq 1 \tag{2.50}$$

Cholesky decomposition needs to determine only one triangular matrix. For large values of n, the cost of computation of solution using Cholesky decomposition is $\sim \dfrac{1}{3}n^3$.

Example 2.13

Solve the problem in Example 2.12 using Cholesky decomposition.

Solution :

Step 1 Determine **L**

We note that the elements of the coefficient matrix are given by

$$a_{i,i} = 4; \ a_{i,j} = 1 \text{ for } 1 \leq i \leq 4, \ 1 \leq j \leq 4$$

We shall verify first that the coefficient matrix is positive definite. We have

$$
\begin{aligned}
\mathbf{x}^T \mathbf{A} \mathbf{x} &= \begin{pmatrix} x_1 & x_2 & x_3 & x_4 \end{pmatrix} \begin{pmatrix} 4 & 1 & 1 & 1 \\ 1 & 4 & 1 & 1 \\ 1 & 1 & 4 & 1 \\ 1 & 1 & 1 & 4 \end{pmatrix} \begin{pmatrix} x_1 \\ x_2 \\ x_3 \\ x_4 \end{pmatrix} \\
&= 4(x_1^2 + x_2^2 + x_3^2 + x_4^2) + 2(x_1 x_2 + x_1 x_3 + x_1 x_4 + x_2 x_4 + x_2 x_3 + x_3 x_4) \\
&= 3(x_1^2 + x_2^2 + x_3^2 + x_4^2) + (x_1 + x_2 + x_3 + x_4)^2
\end{aligned}
$$

This is positive for all **x**. The matrix is seen to be diagonal dominant. Hence the matrix is positive definite. From Expressions 2.49 we get the elements of **L**.

$$l_{1,1} = \sqrt{4} = 2, \ l_{2,1} = l_{3,1} = l_{4,1} = \frac{1}{2} = 0.5, \ l_{2,2} = \sqrt{4 - 0.5^2} = 1.936492$$

$$l_{3,2} = \frac{1 - 0.5 \times 0.5}{1.937942} = 0.387298, \ l_{33} = \sqrt{4 - 0.5^2 - 0.389278^2} = 1.897367$$

$$l_{4,2} = \frac{1 - 0.5 \times 0.5}{1.937942} = 0.387298,$$

$$l_{4,3} = \frac{1 - 0.5 \times 0.5 - 0.387298 \times 0.387298}{1.897367} = 0.316228$$

$$l_{4,4} = \sqrt{4 - 0.5^2 - 0.387298^2 - 0.316228^2} = 1.870829$$

Thus we have (entries are shown rounded to four significant digits after decimals)

$$
\mathbf{A} = \underbrace{\begin{pmatrix} 2 & 0 & 0 & 0 \\ 0.5 & 1.9365 & 0 & 0 \\ 0.5 & 0.3873 & 1.8974 & 0 \\ 0.5 & 0.3873 & 0.3163 & 1.8708 \end{pmatrix}}_{\mathbf{L}} \underbrace{\begin{pmatrix} 2 & 0.5 & 0.5 & 0.5 \\ 0 & 1.9365 & 0.3873 & 0.3873 \\ 0 & 0 & 1.8974 & 0.3162 \\ 0 & 0 & 0 & 1.8708 \end{pmatrix}}_{\mathbf{U} = \mathbf{L}^T}
$$

Calculations are however made including enough number of significant digits after decimals.

Step 2 Forward substitution

First part of the problem consists of solving for vector **y** where

$$
\begin{pmatrix}
2 & 0 & 0 & 0 \\
0.5 & 1.9365 & 0 & 0 \\
0.5 & 0.3873 & 1.8974 & 0 \\
0.5 & 0.3873 & 0.3163 & 1.8708
\end{pmatrix}
\begin{pmatrix}
y_1 \\ y_2 \\ y_3 \\ y_4
\end{pmatrix}
=
\begin{pmatrix}
15.9 \\ 17.7 \\ 13.2 \\ 9.9
\end{pmatrix}
$$

Forward substitution is used to obtain the y's. From the first equation we have

$$
y_1 = \frac{15.9}{2} = 7.95
$$

From the second equation we get

$$
y_2 = \frac{17.7 - 0.5 \times 7.95}{1.9365} = 7.0876
$$

From the third equation we obtain

$$
y_3 = \frac{13.2 - 0.5 \times 7.95 - 0.3873 \times 7.0876}{1.8974} = 3.4153
$$

From the fourth equation we obtain

$$
y_3 = \frac{9.9 - 0.5 \times 7.95 - 0.3873 \times 7.0876 - 0.3163 \times 3.4153}{1.8708} = 1.1225
$$

Step 3 Backward substitution

The second part now consists in solving for vector **x** where

$$
\begin{pmatrix}
2 & 0.5 & 0.5 & 0.5 \\
0 & 1.9365 & 0.3873 & 0.3873 \\
0 & 0 & 1.8974 & 0.3162 \\
0 & 0 & 0 & 1.8708
\end{pmatrix}
\cdot
\begin{pmatrix}
x_1 \\ x_2 \\ x_3 \\ x_4
\end{pmatrix}
=
\begin{pmatrix}
7.95 \\ 7.0876 \\ 3.4153 \\ 1.1225
\end{pmatrix}
$$

Backward substitution is used to obtain the x's. From the fourth equation we have

$$
x_4 = \frac{1.1225}{1.8708} = 0.6
$$

From the third equation we get

$$
x_3 = \frac{3.4153 - 0.3162 \times 0.6}{1.8974} = 1.7
$$

From the second equation we obtain

$$
x_2 = \frac{7.0876 - 0.3873 \times 1.7 - 0.3873 \times 0.6}{1.9365} = 3.2
$$

From the fourth equation we obtain

$$
x_1 = \frac{7.95 - 0.5 \times 3.2 - 0.5 \times 1.7 - 0.5 \times 0.6}{2} = 2.6
$$

Thus the solution vector is given by

$$\mathbf{x} = \begin{pmatrix} 2.6 \\ 3.2 \\ 1.7 \\ 0.6 \end{pmatrix}$$

Program 2.3: *Cholesky decomposition*

```
1  function [X,C] = choleskydecomposition(A,B)
2  %     Input :
3  %              A: coeffieient matrix
4  %              B: right hand Vector
5  %     Output:
6  %              X: solution vector
7  %              C: upper triangular matrix
8  n = size(A,1);    % rank of matrix A
9  C = zeros(n,n);   % initializing C
10 for i = 1:n       % Determining element of C
11     for j = i:n
12         C(i,j) = A(i,j);
13         for k = 1:i-1
14             C(i,j) = C(i,j) - C(k,i)*C(k,j);
15         end
16         if( j==i)
17             C(i,j) = sqrt(C(i,j)) ;
18         else
19             C(i,j) = C(i,j)/C(i,i);
20         end
21     end
22 end
23 Y = C'\B;          %forward  substitution
24 X = C\Y;           %backward substitution
```

Program 2.3 has been applied to Example 2.13

```
A = [4 1 1 1;
     1 4 1 1;
     1 1 4 1;
     1 1 1 4];
B = [15.9; 17.7; 13.2; 9.9];
[x,C] = choleskydecomposition(A,B)
```

The output of the above program is

```
x = [ 2.6000;   3.2000;    1.7000;        0.6000]
C =

    2.0000      0.5000      0.5000      0.5000
         0      1.9365      0.3873      0.3873
         0           0      1.8974      0.3162
         0           0           0      1.8708
```

> **Sparse matrix** is a matrix where majority of the elements are zeros. Such matrices occur frequently in engineering practice. Instead of storing the entire matrix only the nonzero entries can be stored. A simple scheme is to store the row index, column index and the nonzero entry. For example the following matrix
>
> $$\begin{pmatrix} 3 & 1 & 0 \\ 0 & 2 & 0 \\ 0 & 1 & 2 \end{pmatrix} \quad \text{can be rewritten as} \quad \begin{array}{|c|c|c|c|c|c|} \hline i & 1 & 1 & 2 & 3 & 3 \\ \hline j & 1 & 2 & 2 & 2 & 3 \\ \hline a_{i,j} & 3 & 1 & 2 & 1 & 2 \\ \hline \end{array}$$
>
> The computational effort for treating sparse matrices can be reduced largely by employing special techniques, especially for large matrices.
> We shall consider a special form of sparse matrix, tridiagonal matrix and discuss methods to solve the same.

2.6 Tridiagonal matrix algorithm

2.6.1 Cholesky decomposition of a symmetric tri-diagonal matrix

A symmetric tridiagonal matrix is a special case that is very easy to handle using the Cholesky decomposition. The coefficient matrix is of form

$$\mathbf{A} = \begin{pmatrix} a_1 & a_2' & 0 & 0 & \cdots & \cdots & \cdots & 0 \\ a_2' & a_2 & a_3' & 0 & \cdots & \cdots & \cdots & 0 \\ 0 & a_3' & a_3 & a_4' & \cdots & \cdots & \cdots & 0 \\ \cdots & \cdots & \cdots & \cdots & \cdots & \cdots & \cdots & \cdots \\ \cdots & \cdots & \cdots & \cdots & \cdots & a_{n-1}' & a_{n-1} & a_n' \\ 0 & 0 & 0 & \cdots & \cdots & 0 & a_n' & a_n \end{pmatrix} \tag{2.51}$$

The elements that are zero remain so and hence it is necessary to store only single subscripted a and a'. After decomposition we again have only single subscripted d (along the diagonal) and l along either the lower or upper diagonal. Thus the matrix $\mathbf{L}$ is of form

$$\mathbf{L} = \begin{pmatrix} d_1 & 0 & 0 & 0 & \cdots & \cdots & \cdots & 0 \\ l_2 & d_2 & 0 & 0 & \cdots & \cdots & \cdots & 0 \\ 0 & l_3 & d_3 & 0 & \cdots & \cdots & \cdots & 0 \\ \cdots & \cdots & \cdots & \cdots & \cdots & \cdots & \cdots & \cdots \\ \cdots & \cdots & \cdots & \cdots & \cdots & l_{n-1} & d_{n-1} & 0 \\ 0 & 0 & 0 & \cdots & \cdots & 0 & l_n & d_n \end{pmatrix} \tag{2.52}$$

Expressions 2.50 reduce to the following.

$$d_1 = \sqrt{a_1} \,;\, l_i = \frac{a_i'}{d_{i-1}}, \, d_i = \sqrt{a_i - l_i^2}, \, i > 1 \tag{2.53}$$

Example 2.14
Solve the following set of equations by Cholesky decomposition.

$$4x_1 + x_2 \quad = \quad 15.9$$

$$x_1 + 4x_2 + x_3 = 17.7$$
$$x_2 + 4x_3 + x_4 = 13.2$$
$$x_3 + 4x_4 = 9.9$$

Solution :

Step 1 The coefficient matrix is a symmetric tridiagonal matrix with $a_i = 4$, $1 \le i \le 4$, $b_i = 1$, $2 \le i \le 4$. The elements required are calculated using Expressions 2.53.

$$d_1 = \sqrt{4} = 2 \qquad l_2 = \frac{1}{2} = 0.5$$

$$d_2 = \sqrt{4 - 0.5^2} = 1.9365 \qquad l_3 = \frac{1}{1.9365} = 0.5164$$

$$d_3 = \sqrt{4 - 0.5164^2} = 1.9322 \qquad l_4 = \frac{1}{1.9322} = 0.5175$$

$$d_4 = \sqrt{4 - 0.5175^2} = 1.9319$$

Thus we have

$$\begin{pmatrix} 2 & 0 & 0 & 0 \\ 0.5 & 1.9365 & 0 & 0 \\ 0 & 0.5164 & 1.9322 & 0 \\ 0 & 0 & 0.5175 & 1.9319 \end{pmatrix} \cdot \begin{pmatrix} y_1 \\ y_2 \\ y_3 \\ y_4 \end{pmatrix} = \begin{pmatrix} 15.9 \\ 17.7 \\ 13.2 \\ 9.9 \end{pmatrix}$$

Step 2 Forward substitution yields the vector **y** as

$$\mathbf{y}^T = \begin{pmatrix} 7.9500 & 7.0876 & 4.9374 & 3.8018 \end{pmatrix}$$

Now we have

$$\begin{pmatrix} 2 & 0.5 & 0 & 0 \\ 0 & 1.9365 & 0.5164 & 0 \\ 0 & 0 & 1.9322 & 05175 \\ 0 & 0 & 0 & 1.9319 \end{pmatrix} \cdot \begin{pmatrix} x_1 \\ x_2 \\ x_3 \\ x_4 \end{pmatrix} = \begin{pmatrix} 7.9500 \\ 7.0876 \\ 4.9374 \\ 3.8018 \end{pmatrix}$$

Note that **L** is a bidiagonal matrix.

Step 3 Backward substitution now leads to the solution vector

$$\mathbf{x}^T = \begin{pmatrix} 3.1952 & 3.1191 & 2.0282 & 1.9679 \end{pmatrix}$$

2.6.2   General case of a tri-diagonal matrix and the TDMA

In this case the coefficient matrix is of form

$$\mathbf{A} = \begin{pmatrix} a_1 & -a_1' & 0 & 0 & \cdots & \cdots & \cdots & 0 \\ -a_2'' & a_2 & -a_2' & 0 & \cdots & \cdots & \cdots & 0 \\ 0 & -a_3'' & a_3 & -a_3' & \cdots & \cdots & \cdots & 0 \\ \cdots & \cdots & \cdots & \cdots & \cdots & \cdots & \cdots & \cdots \\ \cdots & \cdots & \cdots & \cdots & \cdots & -a_{n-1}'' & a_{n-1} & -a_{n-1}' \\ 0 & 0 & 0 & \cdots & \cdots & 0 & -a_n'' & a_n \end{pmatrix} \qquad (2.54)$$

TDMA or Tridiagonal Matrix Algorithm (also known as Thomas algorithm) aims to convert the coefficient matrix into a bidiagonal matrix with nonzero elements only along the main and upper diagonals. All other elements are zero. Obviously the reduction of coefficient matrix to bidiagonal form requires us to manipulate **b** also by considering the $n \times n + 1$ augmented matrix where the last column contains the elements of vector **b**. Consider hence the augmented matrix given by

$$\mathbf{A'} = \begin{pmatrix} a_1 & -a'_1 & 0 & 0 & \cdots & \cdots & \cdots & 0 & \vdots & b_1 \\ -a''_2 & a_2 & -a'_2 & 0 & \cdots & \cdots & \cdots & 0 & \vdots & b_2 \\ 0 & -a''_3 & a_3 & -a'_3 & \cdots & \cdots & \cdots & 0 & \vdots & \\ \cdots & \cdots & \cdots & \cdots & \cdots & \cdots & \cdots & \cdots & \vdots & \\ \cdots & \cdots & \cdots & \cdots & \cdots & -a'_{n-1} & a_{n-1} & -a'_{n-1} & \vdots & b_{n-1} \\ 0 & 0 & 0 & \cdots & \cdots & 0 & -a''_n & a_n & \vdots & b_n \end{pmatrix} \quad (2.55)$$

Eventually the coefficient matrix has to become bidiagonal and hence we should have

$$\mathbf{A'} = \begin{pmatrix} d_1 & u_1 & 0 & 0 & \cdots & \cdots & \cdots & 0 & \vdots & b'_1 \\ 0 & d_2 & u_2 & 0 & \cdots & \cdots & \cdots & 0 & \vdots & b'_2 \\ 0 & 0 & d_3 & u_3 & \cdots & \cdots & \cdots & 0 & \vdots & \\ \cdots & \cdots & \cdots & \cdots & \cdots & \cdots & \cdots & \cdots & \vdots & \\ \cdots & \cdots & \cdots & \cdots & \cdots & 0 & d_{n-1} & u_{n-1} & \vdots & b'_{n-1} \\ 0 & 0 & 0 & \cdots & \cdots & 0 & 0 & d_n & \vdots & b'_n \end{pmatrix} \quad (2.56)$$

Further we may divide each row by the corresponding diagonal element d_i to get

$$\mathbf{A'} = \begin{pmatrix} 1 & p_1 & 0 & 0 & \cdots & \cdots & \cdots & 0 & \vdots & q_1 \\ 0 & 1 & p_2 & 0 & \cdots & \cdots & \cdots & 0 & \vdots & q_2 \\ 0 & 0 & 1 & p_3 & \cdots & \cdots & \cdots & 0 & \vdots & \\ \cdots & \cdots & \cdots & \cdots & \cdots & \cdots & \cdots & \cdots & \vdots & \\ \cdots & \cdots & \cdots & \cdots & \cdots & 0 & 1 & p_{n-1} & \vdots & q_{n-1} \\ 0 & 0 & 0 & \cdots & \cdots & 0 & 0 & 1 & \vdots & q_n \end{pmatrix} \quad (2.57)$$

Here $p_i = \dfrac{u_i}{d_i}$ and $q_i = \dfrac{b'_i}{d_i}$. The equations are now in a form suitable for solution by back substitution. We have $x_n = q_n$ and in general

$$x_i = p_i x_{i+1} + q_i \quad (2.58)$$

The corresponding equation in the original set would have been

$$a_i x_i = a'_i x_{i+1} + a''_i x_{i-1} + b_i \quad (2.59)$$

Using Equation 2.58 we also have

$$x_{i-1} = p_{i-1} x_i + q_{i-1} \quad (2.60)$$

Substitute this in Equation 2.59 to get

$$a_i x_i = a'_i x_{i+1} + a''_i (p_{i-1} x_i + q_{i-1}) + b_i$$

or on rearrangement

$$x_i = \frac{a'_i}{a_i - a''_i p_{i-1}} x_{i+1} + \frac{b_i + q_{i-1}}{a_i - a''_i p_{i-1}} \tag{2.61}$$

Comparing this with Equation 2.58 we get the following recurrence relations:

$$p_i = \frac{a'_i}{a_i - a''_i p_{i-1}}$$

$$q_i = \frac{b_i + q_{i-1}}{a_i - a''_i p_{i-1}} \tag{2.62}$$

Note also that $p_1 = \dfrac{u_1}{d_1} = \dfrac{a'_1}{a_1}$ and $q_1 = \dfrac{b'_1}{d_1} = \dfrac{b_1}{a_1}$. Calculation of the p's and q's proceed from $i = 1$ to $i = n$ using the above recurrence relations.

Example 2.15

Solve the tridiagonal system of equations given by

$$\begin{pmatrix} 1 & 0 & 0 & 0 & 0 & 0 \\ -0.83 & 2.04 & -1.08 & 0 & 0 & 0 \\ 0 & -0.86 & 2.04 & -1.07 & 0 & 0 \\ 0 & 0 & -0.88 & 2.04 & -1.06 & 0 \\ 0 & 0 & 0 & -0.89 & 2.04 & -1.06 \\ 0 & 0 & 0 & 0 & 0 & 1 \end{pmatrix} \begin{pmatrix} x_1 \\ x_2 \\ x_3 \\ x_4 \\ x_5 \\ x_6 \end{pmatrix} = \begin{pmatrix} 100 \\ 0 \\ 0 \\ 0 \\ 0 \\ 60 \end{pmatrix}$$

using TDMA

Solution :

A spreadsheet was used to solve the problem. Following table shows the results of the calculation. The first five columns are based on the given data. Since zero elements in the coefficient matrix do not change during TDMA it is necessary to only use 3 single subscript arrays as shown in the table. p's and q's are calculated using recurrence relations 2.62. The solution set is obtained by back substitution and hence the calculations in this column proceed upwards starting from the last row. We have rounded the p's and q's to 4 digits after decimals and the solution set is given with only one digit after the decimal point.

i	a	a'	a''	b	p	q	x
1	1	0	0	100	0	100	100
2	2.04	1.08	0.83	0	0.5294	40.6863	71.5
3	2.04	1.07	0.86	0	0.6752	22.0799	58.1
4	2.04	1.06	0.88	0	0.7331	13.4390	53.4
5	2.04	1.06	0.89	0	0.7640	8.6203	54.5
6	1	0	0	60	0	60	60

Alternate solution: Since $x_1 = 100$ and $x_6 = 60$ are easily obtained from the given equations we may solve for x_2 to x_4 only. Again TDMA may be made use of. The corresponding spreadsheet is shown below.

i	a	a'	a''	b	p	q	x
2	2.04	1.08	0	83	0.5294	40.6863	71.5
3	2.04	1.07	0.86	0	0.6752	22.0799	58.1
4	2.04	1.06	0.88	0	0.7331	13.4390	53.4
5	2.04	0	0.89	63.6	0	54.4582	54.5

Cost of computation of matrix using TDMA
Number of computations for determining p_1 and $q_1 = 2$
Total number of operations for determining $p_i = 3$
Total number of operations for determining $q_i = 4$
Number of pairs of p_i and q_i to be determined $= n - 1$
Total number of operations for determining p and q vectors $= 7(n-1)+2$
Total number of operations for performing back substitution $= 2(n-1)$
Total number of computations for performing TDMA $= 9n - 7$
For large matrices, the number of computations for TDMA ($\sim O(n)$) is much smaller than Gauss elimination and LU decomposition methods ($\sim O(n^3)$).
TDMA is Gauss Elimination method applied to a tridiagonal matrix.

MATLAB program for TDMA is given below

Program 2.4: TDMA

```
1  function x = tdma(C,A,B,D)
2  %Input   C = a''
3  %        A = a
4  %        B = a'
5  %        D = b
6  %Output  x = output vector
7  N = length(A);                    %number of unknowns
8  p(1) = B(1)/A(1);
9  q(1) = D(1)/A(1);
10 x = zeros(N,1);                   %initializing output vector
11 for I=2:N                         %calculating p and q
12     p(I)=B(I)/(A(I)-C(I)*p(I-1));
13     p(I)=(D(I)+C(I)*q(I-1))/ (A(I)-C(I)*p(I-1));
14 end
15
16 x(N)= q(N);
17
18 for I=N-1:-1:1                    %calculating output vector
19     x(I)=p(I)*x(I+1)+q(I);
20 end
```

The above program has been applied to tri diagonal matrix given in Example 2.15

```
a   = [1 ;   2.04; 2.04 ; 2.04 ; 2.04 ; 1];
a1 = [0 ;  1.08 ; 1.07 ; 1.06 ; 1.06 ; 0];
a2 = [0 ;  0.83 ; 0.86 ; 0.88 ; 0.89 ; 0];
B   =   [100;0;0;0;0;60];
x  = tdma(a2,a,a1,B)
```

The output of the above program is

```
x = [100.0000;      71.4515;      58.1121;      53.3649;
                    54.4582;      60.0000]
```

2.7   QR Factorization

QR Factorization is a very useful method especially when the coefficient matrix is ill-conditioned. This factorization is also useful in evaluating *eigenvalues* as we shall see later. It is possible to represent the coefficient matrix as

$$\mathbf{A} = \mathbf{QR}$$

where $\mathbf{R}$ is an upper triangular matrix and $\mathbf{Q}$ is an orthogonal matrix that has the important property that $\mathbf{Q}^T\mathbf{Q} = \mathbf{I}_n$ i.e. $\mathbf{Q}^{-1} = \mathbf{Q}^T$. Using the orthogonality property of $\mathbf{Q}$, we easily see that by pre-multiplying matrix $\mathbf{A}$ with $\mathbf{Q}^T$ we get

$$\mathbf{Q}^T\mathbf{A} = \mathbf{Q}^T\mathbf{QR} = \mathbf{R} \tag{2.63}$$

The given equation set $\mathbf{Ax} = \mathbf{b}$ may hence be written in the form

$$\mathbf{Ax} = \mathbf{QRx} = \mathbf{b}$$

Pre-multiplying by $\mathbf{Q}^T$ we then get

$$\mathbf{Q}^T\mathbf{QRx} = \mathbf{Rx} = \mathbf{Q}^T\mathbf{b}$$

which may be solved for $\mathbf{x}$ by back substitution.

QR factorization method differs from LU decomposition in that the Q matrix is orthogonal where as L matrix, in general, is not an orthogonal matrix.

QR factorization may be accomplished by using Gram-Schmidt method, Householder transformation or Givens rotations. These methods are presented below.

2.7.1   Gram-Schmidt Method

Consider the $n \times n$ coefficient matrix $\mathbf{A}$. It may be represented as a set of $1 \times n$ column vectors.

$$\mathbf{A} = \begin{pmatrix} \mathbf{a}_1 & \mathbf{a}_2 & \mathbf{a}_3 & \cdots & \mathbf{a}_i & \cdots & \mathbf{a}_{n-1} & \mathbf{a}_n \end{pmatrix} \tag{2.64}$$

Note that the element $\mathbf{a_i}$ stands for the column vector such that

$$\mathbf{a_i}^T = \left(\begin{array}{ccccccc} a_{1,i} & a_{2,i} & a_{3,i} & \cdots & a_{i,i} & \cdots & a_{n-1,i} & a_{n,i} \end{array} \right) \tag{2.65}$$

In general the column vectors are not orthogonal and hence the dot product of any two column vectors is nonzero. We would like to convert the column vectors to an orthogonal set of column vectors to obtain a matrix

$$\left(\begin{array}{ccccccc} \mathbf{v_1} & \mathbf{v_2} & \mathbf{v_3} & \cdots & \mathbf{v_i} & \cdots & \mathbf{v_{n-1}} & \mathbf{v_n} \end{array} \right) \tag{2.66}$$

Step 1 We will retain the column vector $\mathbf{a_1}$ without any modification and label it as $\mathbf{v_1}$. Consider now the column vector $\mathbf{a_2}$. We would like to modify this vector to $\mathbf{v_2}$ such that $\mathbf{v_1}$ and $\mathbf{v_2}$ are orthogonal to each other. This may be accomplished by the following manipulation.

$$\mathbf{v_2} = \mathbf{a_2} - c_1 \mathbf{v_1} \tag{2.67}$$

where c_1 is a constant that is to be found by requiring that the vectors $\mathbf{v_1}$ and $\mathbf{v_2}$ be orthogonal. Thus we have

$$\mathbf{v_1}^T \mathbf{v_2} = \mathbf{v_1}^T (\mathbf{a_2} - c_1 \mathbf{v_1}) = 0$$

Thus we have

$$c_1 = \frac{\mathbf{v_1}^T \mathbf{a_2}}{\mathbf{v_1}^T \mathbf{v_1}} \tag{2.68}$$

Step 2 Next we consider the column vector $\mathbf{a_3}$. $\mathbf{v_3}$ has to be determined such that $\mathbf{v_1}^T \mathbf{v_3} = 0$ and $\mathbf{v_2}^T \mathbf{v_3} = 0$. The required manipulation is given by

$$\mathbf{v_3} = \mathbf{a_3} - c_1 \mathbf{v_1} - c_2 \mathbf{v_2}$$

where c_1 and c_2 are two constants determined by requiring that $\mathbf{v_1}^T \mathbf{v_3} = 0$ and $\mathbf{v_2}^T \mathbf{v_3} = 0$. It is easily verified then that

$$c_1 = \frac{\mathbf{v_1}^T \mathbf{a_3}}{\mathbf{v_1}^T \mathbf{v_1}} \text{ and } c_2 = \frac{\mathbf{v_2}^T \mathbf{a_3}}{\mathbf{v_2}^T \mathbf{v_2}}$$

Step 3 We are now ready to generalize the above to the kth column vector. We have

$$\mathbf{v_k} = \mathbf{a_k} - \sum_{j=1}^{k-1} c_j \mathbf{v_j} \tag{2.69}$$

where

$$c_j = \frac{\mathbf{v_j}^T \mathbf{a_k}}{\mathbf{v_j}^T \mathbf{v_j}}, \; 1 \le j \le k-1 \tag{2.70}$$

Step 4 We may finally normalize the column vectors by dividing each column vector by its own magnitude ($\sqrt{\mathbf{v}_j^T \mathbf{v}_j}$) to get the matrix $\mathbf{Q}$.

$$\mathbf{Q} = \begin{pmatrix} \mathbf{q}_1 & \mathbf{q}_2 & \mathbf{q}_3 & \cdots & \mathbf{q}_i & \cdots & \mathbf{q}_{n-1} & \mathbf{q}_n \end{pmatrix} \tag{2.71}$$

where

$$\mathbf{q}_j = \frac{\mathbf{v}_j}{\sqrt{\mathbf{v}_j^T \mathbf{v}_j}} \tag{2.72}$$

Example 2.16

Solve the following set of equations by QR factorization.

$$\begin{pmatrix} 1 & 2 & 1 \\ 0 & 1 & -1 \\ 1 & 0 & 1 \end{pmatrix} \begin{pmatrix} x_1 \\ x_2 \\ x_3 \end{pmatrix} = \begin{pmatrix} 2 \\ 1.6 \\ 3.4 \end{pmatrix}$$

Solution :

Step 1 Determining $\mathbf{v}_2$

The column vector $\mathbf{v}_1 = \mathbf{a}_1$ and hence

$$\mathbf{v}_1^T = \begin{pmatrix} 1 & 0 & 1 \end{pmatrix}$$

We have

$$\mathbf{v}_1^T \mathbf{v}_1 = \begin{pmatrix} 1 & 0 & 1 \end{pmatrix} \begin{pmatrix} 1 \\ 0 \\ 1 \end{pmatrix} = 1^2 + 0^2 + 1^2 = 2$$

We also have

$$\mathbf{v}_1^T \mathbf{a}_2 = \begin{pmatrix} 1 & 0 & 1 \end{pmatrix} \begin{pmatrix} 2 \\ 1 \\ 0 \end{pmatrix} = 1 \times 2 + 0 \times 1 + 1 \times 0 = 2$$

Hence

$$c_1 = \frac{\mathbf{v}_1^T \mathbf{a}_2}{\mathbf{v}_1^T \mathbf{v}_1} = \frac{2}{2} = 1$$

Column vector $\mathbf{v}_2$ is then given by

$$\mathbf{v}_2 = \mathbf{a}_2 - c_1 \mathbf{v}_1 = \begin{pmatrix} 2 \\ 1 \\ 0 \end{pmatrix} - \begin{pmatrix} 1 \\ 0 \\ 1 \end{pmatrix} = \begin{pmatrix} 1 \\ 1 \\ -1 \end{pmatrix}$$

Step 2 Now we look at column 3. We have the following (the reader may verify):

$$\mathbf{v}_1^T \mathbf{a}_3 = 2; \quad \mathbf{v}_2^T \mathbf{a}_3 = -1; \quad \mathbf{v}_2^T \mathbf{v}_2 = 3$$

With these we then have

$$c_1 = \frac{\mathbf{v}_1^T \mathbf{a}_3}{\mathbf{v}_1^T \mathbf{v}_1} = \frac{2}{2} = 1; \quad c_2 = \frac{\mathbf{v}_2^T \mathbf{a}_3}{\mathbf{v}_2^T \mathbf{v}_2} = \frac{-1}{3}$$

With these we get

$$\mathbf{v}_3 = \mathbf{a}_3 - c_1\mathbf{v}_1 - c_2\mathbf{v}_2 = \begin{pmatrix} 1 \\ -1 \\ 1 \end{pmatrix} - \begin{pmatrix} 1 \\ 0 \\ 1 \end{pmatrix} + \frac{1}{3}\begin{pmatrix} 1 \\ 1 \\ -1 \end{pmatrix} = \begin{pmatrix} \frac{1}{3} \\ \frac{2}{3} \\ -\frac{2}{3} \\ -\frac{1}{3} \end{pmatrix}$$

Step 3 Normalizing the column vectors by dividing elements in each column by the magnitude of the corresponding column vector we obtain $\mathbf{Q}$.

$$\mathbf{Q} = \begin{pmatrix} 0.7071 & 0.5774 & 0.4082 \\ 0.0000 & 0.5774 & -0.8165 \\ 0.7071 & -0.5774 & -0.4082 \end{pmatrix}$$

We have rounded all the entries to 4 digits after decimals even though the calculations are made using available machine precision. Upper triangle matrix $\mathbf{R}$ may now be obtained as

$$\mathbf{R} = \mathbf{Q}^T\mathbf{A} = \begin{pmatrix} 0.7071 & 0.0000 & 0.7071 \\ 0.5774 & 0.5774 & -0.5774 \\ 0.4082 & -0.8165 & -0.4082 \end{pmatrix}\begin{pmatrix} 1 & 2 & 1 \\ 0 & 1 & -1 \\ 1 & 0 & 1 \end{pmatrix}$$

$$= \begin{pmatrix} 1.4142 & 1.4142 & 1.4142 \\ 0.0000 & 1.7321 & -0.5774 \\ 0.0000 & 0.0000 & 0.8165 \end{pmatrix}$$

Step 4 The right hand matrix $\mathbf{b}$ is to be replaced by $\mathbf{b}'$ given by

$$\mathbf{b}' = \mathbf{Q}^T\mathbf{b} = \begin{pmatrix} 0.7071 & 0.0000 & 0.7071 \\ 0.5774 & 0.5774 & -0.5774 \\ 0.4082 & -0.8165 & -0.4082 \end{pmatrix}\begin{pmatrix} 2 \\ 1.6 \\ 3.4 \end{pmatrix} = \begin{pmatrix} 3.8184 \\ 0.1155 \\ -1.8779 \end{pmatrix}$$

Step 5 The solution to the given set of equations is obtained by solving the system $\mathbf{Rx} = \mathbf{b}'$ by back substitution.

$$\begin{pmatrix} 1.4142 & 1.4142 & 1.4142 \\ 0.0000 & 1.7321 & -0.5774 \\ 0.0000 & 0.0000 & 0.8165 \end{pmatrix}\begin{pmatrix} x_1 \\ x_2 \\ x_3 \end{pmatrix} = \begin{pmatrix} 3.8184 \\ 0.1155 \\ -1.8779 \end{pmatrix}$$

From the last equation we get

$$x_3 = \frac{-1.8779}{0.8165} = -2.3$$

From the second equation we have

$$x_2 = \frac{0.1155 - (-0.5774) \times (-2.3)}{1.7321} = -0.7$$

Finally from the first equation we have

$$x_1 = \frac{3.8184 - 1.4142 \times (-0.7) - 1.4142 \times (-2.3)}{1.4142} = 5.7$$

The solution vector is then written down as

$$\mathbf{x}^T = \begin{pmatrix} 5.7 & -0.7 & -2.3 \end{pmatrix}$$

2.7.2   Householder transformation and QR factorization

Householder method of QR factorization[7] uses a sequence of orthogonal matrices known as Householder matrices to transform the given matrix $\mathbf{A}$ to QR form. The idea of the method is to reduce the coefficient matrix to upper triangular form by multiplying the matrix by a series of Householder matrices.

$$\mathbf{H_n H_{n-1} \ldots H_1 A = U}$$

As Householder matrices are orthogonal, the matrix $\mathbf{H_n H_{n-1} \ldots H_1}$ is also orthogonal. The upper triangular matrix $\mathbf{U}$ thus obtained is $\mathbf{R}$ and thus

$$\mathbf{(H_n H_{n-1} \ldots H_1)^{-1} = Q = H_1^T H_2^T \ldots H_n^T}$$

Consider a Householder matrix $\mathbf{H}_1$ defined as

$$\mathbf{H}_1 = \mathbf{I} - 2\mathbf{uu}^T \tag{2.73}$$

where $\mathbf{I}$ is the $n \times n$ identity matrix and $\mathbf{u}$ is a $n \times 1$ unit vector which will be determined below. $\mathbf{H}_1$ is an orthogonal matrix such that $\mathbf{H}_1^T = \mathbf{H}_1^{-1}$. We verify this below.

$$\begin{aligned}
\mathbf{H}_1 \mathbf{H}_1^T &= (\mathbf{I} - 2\mathbf{uu}^T)(\mathbf{I} - 2\mathbf{uu}^T)^T \\
&= (\mathbf{I}^T - 2(\mathbf{uu}^T)^T)(\mathbf{I} - 2\mathbf{uu}^T) \\
&= (\mathbf{I} - 2\mathbf{uu}^T)(\mathbf{I} - 2\mathbf{uu}^T) \\
&= \mathbf{I} - 4\mathbf{uu}^T + 4\mathbf{uu}^T\mathbf{uu}^T \\
&= \mathbf{I} - 4\mathbf{uu}^T + 4\mathbf{u}(\mathbf{u}^T\mathbf{u})\mathbf{u}^T \\
&= \mathbf{I} - 4\mathbf{uu}^T + 4\mathbf{uu}^T \\
&= \mathbf{I} \tag{2.74}
\end{aligned}$$

In arriving at the above we have made use of the following:

$$(\mathbf{uu}^T)^T = (\mathbf{u}^T)^T\mathbf{u}^T = \mathbf{uu}^T \tag{2.75}$$

$$\mathbf{u}^T\mathbf{u} = 1 \tag{2.76}$$

Let us understand how the Householder transformation achieves QR decomposition. **Step 1** Consider now an $n \times n$ coefficient matrix $\mathbf{A}$.

$$\begin{pmatrix}
a_{1,1} & a_{1,2} & a_{1,3} & \cdots & \cdots & a_{1,n} \\
a_{2,1} & a_{2,2} & a_{2,3} & \cdots & \cdots & a_{2,n} \\
a_{3,1} & a_{3,2} & a_{3,3} & \cdots & \cdots & a_{3,n} \\
\cdots & \cdots & \cdots & \cdots & \cdots & \cdots \\
a_{i,1} & a_{i,2} & a_{i,3} & \cdots & \cdots & a_{i,n} \\
\cdots & \cdots & \cdots & \cdots & \cdots & \cdots \\
a_{n,1} & a_{n,2} & a_{n,3} & \cdots & \cdots & a_{n,n}
\end{pmatrix}$$

---

[7]After Alston S. Householder (1904-1993)- American mathematician

To convert the coefficient matrix to an upper triangular matrix, all the elements below the diagonal elements have to be zero. Let us consider the first column. All the elements below $a_{1,1}$ have to become zero. Let vector $\mathbf{x}$ represent the first column of the coefficient matrix.

$$\mathbf{x}^T = \begin{pmatrix} a_{1,1} & a_{2,1} & \cdots & a_{i,1} & \cdots a_{n,1} \end{pmatrix} \tag{2.77}$$

We desire that this vector be transformed, on pre-multiplication by $\mathbf{H}_1$, to $\hat{x}_1 \mathbf{e}_1$ where $\mathbf{e}_1$ is the unit vector defined as

$$\mathbf{e}_1^T = \begin{pmatrix} 1 & 0 & 0 \cdots & 0 \end{pmatrix} \tag{2.78}$$

and $\hat{x}_1$ is a non-zero number i.e. all the elements below the pivot element are zero. Thus we should have

$$\mathbf{H}_1 \mathbf{x} = \hat{x}_1 \mathbf{e}_1 \tag{2.79}$$

On pre-multiplication by $(\mathbf{H}_1 \mathbf{x})^T$ Equation 2.79 gives

$$(\mathbf{H}_1 \mathbf{x})^T \mathbf{H}_1 \mathbf{x} = \mathbf{x}^T \mathbf{H}_1^T \mathbf{H}_1 \mathbf{x} = \mathbf{x}^T \mathbf{x} = (\hat{x}_1 \mathbf{e}_1)^T \hat{x}_1 \mathbf{e}_1 = \hat{x}_1^2 \tag{2.80}$$

The above follows by the fact that $\mathbf{H}_1$ satisfies Equation 2.74 and that $\mathbf{e}_1^T \mathbf{e}_1 = 1$. Thus the factor $\hat{x}_1$ is given by

$$\hat{x}_1 = \sqrt{\mathbf{x}^T \mathbf{x}} \tag{2.81}$$

Pre-multiplication of Equation 2.79 by $\mathbf{H}_1^T$ gives

$$\mathbf{H}_1^T \mathbf{H}_1 \mathbf{x} = \mathbf{x} = \mathbf{H}_1^T \hat{x}_1 \mathbf{e}_1 = \hat{x}_1 \mathbf{H}_1^T \mathbf{e}_1 = \hat{x}_1 (\mathbf{I} - 2\mathbf{u}\mathbf{u}^T)^T \mathbf{e}_1 = \hat{x}_1 (\mathbf{e}_1 - 2\mathbf{u}\mathbf{u}^T \mathbf{e}_1) \tag{2.82}$$

Expanding the last term on the right side of above equation we get the following:

$$x_1 = \hat{x}_1 (1 - 2u_1^2); \quad x_i = \hat{x}_1 (0 - 2u_1 u_i) \text{ for } i \neq 1 \tag{2.83}$$

Hence we get

$$u_1 = \sqrt{\frac{\hat{x}_1 - x_1}{2\hat{x}_1}} \tag{2.84}$$

In order that u_1 has the largest value possible, we make the difference $\hat{x}_1 - x_1$ take on the largest value by choosing $\hat{x}_1$ as $-\text{sign}\{x_1\}\sqrt{\mathbf{x}^T \mathbf{x}}$. We also have

$$u_i = \frac{x_i}{-2u_1 \hat{x}_1} \tag{2.85}$$

Pre-multiplication of the coefficient matrix by $\mathbf{H}_1$ will achieve the following when $n \times 1$ column vector $\mathbf{a}_1$ is used to evaluate the u's:

$$\begin{pmatrix} a_{1,1} & a_{1,2} & a_{1,3} & \cdots & \cdots & a_{1,n} \\ a_{2,1} & a_{2,2} & a_{2,3} & \cdots & \cdots & a_{2,n} \\ a_{3,1} & a_{3,2} & a_{3,3} & \cdots & \cdots & a_{3,n} \\ \cdots & \cdots & \cdots & \cdots & \cdots & \cdots \\ a_{i,1} & a_{i,2} & a_{i,3} & \cdots & \cdots & a_{i,n} \\ \cdots & \cdots & \cdots & \cdots & \cdots & \cdots \\ a_{n,1} & a_{n,2} & a_{n,3} & \cdots & \cdots & a_{n,n} \end{pmatrix} \Rightarrow \begin{pmatrix} \hat{a}_{1,1} & \hat{a}_{1,2} & \hat{a}_{1,3} & \cdots & \cdots & \hat{a}_{1,n} \\ 0 & a_{2,2}^i & a_{2,3}^i & \cdots & \cdots & a_{2,n}^i \\ 0 & a_{3,2}^i & a_{3,3}^i & \cdots & \cdots & a_{3,n}^i \\ \cdots & \cdots & \cdots & \cdots & \cdots & \cdots \\ 0 & a_{i,2}^i & a_{i,3}^i & \cdots & \cdots & a_{i,n}^i \\ \cdots & \cdots & \cdots & \cdots & \cdots & \cdots \\ 0 & a_{n,2}^i & a_{n,3}^i & \cdots & \cdots & a_{n,n}^i \end{pmatrix} \tag{2.86}$$

Thus all the elements in the first column except the pivot have become zero. We are already making our progress towards an upper triangular form. The elements shown with a - ˆ - will remain unchanged as we proceed further.

Step 2 A second Householder transformation is used to put zeros in all the elements in the second column below the pivot. For this consider the $n-1 \times n-1$ matrix obtained by removing the first row and column of the matrix shown on the right in Equation 2.86. We use the $n-1 \times 1$ vector given by

$$\mathbf{x}^T = \begin{pmatrix} a_{2,2}^i & a_{3,2}^i & \cdots & a_{i,2}^i & \cdots a_{n,2}^i \end{pmatrix} \tag{2.87}$$

We construct a Householder matrix given by $\mathbf{H}_2' = \mathbf{I} - 2\mathbf{u}\mathbf{u}^T$ where the elements of $\mathbf{u}$ are obtained such that the vector $\mathbf{x}$ is transformed to a unit vector $\hat{x}_2\mathbf{e}_2$ where $\mathbf{e}_2$ is $n-1 \times 1$ unit vector. Note that $\mathbf{I}$ in the above is the $n-1 \times n-1$ identity matrix. It is clear that the analysis presented earlier can be used directly now to get the vector $\mathbf{u}$. By adding the row and column that were removed earlier, we construct a matrix $\mathbf{H}_2 = \mathbf{I} - 2\mathbf{u}'\mathbf{u}'^T$ where $\mathbf{I}$ is the $n \times n$ identity matrix and $\mathbf{u}'$ is $n \times 1$ vector given by

$$\mathbf{u}'^T = \begin{pmatrix} 0 & u_1 & u_2 & \cdots & u_{n-1} \end{pmatrix} \tag{2.88}$$

Premultiplication of $\mathbf{H}_1\mathbf{A}$ by $\mathbf{H}_2$ should lead to

$$\begin{pmatrix}
\hat{a}_{1,1} & \hat{a}_{1,2} & \hat{a}_{1,3} & \cdots & \cdots & \hat{a}_{1,n} \\
0 & \hat{a}_{2,2} & \hat{a}_{2,3} & \cdots & \cdots & \hat{a}_{2,n} \\
0 & 0 & a_{3,3}^{ii} & \cdots & \cdots & a_{3,n}^{ii} \\
\cdots & \cdots & \cdots & \cdots & \cdots & \cdots \\
0 & 0 & a_{i,3}^{ii} & \cdots & \cdots & a_{i,n}^{ii} \\
\cdots & \cdots & \cdots & \cdots & \cdots & \cdots \\
0 & 0 & a_{n,3}^{ii} & \cdots & \cdots & a_{n,n}^{ii}
\end{pmatrix} \tag{2.89}$$

Step 3 The process above may be repeated with matrices obtained by removing two rows and two columns, three rows and three columns and so on till we are able to transform the coefficient matrix to upper triangle form $\mathbf{R}$.

$$\mathbf{R} = \begin{pmatrix}
\hat{a}_{1,1} & \hat{a}_{1,2} & \hat{a}_{1,3} & \cdots & \cdots & \hat{a}_{1,n} \\
0 & \hat{a}_{2,2} & \hat{a}_{2,3} & \cdots & \cdots & \hat{a}_{2,n} \\
0 & 0 & \hat{a}_{3,3} & \cdots & \cdots & \hat{a}_{3,n} \\
\cdots & \cdots & \cdots & \cdots & \cdots & \cdots \\
\cdots & \cdots & \cdots & \cdots & \cdots & \cdots \\
0 & 0 & 0 & \cdots & 0 & \hat{a}_{n,n}
\end{pmatrix} \tag{2.90}$$

It is also clear that $\mathbf{H}_{n-1}\mathbf{H}_{n-2}\ldots\mathbf{H}_2\mathbf{H}_1\mathbf{A} = \mathbf{R}$ and hence $\mathbf{Q}$ should be given by

$$\mathbf{Q} = (\mathbf{H}_{n-1}\mathbf{H}_{n-2}\ldots\mathbf{H}_2\mathbf{H}_1)^{-1} = \mathbf{H}_1^T\mathbf{H}_2^T\ldots\mathbf{H}_{n-1}^T \tag{2.91}$$

In summary the sequence of operations of householder transformation are:

- Start with pivot element $a_{1,1}$. Determine H_1 such that all elements below $a_{1,1}$ are zero.
- Move to the next pivot element $a_{2,2}$, consider the $(n-1) \times (n-1)$ matrix obtained by deleting the first row and first column and determine the second Householder matrix.
- Repeat the process to reduce the coefficient matrix into the product of an upper triangular matrix $\mathbf{R}$ and the orthogonal matrix $\mathbf{Q}$.
- Obtain the solution by back substitution.

Example 2.17

Solve the following set of linear equations

$$\begin{pmatrix} 2 & 1 & 1 & 4 \\ 0 & 2 & -1 & 0 \\ 1 & -4 & 4 & 7 \\ 2 & -3 & 2 & -5 \end{pmatrix} \begin{pmatrix} x_1 \\ x_2 \\ x_3 \\ x_4 \end{pmatrix} = \begin{pmatrix} 2 \\ 5 \\ 4 \\ 9 \end{pmatrix}$$

by QR factorization using Householder transformations followed by back substitution.

Solution :

Step 1 We show in detail the use of Householder transformation for putting zeros in the first column below the pivot. For this consider the column vector $\mathbf{a}_1$ given by

$$\mathbf{a}_1^T = \begin{pmatrix} 2 & 0 & 1 & 2 \end{pmatrix}$$

We then have

$$\mathbf{a}_1^T \mathbf{a}_1 = 2^2 + 0^2 + 1^2 + 2^2 = 9$$

$$\hat{a}_1 = -\text{sign}(a_{1,1})\sqrt{\mathbf{a}_1^T \mathbf{a}_1} = -\text{sign}(2)\sqrt{9} = -3$$

The unit vector $\mathbf{u}$ may now be constructed using Equations 2.84 and 2.85.

$$u_1 = \sqrt{\frac{\hat{a}_1 - a_1}{2\hat{a}_1}} = \sqrt{\frac{-3-2}{2 \times (-3)}} = 0.912871$$

$$u_2 = \frac{a_{21}}{(-2\hat{a}_1 u_1)} = \frac{0}{(-2 \times (-3)) \times 0.912871} = 0$$

$$u_3 = \frac{a_{31}}{(-2\hat{a}_1 u_1)} = \frac{1}{(-2 \times (-3)) \times 0.912871} = 0.182574$$

$$u_4 = \frac{a_{41}}{(-2\hat{a}_1 u_1)} = \frac{2}{(-2 \times (-3)) \times 0.912871} = 0.365148$$

The Householder matrix may now be written down as

$$H_1 = \mathbf{I} - 2\mathbf{u}\mathbf{u}^T$$

$$
= \begin{pmatrix} 1 & 0 & 0 & 0 \\ 0 & 1 & 0 & 0 \\ 0 & 0 & 1 & 0 \\ 0 & 0 & 0 & 1 \end{pmatrix} - 2 \begin{pmatrix} 0.912871 \\ 0 \\ 0.182574 \\ 0.365148 \end{pmatrix} \begin{pmatrix} 0.912871 \\ 0 \\ 0.182574 \\ 0.365148 \end{pmatrix}^{T}
$$

$$
= \begin{pmatrix} -0.666667 & 0 & -0.333333 & -0.666667 \\ 0 & 1 & 0 & 0 \\ -0.333333 & 0 & 0.933333 & -0.133333 \\ -0.666667 & 0 & -0.133333 & 0.733333 \end{pmatrix}
$$

Pre-multiplying $\mathbf{A}$ by $\mathbf{H}_1$ we get the following:

$$
\mathbf{H}_1\mathbf{A} = \begin{pmatrix} -3 & 2.666667 & -3.333333 & -1.666667 \\ 0 & 2 & -1 & 0 \\ 0 & -3.666667 & 3.133333 & 5.866667 \\ 0 & -2.3333337 & 0.266667 & -7.266667 \end{pmatrix}
$$

At this stage the first column has been brought to the desired form.

Step 2 The next step involves the second column of $\mathbf{H}_1\mathbf{A}$ consisting of the vector

$$
\mathbf{a}_2' = \begin{pmatrix} 2 & -3.666667 & -2.3333337 \end{pmatrix}^{T}
$$

The calculations follow a procedure similar to that used above to lead to $\mathbf{H}_2$ given by

$$
\mathbf{H}_2 = \begin{pmatrix} 1 & 0 & 0 & 0 \\ 0 & -0.418040 & 0.766406 & 0.487713 \\ 0 & 0.766406 & 0.585781 & -0.263594 \\ 0 & 0.487713 & -0.263594 & 0.832259 \end{pmatrix}
$$

At this stage we also have

$$
\mathbf{H}_2\mathbf{H}_1\mathbf{A} = \begin{pmatrix} -3 & 2.666667 & -3.333333 & -1.666667 \\ 0 & -4.784233 & 2.949503 & 0.952202 \\ 0 & 0 & 0.998750 & 5.352031 \\ 0 & 0 & -1.091705 & -7.594162 \end{pmatrix}
$$

Step 3 Lastly we manipulate the column vector given by

$$
\mathbf{a}_3' = \begin{pmatrix} 0.998750 & -1.091705 \end{pmatrix}^{T}
$$

We obtain $\mathbf{H}_3$ as

$$
\mathbf{H}_3 = \begin{pmatrix} 1 & 0 & 0 & 0 \\ 0 & 1 & 0 & 0 \\ 0 & 0 & -0.674997 & 0.737820 \\ 0 & 0 & 0.737820 & 0.674997 \end{pmatrix}
$$

At this stage we have

$$
\mathbf{H}_3\mathbf{H}_2\mathbf{H}_1\mathbf{A} = \mathbf{R} = \begin{pmatrix} -3 & 2.666667 & -3.333333 & -1.666667 \\ 0 & -4.784233 & 2.949503 & 0.952202 \\ 0 & 0 & -1.479635 & -9.215732 \\ 0 & 0 & 0 & -1.177204 \end{pmatrix}
$$

which is the upper triangle matrix we have been seeking to obtain. The given equations will then be replaced by $\mathbf{Rx} = \mathbf{H}_3\mathbf{H}_2\mathbf{H}_1\mathbf{b}$. Thus we get

$$
\begin{pmatrix}
-3 & 2.666667 & -3.333333 & -1.666667 \\
0 & -4.784233 & 2.949503 & 0.952202 \\
0 & 0 & -1.479635 & -9.215732 \\
0 & 0 & 0 & -1.177204
\end{pmatrix}
\begin{pmatrix}
x_1 \\
x_2 \\
x_3 \\
x_4
\end{pmatrix}
=
\begin{pmatrix}
-8.666667 \\
1.648935 \\
1.860207 \\
6.686519
\end{pmatrix}
$$

Step 4 The above set is solved easily by back substitution to get

$$
\mathbf{x}^T = \begin{pmatrix} -14.48 & 19.56 & 34.12 & -5.68 \end{pmatrix}
$$

Note that all the intermediate calculations show numbers rounded to six digits after the decimal point. However, the final solution vector is *not* rounded.

Cost of computation of QR factorization using Householder algorithm
Cost of determining the Householder matrix of rank i: $\sim O(i^2)$
Cost of determining all the Householder matrices would be : $\sim O(n^3)$
Cost of QR factorization using Householder method: $\sim O(\frac{4}{3}n^3)$

MATLAB program for determining Q and R matrices using Householder algorithm is given below

Program 2.5: *Householder Algorithm for QR factorization*

```
1  function [Q,R] = qrhouseholder(A)
2  % Input  A = coefficient matrix
3  % Output Q = orthogonal matrix
4  %        R = upper triangular matrix
5  n = size(A,1);              %rank of matrix A
6  QT = eye(n);                %initializing transpose of Q
7                              %as identity matrix of Rank n
8  R = A;                      %initializing R matrix
9  for i = 1:n-1               %computing Householder matrices
10                             % Hi to Hn-1
11     x = A(i:n,i);
12     x1 = -sign(A(i,i))*sqrt(x'*x);
13     u = zeros(n-i+1,1);             %initializing vector u
14     u(1) = sqrt((x1-x(1))/(2*x1)); %computing vector u
15     for j=2:n-i+1
16         u(j) = x(j)/(-2*u(1)*x1);
17     end
18 %Determine householder matrix Hi
19     u2 = 2*u*u';
20     H = eye(n);
21     H(i:n,i:n) = H(i:n,i:n) - u2;      % Hi
22     QT(i:n,:) = H(i:n,i:n)*QT(i:n,:);  %updating transpose of Q
23     A(i:n,:) = H(i:n,i:n)*A(i:n,:);    %updating A
24 end
25 Q = QT';                    %Q = Q'
26 R = A;
```

Use the program to solve the system of equations in Example 2.17.

```
A = [ 2  1  1  4;0  2 -1  0; 1 -4  4  7; 2 -3  2 -5];
B = [2; 5; 4; 9];
[Q,R] = qrhouseholder(A);
B1 = Q'*B;
x  = R\B1;
```

The output of the program is

```
Q =
   -0.6667      -0.5806      -0.3314      -0.3296
        0       -0.4180      -0.1575       0.8947
   -0.3333       0.6503      -0.6562       0.1884
   -0.6667       0.2555       0.6594       0.2354
R =
   -3.0000       2.6667      -3.3333      -1.6667
    0.0000      -4.7842       2.9495       0.9522
    0.0000       0.0000      -1.4796      -9.2157
    0.0000      -0.0000            0      -1.1772
x = [-14.4800;   19.5600;   34.1200;    -5.6800]
```

Alternatively, MATLAB provides several intrinsic functions to perform QR factorization. One such function is

```
[Q,R] = qr(A)
```

2.7.3   Givens rotation and QR factorization

Givens rotation method is similar to Householder algorithm where a number of orthogonal matrices known as the Givens matrices[8] multiply the coefficient matrix to reduce it to upper triangular form $\mathbf{R}$. The product of the Givens matrices are related to the orthogonal matrix $\mathbf{Q}$.

Givens matrix is a 2×2 matrix given by

$$\mathbf{G} = \begin{pmatrix} \cos\theta & \sin\theta \\ -sin\theta & \cos\theta \end{pmatrix} \tag{2.92}$$

where θ is any angle. Givens matrix is orthogonal since $\mathbf{G}^T\mathbf{G} = \mathbf{I}_2$.
A vector $\mathbf{x}^T = \begin{pmatrix} x_1 & x_2 \end{pmatrix}$ will then be rotated by an angle equal to θ and transformed to $\mathbf{x}'^T = \begin{pmatrix} (x_1\cos\theta + x_2\sin\theta) & (-x_1\sin\theta + x_2\cos\theta) \end{pmatrix}$. However there is no change in the magnitude of the vector i.e. $||\mathbf{x}|| = ||\mathbf{x}'||$. Consider now a matrix of the form (example given here is a 6×6 matrix)

$$\mathbf{G}(3,5,\theta) = \begin{pmatrix} 1 & 0 & 0 & 0 & 0 & 0 \\ 0 & 1 & 0 & 0 & 0 & 0 \\ 0 & 0 & \cos\theta & 0 & \sin\theta & 0 \\ 0 & 0 & 0 & 1 & 0 & 0 \\ 0 & 0 & -sin\theta & 0 & \cos\theta & 0 \\ 0 & 0 & 0 & 0 & 0 & 1 \end{pmatrix} \tag{2.93}$$

---

[8]Named after James Wallace Givens (1910-1993)- American mathematician

Note that this matrix has been obtained by modifying $\mathbf{I}_6$ by replacing suitable diagonal elements by $\cos\theta$ and correspondingly the suitable non-diagonal elements by $\sin\theta$ or $-\sin\theta$. In the present case $i = 3$ and $j = 5$ represent the two rows that are involved. If we premultiply a vector $\mathbf{x}$ by $\mathbf{G}(i,j,\theta)$ only the components x_i and x_j will be affected and effectively $\mathbf{x}^T = \begin{bmatrix} x_1 & x_2 & x_3 & x_4 & x_5 & x_6 \end{bmatrix}$ is transformed to $\mathbf{x}'^T = \begin{bmatrix} x_1 & x_2 & (x_3\cos\theta + x_5\sin\theta) & x_4 & (-x_3\sin\theta + x_5\cos\theta) & x_6 \end{bmatrix}$, in the present case. We note that there is no change in the magnitude of the vector since $\mathbf{G}(3,5,\theta)$ or in general $\mathbf{G}(i,j,\theta)$ is orthogonal.

With this background we are ready to look at Givens rotation and QR factorization. Consider a simple 2×2 matrix given by

$$\mathbf{A} = \begin{pmatrix} a_{1,1} & a_{1,2} \\ a_{2,1} & a_{2,2} \end{pmatrix} \tag{2.94}$$

If we premultiply $\mathbf{A}$ by the Givens matrix $\mathbf{G}$ (Equation 2.92) we get

$$\begin{pmatrix} \cos\theta & \sin\theta \\ -\sin\theta & \cos\theta \end{pmatrix} \begin{pmatrix} a_{1,1} & a_{1,2} \\ a_{2,1} & a_{2,2} \end{pmatrix} =$$
$$\begin{pmatrix} a_{1,1}\cos\theta + a_{2,1}\sin\theta & a_{1,2}\cos\theta + a_{2,2}\sin\theta \\ -a_{1,1}\sin\theta + a_{2,1}\cos\theta & -a_{1,2}\sin\theta + a_{2,2}\cos\theta \end{pmatrix} \tag{2.95}$$

We at once see that the resultant matrix becomes upper triangular if θ is chosen such that $-a_{1,1}\sin\theta + a_{2,1}\cos\theta = 0$ or $\tan\theta = \dfrac{a_{2,1}}{a_{1,1}}$.

Step 1 Consider now a $n \times n$ matrix $\mathbf{A}$ which needs to be brought to **QR** form. Assume that all matrix elements are nonzero and hence we would like to make $a_{n,1}$ zero by premultiplying $\mathbf{A}$ by a suitable Givens matrix $\mathbf{G}_1$. We need to manipulate only the rows $n-1$ and n and hence $\mathbf{G}_1 = \mathbf{G}(n-1,n,\theta_1)$ where we choose θ_1 such that $\tan\theta_1 = \dfrac{a_{n,1}}{a_{n-1,1}}$. To make matters simple let us again consider $n = 6$. We thus need $\mathbf{G}(5,6,\theta_1)$ with $\tan\theta_1 = \dfrac{a_{6,1}}{a_{5,1}}$ to put a zero in place of $a_{6,1}$. Thus we have

$$\mathbf{G}(5,6,\theta_1) = \begin{pmatrix} 1 & 0 & 0 & 0 & 0 & 0 \\ 0 & 1 & 0 & 0 & 0 & 0 \\ 0 & 0 & 1 & 0 & 0 & 0 \\ 0 & 0 & 0 & 1 & 0 & 0 \\ 0 & 0 & 0 & 0 & \cos\theta_1 & \sin\theta_1 \\ 0 & 0 & 0 & 0 & -\sin\theta_1 & \cos\theta_1 \end{pmatrix}$$

This operation will put a zero in place of a_{61} and also modify all the elements on the 5th and 6th rows. All other elements of $\mathbf{A}$ remain the same. Hence we would expect to see the following.

$$\mathbf{G}(5,6,\theta_1)\mathbf{A} = \begin{pmatrix} a_{1,1} & a_{1,2} & a_{1,3} & a_{1,4} & a_{1,5} & a_{1,6} \\ a_{2,1} & a_{2,2} & a_{2,3} & a_{2,4} & a_{2,5} & a_{2,6} \\ a_{3,1} & a_{3,2} & a_{3,3} & a_{3,4} & a_{3,5} & a_{3,6} \\ a_{4,1} & a_{4,2} & a_{4,3} & a_{4,4} & a_{4,5} & a_{4,6} \\ a_{5,1}^1 & a_{5,2}^1 & a_{5,3}^1 & a_{5,4}^1 & a_{5,5}^1 & a_{5,6}^1 \\ 0 & a_{6,2}^1 & a_{6,3}^1 & a_{6,4}^1 & a_{6,5}^1 & a_{6,6}^1 \end{pmatrix}$$

In the above superscript 1 indicates that the element has changed because of Givens rotation.

Step 2 In the next step we would like to put a zero in place of $a_{5,1}^1$ and this may be accomplished by pre-multiplying the above by $\mathbf{G}(4,5,\theta_2)$ where $\tan\theta_2 = \dfrac{a_{5,1}^1}{a_{4,1}}$. Then we should have

$$\mathbf{G}(4,5,\theta) = \begin{pmatrix} 1 & 0 & 0 & 0 & 0 & 0 \\ 0 & 1 & 0 & 0 & 0 & 0 \\ 0 & 0 & 1 & 0 & 0 & 0 \\ 0 & 0 & 0 & \cos\theta_2 & \sin\theta_2 & 0 \\ 0 & 0 & 0 & -\sin\theta_2 & \cos\theta_2 & 0 \\ 0 & 0 & 0 & 0 & 0 & 1 \end{pmatrix}$$

and

$$\mathbf{G}(4,5,\theta_2)\mathbf{G}(5,6,\theta)\mathbf{A} = \begin{pmatrix} a_{1,1} & a_{1,2} & a_{1,3} & a_{1,4} & a_{1,5} & a_{1,6} \\ a_{2,1} & a_{2,2} & a_{2,3} & a_{2,4} & a_{2,5} & a_{2,6} \\ a_{3,1} & a_{3,2} & a_{3,3} & a_{3,4} & a_{3,5} & a_{3,6} \\ a_{4,1}^1 & a_{4,2}^1 & a_{4,3}^1 & a_{4,4}^1 & a_{4,5}^1 & a_{4,6}^1 \\ 0 & a_{5,2}^2 & a_{5,3}^2 & a_{5,4}^2 & a_{5,5}^2 & a_{5,6}^2 \\ 0 & a_{6,2}^1 & a_{6,3}^1 & a_{6,4}^1 & a_{6,5}^1 & a_{6,6}^1 \end{pmatrix}$$

In the above entries with subscript ^{2} have undergone a "second" change. We see that eventually we should have

$$\begin{aligned} \mathbf{G}(1,2,\theta_5)\mathbf{G}(2,3,\theta_4)\mathbf{G}(3,4,\theta_3) \\ \mathbf{G}(4,5,\theta_2)\mathbf{G}(5,6,\theta_1)\mathbf{A} \end{aligned} = \begin{pmatrix} a_{1,1}^1 & a_{1,2}^1 & a_{1,3}^1 & a_{1,4}^1 & a_{1,5}^1 & a_{1,6}^1 \\ 0 & a_{2,2}^2 & a_{2,3}^2 & a_{2,4}^2 & a_{2,5}^2 & a_{2,6}^2 \\ 0 & a_{3,2}^2 & a_{3,3}^2 & a_{3,4}^2 & a_{3,5}^2 & a_{3,6}^2 \\ 0 & a_{4,2}^2 & a_{4,3}^2 & a_{4,4}^2 & a_{4,5}^2 & a_{4,6}^2 \\ 0 & a_{5,2}^2 & a_{5,3}^2 & a_{5,4}^2 & a_{5,5}^2 & a_{5,6}^2 \\ 0 & a_{6,2}^1 & a_{6,3}^1 & a_{6,4}^1 & a_{6,5}^1 & a_{6,6}^1 \end{pmatrix} \tag{2.96}$$

Step 3 Now we may consider the $n-1 \times n-1$ matrix obtained by removing the first row and first column of the transformed matrix and repeat a similar process to put zeroes below the element in the first column and first row of this matrix. In the above example of 6×6 matrix we would be considering the 5×5 matrix given by

$$\mathbf{A}^1 = \begin{pmatrix} a_{2,2}^2 & a_{2,3}^2 & a_{2,4}^2 & a_{2,5}^2 & a_{2,6}^2 \\ a_{3,2}^2 & a_{3,3}^2 & a_{3,4}^2 & a_{3,5}^2 & a_{3,6}^2 \\ a_{4,2}^2 & a_{4,3}^2 & a_{4,4}^2 & a_{4,5}^2 & a_{4,6}^2 \\ a_{5,2}^2 & a_{5,3}^2 & a_{5,4}^2 & a_{5,5}^2 & a_{5,6}^2 \\ a_{6,2}^1 & a_{6,3}^1 & a_{6,4}^1 & a_{6,5}^1 & a_{6,6}^1 \end{pmatrix}$$

which would eventually take the form

$$\mathbf{G}(1,2,\theta_5)\mathbf{G}(2,3,\theta_4) \\ \mathbf{G}(3,4,\theta_2)\mathbf{G}(4,5,\theta_1)\mathbf{A}^1 = \left\{ \begin{array}{cccccc} a_{2,2}^3 & a_{2,3}^3 & a_{2,4}^3 & a_{2,5}^3 & a_{2,6}^3 \\ 0 & a_{3,3}^4 & a_{3,4}^4 & a_{3,5}^4 & a_{3,6}^4 \\ 0 & a_{4,3}^4 & a_{4,4}^4 & a_{4,5}^4 & a_{4,6}^4 \\ 0 & a_{5,3}^4 & a_{5,4}^4 & a_{5,5}^4 & a_{5,6}^4 \\ 0 & a_{6,3}^2 & a_{6,4}^2 & a_{6,5}^2 & a_{6,6}^2 \end{array} \right\}$$

(2.97)

Step 4 We proceed sequentially with $n-2 \times n-2,\dots,2 \times 2$ matrices to put zeroes below the diagonal of matrix $\mathbf{A}$ to finally arrive at an upper triangular matrix R given by (in the present case of 6×6 matrix)

$$\mathbf{R} = \left\{ \begin{array}{cccccc} a_{1,1}^1 & a_{1,2}^1 & a_{1,3}^1 & a_{1,4}^1 & a_{1,5}^1 & a_{1,6}^1 \\ 0 & a_{2,2}^3 & a_{2,3}^3 & a_{2,4}^3 & a_{2,5}^3 & a_{2,6}^3 \\ 0 & 0 & a_{3,3}^5 & a_{3,4}^5 & a_{3,5}^5 & a_{3,6}^5 \\ 0 & 0 & 0 & a_{4,4}^7 & a_{4,5}^7 & a_{4,6}^7 \\ 0 & 0 & 0 & 0 & a_{5,5}^9 & a_{5,6}^9 \\ 0 & 0 & 0 & 0 & 0 & a_{6,6}^5 \end{array} \right\}$$

(2.98)

where the superscript indicates the number of changes undergone by an element of $\mathbf{A}$. A MATLAB program to perform one Givens rotation is given below

Program 2.6: *Givens rotation*

```
 1 function [G,R] = givensrotation(A,i,j1,j2)
 2 % Input   A :  coefficient  matrix
 3 %         Q :  orthogonal  matrix
 4 %         i :  column  number  of  elements
 5 %         j1: row  number  of  first  element
 6 %         j2: row  number  of  second  element
 7 % Output  G :  Givens  matrix  (2 x2  matrix)
 8 %         R :  transformed  matrix
 9 n = size(A,1);        % rank  of  matrix  A
10 G = zeros(2);         % initialize  Givens  matrix
11 A1= zeros(2,n);       % extract  two  rows  of  A
12 A1(1,:) = A(j1,:);    % involved  in  matrix  operation
13 A1(2,:) = A(j2,:);
14 R = A;                % initialize  R  matrix
15 theta = atan(A(j2,i)/A(j1,i)); % compute  θ
16 G(1,1) = cos(theta);  % elements  of  Givens  matrix
17 G(1,2) = sin(theta);  % elements  of  Givens  matrix
18 G(2,1) = -sin(theta); % elements  of  Givens  matrix
19 G(2,2) = cos(theta);  % elements  of  Givens  matrix
20 A1 = G*A1;            % perform  matrix  multiplication
21 R(j1,:) = A1(1,:);    % update  R  matrix
22 R(j2,:) = A1(2,:);
```

An example is worked out in detail to demonstrate the above.

Example 2.18

Obtain the upper triangle form of 4×4 matrix shown below by using Givens rotations.

$$\mathbf{A} = \begin{pmatrix} 2 & 1 & 1 & 4 \\ 0 & 2 & -1 & 0 \\ 1 & -4 & 4 & 7 \\ 2 & -3 & 2 & -5 \end{pmatrix}$$

Solution :

Step 1 Element 4,1 is zeroed first. We choose $\tan\theta_1 = \dfrac{2}{1} = 2$ to get $\theta_1 = \tan^{-1}(2) = 1.107149\,\text{rad}$, $\cos\theta_1 = 0.447214$, $\sin\theta_1 = 0.894427$ and hence the Givens matrix is

$$\mathbf{G}(3,4,1.107149) = \begin{pmatrix} 1 & 0 & 0 & 0 \\ 0 & 1 & 0 & 0 \\ 0 & 0 & 0.447214 & 0.894427 \\ 0 & 0 & -0.894427 & 0.447214 \end{pmatrix}$$

We then get

$$\mathbf{A}^1 = \mathbf{G}(3,4,1.107149)\mathbf{A} = \begin{pmatrix} 2 & 1 & 1 & 4 \\ 0 & 2 & -1 & 0 \\ 2.236068 & -4.472136 & 3.577709 & -1.341641 \\ 0 & 2.236068 & -2.683282 & -8.497058 \end{pmatrix}$$

Thus we have managed to put a zero in position 4,1.

Step 2 Next we would like to put a zero in position 3,1. Since element 2,1 is already zero we would like not to affect it. Hence we consider a second Givens matrix $\mathbf{G}(1,3,\theta_2)$ where $\tan\theta_2 = \dfrac{2.236068}{2} = 1.118034$, $\cos\theta_2 = 0.666667$, $\sin\theta_2 = 0.745356$ to get

$$\mathbf{G}(1,3,1.118034) = \begin{pmatrix} 0.666667 & 0 & 0.745356 & 0 \\ 0 & 1 & 0 & 0 \\ 0.745356 & 0 & 0.666667 & 0 \\ 0 & 0 & 0 & 1 \end{pmatrix}$$

Pre-multiplying $\mathbf{A}^1$ by the above matrix we get

$$\mathbf{A}^2 = \mathbf{G}(1,3,1.118034)\mathbf{A}^1 = \begin{pmatrix} 3 & -2.666667 & 3.333333 & 1.666667 \\ 0 & 2 & -1 & 0 \\ 0 & -3.726780 & 1.639783 & -3.875851 \\ 0 & 2.236068 & -2.683282 & -8.497058 \end{pmatrix}$$

Thus we have put zeros below the diagonal element in the first column.

Step 3 For further processing we drop the first row and first column and consider the 3×3 matrix

$$\mathbf{B} = \begin{pmatrix} 2 & -1 & 0 \\ -3.726780 & 1.639783 & -3.875851 \\ 2.236068 & -2.683282 & -8.497058 \end{pmatrix}$$

We put a zero in the position 3,1 of this matrix by using the following Givens matrix with
$\tan\theta_1 = \dfrac{2.236068}{-3.726780} = -0.540420$, $\cos\theta_1 = 0.857493$, $\sin\theta_1 = -0.514496$.

$$\mathbf{G}(2,3,-0.540420) = \begin{pmatrix} 1 & 0 & 0 \\ 0 & 0.857493 & -0.514496 \\ 0 & 0.514496 & 0.857493 \end{pmatrix}$$

Then we have

$$\mathbf{B}^1 = \mathbf{G}(2,3,-0.540420)\mathbf{B} = \begin{pmatrix} 2 & -1 & 0 \\ -4.346135 & 2.786639 & 1.048185 \\ 0 & -1.457233 & -9.280276 \end{pmatrix}$$

Step 4 Now we would like to put a zero in position 2,1 of above matrix. This is accomplished by constructing a Givens matrix $\mathbf{G}(1,2,\theta_2)$ where $\theta_2 = \tan^{-1}\dfrac{-4.346135}{2} = \tan^{-1}(-2.173067) = -1.139510\,\text{rad}$. Thus we have

$$\mathbf{G}(1,2,-1.139510) = \begin{pmatrix} 4.784233 & -2.949503 & -0.952202 \\ 0 & 0.173352 & -0.984860 \\ 0 & 0.984860 & 0.173352 \end{pmatrix}$$

We then get

$$\mathbf{B}^2 = \mathbf{G}(1,2,-1.139510)\mathbf{B}^1 = \begin{pmatrix} 4.784233 & -2.949503 & -0.952202 \\ 0 & 0.256498 & 0.438183 \\ 0 & -1.457233 & -9.280276 \end{pmatrix}$$

Step 5 Now that the second column has zeroes below the diagonal we consider lastly the 2×2 matrix given by
$$\mathbf{C} = \begin{pmatrix} 0.256498 & 0.438183 \\ -1.457233 & -9.280276 \end{pmatrix}$$

This is easily handled by the following Givens matrix with $\theta_1 = \tan^{-1}\dfrac{-1.457233}{0.256498} = \tan^{-1}(-5.681277) = -1.396564$. $\cos\theta_1 = 0.173352$, $\sin\theta_1 = -0.984860$

$$\mathbf{G}(1,2,-1.396564) = \begin{pmatrix} 0.173352 & -0.984860 \\ 0.984860 & 0.173352 \end{pmatrix}$$

We then get
$$\mathbf{C}^1 = \mathbf{G}(1,2,-1.396564)\mathbf{C} = \begin{pmatrix} 1.479635 & 9.215732 \\ 0 & -1.177204 \end{pmatrix}$$

We see that the desired upper triangle matrix is obtained as

$$\mathbf{R} = \begin{pmatrix} 3 & -2.666667 & 3.333333 & 1.666667 \\ 0 & 4.784233 & -2.949503 & -0.952202 \\ 0 & 0 & 1.479635 & 9.215732 \\ 0 & 0 & 0 & -1.177204 \end{pmatrix}$$

Apply Program 2.6 to obtain the upper triangular matrix.

```
A = [ 2   1   1   4
        0   2  -1   0
        1  -4   4   7
        2  -3   2  -5];
[G,R] = givensmatrix(A,1,3,4)
[G,R] = givensmatrix(R,1,1,3)
[G,R] = givensmatrix(R,2,3,4)
[G,R] = givensmatrix(R,2,2,3)
[G,R] = givensmatrix(R,3,3,4)
```

The output is given below

```
R =
        3.0000      -2.6667       3.3333       1.6667
       -0.0000       4.7842      -2.9495      -0.9522
       -0.0000       0.0000       1.4796       9.2157
        0.0000      -0.0000      -0.0000      -1.1772
```

It is noted that the $\mathbf{Q}$ and $\mathbf{R}$ obtained here is not identical to those obtained by Householder transformation. It is possible to have sign changes, with identical magnitude, in the matrices. However, the product of $\mathbf{Q}$ and $\mathbf{R}$ remains equal to $\mathbf{A}$ in both the cases.

We have obtained the upper triangle factor in the above example by the use of a sequence of Givens rotations. We would like to determine the factor $\mathbf{Q}$ now. The required information is easily obtained by using the Givens matrices that were computed above. We see that the following holds in the specific example considered above.

$$\left\{ \begin{array}{l} \underbrace{\mathbf{G}(1,2,-1.396564)}_{2\times 2} \ \underbrace{\mathbf{G}(1,2,-1.139510)\mathbf{G}(2,3,-0.540420)}_{3\times 3} \\ \underbrace{\mathbf{G}(1,3,0.666667)\mathbf{G}(3,4,1.107149)}_{4\times 4} \end{array} \right\} \mathbf{A} = \mathbf{R}$$

We may recast this as

$$\mathbf{A} = \left\{ \begin{array}{l} \underbrace{\mathbf{G}(1,2,-1.396564)}_{2\times 2} \ \underbrace{\mathbf{G}(1,2,-1.139510)\mathbf{G}(2,3,-0.540420)}_{3\times 3} \\ \underbrace{\mathbf{G}(1,3,1.118034)\mathbf{G}(3,4,1.107149)}_{4\times 4} \end{array} \right\}^{T} \mathbf{R} = \mathbf{QR}$$

Note that all the $\mathbf{G}'s$ are orthogonal and hence the inverse and transpose are identical. We thus recognize the term inside the bracket as $\mathbf{Q}$. The reader should note then that

$$\mathbf{Q} = \begin{array}{c} \underbrace{\mathbf{G}^{T}(3,4,1.107149)\mathbf{G}^{T}(1,3,1.118034)}_{4\times 4} \\ \underbrace{\mathbf{G}^{T}(2,3,-0.540420)\mathbf{G}^{T}(1,2,-1.139510)}_{3\times 3} \underbrace{\mathbf{G}^{T}(1,2,-1.396564)}_{2\times 2} \end{array}$$

Note that all matrices are made 4×4 by adding rows and columns, to the left and above, as necessary, those of an identity matrix. The following shows how it is to be done.

Example 2.19

Obtain the **Q** *matrix in Example 2.18.*

Solution :

Transposes of all Givens matrices in the above example are written down first.

4×4 Matrices:

$$\mathbf{G}^T(3,4,1.107149) = \mathbf{G}_1^T = \begin{pmatrix} 1 & 0 & 0 & 0 \\ 0 & 1 & 0 & 0 \\ 0 & 0 & 0.447214 & -0.894427 \\ 0 & 0 & 0.894427 & 0.447214 \end{pmatrix}$$

$$\mathbf{G}^T(1,3,1.118034) = \mathbf{G}_2^T = \begin{pmatrix} 0.666667 & 0 & -0.745356 & 0 \\ 0 & 1 & 0 & 0 \\ 0.745356 & 0 & 0.666667 & 0 \\ 0 & 0 & 0 & 1 \end{pmatrix}$$

3×3 Matrices:

$$\mathbf{G}^T(1,2,-0.540420) = \mathbf{G}_3^T = \begin{pmatrix} 1 & 0 & 0 \\ 0 & 0.857493 & 0.514496 \\ 0 & -0.514496 & 0.857493 \end{pmatrix}$$

$$\mathbf{G}^T(1,2,-1.139510) = \mathbf{G}_4^T = \begin{pmatrix} 0.418040 & 0.908429 & 0 \\ -0.908429 & 0.418040 & 0 \\ 0 & 0 & 1 \end{pmatrix}$$

2×2 matrix: Finally we have

$$\mathbf{G}^T(1,2,-1.396564) = G_5^T = = \begin{pmatrix} 0.173352 & 0.984860 \\ -0.984860 & 0.173352 \end{pmatrix}$$

Conversion to 4×4: By adding columns to the left and rows above, we get:

$$\mathbf{G}_3^T = \begin{pmatrix} \mathbf{1} & \mathbf{0} & \mathbf{0} & \mathbf{0} \\ \mathbf{0} & \mathbf{1} & 0 & 0 \\ \mathbf{0} & \mathbf{0} & 0.857493 & 0.514496 \\ \mathbf{0} & \mathbf{0} & -0.514496 & 0.857493 \end{pmatrix}$$

$$\mathbf{G}_4^T = \begin{pmatrix} \mathbf{1} & \mathbf{0} & \mathbf{0} & \mathbf{0} \\ \mathbf{0} & 0.418040 & 0.908429 & 0 \\ \mathbf{0} & -0.908429 & 0.418040 & 0 \\ \mathbf{0} & \mathbf{0} & \mathbf{0} & 1 \end{pmatrix}$$

$$\mathbf{G}_5^T = \begin{pmatrix} \mathbf{1} & \mathbf{0} & \mathbf{0} & \mathbf{0} \\ \mathbf{0} & \mathbf{1} & \mathbf{0} & \mathbf{0} \\ \mathbf{0} & \mathbf{0} & 0.173352 & 0.984860 \\ \mathbf{0} & \mathbf{0} & -0.984860 & 0.173352 \end{pmatrix}$$

In the above the added entries are shown **bold**. We finally have

$$\mathbf{G}_1^T \mathbf{G}_2^T \mathbf{G}_3^T \mathbf{G}_4^T \mathbf{G}_5^T = \mathbf{Q} = \begin{pmatrix} 0.666667 & 0.580611 & 0.331360 & -0.329617 \\ 0.000000 & 0.418040 & 0.157478 & 0.894675 \\ 0.333333 & -0.650284 & 0.656158 & 0.188353 \\ 0.666667 & -0.255469 & -0.659438 & 0.235441 \end{pmatrix}$$

MATLAB program has been provided below to determine **Q**, **R** matrices using Givens rotation.

Program 2.7: *QR using Givens rotation*

```
1  function [Q,R] = qrgivens(A)
2  n = size(A,1);           % rank of matrix A
3  QT = eye(n);             % initializing transpose of Q matrix
4  R = A;                   % initializing R
5  A1 = zeros(2,n);         % matrix A1 will be used for matrix
                            % multiplication
6  for i=1:n-1
7      for j=n-1:-1:i
8      A1(1,:) = QT(i,:);   % involved in matrix operation
9      A1(2,:) = QT(j+1,:);
10     [G,R] = givensmatrix(R,i,i,j+1); % Givens rotation
11     A1 = G*A1;           % perform matrix multiplication
12     QT(i,:) = A1(1,:);   % update QT
13     QT(j+1,:) = A1(2,:); % update QT
14     end
15 end
16 Q=QT';
```

Cost of computation of QR factorization using Givens rotation

Consider i^{th} column. Matrix multiplication is the most expensive operation involved

Number of computations to be performed for matrix multiplication with Givens matrix for i^{th} column: $6(n-i+1)$

Number of rows for which Givens rotation has to be performed = $n-i$

Total number of operations for determining **R** and $\mathbf{Q} = 2\sum_{i=1}^{n-1} 6(n-i+1)(n-i) \sim \frac{4}{3}n^3$

It is apparent that if **A** is a dense matrix a large number of Givens rotations are needed to obtain its **QR** factorization. A large number of matrix multiplications are involved. Hence the Householder method is advantageous in such situations. However the Givens method is useful if the coefficient matrix is sparse, as for example, when it is tridiagonal. The following example will demonstrate this.

Example 2.20

Solve the follow tridiagonal system of equations by **QR** *factorization using Givens rotations.*

$$\begin{pmatrix} -2.0625 & 1 & 0 & 0 \\ 1 & -2.0625 & 1 & 0 \\ 0 & 1 & -2.0625 & 1 \\ 0 & 0 & 1 & -1.03125 \end{pmatrix} \begin{pmatrix} x_1 \\ x_2 \\ x_3 \\ x_4 \end{pmatrix} = \begin{pmatrix} -1 \\ 0 \\ 0 \\ 0 \end{pmatrix}$$

Solution :

Step 1 The coefficient matrix is a 4×4 square matrix. Because it is sparse only three Givens rotations are required for **QR** factorization. We make element 2,1 zero by choosing $\theta_1 = \tan^{-1}\dfrac{1}{-2.0625} = \tan^{-1}(-0.484848) = -0.451453$ rad. The Givens matrix is given by

$$\mathbf{G}_1 = \mathbf{G}(1,2,-0.451453) = \begin{pmatrix} 0.899814 & -0.436274 & 0 & 0 \\ 0.436274 & 0.899814 & 0 & 0 \\ 0 & 0 & 1 & 0 \\ 0 & 0 & 0 & 1 \end{pmatrix}$$

This transforms the coefficient matrix to

$$\mathbf{A}_1 = \mathbf{G}_1\mathbf{A} = \begin{pmatrix} -2.292140 & 1.799628 & -0.436274 & 0 \\ 0 & -1.419593 & 0.899814 & 0 \\ 0 & 1 & -2.0625 & 1 \\ 0 & 0 & 1 & -1.03125 \end{pmatrix}$$

Step 2 We make element 3,2 of $\mathbf{A}_1$ zero by choosing $\theta_2 = \tan^{-1}\dfrac{1}{-1.419593} = \tan^{-1}(-0.704427) = -0.613691$ rad. The Givens matrix is given by

$$\mathbf{G}_2 = \mathbf{G}(3,2,-0.613691) = \begin{pmatrix} 1 & 0 & 0 & 0 \\ 0 & 0.817528 & -0.575889 & 0 \\ 0 & 0.575889 & 0.817528 & 0 \\ 0 & 0 & 0 & 1 \end{pmatrix}$$

This transforms the coefficient matrix to

$$\mathbf{A}_2 = \mathbf{G}_2\mathbf{A}_1 = \begin{pmatrix} -2.292140 & 1.799628 & -0.436274 & 0 \\ 0 & -1.736446 & 1.923394 & -0.575889 \\ 0 & 0 & -1.167958 & 0.817528 \\ 0 & 0 & 1 & -1.03125 \end{pmatrix}$$

Step 3 We make element 4,3 of $\mathbf{A}_2$ zero by choosing $\theta_3 = \tan^{-1}\dfrac{1}{-1.167958} = \tan^{-1}(-0.856195) = -0.708080$ rad. The Givens matrix is given by

$$\mathbf{G}_3 = \mathbf{G}(4,3,-0.708080) = \begin{pmatrix} 1 & 0 & 0 & 0 \\ 0 & 1 & 0 & 0 \\ 0 & 0 & 0.759612 & -0.650376 \\ 0 & 0 & 0.650376 & 0.759612 \end{pmatrix}$$

This transforms the coefficient matrix to

$$\mathbf{A}_3 = \mathbf{G}_3\mathbf{A}_2 = \mathbf{R} = \begin{pmatrix} -2.292140 & 1.799628 & -0.436274 & 0 \\ 0 & -1.736446 & 1.923394 & -0.575889 \\ 0 & 0 & -1.537572 & 1.291705 \\ 0 & 0 & 0 & -0.251650 \end{pmatrix}$$

Step 4 The transpose of the **Q** is then obtained directly as $\mathbf{Q}^T = \mathbf{G}_3\mathbf{G}_2\mathbf{G}_1$ as

$$\mathbf{Q}^T = \mathbf{G}_3\mathbf{G}_2\mathbf{G}_1 = \begin{pmatrix} 0.899814 & -0.436274 & 0 & 0 \\ 0.356666 & 0.735623 & -0.575889 & 0 \\ 0.190849 & 0.393626 & 0.621004 & -0.650376 \\ 0.163404 & 0.337020 & 0.531701 & 0.759612 \end{pmatrix}$$

Step 5 The **b** matrix is transformed to

$$(\mathbf{Q}^T\mathbf{b})^T = \mathbf{b}_1^T = \begin{pmatrix} -0.899814 & -0.356666 & -0.190849 & -0.163404 \end{pmatrix}$$

and finally we get the following system of equations.

$$\begin{pmatrix} -2.292140 & 1.799628 & -0.436274 & 0 \\ 0 & -1.736446 & 1.923394 & -0.575889 \\ 0 & 0 & -1.537572 & 1.291705 \\ 0 & 0 & 0 & -0.251650 \end{pmatrix} \begin{pmatrix} x_1 \\ x_2 \\ x_3 \\ x_4 \end{pmatrix} = \begin{pmatrix} -0.899814 \\ -0.356666 \\ -0.190849 \\ -0.163404 \end{pmatrix}$$

Step 6 We solve these by back substitution to get

$$\mathbf{x}^T = \begin{pmatrix} 0.839644 & 0.731766 & 0.669623 & 0.649331 \end{pmatrix}$$

2.8 Iterative methods of solution

The methods discussed so far have considered reorganizing the coefficient matrix such that simple backward or forward substitutions are performed to obtain the desired solution. These methods directly give the solution. There are another class of solvers where the solution is determined iteratively. A solution vector is initially assumed and through a series of operations the solution vector is systematically corrected towards the solution. Once the error is within acceptable limits, the iterations are stopped. To make the iterative process stable, a **preconditioner** step (similar to pivoting) can be applied to the system of equations.

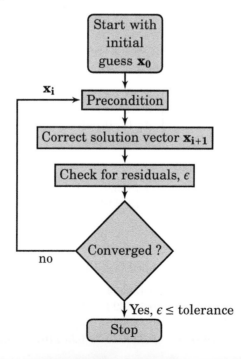

2.8.1 Jacobi and Gauss Seidel methods

Consider the linear system $\mathbf{Ax} = \mathbf{b}$. We may write the coefficient matrix as $\mathbf{A} = \mathbf{L} + \mathbf{D} + \mathbf{U}$ where $\mathbf{L}$ is strict lower triangular[9], $\mathbf{D}$ is diagonal and $\mathbf{U}$ is strict upper triangular [10]. These are respectively given by the following, for a 7×7 matrix (as an example):

$$\mathbf{L} = \begin{pmatrix} 0 & 0 & 0 & 0 & 0 & 0 & 0 \\ a_{2,1} & 0 & 0 & 0 & 0 & 0 & 0 \\ a_{3,1} & a_{3,2} & 0 & 0 & 0 & 0 & 0 \\ a_{4,1} & a_{4,2} & a_{4,3} & 0 & 0 & 0 & 0 \\ a_{5,1} & a_{5,2} & a_{5,3} & a_{5,4} & 0 & 0 & 0 \\ a_{6,1} & a_{6,2} & a_{6,3} & a_{6,4} & a_{6,5} & 0 & 0 \\ a_{7,1} & a_{7,2} & a_{7,3} & a_{7,4} & a_{7,5} & a_{7,6} & 0 \end{pmatrix}$$

$$\mathbf{D} = \begin{pmatrix} a_{1,1} & 0 & 0 & 0 & 0 & 0 & 0 \\ 0 & a_{2,2} & 0 & 0 & 0 & 0 & 0 \\ 0 & 0 & a_{3,3} & 0 & 0 & 0 & 0 \\ 0 & 0 & 0 & a_{4,4} & 0 & 0 & 0 \\ 0 & 0 & 0 & 0 & a_{5,5} & 0 & 0 \\ 0 & 0 & 0 & 0 & 0 & a_{6,6} & 0 \\ 0 & 0 & 0 & 0 & 0 & 0 & a_{7,7} \end{pmatrix}$$

and

$$\mathbf{U} = \begin{pmatrix} 0 & a_{1,2} & a_{1,3} & a_{1,4} & a_{1,5} & a_{1,6} & a_{1,7} \\ 0 & 0 & a_{2,3} & a_{2,4} & a_{2,5} & a_{2,6} & a_{2,7} \\ 0 & 0 & 0 & a_{3,4} & a_{3,5} & a_{3,6} & a_{3,7} \\ 0 & 0 & 0 & 0 & a_{4,5} & a_{4,6} & a_{4,7} \\ 0 & 0 & 0 & 0 & 0 & a_{5,6} & a_{5,7} \\ 0 & 0 & 0 & 0 & 0 & 0 & a_{6,7} \\ 0 & 0 & 0 & 0 & 0 & 0 & 0 \end{pmatrix}$$

We then rewrite the given equation set as

$$\mathbf{Dx} = \mathbf{b} - (\mathbf{L} + \mathbf{U})\mathbf{x}$$

We then write the solution in the form

$$\mathbf{x} = \mathbf{D}^{-1}[\mathbf{b} - (\mathbf{L} + \mathbf{U})\mathbf{x}] \tag{2.99}$$

This way of writing the equation makes it possible to obtain the solution by an iterative process where a guess value $\mathbf{x}^k$ is used on the right hand side to obtain an improved value $\mathbf{x}^{k+1}$ as

$$\mathbf{x}^{k+1} = \mathbf{D}^{-1}[\mathbf{b} - (\mathbf{L} + \mathbf{U})\mathbf{x}^k] \tag{2.100}$$

---

[9] Strict lower triangular matrix has non-zero elements for $j < i$, zero elements for $i \geq j$

[10] Strict upper triangular matrix has non-zero elements for $j > i$, zero elements for $i \leq j$

Noting that $\mathbf{D}^{-1}$ is also diagonal (Section 2.2.2) with elements represented by $\dfrac{1}{a_{i,i}}$ the above may be rewritten in the more familiar form

$$x_i^{k+1} = \frac{b_i - \sum_{j=1}^{i-1} a_{i,j} x_j^k - \sum_{j=i+1}^{n} a_{i,j} x_j^k}{a_{i,i}} \tag{2.101}$$

This constitutes what is known as the Jacobi iteration scheme.[11] Iteration stops when the change $|x_i^{k+1} - x_i^k| \le \varepsilon$ for $1 \le i \le n$ where ε is a prescribed small number.

A variant of the above is the Gauss Seidel[12] scheme which recognizes that x_j^{k+1} for $i < j$ are available during the iteration process and writes the equation set as

$$\mathbf{x}^{k+1} = \mathbf{D}^{-1}[\mathbf{b} - \mathbf{L}\mathbf{x}^{k+1} - \mathbf{U}\mathbf{x}^k] \tag{2.102}$$

The above may be rewritten in the more familiar form

$$x_i^{k+1} = \frac{b_i - \sum_{j=1}^{i-1} a_{i,j} x_j^{k+1} - \sum_{j=i+1}^{n} a_{i,j} x_j^k}{a_{i,i}} \tag{2.103}$$

Another variant is the method of successive under/over-relaxation that is constructed by writing the update as a weighted sum of the old and new values. Thus we would replace Equation 2.103 by

$$\mathbf{x}^{k+1} = (1 - \omega)\mathbf{x}^k + \omega \mathbf{D}^{-1}[\mathbf{L}\mathbf{x}^{k+1} - \mathbf{U}\mathbf{x}^k] \tag{2.104}$$

where ω is known as the relaxation parameter. $\omega < 1$ indicates underrelaxation, $\omega = 1$ represents Gauss Seidel iteration and $\omega > 1$ represents overrelaxation. This may be written in the more convenient form

$$x_i^{k+1} = (1 - \omega)x_i^k + \omega \frac{b_i - \sum_{j=1}^{i-1} a_{i,j} x_j^{k+1} - \sum_{j=i+1}^{n} a_{i,j} x_j^k}{a_{i,i}} \tag{2.105}$$

If the coefficient matrix is diagonal dominant, the above methods will converge. In the case of successive over-relaxation method $\omega < 2$. For determining the optimum value for ω see reference[13].

Program 2.8: *Gauss Seidel for system of linear equations*

```
1  function X  = gaussSeidel(A,B,Xg,relax,tolerance)
2  %    Input :
3  %              A:      coefficient matrix
4  %              B:      right hand Vector
5  %              Xg:     guess for output vector
6  %              relax:  relaxation factor
```

[11]after Carl Gustav Jacob Jacobi, 1804-1851, a German mathematician,
[12]after Philipp Ludwig von Seidel, 1821-1896, a German mathematician
[13]Reid, J. K. "A method for finding the optimum successive over-relaxation parameter." The Computer Journal 9.2 (1966): 200-204.

```
 7 %                tolerance
 8 %    Output :
 9 %            X:        output Vector
10 n = size(A,1);                    % rank of matrix A
11 r = B-A*Xg;                       % residual
12 X = Xg;                           % initialize output vector
13 s = r'*r;                         % initial error
14 count = 0;                        % initialize iteration count
15 while( s > tolerance )            % loop until error < tolerance
16     for i=1:n                         % loop for each variable
17         sum = B(i);
18         for j=1:n                      % inner loop
19             if( i~=j)
20                 sum = sum - A(i,j)*X(j);  % for Gauss Seidel
21                 %sum = sum - A(i,j)*Xg(j); % for Jacobi
22             end
23         end
24         X(i) = (1-relax)*Xg(i)+relax*sum/A(i,i);
25                                     % update output vector
26     end
27     r = B-A*X;                       % update residual
28     s = r'*r;                        % update error
29     Xg = X;
30     count = count + 1;              % update iteration count
31 end
```

Example 2.21

Figure 2.4 shows a resistance network connected to a battery that supplies 9 V across its terminals. It is desired to determine the voltages with respect to ground at the nodes labeled 2 -5 using Gauss Seidel iteration.

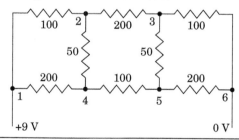

Figure 2.4: *Resistance network for Example 2.21 (All resistances are in Ω)*

Solution :

Step 1 Determining nodal equations Nodal equations are obtained by applying Kirchoff's current law to the respective nodes. The current law states that the sum of all the currents entering a node must be zero. Ohms law is used to calculate the currents. Consider node 2 as an example. We take $v_1 = 9\,V$ and get

$$\frac{9-v_2}{100} + \frac{v_3-v_2}{200} + \frac{v_4-v_2}{50} = 0$$

This may be rewritten as

$$-0.035v_2 + 0.005v_3 + 0.02v_4 = -0.09$$

Similarly all the other nodal equations may be derived to get the following set of equations.

$$
\begin{pmatrix}
-0.035 & 0.005 & 0.02 & 0 \\
0.005 & -0.035 & 0 & 0.02 \\
0.02 & 0 & -0.035 & 0.01 \\
0 & 0.02 & 0.01 & -0.035
\end{pmatrix}
\begin{pmatrix}
v_2 \\
v_3 \\
v_4 \\
v_5
\end{pmatrix}
=
\begin{pmatrix}
-0.09 \\
0 \\
-0.045 \\
0
\end{pmatrix}
$$

Note that the set of equations are diagonal dominant. These equations may be recast in a form suitable for Gauss Seidel iteration. Thus

$$
v_2^{k+1} = \frac{-0.09 - 0.005 v_3^k - 0.02 v_4^k}{-0.035}
$$

$$
v_3^{k+1} = \frac{-0.005 v_2^{k+1} - 0.02 v_5^k}{-0.035}
$$

$$
v_4^{k+1} = \frac{-0.045 - 0.02 v_2^{k+1} - 0.01 v_5^k}{-0.035}
$$

$$
v_5^{k+1} = \frac{-0.02 v_3^{k+1} - 0.01 v_4^{k+1}}{-0.035}
$$

Step 2 We start Gauss Seidel iteration with the initial set given by

$$
\mathbf{v}^T = \begin{pmatrix} 6 & 3 & 6 & 3 \end{pmatrix}
$$

where all nodal voltages are in V. The calculations have been done using a spreadsheet and the results are shown in the following table. We track the change in each nodal voltage between iterations and stop the calculations when the change is less than a mV.

Iteration No.	0	1	2	3	4
v_2	6	6.429	6.271	6.257	6.260
v_3	3	2.633	2.705	2.720	2.725
v_4	6	5.816	5.774	5.774	5.778
v_5	3	3.166	3.195	3.204	3.208
Change					
Δv_2		0.429	-0.157	-0.014	0.002
Δv_3		-0.367	0.072	0.015	0.005
Δv_4		-0.184	-0.042	0.000	0.004
Δv_5		0.166	0.029	0.009	0.004

Iteration No.	5	6	7	8
v_2	6.262	6.264	6.266	6.267
v_3	2.728	2.729	2.730	2.731
v_4	5.781	5.783	5.784	5.785
v_5	3.210	3.212	3.213	3.213
Change				
Δv_2	0.003	0.002	0.001	0.000
Δv_3	0.003	0.002	0.001	0.000
Δv_4	0.003	0.002	0.001	0.000
Δv_5	0.002	0.001	0.001	0.000

The solution set is thus given by

$$\mathbf{v}^T = \begin{pmatrix} 6.267 & 2.731 & 5.785 & 3.213 \end{pmatrix}$$

Cost of computation of <u>one iteration</u> of Gauss Seidel/Jacobi method
Number of operations for one unknown variable = addition '+' = n
multiplication '×' = n-1
division '/' = 1
Total number of operations for one unknown variables = $2n$
Total number of operations for n unknown variables = $2n^2$

Example 2.22

The following sparse but not tridiagonal set of equations need to be solved by iterative methods.

$$\begin{pmatrix} 4 & -1 & 0 & 0 & -1 & 0 & 0 & 0 \\ -1 & 4 & -1 & 0 & 0 & -1 & 0 & 0 \\ 0 & -1 & 4 & -1 & 0 & 0 & -1 & 0 \\ 0 & 0 & -2 & 4 & 0 & 0 & 0 & -1 \\ -1 & 0 & 0 & 0 & 4 & -1 & 0 & 0 \\ 0 & -1 & 0 & 0 & -1 & 4 & -1 & 0 \\ 0 & 0 & -1 & 0 & 0 & -1 & 4 & -1 \\ 0 & 0 & 0 & -1 & 0 & 0 & -2 & 4 \end{pmatrix} \begin{pmatrix} x_1 \\ x_2 \\ x_3 \\ x_4 \\ x_5 \\ x_6 \\ x_7 \\ x_8 \end{pmatrix} = \begin{pmatrix} 150 \\ 100 \\ 100 \\ 100 \\ 100 \\ 50 \\ 50 \\ 50 \end{pmatrix}$$

Compare the convergence rates of Jacobi, Gauss Seidel and SOR techniques in this case.

Solution :

Case 1: Jacobi iterations
We write below the equations for performing Jacobi iterations. Values before performing an iteration are identified as 'old' and those after performing an iteration are identified as 'new'.

$$x_1^{new} = \frac{x_2^{old} + x_5^{old} + 150}{4} \qquad x_2^{new} = \frac{x_1^{old} + x_3^{old} + x_6^{old} + 100}{4}$$

$$x_3^{new} = \frac{x_2^{old} + x_4^{old} + x_7^{old} + 100}{4} \qquad x_4^{new} = \frac{2x_3^{old} + x_8^{old} + 150}{4}$$

$$x_5^{new} = \frac{x_1^{old} + x_6^{old} + 100}{4} \qquad x_6^{new} = \frac{x_2^{old} + x_5^{old} + x_7^{old} + 50}{4}$$

$$x_7^{new} = \frac{x_3^{old} + x_6^{old} + x_8^{old} + 50}{4} \qquad x_8^{new} = \frac{x_4^{old} + 2x_7^{old} + 150}{4}$$

We initialize the vector as $\mathbf{x}^{T,old} = \begin{pmatrix} 80 & 80 & 80 & 80 & 60 & 60 & 60 & 60 \end{pmatrix}$. Tolerance is set as $\epsilon = 0.001$. Convergence is achieved after 21 iterations to the solution vector given by $\mathbf{x}^{T,new} = \begin{pmatrix} 72.019 & 79.246 & 81.658 & 82.237 & 58.831 & 63.307 & 65.149 & 65.634 \end{pmatrix}$.

Case 2: Gauss Seidel iterations

We write below the equations used for performing Gauss Seidel iterations.

$$x_1^{new} = \frac{x_2^{old} + x_5^{old} + 150}{4} \qquad x_2^{new} = \frac{x_1^{new} + x_3^{old} + x_6^{old} + 100}{4}$$

$$x_3^{new} = \frac{x_2^{new} + x_4^{old} + x_7^{old} + 100}{4} \qquad x_4^{new} = \frac{2x_3^{new} + x_8^{old} + 150}{4}$$

$$x_5^{new} = \frac{x_1^{new} + x_6^{old} + 100}{4} \qquad x_6^{new} = \frac{x_2^{new} + x_5^{new} + x_7^{old} + 50}{4}$$

$$x_7^{new} = \frac{x_3^{new} + x_6^{new} + x_8^{old} + 50}{4} \qquad x_8^{new} = \frac{x_4^{new} + 2x_7^{new} + 150}{4}$$

We initialize the vector as in Case 1. Tolerance is again set as $\epsilon = 0.001$. Convergence is achieved after 14 iterations to the solution vector given earlier in Case 1.

Case 3: Iteration with overrelaxation

We choose a relaxation parameter of $\omega = 1.2$ and perform iterations to obtain the solution. The initial vector is the same as that used in the above two cases. Tolerance is again set as $\epsilon = 0.001$. Convergence is achieved after 11 iterations to the solution vector given earlier in Case 1. Figure 2.5 compares the rate of change of x_1 with iterations for the three schemes.

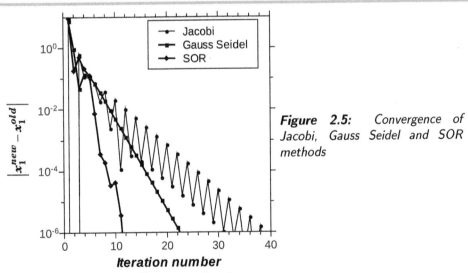

Figure 2.5: *Convergence of Jacobi, Gauss Seidel and SOR methods*

2.8.2 Conjugate Gradient method

In many engineering applications we need to solve a large set of equations where the coefficient matrix is symmetric, sparse, diagonally dominant and hence positive definite. The given equations are equivalent to the determination of the optimum of a quadratic form[14] as will be shown below.

---

[14]Quadratic form is a polynomial of degree 2. For example, in two dimensions, $f(x_1, x_2) = 2x_1^2 + 2x_1x_2 + 3x_2^2 - 3x_1 + 9x_2$ is a quadratic form. This may be written in matrix notation indicated by Equation 2.106 where $\mathbf{x}^T = [x_1 \ x_2]$, $\mathbf{A} = \begin{pmatrix} 4 & 2 \\ 2 & 6 \end{pmatrix}$, $\mathbf{b}^T = [3 \ -9]$ and $\mathbf{c}$ is the null vector.

Consider a quadratic form given by

$$f(\mathbf{x}) = \frac{1}{2}\mathbf{x}^T\mathbf{A}\mathbf{x} - \mathbf{b}^T\mathbf{x} + \mathbf{c} \tag{2.106}$$

where $\mathbf{A}$ is an $n \times n$ symmetric matrix, $\mathbf{b}$ is a vector, $\mathbf{x}$ is the vector that needs to be determined and $\mathbf{c}$ is a constant vector. If we differentiate the first term on the right hand side with respect to $\mathbf{x}$ we get

$$\frac{d}{d\mathbf{x}}\left(\frac{1}{2}\mathbf{x}^T\mathbf{A}\mathbf{x}\right) = \frac{1}{2}(\mathbf{A}\mathbf{x} + \mathbf{A}^T\mathbf{x}) = \mathbf{A}\mathbf{x}$$

since $\mathbf{A}^T = \mathbf{A}$ for a symmetric matrix. Differentiate Equation 2.106 with respect to $\mathbf{x}$ to get

$$\frac{df(\mathbf{x})}{d\mathbf{x}} = f'(\mathbf{x}) = \mathbf{A}\mathbf{x} - \mathbf{b} \tag{2.107}$$

If the derivative given by Equation 2.107 vanishes the given set of linear equations are satisfied and the quadratic form given by Equation 2.106 has an optimum. Since we assumed the coefficient matrix to be positive definite the optimum will be a minimum of the quadratic form. Hence the solution represents a *critical point*[15] of the quadratic form. Therefore it is possible to solve such systems by the optimization algorithms such as Steepest Descent or the Conjugate Gradient (CG) method.

Prelude to the CG method: Steepest Descent method (SD) Before moving on to the CG we will introduce the SD that is easy to understand.

Step 1 Determine direction of descent
We start from a guess point $\mathbf{x}_i$ and slide down to the bottom of the paraboloid that is represented by the quadratic form. It certainly is not possible to move to the bottom in a single step since the direction of slide has to be the right one. All we can do once we have chosen a starting point is to move in a direction along which the quadratic form decreases most rapidly. Obviously the direction of choice is opposite the direction of $f'(\mathbf{x}_i)$ i.e. $-f'(\mathbf{x}_i) = \mathbf{b} - \mathbf{A}\mathbf{x}_i$. Of course if $f'(\mathbf{x}_i) = 0$ we are already at the critical point $\mathbf{x}^*$. However it may not be so and hence we represent the error as $\mathbf{e}_i = \mathbf{x}_i - \mathbf{x}^*$.
The residual $\mathbf{r}_i$ is defined as

$$\mathbf{r}_i = \mathbf{b} - \mathbf{A}\mathbf{x}_i = \mathbf{b} - \mathbf{A}\mathbf{x}^* - \mathbf{A}\mathbf{e}_i = -\mathbf{A}\mathbf{e}_i = -f'(\mathbf{x}_i) \tag{2.108}$$

Hence $\mathbf{r}_i$ is the direction of steepest descent.

Step 2 Determine the new point The direction of descent is known, but what should be the step by which we should move? Let the new point be $\mathbf{x}_{i+1} = \mathbf{x}_i + \alpha_i\mathbf{r}_i$. α_i is the 'step size', a scalar which has to be selected such that the error is minimum. Hence, the derivative of the function with respect to α_i must be zero which gives

$$\frac{df(\mathbf{x}_{i+1})}{d\alpha_i} = f'(\mathbf{x}_{i+1})\frac{d\mathbf{x}_{i+1}}{d\alpha_i} = -\mathbf{r}_{i+1}^T\mathbf{r}_i = 0 \tag{2.109}$$

---

[15]Critical point represents the location of an optimum. More on this later.

The above has used the defining equations for $\mathbf{r}_i$, $\mathbf{x}_{i+1}$, and rules of matrix multiplication to indicate that $\mathbf{r}_i$ and $\mathbf{r}_{i+1}$ are orthogonal. The above may further be rewritten as

$$\mathbf{r}_{i+1}^T \mathbf{r}_i = [\mathbf{b} - \mathbf{A}\mathbf{x}_{i+1}]^T \mathbf{r}_i = [\mathbf{b} - \mathbf{A}(\mathbf{x}_i + \alpha_i \mathbf{r}_i)]^T \mathbf{r}_i = \mathbf{r}_i^T \mathbf{r}_i - \alpha_i (\mathbf{A}\mathbf{r}_i)^T \mathbf{r}_i = 0$$

and solved for α_i to get

$$\alpha_i = \frac{\mathbf{r}_i^T \mathbf{r}_i}{(\mathbf{A}\mathbf{r}_i)^T \mathbf{r}_i} = \frac{\mathbf{r}_i^T \mathbf{r}_i}{\mathbf{r}_i^T \mathbf{A}\mathbf{r}_i} \tag{2.110}$$

We check if the solution has converged. We terminate the iterations if yes otherwise we continue the iterations.

In practice we reduce the number of arithmetical operations by writing $\mathbf{x}_{i+1} = \mathbf{x}_i + \alpha_i \mathbf{r}_i$ after pre-multiplication by $\mathbf{A}$ as

$$\mathbf{A}\mathbf{x}_{i+1} = \mathbf{A}\mathbf{x}_i + \alpha_i \mathbf{A}\mathbf{r}_i \quad \text{or} \quad \mathbf{r}_{i+1} = \mathbf{r}_i - \alpha_i \mathbf{A}\mathbf{r}_i \tag{2.111}$$

We present an example below on the use of SD method before moving on to the CG method.

Example 2.23

Solve the following linear system of equations by the Steepest Descent method.

$$\begin{pmatrix} 0.00425 & -0.00100 & -0.00125 \\ -0.00100 & 0.00467 & -0.00200 \\ -0.00125 & -0.00200 & 0.00425 \end{pmatrix} \begin{pmatrix} x_1 \\ x_2 \\ x_3 \end{pmatrix} = \begin{pmatrix} 0.018 \\ 0 \\ 0 \end{pmatrix}$$

Solution :

Step 1 Determine the direction of descent
We start from an arbitrarily chosen point given by $\mathbf{x}_0^T = [1 \quad 1 \quad 1]$. We calculate the residual at this point as

$$
\begin{aligned}
r_{10} &= 0.018 - 0.00425 + 0.00100 + 0.00125 = 0.01600 \\
r_{20} &= 0.00100 - 0.00467 + 0.00200 = -0.00167 \\
r_{30} &= 0.00125 + 0.00200 - 0.00425 = -0.00100
\end{aligned}
$$

$$\mathbf{r}_0^T = \begin{pmatrix} 0.01600 & -0.00167 & -0.00100 \end{pmatrix}$$

Step 2 Determine step size and update to a new point
Then we have

$$\mathbf{r}_0^T \mathbf{r}_0 = [0.01600 \quad -0.00167 \quad -0.001] \begin{pmatrix} 0.01600 \\ -0.00167 \\ -0.00100 \end{pmatrix} = 2.59778 \times 10^{-4}$$

We also have

$$\mathbf{A}\mathbf{r}_0 = \begin{pmatrix} 0.00425 & -0.00100 & -0.00125 \\ -0.00100 & 0.00467 & -0.00200 \\ -0.00125 & -0.00200 & 0.00425 \end{pmatrix} \begin{pmatrix} 0.01600 \\ -0.00167 \\ -0.00100 \end{pmatrix} = \begin{pmatrix} 0.00007 \\ -0.00002 \\ -0.00002 \end{pmatrix}$$

Hence

$$\mathbf{r}_0^T \mathbf{A} \mathbf{r}_0 == \left(0.01600 \quad -0.00167 \quad -0.001 \right) \begin{pmatrix} 0.00007 \\ -0.00002 \\ -0.00002 \end{pmatrix} = 1.11919 \times 10^{-6}$$

With these the step size parameter is obtained as

$$\alpha_0 = \frac{2.59778 \times 10^{-4}}{1.11919 \times 10^{-6}} = 217.95639$$

The next point for search is then given by

$$\mathbf{x}_1 = \begin{pmatrix} 1 \\ 1 \\ 1 \end{pmatrix} + 217.95369 \begin{pmatrix} 0.00007 \\ -0.00002 \\ -0.00002 \end{pmatrix} = \begin{pmatrix} 4.48730 \\ 0.63674 \\ 0.78204 \end{pmatrix}$$

Step 3 Calculate new residuals
We may now calculate the new residuals as

$$
\begin{aligned}
r_{11} &= 0.018 - 0.00425 \times 4.48730 + 0.00100 \times 0.63674 + 0.00125 \times 0.78204 \\
&= 0.00054 \\
r_{21} &= 0.00100 \times 4.48730 - 0.00467 \times 0.63674 + 0.00200 \times 0.78204 = 0.00038 \\
r_{31} &= 0.00125 \times 4.48730 + 0.00200 \times 0.63674 - 0.00425 \times 0.78204 = 0.00356
\end{aligned}
$$

Hence

$$\mathbf{r}_1^T \mathbf{r}_1 = \left(0.00054 \quad 0.00038 \quad 0.00356 \right) \begin{pmatrix} 0.00054 \\ 0.00038 \\ 0.00356 \end{pmatrix} = 2.24471 \times 10^{-5}$$

We thus see that the magnitude of the residual vector is smaller than that with which we started.

We have completed one iteration now. We may continue this process till the residuals converge to zero with desired tolerance or we may look at the change in the vector $\mathbf{x}$ i.e. $\mathbf{x}_{i+1} - \mathbf{x}_i$ to decide when to stop. After 13 iterations we see that the change in each component of the vector is smaller than 10^{-3} and we may stop the iteration process. Thus the solution may be written down as

$$\mathbf{x}^T = [5.615 \quad 2.393 \quad 2.777]$$

Progressive convergence to the desired solution is shown graphically in Figure 2.6. Though the first few iterations show quick approach towards the solution convergence is slow after that.

The reason for slow convergence of the SD method may be explained by looking at a simple case of a quadratic form in two dimensions. Consider the solution of two equations in two unknowns given by

$$\begin{pmatrix} 4 & 2 \\ 2 & 6 \end{pmatrix} \begin{pmatrix} x_1 \\ x_2 \end{pmatrix} = \begin{pmatrix} 3 \\ -9 \end{pmatrix} \tag{2.112}$$

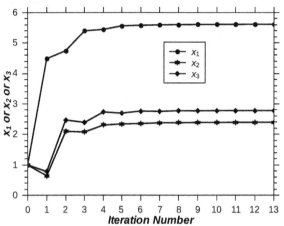

Figure 2.6: Convergence of solution in Example 2.23

The corresponding quadratic form is seen to be

$$f(x_1, x_2) = 2x_1^2 + 2x_1 x_2 + 3x_2^2 - 3x_1 + 9x_2 \tag{2.113}$$

We start the Steepest Descent procedure for minimizing the above quadratic form and obtain the minimum after a large number of iterations, as shown in Figure 2.7. Convergence is seen to involve a staircase pattern with 90° turns. The steps become

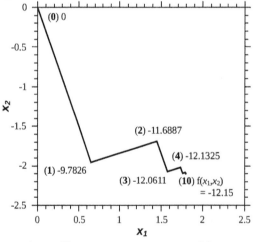

Figure 2.7: Convergence of the Steepest Descent method for the quadratic form given in Equation 2.113

smaller and smaller as we progress and hence we need a large number of iterations. In the present case we have stopped the calculations after 10 iterations to get a minimum value of $f_{min} = -12.15$. Corresponding to this the root is given by $x_1 = 1.8$, $x_2 = -2.1$. It is seen that the minimum could have been easily reached by taking a step along direction 1 - 10 after the first iteration. The CG method does exactly this and reaches the minimum in just 2 steps as will be shown later. We develop the CG method now.

CG Method

The CG method chooses the new search direction after each iteration by requiring it to be **A**-conjugate with respect the initial search direction. Two directions $\mathbf{d}_i$ and $\mathbf{d}_j$ are **A**-conjugate if $\mathbf{d}_i^T \mathbf{A} \mathbf{d}_j = 0$.

Let us assume that we have reached $\mathbf{x}_i$ after i iteration steps. We choose the next point using the expression

$$\mathbf{x}_{i+1} = \mathbf{x}_i + \alpha_i \mathbf{d}_i \qquad (2.114)$$

The value of α_i is chosen so as to make the derivative $\dfrac{df(\mathbf{x}_{i+1})}{d\alpha_i} = 0$. Thus we have

$$\frac{df(\mathbf{x}_{i+1})}{d\alpha_i} = f'(\mathbf{x}_{i+1})\frac{d\mathbf{x}_{i+1}}{d\alpha_i} = -\mathbf{r}_{i+1}^T \mathbf{d}_i = 0$$

With $\mathbf{r}_{i+1} = \mathbf{b} - \mathbf{A}\mathbf{x}_{i+1} = \mathbf{b} - \mathbf{A}\mathbf{x}_i - \alpha_i \mathbf{A}\mathbf{d}_i = \mathbf{r}_i - \alpha_i \mathbf{A}\mathbf{d}_i$ we have

$$\mathbf{r}_{i+1}^T \mathbf{d}_i = \mathbf{r}_i^T \mathbf{d}_i - \alpha_i \mathbf{d}_i^T \mathbf{A}\mathbf{d}_i = 0$$

or solving for α_i we have

$$\alpha_i = \frac{\mathbf{r}_i^T \mathbf{d}_i}{\mathbf{d}_i^T \mathbf{A}\mathbf{d}_i} \qquad (2.115)$$

Let us now seek for the next direction of search

$$\mathbf{d}_{i+1} = \mathbf{r}_{i+1} + \beta_i \mathbf{d}_i \qquad (2.116)$$

where β_i is to be determined using the condition that $\mathbf{d}_i$ and $\mathbf{d}_{i+1}$ are $\mathbf{A}$ conjugate. We then have

$$\mathbf{d}_{i+1}^T \mathbf{A}\mathbf{d}_i = \left(\mathbf{r}_{i+1} + \beta_i \mathbf{d}_i\right)^T \mathbf{A}\mathbf{d}_i = 0$$

Solving for β_i we get

$$\beta_i = -\frac{\mathbf{r}_{i+1}^T \mathbf{A}\mathbf{d}_i}{\mathbf{d}_i^T \mathbf{A}\mathbf{d}_i} \qquad (2.117)$$

It is obvious that the Conjugate Gradient method may be applied only after the first step. It is usual to choose the first step as a steepest descent step.

Summary of Conjugate gradient method

Step 1 Make an initial guess, $\mathbf{x}_0$. Perform steepest descent to find the first direction of descent ($\mathbf{d}_0$). In the first iteration, the direction of descent and the residual $\mathbf{r}_0$ are same.

$$\mathbf{r}_0 = \mathbf{d}_0 = \mathbf{b} - \mathbf{A}\mathbf{x}_0$$

Step 2 Determine the optimum step size α_1

$$\alpha_1 = \frac{\mathbf{r}_0^T \mathbf{d}_0}{\mathbf{d}_0^T \mathbf{A}\mathbf{d}_0} = \frac{\mathbf{r}_0^T \mathbf{r}_0}{\mathbf{d}_0^T \mathbf{A}\mathbf{r}_0}$$

Step 3 Update the point to $\mathbf{x}_1$.

$$\mathbf{x}_1 = \mathbf{x}_0 + \alpha_1 \mathbf{d}_0$$

Step 4 Evaluate new residual

$$\mathbf{r}_1 = \mathbf{r}_0 - \alpha_1 \mathbf{A}\mathbf{d}_0$$

If residual $\mathbf{r}_1^T \mathbf{r}_1$ is less than tolerance stop iteration. Otherwise go to step 5.

Step 5 Determining new direction for descent
Determine β_1

$$\beta_1 = -\frac{\mathbf{r}_1^T \mathbf{A} \mathbf{d}_0}{\mathbf{d}_0^T \mathbf{A} \mathbf{d}_0}$$

Update the direction of descent

$$\mathbf{d}_1 = \mathbf{r}_1 + \beta_1 \mathbf{d}_0$$

Step 6 Repeat steps 2-5 until convergence is reached

Consider again the quadratic form represented by Equation 2.113. The starting values are taken as $\mathbf{x}_0^T = [0\ \ 0]$ as we did earlier by the SD method.

The matrix elements for the present case are shown below.

A		b
4	2	3
2	6	-9

The steepest descent method uses the residuals to calculate the step size and the next point $\mathbf{x}_1$ as shown below.

$\mathbf{x}_0$	$\mathbf{r}_0$	$\mathbf{A}\mathbf{r}_0$	$\mathbf{r}_0^T \mathbf{A}\mathbf{r}_0$	α_0	$\mathbf{x}_1$
0	3	-6	414	0.2174	0.6522
0	-9	-48			-1.9565

We set $\mathbf{d}_0 = \mathbf{r}_0$ and seek the next trial direction $\mathbf{d}_1$ using the CG method.

$\mathbf{d}_0$	$\mathbf{d}_0^T \mathbf{A}\mathbf{d}_0$	$\mathbf{x}_1$	$\mathbf{r}_1$	$\mathbf{r}_1^T \mathbf{A}\mathbf{d}_0$	β_0	$\mathbf{d}_1$
3	414	0.6522	4.3043	-94.6957	0.2287	4.9905
-9		-1.9565	1.4348			-0.6238

Using the above values the next point may be calculated.

$\mathbf{x}_1$	$\mathbf{r}_1$	$\mathbf{A}\mathbf{r}_1$	$\mathbf{r}_1^T \mathbf{A}\mathbf{r}_1$	α_1	$\mathbf{x}_2$
0.6522	4.3043	20.0870	111.1645	0.2300	1.8
-1.9565	1.4348	17.2174			-2.1

The residuals are zero at $\mathbf{x}_2$ and hence we terminate the calculations. The solution has been obtained after just two iterations! The convergence to the minimum is shown in Figure 2.8 as a 3D plot. The vector is a point on the surface and we see that the point moves to the minimum value in two iterations. The SD method converges slowly as indicated in the same figure. A MATLAB code for CG method is given below.

Program 2.9: Conjugate gradient method for system of linear equations

```
1  function [x] = cg(A,b,xguess,tolerance)
2  % Input A          : coeffieient matrix
3  %        b          : right hand vector
```

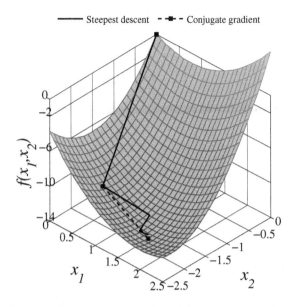

— Steepest descent   - ■ - Conjugate gradient

Figure 2.8: *Convergence of the Conjugate Gradient and Steepest descent methods for the quadratic form given in Equation 2.113*

```
 4 %           xguess   : initial guess
 5 %           tolerance: tolerance for convergence
 6 % Output x      : final output vector
 7
 8 r = b-A*xguess;              % calculate initial residual
 9 d = r;                        % initial direction of descent
10 s = r'*r;                     % calculate initial error
11 count = 0;                    % initialize iteration count
12 while (s >= tolerance)        % loop until error ≤ tolerance
13     alpha = r'*d/(d'*A*d);    % calculate step size alpha
14     xguess = xguess + alpha*d; % update x
15     r = r - alpha*A*d;        % calculate new residual
16     beta = -r'*A*d/(d'*A*d);  % calculate beta
17     d = r+ beta*d;           % update new direction CG step
18 %    d = r;                   % SD step
19 %    uncomment SD step if SD is desired
20
21     s = r'*r;                 % calculate error
22     count = count + 1;        % update iteration
23 end
24 x = xguess;                   % update final vector
```

Computation cost of Conjugate Gradient method for <u>one iteration</u>
Conjugate gradient method involves:
Matrix multiplication : 1 ($\mathbf{Ad}_0$) :$\sim O(n^2)$
Dot products : 3 :$\sim O(n)$
Hence, for large values of n, the cost of computation of CG method is $\sim O(n^2)$.

Example 2.24

Consider Example 2.21 again but solve using the Conjugate Gradient method. Use the same starting values.

Solution :

Step 1: Steepest Descent method. We start with the initial guess $\mathbf{x}_0 = [6\ 3\ 6\ 3]$ and the calculations as shown below in tabular form leads to the next point $\mathbf{x}_1$. Note that we set $\mathbf{d}_0 = \mathbf{r}_0$ in this step.

$\mathbf{x}_0$	$\mathbf{r}_0$	$\mathbf{d}_0$	α_0	$\mathbf{x}_1$	β_0	$\mathbf{d}_1$
6	0.015	0.015	16	6.24000	0.00160	0.00062
3	-0.015	-0.015		2.76000		-0.00062
6	-0.015	-0.015		5.76000		0.00058
3	0.015	0.015		3.24000		-0.00058

Step 2: The next step is the first Conjugate Gradient Step which takes us from $\mathbf{x}_1$ to $\mathbf{x}_2$ as shown in the tabulation below. Note that $\mathbf{d}_1$ *differs* from $\mathbf{r}_1$.

$\mathbf{x}_1$	$\mathbf{r}_1$	$\mathbf{d}_1$	α_1	$\mathbf{x}_2$	β_1	$\mathbf{d}_2$
6.24	0.00060	0.00062	44.64286	6.26786	0.00446	0.00007
2.76	-0.00060	-0.00062		2.73214		-0.00007
5.76	0.00060	0.00058		5.78571		-0.00007
3.24	-0.00060	-0.00058		3.21429		0.00007

Step 3: We take a second CG step now. We at once see that $\mathbf{r}_2 = 0$ and hence there is no change in the solution i.e. $\mathbf{x}_2 = \mathbf{x}_3$ and hence the method has converged to the solution, essentially in two iteration steps.

$\mathbf{x}_2$	$\mathbf{r}_2$	$\mathbf{d}_2$	α_2	$\mathbf{x}_3$	β_2	$\mathbf{d}_3$
6.26786	0	0.00007	0	6.26786	0	0
2.73214	0	-0.00007		2.73214		0
5.78571	0	-0.00007		5.78571		0
3.21429	0	0.00007		3.21429		0

The procedure has converged after one SD and one CG iterations. The convergence can be followed by looking at the magnitude of the square of the residual given by $\mathbf{r}^T\mathbf{r}$. We have $\mathbf{r}_0^T\mathbf{r}_0 = 0.0009$, $\mathbf{r}_1^T\mathbf{r}_1 = 0.00000144$ and $\mathbf{r}_2^T\mathbf{r}_2 = 0$.

Use Program 2.9 to solve the Example.

```
A = [-0.035   0.005    0.02      0;
      0.005  -0.035       0   0 02;
      0.02        0  -0.035   0.01;
         0    0.02    0.01   -0.035];

B = [-0.09; 0; -0.045; 0];

xguess = [6; 3; 6; 3];
x = cg(A,B,xguess,1e-6)
```

The output of the program is given below.

```
x  =   [6.2679;       2.7321;       5.7857;       3.2143]
```

Example 2.25

Compare the convergence rates of Gauss Seidel, SOR and CG method for solving the system of equations

$$\begin{pmatrix} 4 & -1 & -3/4 & -1/2 & -1/4 & -1/8 & -1/16 & -1/32 \\ -1 & 4 & -1 & -1/2 & -1/4 & -1/8 & -1/16 & -1/32 \\ -3/4 & -1 & 4 & -1 & -1/4 & -1/8 & -1/16 & -1/32 \\ -1/2 & -1/2 & -1 & 4 & -1 & -1/8 & -1/16 & -1/32 \\ -1/4 & -1/4 & -1/4 & -1 & 4 & -1 & -1/16 & -1/32 \\ -1/8 & -1/8 & -1/8 & -1/8 & -1 & 4 & -1 & 0 \\ -1/16 & -1/16 & -1/16 & -1/16 & -1/16 & -1 & 4 & -1 \\ -1/32 & -1/32 & -1/32 & -1/32 & -1/32 & 0 & -1 & 4 \end{pmatrix} \begin{pmatrix} x_1 \\ x_2 \\ x_3 \\ x_4 \\ x_5 \\ x_6 \\ x_7 \\ x_8 \end{pmatrix} = \begin{pmatrix} 150 \\ 100 \\ 100 \\ 100 \\ 100 \\ 50 \\ 50 \\ 50 \end{pmatrix}$$

Also compare the solutions obtained using Gauss elimination and Cholesky decomposition.

Solution :

MATLAB Programs 2.9 and 2.8 have been used to solve the system of linear equations

```
A= [4.0000      -1.0000      -0.7500     -0.5000      -0.2500   ...
                 -0.1250     -0.0625     -0.03125
    -1.0000      4.0000      -1.0000     -0.5000      -0.2500   ...
                 -0.1250     -0.0625     -0.03125
    -0.7500     -1.0000       4.0000      -1.0000      -0.2500   ...
                 -0.1250     -0.0625     -0.03125
    -0.5000     -0.5000      -1.0000      4.0000      -1.0000   ...
                 -0.1250     -0.0625     -0.03125
    -0.2500     -0.2500      -0.2500     -1.0000       4.0000   ...
                 -1.0000     -0.0625     -0.03125
    -0.1250     -0.1250      -0.1250     -0.1250      -1.0000   ...
                  4.0000     -1.0000      0.0000
    -0.0625     -0.0625      -0.0625     -0.0625      -0.0625 ...
                 -1.0000      4.0000     -1.0000
    -0.03125     -0.03125      -0.03125    -0.03125     -0.03125 ...
                  0.0000     -1.0000      4.0000 ];

B = [150; 100; 100; 100; 100; 50; 50; 50];

xguess = [80; 80; 80; 80; 60; 60; 60; 60];
xcg  = cg(A,B,xguess,1e-6); % Conjugate  Gradient
xgs  = gaussSeidel(A,B,xguess,1.0,1e-6); % Gauss  Seidel
xgsr = gaussSeidel(A,B,xguess,1.2,1e-6); %SOR
[xcholesky,C] = choleskydecomposition(A,B); %Cholesky
xgausselim = gausselimination(A,B);     % Gauss  elimination
```

All the three iterative methods produce the same output vector

```
xcg = [99.6684;          94.6375;          99.3832;          97.2898;
   82.3596;      55.3859;          40.3113      26.2758]
```

However, the rate of convergence of Conjugate Gradient method was highest followed by SOR and Gauss Seidel. Figure 2.9 indicates the rate of convergence of the three methods. There is a good agreement between the direct methods and the iterative methods for the

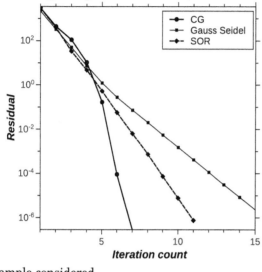

Figure 2.9: *Convergence of different iteration schemes*

example considered.

The above example clearly indicates that the CG method converges faster than all the other methods. When the number of equations to be solved is very large, this will be a great advantage, which may not be apparent from the present 8×8 case where the number of iterations has come down by a factor of 2 with respect to the Gauss Seidel scheme. Also note that the SOR method needs to be used with caution so that it does not diverge. In some cases encountered in solving large systems in engineering, it is actually necessary to use under-relaxation to see that the solution converges.

The MATLAB codes presented in the chapter have not been optimized for computational time. However, subroutines available in libraries such as LAPACK and routines available in MATLAB are optimized so that they are efficient in terms of computational time requirements. When solving large scale finite difference and finite element programs such as in structural analysis, computational fluid dynamics it is necessary to optimize the codes.

Concluding remarks

In the present chapter we have considered several techniques for the solution of a set of linear equations. The emphasis has been on the application of these methods and this has been done via several examples. Many a time the same example has been considered by different methods to bring out the relative performance of the methods. It is expected that the reader would himself come to the conclusions regarding which method to use in a particular application.

On the whole iterative methods score over other methods in the solution of a set of large number of equations especially when the coefficient matrix is sparse and diagonal dominant. In engineering practice many important field problems do lead such sparse equations.

However, when the matrix is dense non iterative methods such as Gauss elimination, LU and QR factorizations may be more useful. QR factorization is more computationally expensive than Gauss elimination. However, QR factorization is more suitable for ill conditioned matrices.

In view of all this MATLAB uses the most suitable method for solution of the equation.

2.A MATLAB routines related to Chapter 2

MATLAB routine	Function
rank(A)	rank of matrix **A**
rref(A)	returns the reduced row echelon form of matrix **A**
zeros	returns a matrix whose elements are zeros
eye(n)	returns an $n \times n$ identity matrix
ones	returns a matrix whose elements are one
norm(A)	returns norm of matrix A
triu(A)	returns upper triangular part of matrix **A**
tril(A)	returns lower triangular part of matrix **A**
diag(A)	returns a vector containing the diagonal elements of matrix **A**
inv(A)	returns inverse of matrix **A**
cond(A)	returns condition number of matrix **A**
lu(A)	returns **L** and **U** matrices of **A**
chol(A)	returns Cholesky matrix of **A**
qr(A)	returns **Q** and **R** matrices of **A**
mldivide(A,b)	equivalent to **A\b**
linsolve(A,b)	solves a system of linear equations **Ax = b**

2.B Suggested reading

1. **E. Kreyszig** *Advanced engineering mathematics* Wiley-India, 2007
2. **S. Lipschutz** and **Lipson, M.** *Schaum's outline of theory and problems of linear algebra* Schaum's Outline Series, 2001
3. **G. Strang** *Introduction to linear algebra* Wellesley Cambridge Press, 2000
4. **C. Meyer** *Matrix analysis and applied linear algebra* Society for Industrial and Applied Mathematics, 2000
5. **G. Golub** and **Van Loan, C.** *Matrix computations* Johns Hopkins Univ Pr, 1996
6. **E. Anderson** and others *LAPACK Users' guide* Society for Industrial and Applied Mathematics, 1999

Chapter 3

Computation of eigenvalues

This chapter deals with eigenvalues and eigenvectors of matrices. The material here is a sequel to Chapter 2 dealing with the solution of linear equations. Eigenvalues are very important since many engineering problems naturally lead to eigenvalue problems. When the size of a matrix is large special numerical methods are necessary for obtaining eigenvalues and eigenvectors.

3.1   Examples of eigenvalues

Vector **x** is an eigenvector of a matrix **A** if the following equation is satisfied

$$\mathbf{A}\mathbf{x} = \lambda \mathbf{x} \qquad (3.1)$$

Scalar λ is known as eigenvalue of the eigenvector. System of equations which can be reduced to the above form are classified as eigenvalue problems.

One might wonder about the importance of eigenvalues in practical engineering problems. Actually there are several practical examples which reduce to eigenvalue problems. Eigenvalues form an important foundation for quantum mechanics. Eigenvalues are employed in data analysis tools such as "principal component analysis" (reader may refer to advanced books on linear algebra for this topic). The most remarkable example of eigenvalue use for data analysis is the algorithm behind GOOGLE's search engine. Before treating eigenvalue problems mathematically, we try to understand what eigenvalues represent in physical systems by presenting a number of examples taken from different disciplines.

3.1.1   Eigenvalue problem in geometry

Consider a non-singular $n \times n$ square matrix **A**. The product **Ax** where **x** is a $n \times 1$ vector yields a $n \times 1$ vector **x**$'$ whose magnitude and direction are, in general, different. Vector **x** is said to be transformed to vector **x**$'$ by the *linear transformation* $\mathbf{A}\mathbf{x} = \mathbf{x}'$.

Consider a simple 2×2 matrix **A** given by

$$\mathbf{A} = \left(\begin{array}{rr} 1 & 2 \\ -1 & 3 \end{array} \right)$$

A vector $\mathbf{x}^T = \left(\begin{array}{cc} 1 & 1 \end{array} \right)$ is transformed to $\mathbf{x}'^T = \left(\begin{array}{cc} 3 & 2 \end{array} \right)$ by the transformation. Vector **x** makes an angle of $45°$ with the horizontal axis while vector **x**$'$ makes an angle of $33.7°$ with the horizontal axis. The magnitude of **x** is 1.4142 while the magnitude of **x**$'$ is 3.6056. Thus the transformation magnifies the vector by a factor of 2.5495 and rotates it clockwise by $11.31°$(see Figure 3.1).

The relation between **x** and **x**$'$ may, in general, be written in the form

$$\underbrace{\left(\begin{array}{ccccc} \mathbf{a}_1 & \mathbf{a}_2 & \cdots & \cdots & \mathbf{a}_n \end{array} \right)}_{\text{Matrix of column vectors}} \underbrace{\left(\begin{array}{c} \mathbf{x} \end{array} \right)}_{\text{Given column vector}} = \underbrace{\left(\begin{array}{c} \mathbf{x}' \end{array} \right)}_{\text{Transformed column vector}} \qquad (3.2)$$

Column vectors $\mathbf{a}_1 \ \mathbf{a}_2 \ \cdots \ \cdots \ \mathbf{a}_n$, assumed to be linearly independent, define the *column space* i.e. any arbitrary vector in the column space may be represented as a linear combination of the column vectors that define the column space. The column vector **x**$'$ is hence obtained as a linear combination of vectors in this column space. However there

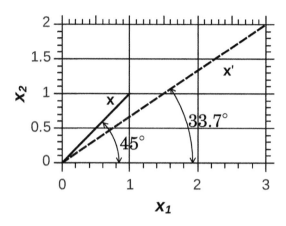

Figure 3.1: *Vector* **x** *and transformed vector* **x'** *due to linear transformation* **x'** *=* **Ax**

are certain vectors, known as eigenvectors that show only a change of magnitude but no change in direction i.e. **x** and **x'** are parallel to each other. This may be expressed by the matrix equation

$$\mathbf{Ax} = \lambda \mathbf{x} \tag{3.3}$$

where λ is a scalar known as the eigenvalue. Obviously we are looking for a non trivial solution to this equation. We may rewrite the above equation as

$$(\mathbf{A} - \lambda \mathbf{I})\mathbf{x} = 0 \tag{3.4}$$

The nontrivial solutions (vectors) define the *null space*. Non-trivial solution is possible if the matrix $\mathbf{A} - \lambda \mathbf{I}$ is singular. This requires that the determinant of the above matrix vanish. Thus we have to solve the equation

$$\begin{vmatrix} (a_{1,1}-\lambda) & a_{1,2} & a_{1,3} & \cdots & \cdots & a_{1,n} \\ a_{2,1} & (a_{2,2}-\lambda) & a_{23} & \cdots & \cdots & a_{2,n} \\ a_{3,1} & a_{3,2} & (a_{3,3}-\lambda) & \cdots & \cdots & a_{3,n} \\ \cdots & \cdots & \cdots & \cdots & \cdots & \cdots \\ \cdots & \cdots & \cdots & \cdots & \cdots & \cdots \\ a_{n,1} & a_{n,2} & a_{n,3} & \cdots & \cdots & (a_{n,n}-\lambda) \end{vmatrix} = 0 \tag{3.5}$$

to obtain the eigenvalues. When expanded the above determinant leads to a polynomial (known as the characteristic polynomial) of degree n in λ. The roots of the characteristic polynomial represent the eigenvalues of $\mathbf{A}$. In principle all eigenvalues may be obtained by obtaining all the roots of this polynomial. Since root finding is very difficult alternate methods of obtaining eigenvalues are more or less universally employed in practice. However, in this introductory part we shall demonstrate many important underlying principles and concepts by working with square matrices of small n (for example, $n = 2$) and obtaining the eigenvalues by *actually* solving for the roots of characteristic polynomials.

3.1.2   Solution of a set of ordinary differential equations (ODE)

We consider a set of ordinary differential equations, such as, the set of two equations given below.

$$\frac{dy_1}{dx} = ay_1 + by_2$$

$$\frac{dy_2}{dx} = cy_1 + dy_2 \tag{3.6}$$

Using matrix notation it is possible to recast the above set as

$$\frac{d\mathbf{y}}{dx} = \mathbf{Ay} \tag{3.7}$$

where

$$\mathbf{y}^T = \begin{pmatrix} y_1 & y_2 \end{pmatrix} \tag{3.8}$$

and

$$\mathbf{A} = \begin{pmatrix} a & b \\ c & d \end{pmatrix} \tag{3.9}$$

where the elements $a - d$ are assumed to be real numbers. The reader may recall that the solution of an ODE with constant coefficients is of exponential form. Thus we may look for a solution to the set of ODE above in the form $\mathbf{y} = \mathbf{y}_0 e^{\lambda x}$ where $\mathbf{y}_0$ is determined using the specified initial values. We then see that $\frac{d\mathbf{y}}{dx} = \mathbf{y}_0 \lambda e^{\lambda x} = \lambda \mathbf{y}$. This in Equation 3.7 immediately leads to the eigenvalue problem

$$\lambda \mathbf{y} = \mathbf{Ay} \tag{3.10}$$

It is obvious that the above may be extended to the case where there are n ODEs in the given set.

We consider now a second order ODE with constant coefficients such as

$$\frac{d^2 y}{dx^2} + a\frac{dy}{dx} + by = 0 \tag{3.11}$$

where a and b are constants. The problem statement is complete once the initial values are specified. Let $\frac{dy}{dx} = z$. The above equation may then be written down as the following set of two equations.

$$\frac{dy}{dx} = z$$

$$\frac{dz}{dx} = -by - az \tag{3.12}$$

or in the matrix form

$$\frac{d\mathbf{y}}{dx} = \begin{pmatrix} 0 & 1 \\ -b & -a \end{pmatrix} \mathbf{y} \tag{3.13}$$

where $\mathbf{y}^T = \begin{pmatrix} y & z \end{pmatrix}$. We see that the above is again in the form $\frac{d\mathbf{y}}{dx} = \mathbf{Ay}$. This case will also lead, as before, to an eigenvalue problem.

3.1.3 Standing waves on a string

Consider a thin wire of length L suspended between two rigid supports under tension. The wire is disturbed slightly from its equilibrium position (no change in the tension because of small displacement) to create standing waves in the string. It is intended to study the shape of the string as a function of time. The governing equation of motion of the string is easily shown to be[1]

$$a^2 \frac{\partial^2 u}{\partial x^2} = \frac{\partial^2 y}{\partial t^2} \tag{3.14}$$

where y is the displacement of the string from position of equilibrium and a is the speed of propagation of wave in the string. The equation is valid for small amplitudes. If we apply separation of variables to the above governing equation where displacement $u(x,t) = v(x)w(t)$, we obtain a set of ordinary differential equations such as

$$\frac{d^2 v}{dx^2} = -\frac{\lambda^2}{a^2} v \tag{3.15}$$

$$\frac{d^2 w}{dt^2} = -\lambda^2 w \tag{3.16}$$

where λ is a constant. The first equation is related to the shape of the string and the second equation gives the temporal response of the displacement. Solution to these equations describe the vibrations of the string.

What are the possible shapes the string can assume? As the two ends of the string are fixed, there would be no deflection at the two ends and the deflection can occur in the rest of the string. It can be realized that the solution to first of Equation 3.15 would take up the form $\sin\left(\frac{\lambda x}{a}\right)$. Similarly, Equation 3.16 is a sinusoidal function of time, $\sin(\lambda t)$. It is surprising that there can be an infinite number of λ values that can satisfy both the governing equations and boundary conditions. The general solution to the equation would thus become

$$u(x) = \sum_{n=1}^{\infty} A_n \sin \frac{\lambda_n x}{a} \sin \lambda_n t \tag{3.17}$$

where $\lambda_n = \frac{n a \pi}{L}$, n is an integer and A_n is the weightage for the nth harmonic. This means the shape of the string can be represented by a weighted sum of sinusoidal waves. Let us understand the significance of the above deduction using Figure 3.2. The figure shows the fundamental waveform and the first three overtones which satisfy the governing equations and boundary conditions. The deflection in the string is a weighted sum of the fundamental mode and the overtones. The weights for each waveform A_n are determined by the governing equations, initial conditions and boundary conditions. It is to be noted that, each waveform is independent of all others i.e. a waveform cannot be represented as a weighted sum of other waveforms. Also, it can be shown that each waveform is orthogonal to all other waveforms. The waveforms represent a system of eigenvectors and λ_n are the eigenvalues of the corresponding eigenvectors.

---

[1]See Chapter 15

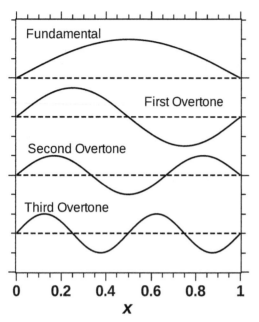

Figure 3.2: *Vibration in a string showing the fundamental and first three overtones*

3.1.4 Resonance

The physics of musical instruments lies in the resonance properties of the instrument and can be explained completely by eigenvalues. The principle discussed in the previous section becomes central to the design of several musical instruments such as guitar, cello, piano and flute. When a vibrating tuning fork is brought in the vicinity of the wire, the wire would also vibrate, if the characteristic frequency of the tuning fork and the wire are the same and this is known as resonance. The characteristic frequency of a string depends on the tension and density of the string. By changing the tension in the string one can change the fundamental frequency of vibration. Musicians tune their instruments by changing the tension in the strings.

Knowledge of characteristic frequency would be important for design of structures such as beams, trusses and bridges. Eigenvalues represent the fundamental modes of vibration of structures such as beam, truss and bridges. They are indicators of when the structures might experience destructive vibrations due to forces of wind, water, earthquake etc.

3.1.5 Natural frequency of a spring mass system

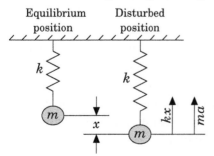

Figure 3.3: *An undamped spring mass system*

Consider an undamped spring mass system as shown in Figure 3.3. The mass is displaced by a small amount from the equilibrium position. The spring would tend to restore the mass to its original position and the mass moves towards the equilibrium position. However, the mass approaches the equilibrium position with certain velocity and would pass the equilibrium position. If the spring mass system is assumed to be ideal, the mass would oscillate about the equilibrium position. Such a motion is known as simple harmonic motion. Another common example of a simple harmonic motion is that executed by a pendulum.

Let us determine the frequency of the oscillations of an undamped spring mass system. Force balance applied to the spring mass system is as follows:

$$m\,\mathbf{a} = -k\,\mathbf{x} \longrightarrow m\,\frac{d^2\mathbf{x}}{dt^2} = -k\,\mathbf{x} \tag{3.18}$$

Observing the equations carefully, it can be deduced that the displacement is sinusoidal i.e. $\mathbf{x} \sim \sin(\omega t)$, where ω is the natural frequency of oscillations of the spring mass system. The frequency of the oscillations is determined by substituting $\sin(\omega t)$ in Equation 3.18 which would yield

$$\omega^2 \sin(\omega t) = \frac{k}{m}\sin(\omega t) \longrightarrow \omega = \sqrt{\frac{k}{m}} \tag{3.19}$$

For a simple spring mass system considered, there is only one natural frequency. Let us

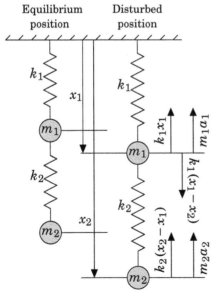

Figure 3.4: *Two Spring mass system*

consider the system shown in Figure 3.4 where two masses are connected by a spring and the system is fixed to a rigid support via another spring. There are two degrees of freedom i.e. the dynamic state of the system requires two coordinates $x_1(t)$ and $x_2(t)$, both functions of time t, to describe. When the spring mass system is displaced from the equilibrium position, the system performs a simple harmonic motion with displacement being sinusoidal with respect to time. Assembling the force equations for the two spring mass systems (with $\ddot{x} = \dfrac{d^2x}{dt^2}$)

$$
\begin{aligned}
m_1\ddot{x}_1 &= -(k_1 + k_2)x_1 + k_2 x_2 \\
m_2\ddot{x}_2 &= k_2 x_1 - k_2 x_2
\end{aligned}
\tag{3.20}
$$

Representing the two equations in matrix form, we have

$$
\begin{pmatrix} m_1 & 0 \\ 0 & m_2 \end{pmatrix}
\begin{pmatrix} \ddot{x}_1 \\ \ddot{x}_2 \end{pmatrix}
=
\begin{pmatrix} -(k_1+k_2) & k_2 \\ k_2 & -k_2 \end{pmatrix}
\begin{pmatrix} x_1 \\ x_2 \end{pmatrix}
\tag{3.21}
$$

As the displacement is sinusoidal, it can be written as $\mathbf{x} = \mathbf{y}e^{j\omega t}$ and the matrix equations become

$$
-\omega^2 \begin{pmatrix} m_1 & 0 \\ 0 & m_2 \end{pmatrix}
\begin{pmatrix} y_1 \\ y_2 \end{pmatrix}
=
\begin{pmatrix} -(k_1+k_2) & k_2 \\ k_2 & -k_2 \end{pmatrix}
\begin{pmatrix} y_1 \\ y_2 \end{pmatrix}
\tag{3.22}
$$

The equations can be further simplified to the following form[2]

$$
-\omega^2 \begin{pmatrix} y_1 \\ y_2 \end{pmatrix}
=
\begin{pmatrix} -\dfrac{k_1+k_2}{m_1} & \dfrac{k_2}{m_1} \\[2ex] \dfrac{k_2}{m_2} & -\dfrac{k_2}{m_2} \end{pmatrix}
\begin{pmatrix} y_1 \\ y_2 \end{pmatrix}
\tag{3.23}
$$

which assumes the eigenvalue form introduced in Equation 3.1. From the above equation it is evident that the problem of determining natural frequencies reduces to that of an eigenvalue problem where ω^2 is the eigenvalue.

Equation 3.22 represents a **generalized eigenvalue problem**. Generalized eigenvalue problem is of the form

$$
\mathbf{Ax} = \lambda \mathbf{Bx}
\tag{3.24}
$$

where $\mathbf{A}$ and $\mathbf{B}$ are matrices. Generalized eigenvalue problem can be simplified by pre-multiplying by $\mathbf{B}^{-1}$ on both the sides to get

$$
\mathbf{B}^{-1}\mathbf{Ax} = \lambda \mathbf{B}^{-1}\mathbf{Bx} = \lambda \mathbf{x}
\tag{3.25}
$$

However, this simplification is ruled out when matrix $\mathbf{B}$ is singular and special treatment would be required. Such cases will not be covered in the present book.

When there are n spring masses, there would be n fundamental frequencies for the system which are nothing but eigenvalues of the system. Several complex systems can be represented by equivalent spring mass systems.

3.2   Preliminaries on eigenvalues

Having seen how eigenvalue problems originate, we are now ready to discuss the mathematics behind them.

---

[2]This uses the inverse of the diagonal matrix $\begin{pmatrix} m_1 & 0 \\ 0 & m_2 \end{pmatrix}$ given by $\begin{pmatrix} 1/m_1 & 0 \\ 0 & 1/m_2 \end{pmatrix}$

3.2.1 Some important points

- The eigenvalues of a diagonal matrix and a triangular matrix are the diagonal elements of the matrix.

$$\Lambda = \begin{pmatrix} d_{1,1} & 0 & 0 \\ 0 & d_{2,2} & 0 \\ 0 & 0 & d_{3,3} \end{pmatrix}$$

$d_{1,1}$, $d_{2,2}$ and $d_{3,3}$ are the eigenvalues of the above diagonal matrix.

- Trace of a matrix is defined as the sum of the diagonal elements of the matrix. The trace of a matrix is also equal to sum of all the eigenvalues of the matrix.

$$Tr = \sum_{i=1}^{n} d_{i,i} = \sum_{i=1}^{n} \lambda_i \tag{3.26}$$

3.2.2 Similarity transformation

Matrices $\mathbf{A}$ and $\mathbf{B}$ are *similar* if there exists a nonsingular matrix $\mathbf{P}$ such that the following condition is satisfied.

$$\mathbf{B} = \mathbf{P}^{-1}\mathbf{A}\mathbf{P} \tag{3.27}$$

Let $\mathbf{y}$ be another vector such that $\mathbf{y} = \mathbf{P}^{-1}\mathbf{x}$. Then

$$\mathbf{B}\mathbf{y} = \mathbf{P}^{-1}\mathbf{A}\mathbf{P}\,\mathbf{P}^{-1}\mathbf{x} = \mathbf{P}^{-1}\mathbf{A}\mathbf{x} \tag{3.28}$$

If λ is the eigenvalue of $\mathbf{A}$, Equation 3.28 becomes

$$\mathbf{B}\mathbf{y} = \mathbf{P}^{-1}\mathbf{A}\mathbf{x} = \mathbf{P}^{-1}\lambda\mathbf{x} = \lambda\mathbf{P}^{-1}\mathbf{x} = \lambda\mathbf{y} \tag{3.29}$$

Therefore, similar matrices have the same **eigenvalues**, characteristic polynomial and trace. However, the eigenvectors of similar matrices would not be same. If $\mathbf{P}$ is orthogonal, $\mathbf{P}^{-1} = \mathbf{P}^T$ and the similarity transformation becomes $\mathbf{B} = \mathbf{P}^T\mathbf{A}\mathbf{P}$.

Determining eigenvalues using analytical methods become difficult as the rank of the matrix increases. However, if it is possible to determine either diagonal or triangular similar matrices, the evaluation of eigenvalues could become easier. The operation of converting a matrix into a similar triangular matrix is known as **Schur decomposition**[3].

$$\mathbf{U} = \mathbf{Q}^{-1}\mathbf{A}\mathbf{Q} \tag{3.30}$$

$\mathbf{A}$ is a real[4] matrix, $\mathbf{Q}$ is an orthogonal matrix and $\mathbf{U}$ is an upper triangular matrix, which is called as **Schur form** of $\mathbf{A}$. QR iteration method is based on converting the matrix into a similar upper triangular matrix of Schur form. Householder and Givens matrices, discussed in Chapter 2, are orthogonal matrices and can be used to transform the matrix into an upper triangular form i.e. **Schur form**. Under certain conditions, the Schur form is a diagonal matrix Λ.

---

[3]After Issai Schur, 1875-1941, German mathematician

[4]If $\mathbf{A}$ is a complex matrix, $\mathbf{Q}$ will be a **unitary** complex matrix which is analogous to orthogonal real matrices.

3.2.3   More about the 2×2 case

Consider the 2×2 real matrix given by the following equation.

$$\mathbf{A} = \begin{pmatrix} a & b \\ c & d \end{pmatrix} \tag{3.31}$$

The characteristic polynomial may be written as

$$\lambda^2 - (a+d)\lambda + (ad - bc) = 0 \tag{3.32}$$

Note that the coefficient of λ is the negative of Trace of the matrix given by the sum of the diagonal elements and the constant term in the characteristic equation is the determinant of $\mathbf{A}$. The two roots of the quadratic may be written down as

$$\lambda = \frac{(a+d) \pm \sqrt{(a+d)^2 - 4(ad - bc)}}{2} \tag{3.33}$$

The discriminant, the term under the square root determines the nature of the roots. If the discriminant is positive the roots are real and distinct. If the discriminant is zero we have the case of two repeated roots. If the discriminant is negative we have two complex roots which are complex conjugates of each other.

In the special case of $b = c$, the matrix is symmetric and the discriminant reduces to

$$\text{Dis} = (a+d)^2 - 4(ad - b^2) = (a-d)^2 + 4b^2$$

which is positive. In this case the roots are *real* and *distinct*.

Example of symmetric matrix

Consider a symmetric coefficient matrix given by

$$\mathbf{A} = \begin{pmatrix} 1 & 1 \\ 1 & 3 \end{pmatrix}$$

This corresponds to the case where $a = b = c = 1$ and $d = 3$ in Equation 3.31. Thus all the elements of the matrix are real. The characteristic polynomial for this case is

$$(1-\lambda)(3-\lambda) - 1 = \lambda^2 - 4\lambda + 2 = 0$$

This has roots given by $\lambda = 2 \pm \sqrt{2}$ and hence the eigenvalues are $\lambda_1 = 3.4142$ and $\lambda_2 = 0.5858$.

We pursue this case further to bring out some facts which will then be generalized. Choose a unit vector given by $\mathbf{x}^T = \begin{pmatrix} 0.7071 & 0.7071 \end{pmatrix}$. This is a vector that makes $45°$ with the horizontal axis. The product $\mathbf{Ax}$ yields the vector $\mathbf{x}'^T = \begin{pmatrix} 1.4142 & 2.8284 \end{pmatrix}$. This

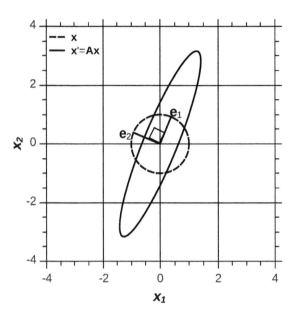

Figure 3.5: *Transformation of unit circle to an ellipse by the transformation $\mathbf{x}' = \mathbf{A}\mathbf{x}$*

vector makes an angle with the horizontal axis of $\theta' = \tan^{-1}\dfrac{2.8284}{1.4142} = 63.4°$. The vector has thus undergone a net rotation of $63.4 - 45 = 18.4°$ in the counterclockwise direction. The magnitude of the vector $\mathbf{x}'$ is given by $\sqrt{1.4142^2 + 2.8284^2} = 3.1623$.

Consider now a vector given by $\mathbf{x}^T = \begin{pmatrix} 0.3827 & 0.9239 \end{pmatrix}$ which makes an angle of $67.5°$ with the horizontal axis. This is transformed to the vector $\mathbf{x}'^T = \begin{pmatrix} 1.3066 & 3.1543 \end{pmatrix}$ which makes exactly the same angle with the horizontal axis. Hence in this case the net rotation is zero. However the vector has been magnified by a factor equal to 3.4142. Thus we identify $\mathbf{x}^T = \begin{pmatrix} 0.3827 & 0.9239 \end{pmatrix}$ as one of the eigenvectors and the corresponding eigenvalue as $\lambda_1 = 3.4142$. This is in agreement with the earlier derivations. As the vector $\mathbf{x}$ sweeps a unit circle the vector $\mathbf{x}'$ sweeps an ellipse with semi-major axis equal to $\lambda_1 = 3.4142$ and semi-minor axis equal to $\lambda_2 = 0.5858$ (see Figure 3.5). In this figure the eigenvectors shown as $\mathbf{e}_1$ and $\mathbf{e}_2$ are oriented along the major and minor axes of the ellipse and are orthogonal.

Consider now the matrix that consists of eigenvectors as its columns. In the present case the matrix is denoted as $\mathbf{Q}$ and is given by

$$\mathbf{Q} = \begin{pmatrix} 0.3827 & -0.9239 \\ 0.9239 & 0.3827 \end{pmatrix}$$

First column is the first eigenvector while the second column is the second eigenvector of $\mathbf{A}$ (the two eigenvectors are orthogonal). The transpose of this vector is given by

$$\mathbf{Q}^T = \begin{pmatrix} 0.3827 & 0.9239 \\ -0.9239 & 0.3827 \end{pmatrix}$$

If we construct a diagonal matrix Λ given by

$$\Lambda = \begin{pmatrix} 3.4142 & 0 \\ 0 & 0.5858 \end{pmatrix}$$

it may be verified that

$$\mathbf{Q}\Lambda\mathbf{Q}^T = \mathbf{A}$$

The matrix $\mathbf{A}$ is said to be diagonalizable. The diagonal elements of Λ represent the eigenvalues of the symmetric matrix $\mathbf{A}$. Note that for a symmetric matrix $\mathbf{Q}^T = \mathbf{Q}^{-1}$. Hence, in general, for a symmetric square matrix we may write

$$\mathbf{A} = \mathbf{Q}\Lambda\mathbf{Q}^{-1} \text{ or } \mathbf{A}\mathbf{Q} = \mathbf{Q}\Lambda \tag{3.34}$$

as long as $\mathbf{A}$ has n *independent* eigenvectors. In the present case the two eigenvalues are distinct and hence the eigenvectors are linearly independent. Matrix Λ represents the **Schur** form of $\mathbf{A}$.

Asymmetric matrix: real and distinct eigenvalues

Consider now an asymmetric 2×2 matrix.

$$\mathbf{A} = \begin{pmatrix} 2 & 2 \\ 1 & 2 \end{pmatrix}$$

The eigenvalues may easily be deduced as $\lambda_1 = 3.4142$ and $\lambda_2 = 0.5858$. The eigenvector $\mathbf{e}_1$ corresponding to the first eigenvalue is obtained by requiring $(\mathbf{A} - \lambda\mathbf{I})\mathbf{x}_1 = \mathbf{O}$ where $\mathbf{O}$ is the null vector. This is written in expanded form as

$$\begin{pmatrix} (2-3.4142) & 2 \\ 1 & (2-3.4142) \end{pmatrix} \begin{pmatrix} x_1 \\ y_1 \end{pmatrix} = \begin{pmatrix} 0 \\ 0 \end{pmatrix}$$

Choosing $y_1 = 1$ we get $x_1 = 1.4142$ or in the normalized form $\mathbf{e}_1^T = \begin{pmatrix} 0.8165 & 0.5774 \end{pmatrix}$. We can easily find a unit vector orthogonal to this as $\mathbf{v}_2 = \begin{pmatrix} -0.5774 & 0.8165 \end{pmatrix}$. We form a matrix $\mathbf{Q}$ with these two vectors for columns as

$$\mathbf{Q} = \begin{pmatrix} 0.8165 & -0.5774 \\ 0.5774 & 0.8165 \end{pmatrix}$$

The reader may easily verify then that

$$\mathbf{Q}^{-1}\mathbf{A}\mathbf{Q} = \begin{pmatrix} 0.5858 & 1 \\ 0 & 3.4142 \end{pmatrix}$$

The Schur form is thus seen to be upper triangular. The two eigenvalues appear along the diagonal. Note that only one column of $\mathbf{Q}$ is an eigenvector in this case. The second eigenvector turns out to be $\mathbf{e}_2^T = \begin{pmatrix} -0.8165 & 0.5774 \end{pmatrix}$ which certainly is not orthogonal to $\mathbf{e}_1$. Figure 3.6 clearly shows that the two eigenvectors are not along the major and minor axes of the ellipse. Also the angle between the two eigenvectors is not equal to $90°$.

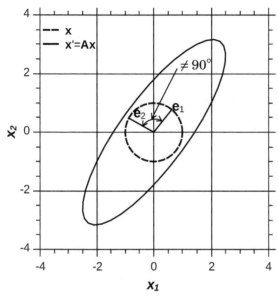

Figure 3.6: *Eigenvectors in the case of an asymmetric matrix*

Asymmetric matrix: real repeated eigenvalues

Consider another 2×2 asymmetric matrix given by

$$\mathbf{A} = \begin{pmatrix} 2 & 1 \\ -0.25 & 3 \end{pmatrix}$$

$\text{Tr} = 2+3 = 5$ and $\text{Det} = 2 \times 3 + 1 \times 0.25 = 6.25$ for this matrix. Then $\text{Dis} = 5^2 - 4 \times 6.25 = 0$ and hence matrix $\mathbf{A}$ has eigenvalue $\lambda = 2.5$ with *algebraic* multiplicity of 2 (i.e. the eigenvalue of 2.5 repeats itself). Now we can calculate one eigenvector by solving the following.

$$\mathbf{A} - 2.5\mathbf{I} = \begin{pmatrix} (2-2.5) & 1 \\ -0.25 & (3-2.5) \end{pmatrix} \begin{pmatrix} x_1 \\ x_2 \end{pmatrix} = \begin{pmatrix} 0 \\ 0 \end{pmatrix}$$

The solution is such that $x_1 = 2$, $x_2 = 1$ or $\mathbf{e}_1^T = \begin{pmatrix} 2 & 1 \end{pmatrix}$. It is clear that there is only one eigenvector that is possible in this case. We say that geometric multiplicity is one (it is the same as the number of independent eigenvectors that are possible). In view of this the matrix is referred to as being defective. We look for a second independent eigenvector by seeking what is known as a *generalized* eigenvector.

Aside

Consider a set of two ordinary differential equations $\mathbf{y}' = \mathbf{A}\mathbf{y}$ that have repeated eigenvalues, say λ with algebraic multiplicity 2 and geometric multiplicity 1. We know that one solution to the set of equations is given by $\mathbf{y} = \mathbf{e}_1 e^{\lambda t}$. In order to get a second independent solution we seek a solution in the form $\mathbf{y} = \mathbf{e}_2 e^{\lambda t} + \mathbf{e}_1 t e^{\lambda t}$. We now have, by substituting these in the given set of ordinary differential equations, the relation

$$\mathbf{e}_2 \lambda e^{\lambda t} + \mathbf{e}_1 e^{\lambda t} + \mathbf{e}_1 t \lambda e^{\lambda t} = \mathbf{A}(\mathbf{e}_2 e^{\lambda t} + \mathbf{e}_1 t e^{\lambda t})$$

The above requires that the following two relations hold:

$$\mathbf{A}\mathbf{e}_1 = \lambda\mathbf{e}_1$$
$$\mathbf{A}\mathbf{e}_2 - \lambda\mathbf{e}_2 = \mathbf{e}_1$$

Here $\mathbf{e}_2$ is the generalized eigenvector that is being sought. The above may be generalized to the case where the algebraic multiplicity of an eigenvalue is k. The generalized eigenvectors are such that we have

$$\mathbf{y} = \mathbf{e}_k e^{\lambda t} + t\mathbf{e}_{k-1}e^{\lambda t} + \frac{t^2}{2!}\mathbf{e}_{k-2}e^{\lambda t} + \cdots\cdots + \frac{t^k}{k!}\mathbf{e}_1 e^{\lambda t}$$

Substitute this in the set of ODE and observe that the following will hold:

$$\mathbf{A}\mathbf{e}_1 = \lambda\mathbf{e}_1$$
$$\mathbf{A}\mathbf{e}_2 = \mathbf{e}_1 + \lambda\mathbf{e}_2$$
$$\cdots \quad \cdots \quad \cdots$$
$$\mathbf{A}\mathbf{e}_i = \mathbf{e}_{i-1} + \lambda\mathbf{e}_i$$
$$\cdots \quad \cdots \quad \cdots$$
$$\mathbf{A}\mathbf{e}_k = \mathbf{e}_{k-1} + \lambda\mathbf{e}_k$$

We thus obtain a chain of vectors that may be used to populate the columns of $\mathbf{X}$ to proceed with the Schur form.

Back to example

Substituting for $\mathbf{e}_1$, we have to solve the system of equations given by

$$\begin{pmatrix} (2-2.5) & 1 \\ -0.25 & (3-2.5) \end{pmatrix} \begin{pmatrix} x_1 \\ x_2 \end{pmatrix} = \begin{pmatrix} 2 \\ 1 \end{pmatrix}$$

The reader may verify that this system has a solution given by $x_2 = 2 + \dfrac{x_1}{2}$ or choosing $x_1 = 0$ we get $x_2 = 2$ or $\mathbf{e}_2^T = \begin{pmatrix} 0 & 2 \end{pmatrix}$.

Using $\mathbf{e}_1$ and $\mathbf{e}_2$ as the two columns of a square matrix, we have

$$\mathbf{X} = \begin{pmatrix} 2 & 0 \\ 1 & 2 \end{pmatrix}$$

We may then show that

$$\mathbf{X}^{-1}\mathbf{A}\mathbf{X} = \begin{pmatrix} 2.50 & 1 \\ 0 & 2.50 \end{pmatrix}$$

The repeated eigenvalues appear along the diagonal. The above Schur form that is non-diagonal is referred to as the Jordan form. More about this form later.

Asymmetric matrix: complex eigenvalues

The last possibility is that the matrix has complex eigenvalues, the discriminant being negative. In this case the eigenvalues appear in pairs with the two being complex conjugates of each other. Before proceeding further we introduce some definitions which are required while dealing with complex matrices.

Let $\mathbf{A}$ be a matrix whose elements may be complex. A complex number consists of a real part and an imaginary part and is represented by $z = x + jy$ where $j = \sqrt{-1}$ is a pure imaginary number. The complex conjugates is represented as $\bar{z}$ and is given by $\bar{z} = x - jy$. Square of the magnitude of a complex number is given by $r^2 = z\bar{z} = x^2 + y^2$. The angle made by the complex number with the real axis (x axis) is known as the argument of z and is given by $\theta = \arg(z) = \tan^{-1}\dfrac{y}{x}$. A complex number may also be represented in the following forms:

$$z = re^{j\theta} \text{ or } z = r(\cos\theta + j\sin\theta)$$

Since $\mathbf{A}$ has complex elements, we define a complex conjugate $\bar{\mathbf{A}}$ as a matrix whose elements are the complex conjugate of elements of $\mathbf{A}$. The transpose of $\bar{\mathbf{A}}$ is obtained by transposing the elements of $\bar{\mathbf{A}}$ and is represented by the symbol $\mathbf{A}^*$. Matrix $\mathbf{A}$ is said to be unitary if $\mathbf{A}\mathbf{A}^* = \mathbf{I}$. Thus $\mathbf{A}^*$ plays the role of $\mathbf{A}^{-1}$ for a unitary matrix.

With this background consider a 2×2 matrix given by

$$\mathbf{A} = \begin{pmatrix} 0 & 1 \\ -1 & 0 \end{pmatrix}$$

The characteristic polynomial of this matrix is easily seen to be $\lambda^2 + 1 = 0$ which leads to the two eigenvalues given by $\lambda_1 = j$ and $\lambda_2 = -j$. We see that the two eigenvalues are complex conjugates of each other. The eigenvector corresponding to first of these eigenvalues (i.e. $\lambda_1 = j$) requires that

$$\begin{pmatrix} -j & 1 \\ -1 & -j \end{pmatrix} \begin{pmatrix} x \\ y \end{pmatrix} = \begin{pmatrix} 0 \\ 0 \end{pmatrix}$$

These equations are satisfied if $y = jx$. Normalized eigenvector is thus seen to be $\mathbf{e}_1^T = \begin{pmatrix} \dfrac{1}{\sqrt{2}} & \dfrac{j}{\sqrt{2}} \end{pmatrix}$. We may now compute a second vector perpendicular to this as $\mathbf{v}_2^T = \begin{pmatrix} \dfrac{-1}{\sqrt{2}} & \dfrac{j}{\sqrt{2}} \end{pmatrix}$.[5] With these as the columns we may write a unitary matrix $\mathbf{X}$ as

$$\mathbf{X} = \begin{pmatrix} \dfrac{1}{\sqrt{2}} & -\dfrac{1}{\sqrt{2}} \\ \dfrac{j}{\sqrt{2}} & \dfrac{j}{\sqrt{2}} \end{pmatrix}$$

---

[5]Note that two complex numbers are perpendicular if $z\bar{z} = 0$ Hence two vectors are perpendicular to each other if $\mathbf{e}_1^T \bar{\mathbf{v}}_2 = 0$

We see at once that

$$\mathbf{X}^* = \begin{pmatrix} \dfrac{1}{\sqrt{2}} & -\dfrac{j}{\sqrt{2}} \\ -\dfrac{1}{\sqrt{2}} & -\dfrac{j}{\sqrt{2}} \end{pmatrix}$$

We then have

$$\mathbf{X}^*\mathbf{A}\mathbf{X} = \begin{pmatrix} j & 0 \\ 0 & -j \end{pmatrix}$$

which is diagonal and has the eigenvalues along the diagonal. This is the Schur form of the matrix.

3.3   Analytical evaluation of eigenvalues and eigenvectors in simple cases

Having looked at simple 2×2 cases we now look at cases with $n > 2$. Analytical evaluation is on the same lines as done above in the case of 2×2 matrices. The goal of the analysis is to obtain all the eigenvalue-eigenvector pairs by analysis. The first example considers a symmetric matrix that has distinct eigenvalues.

Example 3.1

Obtain all the eigenpairs (eigenvalues and the corresponding eigenvectors) of the 3×3 symmetric square matrix

$$\mathbf{A} = \begin{pmatrix} 1 & 0 & 1 \\ 0 & 2 & 2 \\ 1 & 2 & 3 \end{pmatrix}$$

Also obtain the Schur form of this matrix.

Solution :

Step 1 The characteristic polynomial of $\mathbf{A}$ is obtained by setting the determinant of $\mathbf{A} - \lambda\mathbf{I} = 0$ as

$$(1-\lambda)[(2-\lambda)(3-\lambda)-4] - (2-\lambda) = -6\lambda + 6\lambda^2 - \lambda^3 = 0$$

The eigenvalues are given by $\lambda_1 = 0$, $\lambda_2 = 3+\sqrt{3} = 4.7321$ and $\lambda_3 = 3-\sqrt{3} = 1.2679$. The three eigenvectors are obtained by solving the homogeneous equations that result on substituting the respective eigenvalues in the equation defining the eigenvectors (Equation 3.3).

Step 2 Eigenvector corresponding to $\lambda_1 = 0$:
The eigenvector is a solution of $\mathbf{A}\mathbf{x}_1 = 0$. The components of the eigenvector satisfy the following three equations:

$$\begin{aligned} x_1 + 0y_1 + z_1 &= 0 \\ 0x_1 + 2y_1 + 2z_1 &= 0 \\ x_1 + 2y_1 + 3z_1 &= 0 \end{aligned}$$

Setting $z_1 = 1$ we see that $x_1 = y_1 = -1$ satisfy all the equations. A unit vector may now be constructed as $\mathbf{e}_1$. In fact we can make the eigenvector a unit vector by dividing each of the components by the magnitude of the vector. Thus the eigenvector corresponding to $\lambda_1 = 0$ is $\mathbf{e}_1^T = \begin{pmatrix} -0.5774 & -0.5774 & 0.5774 \end{pmatrix}$.

Step 3 Eigenvector corresponding to $\lambda_2 = 4.7321$:
The eigenvector is a solution of $(\mathbf{A} - 4.7321\mathbf{I})\mathbf{x}_2 = 0$. The components of the eigenvector satisfy the following three equations:

$$
\begin{aligned}
(1 - 4.7321)x_2 + 0y_2 + z_2 &= 0 \\
0x_2 + (2 - 4.7321)y_2 + 2z_2 &= 0 \\
x_2 + 2y_2 + (3 - 4.7321)z_2 &= 0
\end{aligned}
$$

Setting $z_2 = 1$ we see that $x_2 = 0.2679$ and $y_2 = 0.7321$ satisfy the equations. A unit vector may now be constructed as $\mathbf{e}_2$. Thus the eigenvector corresponding to $\lambda_2 = 4.7321$ is $\mathbf{e}_2^T =$
$\begin{pmatrix} 0.2113 & 0.5774 & 0.7887 \end{pmatrix}$.

Step 4 Eigenvector corresponding to $\lambda_3 = 1.2679$:
The eigenvector is a solution of $(\mathbf{A} - 1.2679\mathbf{I})\mathbf{x}_3 = 0$. The components of the eigenvector satisfy the following three equations:

$$
\begin{aligned}
(1 - 1.2679)x_3 + z_3 &= 0 \\
(2 - 1.2679)y_3 + 2z_3 &= 0 \\
x_3 + 2y_3 + (3 - 1.2679)z_3 &= 0
\end{aligned}
$$

Setting $z_3 = 1$ we see that $x_3 = 3.7321$ and $y_3 = -2.7321$ satisfy the equations. A unit vector may now be constructed as $\mathbf{e}_3$. Thus the eigenvector corresponding to $\lambda_3 = 1.2679$ is $\mathbf{e}_3^T =$
$\begin{pmatrix} 0.7887 & -0.5774 & 0.2113 \end{pmatrix}$.

Step 5 The reader may verify that the three eigenvectors are mutually orthogonal. Hence we may use these as the columns of a square matrix to derive the Schur form of $\mathbf{A}$. Thus

$$
\mathbf{U} = \begin{pmatrix}
-0.5774 & 0.2113 & 0.7887 \\
-0.5774 & 0.5774 & -0.5774 \\
0.5774 & 0.7887 & 0.2113
\end{pmatrix}
$$

We at once obtain the Schur form of $\mathbf{A}$ as

$$
\mathbf{U}^T \mathbf{A} \mathbf{U} = \begin{pmatrix}
0 & 0 & 0 \\
0 & 4.7321 & 0 \\
0 & 0 & 1.2679
\end{pmatrix}
$$

The eigenvalues appear along the diagonal.

The next example considers the case of a matrix with repeated eigenvalues. However all the eigenvalues are real.

Example 3.2
Determine all the eigenpairs pairs of the matrix

$$
\mathbf{A} = \begin{pmatrix}
1 & 1 & -1 \\
0 & 0 & 2 \\
0 & -1 & 3
\end{pmatrix}
$$

Obtain also the Schur form of the matrix.

Solution :

Step 1 The characteristic polynomial is obtained as $(1 - \lambda)[-\lambda(3 - \lambda) + 2] = (1 - \lambda)^2(2 - \lambda) = 0$ with $\lambda = 1$ (algebraic multiplicity $=2$) and $\lambda = 2$ (algebraic multiplicity $=1$) as the eigenvalues.

Step 2 Let us first obtain one of the eigenvectors corresponding to $\lambda = 1$. We have $(\mathbf{A} - \mathbf{I})\mathbf{x} = \mathbf{x}$ which yields the following three equations.

$$
\begin{aligned}
0x + y - z &= 0 \\
0x - 1y + 2z &= 0 \\
0x - 1y + 2z &= 0
\end{aligned}
$$

Only one possible solution for these three equations is given by $x = 1$, $y = z = 0$. Hence $\mathbf{e}_1^T = \begin{pmatrix} 1 & 0 & 0 \end{pmatrix}$.

Step 3 A second generalized eigenvector is obtained by the method that was earlier presented in a previous section. For this we use the equations given below.

$$
\begin{aligned}
0x + y - z &= 1 \\
0x - 1y + 2z &= 0 \\
0x - 1y + 2z &= 0
\end{aligned}
$$

The second and third of the above require $y = 2z$. This in the first gives $z = 1$. Hence a second eigenvector corresponding to $\lambda = 1$ is given by $\mathbf{e}_2^T = \begin{pmatrix} 0 & 2 & 1 \end{pmatrix}$.

Step 4 Lastly the eigenvector corresponding to $\lambda = 2$ requires the following to hold.

$$
\begin{aligned}
-x + y - z &= 0 \\
0x - 2y + 2z &= 0 \\
0x - 1y + z &= 0
\end{aligned}
$$

These are satisfied if $y = z$ and $x = 0$. Thus the required eigenvector is $\mathbf{e}_3^T = \begin{pmatrix} 0 & 1 & 1 \end{pmatrix}$.

Step 5 We choose to construct a matrix $\mathbf{X}$ with the above three vectors as its columns. Thus

$$
\mathbf{X} = \begin{pmatrix} 0 & 1 & 0 \\ 1 & 0 & 2 \\ 1 & 0 & 1 \end{pmatrix}
$$

The inverse of this matrix is easily shown to be

$$
\mathbf{X}^{-1} = \begin{pmatrix} 0 & -1 & 2 \\ 1 & 0 & 0 \\ 0 & 1 & -1 \end{pmatrix}
$$

The Schur form of the given matrix is then obtained as

$$
\mathbf{X}^{-1}\mathbf{A}\mathbf{X} = \left(\begin{array}{c|cc} 2 & 0 & 0 \\ \hline 0 & 1 & 1 \\ 0 & 0 & 1 \end{array} \right)
$$

The Schur form has turned out with an extra 1 appearing in the third column leading to a bi-diagonal structure. This matrix is said to be in **Jordan**[6] form. The 3×3 matrix is said to be in block diagonal form with one 1×1 block and a second 2×2 block.

---

[6]Named after Marie Ennemond Camille Jordan 1838-1922, French mathematician

The 1×1 block contains a single element 2 while the 2×2 block is in the bidiagonal form given by $\begin{pmatrix} 1 & 1 \\ 0 & 1 \end{pmatrix}$.

Jordan matrix

In general, the Jordan form of a matrix will appear as below:

$$\mathbf{J} = \begin{pmatrix} \mathbf{J}_1 & & & & \\ & \mathbf{J}_2 & & & \\ & & \ddots & & \\ & & & \ddots & \\ & & & & \mathbf{J}_k \end{pmatrix}$$

The diagonal entries are Jordan blocks while all other entries are zero (left blank in the above). The ith Jordan block is of the form

$$\mathbf{J}_i = \begin{pmatrix} \lambda_i & 1 & & & \\ & \lambda_i & 1 & & \\ & & \ddots & \ddots & \\ & & & \ddots & 1 \\ & & & & \lambda_i \end{pmatrix}$$

n_i, the size of the Jordan block $\mathbf{J}_i$, represents the algebraic multiplicity of the eigenvalue λ_i. Jordan form may be obtained by evaluating the chain of generalized eigenvectors as indicated earlier, to populate the matrix $\mathbf{X}$.

Example 3.3

Obtain the Jordan form of the following matrix.

$$\mathbf{A} = \begin{pmatrix} 0 & 1 & 0 \\ 0 & 0 & 1 \\ 1 & -3 & 3 \end{pmatrix}$$

Solution :

Step 1 The characteristic polynomial of this matrix is obtained as

$$-\lambda[-\lambda(3-\lambda)+3]+1 = -\lambda^3 + 3\lambda^2 - 3\lambda + 1 = -(1-\lambda)^3 = 0$$

The matrix has eigenvalue of $\lambda = 1$ with algebraic multiplicity of 3.

Step 2 One eigenvector is certainly obtainable and satisfies the following equations:

$$\begin{aligned} -x_1 + x_2 + 0x_3 &= 0 \\ 0x_1 - x_2 + x_3 &= 0 \\ x_1 - 3x_2 + 2x_3 &= 0 \end{aligned}$$

We see that $x_1 = x_2 = x_3$ is the desired solution. We may choose each one of these to be 1 to get $\mathbf{v}_1^T = \begin{pmatrix} 1 & 1 & 1 \end{pmatrix}$. This is the only possible eigenvector and hence geometric multiplicity is 1.

Step 3 We need to determine two generalized eigenvectors in order to construct the Schur form of the matrix. The first generalized eigenvector is obtained by using the condition $\mathbf{Av_2} - \lambda\mathbf{v_2} = \mathbf{v_1}$. Thus we should have

$$
\begin{aligned}
-x_1 + x_2 + 0x_3 &= 1 \\
0x_1 - x_2 + x_3 &= 1 \\
x_1 - 3x_2 + 2x_3 &= 1
\end{aligned}
$$

One may verify that $x_1 = x_2 - 1$ and $x_3 = x_2 + 1$ satisfy these three equations. Choosing $x_2 = 0$ we get $x_1 = -1$ and $x_3 = 1$. Thus the first generalized eigenvector is $\mathbf{v}_2^T = \begin{pmatrix} -1 & 0 & 1 \end{pmatrix}$.

Step 4 The second generalized eigenvector has to satisfy the condition $\mathbf{Av_3} - \lambda\mathbf{v_3} = \mathbf{v_2}$. Thus we have

$$
\begin{aligned}
-x_1 + x_2 + 0x_3 &= -1 \\
0x_1 - x_2 + x_3 &= 0 \\
x_1 - 3x_2 + 2x_3 &= 1
\end{aligned}
$$

These equations are satisfied by $x_1 = x_2 + 1$ and $x_3 = x_2$. Thus $x_1 = 1$, $x_2 = 0$ and $x_3 = 0$ satisfy this. Thus the second generalized eigenvector is $\mathbf{v}_3^T = \begin{pmatrix} 1 & 0 & 0 \end{pmatrix}$.

Step 5 Matrix $\mathbf{X}$ uses these three vectors as columns and hence we have

$$
\mathbf{X} = \begin{pmatrix} 1 & -1 & 1 \\ 1 & 0 & 0 \\ 1 & 1 & 0 \end{pmatrix}
$$

This matrix is easily inverted to get

$$
\mathbf{X}^{-1} = \begin{pmatrix} 0 & 1 & 0 \\ 0 & -1 & 1 \\ 1 & -2 & 1 \end{pmatrix}
$$

The Schur form is obtained as

$$
\mathbf{X}^{-1}\mathbf{AX} = \begin{pmatrix} 1 & 1 & 0 \\ 0 & 1 & 1 \\ 0 & 0 & 1 \end{pmatrix}
$$

The Schur form, in this case, is in the Jordan block form. The diagonal elements are all equal to 1 corresponding to the multiplicity of eigenvalues. The super diagonal has two 1's thus leading to Jordan block form.

Next example considers a matrix which has repeated eigenvalues but with all independent eigenvectors. The Schur form turns out to be diagonal.

Example 3.4

Determine all the eigenpairs pairs of the matrix

$$
\mathbf{A} = \begin{pmatrix} -3 & 0 & 0 & 0 \\ 0 & -3 & 0 & 0 \\ 12 & 6 & 2 & 0 \\ 0 & 0 & 0 & 2 \end{pmatrix}
$$

Solution :

Step 1 It is easily seen that the eigenvalues are given by $\lambda_1 = 2$ and $\lambda_2 = -3$ both with algebraic multiplicity of 2. The eigenvectors are now evaluated.

Step 2 Eigenvectors for $\lambda_1 = 2$:

Forming $\mathbf{A} - 2\mathbf{I}$ we get the following three[7] equations:

$$
\begin{aligned}
-5x_1 &= 0 \\
-5x_2 &= 0 \\
2x_1 + 6x_2 &= 0
\end{aligned}
$$

Both x_1 and x_2 have to be zero and x_3 and x_4 may have arbitrary values. The obvious choices lead to independent eigen vectors given by $\mathbf{e}_1^T = \begin{pmatrix} 0 & 0 & 0 & 1 \end{pmatrix}$ and $\mathbf{e}_2^T = \begin{pmatrix} 0 & 0 & 1 & 0 \end{pmatrix}$.

Step 3 Eigenvectors for $\lambda_2 = -3$:

Forming $\mathbf{A} + 3\mathbf{I}$ we get the following two equations:

$$
\begin{aligned}
12x_1 + 6x_2 + 5x_3 &= 0 \\
5x_4 &= 0
\end{aligned}
$$

x_4 has to be zero. If we set $x_3 = 0$ we have $x_2 = -2x_1$. We may conveniently choose $x_1 = 1$ to get $x_2 = -2$. Thus we have $\mathbf{e}_3^T = \begin{pmatrix} 1 & -2 & 0 & 0 \end{pmatrix}$. Alternately we may choose $x_1 = 0$ to get $x_3 = -\dfrac{6}{5}x_2$. Choosing $x_2 = 1$ we get $x_3 = -1.2$. Hence we have another independent eigenvector given by $\mathbf{e}_4^T = \begin{pmatrix} 0 & 1 & -1.2 & 0 \end{pmatrix}$.

Step 4 Using these vectors as the basis we may construct a matrix $\mathbf{X}$ as

$$
\mathbf{X} = \begin{pmatrix}
0 & 0 & 1 & 0 \\
0 & 0 & -2 & 1 \\
0 & 1 & 0 & -1.2 \\
1 & 0 & 0 & 0
\end{pmatrix}
$$

Inverse of this matrix may be computed and shown to be

$$
\mathbf{X}^{-1} = \begin{pmatrix}
0 & 0 & 0 & 1 \\
-2.4 & 1.2 & 1 & 0 \\
1 & 0 & 0 & 0 \\
-2 & 1 & 0 & 0
\end{pmatrix}
$$

The Schur form of $\mathbf{A}$ is then obtained as

$$
\mathbf{X}^{-1}\mathbf{A}\mathbf{X} = \begin{pmatrix}
2 & 0 & 0 & 0 \\
0 & 2 & 0 & 0 \\
0 & 0 & -3 & 0 \\
0 & 0 & 0 & -3
\end{pmatrix}
$$

which is a diagonal matrix. This example shows that the matrix is diagonalizable as long as all the eigenvectors are linearly independent i.e. geometric multiplicity is the same as the algebraic multiplicity.

---

[7]The fourth equation will be an identity $0 = 0$

Computation of eigenvalues and eigenvectors
Analytical methods presented in the previous section are useful only for small matrices, possibly $n \leq 4$. Obtaining the characteristic polynomial is itself a major problem for a large system. Solution of the characteristic polynomial is difficult for large n (say $n \geq 5$) and hence we take recourse to iterative methods of computing the eigenvectors. If our goal is to obtain only the dominant eigenvalue it may be done by the Power method or the Rayleigh quotient iteration. If all the eigenvalues are desired we use the **QR** iteration. Acceleration of convergence of these methods requires some variants such as the shift method. These will be considered in the following sections.

3.4   Power method

Assume that a $n \times n$ matrix $\mathbf{A}$ has n independent eigenvectors $\mathbf{e}_i$, $i = 1, 2, 3 \ldots n$. Any normalized vector $\mathbf{x}^{(0)}$ (for a normalized vector $\mathbf{x}^T \mathbf{x} = ||\mathbf{x}|| = 1$) may be represented as a linear combination of the n eigenvectors as $\mathbf{x} = \sum_{i=1}^{n} c_i \mathbf{e}_i$ where c_i are n scalar constants.

We shall assume that all the eigenvectors are also normalized. If we premultiply $\mathbf{x}^{(0)}$ by $\mathbf{A}$ we get

$$\mathbf{A}\mathbf{x}^{(0)} = \sum_{i=1}^{n} c_i \mathbf{A}\mathbf{e}_i = \sum_{i=1}^{n} c_i \lambda_i \mathbf{e}_i$$

The above result follows from the fact that $\mathbf{A}\mathbf{e}_i = \lambda_i \mathbf{e}_i$ since $\mathbf{e}_i$ is an eigenvector of matrix $\mathbf{A}$. We represent $\mathbf{A}\mathbf{x}^{(0)}$ as $\mathbf{y}^{(0)}$ and define a second normalized vector $\mathbf{x}^{(1)} = \dfrac{\mathbf{y}^{(0)}}{||\mathbf{y}^{(0)}||}$. Hence we have

$$\mathbf{x}^{(1)} = \frac{\lambda_1}{||\mathbf{y}^{(0)}||}\left(c_1\mathbf{e}_1 + c_2\left(\frac{\lambda_2}{\lambda_1}\right)\mathbf{e}_2 + \ldots + c_n\left(\frac{\lambda_n}{\lambda_1}\right)\mathbf{e}_n\right) \tag{3.35}$$

If we multiply $\mathbf{x}^{(1)}$ by $\mathbf{A}$ again[8] we get

$$\mathbf{A}\mathbf{x}^{(1)} = \frac{\lambda_1^2}{||\mathbf{y}^{(0)}||}\left(c_1\mathbf{e}_1 + c_2\left(\frac{\lambda_2}{\lambda_1}\right)^2\mathbf{e}_2 + \ldots + c_n\left(\frac{\lambda_n}{\lambda_1}\right)^2\mathbf{e}_n\right)$$

We set $\mathbf{A}\mathbf{x}^{(1)}$ as $\mathbf{y}^{(1)}$ and normalize it to obtain

$$\mathbf{x}^{(2)} = \frac{\lambda_1^2}{||\mathbf{y}^{(0)}||\,||\mathbf{y}^{(1)}||}\left(c_1\mathbf{e}_1 + c_2\left(\frac{\lambda_2}{\lambda_1}\right)^2\mathbf{e}_2 + \ldots + c_n\left(\frac{\lambda_n}{\lambda_1}\right)^2\mathbf{e}_n\right) \tag{3.36}$$

The above may be generalized by repeating the process of multiplication by $\mathbf{A}$ and normalization as

$$\mathbf{x}^{(k)} = \frac{\lambda_1^k}{||\mathbf{y}^{(0)}||\,||\mathbf{y}^{(1)}||\ldots||\mathbf{y}^{(k-1)}||}\left(c_1\mathbf{e}_1 + c_2\left(\frac{\lambda_2}{\lambda_1}\right)^k\mathbf{e}_2 + \ldots + c_n\left(\frac{\lambda_n}{\lambda_1}\right)^k\mathbf{e}_n\right) \tag{3.37}$$

---

[8]Repeated multiplication of $\mathbf{x}^{(0)}$ by $\mathbf{A}$ is like multiplying $\mathbf{x}^{(0)}$ by different powers of $\mathbf{A}$ and hence the name of the method

Let us assume that the eigenvalues are such that $|\lambda_1| > |\lambda_2| \ldots > |\lambda_n|$. As $k \to \infty$ the ratio $\left(\dfrac{\lambda_j}{\lambda_1}\right)^k \to 0$ for $j \neq 1$ and hence we have

$$\lim_{k \to \infty} \mathbf{x}^{(k)} = \lim_{k \to \infty} \frac{c_1 \lambda_1^k}{||\mathbf{y}^{(0)}|| \, ||\mathbf{y}^{(1)}|| \ldots ||\mathbf{y}^{(k-1)}||} \mathbf{e}_1 \tag{3.38}$$

We also have

$$\lim_{k \to \infty} \mathbf{x}^{(k+1)} = \lim_{k \to \infty} \frac{c_1 \lambda_1^{k+1}}{||\mathbf{y}^{(0)}|| \, ||\mathbf{y}^{(1)}|| \ldots ||\mathbf{y}^{(k)}||} \mathbf{e}_1 \tag{3.39}$$

Since both $\mathbf{x}^{(k)}$ and $\mathbf{e}_1$ are normalized vectors it is obvious that the former tends to the latter as $k \to \infty$. Thus the coefficients of $\mathbf{e}_1$ are equal in the above two expressions as $k \to \infty$ and hence we see that $||y^k|| \to \lambda_1$ as $k \to \infty$. General procedure for Power method is given below:

- Start with a guess for eigenvector
- Evaluate $\mathbf{y}_{i+1} = \mathbf{A}\mathbf{x}_i$. The norm of $\mathbf{x}_i$ is equal to the magnitude of the eigenvalue λ_{i+1}.
- Normalize $\mathbf{y_{i+1}}$ to determine the new eigenvector, $\mathbf{x}_{i+1}$.
- Continue iterations until the eigenvalue and eigenvector converge.

MATLAB program for determining the dominant eigenvalue using power method is given below.

Program 3.1: *Power method*

```
1  function [e,x] = powermethod(A,xguess,tolerance)
2  % Input   A         :  coefficient matrix
3  %         xguess    :  guess for eigen vector
4  %         tolerance :  tolerance for convergence
5  % Output  e         :  eigenvalue
6  %         x         :  eigenvector
7  x = xguess;                    % initialize eigenvector
8  e = 0;                         % initialize eigenvalue
9  count = 0;                     % initialize iteration count
10 res = 100;                     % initialize residual
11 while ( res > tolerance)       % loop until convergence
12     x = A*x;                   % update λx
13     x = x/sqrt(x'*x );         % normalize eigenvector
14     e = x'*A*x;                % calculate eigenvalue
15     res = abs(A*x-e*x);        % calculate residual
16 end
```

Example 3.5

Obtain the dominant eigenvalue of the symmetric square matrix given below by the Power Method.

$$\mathbf{A} = \begin{pmatrix} 1 & 1 & 1 \\ 1 & 2 & 2 \\ 1 & 2 & 3 \end{pmatrix}$$

Solution :

Step 1 We start the Power method with $\mathbf{x}^{(0)T} = \left(\dfrac{1}{\sqrt{3}} \ \dfrac{1}{\sqrt{3}} \ \dfrac{1}{\sqrt{3}} \right)$. $\mathbf{Ax}^{(0)}$ is then obtained as

$$\mathbf{Ax}^{(0)} = \begin{pmatrix} 1 & 1 & 1 \\ 1 & 2 & 2 \\ 1 & 2 & 3 \end{pmatrix} \begin{pmatrix} \dfrac{1}{\sqrt{3}} \\ \dfrac{1}{\sqrt{3}} \\ \dfrac{1}{\sqrt{3}} \end{pmatrix} = \begin{pmatrix} 1.732051 \\ 2.886751 \\ 3.464102 \end{pmatrix}$$

On normalizing this vector we get $\mathbf{x}^{(1)T} = \left(0.358569\ 0.597614\ 0.717137 \right)$ We also have $\|\mathbf{y}^{(0)}\| = 4.830459$.

Step 2 We need to proceed at least one more iteration step before we can check for convergence of the Power method. This is done by using $\mathbf{x}^{(1)}$ in place of $\mathbf{x}^{(0)}$ and repeating the above calculations. We get

$$\mathbf{Ax}^{(1)} = \begin{pmatrix} 1.673320 \\ 2.988072 \\ 3.705209 \end{pmatrix}$$

and $\|\mathbf{y}^{(1)}\| = 5.045507$. Since the change is large it is necessary to continue the iteration process.

Step 3 Further calculations are presented below in tabular form.

Iteration No.	1	2	3	4	5
$\|\mathbf{y}\|$	4.830459	5.045508	5.048864	5.048917	5.048917
Change	......	0.215048	0.003356	5.309959E-5	8.562776E-7

The dominant eigenvalue value is thus given by $\lambda_1 = 5.048917$ which is accurate to six digits after decimals. The convergence of the eigenvector is shown in the following table.

$\mathbf{x}^{(0)}$	$\mathbf{x}^{(1)}$	$\mathbf{x}^{(2)}$	$\mathbf{x}^{(3)}$	$\mathbf{x}^{(4)}$	$\mathbf{x}^{(5)}$
0.577350	0.358569	0.331646	0.328436	0.328042	0.327992
0.577350	0.597614	0.592224	0.591185	0.591033	0.591012
0.577350	0.717137	0.734358	0.736635	0.736932	0.736971

We use Program 3.1 to Example 3.5 and get the following:

```
A =[1 1 1
    1 2 2
    1 2 3];
xguess = 0.5774*[ 1; 1; 1];
[e,x] = powermethod(A,xguess,1e-6)
```

Output of the program:

```
e = 5.0489
x = [0.3280;     0.5910;     0.7370]
```

3.4.1 Inverse Power Method

The Power method yields only the dominant eigenvalue. The smallest eigenvalue may be obtained by the inverse Power Method. Premultiply Equation 3.3 by $\mathbf{A}^{-1}$ to get

$$\mathbf{A}^{-1}\mathbf{A}\mathbf{x} = \mathbf{x} = \lambda\mathbf{A}^{-1}\mathbf{x}$$

Rearrange this as

$$\mathbf{A}^{-1}\mathbf{x} = \frac{1}{\lambda}\mathbf{x} = \lambda'\mathbf{x} \tag{3.40}$$

We thus see that $\lambda' = \dfrac{1}{\lambda}$ is the eigenvalue of $\mathbf{A}^{-1}$. Hence the reciprocal of the smallest eigenvalue of $\mathbf{A}$ will be the dominant eigenvalue of $\mathbf{A}^{-1}$. The latter may be obtained by using the Power method and the least eigenvalue of $\mathbf{A}$ may then be obtained as $\lambda = \dfrac{1}{\lambda'}$.

Example 3.6
Obtain the smallest eigenvalue of the symmetric square matrix given below

$$\mathbf{A} = \begin{pmatrix} 1 & 1 & 1 \\ 1 & 2 & 2 \\ 1 & 2 & 3 \end{pmatrix}$$

by the Inverse Power Method.

Solution :

Step 1 $\mathbf{A}^{-1}$ may be obtained by any of the methods previously dealt with. We can show that

$$\mathbf{A}^{-1} = \begin{pmatrix} 2 & -1 & 0 \\ -1 & 2 & -1 \\ 0 & -1 & 1 \end{pmatrix}$$

Step 2 The Power Method is used now to get λ' of the inverse matrix by following the steps indicated in Example3.5. Table below shows the convergence of the eigenvalue.

Iteration no.	$\|\mathbf{y}\|$	Change	Iteration no.	$\|\mathbf{y}\|$	Change
	0.57735		6	3.24531	0.00370
1	2.23607	1.65872	7	3.24670	0.00139
2	2.89828	0.66221	8	3.24692	0.00022
3	3.15096	0.25269	9	3.24697	5E-05
4	3.22386	0.07290	10	3.24698	1E-05
5	3.24162	0.01775			

Step 3 We have stopped with an error of 1 unit in the 5th digit after decimals. The desired least eigenvalue of **A** is than given by $\lambda_3 = \dfrac{1}{3.24698} = 0.30798$. The convergence of the eigenvector is shown in the following table.

$\mathbf{x}^{(0)}$	$\mathbf{x}^{(1)}$	$\mathbf{x}^{(2)}$	$\mathbf{x}^{(3)}$	$\mathbf{x}^{(4)}$	$\mathbf{x}^{(5)}$
0.57735	1.00000	0.89443	0.77152	0.68558	0.63798
0.57735	0.00000	-0.44721	-0.61721	-0.68558	-0.71393
0.57735	0.00000	0.00000	0.15430	0.24485	0.28861

$\mathbf{x}^{(6)}$	$\mathbf{x}^{(7)}$	$\mathbf{x}^{(8)}$	$\mathbf{x}^{(9)}$	$\mathbf{x}^{(10)}$
0.61385	0.60203	0.59630	0.59355	0.59223
-0.72632	-0.73196	-0.73459	-0.73584	-0.73643
0.30927	0.31906	0.32372	0.32594	0.32701

We use Program 3.1 to determine the smallest eigenvalue.

```
A1 = inv(A);
[e,x] = powermethod(A1,xguess,1e-6)
e = 1/e;
```

The output of the program is given below.

```
e = 0.3080
x = [0.5911;      -0.7369;      0.3279]
```

Note that a finer tolerance has yielded a better value.

3.4.2   Inverse Power Method with Shift

The Power method yields the dominant eigenvalue of a matrix. If one is interested in another eigenvalue it is possible to obtain it by using the Inverse Power method with shift. Subtract $c\mathbf{I}$ from **A** to get matrix $\mathbf{B} = \mathbf{A} - c\mathbf{I}$. Here c is a non-zero scalar and we expect one of the eigenvalues to be close to this. Consider now the eigenvalues of **B**. We have

$$\mathbf{Bx} = \lambda'\mathbf{x}$$

or

$$(\mathbf{A} - c\mathbf{I})\mathbf{x} = (\lambda - c)\mathbf{x} = \lambda'\mathbf{x}$$

Hence we have $\lambda' = \lambda - c$ i.e. the eigenvalue of **B** is shifted by c from the eigenvalue of **A**. If we use the Inverse Power method now it is clear that the reciprocal of the above eigenvalue will be the dominant eigenvalue which is obtained easily by the Power Method. We have obtained the biggest and the smallest eigenvalues of the same 3×3 square matrix in Examples 3.5 and 3.6. We now obtain the third eigenvalue by use of Inverse Power Method with Shift.

Example 3.7

Obtain the second eigenvalue of the symmetric square matrix given below by the Inverse Power Method with Shift.

$$\mathbf{A} = \begin{pmatrix} 1 & 1 & 1 \\ 1 & 2 & 2 \\ 1 & 2 & 3 \end{pmatrix}$$

Solution :

Step 1 From the previous two examples we know that $\lambda_1 = 5.04892$ and $\lambda_3 = 0.30798$. Let us use a shift of $c = 0.6$ hoping that the second eigenvalue is close to this value. We thus consider $\mathbf{B} = \mathbf{A} - 0.6\mathbf{I}$ as the matrix whose smallest eigenvalue we shall determine by the Inverse Power Method. We thus have

$$\mathbf{B} = \begin{pmatrix} 0.4 & 1 & 1 \\ 1 & 1.4 & 2 \\ 1 & 2 & 2.4 \end{pmatrix}$$

Step 2 $\mathbf{B}^{-1}$ may be obtained by any of the methods previously dealt with. We can show that

$$\mathbf{B}^{-1} = \begin{pmatrix} 11.42857 & 7.14286 & -10.71429 \\ 7.14286 & 0.71429 & -3.57143 \\ -10.71429 & -3.57143 & 7.85714 \end{pmatrix}$$

Step 3 The Power Method is used now to get λ' of the inverse matrix by following the steps indicated in Example 3.5. The following table shows the convergence of the eigenvalue.

$\|\mathbf{y}\|$	Change
6.36209	
23.15004	16.78795
23.19884	0.04880
23.19961	0.00078
23.19963	0.00002
23.19963	0.00000

Step 4 We have stopped with an error of 1 unit in the 6th digit after decimals. The desired least eigenvalue of $\mathbf{A}$ is than given by $\lambda_3 = \dfrac{1}{23.19963} + 0.6 = 0.64310$. Convergence of the eigenvector is shown in the table below.

$\mathbf{x}^{(0)}$	$\mathbf{x}^{(1)}$	$\mathbf{x}^{(2)}$	$\mathbf{x}^{(3)}$	$\mathbf{x}^{(4)}$	$\mathbf{x}^{(5)}$
0.57735	0.71302	0.74200	0.73625	0.73708	0.73696
0.57735	0.38892	0.32200	0.32890	0.32785	0.32801
0.57735	-0.58338	-0.58800	-0.59141	-0.59095	-0.59102

We may also use Program 3.1 to determine the eigenvalue with a shift.

```
A = [ 1 1 1;
      1 2 2;
      1 2 3];
xguess = 0.5774*[ 1;1;1];
A1 = A - 0.6*eye(3);% shift =0.6
A1 = inv(A1);% A1 is the inverse of the given matrix after
                %shift
[e,x] = powermethod(A1,xguess,1e-6);
e = 1/e+0.6;
```

The output of the program is

```
e  =   0.6431
```

To summarize, it is possible to obtain all the eigenvalues of a square matrix by the Power Method and its variants viz. Inverse Power Method and the Inverse Power Method with Shift.

Convergence of the Power Method depends on the ratio $\dfrac{|\lambda_2|}{|\lambda_1|}$ being less than 1. Of course the smaller it is compared to one the faster the convergence. After each iteration the convergence is linear with this ratio. We shall look at a variant of the Power Method that converges faster. It is known as the Rayleigh Quotient Method.

3.5   Rayleigh Quotient Iteration

Premultiply Equation 3.3 by $\mathbf{x}^T$ and rearrange to get $\lambda = \dfrac{\mathbf{x}^T \mathbf{A} \mathbf{x}}{\mathbf{x}^T \mathbf{x}}$. Assume that we have an approximate normalized eigenvector given by $\mathbf{x}^{(i)}$. We see then that $R^i = \dfrac{\mathbf{x}^{(i)T} \mathbf{A} \mathbf{x}^{(i)}}{\mathbf{x}^{(i)T} \mathbf{x}^{(i)}} = \mathbf{x}^{(i)T} \mathbf{A} \mathbf{x}^{(i)}$, known as the Rayleigh Quotient is an approximation to the eigenvalue of the matrix $\mathbf{A}$. The Rayleigh Quotient Iteration method of determining the eigenvalue of a matrix is based on looking for convergence of the Rayleigh Quotient as iterations proceed. We start or continue the iteration with an assumed starting approximation to the eigenvector $\mathbf{x}^{(i)}$ and then calculate the Rayleigh Quotient. We now form a new matrix by introducing a shift equal to R^i as $\mathbf{B} = \mathbf{A} - R^i \mathbf{I}$. Now calculate a vector $\mathbf{y}^{(i)} = \mathbf{B}^{(-1)} \mathbf{x}^i$ and normalize it to get $\mathbf{x}^{(i+1)} = \dfrac{\mathbf{y}^{(i)}}{\|\mathbf{y}^{(i)}\|}$. Calculate the new value of Rayleigh Quotient given by $R^{i+1} = \mathbf{x}^{(i+1)T} \mathbf{A} \mathbf{x}^{(i+1)}$. Continue the iteration with $\mathbf{x}^{(i+1)}$. Since the Rayleigh Quotient tends to the eigenvalue as $i \to \infty$ matrix $\mathbf{A}$ with shift R^i should become singular as $i \to \infty$. Iterations stop when indeed this happens.

General procedure for Rayleigh quotient method

- Start with a guess for normalized eigenvector $\mathbf{x}_i$.
- Evaluate Rayleigh quotient $\mathbf{R}_i = \mathbf{x}_i^T \mathbf{A} \mathbf{x}_i$.
- Evaluate new eigenvector. $\mathbf{y}_{i+1} = (\mathbf{A} - \mathbf{I} R_i)^{-1} \mathbf{x}_i$.
- Normalize $\mathbf{y}_{i+1}$ to determine eigenvector $\mathbf{x}_{i+1}$
- Continue iterations until $(\mathbf{A} - \mathbf{I} R_i)$ becomes singular. This can be done by checking the condition number of the matrix.

Rayleigh Quotient Iteration converges fast and does so to different eigenvalues depending on the starting value of the eigenvector.

Example 3.8

Obtain the first eigenvalue of the symmetric square matrix given below by Rayleigh Quotient Iteration.

$$\mathbf{A} = \begin{pmatrix} 1 & 1 & 1 \\ 1 & 2 & 2 \\ 1 & 2 & 3 \end{pmatrix}$$

Solution :

Step 1 Note that the matrix being considered here is the same as that considered in previous three examples. We already know what the eigenvalues are. We start with normalized vector $\mathbf{x}^{(0)T} = \left(\dfrac{1}{\sqrt{3}} \quad \dfrac{1}{\sqrt{3}} \quad \dfrac{1}{\sqrt{3}} \right) = (0.577350 \quad 0.577350 \quad 0.577350)$ to start the iteration process. We calculate the Rayleigh Quotient as

$$R^0 = (0.577350\ 0.577350\ 0.577350) \begin{pmatrix} 1 & 1 & 1 \\ 1 & 2 & 2 \\ 1 & 2 & 3 \end{pmatrix} \begin{pmatrix} 0.577350 \\ 0.577350 \\ 0.577350 \end{pmatrix} = 4.666667$$

Now impose a shift of R^0 on $\mathbf{A}$ to get

$$\mathbf{A} - 4.666667\mathbf{I} = \begin{pmatrix} -3.666667 & 1.000000 & 1.000000 \\ 1.000000 & -2.666667 & 2.000000 \\ 1.000000 & 2.000000 & -1.666667 \end{pmatrix}$$

Invert this matrix using any method described earlier and premultiply $\mathbf{x}^{(0)}$ with it to get

$$\mathbf{y}^{(1)T} = \left(0.755978\ 1.473679\ 1.875591 \right)$$

Normalize this to get

$$\mathbf{x}^{(1)T} = \left(0.302124\ 0.588950\ 0.749573 \right)$$

Step 2 We now replace $\mathbf{x}^{(0)}$ with $\mathbf{x}^{(1)}$ and run through the calculation to get $R^1 = 5.045222$ and the next approximation to the eigenvector as

$$\mathbf{x}^{(2)T} = \left(0.328007\ 0.591011\ 0.736965 \right)$$

The convergence is seen to be rapid and is summarized below as a table.

Iteration No.	1	2	3	4
R	4.666667	5.045223	5.048917	5.048917
Change	...	0.378556	0.003695	0.000000

Essentially the R value converges to the first eigenvalue in just two iterations. The eigenvector converges to

$$\mathbf{x} = \begin{pmatrix} 0.327985 \\ 0.591009 \\ 0.736976 \end{pmatrix}$$

Program 3.2: *Rayleigh quotient*

```
 1 function [R,x] = rayleighcoefficient(A,xguess,tolerance)
 2 % Input    A           : coefficient matrix
 3 %          xguess       : guess for eigen vector
 4 %          tolerance : tolerance for convergence
 5 % Output e              : eigenvalue
 6 %          x            : eigenvector
 7 n = size(A,1);
 8 x = xguess;                      % initialize eigenvector
 9 count = 0;                       % initialize iteration count
10 R =   x'*A*x;                    % initialize Rayleigh coefficient
11 I = eye(n);                      % idendity matrix of rank n
12 B = A-R*I;                       % initialize  matrix B
13 res = abs(B*x);                  % initialize residual
14 while ( res > tolerance) % loop until convergence
15     x =   B\x;                   % update new eigenvector
16     x = x/sqrt(x'*x );           % normalize eigenvector
17     R =   x'*A*x;                % update eigenvalue
18     B = A-R*I;
19     res = abs(B*x);              % calculate residual
20     count = count + 1;           % update iteration count
21 end
```

3.5.1 Deflation of a Matrix

Consider a square matrix $\mathbf{A}$ whose dominant eigenvalue λ_1 has been determined, as for example, by the power method. Let the corresponding normalized eigenvector be $\mathbf{e}_1$. We form a matrix given by $\lambda_1 \mathbf{e}_1 \mathbf{e}_1^T$. It may be verified easily that the matrix $\mathbf{B} = \mathbf{A} - \lambda_1 \mathbf{e}_1 \mathbf{e}_1^T$ has one of the eigenvalues equal to zero, the one that corresponded to the dominant eigenvalue of the matrix $\mathbf{A}$. The next lower eigenvalue of matrix $\mathbf{A}$ becomes the dominant eigenvalue of $\mathbf{B}$ and may be determined by the application of the power method or the much faster Rayleigh quotient iteration.

Example 3.9

The dominant eigenvalue and eigenvector of matrix in Example 3.8 may be assumed known. Determine a second eigenvalue eigenvector pair by deflation using Rayleigh quotient iteration.

Solution :

Step 1 From earlier result in Example 3.8 the dominant eigenvalue eigenvector pair are given by $\lambda_1 = 5.048917$ and $\mathbf{e}_1^T = \begin{pmatrix} 0.327985 & 0.591009 & 0.736976 \end{pmatrix}$. We modify the matrix $\mathbf{A}$ of Example 3.8 by subtracting $\lambda_1 \mathbf{e}_1 \mathbf{e}_1^T$ from it to get

$$\mathbf{B} = \begin{pmatrix} 0.456866 & 0.021306 & -0.220411 \\ 0.021306 & 0.236455 & -0.199105 \\ -0.220411 & -0.199105 & 0.257762 \end{pmatrix}$$

Step 2 We now use Rayleigh quotient iteration to obtain a second eigenvalue eigenvector pair. We start the iteration with unit vector

$$\mathbf{x}_0^T = \begin{pmatrix} 0.577350 & 0.577350 & -0.577350 \end{pmatrix}$$

The Rayleigh quotient will then be obtained as $\mathbf{x}_0^T \mathbf{B} \mathbf{x}_0 = 0.610909$. We now form the vector $\mathbf{y}_1 = \mathbf{B}^{-1}\mathbf{x}_0$ and normalize it to get

$$\mathbf{x}_1 = \begin{pmatrix} 0.752693 & 0.302000 & -0.585021 \end{pmatrix}$$

Step 3 A new value of Rayleigh quotient is evaluated as $\mathbf{x}_1^T \mathbf{B} \mathbf{x}_1 = 0.642773$. The iterative process is continued till **B** becomes singular. In the present case convergence is obtained in three iterations to get

$$\lambda_2 = 0.643104$$
$$\mathbf{e}_2^T = \begin{pmatrix} 0.736976 & 0.327985 & -0.591009 \end{pmatrix}$$

The following program demonstrates estimation of eigenvalues for matrix in Example 3.8 using power method combined with deflation.

```
A = [    1 1 1;
         1 2 2;
         1 2 3];
xguess = 0.5774*[ 1;1;1];
[e1,x1] = powermethod(A,xguess,1e-6);
A1 = A-e1*x1*x1';
[e2,x2] = powermethod(A1,xguess,1e-6);
A2 = A1-e2*x2*x2';
[e3,x3] = powermethod(A2,xguess,1e-6);
```

The output of the program

```
e1 =    5.0489
e2 =    0.6431
e3 =    0.3080
```

Thus all three eigenvalues have been obtained by the use of power method with deflation. MATLAB provides several built in functions to estimate eigenvalues. One such function has been shown below. The output of the same applied to matrix in Example 3.8 is given below.

```
e = eig(A)
e = [0.3080;    0.6431;    5.0489]
```

3.6 Eigenvalue eigenvector pair by QR iteration

In Chapter 2 we have presented a method for performing **QR** factorization of a matrix **A**. Here we shall show that such a factorization may be used to obtain all the eigenvalue eigenvector pairs. Recall that a square matrix $\mathbf{A}$ is brought to the form $\mathbf{A} = \mathbf{QR}$ by the use of either Householder transformations or Givens rotations. Let us consider the following.

$$\mathbf{A} = \mathbf{Q}_1 \mathbf{R}_1$$

Premultiply by $\mathbf{Q}_1^T$ the above to get

$$\mathbf{Q}_1^T \mathbf{A} = \mathbf{Q}_1^T \mathbf{Q}_1 \mathbf{R}_1 = \mathbf{R}_1$$

Now consider $\mathbf{A}_1 = \mathbf{R}_1 \mathbf{Q}_1$ and recast it as $\mathbf{A}_1 = \mathbf{Q}_1^T \mathbf{A} \mathbf{Q}_1$. Matrix $\mathbf{A}_1$ is similar to matrix $\mathbf{A}$. Similar matrices have the same eigenvalues as shown earlier.

We may now generalize the above by defining an iteration process given by

$$\mathbf{A}_{k-1} = \mathbf{Q}_k \mathbf{R}_k; \ \mathbf{A}_k = \mathbf{R}_k \mathbf{Q}_k = \mathbf{Q}_k^T \mathbf{A}_{k-1} \mathbf{Q}_k \tag{3.41}$$

Thus we have a set of similar matrices given by $\mathbf{A}, \mathbf{A}_1, \cdots\cdots, \mathbf{A}_k$. As $k \to \infty$ $\mathbf{A}_k$ tends to a diagonal matrix, in case the matrix $\mathbf{A}$ is real symmetric. Because the diagonal matrix is similar to $\mathbf{A}$ the diagonal elements of $\mathbf{A}_k$ represent the eigenvalues of $\mathbf{A}$. We also see that

$$
\begin{aligned}
\mathbf{A}_k &= \mathbf{Q}_k^T \mathbf{A}_{k-1} \mathbf{Q}_k = \mathbf{Q}_k^T \mathbf{Q}_{k-1}^T \mathbf{A}_{k-2} \mathbf{Q}_{k-1} \mathbf{Q}_k \\
&= \ldots = \mathbf{Q}_k^T \cdots\cdots \mathbf{Q}_1^T \mathbf{A} \mathbf{Q}_1 \ldots \mathbf{Q}_k
\end{aligned}
$$

The above may be rewritten after setting $\mathbf{A}_k = \Lambda$, a diagonal matrix containing eigenvalues along the diagonal, as

$$\mathbf{A} \mathbf{Q}_1 \ldots \mathbf{Q}_k = \Lambda \mathbf{Q}_1 \ldots \mathbf{Q}_k \tag{3.42}$$

At once it is clear that the eigenvectors are the columns of the matrix product $\mathbf{Q}_1 \ldots \mathbf{Q}_k$.

The general procedure for QR iteration is

- Perform QR factorization of the matrix $\mathbf{A}_i$ to determine matrices $\mathbf{Q}_i$ and $\mathbf{R}_i$
- Determine new matrix $\mathbf{A}_{i+1} = \mathbf{R}_i \mathbf{Q}_i$
- Continue the process until $\mathbf{A}_{i+1}$ becomes a diagonal matrix or in general an upper triangular matrix.

Example 3.10

Determine all the eigenpairs pairs of the symmetric matrix

$$\mathbf{A} = \begin{pmatrix} 2.5 & 3 & 4 \\ 3 & 3.5 & 5 \\ 4 & 5 & 5.5 \end{pmatrix}$$

by $\mathbf{QR}$ *iteration.*

Solution :

Step 1 We show one iteration in detail to indicate how the method works. Using Givens rotations we may bring the coefficient matrix to upper triangular form

$$\mathbf{R}_0 = \begin{bmatrix} 5.59017 & 6.79765 & 8.40762 \\ 0 & 0.20494 & -0.74168 \\ 0 & 0 & 0.10911 \end{bmatrix}$$

Correspondingly we also have

$$\mathbf{Q}_0 = \begin{bmatrix} 0.44721 & -0.19518 & 0.87287 \\ 0.53666 & -0.72217 & -0.43644 \\ 0.71554 & 0.66361 & -0.21822 \end{bmatrix}$$

The reader may verify that we get back **A** by multiplication of matrices $\mathbf{Q}_0$ and $\mathbf{R}_0$.

Step 2 We now calculate

$$\mathbf{A}_1 = \mathbf{R}_0\mathbf{Q}_0 = \begin{bmatrix} 12.16400 & -0.42072 & 0.07807 \\ -0.42072 & -0.64019 & 0.07241 \\ 0.07807 & 0.07241 & -0.02381 \end{bmatrix}$$

This completes one iteration. We see that the off diagonal elements have already become smaller. Also notice that the transformed matrix preserves symmetry.

Step 3 The convergence of the iteration process is slow since it takes 6 iterations to get the following to 5 significant digits after decimals.

$$\mathbf{A}_6 = \mathbf{R}_5\mathbf{Q}_5 = \begin{bmatrix} 12.17828 & 0 & 0 \\ 0 & -0.66279 & 0 \\ 0 & 0 & -0.01549 \end{bmatrix} = \Lambda$$

We also obtain the three eigenvectors as the columns of the matrix given below. The matrix has been obtained by multiplying the six **Q**'s.

$$\mathbf{E} = \begin{bmatrix} 0.45877 & -0.28059 & 0.84309 \\ 0.55732 & -0.64812 & -0.51897 \\ 0.69205 & 0.70796 & -0.14096 \end{bmatrix}$$

It may be verified that the columns of the **E** matrix are mutually orthogonal.

Program 3.3: *QR iteration for eigenvalues*

```
1  function B = eigenqr(A)
2  % Input   A : coefficient matrix
3  % Output B : Matrix with eigenvalues along the diagonal
4  res = norm(tril(A,-1));        % initialize residual
5  % tril : MATLAB function to extract lower triangular matrix
6  % norm : MATLAB function to determine norm of a matrix
7  count = 0;                     % initialize iteration count
8  while( res > 1e-6)             % loop until convergence
9      [Q,R] = householder(A);   % estimate Q and R matrices
10     A = R*Q;                   % update A;
11     res = norm(tril(A,-1));    % update residual
12     count = count + 1;         % update iteration count
13 end
14 B = A;
```

The program has been applied to matrix in Example 3.10

```
A   = [ 2.5     3      4
          3    3.5     5
          4     5     5.5];
e = eigenqr(A);
```

The output of the program is as follows

```
e =
      12.1783     -0.0000      0.0000
      -0.0000     -0.6628      0.0000
       0.0000      0.0000     -0.0155
```

3.7 Modification of QR iteration for faster convergence

Since the basic **QR** iteration is slow and involves excessive calculations, especially when the matrix is dense, we explore the possibility of improving **QR** iteration for practical applications. The first trick that is employed is to reduce the given matrix to Upper Hessenberg form.[9] The Hessenberg matrix is similar to the given matrix and has zero elements for $i > j + 1$. This means that the matrix has only one non zero sub-diagonal. Hessenberg matrices have the important property that they remain Hessenberg under transformations that are used to convert the matrix to the Schur form. The second technique that is used is to introduce shift to accelerate convergence. Both these will be dealt with in what follows.

3.7.1 Upper Hessenberg form

Consider a given square matrix **A** of size $n \times n$. Schematic of the matrix is as shown below.

$$\mathbf{A} = \begin{pmatrix} * & * & * & * & * & * \\ * & * & * & * & * & * \\ * & * & * & * & * & * \\ * & * & * & * & * & * \\ * & * & * & * & * & * \\ * & * & * & * & * & * \end{pmatrix}$$

where $*$ represents, in general, a non-zero entry. Were it to be converted to the Hessenberg form ($\mathbf{A}_\mathrm{H}$) we should have the following schematic for the matrix.

$$\mathbf{A}_\mathrm{H} = \begin{pmatrix} * & * & * & * & * & * \\ * & * & * & * & * & * \\ 0 & * & * & * & * & * \\ 0 & 0 & * & * & * & * \\ 0 & 0 & 0 & * & * & * \\ 0 & 0 & 0 & 0 & * & * \end{pmatrix}$$

---

[9]Named after Karl Adolf Hessenberg, 1904-1959, German mathematician and engineer

The entries are zero for row numbers greater than column number +1. We convert **A** to the Hessenberg form (**A**$_\mathrm{H}$) by a succession of similarity transformations of form **GAG**^{T}, using Givens rotations or Householder transformations. These transformations preserve the eigenvalues and hence the eigenvalues of **A**$_\mathrm{H}$ are the same as the eigenvalues of **A**. **QR** of **A**$_\mathrm{H}$ involves much smaller number of calculations and hence there is a gain in the speed of obtaining the eigenvalues. Note that the zero entries in **A**$_\mathrm{H}$ do not change during **QR** iterations and hence this advantage. In case the given matrix is symmetric, the **A**$_\mathrm{H}$ matrix will turn out to be also symmetric and hence tri-diagonal.

$$\mathbf{A_H} = \begin{pmatrix} * & * & 0 & 0 & 0 & 0 \\ * & * & * & 0 & 0 & 0 \\ 0 & * & * & * & 0 & 0 \\ 0 & 0 & * & * & * & 0 \\ 0 & 0 & 0 & * & * & * \\ 0 & 0 & 0 & 0 & * & * \end{pmatrix}$$

It is seen that the **QR** factorization of **A**$_\mathrm{H}$ requires just $n-1$ Givens rotations. The following MATLAB program demonstrates the use of Householder matrix to covert the matrix into Hessenberg form.

Program 3.4: *Hessenberg matrix using Householder transformation*

```
 1 function H = hessenberg(A)
 2 n = size(A,1);                      % rank  of  matrix  A
 3 for i = 1:n-2
 4     x = A(i+1:n,i);
 5     x1 = -sign(A(i+1,i))*sqrt(x'*x);
 6     u = zeros(n-i,1);   % initializing  vector  u
 7     u(1) = sqrt((x1-x(1))/(2*x1)); %computing  vector  u
 8     for j=2:n-i
 9         u(j) = x(j)/(-2*u(1)*x1);
10     end
11 %determine  householder  matrix  for  current  column
12     H = eye(n);
13     H(i+1:n,i+1:n) = H(i+1:n,i+1:n)  -  2*u*u';
14     A = H'*A*H;          % similarity  transformation
15 end
16 H = A;                   % Hessenberg  matrix
```

Example 3.11

Transform the following symmetric matrix to Upper Hessenberg form.

$$\mathbf{A} = \begin{pmatrix} 1 & 2 & 3 & 4 \\ 2 & 1 & 2 & 2 \\ 3 & 2 & 1 & 3 \\ 4 & 2 & 3 & 1 \end{pmatrix}$$

Solution :

Step 1 Since the given matrix is a 4×4 matrix Upper Hessenberg form will involve three ($n = 4$, $n - 1 = 3$) similarity transformations to put three zeroes in appropriate places in matrix $\mathbf{A}$. The first transformation uses the Givens rotation $\mathbf{G}_1 = \mathbf{G}(3,4,\theta)$ where $\theta = \tan^{-1}\left(\dfrac{4}{3}\right) = 0.9273$ rad. We thus have

$$\mathbf{G}_1 = \begin{pmatrix} 1 & 0 & 0 & 0 \\ 0 & 1 & 0 & 0 \\ 0 & 0 & 0.6 & 0.8 \\ 0 & 0 & -0.8 & 0.6 \end{pmatrix}$$

The first similarity transformation $\mathbf{G}_1\mathbf{A}\mathbf{G}_1^T$ then gives matrix $\mathbf{B}$ as

$$\mathbf{B} = \mathbf{G}_1\mathbf{A}\mathbf{G}_1^T = \begin{pmatrix} 1 & 2 & 5 & 0 \\ 2 & 1 & 2.8 & -0.4 \\ 5 & 2.8 & 3.88 & -0.84 \\ 0 & -0.4 & -0.84 & -1.88 \end{pmatrix}$$

Step 2 The second transformation uses the Givens rotation $\mathbf{G}_2 = \mathbf{G}(2,3,\theta)$ where $\theta = \tan^{-1}\left(\dfrac{5}{2}\right) = 1.1903$ rad. We thus have

$$\mathbf{G}_2 = \begin{pmatrix} 1 & 0 & 0 & 0 \\ 0 & 0.3714 & 0.9285 & 0 \\ 0 & -0.9285 & 0.3714 & 0 \\ 0 & 0 & 0 & 1 \end{pmatrix}$$

The second similarity transformation $\mathbf{G}_2\mathbf{B}\mathbf{G}_2^T$ then gives matrix $\mathbf{C}$ as

$$\mathbf{C} = \mathbf{G}_2\mathbf{B}\mathbf{G}_2^T = \begin{pmatrix} 1 & 5.3852 & 0 & 0 \\ 5.3852 & 5.4138 & -1.0345 & -0.9285 \\ 0 & -1.0345 & -0.5338 & 0.0594 \\ 0 & -0.9285 & 0.0594 & -1.88 \end{pmatrix}$$

The first column of $\mathbf{A}$ has the desired form now.

Step 3 We now take care of the second column. The third transformation uses the Givens rotation $\mathbf{G}_3 = \mathbf{G}(3,4,\theta)$ where $\theta = \tan^{-1}\left(\dfrac{-0.9285}{-1.0345}\right) = 0.7314$ rad. We thus have

$$\mathbf{G}_3 = \begin{pmatrix} 1 & 0 & 0 & 0 \\ 0 & 1 & 0 & 0 \\ 0 & 0 & 0.7442 & 0.6679 \\ 0 & 0 & -0.6679 & 0.7442 \end{pmatrix}$$

The third similarity transformation $\mathbf{G}_3\mathbf{C}\mathbf{G}_3^T$ then gives matrix $\mathbf{A}_\mathrm{H}$ as

$$\mathbf{A}_\mathrm{H} = \mathbf{G}_3\mathbf{B}\mathbf{G}_3^T = \begin{pmatrix} 1 & 5.3852 & 0 & 0 \\ 5.3852 & 5.4138 & -1.3900 & 0 \\ 0 & -1.3900 & -1.0753 & -0.6628 \\ 0 & 0 & -0.6628 & -1.3385 \end{pmatrix}$$

$\mathbf{A}_\mathrm{H}$ is in Upper Hessenberg form and is similar to $\mathbf{A}$. Since $\mathbf{A}$ is symmetric $\mathbf{A}_\mathrm{H}$ is also symmetric and hence is in tri-diagonal form.

We use Program 3.4 to determine the Hessenberg matrix.

```
A = [1       2       3       4
     2       1       2       2
     3       2       1       3
     4       2       3       1];
H = hessenberg(A);
```

The output of the program is given below

```
H =
       1.0000     -5.3852      0.0000     -0.0000
      -5.3852      5.4138      1.3900     -0.0000
       0.0000      1.3900     -1.0753      0.6628
      -0.0000     -0.0000      0.6628     -1.3385
```

Checking if the eigenvalues of the Hessenberg matrix and the original matrix are same.

```
e1 = eig(A);    e2 = eig(H);
e1 = [-3.0000;    -1.7115;    -0.4466;    9.1581]
e2 = [ 9.1581;    -3.0000;    -0.4466;    -1.7115]
```

Number of operations required for QR factorization

Operations	QR factorization	
	complete matrix	Hessenberg matrix
Number of Givens rotation required	$\dfrac{n^2}{2} - \dfrac{n}{2}$	n
Number of operations required for one matrix multiplication in i_{th} column	$6(n-i+1)$	
Total number of operations	$\displaystyle\sum_{i}^{n-1} 6(n-i)(n-i+1)$ $\sim O(n^3)$	$\displaystyle\sum_{i}^{n-1} 6(n - i + 1)$ $\sim O(n^2)$

Number of operations for one QR iteration on $n \times n$ matrix $\sim O(n^3)$
Number of operations for converting matrix to Hessenberg form: $\sim O(n^3)$
Number of operations for one QR iteration on Hessenberg matrix $\sim O(n^2)$

The number of operations required for converting to Hessenberg is of the same order as that of one QR iteration of complete matrix. But the number of computations for QR iteration of Hessenberg matrix is one order of magnitude smaller than QR iteration of complete matrix.

A MATLAB program has been provided to determine Q and R matrices of a Hessenberg matrix using Givens rotations

Program 3.5: QR factorization for a Hessenberg matrix using Givens rotation

```
1 function [Q,R] = qrhessenberg(A)
2 % Input    A : Hessenberg matrix
3 % Output   Q : orthogonal matrix
4 %          R : upper Triangular matrix
5 n = size(A,1);           % rank of matrix A
```

```
6  QT = eye(n);              % initializing  transpose  of Q
7  R = A;
8  for  i=1:n-1
9      [G,R] = givensmatrix(R,i,i,i+1);
10     QT(i:i+1,:) = G*QT(i:i+1,:); % matrix  multiplication
11 end
12 Q=QT'
```

Example 3.12

Obtain all the eigenvalues of matrix in Example 3.11 starting with the Upper Hessenberg matrix derived there.

Solution :

Since the Upper Hessenberg form is preserved during Givens rotations, we need to apply only three Givens rotations to convert the Upper Hessenberg form obtained in the previous example to $\mathbf{QR}$ form. Matrix $\mathbf{A}$ being a 4×4 square matrix would have required 6 Givens rotations to complete $\mathbf{QR}$ factorization and hence we have saved the labor of applying 3 Givens rotations in each $\mathbf{QR}$ iteration. Again we shall show one $\mathbf{QR}$ iteration and then the final result.

Step 1 First Givens rotation will put a zero at position (2,1) of $\mathbf{A_H}$. We see that the required Givens matrix is $\mathbf{G_1} = \mathbf{G}(1, 2, -1.3872 \text{ rad})$ and is given by

$$\mathbf{G_1} = \left(\begin{array}{cccc} 0.1826 & -0.9832 & 0 & 0 \\ 0.9832 & 0.1826 & 0 & 0 \\ 0 & 0 & 1 & 0 \\ 0 & 0 & 0 & 1 \end{array} \right)$$

Premultiply $\mathbf{A_H}$ by $\mathbf{G_1}$ to get

$$\mathbf{B} = \mathbf{G_1 A_H} = \left(\begin{array}{cccc} 5.4772 & -6.3060 & -1.3667 & 0 \\ 0 & -4.3062 & 0.2538 & 0 \\ 0 & 1.3900 & -1.0753 & 0.6628 \\ 0 & 0 & 0.6628 & -1.3385 \end{array} \right)$$

Step 2 Second Givens rotation will put a zero at position (3,2) of $\mathbf{B}$. We see that the required Givens matrix is $\mathbf{G_2} = \mathbf{G}(2, 3, -0.3122 \text{ rad})$ and is given by

$$\mathbf{G_2} = \left(\begin{array}{cccc} 1 & 0 & 0 & 0 \\ 0 & 0.9516 & -0.3072 & 0 \\ 0 & 0.3072 & 0.9516 & 0 \\ 0 & 0 & 0 & 1 \end{array} \right)$$

Premultiply $\mathbf{B}$ by $\mathbf{G_2}$ to get

$$\mathbf{C} = \mathbf{G_2 B} = \left(\begin{array}{cccc} 5.4772 & -6.3060 & -1.3667 & 0 \\ 0 & -4.5250 & 0.5718 & -0.2036 \\ 0 & 0 & -0.9454 & 0.6307 \\ 0 & 0 & 0.6628 & -1.3385 \end{array} \right)$$

Step 3 The third Givens rotation will put a zero at position (4,3) of **C**. We see that the required Givens matrix is $\mathbf{G}_3 = \mathbf{G}(3,4,-0.6115 \text{ rad})$ and is given by

$$\mathbf{G}_3 = \begin{pmatrix} 1 & 0 & 0 & 0 \\ 0 & 1 & 0 & 0 \\ 0 & 0 & 0.8188 & -0.5741 \\ 0 & 0 & 0.5741 & 0.8188 \end{pmatrix}$$

Premultiply **C** by $\mathbf{G}_3$ to get

$$\mathbf{R} = \mathbf{G}_3\mathbf{C} = \begin{pmatrix} 5.4772 & -6.3060 & -1.3667 & 0 \\ 0 & -4.5250 & 0.5718 & -0.2036 \\ 0 & 0 & -1.1546 & 1.2848 \\ 0 & 0 & 0 & -0.7339 \end{pmatrix}$$

The matrix **Q** is obtained as

$$\mathbf{Q} = \mathbf{G}_1^T\mathbf{G}_2^T\mathbf{G}_3^T = \begin{pmatrix} 0.1826 & 0.9357 & 0.2473 & 0.1734 \\ -0.9832 & 0.1737 & 0.0459 & 0.0322 \\ 0 & -0.3072 & 0.7792 & 0.5463 \\ 0 & 0 & -0.5741 & 0.8188 \end{pmatrix}$$

Step 4 Finally we form the product **RQ** to get

$$\mathbf{RQ} = \begin{pmatrix} 7.2 & 4.4490 & 0 & 0 \\ 4.4490 & -0.9619 & 0.3547 & 0 \\ 0 & 0.3547 & -1.6372 & 0.4213 \\ 0 & 0 & 0.4213 & -0.6009 \end{pmatrix}$$

which again is in Upper Hessenberg form. Off diagonal terms are already smaller after one iteration. We replace the matrix $\mathbf{A}_H$ by **RQ** and repeat the process till all off diagonal terms are zero to the desired number of digits after the decimal point. Even though we have shown only 4 significant digits in the above, calculations have been done using the available precision of the computer. The converged matrix is in diagonal form and is given by

$$\mathbf{\Lambda} = \begin{pmatrix} 9.158122 & 0 & 0 & 0 \\ 0 & -3.000000 & 0 & 0 \\ 0 & 0 & -1.711536 & 0 \\ 0 & 0 & 0 & -0.446587 \end{pmatrix}$$

where we have shown 6 digits after decimals. The eigenvalues are the entries along the diagonal now. They are seen to be real and distinct.

3.7.2   QR iteration with shift

Wilkinson shift

Consider a Hessenberg matrix **A** and assume that all its eigenvalues are real. We can accelerate convergence of the **QR** iteration method if we introduce a shift σ (a real scalar) before performing **QR** factorization such that

$$\mathbf{A}_{k-1} - \sigma\mathbf{I} = \mathbf{Q}_k\mathbf{R}_k \tag{3.43}$$

Then we obtain

$$\mathbf{A}_k = \mathbf{R}_k \mathbf{Q}_k + \sigma \mathbf{I} \tag{3.44}$$

The shift itself may be selected as the element $a_{n,n}$, or alternatively, as Wilkinson shift that is obtained as follows. Consider the eigenvalues of the trailing 2×2 matrix given by the elements within the box in Equation 3.45.

$$
\begin{pmatrix}
* & * & * & * & * & * & * \\
* & * & * & * & * & * & * \\
0 & * & * & * & * & * & * \\
0 & 0 & * & * & * & * & * \\
0 & 0 & 0 & * & * & * & * \\
0 & 0 & 0 & 0 & * & * & * \\
0 & 0 & 0 & 0 & 0 & * & *
\end{pmatrix}
\qquad
\begin{pmatrix}
a_{n-1,n-1} & a_{n-1,n} \\
a_{n,n-1} & a_{n,n}
\end{pmatrix}
\tag{3.45}
$$

The two eigenvalues of this matrix may be easily obtained as the roots of a quadratic equation. We choose the eigenvalue that is closer to $a_{n,n}$ as the shift. As the iteration proceeds the sub diagonal element $(a_{n,n-1})$ next to $a_{n,n}$ becomes almost zero. This means that $a_{n,n}$ has converged to an eigenvalue.

$$
\begin{pmatrix}
* & * & * & * & * & * & * \\
* & * & * & * & * & * & * \\
0 & * & * & * & * & * & * \\
0 & 0 & * & * & * & * & * \\
0 & 0 & 0 & * & * & * & * \\
0 & 0 & 0 & 0 & * & * & * \\
0 & 0 & 0 & 0 & 0 & 0 & *
\end{pmatrix}
$$

Now the last row and last column are removed and the eigenvalues are calculated for the resultant $n-1 \times n-1$ matrix. The above procedure of eliminating rows and columns is known as deflation.

Deflation

Another strategy in speeding up the **QR** iteration is to *deflate* the matrix when possible. Consider the following Hessenberg matrix.

$$
\mathbf{A}_H =
\begin{pmatrix}
* & * & * & * & * & * & * \\
* & * & * & * & * & * & * \\
0 & * & * & * & * & * & * \\
0 & 0 & * & * & * & * & * \\
0 & 0 & 0 & \textcircled{0} & * & * & * \\
0 & 0 & 0 & 0 & * & * & * \\
0 & 0 & 0 & 0 & 0 & * & *
\end{pmatrix}
$$

When one of the lower sub diagonal elements is smaller than a specified tolerance, the eigenvalue problem can be simplified further. The large matrix can be broken into two smaller matrices. The eigenvalues of these smaller matrices are the eigenvalue of the large matrix. The computation time for smaller matrices would be much smaller than that for a single large matrix.

Summary of **QR** iteration with Wilkinson shift:

- Determine eigenvalues of the 2×2 matrix at the right bottom of the original matrix. The eigenvalue which is closer to the diagonal element of the last row $a_{n,n}$ is chosen as the shift parameter (σ).
- Perform **QR** factorization of matrix $\mathbf{A} - \sigma\mathbf{I}$.
  thus $\mathbf{Q}_k\mathbf{R}_k = \mathbf{A}_k - \sigma_k\mathbf{I}$
- Update matrix $\mathbf{A}_{k+1} = \mathbf{Q}_k\mathbf{R}_k + \sigma_k\mathbf{I}$
- If $a_{n,n-1} \approx 0$, $a_{n,n}$ is an eigenvalue of matrix $\mathbf{A}$. Deflate the matrix i.e. remove last row and last column. Continue the procedure for $n-1 \times n-1$ matrix.

Next example considers a real symmetric 3×3 square matrix to demonstrate the above method.

Example 3.13

Use Wilkinson shift to obtain the eigenvalues of the following symmetric matrix.

$$\mathbf{A} = \begin{pmatrix} 1 & 1 & 1 \\ 1 & 2 & 2 \\ 1 & 2 & 3 \end{pmatrix}$$

Solution :

Step 1 We may easily convert the given matrix to Hessenberg form (reader may do this part) to get

$$\mathbf{A} = \begin{pmatrix} 1 & 1.4142 & 0 \\ 1.4142 & 4.5 & 0.5 \\ 0 & 0.5 & 0.5 \end{pmatrix}$$

Now consider the trailing 2×2 matrix (elements shown boldface). The eigenvalues of this matrix are the solution of the quadratic $\lambda^2 - \text{Tr}\lambda + \text{Det} = 0$ where Tr and Det are the trace and determinant respectively. In the present case Tr $= 4.5 + 0.5 = 5$ and Det $= 4.5 \times 0.5 - 0.5^2 = 2$. The discriminant is $5^2 - 4 \times 2 = \sqrt{17}$. The two eigenvalues are then given by $\dfrac{5 + \sqrt{17}}{2} = 4.5616$ and $\dfrac{5 - \sqrt{17}}{2} = 0.4384$. The latter is close to $a_{3,3}$ and hence the shift is chosen as $\sigma_1 = 0.4384$.

Step 2 With this shift the matrix we would like to work with becomes

$$\mathbf{A} - \sigma_1\mathbf{I} = \mathbf{B} = \begin{pmatrix} 0.5616 & 1.4142 & 0 \\ 1.4142 & 4.0616 & 0.5 \\ 0 & 0.5 & 0.0616 \end{pmatrix}$$

This matrix is converted to upper triangular form by the use of Givens rotations. The reader may verify that the Givens rotations are such that

$$\mathbf{G}_1 = \begin{pmatrix} 0.3690 & 0.9294 & 0 \\ -0.9294 & 0.3690 & 0 \\ 0 & 0 & 1 \end{pmatrix}$$

and

$$G_2 = \begin{pmatrix} 1 & 1 & 0 \\ 0 & 0.3462 & 0.9382 \\ 0 & -0.9382 & 0.3462 \end{pmatrix}$$

With these we also have

$$R_1 = G_2 G_1 B = \begin{pmatrix} 1.5216 & 4.2968 & 0.4647 \\ 0 & 0.5330 & 0.1216 \\ 0 & 0.0000 & -0.1518 \end{pmatrix}$$

and

$$Q_1 = G_1^T G_2^T = \begin{pmatrix} 0.3690 & -0.3218 & 0.8719 \\ 0.9294 & 0.1278 & -0.3462 \\ 0 & 0.9382 & 0.3462 \end{pmatrix}$$

Step 3 Now we calculate the next iterate as

$$R_1 Q_1 + \sigma I = \begin{pmatrix} 4.9935 & 0.4953 & 0 \\ 0.4953 & 0.6207 & -0.1424 \\ 0 & -0.1424 & 0.3859 \end{pmatrix}$$

Step 4 We note two things. 1) The Hessenberg form is retained 2) The off diagonal terms show a big reduction after just one iteration. In fact the process converges, after just three more iterations to the following:[10]

$$R_4 Q_4 + \sigma_3 I = \begin{pmatrix} 5.048917E+00 & 1.505822E-04 & 6.989134E-16 \\ 1.505822E-04 & 6.431041E-01 & 4.231264E-16 \\ -5.313714E-17 & 2.387616E-20 & \mathbf{3.079785E-01} \end{pmatrix}$$

Assuming that element $(3,2)$ is close enough to zero, the bold entry gives one of the eigenvalues of A i.e. $\lambda_1 = 0.307979$. The matrix has been brought to **reduced** Hessenberg form.[11] Now we may *deflate* the matrix by removing the third row and third column to get the following 2×2 matrix.

$$\begin{pmatrix} 5.048916E+00 & 1.505822E-04 \\ 1.505822E-04 & 6.431041E-01 \end{pmatrix}$$

Step 5 The other two eigenvalues are then obtained by solving the characteristic equation of this matrix which is a quadratic equation. This calculation is similar to that used in obtaining the Wilkinson shift and hence will not be repeated. We thus have the other two eigenvalues as $\lambda_2 = \mathbf{0.643104}$ and $\lambda_3 = \mathbf{5.048917}$. Thus we have managed to obtain all the three eigenvalues of the given matrix.

---

[10]As and when required results are given with available precision even though numbers are generally shown rounded to four or six digits

[11]Unreduced Hessenberg form has all nonzero elements for $i > j+1$ while the reduced Hessenberg form has one zero element for $i > j+1$ i.e. $n, n-1$

Complex eigenvalues and double shift (Francis shift)

Now we consider a real square asymmetric matrix that may have complex eigenvalues. Above method that uses Wilkinson shift does not work in this case. Of course one does not know whether the matrix has complex eigenvalues by simply looking at the matrix. We convert the given matrix to Hessenberg form and hence what follows will concern a matrix in Hessenberg form. We may start the iteration process by using Wilkinson shift. It is possible that after a few iterations the 2×2 matrix at the right bottom of the matrix has eigenvalues that are complex. This is when the double shift strategy also known as Francis QR step comes in.

Let the matrix at this stage be identified by $\mathbf{A}_0$. Let σ and $\bar{\sigma}$ be the eigenvalues of the trailing 2×2 matrix. These are the roots of the quadratic equation that represents the characteristic equation of this matrix. Apply shift σ and obtain the $\mathbf{Q}_1 \mathbf{R}_1$ factorization of $\mathbf{A}_0$. Thus we have, at this stage the following.

$$
\begin{aligned}
\mathbf{A}_0 - \sigma \mathbf{I} &= \mathbf{Q}_1 \mathbf{R}_1 \ldots\ldots(a) \\
\mathbf{A}_1 &= \mathbf{R}_1 \mathbf{Q}_1 + \sigma \mathbf{I} \ldots(b)
\end{aligned}
\tag{3.46}
$$

Now introduce a second shift $\bar{\sigma}$ and again perform **QR** factorization of $\mathbf{A}_1$. Thus we have

$$
\begin{aligned}
\mathbf{A}_1 - \bar{\sigma} \mathbf{I} &= \mathbf{Q}_2 \mathbf{R}_2 \ldots\ldots(a) \\
\mathbf{A}_2 &= \mathbf{R}_2 \mathbf{Q}_2 + \bar{\sigma} \mathbf{I} \ldots(b)
\end{aligned}
\tag{3.47}
$$

Since the shifts are complex numbers it appears we have to work with complex arithmetic if we want to perform the above steps. However, as shown in what follows, the above two shift process may be replaced by an *implicit* double shift process that uses only real arithmetic. Substitute Equation 3.46(b) in Equation 3.47(a) to get

$$
\mathbf{R}_1 \mathbf{Q}_1 + \sigma \mathbf{I} - \bar{\sigma} \mathbf{I} = \mathbf{Q}_2 \mathbf{R}_2
$$

which may be rearranged as

$$
\mathbf{R}_1 \mathbf{Q}_1 + (\sigma - \bar{\sigma}) \mathbf{I} = \mathbf{Q}_2 \mathbf{R}_2
\tag{3.48}
$$

Premultiply the above by $\mathbf{Q}_1$ and postmultiply by $\mathbf{R}_1$ to obtain

$$
\mathbf{Q}_1 (\mathbf{R}_1 \mathbf{Q}_1) \mathbf{R}_1 + (\sigma - \bar{\sigma}) \mathbf{Q}_1 \mathbf{R}_1 = (\mathbf{Q}_1 \mathbf{Q}_2)(\mathbf{R}_2 \mathbf{R}_1)
$$

The left hand side of the above equation may be rewritten as

$$
\mathbf{Q}_1 \mathbf{R}_1 [\mathbf{Q}_1 \mathbf{R}_1 + (\sigma - \bar{\sigma}) \mathbf{I}] = (\mathbf{A}_0 - \sigma \mathbf{I})(\mathbf{A}_0 - \bar{\sigma} \mathbf{I})
\tag{3.49}
$$

where we have made use of Equation 3.46(a). Thus we have the important result

$$
(\mathbf{A}_0 - \sigma \mathbf{I})(\mathbf{A}_0 - \bar{\sigma} \mathbf{I}) = (\mathbf{Q}_1 \mathbf{Q}_2)(\mathbf{R}_2 \mathbf{R}_1) = \mathbf{Z} \mathbf{R}
\tag{3.50}
$$

where we have introduced the notation $\mathbf{Z} = \mathbf{Q}_1 \mathbf{Q}_2$ and $\mathbf{R} = \mathbf{R}_2 \mathbf{R}_1$. We may easily see that the left hand side is given by

$$(\mathbf{A}_0 - \sigma \mathbf{I})(\mathbf{A}_0 - \bar{\sigma} \mathbf{I}) = \mathbf{A}_0^2 - (\sigma + \bar{\sigma})\mathbf{A}_0 + \sigma \bar{\sigma} \mathbf{I}$$

Note that $\sigma + \bar{\sigma} = \text{Tr}$ and $\sigma \bar{\sigma} = \text{Det}$ are the trace and determinant respectively of the 2×2 matrix. Both of these are real quantities. Thus we may write the above as

$$(\mathbf{A}_0 - \sigma \mathbf{I})(\mathbf{A}_0 - \bar{\sigma} \mathbf{I}) = \mathbf{A}_0^2 - \text{Tr}\mathbf{A}_0 + \text{Det}\mathbf{I} = \mathbf{M} \tag{3.51}$$

Thus the left hand side turns out be real and hence we conclude that $\mathbf{Z}$ and $\mathbf{A}$ are real. Thus the factorization $\mathbf{M} = \mathbf{ZR}$ yields the factorization of $\mathbf{A}_0$ directly and does not involve complex arithmetic. The shifts are implicit and hence the method is referred to implicit double shift method. As the last step we compute

$$\mathbf{A}_2 = \mathbf{Z}^T \mathbf{A}_0 \mathbf{Z} \tag{3.52}$$

to complete one iteration cycle using implicit double shift. As the iteration proceeds the sub diagonal element $(a_{n-1,n-2})$ next to $a_{n-1,n-1}$ becomes almost zero. This means that the eigenvalues of the 2×2 trailing submatrix are also eigenvalues of matrix $\mathbf{A}$.

$$\begin{pmatrix} * & * & * & * & * & * & * \\ * & * & * & * & * & * & * \\ 0 & * & * & * & * & * & * \\ 0 & 0 & * & * & * & * & * \\ 0 & 0 & 0 & * & * & * & * \\ 0 & 0 & 0 & 0 & \textcircled{0} & * & * \\ 0 & 0 & 0 & 0 & 0 & * & * \end{pmatrix} \qquad \begin{pmatrix} a_{n-1,n-1} & a_{n-1,n} \\ a_{n,n-1} & a_{n,n} \end{pmatrix} \tag{3.53}$$

Now the last two rows and last two columns are removed and the procedure is performed on resultant $n - 2 \times n - 2$ matrix.

Summary of QR iteration with shifts (Francis or Wilkinson):

- Calculate the eigenvalues $(\sigma_k, \bar{\sigma}_k$ of 2×2 submatrix at the right bottom of the original matrix $\mathbf{A}_k$. If the eigenvalues are complex use Francis shift else use Wilkinson shift (discussed earlier).
- Determine matrix $\mathbf{M}_k = \mathbf{A}_k^2 - (\sigma_k + \bar{\sigma}_k)\mathbf{A}_k + \sigma_k \bar{\sigma}_k \mathbf{I}$
- Perform $\mathbf{QR}$ factorization of $\mathbf{M}_k$
  $$\mathbf{Z}_k \mathbf{R}_k = \mathbf{M}_k$$
- Update matrix $\mathbf{A}_{k+1} = \mathbf{Z}_k^T \mathbf{A}_k \mathbf{Z}_k$
- If $a_{n-1,n-2} \approx 0$, the eigenvalues of the 2×2 submatrix at the right bottom are also the eigenvalues of $\mathbf{A}$. Deflate the matrix to $n - 2 \times n - 2$ matrix i.e. remove last two rows and two columns. Continue the iterations for the deflated matrix.

We work out an example in full detail to bring out all the important aspects involved in the above procedure.

Example 3.14

Obtain all the eigenvalues of the following Hessenberg matrix.

$$\mathbf{A} = \begin{pmatrix} 2 & -1 & 1 & 0.5 \\ -1 & 2 & 1 & 0.4 \\ 0 & -1 & 2 & 0.3 \\ 0 & 0 & -0.5 & 1 \end{pmatrix}$$

You may expect complex eigenvalues in this case.

Solution :

Consider the trailing 2×2 matrix of **A** given by

$$\begin{pmatrix} 2 & 0.3 \\ -0.5 & 1 \end{pmatrix}$$

The eigenvalues of this matrix are obtained (both are real) by solving characteristic polynomial (quadratic) as $\sigma_1 = \dfrac{3 + \sqrt{0.4}}{2} = 1.8162$ and $\sigma_2 = \dfrac{3 - \sqrt{0.4}}{2} = 1.1838$. The latter being the one closer to $a_{4,4} = 1$ is used as the Wilkinson shift to obtain the first **QR** factorization of **A**, viz.

$$\mathbf{A}_1 = \begin{pmatrix} 2.9797 & -0.0877 & -0.2083 & 0.0968 \\ 0.8002 & 2.1264 & 1.0218 & 0.0497 \\ 0.0000 & -1.2211 & 0.6667 & -0.4009 \\ 0.0000 & 0.0000 & 0.0187 & 1.2272 \end{pmatrix}$$

The eigenvalues of the trailing 2×2 matrix of $\mathbf{A}_1$ are also real and the one that is closer to 1.2272 i.e. $\sigma = 1.2134$ is used as the Wilkinson shift. The following is obtained after a second iteration.

$$\mathbf{A}_2 = \begin{pmatrix} 3.1022 & -0.4959 & -0.2933 & 0.1195 \\ 0.6182 & 1.1606 & 1.7225 & 0.2671 \\ 0.0000 & -0.4177 & 1.5017 & -0.2379 \\ 0.0000 & 0.0000 & 0.0008 & 1.2355 \end{pmatrix}$$

After third iteration, the matrix becomes

$$\mathbf{A}_3 = \begin{pmatrix} 2.9467 & -0.4729 & 1.0141 & 0.1979 \\ 0.1340 & 1.2367 & 0.4126 & 0.2757 \\ 0.0000 & -1.7113 & 1.5812 & 0.1645 \\ 0.0000 & 0.0000 & 0.0000 & 1.2354 \end{pmatrix}$$

The element at (4,3) has become zero and hence $\lambda_1 = 1.235393$ and we may deflate the matrix by dropping the last row and column. The other three eigenvalues will use the following 3×3 matrix.

$$\mathbf{B} = \begin{pmatrix} 2.9467 & -0.4729 & -1.0141 \\ 0.1340 & 1.2367 & 0.4126 \\ 0.0000 & -1.7113 & 1.5812 \end{pmatrix}$$

At this stage we see that the eigenvalues of the trailing 2×2 matrix are complex and hence we initiate the implicit double shift now. We calculate the trace and determinant as Tr $= 1.2367 + 1.5812 = 2.8179$ and Det $= 1.12367 \times 1.5812 + 0.4126 \times 1.7113 = 2.6616$.

The discriminant $\text{Tr}^2 - 4\text{Det} = 2.8179^2 - 4 \times 2.6616 = -2.7057$ is negative showing that the eigenvalues are complex. We construct matrix $\mathbf{M}$ as

$$\mathbf{M} = \begin{pmatrix} 2.9778 & 1.0897 & -1.9292 \\ 0.1830 & -0.0634 & -0.1359 \\ -0.2293 & 0.0000 & 0.0000 \end{pmatrix}$$

Note that $\mathbf{M}$ has non-zero elements along two sub-diagonals and hence is *not* in Hessenberg form. We now obtain the $\mathbf{ZR}$ using Givens rotations as required. At the end of one iteration we get the following.

$$\mathbf{R} = \begin{pmatrix} 2.9922 & 1.0806 & -1.9282 \\ 0 & -0.1543 & 0.0652 \\ 0 & 0 & -0.1342 \end{pmatrix}$$

and

$$\mathbf{Z} = \begin{pmatrix} 0.9952 & -0.0928 & 0.0315 \\ 0.0611 & 0.8387 & 0.5411 \\ -0.0766 & -0.5366 & 0.8404 \end{pmatrix}$$

Finally we also have

$$\mathbf{Z}^T\mathbf{AZ} = \begin{pmatrix} 2.9951 & 0.0986 & -0.9790 \\ -0.0069 & 1.9110 & 0.7338 \\ 0.0000 & -1.4884 & 0.8585 \end{pmatrix}$$

This completes one implicit double shift iteration. Two more iterations are required to get the following transformed $\mathbf{A}$:

$$\mathbf{A}' = \begin{pmatrix} \mathbf{2.9917} & 0.9898 & -0.0867 \\ 0.0000 & 1.0364 & 1.6542 \\ 0.0000 & -0.5640 & 1.7365 \end{pmatrix}$$

The second real eigenvalue is thus given by $\lambda_2 = 2.9917$. The two complex eigenvalues are those of the deflated matrix

$$\mathbf{C} = \begin{pmatrix} 1.0364 & 1.6542 \\ -0.5640 & 1.7365 \end{pmatrix}$$

The two complex eigenvalues are easily obtained as the roots of the quadratic characteristic polynomial of $\mathbf{C}$. These may be shown to be given by $\lambda_3 = 1.38645 + j0.9003$ and $\lambda_4 = 1.3865 - j0.9003$.

The following MATLAB program incorporates QR iteration with shifts.

***Program 3.6:** QR iteration with shifts*

```
1 function eigen = eigenqrshift(A)
2 % Input   A : coefficient matrix
3 % Output B : Matrix with eigenvalues at diagonal
4 n = size(A,1);
5 B = hess(A);      % B = hessenberg matrix of A
6 n1 = n;           % initialize size of matrix B
7 count = 0;        % initialize iteration count
8 eigen(1:n,1) = 0; % initialize eigenvalues
```

```
 9 while 1                % loop until  convergence
10     A1 = B(n1-1:n1,n1-1:n1);  % 2x2 submatrix
11     Tr = trace(A1);          % trace
12     D  = det(A1);            % determinant
13     count = count + 1;       % update iteration count
14     if( Tr^2-4*D < 0 )       % Francis double shift
15         M = B*B - Tr*B + D*eye(n1); % determine M
16         [Z,R] = qrgivens(M);  % estimate Q and R matrices
17         B = Z'*B*Z;           % update B;
18     else                     % Wilkinson  shift
19         e = eig(A1);         % eigenvalue of 2x2 submatrix
20         if( abs(e(1)-A1(2,2)) <= abs(e(2)-A1(2,2)) )
21             mu = e(1);       % select Wilkinson shift
22         else
23             mu = e(2);
24         end
25         [Q,R] = qrhessenberg(B-mu*eye(n1)); % QR factorization
26         B = R*Q+mu*eye(n1);   % update B
27     end
28     if(abs(B(n1-1,n1-2)) < 1e-6) % check for deflation
29         e = eig(B(n1-1:n1,n1-1:n1)); % (Francis step)
30         eigen(n-n1+1) = e(1);      % Update eigenvalues
31         eigen(n-n1+2) = e(2);
32         B1 = B(1:n1-2,1:n1-2);     % deflate matrix B
33         B  = B1;
34         n1 = n1-2;                 % update rank of B
35     elseif(abs(B(n1,n1-1)) < 1e-6 ) % check for deflation
36         eigen(n-n1+1) = B(n1,n1);  % (Wilkinson step)
37         B1 = B(1:n1-1,1:n1-1);     % deflate B
38         B  = B1;
39         n1 = n1-1;                 % update rank of B
40     end
41     if(n1 == 1)                    % if rank of B is 1 or 2
42         eigen(n)= B(1,1);          % update eigenvalues
43         break                      % break the loop
44     elseif(n1 == 2)
45         e = eig(B);
46         eigen(n-1) = e(1);
47         eigen(n)   = e(2);
48         break                      % break the loop
49     end
50 end
```

Example 3.15

Determine the eigenvalues of the matrix

$$\mathbf{A} = \begin{pmatrix} 2 & 1 & 1 & 4 & 2 \\ 0 & 2 & -1 & 0 & 5 \\ 1 & -4 & 4 & 7 & 3 \\ 2 & -3 & 2 & -5 & 4 \\ 2 & 1 & 1 & 3 & 6 \end{pmatrix}$$

Compare different methods in terms of convergence.

Solution :

MATLAB Programs 3.3 and 3.6 have been used to compare different methods. The following table lists the performance of different methods. It has to be noted that same convergence criteria (sub diagonal element less than 10^{-6}) has been adopted for all the methods.

Method	Number of iterations
Simple QR	157
QR + Hessenberg	137
QR + shift	9
QR + Hessenberg + shift	8

The detailed output of the Program 3.6 after every iteration is given now.

```
% Hesenberg matrix
2        -4.3333    0.7633    1.4223    -0.7853
-3        6.8889    2.4394    3.9965    -2.9361
0         8.1027   -3.1078   -1.1834    2.283
0         0        -2.4375   0.914     5.3054
0         0         0        4.5649    2.3049
% Matrix after iteration 1
0.1066   1.9124    2.0455    3.4806    0.7168
4.6687   0.4972   -8.5003    1.4168    4.2339
0       -4.8981    2.7393   -6.1354    0.0715
0        0        -2.7703   -0.9211   -0.5916
0        0         0        0.0019    6.5781
% Matrix after iteration 2
-2.8823   4.9107    1.1637    4.7719   -1.8949
3.6315   -1.6656    2.2611   -1.7928    2.4227
0         2.7613    6.2293    2.9841   -1.3618
0         0         7.5645    0.7411   -2.7349
0         0         0         0         6.5775    %λ₁=6.5775
% Matrix after iteration 3
-3.69    -3.9371   2.2168    1.8185
-5.425    1.8714   6.6009    0.1199
0         4.8792   5.3634   -5.015
0         0        1.0507   -1.1222
% Matrix after iteration 6
-0.7835  10.1254   0.1236   -0.6807
6.6472    2.8967   -2.751   -4.8846
0         0.0057   0.3833   -2.4842
0         0         0       -0.0741    %λ₂=-0.0741
% Matrix after iteration 8
1.4169   10.3173  -0.2321
6.8376    0.6961   -2.7477
0         0        0.3835               %λ₃=0.3835
% the eigenvalues
eigen = [6.5775;    -0.0741;    0.3835;    9.4634;    -7.3504]
```

Example 3.16

Determine the eigenvalues of the matrix

$$
\begin{pmatrix}
2 & 1 & 1 & 5 & 3 \\
0 & 2 & -1 & 0 & 8 \\
1 & -4 & 4 & 7 & 1 \\
2 & -3 & 2 & -5 & 4 \\
5 & 1 & 1 & 3 & 6
\end{pmatrix}
$$

Solution :

MATLAB Program 3.6 has been used to determine the eigenvalues. All the eigenvalues are determined in 6 iterations. The detailed output of the program has been given below.

```
% Hessenberg matrix
2.0000   -4.746   -0.759   -1.725   -3.148
-5.477    7.733   -0.447    0.705   -1.210
     0    7.678    0.162   -1.071    0.640
     0        0   -3.648    6.164   -0.993
     0        0        0   -5.229   -7.060
% Matrix after 3 iterations
11.401    3.698   -0.659    3.689   -0.2229
-2.11     4.797    1.986   -2.16    -2.9475
     0    5.027    2.598    1.908   -1.3652
     0        0   -1.435   -2.852    4.8469
     0        0        0        0   -6.9449    % λ₁ = -6.9449
% Matrix after 6 iterations
10.294    3.206   -1.5710   4.721
-0.461    7.877   -2.6835   3.697
     0        0    0.4810   1.136               % λ₂ = -1.1135 + 0.6249i
     0        0   -2.5821  -2.708               % λ₃ = -1.1135 - 0.6249i
eigen =
  -6.9449
  -1.1135 + 0.6249i    % 'i' refers to pure imaginary
  -1.1135 - 0.6249i    % number in MATLAB
   9.0860 + 0.1396i
   9.0860 - 0.1396i
```

Concluding remarks

In this chapter we have discussed different types of problems that lead to eigenvalues. We have also given physical as well as geometric interpretation of eigenpairs. Since analytical methods are not always feasible numerical determination of eigenvalues are necessary. When the largest (smallest) eigenvalue alone is required, it is possible to use the power method or its variants.

However, if all eigenvalues are required it is advisable to convert the matrix to Hessenberg form and use QR factorization method with Francis or Wilkinson shift.

3.A   MATLAB routines related to Chapter 3

MATLAB routine	Function
poly(A)	returns characteristic polynomial of matrix **A**
eig(A)	returns eigenvalues of matrix **A**
eig(A,B)	returns eigenvalues of generalized matrix **A,B**
eigs(A)	returns dominant eigenvalues (first 6) of matrix **A**
eigs(A,B)	returns dominant eigenvalues (first 6) of generalized matrix **A,B**
hess(A)	reduces matrix **A** to Hessenberg form
schur(A)	determines Schur form of matrix **A**
qz(A,B)	performs **QZ** decomposition of generalized matrix system **A,B**

3.B   Suggested reading

1. **E. Kreyszig** *Advanced engineering mathematics* Wiley-India, 2007
2. **S. Lipschutz** and **M. Lipson** *Schaum's outline of theory and problems of linear algebra* Schaum's Outline Series, 2001
3. **G. Strang** *Introduction to linear algebra* Wellesley Cambridge Press, 2000
4. **C. Meyer** *Matrix analysis and applied linear algebra* Society for Industrial and Applied Mathematics, 2000
5. **G. Golub** and **C. Van Loan** *Matrix computations* Johns Hopkins Univ Pr, 1996
6. **E. Anderson** and **others** *LAPACK Users' guide* Society for Industrial and Applied Mathematics, 1999

Chapter 4

Solution of algebraic equations

Algebraic equations are routinely encountered while modeling problems in various areas of engineering. Finding the solution of such equations is also referred to as "root" finding or finding the "zeroes" of the equation. Sometimes equations occur as simultaneous equations and the solution of such equations is extremely important in engineering applications. Such solutions represent operating point(s) of an engineering system, as will be made clear later on. Several methods of root finding listed below will be considered here.

- *Bisection method*
- *Fixed point iteration method*
- *Gauss Seidel iteration method for a set of nonlinear equations*
- *Newton Raphson[a] method*
- *Secant method*
- *Regula falsi method*

We also consider optimization problems since they are essentially root finding problems.

---

[a]after Isaac Newton, 1642-1727, an English physicist and mathematician and Joseph Raphson, 1648-1715, an English mathematician known best for the Newton Raphson method

4.1 Univariate nonlinear equation

Non-linear equation of a single variable is referred to as univariate non-linear equation. Many interesting physical problems lead to such equations. Roots or zeroes of such equations represent interesting features of the functions. These are useful in graphing the function as well as in understanding the behavior of physical systems. Hence root finding is an important problem from numerical analysis point of view since analytical methods are not always adequate.

4.1.1 Plotting graph: the simplest method

The simplest way to localize the roots of a given function is to make a plot. We demonstrate a simple method for locating the approximate position of the root by an example.

Example 4.1

Locate the zeroes of the polynomial $y(x) = \dfrac{1}{8}(35x^4 - 30x^2 + 3)$ defined in the interval $-1 \le x \le 1$.

Background :
The function $y(x)$ defined in Example 4.5 is fourth order **Legendre polynomial** $P_4(x)$. Legendre polynomials are solutions to Legendre differential equation and are associated with the solution of Laplace equation[1] in spherical coordinates. Legendre polynomials form an orthogonal set and are used frequently in analysis. Similar to Fourier series (based on circular functions) used for approximating functions in problems involving Cartesian coordinate system, Legendre polynomials are used as Fourier Legendre series for approximating functions in spherical coordinates. The roots of the Legendre polynomials also represent the nodes in Gauss Quadrature (considered in a later chapter).

Solution :
We locate the root by plotting the function in MATLAB

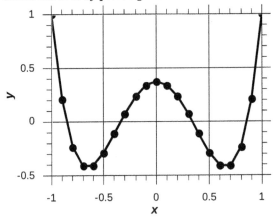

Figure 4.1: Plot of $y = f(x)$

---

[1]after Pierre-Simon, marquis de Laplace, 1749-1827, French mathematician and astronomer

```
%MATLAB Program for plotting a function
x = [-1:0.1:1];                    % Creating array of x
                                   % with 0.1 spacing
y = (35*x.^4 - 30*x.^2 + 3)/8;     % Defining y
plot(x,y);                         % plotting x and y
```

By plotting, it is seen that $y(x)$ has four real roots between the points $(-0.9, -0.8)$, $(-0.4, -0.3)$, $(0.3, 0.4)$ and $(0.8, 0.9)$.

The location of root can be further improved by refining the grid near the location of roots. Once the root has been located approximately, methods such as **bisection method**, **fixed point iteration** and **Newton Raphson** can be applied to obtain a more accurate estimates for the root.

4.1.2 Bracketing methods

By plotting, we can easily find the points on either side of a root. The bounds enclosing the roots can be further reduced by applying the bracketing methods. The simplest of these method is the bisection method.

Bisection method

In the bisection method the interval is halved after every iteration. The method may be explained with reference to Figure 4.2.

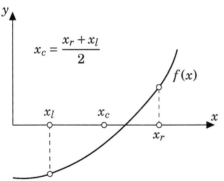

Figure 4.2: *Bisection method for determining the root of an equation*

Step 1 Calculation begins with the choice of two points, one to the left of the root (x_l) and the second to the right of the root (x_r). Thus the function changes sign between the two chosen points, i.e. the function crosses the x axis at some location between the two chosen points. Hence $f(x_l) \times f(x_r) < 0$.

Step 2 As a better choice for the root we try a point mid way between x_l and x_r i.e $x_c = \dfrac{x_l + x_r}{2}$.

Step 3

(a) If $f(x_c) = 0$, x_c is the root.

(b) If $f(x_c) \times f(x_r)$ is positive the mid point is on the same side of the root as x_r and we replace x_r by x_c and continue the process.

(c) Otherwise we replace x_l by x_c and continue the process. At any stage of the iteration process the root is somewhere between the current values of x_l and x_r.

Example 4.2

Determine one of the roots of the quadratic equation $x^2 - 3x - 2 = 0$ by the bisection method.

Solution :

The quadratic equation may easily be solved leading to a closed form solution.   This procedure is familiar to the reader and leads to the following two roots:

$$x_1 = \mathbf{3.561553}\,;\, x_2 = -\mathbf{0.561553}$$

First root We note that the first root is between **3** and **4**. We may start with these two values as the starting values for the use of the bisection method. The calculation proceeds as indicated in the following table.

Spreadsheet for the bisection method of Example 4.2

x_l	x_c	x_r	f_l	f_c	f_r	S_1	S_2
3	3.5	4	-2	-0.25	2	+	-
3.5	3.75	4	-0.25	0.8125	2	-	+
3.5	3.625	3.75	-0.25	0.2656	0.8125	-	+
3.5	3.5625	3.625	-0.25	0.0039	0.2656	-	+
3.5	3.5313	3.5625	-0.25	-0.1240	0.0039	+	-
3.5313	3.5469	3.5625	-0.1240	-0.0603	0.0039	+	-
3.5469	**3.5547**	3.5625	-0.0603	-0.0283	0.0039	+	-
...	...	...	...	...	...	...	...

$f_r = f(x_r)$ etc., $S_1 = \text{SIGN}(f_l \times f_c)$ and $S_2 = \text{SIGN}(f_c \times f_r)$

The calculation may be continued till we get a value acceptable to us. The best value, at any stage is given by the value in the second column labeled x_c.
Second root lies between **-1** and **0**. A MATLAB code has been written to perform the iterations.

```
xr = 0;                          % Input: first bound
xl = -1;                         % Input: second bound
tolerance = 1e-4;                % tolerance for stopping iteration
f = inline('x^2-3*x-2');         % definition of function "inline"
fr = f(xr);                      % evaluate function at first bound
fl = f(xl);                      % evaluate function at second bound
xc  = (xr+xl)/2;                 % calculate mid point
residual = abs(xr-xl)/2;         % residual (half of the range)
count = 1;                       % initialize iteration number
while ( residual >= tolerance)   % loop for calculating root
    fc = f(xc);                  % evaluate function at mid point
    if( abs(fc) <= tolerance)    % if fc≈0
        break;                   % break loop
    elseif(f(xc)*f(xr) >= 0 )    % check for position of root
        xr = xc;                 % update xr
        fr = fc;                 % update fr
    else
        xl = xc;                 % update xl
```

```
        fl  = fc;                % update  fl
    end
    xc   = (xr+xl)/2;            % update  center  point
    residual = residual/2;       % update  residual
    count = count + 1;           % increment iteration count
end                              % end of loop
% the  root  of  the  algebraic  equation  is  xc
```

The second root of the quadratic equation obtained after **14** iterations is

```
xc = -0.5616
```

Notes on bracketing methods

1. Bracketing methods always converge to a root for a unimodal function that has only one root in the initial range.
2. Other bracketing methods such as **Golden search** and **Fibonacci search** method may be used to improve the speed of convergence.
3. Bracketing methods can be followed up with **fixed point iteration methods** to obtain more accurate roots.

4.1.3   Fixed point iteration method

The method is so named after the fact that it converges towards the root, also referred to as the **fixed point**, by an iterative process, starting from an initial guess value, possibly in the **close vicinity** of the fixed point.

Consider the solution of a non-linear equation $f(x) = 0$. We recast, by a suitable manipulation, this equation in the form

$$x = g(x) \tag{4.1}$$

We may interpret the above as equivalent to finding the point of intersection of line $y_1 = x$ and the curve $y_2 = g(x)$. The determination of the fixed point itself is by an iteration process.

Step 1 We start the iteration with a guess value $x = x_g$.

Step 2 Equation 4.1 is used to get a better value (closer to the intersection point and hence the root of the equation) as $x_b = g(x_g)$.

Step 3

    a If the difference $|x_b - x_g| \leq \varepsilon$ where ε is a specified tolerance, the iteration stops.

    b Otherwise set $x_g = x_b$ and continue the iteration process.

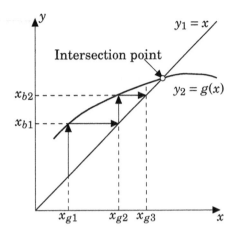

Figure 4.3: *Approach towards the root by the fixed point iteration method*

Geometrical interpretation of the fixed point iteration method is presented using Figure 4.3 as an illustration. The figure shows the line $y_1 = x$ passing through the origin and the curve $y_2 = g(x)$. The line will make $45°$ with the x axis if the the y and x axes use identical scales. We start the iteration by choosing $x_g = x_{g1}$ as shown. The better value x_{b1} is obtained as the corresponding ordinate on the curve. We see that we have already moved closer to the intersection point. We now set $x_{g2} = x_{b1}$ and move to point x_{b2}. This process is continued till we approach the intersection point within the specified tolerance.

Example 4.3

Determine cube root of 2 using the fixed point iteration method.

Solution :

The problem is equivalent to obtaining the real root of equation

$$f(x) = x^3 - 2 = 0$$

We may recast the equation in the form

$$f(x) = x^3 - 2 = x^3 + x^3 - x^3 - 2 = 2x^3 - x^3 - 2 = 0$$

or

$$2x^3 = x^3 + 2 \ or \ x = \frac{x^3 + 2}{2x^2} = \frac{x}{2} + \frac{1}{x^2}$$

The last part is interpreted as being in the form $x = g(x)$ where $g(x) = \frac{x}{2} + \frac{1}{x^2}$, useful for the application of the fixed point iteration method.

Table below shows the spreadsheet that is used for obtaining the root by fixed point iteration method. The root has converged to 5 significant digits after the decimal point in 16 iterations. The convergence of the method is also shown graphically in Figure 4.4.

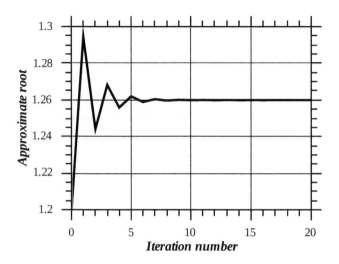

Figure 4.4: *Convergence of the fixed point iteration method for Example 4.3*

Spreadsheet for the fixed point iteration method of Example 4.3

Trial No.	x_g	x_b	$x_b - x_g$	Trial No.	x_g	x_b	$x_b - x_g$
1	1.2	1.2944	0.0944	7	1.2589	1.2604	0.0015
2	1.2944	1.244	-0.0504	8	1.2604	1.2597	-0.0008
3	1.244	1.2682	0.0242	9	1.2597	1.2601	0.0004
4	1.2682	1.2559	-0.0123	10	1.2601	1.2599	-0.0002
5	1.2559	1.262	0.0061	11	1.2599	1.26	0.0001
6	1.262	1.2589	-0.0031	12	1.26	1.2599	-0.0001

Example 4.4

Find the roots of the transcendental equation $f(x) = 1 - 3x + 0.5xe^x = 0$ by the fixed point iteration method.

Solution :

The above transcendental equation can be rewritten in the form $x = g(x)$ in the following three possible ways

$$x = g_1(x) = \frac{1}{3} + \frac{xe^x}{6}; \quad x = g_2(x) = \ln\left[\frac{3x-1}{0.5x}\right]; \quad x = g_3(x) = (6x-2)e^{-x}$$

On plotting the above three functions (Figure 4.5), we can see that these curves intersect the straight line $y = x$ at two points **A** (near $x = 0.5$) and **B** (near $x = 1.5$).

Let us apply fixed point iteration method to each of the above forms of equations and obtain the two roots. $g_1(x)$: Let us start the calculation with a guess value of $x_g = 1$. We obtain the root as $x = 0.4516$ correct to four places after the decimal point. Spreadsheet 1 presented below shows convergence details.

In order to obtain the root at **B**, we take the guess value as $x_g = 1.6$. However, the fixed point iteration diverges away from the solution. No matter what the guess value we take for this form of equation, we either end up at root **A** or solution diverges.

Spreadsheet-1 for the fixed point iteration method in Example 4.4

Trial No.	x_g	x_b	$x_b - x_g$	Trial No.	x_g	x_b	$x_b - x_g$
1	1	0.7864	-0.2136	7	0.4560	0.4533	-0.0028
2	0.7864	0.6211	-0.1653	8	0.4533	0.4522	-0.0011
3	0.6211	0.5260	-0.0951	9	0.4522	0.4518	-0.0004
4	0.5260	0.4817	-0.0443	10	0.4518	0.4516	-0.0002
5	0.4817	0.4633	-0.0184	11	0.4516	0.4516	-0.0001
6	0.4633	0.4560	-0.0072	12	0.4516	0.4516	0.0000

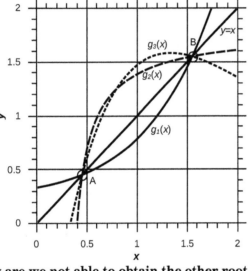

Figure 4.5: *Plot of the transcendental function in different ways*

Why are we not able to obtain the other root with $g_1(x)$?

$g_2(x)$ **:** Let us try to obtain the roots with second form of equation. We start the calculation with a guess value of $x_g = 3$. MATLAB program has been provided below for determining the root of the above equation. Readers are encouraged to change the function and initial guess values and observe the convergence/divergence of fixed point iteration.

```
% Inputs
xg = 3;  % initial guess of root reader may change the value
tolerance = 1e-4;   % tolerance for stopping iteration
% define function   x = g(x) reader may change function
g = inline('log((3*xg-1)/(0.5*xg))'); %  the function
count = 1;  % initialize iteration number
while 1                              % loop for calculating root
    xb = g(xg);                      % new value of root
    residual = abs(xb-xg);           % calculate residual
    count = count + 1;               % increment iteration count
    xg = xb;                         % update value of root
    if (residual < tolerance)        % check for convergence
        break
    end
end
% the root of the equation is xb
```

The second point converges to $x = \mathbf{1.5495}$ correct to four places after the decimal point. The spreadsheet is presented in tabular form below.

Spreadsheet-2 for the fixed point iteration method in Example 4.4

Trial No.	x_g	x_b	$x_b - x_g$	Trial No.	x_g	x_b	$x_b - x_g$
1	3	1.6740	1.3260	5	1.5502	1.5496	0.0005
2	1.6740	1.5697	0.1043	6	1.5496	1.5496	0.0001
3	1.5697	1.5531	0.0167	7	1.5496	1.5495	0.0000
4	1.5531	1.5502	0.0029	8	1.5495	1.5495	0.0000

However, we fail to obtain root at **A** using $g_2(x)$ with any guess values. It is left to the reader to probe the convergence/divergence of $x = g_3(x)$.

Convergence of fixed point iteration How do we know whether a solution converges or diverges, and which of the roots is obtained by a given formulation?

An iterative scheme is said to converge when the ratio of errors of successive iterations is less than 1.

$$\frac{E_{n+1}}{E_n} = \left| \frac{x_{n+1} - x_n}{x_n - x_{n-1}} \right| \le 1$$

It is easy to see that the iteration will converge to the root if $\left| \dfrac{dg}{dx} \right|_{x_b} \le 1$, i.e. if the improved value moves closer to the straight line $y = x$. However the process sometimes converges even if the initial guess value does not satisfy this condition.

The speed of convergence of a given method can be determined from the relationship between two successive errors $|x_{n+1} - x_n| = C |x_n - x_{n-1}|^k$ for large n, where C is a proportionality constant and k is known as the **order** of convergence.

The rate of convergence of fixed point iteration is first order i.e. $|x_{n+1} - x_n| = C |x_n - x_{n-1}|$.

In general, the order of convergence of any numerical scheme can be found by plotting E_n and E_{n+1} on a Log-Log graph and finding the slope of the resultant curve. Such a plot is presented in Example 4.7.

4.1.4 Newton Raphson method

Newton Raphson method is also a **fixed point iteration method**. However, the formulation of the alternate form given by Equation 4.1 is based on the use of Taylor expansion around the guess point, that includes the first derivative term. We write the value of function at x_b as

$$f(x_b) = f(x_g) + \frac{df}{dx}\bigg|_{x_g} (x_b - x_g) + \text{Higher order terms} \qquad (4.2)$$

If we ignore the higher order terms, set the left side to zero, we get, on rearrangement

$$x_b = \underbrace{x_g - \frac{f(x_g)}{f^{(1)}(x_g)}}_{g(x_g)} \qquad (4.3)$$

where we have also introduced the notation $f^{(1)} = \dfrac{df}{dx}$. We notice that Equation 4.3 is in the form of Equation 4.1 with $g(x) = x - \dfrac{f(x)}{f^{(1)}(x)}$. A geometric derivation of the Newton

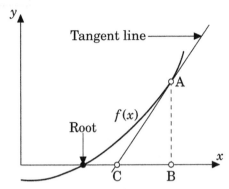

Figure 4.6: *Geometric derivation of the Newton Raphson method*

Raphson method is also possible. Referring to Figure 4.6 we choose the point C to be the point at which tangent drawn to the curve at A cuts the x axis. From the triangle ABC we than have

$$\text{Slope of tangent} = f^{(1)}(x_g) = \frac{AB}{BC} = \frac{f(x_g) - 0}{x_g - x_b} \tag{4.4}$$

On rearrangement the above reduces to the expression given by Equation 4.3. The iterative procedure followed for Newton Raphson method is similar to that for the fixed point iteration method.

Example 4.5

Solve for one of the roots of the cubic $x^3 - 6x^2 + 8x + 0.8 = 0$ by the Newton Raphson method.

Solution :

By making a plot of the function it is seen that all the three roots are real (see Figure 4.1).[2]

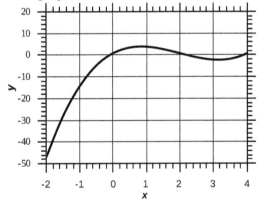

Figure 4.7: *Plot of the cubic in Example 4.5*

The derivative required for the Newton Raphson method is given by $f^{(1)}(x) = 3x^2 - 12x + 8$ and the Newton Raphson uses the algorithm

$$x_b = x_g - \frac{x_g^3 - 6x_g^2 + 8x_g + 0.8}{3x_g^2 - 12x_g + 8} \tag{4.5}$$

---

[2]See "Solving Cubic Equations" by John Kennedy, available for download at http://homepage.smc.edu/kennedy_john/PAPERS.HTM, that explains how one can determine the nature of the roots of the cubic.

Depending on the starting guess value the Newton Raphson method yields one of the three roots. For example, if we start with value $x_g = -0.5$ the method yields the root $x = -0.093361$ after just 5 iterations. A MATLAB code has been presented below for this choice of the starting value. The reader is encouraged to change the values of initial guess in the MATLAB program and check for convergence/divergence of the iteration scheme.

```
xg = -0.5;                 % initial guess
tolerance = 1e-4;          % specified tolerance for convergence
f = inline('x^3-6*x^2+8*x+0.8'); % define function
f1 = inline('3*x^2-12*x+8');     % define first derivative
count = 0;                 % initialize iterative count
while 1                    % loop for calculating root
    xb = xg - f(xg)/f1(xg);    % update root
    residual = abs(xb-xg);     % calculate residual
    count = count + 1;         % increment iteration count
    xg = xb;                   % update old value of root
    if (residual < tolerance)  % check for convergence
        break                  % if converged break loop
    end
end
```

Spreadsheet for Newton Raphson method in Example 4.5

Trial No.	x_g	x_b	$f(x_b)$	% Change
1	-0.5	-0.172881	-0.767546	-65.423729
2	-0.172881	-0.097367	-0.036741	-43.679852
3	-0.097367	-0.093372	-0.000100	-4.103030
4	-0.093372	-0.093361	-0.000000	-0.011751
5	-0.093361	-0.093361	0	-0.000000

Convergence of Newton Raphson method Convergence of Newton Raphson method to the root is assured when

$$\frac{1}{2}\left|\frac{f^{(2)}(x)}{f^{(1)}(x)}(x_n - x_{root})\right| \le 1$$

where x_n is the nth iterate and x_{root} is the exact root. The above condition is based on the second derivative term that was neglected in the Taylor expansion of the function around the guess value for the root. The following points are important to ensure convergence of NR method

- The initial guess for NR method must be sufficiently close to the root. Otherwise, the solution may diverge away from exact solution.
- NR may fail when the first derivative is very small. As $f^{(1)}(x) \to 0$, we may have division by zero error.

Under certain conditions, the speed of convergence of NR method is quadratic i.e. $|x_{n+1} - x_n| = C|x_n - x_{n-1}|^2$.

Newton Raphson and multiple roots

Consider the function $f(x) = x^3 - 2x^2 + x$. The roots of the equation are 0, 1 and 1 i.e. the root at $x = 1$ has multiplicity of 2. the first derivative at $x = 1$ is equal to zero, which means $f(x)/f^{(1)}(x)$ is not defined at $x = 1$. As the first derivative is zero at the root, the convergence is poor. Starting the iteration from $x = 2$, it takes 20 iterations to converge within six decimal accuracy. In general, when a root displays multiplicity as in the function considered, NR method reduces to a first order method. However, the convergence of NR method can be accelerated by modifying Equation 4.3

$$x_b = x_g - M \frac{f(x_g)}{f^{(1)}(x_g)} \tag{4.6}$$

where M is the multiplicity of the root. Applying above equation to $f(x) = x^3 - 2x^2 + x$, we achieve convergence within six decimals accuracy in only 5 iterations!

4.1.5   Secant method

Newton Raphson method requires calculation of function and its derivative at the guess value. But, there can be instances where calculation of derivatives are not as simple as in the examples discussed earlier. The secant method uses only function calculations, avoiding the calculation of derivatives. The procedure starts with two points on the curve say x_{i-1} and x_i as shown in Figure 4.8.

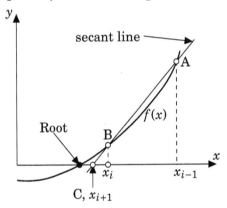

Figure 4.8:   *Geometry for the Secant method*

The secant line is a straight line that passes through these two points. It is extended to cut the x axis at point x_{i+1}. This point is expected to be closer to the root than the two points we started with. From geometry, we have

$$\frac{f(x_{i-1}) - f(x_i)}{x_{i-1} - x_i} = \frac{f(x_i) - 0}{x_i - x_{i+1}}$$

This may be rearranged as

$$x_{i+1} = x_i - \frac{f(x_i)(x_{i-1} - x_i)}{f(x_{i-1}) - f(x_i)} \tag{4.7}$$

The above equation can be interpreted as Newton Raphson method with derivative $f^{(1)}(x)$ evaluated using forward differences (to be described in a later chapter). Iteration proceeds by replacing x_i by x_{i+1} and x_{i-1} by x_i. Iteration stops when a stopping condition is satisfied.

Example 4.6

Solve the cubic $x^3 - 6x^2 + 8x + 0.8 = 0$ of Example 4.5 by the Secant method.

Solution :

We start the application of the secant method by using $x_1 = 2.5$ and $x_2 = 2$. Using the secant method x_3 is obtained as 2.213333. We now replace x_1 by x_2 and x_2 by x_3 and continue the process as indicated in table below. The converged value of the root is 2.202063, accurate to 6 significant digits after the decimal point.

The MATLAB code and the spreadsheet for secant method have been provided below.

```
% Input
x1 = 2.5;                   % first guess value
x2 = 2.0;                   % second guess value
tolerance = 1e-4;           % tolerance for convergence
f = inline('x^3-6*x^2+8*x+0.8');       % define function
f1 = f(x1);                 % evaluate function at first point
f2 = f(x2);                 % evaluate function at second point
count = 0;                  % initialize iteration count
residual = abs(x1-x2)       % initialize residual
while (residual > tolerance)
    xb = x1 - f1*(x2-x1)/(f2-f1);   % evaluate new point
    residual = abs(xb-x1);          % evaluate residual
    count = count + 1;              % update iteration count
    x2 = x1;                        % update x2
    f2 = f1;
    x1 = xb;                        % update x1
    f1 = f(x1);                     % function at first point
end
```

Spreadsheet for secant method in Example 4.6

Trial No.	x_{n-1}	x_n	$f(x_{n-1})$	$f(x_n)$	x_{n+1}
1	2.5	2	-1.075	0.8	2.213333
2	2	2.213333	0.8	-0.043624	2.202302
3	2.213333	2.202302	-0.043624	-0.000927	2.202062
4	2.202302	2.202062	-0.000928	0.000002	2.202063
5	2.202062	2.202063	0.000002	-0.000000	2.202063
6	2.202063	2.202063	-0.000000	0.000000	2.202063

Convergence of Secant method: The order of convergence of the Secant method, under special circumstances is around 1.618 (known as the golden ratio). Convergence of secant method will be slower for functions with multiple roots. All points discussed for convergence of NR method are also applicable for convergence of secant method. Convergence of secant method will be poor when the slope of the secant near the root is very small. In case of multiple roots, one can apply the same acceleration technique as prescribed for NR method.

Example 4.7

Obtain a root of equation $x^{2.1} - 0.5x - 3 = 0$ by the fixed point iteration method, Newton

Raphson method and secant method. Compare the three methods from the view point of convergence. Use a starting value of $x_g = 2$ in each case.

Solution :

The fixed point iteration scheme uses the following for carrying out the iterations.

$$x_b = (0.5x_g + 3)^{\frac{1}{2.1}} \tag{4.8}$$

Table below shows the spreadsheet for this case.

Spreadsheet for fixed point iteration method in Example 4.7

Trial No.	x_b	x_g	E_n	E_n/E_{n-1}
1	2.000000	1.935064	0.064936	
2	1.935064	1.927568	0.007495	0.115428
3	1.927568	1.926701	0.000867	0.115704
4	1.926701	1.926600	0.000100	0.115736
5	1.926600	1.926589	0.000012	0.115739
6	1.926589	1.926587	0.000001	0.115740
7	1.926587	1.926587	0.000000	0.115740

The last column of the table indicates first order convergence of the fixed point iteration method.

We set up the Newton Raphson iteration scheme by noting that $f^{(1)}(x) = 2.1x^{1.1} - 0.5$ and hence

$$x_b = x_g - \frac{x_g^{2.1} - 0.5x_g - 3}{2.1x^{1.1} - 0.5} \tag{4.9}$$

Table below shows the spreadsheet for this case.

Spreadsheet for fixed point iteration method in Example 4.7

Trial No.	x_b	x_g	E_n	E_n/E_{n-1}^2
1	2.000000	1.928253	0.071747	
2	1.928253	1.926588	0.001664	0.323316
3	1.926588	1.926587	0.000001	0.322863
4	1.926587	1.926587	0.000000	0.322586

The last column of the table indicates second order convergence of the Newton Raphson method.

The root of the equation has also been found by the Secant method with initial values of 2 and 2.5 and the results have been tabulated in the spreadsheet below.

Spreadsheet for Secant method in Example 4.7

Trial No.	x_{n-1}	x_n	$f(x_{n-1})$	$f(x_n)$	x_{n+1}	E_n
1	2.500000	2.000000	2.599739	0.287094	1.937930	0.062070
2	2.000000	1.937930	0.287094	0.043486	1.926849	0.011080
3	1.937930	1.926849	0.043486	0.001001	1.926588	0.000261
4	1.926849	1.926588	0.001001	0.000004	1.926587	0.000001
5	1.926588	1.926587	0.000004	0.000000	1.926587	0.000000

We have compared the convergence characteristics of all the three methods in Figure 4.9. The slope of the convergence curve for Secant method is 1.530, which is between that for fixed point iteration method and Newton Raphson method. The Secant method is faster than fixed point iteration method but slower than the Newton Raphson method.

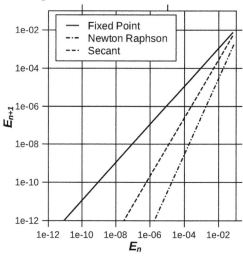

Figure 4.9: *Convergence characteristics of Fixed point iteration, Newton Raphson and Secant methods*

4.1.6 Regula Falsi method

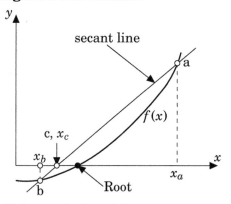

Figure 4.10: *Regula Falsi method*

Regula Falsi method or false position method is a cross between bracketing method and secant method. The main advantage of this method is that convergence is always guaranteed. The calculation starts similar to bisection method, where two guess points x_a and x_b are chosen such that the root is *bracketed* by the points. The new guess point x_c is where the secant line joining the two points intersects the x axis as shown in Figure 4.10. The new set of points for next iteration are chosen as in the case of the bisection method.

(a) If $f(x_c) = 0$, x_c is the root.

(b) If $f(x_a) \times f(x_c)$ is positive, the new guess point is on the same side of the root as x_a. Hence the points for the next iteration are x_c and x_b.

(c) Otherwise the new set of points are x_c and x_a. At any stage of the iteration process the root is bracketed by the two points. The iteration is continued until $|x_a - x_b| \le$ tolerance.

Being a bracketing method, convergence is guaranteed. However, the rate of convergence would depend on the nature of the function. The method is generally slower than the secant method. For certain functions, regula falsi converges faster than the bisection method. For the function $f(x) = x^2 - 3x - 2$ (Example 4.2), regula falsi requires 6 iteration steps for convergence where as bisection method takes 14 iterations.

However, there are certain types of functions for which the method converges rather very slowly. Now let us consider the function $x^5 - 2 = 0$ whose root is equal to 1.1487. Taking the initial interval as $x = 1$ and 2, the results from bisection method and regula falsi method have been summarized in the table on next page.

Bisection method converges in 14 iterations where as regula falsi method takes 34 iterations for convergence. NR method takes 4 iterations where as secant method takes 6 iterations for converged solution. Notice that for regula falsi method, one of the guess points is anchored at 2 for all the iterations and the step size does not go to zero when converged. (in fact the size of the interval is significant). Unlike the bisection method, the reduction in interval for regula falsi is not consistent. The reason for poor convergence of regula falsi method when applied to above example is that the function is concave around the root. In general convergence of regula falsi method would be poor when the order of the nonlinear equation is large.

Bisection method				Regula falsi method			
Iteration	x_a	x_b	$|x_b - x_a|$	Iteration	x_a	x_b	$|x_b - x_a|$
1	1	2	1	1	1	2	1
2	1	1.5	0.5	2	1.03226	2	0.96774
3	1	1.25	0.25	3	1.05825	2	0.94175
4	1.125	1.25	0.125	4	1.07891	2	0.92109
...	...	...	...	...	...	...	...
13	1.14868	1.14892	0.00024	33	1.14867	2	0.85132
14	1.14868	1.14880	0.00012	34	1.14868	2	0.85132

The methods that have been discussed to determine roots of univariate functions have to be applied with caution. The application of these methods depends on the nature of the function. It is difficult to define a well defined function as each method is bound by its own limitations. For iterative methods such as fixed point iteration, NR and secant method, convergence is guaranteed under certain conditions. In such a case, initial guess is important for ensuing convergence. On the other hand, bracketing methods guarantee convergence.

A strategy for ensuring converged solutions is to use bracketing methods to reduce the bound of the solution followed by iterative methods. In practice, it would be better to use a combination of different methods (*hybrid methods*) to determine the roots. The hybrid method automatically decides the best method to determine the solution and hence guarantees convergence of solution to the root. Regula Falsi method is one such method, where secant method and bisection method were combined together (However, regula falsi has its share of limitations).

MATLAB has an inbuilt function **fzero** to determine the roots of a univariate nonlinear equation. The algorithm behind this function is Brent's method to determine roots. Brent's method is a hybrid algorithm which uses bisection method, secant method and quadratic search method (discussed in optimization) to determine the root.

4.2 Multivariate non linear equations

4.2.1 Gauss Seidel iteration

Fixed point iterative method discussed above may be extended to solve a set of nonlinear equations involving more than one variable. This is possible by the use of the Gauss Seidel iteration method, also called 'simulation' in engineering literature. Recall that the Gauss Seidel method has earlier been used for the solution of a set of linear equations in Section 2.8.1. As an example consider a set of two nonlinear equations given by $f_1(x_1, x_2) = 0$ and $f_2(x_1, x_2) = 0$. We rewrite these equations in the following form:

$$\begin{aligned} x_1 &= g_1(x_1, x_2) \\ x_2 &= g_2(x_1, x_2) \end{aligned} \tag{4.10}$$

The solution starts with guess values for the two unknowns. We substitute these in the first equation to get an update for x_1. This updated value and the initial guess value for x_2 are used in the second equation to get an update for x_2. This process is repeated till convergence.

Example 4.8

Find the point of intersection between the curve $x^2 - y^2 = 2$ and the curve $xy = 2$. Confine your solution to the first quadrant only. Use the Gauss Seidel method and start with the guess set $x_0 = 1, y_0 = 1$.

Solution :

Step 1 Equation of the first curve is used to solve for x to get

$$x_{i+1} = \sqrt{2 + y_i^2}$$

where subscript i stands for iteration count.

Step 2 Equation for the second curve is solved for y to get

$$y_{i+1} = \frac{2}{x_{i+1}}$$

For example, the very first iteration would yield

$$x_1 = \sqrt{2 + y_0^2} = \sqrt{2 + 1^2} = 1.732051$$

$$y_1 = \frac{2}{x_1} = \frac{2}{1.732051} = 1.1547005$$

When this process is continued we obtain the values shown in the following table. It is seen that the root converges to 6 digits after decimals in **15** iterations.

Spreadsheet for the Gauss Seidel iteration method in Example 4.8

Iteration Number	x_g	y_g	Iteration Number	x_g	y_g
0	1	1	8	1.798990	1.111735
1	1.732051	1.154701	9	1.798876	1.111805
2	1.825742	1.095445	10	1.798919	1.111779
3	1.788854	1.118034	11	1.798903	1.111789
4	1.802776	1.109400	12	1.798909	1.111785
5	1.797434	1.112697	13	1.798907	1.111786
6	1.799471	1.111438	14	1.798908	1.111786
7	1.798692	1.111919	15	**1.798908**	**1.111786**

Convergence is also shown graphically by plotting the x, y pairs as in Figure 4.11.

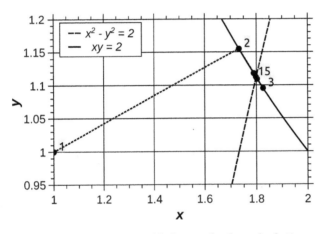

Figure 4.11: Convergence of root in Example 4.8. Numbers indicate the position of root after that many iterations

A Matlab program is presented below to do the calculations.

```
% Inputs
x = 1.0;                    % guess value of x
y = 1.0;                    % guess value of y
tolerance = 1e-4;           % specified tolerance for convergence
count = 0;                  % initialize iteration count
while 1
    xnew = sqrt(2+y^2);     % calculate new value for x
    ynew = 2/xnew;          % calculate new value for y
    residualx = abs(x-xnew); % calculate change in x
    residualy = abs(y-ynew); % calculate change in y
    x = xnew;               % update values of x
    y = ynew;               % update values of y
    count = count + 1;      % increment iteration count
    if( max([residualx; residualy]) < tolerance )
        break               % if converged, break loop
    end
end
```

4.2.2   Newton Raphson method

The Newton Raphson method is easily extended to the solution of set of nonlinear equations. We shall illustrate the procedure for a set of two non-linear equations and then generalize to a larger system of equations.

Let it be required to determine the solution of the two equations given by

$$f_1(x_1, x_2) = 0; \quad f_2(x_1, x_2) = 0 \tag{4.11}$$

These two equations represent curves in x_1, x_2 plane and the root required represents the point of intersection between the two curves. We start with a guess point x_{1g}, x_{2g} and Taylor expand the two functions around the guess point to get

$$
\begin{aligned}
f_1(x_{1b}, x_{2b}) &= f_1(x_{1g}, x_{2g}) + \left.\frac{\partial f_1}{\partial x_1}\right|_{x_{1g}, x_{2g}} (x_{1b} - x_{1g}) + \left.\frac{\partial f_1}{\partial x_2}\right|_{x_{1g}, x_{2g}} (x_{2b} - x_{2g}) + O(\mathbf{x}^T \mathbf{x}) \\
f_2(x_{1b}, x_{2b}) &= f_2(x_{1g}, x_{2g}) + \left.\frac{\partial f_2}{\partial x_1}\right|_{x_{1g}, x_{2g}} (x_{1b} - x_{1g}) + \left.\frac{\partial f_2}{\partial x_2}\right|_{x_{1g}, x_{2g}} (x_{2b} - x_{2g}) + O(\mathbf{x}^T \mathbf{x})
\end{aligned}
\tag{4.12}
$$

Introduce the notation $f_{1g} = f_1(x_{1g}, x_{2g})$, $f_{2g} = f_2(x_{1g}, x_{2g})$, $\left.\frac{\partial f_1}{\partial x_1}\right|_g = \left.\frac{\partial f_1}{\partial x_1}\right|_{x_{1g}, x_{2g}}$, $\left.\frac{\partial f_1}{\partial x_2}\right|_g = \left.\frac{\partial f_1}{\partial x_1}\right|_{x_{1g}, x_{2g}}$, $\left.\frac{\partial f_2}{\partial x_1}\right|_g = \left.\frac{\partial f_2}{\partial x_1}\right|_{x_{1g}, x_{2g}}$, $\left.\frac{\partial f_2}{\partial x_2}\right|_g = \left.\frac{\partial f_2}{\partial x_2}\right|_{x_{1g}, x_{2g}}$, $\Delta x_1 = x_{1b} - x_{1g}$ and $\Delta x_2 = x_{2b} - x_{2g}$.

Ignoring higher order terms, we set both $f_1(x_{1b}, x_{2b})$ and $f_2(x_{1b}, x_{2b})$ to zero, to rewrite Equation 4.12 as

$$
\begin{aligned}
\left.\frac{\partial f_1}{\partial x_1}\right|_g \Delta x_1 + \left.\frac{\partial f_1}{\partial x_2}\right|_g \Delta x_2 &= -f_{1g} \\
\left.\frac{\partial f_2}{\partial x_1}\right|_g \Delta x_1 + \left.\frac{\partial f_2}{\partial x_2}\right|_g \Delta x_2 &= -f_{2g}
\end{aligned}
\tag{4.13}
$$

which may also be recast in the matrix form as

$$
\underbrace{\begin{pmatrix} \left.\dfrac{\partial f_1}{\partial x_1}\right|_g & \left.\dfrac{\partial f_1}{\partial x_2}\right|_g \\[2mm] \left.\dfrac{\partial f_2}{\partial x_1}\right|_g & \left.\dfrac{\partial f_2}{\partial x_2}\right|_g \end{pmatrix}}_{\text{Jacobian matrix}} \underbrace{\begin{pmatrix} \Delta x_1 \\ \Delta x_2 \end{pmatrix}}_{\text{step vector}} = \begin{pmatrix} -f_{1g} \\ -f_{2g} \end{pmatrix}
\tag{4.14}
$$

As indicated above, the square matrix involving the first derivatives is known as the Jacobian matrix. The NR method has replaced the original set of non-linear equations by a set of linear equations. Newton Raphson method is hence a **linearization** scheme which gives a set of linear equations for the step size.

We solve for Δx_1 and Δx_2 starting from an initial guess value and use an iterative solution till each of Δx_1 and Δx_2 is less than or equal to a specified tolerance.

The above procedure may be extended easily to n equations where $n > 2$. We would obtain a system of simultaneous equations of the form

$$\mathbf{J}\Delta\mathbf{x} = -\mathbf{f}$$

where $\mathbf{J}$ is the Jacobian matrix, $\Delta\mathbf{x}$ denotes the change in the value of the vector of roots (step size). The above system of equations can be solved by methods already familiar to us.

Example 4.9

Find an intersection point between the circle $x^2 + y^2 = 4$ and the curve $e^x + y = 1$ using the Newton Raphson method.

Solution :

There are two points of intersection between the circle and the curve as indicated in Figure 4.12. The first one is in the second quadrant while the second is in the fourth quadrant.

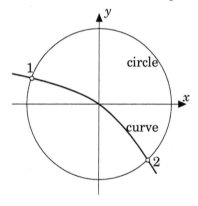

Figure 4.12: *Intersection points between circle and curve of Example 4.9*

The two equations given in the example are written as

$$f_1(x, y) = x^2 + y^2 - 4 = 0; \qquad f_2(x, y) = e^x + y - 1 = 0$$

The partial derivatives required in the Newton Raphson method are given by

$$\frac{\partial f_1}{\partial x} = 2x \quad ; \quad \frac{\partial f_1}{\partial y} = 2y$$

$$\frac{\partial f_2}{\partial x} = e^x \quad ; \quad \frac{\partial f_2}{\partial y} = 1$$

We start with guess values of $x_g = -1, y_g = 0.5$ (a point in the second quadrant) and show below the calculations involved in one iteration. The Jacobian matrix is given by

$$\mathbf{J} = \begin{pmatrix} -2 & 1 \\ 0.367879 & 1 \end{pmatrix}$$

The function values at the guess point are

$$f_1(-1, 0.5) = -2.75; \, f_2(-1, 0.5) = -0.132120$$

Using Kramer's rule the increments in x and y are obtained as

$$\Delta x = \frac{2.75 \times 1 - 0.132120 \times 1}{-2 - 0.367879} = -1.105580$$

$$\Delta y = \frac{0.132120 \times -2 - 2.75 \times 0.367879}{-2 - 0.367879} = 0.538840$$

The new point is obtained as

$$x_b = -1 - 1.105580 = -2.105580; \ y_b = 0.5 + 0.538840 = 1.038840$$

The process converges after two more iterations to the point $-1.816260, 0.837368$. We present below the MATLAB code to obtain the other root.

```
% Inputs
xg = 1;                    % initial guess value of x
yg = -0.5;                 % initial guess value of y
tolerance = 1e-4;          % specified tolerance for convergence
count = 0;                 % initialize iteration count
residual = 100;            % initialize residual
while (residual > tolerance)
    f1 = xg^2+yg^2-4;      % evaluate first function
    f2 = exp(xg)+yg-1;     % evaluate second function
    f1x= 2*xg;             % evaluate first derivatives
    f1y= 2*yg;             % evaluate first derivatives
    f2x= exp(xg);          % evaluate first derivatives
    f2y= 1;                % evaluate first derivatives
    % calculate dx and dy using Cramers rule
    dx = -(f1*f2y-f2*f1y)/(f1x*f2y-f2x*f1y);
    dy = -(f1x*f2-f2x*f1)/(f1x*f2y-f2x*f1y);
    residual = max(abs(dx),abs(dy)); % evaluate residual
    count = count + 1;     % increment iteration count
    xg = xg+dx;            % update x
    yg = yg+dy;            % update y
end
```

The second root obtained in 6 iterations converges to the point $1.0042, -1.7296$.

4.3 Root finding and optimization

Root finding of an equation may also happen naturally when we want to determine the maximum or minimum of a function. From calculus we know that the maximum of a function of one variable occurs at a point where $f^{(1)}(x) = 0$ and $f^{(2)}(x) < 0$. Similarly, minimum of a function would occur where $f^{(1)}(x) = 0$ and $f^{(2)}(x) > 0$. Naturally the methods discussed so far can be adapted for optimization. Let us consider a simple example to demonstrate this.

Example 4.10

Obtain the maximum of the function $f(x) = \dfrac{x^5}{e^x - 1}$.

Background :

A note on Planck's distribution: Function given in this example is related to the Planck distribution function describing the variation of black body emissive power with wavelength. x is equivalent to the reciprocal of product of wavelength of emission λ and temperature of black body T. The solution to the exercise is related to the Wein's displacement law which states that the wavelength at which emission from a black body is maximum is inversely proportional to the temperature of the black body. The appropriate relationship is given by

$$\lambda_{max} = \frac{2897.8}{T} \mu m\, K.$$

Solution :

The derivative of the function is given by

$$f^{(1)}(x) = \frac{5x^4}{e^x - 1} - \frac{x^5 e^x}{(e^x - 1)^2} \tag{4.15}$$

It is evident that the corresponding root finding problem may be written as

$$g(x) = 5(e^x - 1) - xe^x = 0 \tag{4.16}$$

The Newton Raphson method requires the derivative given by

$$g^{(1)}(x) = 5e^x - e^x - xe^x = 4e^x - xe^x \tag{4.17}$$

Newton Raphson method then uses the following for iteration:

$$x_b = x_g - \frac{5(e^{x_g} - 1) - x_g e^{x_g}}{4e^{x_g} - x_g e^{x_g}} \tag{4.18}$$

We start the iteration process with a guess value of $x_g = 5$ and obtain the converged value of $x_b = 4.965114$, correct to 6 significant digits after the decimal point, after just 4 iterations (see table below).

Spreadsheet for Newton Raphson method in Example 4.10

Trial No.	x_g	x_b	$g(x_b)$	|% Change|
1	5	4.966310	-0.165643	0.673795
2	4.966310	4.965116	-0.0002013	0.024054
3	4.965116	4.965114	-0.0000003	0.000029
4	4.965114	4.965114	0	0.000000

We now calculate the second derivative $g^{(2)}(4.965114)$ as

$$g^{(2)}(4.965114) = (3 - 4.965114)e^{4.965114} = -281.649843$$

The second derivative is negative and hence the root determined above represents the point of maximum for the original function. The maximum value of the function may be determined as

$$f(4.965114) = \frac{4.965114^5}{e^{4.965114} - 1} = 21.201436$$

4.3.1 Search methods of optimization: Univariate case

Even though Newton Raphson method is useful, as shown above, for optimizing a function of a single variable, the procedure may not yield a global optimum. There may be several optima that may be obtained by using the Newton Raphson method, starting with different initial guess values. The one that yields a global optimum may be obtained by sorting these.

Many search methods are available[3] that need only function calculation and not the calculation of the derivative. We have already discussed Bisection method for root finding. The same can be used for finding the maxima and minima. One technique which is very useful and that is based on interpolation, known as the quadratic interpolation method, will be described here.

Quadratic interpolation method

The idea of quadratic interpolation or quadratic search method is to assume the function to be quadratic and determine the minimum/maximum of the quadratic function. We start the search initially with three points x_1, x_2 and x_3 which are hopefully close to the optimum. We calculate three function values at these three points, say $f(x_1) = f_1$, $f(x_2) = f_2$ and $f(x_3) = f_3$. We may easily fit a quadratic through these three function values.

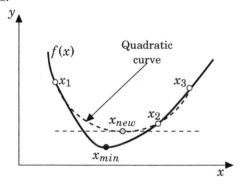

Figure 4.13: *Functions values for quadratic interpolation*

A typical example of a fictitious function passing through three points is shown in Figure 4.13. A quadratic function of the form $q(x) = a + bx + cx^2$ defined in the interval $x_1 < x < x_3$ that passes through the three points satisfies the following equations.

$$f_1 = a + bx_1 + cx_1^2; \quad f_2 = a + bx_2 + cx_2^2; \quad f_3 = a + bx_3 + cx_3^2 \tag{4.19}$$

The coefficients of the quadratic equation are determined by solving the following matrix equation

$$\begin{pmatrix} 1 & x_1 & x_1^2 \\ 1 & x_2 & x_2^2 \\ 1 & x_3 & x_3^2 \end{pmatrix} \begin{pmatrix} a \\ b \\ c \end{pmatrix} = \begin{pmatrix} f_1 \\ f_2 \\ f_3 \end{pmatrix} \tag{4.20}$$

---

[3]See, for example, S.S. Rao, Optimization-Theory and Applications (Second Edition), Wiley Eastern Ltd.,New Delhi, 1984

The minimum of the quadratic is at the point where the first derivative of the function becomes zero which is nothing but $b + 2cx^* = 0$ or $x^* = -\dfrac{b}{2c}$. The procedure may be deemed to have converged if $|f(x^*) - f(x_1)| \leq \epsilon$ where ϵ is a chosen tolerance. Otherwise we shift x_1 to x^*, x_2 to x_1 and x_3 to x_2 and repeat the process till convergence.

The general procedure of a quadratic interpolation scheme is

- Start with three guess points x_1, x_2 and x_3. The initial points may be equi-spaced say $x_2 = x_1 + h$ and $x_3 = x_1 + 2h$
- Determine b and c and determine the root $x^* = -\dfrac{b}{2c}$
- Check if the iteration has converged i.e. $|x^* - x_1| < \epsilon$
- Update $x_3 = x_2$, $x_2 = x_1$ and $x_1 = x^*$

The procedure is similar in character to the secant method where a straight line is used to determine the roots of the equation instead of a quadratic curve. A simple example is considered below.

Example 4.11

Find the minimum value of the function $f(x) = x^6 - 5x^4 - 20x + 5$ by univariate search based on quadratic interpolation. Start from $x = 1$ and a suitable h.

Solution :

MATLAB program for Example has been given below.

```
% Initialize points
x3 = 1;    x2 = 1 + 0.75;   x1 = 1 + 2*0.75;
% define objective function f(x)
f = inline('x^6-5*x^4-20*x+5');
% evaluate function at points
f3 = f(x3); f2 = f(x2);   f1 = f(x1);
tol = 1e-6;    % initialize tolerance
res = 0.75;    % initialize residual
count = 0;     % initialize iteration number
while(res > tol) % iterate until convergence
    % evaluate b and c
    b = -(x1+x2)*f3/((x3-x1)*(x3-x2)) -  ...
        (x3+x2)*f1/((x1-x3)*(x1-x2)) - ...
        (x3+x1)*f2/((x2-x1)*(x2-x3));
    % In MATLAB, when a command is too long to fit in a line,
    % we can use ... to break the command over several lines
    c = f3/((x3-x1)*(x3-x2)) + f1/((x1-x3)*(x1-x2)) ...
        + f2/((x2-x1)*(x2-x3));
    xnew = -0.5*b/c; % evaluate new optimum point
    res = abs(xnew-x1); % update residual
             % update points and functions
    x3 = x2;   x2 = x1; f3 = f2; f2 = f1;
    x1 = xnew; f1 = f(x1);
    count = count + 1; % update iteration count
end
```

As indicated we start the iteration process from $x = 1$. A few trials indicate that the step size may be chosen as $h = 0.75$ to commence the quadratic interpolation process. The state of affairs at the end of one iteration is as indicated in the following extract of a spreadsheet. Note that the first derivative calculated as $f'(x) = 6x^5 - 20x^3 - 20$ is also indicated in the spreadsheet.

Start iteration		Step size =	0.75
from $x =$	1		
x	$f(x)$	Quadratic interpolation:	
1	-19.0000	$b = $ -237.3145	
1.75	-48.1716	$c = $ 72.1523	
2.5	3.8281	$x^* = $	1.6445
$f'(x^*) = $	-36.7809	$|x^* - x_1| = 0.8555$	

The points are updated with $x_1 = 1.6445$ and retaining the other two points. The second iteration is as follows

Iteration 2			
$x_1 = $			
x	$f(x)$	Quadratic interpolation:	
1.75	-48.1716	$b = $	-439.5631
2.5	3.8281	$c = $	119.7403
1.6445	-44.6808	x^*	= 1.8355
$f'(x^*) = $	-18.6764	$|x^* - x_1| = $	0.1909

We see that the derivative has already decreased substantially and it appears that we are close to the minimum of the function. The next iteration has been summarized as follows

Iteration 3			
from $x_1 = $	1.8355		
x	$f(x)$	Quadratic interpolation:	
2.5	3.8281	$b = $	-477.9471
1.6445	-44.6808	$c = $	129.0016
1.8355	-50.2218	x^*	= 1.8525
$f'(x^*) = $	-16.2480	$|x^* - x_1| = $	0.0170

The solution converges to minimum of -51.3165 at $x_1 = 1.9457$ in 10 iterations.

4.4 Multidimensional unconstrained optimization

Above we have considered maxima/minima of a function in one variable as a problem in root finding (see Example 4.10). Optimization of a function of more than one variable should also be possible using root finding but in more than one dimension. Optimization is a topic of many books and hence only a few simple cases will be considered here, as an introduction to the topic. Calculus methods are appropriate in simple cases while computational schemes become important in more difficult cases.

4.4.1   Calculus based Newton method

Consider, as an example, the optimization (finding minimum/maximum) of a function $f(x_1, x_2)$ in two variables. In an unconstrained optimization case we just have to find the coordinates x_1^*, x_2^* (critical point) such that $f(x_1^*, x_2^*)$ has a maximum/minimum value.

The condition that is to be satisfied by the function at the optimum point is that the gradient be zero. Thus both derivatives $\dfrac{\partial f}{\partial x_1}$ and $\dfrac{\partial f}{\partial x_2}$ should be zero at the critical point x_1^*, x_2^*. The second condition that has to be satisfied, if the function has a minimum at this point, is that the Hessian matrix be positive definite. Hessian matrix is the matrix that is defined, in the case of a function of two variables, as

$$\mathbf{H} = \begin{pmatrix} \dfrac{\partial^2 f}{\partial x_1^2} & \dfrac{\partial^2 f}{\partial x_1 \partial x_2} \\[4mm] \dfrac{\partial^2 f}{\partial x_1 \partial x_2} & \dfrac{\partial^2 f}{\partial x_2^2} \end{pmatrix} \tag{4.21}$$

The Hessian matrix is positive definite if both its eigenvalues are positive at the optimum (or critical) point. In the above case of a function of two variables it may be shown that this condition is satisfied if $\dfrac{\partial^2 f}{\partial x_1^2} \cdot \dfrac{\partial^2 f}{\partial x_2^2} - \left(\dfrac{\partial^2 f}{\partial x_1 \partial x_2} \right)^2 > 0$ (Determinant of the Hessian - also the discriminant of the characteristic equation - a quadratic, in this case) and $\dfrac{\partial^2 f}{\partial x_1^2} > 0$. If the critical point is a maximum the Hessian matrix should be negative definite at the critical point. The condition to be satisfied is that $\dfrac{\partial^2 f}{\partial x_1^2} < 0$ and the discriminant be positive.

The Newton method is simply the application of Newton Raphson to determine the optimum point by iteration. For this purpose the gradient vector with its two components have to be set to zero and hence is equivalent to solving a set of two simultaneous equations $\dfrac{\partial f}{\partial x_1} = 0$ and $\dfrac{\partial f}{\partial x_2} = 0$. The equations to be solved are

$$\Delta \mathbf{x} = -\mathbf{H}^{-1} \nabla f(x) \tag{4.22}$$

We will show this by a simple example.

Example 4.12

Obtain the minimum of the function $f(x_1, x_2) = 3x_1^2 + 4x_2^2 - 5x_1x_2 - 8x_1$ and its location.

Solution :

Note that the function $f(x_1, x_2)$ is a quadratic form. Following expressions may be derived for the function $f(x_1, x_2)$ specified in the problem.

$$\frac{\partial f}{\partial x_1} = 6x_1 - 5x_2 - 8; \qquad \frac{\partial f}{\partial x_2} = 8x_2 - 5x_1$$

$$\frac{\partial^2 f}{\partial x_1^2} = 6; \quad \frac{\partial^2 f}{\partial x_2^2} = 8; \quad \text{and} \quad \frac{\partial^2 f}{\partial x_1 \partial x_2} = -5$$

We may now verify that the determinant of the Hessian (Δ) is given by $\Delta = 6 \times 8 - 5 \times 5 = 23 > 0$. Also $\frac{\partial^2 f}{\partial x_1^2} = 6 > 0$. Both are independent of x_1, x_2. The two components of the gradient vanish when

$$6x_1 - 5x_2 - 8 = 0$$
$$8x_2 - 5x_1 = 0$$

These two equations are easily solved to get $x_1^* = 2.7826$ and $x_2^* = 1.7391$. This point must be a point of minimum. The minimum value of the function is then given by

$$f(2.7826, 1.7391) = -11.1304$$

We consider now a slightly more difficult problem that requires the use of Newton method.

Example 4.13

Obtain the minimum of the function $f(x_1, x_2) = (x_1^2 - x_2)^2 + (1 - x_1)^2 + x_1 e^{x_2}$ and its location.

Solution :

We introduce the following notation for the sake of simplicity.

$$\frac{\partial f}{\partial x_1} = f_1; \quad \frac{\partial f}{\partial x_2} = f_2$$

$$\frac{\partial^2 f}{\partial x_1^2} = f_{11}; \quad \frac{\partial^2 f}{\partial x_2^2} = f_{22}; \quad \frac{\partial^2 f}{\partial x_1 \partial x_2} = f_{12}$$

For the given function we have the following:

$$f_1 = 4(x_1^2 - x_2)x_1 - 2(1 - x_1) + e^{x_2}; \quad f_2 = -2(x_1^2 - x_2) + x_1 e^{x_2}$$

$$f_{11} = 12x_1^2 - 4x_2 + 2; \quad f_{22} = 2 + x_1 e^{x_2}; \quad f_{12} = -4x_1 + e^{x_2}$$

The Newton Raphson procedure requires that we start from a guess point (x_{1g}, x_{2g}) and take steps Δx_1 and Δx_2 that are solutions of the following linear equations.

$$\begin{pmatrix} f_{11} & f_{12} \\ f_{12} & f_{22} \end{pmatrix} \begin{pmatrix} \Delta x_1 \\ \Delta x_2 \end{pmatrix} + \begin{pmatrix} f_1 \\ f_2 \end{pmatrix} = \begin{pmatrix} 0 \\ 0 \end{pmatrix}$$

where all the functions and derivatives are evaluated at the guess point. We shall start with the guess point $(0.5, 0.5)$. Function and derivative values are evaluated at this point. Function $f(0.5, 0.5)$ is evaluated as 1.1369. The set of equations to be solved turn out to be

$$\begin{pmatrix} 3.0000 & -0.3513 \\ -0.3513 & 2.8244 \end{pmatrix} \begin{pmatrix} \Delta x_1 \\ \Delta x_2 \end{pmatrix} + \begin{pmatrix} 0.1487 \\ 1.3244 \end{pmatrix} = \begin{pmatrix} 0 \\ 0 \end{pmatrix}$$

The solution is obtained by Cramer's rule as $\Delta x_1 = -0.1060$ and $\Delta x_2 = -0.4821$ and a point closer to the optimum is obtained as $0.3940, 0.0179$. At the end of the above iteration the function value has reduced to 0.7872. Two more iterations are required for convergence to the critical point $(0.3799, -0.0384)$ where the function value is 0.7835. Convergence of the method is brought home through the following table. We note that the optimum is a minimum since all the conditions for a minimum are satisfied.

Iteration No.	1	2	3	4
x_1	0.500000	0.393977	0.379831	**0.379946**
x_2	0.500000	0.017907	-0.038053	**-0.038449**
f_{11}	3.000000	3.790982	3.883468	3.886100
f_{12}	-0.351279	-0.557838	-0.556661	-0.557502
f_{22}	2.824361	2.401095	2.365649	2.365615
f_1	0.148721	0.022410	0.000667	3.18E-07
f_2	1.324361	0.126474	0.001000	-4.17E-08
Δx_1	-0.106023	-0.014146	0.000115	-8.20E-08
Δx_2	-0.482093	-0.055960	-0.000396	-1.72E-09
$f(x_1, x_2)$	1.136861	0.787214	0.783501	**0.783500**

Bold entries - converged solution

The procedure involved in Newton's method for optimization is simple. However, the method requires determination of the inverse of the Hessian matrix. As the number of variables increases, evaluation of inverse of Hessian matrix may become computationally intensive as has already been discussed earlier. Alternatively, there are methods known as **Quasi Newton methods** where the inverse of the Hessian matrix is approximated using the information about derivatives. This also reduces the number of computations required.

The method is bound to fail when the Hessian matrix is poorly conditioned i.e. $|H| \approx 0$. The solution under such conditions would diverge. There are some methods such as **Levenberg Marquadt** method where an extra term is added to the inverse of the Hessian matrix. Levenberg Marquadt method is a widely used algorithm in many optimization packages. The method can be considered to be weighted sum of Newton methods and Gradient descent methods. Examples of Gradient descent methods are steepest descent and conjugate gradient which will be discussed in the following section.

4.4.2   Gradient descent search methods

Search methods avoid the calculation and evaluation of the Hessian. We have already presented in Chapter 2 the method of steepest descent and the conjugate gradient method for solving a set of linear equations, by minimizing the residuals. In the present case the equations are non-linear and require some modification of the methods. These will be considered here.

Steepest descent method

As we have seen earlier, in the steepest descent method (for locating a minimum) or the steepest ascent method (for locating a maximum) the search direction from a specified starting point is either opposite the gradient vector (minimum) or along the gradient vector (maximum). When the direction of search is known we determine the optimum step along the search direction such that the given function takes on its minimum (or maximum) value, by univariate search using a method such as the quadratic interpolation method discussed earlier. This method uses only the function calculations and not the Hessian which would necessarily require the evaluation of second derivatives of the function. Also no matrix inversion is required. If the minimum (or maximum) thus determined satisfies the conditions for a minimum (or maximum) we stop the calculation. Otherwise we switch the starting point to the new point and search again in a descent (or ascent) direction as may be. The general algorithm of steepest descent for nonlinear equations is given below.

	Operation	Nonlinear SD	Linear SD
1	Direction of descent	$\mathbf{r}_k = -\nabla f(\mathbf{x}_k)$	$f(\mathbf{x}_k) = \mathbf{x}_k^T \mathbf{A}\mathbf{x}_k - \mathbf{b}^T\mathbf{x} - c$ Hence $\mathbf{r}_k = \mathbf{b} - \mathbf{A}\mathbf{x}_k$
2	Determine α_k : $\dfrac{df(\mathbf{x}_k)}{d\alpha_k} = 0$	α_k estimated using line search methods	$\alpha_k = \dfrac{\mathbf{r}_k^T \mathbf{r}_k}{\mathbf{r}_k^T \mathbf{A}\mathbf{r}_k}$
3	Update $\mathbf{x}_{k+1}$	$\mathbf{x}_{k+1} = \mathbf{x}_k + \alpha_k \mathbf{r}_k$	

Example 4.14

Obtain a maximum of the function $f(x,y) = \dfrac{xy(10-xy)}{(x+2y)}$ *by the steepest ascent method. Start the iteration with* $(x_0, y_0) = (1,1)$.

Solution :

The first partial derivatives required to determine the gradient vector are obtained by differentiation of $f(x,y)$ as

$$f_x = \frac{(10y - 2xy^2)}{(x+2y)} - \frac{(10xy - x^2y^2)}{(x+2y)^2}$$

$$f_y = \frac{(10x - 2x^2y)}{(x+2y)} - \frac{2(10xy - x^2y^2)}{(x+2y)^2}$$

We start with $x_0 = 1$ and $y_0 = 1$ and calculate various quantities of interest as below:

x_0	y_0	$f(x_0, y_0) = f(0)$	$f_x(x_0, y_0)$	$f_y(x_0, y_0)$	$|\mathbf{r}|(x_0, y_0)$
1	1	3	1.66667	0.66667	1.79505

The last column shows the magnitude of the gradient vector which is represented by $\mathbf{r}$. We know now the direction of ascent and hence are in a position to perform a univariate search using quadratic interpolation that is familiar to us from an earlier section. After a few trials we hit upon a step size of $h = 0.6$ for which the appropriate calculations are displayed below.

h	$f(x_0 + hf_x, y_0 + hf_y)$ $= f(h)$	$f(x_0 + 2hf_x, y_0 + 2hf_y)$ $= f(2h)$	b	c	h_{opt}
0.6	4.2	3.76364	3.36364	-2.27273	0.74

Having thus obtained the optimum step size (h_{opt}) the candidate point for the maximum is given by $x_1 = x_0 + h_{opt}f_x(x_0, y_0) = 2.12105$ and $y_1 = y_0 + h_{opt}f_y(x_0, y_0) = 1.48947$. We now give the state of affairs at this point below.

x_1	y_1	$f(x_1, y_1) = f(0)$	$f_x(x_1, y_1)$	$f_y(x_1, y_1)$	$|\mathbf{r}|(x_1, y_1)$
2.12105	1.48947	4.23758	0.24430	-0.13069	0.27706

After one iteration we see a substantial increase in the value of the function and also a substantial decrease in the magnitude of the gradient vector. More iterations are required for an acceptable solution to the problem. The following table summarizes the result of such an iteration process.

x	y	$f(x, y)$	f_x	f_y	h_{opt}	$|\mathbf{r}|$
1	1	3	1.66667	0.66667	0.74	1.79505
2.12105	1.48947	4.23758	0.24430	-0.13069	1.64372	0.27706
2.52261	1.27466	4.30121	0.04892	0.07905	0.38876	0.09296
2.54163	1.30539	4.30289	0.01726	-0.01064	1.64289	0.02027
2.56998	1.28791	4.30323	0.00974	0.01565	0.38331	0.01843
2.57372	1.29391	4.30330	0.00348	-0.00217	1.63957	0.00410
2.57942	1.29035	4.30331	0.00207	0.00332	0.38235	0.00391
2.58021	1.29162	4.30331	0.00074	-0.00047	1.63911	0.00088
2.58143	1.29086	4.30331	0.00045	0.00072	0.38210	0.00085
2.58160	1.29113	4.30331	0.00016	-0.00010	1.63961	0.00019
2.58187	1.29096	4.30331	0.00010	0.00016	0.38204	0.00018
2.58191	1.29102	4.30331	0.00003	-0.00002	1.63972	0.00004

Conjugate gradient method

The conjugate gradient method is similar to that presented for the solution of linear equations that involved the optimization of a quadratic form. Since the optimization problem is nonlinear in nature involving non-quadratic form the method presented earlier is modified by working with the gradient rather than the residual. First step is as usual the steepest descent (or ascent) step.

Iteration 1 Let the steepest descent step be given by $\mathbf{d}_0 = -\nabla f(\mathbf{x}_0) = \mathbf{r}_0$ where $\mathbf{x}_0$ is the initial vector from where we start the iteration. Univariate search is now made using a method such as the quadratic interpolation method to determine the step size $\alpha_0 = h_{opt}$ such that error is minimum i.e. $\dfrac{df(\mathbf{x}_0 + \alpha_0 \mathbf{d}_0)}{d\alpha_0} = 0$. The next trial vector will be $\mathbf{x}_1 = \mathbf{x}_0 + \alpha_0 \mathbf{d}_0$ or $\mathbf{d}_0 = \dfrac{\mathbf{x}_1 - \mathbf{x}_0}{\alpha_0}$.

Now calculate the gradient using vector $\mathbf{x}_1$ as $\nabla f(\mathbf{x}_1) = -\mathbf{r}_1$. Let the conjugate direction we are looking for be given by $\mathbf{d}_1$ such that $\mathbf{d}_1 = -\nabla f(\mathbf{x}_1) + \beta_0 \mathbf{d}_0 = \mathbf{r}_1 + \beta_0 \mathbf{d}_0$. Also let the Hessian be approximated by finite differences such that $\mathbf{H}_1(\mathbf{x}_1 - \mathbf{x}_0) = \nabla f(\mathbf{x}_1) - \nabla f(\mathbf{x}_0) = -\dfrac{\mathbf{r}_1 - \mathbf{r}_0}{\alpha_0}$. Directions $\mathbf{d}_1$ and $\mathbf{d}_0$ are $\mathbf{H}_1$ conjugate and hence we should have

$$\mathbf{d}_1^T \mathbf{H}_1 \mathbf{d}_0 = -[\mathbf{r}_1 + \beta_0 \mathbf{d}_0]^T \left[\frac{\mathbf{r}_1 - \mathbf{r}_0}{\alpha_0} \right] = 0$$

Solving for β_0 we get

$$\beta_0 = -\frac{\mathbf{r}_1^T [\mathbf{r}_1 - \mathbf{r}_0]}{\mathbf{d}_0^T [\mathbf{r}_1 - \mathbf{r}_0]}$$

Iteration 2 We now determine an optimum step size α_1 along direction $\mathbf{d}_1$ that gives a minimum for the function. The next point is then reached as $\mathbf{x}_2 = \mathbf{x}_1 + \alpha_1 \mathbf{d}_1$ or $\mathbf{d}_1 = \dfrac{\mathbf{x}_2 - \mathbf{x}_1}{\alpha_1}$. Let the direction of search now be taken as $\mathbf{d}_2 = \mathbf{r}_2 + \beta_1 \mathbf{d}_1$ such that $\mathbf{d}_2$ is $\mathbf{H}_2$ conjugate to $\mathbf{d}_1$. Here $\mathbf{H}_2 \mathbf{d}_1$ is again approximated as $\mathbf{H}_2 \mathbf{d}_1 = -\dfrac{\mathbf{r}_2 - \mathbf{r}_1}{\alpha_1}$. Directions $\mathbf{d}_2$ and $\mathbf{d}_1$ are $\mathbf{H}_2$ conjugate and hence we should have

$$\mathbf{d}_2^T \mathbf{H}_2 \mathbf{d}_1 = -[\mathbf{r}_2 + \beta_1 \mathbf{d}_1]^T \left[\frac{\mathbf{r}_2 - \mathbf{r}_1}{\alpha_1} \right] = 0$$

Solving for β_1 we get

$$\beta_1 = -\frac{\mathbf{r}_2^T [\mathbf{r}_2 - \mathbf{r}_1]}{\mathbf{d}_1^T [\mathbf{r}_2 - \mathbf{r}_1]}$$

With this we are in a position to generalize the iteration process. Assume that we have to take a step starting from $\mathbf{x}_k$ along the direction $\mathbf{d}_k$ that gives a minimum. The step size is given by α_k such that

$$\mathbf{x}_{k+1} = \mathbf{x}_k + \alpha_k \mathbf{d}_k \text{ or } \mathbf{d}_k = \frac{\mathbf{x}_{k+1} - \mathbf{x}_k}{\alpha_k} \tag{4.23}$$

Direction of search is then given by

$$\mathbf{d}_{k+1} = \mathbf{r}_{k+1} + \beta_k \mathbf{d}_k \tag{4.24}$$

where

$$\beta_k = -\frac{\mathbf{r}_{k+1}^T [\mathbf{r}_{k+1} - \mathbf{r}_k]}{\mathbf{d}_k^T [\mathbf{r}_{k+1} - \mathbf{r}_k]} \tag{4.25}$$

The above represents the Hestenes-Stiefel formula. In practice two alternate formulae are used. The first one known as the Fletcher-Reeve formula is given by

$$\beta_k = \frac{\mathbf{r}_{k+1}^T \mathbf{r}_{k+1}}{\mathbf{r}_k^T \mathbf{r}_k} \tag{4.26}$$

The second one known as Polak-Rebiere formula is given by

$$\beta_k = \frac{\mathbf{r}_{k+1}^T [\mathbf{r}_{k+1} - \mathbf{r}_k]}{\mathbf{r}_k^T \mathbf{r}_k} \tag{4.27}$$

At the end of every n iterations (n is the size of vector) a steepest descent step is introduced. The general algorithm of nonlinear CG has been given below.

	Operation	Nonlinear CG	Linear CG
1	Determine residual $\mathbf{r}$	$\mathbf{r}_k = -\nabla f(\mathbf{x}_k)$	$f(\mathbf{x}_k) = \mathbf{x}_k^T \mathbf{A} \mathbf{x}_k - \mathbf{b}^T \mathbf{x} - c$ Hence $\mathbf{r}_k = \mathbf{b} - \mathbf{A}\mathbf{x}_k$
2	Determine $\alpha_k : \dfrac{df(\mathbf{x}_k)}{d\alpha_k} = 0$	α_k estimated using line search methods	$\alpha_k = \dfrac{\mathbf{r}_k^T \mathbf{r}_k}{\mathbf{r}_k^T \mathbf{A} \mathbf{r}_k}$
3	Update $\mathbf{x}_{k+1}$	$\mathbf{x}_{k+1} = \mathbf{x}_k + \alpha_k \mathbf{d}_k$	

$\cdots$ Continued on next page

continued from previous page⋯

	Operation	Nonlinear CG	Linear CG
4	Update $\mathbf{r}$	$\mathbf{r}_{k+1} = -\nabla f(\mathbf{x}_{k+1})$	$\mathbf{r}_{k+1} = \mathbf{b} - \mathbf{A}\mathbf{x}_{k+1}$
5	Determine β_k	Fletcher Reeve $$\beta_k = \frac{\mathbf{r}_{k+1}^T \mathbf{r}_{k+1}}{\mathbf{r}_k^T \mathbf{r}_k}$$	$$\beta_k = -\frac{\mathbf{r}_{k+1}^T \mathbf{A}\mathbf{d}_k}{\mathbf{d}_k^T \mathbf{A}\mathbf{d}_k}$$
6	Update direction of descent $\mathbf{d}_{k+1}$	$\mathbf{d}_{k+1} = \mathbf{r}_{k+1} + \beta_k \mathbf{d}_k$ For first iteration ($k = 0$): $\mathbf{d}_0 = -\mathbf{r}_0$	

A MATLAB program has been provided which calculates the optimum point using CG method.

Program 4.1: *Nonlinear Conjugate gradient method*

```
1  function x = conjugategradient(xguess,tol,yfunction,grady)
2  %   Inputs      yfunction: name of function whose minima is desired
3  %               grady    : name of function which calculates ∇f
4  %   Both yfunction and grady have to be provided by the user
5  %               xguess   : guess vector
6  %               tol      : tolerance for convergence
7  %   Outputs     x        : optimum vector
8  x = xguess;                    % initialize x
9  sp = grady(x);                 % initialize ∇f
10 s = sp'*sp;                    % initialize residual
11 count = 0;                     % initialize iteration count
12 dp = -sp;                      % initialize direction of descent
13 a = 1;                         % initialize α
14 while ( abs(s) > tol ) % iteration loop
15 % determine optimum α and update x using Quadratic search
16 [x,a]= cgquadsearch(x,dp/sqrt(dp'*dp),a,0.1*a,tol,yfunction);
17 spnew = grady(x);              % update ∇f
18 dp1 = dp*(spnew'*spnew)/(sp'*sp); % Fletcher-Reeve step
19 dp = -spnew;                   % SD step
20 dp = dp + dp1;                 % CG step
21 % commenting CG step will result in SD
22 sp = spnew;
23 s = sp'*sp;                    % update residual
24 count = count +1;             % update iteration count
25 end
```

Function cgquadsearch is called by Function conjugategradient to determine α. cgquadsearch incorporates quadratic interpolation algorithm.

Program 4.2: *Quadratic interpolation search method called by conjugate gradient*

```
1  function [x1,a1] = cgquadsearch(x,dp,a,h,tol,yfunction)
2  %   Inputs      x    : vector
3  %               dp   : direction of descent
4  %               a    : initial α
5  %               h    : initial step size
6  %               tol  : tolerance for convergence
```

```
 7  %                        yfunction : function name of objective function
 8  %   Outputs      x1 : updated vector
 9  %                a1 : optimum α
10  a3 = a; a2 = a + h; a1 = a + 2*h;  % Initialize points
11  f3 = yfunction(x+a3*dp);      % evaluate function at points
12  f2 = yfunction(x+a2*dp); f1 = yfunction(x+a1*dp);
13  res = abs(h);                 % initialize residual
14  count = 0;                    % initialize iteration count
15  while(res > tol)              % iterate until convergence
16      % evaluate b and c
17      b = -(a1+a2)*f3/((a3-a1)*(a3-a2)) ...
18          - (a3+a2)*f1/((a1-a3)*(a1-a2)) ...
19          - (a3+a1)*f2/((a2-a1)*(a2-a3));
20      c = f3/((a3-a1)*(a3-a2)) + f1/((a1-a3)*(a1-a2)) ...
21          + f2/((a2-a1)*(a2-a3));
22      anew = -0.5*b/c;          % evaluate new optimum point
23      res = abs(anew-a1);       % update residual
24      a3 = a2;                  % update points and functions
25      a2 = a1; f3 = f2; f2 = f1; a1 = anew;
26      f1 = yfunction(x+a1*dp);% update function at new point
27      count = count + 1;        % update iteration count
28  end
29  x1 = x+a1*dp;                 % update point
```

Example 4.15

Determine the minimum of the function $f(x,y) = (x^2 - y)^2 + (1-x)^2 + xy$ using the method of Fletcher Reeves. Use the origin to start the iteration.

Solution :

The first step is the steepest descent step that moves in a direction that is opposite the gradient of the function. The components of the gradient vector are given by

$$f_x(x,y) = 4x(x^2 - y) - 2(1-x) + y; \qquad f_y(x,y) = -2(x^2 - y) + x$$

The starting point is given as the origin and hence $x_0 = 0$, $y_0 = 0$ is the starting point at which the components of the gradient are $f_x(0,0) = -2$ and $f_y(0,0) = 0$. The magnitude of the gradient vector is seen to be $||\mathbf{r}_0|| = \sqrt{(-2)^2 + 0^2} = 2$. The direction of descent is thus given by $\mathbf{d}_0^T = \begin{pmatrix} 2 & 0 \end{pmatrix}$. We now look for a step size that makes the function a minimum, as we move in this direction. Quadratic interpolation method is made use of. After a few trials we find that $h = 0.2$ will be suitable for conducting the quadratic interpolation. We tabulate the results of quadratic interpolation below.

Guess		$x_0 =$	0	h =	0.2
Point :		$y_0=$	0		
	Quadratic interpolation:				
$f(x_0,y_0)$		1		b =	-4.768
$f(x_0 + d_{01}h, y_0 + d_{02}h)$		0.3856		c =	8.48
$f(x_0 + 2d_{01}h, y_0 + 2d_{02}h)$		0.4496		h_{opt}	0.2811
$d_{01} = 2$ and $d_{02} = 0$ are components of direction vector $\mathbf{d}_0$					

Thus the optimum step size is taken as $\alpha_0 = 0.2811$ to reach the point $x_1 = 0 + 2 \times 0.2811 = 0.5623$ and $y_1 = 0 + 0 \times 0.2811 = 0$. The components of the gradient vector at $\mathbf{x}_1$ are calculated as $f_x(0.5623, 0) = -0.1645$ and $f_y(0.5623, 0) = -0.0700$. Magnitude of the gradient vector is given by $\|\mathbf{r}_1\| = \sqrt{(-0.1645)^2 + (-0.0700)^2} = 0.1787$. There is thus a significant reduction in the gradient after just one iteration step. The direction of search now uses the Fletcher-Reeve formula.

$$d_{11} = 0.1645 + \left(\frac{0.1787}{2}\right)^2 \times (2) = 0.1804$$

$$d_{12} = 0.0700 + \left(\frac{0.1787}{2}\right)^2 \times (0) = 0.0700$$

A univariate search using quadratic interpolation is again used now to determine the optimum step size. This step is summarized as a table below.

Point		$x_1 =$	0.5623	h =	0.2
		$y_1 =$	0		
		Quadratic interpolation:			
$f(x_1, y_1)$		0.2916		b =	-0.0353
$f(x_1 + d_{11}h, y_0 + d_{12}h)$		0.2881		c =	0.0889
$f(x_1 + 2d_{11}h, y_0 + 2d_{12}h)$		0.2917		h_{opt}	0.1986
$d_{11} = 01804$ and $d_{12} = 0.0700$ are components of direction vector $\mathbf{d}_1$					

Thus the optimum step is taken as $\alpha_1 = 0.1986$ to reach the point $x_2 = 0.5623 + 0.1804 \times 0.1986 = 0.5981$ and $y_2 = 0 + 0.0700 \times 0.1986 = 0.0139$. The components of the gradient vector at $\mathbf{x}_2$ are calculated as $f_x(0.5981, 0.0139) = -0.0326$ and $f_y(0.5981, 0.0139) = 0.0895$. Magnitude of the gradient vector is given by $\|\mathbf{r}_2\| = \sqrt{(-0.0316)^2 + (0.0895)^2} = 0.0953$. The calculations proceed with a few more iterations using the Fletcher-Reeve formula. The results are summarized as a table.

i	x_i	y_i	d_{i1}	d_{i2}	h_{opt}	$\|\mathbf{r}_i\|$
0	0.00000	0.00000	2.00000	0.00000	0.28113	2.00000
1	0.56226	0.00000	0.18043	0.07002	0.19860	0.17874
2	0.59810	0.01391	0.01865	0.10944	0.45188	0.09530
3	0.60652	0.06336	-0.01470	0.00536	0.14710	0.01539
4	0.60436	0.06415	-0.00110	-0.00203	0.55770	0.00228
5	0.60375	0.06301	0.00086	-0.00173	0.17007	0.00158
6	0.60389	0.06272	0.00008	0.00003	0.22103	0.00009
7	0.60391	0.06273	0.00001	0.00006	0.43575	0.00005
8	0.60391	0.06275	-0.00001	0.00000	0.14765	0.00001

Depending on the required precision we may stop the iteration by observing the last column that gives the magnitude of the gradient vector. Convergence is initially very rapid and at $i = 4$ we have reached close to the optimum point. However, we have continued the iterations till the gradient has reduced to 0.00001. Convergence to the minimum is also shown graphically by the surface plot shown in Figure 4.14. Points, as search progresses

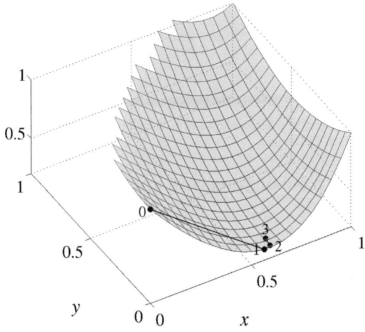

Figure 4.14: *Convergence of Fletcher Reeves to the minimum in Example 4.15*

are numbered in ascending order and are joined by lines. Iterates beyond $i = 3$ do not resolve in this figure.

We perform appropriate test to see that the solution obtained above corresponds to a minimum. For this purpose the second derivatives of $f(x, y)$ are calculated at the critical point as follows:

$$
\begin{aligned}
f_{xx}(0.6039, 0.0628) &= 12 \times 0.6039^2 - 4 \times 0.0628 + 2 = 6.1255 \\
f_{yy}(0.6039, 0.0628) &= 2 \\
f_{xy}(0.6039, 0.0628) &= -4 \times 0.6039 + 1 = -1.4157
\end{aligned}
$$

We note that $f_{xx} > 0$ and $f_{xx}f_{yy} - f_{xy}^2 = 6.1255 \times 2 - (-1.4157)^2 = 10.2470 > 0$ and hence the point $(0.6039, 0.0628)$ represents a minimum of the function.

Program 4.1 has been used to determine the optimum point. Two functions defining the objective function and its gradient, **EX319** and **EX319quad** respectively, have been provided.

```
function y = EX319(x)
y = (x(1)^2-x(2))^2 + (1-x(1))^2 + x(1)*x(2);     % x(1) = x
                                                  % x(2) = y

function sp = EX319grad(x)
sp(1,1) = 4*x(1)*(x(1)^2-x(2))-2*(1-x(1))+x(2);
sp(2,1) = -2*(x(1)^2-x(2))+x(1);
```

The output of the program is given below

```
x = conjugategradient([0;0],1e-6,@EX319,@EX319grad)
% External functions EX319 and EX319quad have
%                        to be prefixed with '@'
x =
    0.6039
    0.0625
```

The conjugate gradient method converges in 5 iterations where as the steepest descent method converges in 8 iterations.

4.5   Examples from engineering applications

In this section we consider typical engineering applications of the methods that have been presented earlier. We give a brief description of the genesis of the problem, the necessary background information needed to formulate the problem. The solution is presented by one of the many methods that have been presented before.

4.5.1   Operating point of a fan-duct system

In air handling applications we come across problems that require the choice of a fan to sustain a flow through a duct system. The fan is characterized by its own characteristics that is obtained either theoretically or experimentally. A relation between the pressure developed p and the volume flow rate of air Q is obtained to represent the fan characteristic. The air handling system uses a fan coupled to the duct system. The duct system produces a pressure drop when we drive air through it. The pressure drop may be due to losses at entry, exit, bends and also due to friction in the duct system. The pressure drop is represented as a function of the volume flow rate to characterize the behavior of the duct. When coupled to each other the fan and duct system will perform satisfactorily if the fan is able to sustain the desired flow rate in the duct system. This requires that the pressure developed by the fan be just equal to the pressure loss in the duct system for a given or specified air volume flow rate.

Example 4.16

Flow rate Q (m^3/h) and head developed Δp (Pa) by a fan are related by the equation $Q = 20 - 0.0001\Delta p^2$ m^3/h. Head loss in a duct is related to the flow rate by the equation $\Delta p = 100 + 8Q^2$. Determine the operating point if the fan and duct are coupled.

Solution :

The operating point is determined by Q and Δp values that satisfy both the fan and duct characteristics. Thus we have to solve simultaneously the two equations given below.

$$f_1(\Delta p, Q) = \Delta p - 100 - 8Q^2 = 0$$
$$f_2(\Delta p, Q) = 0.0001\Delta p^2 + Q - 20 = 0$$

The Jacobian matrix is given by

$$\frac{\partial f_1}{\partial \Delta p} = 1 \quad ; \quad \frac{\partial f_1}{\partial Q} = -16Q$$

$$\frac{\partial f_2}{\partial \Delta p} = 0.0002\Delta p \quad ; \quad \frac{\partial f_2}{\partial Q} = 1$$

We start the calculation with guess values of $\Delta p_g = 400$ and $Q_g = 4$ and use the Newton Raphson method to obtain converged values of $\Delta p_b = 375.87$ and $Q_b = 5.87$ after just 3 iterations as shown in the spreadsheet shown below. Even though the calculations have been performed with a larger number of digits the table shows values rounded to 2 digits after the decimal point.

Spreadsheet for Newton Raphson
method in Example 4.16

Δp_g	400.00	376.50	375.87
Q_g	6.00	5.88	5.87
$f_1(\Delta p_g, Q_g)$	12.00	-0.11	0.00
$f_2(\Delta p_g, Q_g)$	2.00	0.06	0.00
$f_{1\Delta p}$	1.00	1.00	1.00
f_{1Q}	-96.00	-94.08	-93.96
$f_{2\Delta p}$	0.08	0.08	0.08
f_{2Q}	1.00	1.00	1.00
Δp_b	376.50	375.87	375.87
Q_b	5.88	5.87	5.87

From the point of view of the application it is reasonable to retain only 2 digits after decimals. The operating point of the fan-duct system is $\Delta p = \mathbf{375.87\,Pa}$ and $Q = \mathbf{5.87\,m^3/h}$.

4.5.2   Pumps operating in parallel

Water pumping stations use pumps in parallel to cater for varying mass flow rates. When two pumps run in parallel they have to produce the same head but will provide different flow rates (see Figure 4.15). The example uses simulation or Gauss Seidel method for determining the operating point. Three equations are to be simultaneously satisfied in order to define the operating point for

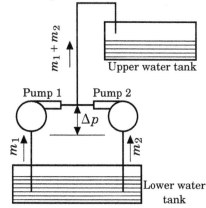

Figure 4.15: *Schematic of two pumps connected in parallel*

two pumps operating in parallel. The first equation represents the pressure difference (or head) required to maintain total flow in the pipe connecting the two tanks. The pressure difference (or head) mass flow rate characteristics of the two individual pumps provide the other two equations. When the two pumps are connected in parallel the head developed by the two pumps must be equal to the head required to maintain the mass flow rate through the connecting pipe line. In Example 4.17 these three equations correspond respectively to Equations 4.28(a) through 4.28(c).

Example 4.17

A water pumping system consists of two parallel pumps drawing water from a lower tank and delivering it to another tank that is at an elevation 40 m (i.e. 392.3 kPa) above the pumps. In addition to overcoming the pressure difference due to the elevation, friction in the pipe is to be overcome by the flow. We are given the pump and friction characteristics in terms of pressure drop, mass flow rates. Consistent units are used and the following equations result:

$$
\begin{array}{lrcll}
\text{Head losses in piping and} & p - 8(m_1 + m_2)^2 & = & 392.3 & (a) \\
\text{due to elevation:} & & & & \\
\text{Characteristic of Pump 1:} & p + 20m_1 + 3.5m_1^2 & = & 810 & (b) \\
\text{Characteristic of Pump 2:} & p + 60m_2 + 20m_2^2 & = & 900 & (c)
\end{array} \qquad (4.28)
$$

where p represents the head developed (in kPa) by either pump when connected in parallel, m_1 and m_2 are the mass flow rates in kg/s through the two pumps respectively. Determine the operating point for the system.

Solution :

Iteration is started with guess value for one of the unknowns, say $m_1 = m_{1,0}$. From Equation 4.28(b) we then have $p = p_0 = 810 - 20m_{1,0} - 3.5m_{1,0}^2$. We use Equation 4.28(c) to solve for m_2.

$$
m_2 = m_{2,0} = \frac{-60 + \sqrt{(60^2 - 4 \times 20 \times (m_{1,0} - 900))}}{2 \times 20}
$$

Now we use Equation 4.28(a) to get a new value for m_1 as

$$
m_{1,1} = \sqrt{\frac{p_0 - 392.3}{8}} - m_{2,0}
$$

We use $m_{1,1}$ and repeat the process to get $p_{1,1}$ and $m_{2,1}$. The procedure is continued until desired convergence has been achieved.

For example, starting with $m_{1,0} = 5$, we get

$$
p_0 = 810 - 20 \times 5 - 3.5 \times 5^2 = 622.5
$$

$$
m_{2,1} = \frac{-60 + \sqrt{[60^2 - 4 \times 20 \times (622.5 - 900)]}}{2 \times 20} = 2.516
$$

We now update $m_{1,0}$ as

$$
m_{1,1} = \sqrt{\frac{622.5 - 392.3}{8}} - 2.516 = 2.849
$$

Gauss Seidel scheme for the current problem shows oscillations that die down slowly and the converged solution (3 digits after decimals) is available at the end of 48 iterations. Convergence of m_1 is shown graphically in Figure 4.16.

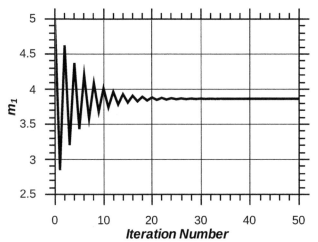

Figure 4.16: Convergence of m_1 in Example 4.17

Spreadsheet for the Gauss Seidel
iteration method in Example 4.17

Iteration Number	m_1 kg/s	p kPa	m_2 kg/s
0	5.000	622.500	2.516
1	2.849	724.616	1.820
2	4.626	642.591	2.389
3	3.205	709.942	1.928
24	3.873	680.043	2.140
48	3.865	680.434	2.137

Alternately we may solve this problem by the Newton Raphson method. We pose the problem in terms of solving simultaneously three equations given by

$$
\begin{array}{llll}
f_1(p, m_1, m_2) &=& p - 8(m_1 + m_2)^2 - 392.3 = 0 & (a) \\
f_2(p, m_1, m_2) &=& p + 20m_1 + 3.5m_1^2 - 810 & (b) \\
f_3(p, m_1, m_2) &=& p + 60m_2 + 20m_2^2 - 900 & (c)
\end{array} \qquad (4.29)
$$

The Jacobian matrix is a 3×3 matrix given by the following:

$$
\mathbf{J} = \left(\begin{array}{ccc}
f_{1p} & f_{1m_1} & f_{1m_2} \\
f_{2p} & f_{2m_1} & f_{2m_2} \\
f_{3p} & f_{3m_1} & f_{3m_2}
\end{array}\right) = \left(\begin{array}{ccc}
1 & -16(m_1 + m_2) & -16(m_1 + m_2) \\
1 & 20 + 7m_1 & 0 \\
1 & 0 & 60 + 40m_2
\end{array}\right) \qquad (4.30)
$$

Newton Raphson method consists in the use of the following iteration scheme.

$$
\left(\begin{array}{c}
\Delta p \\
\Delta m_1 \\
\Delta m_2
\end{array}\right) = - \left(\begin{array}{ccc}
1 & -16(m_1 + m_2) & -16(m_1 + m_2) \\
1 & 20 + 7m_1 & 0 \\
1 & 0 & 60 + 40m_2
\end{array}\right)^{-1} \left(\begin{array}{c}
f_1 \\
f_2 \\
f_3
\end{array}\right)
$$

where the left hand side represents the change in each of the variables from iteration to iteration and the right hand side is calculated based on the previous values of the variables. Iterative solution stops when the left hand side vector becomes smaller or equal to a preset tolerance. In the present case we have used a tolerance on fractional change of 10^{-5} as the stopping criterion.

We start with the initial set $m_1 = 5.000$, $p = 622.500$ and $m_2 = 2.516$. With these the function values are calculated as

$$\begin{aligned}
f_1 &= 622.5 - 8(5 + 2.516)^2 - 392.3 = -221.722 \\
f_2 &= 622.5 + 20 \times 5 + 3.5 \times 5^2 - 810 = 0 \\
f_3 &= 622.5 + 60 \times 2.516 + 20 \times 2.516^2 - 900 = 0.065119
\end{aligned}$$

The Jacobian matrix is evaluated as

$$\mathbf{J} = \begin{pmatrix} 1 & -120.256 & -120.256 \\ 1 & 55 & 0 \\ 1 & 0 & 160.64 \end{pmatrix}$$

The augmented matrix is obtained by combining the Jacobian with the column matrix representing negative function vector. Thus the augmented matrix is written down as

$$\begin{pmatrix} 1 & -120.256 & -120.256 & \vdots & 221.722 \\ 1 & 55 & 0 & \vdots & 0 \\ 1 & 0 & 160.64 & \vdots & 0.065119 \end{pmatrix}$$

Using elementary row operations the above may be transformed to the upper triangular form

$$\begin{pmatrix} 1 & -120.256 & -120.256 & \vdots & 221.722 \\ 0 & 1 & 0.686173 & \vdots & -1.265132 \\ 0 & 0 & 1.649644 & \vdots & -0.579159 \end{pmatrix}$$

The solution may be obtained by back substitution. We thus have

$$\begin{aligned}
\Delta m_2 &= \frac{-0.579159}{1.649644} = -0.351082 \\
\Delta m_1 &= -1.265132 - 0.686173 \times (-0.351082) = -1.024230 \\
\Delta p &= 221.722 - [-120.256 \times (-1.024230 - 0.351082)] = 56.332627
\end{aligned}$$

The new values of the variables are then obtained as

$$\begin{aligned}
m_1 &= 5 - 1.024230 = 3.975770 \\
m_2 &= 2.516 - 0.351082 = 2.164918 \\
p &= 622.5 + 56.332627 = 678.833
\end{aligned}$$

Fractional change in the values of the variables are given by

$$\begin{aligned}
\frac{\Delta p}{p} &= \frac{56.332627}{622.5} = 0.090494 \\
\frac{\Delta m_1}{m_1} &= \frac{-1.024230}{5} = -0.204846 \\
\frac{\Delta m_2}{m_2} &= \frac{-0.351082}{2.516} = 0.13954
\end{aligned}$$

However the convergence is very fast and only three more iterations are required to obtain the solution as seen from the following table.

Convergence of the Newton Raphson solution

Iteration No.	p kPa	m_1 kg/s	m_2 kg/s	$\dfrac{\Delta p}{p}$ ...	$\dfrac{\Delta m_1}{m_1}$ ...	$\dfrac{\Delta m_2}{m_2}$ ...
2	680.427	3.866	2.137	2.35×10^{-3}	-2.77×10^{-2}	-1.28×10^{-2}
3	680.442	3.864	2.137	2.5×10^{-5}	-3.09×10^{-4}	-9.7×10^{-5}
4	680.442	3.864	2.137	3×10^{-7}	-6×10^{-6}	-6×10^{-7}

4.5.3 Operating point of a heat exchanger

The next example we consider is an oil to water heat exchanger operating with a controller that selects the flow rate of coolant (water) based on the outlet temperature of oil.[4] The schematic of the oil to water heat exchanger is shown in Figure 4.17. The oil flow rate and inlet temperature

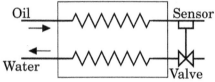

Figure 4.17: *Oil to water heat exchanger with coolant flow control*

are given as m_h and T_{hi}. The specific heat of oil is specified as c_h. The inlet temperature of water is specified as T_{ci}. Controller action is specified as a relation between oil outlet temperature T_{h0} and the mass flow rate of coolant water m_c.

$$f(m_c, T_{ho}) = 0 \qquad (4.31)$$

Specific heat of water is known and given as c_c. The goal of the simulation is to determine the outlet temperature of oil T_{ho}, the mass flow rate of coolant m_c and the outlet temperature of water T_{co}. Three equations are needed for performing the simulation. Equation 4.31 is the first equation. We equate the rate at which heat is given up by oil to that acquired by water to get the second equation.

$$m_c c_c (T_{co} - T_{ci}) = m_h c_h (T_{hi} - T_{ho}) \qquad (4.32)$$

The third equation is obtained by equating any one of the above to the heat transferred in the heat exchanger based on the specified overall heat transfer area product for the heat exchanger $U \cdot A$ and the logarithmic mean temperature difference $LMTD$. Thus

$$m_h c_h (T_{hi} - T_{ho}) = U \cdot A\, LMTD \qquad (4.33)$$

Since the heat exchanger is of the counter current type $LMTD$ is given by

$$LMTD = \frac{(T_{hi} - T_{co}) - (T_{ho} - T_{ci})}{\ln \left[\dfrac{T_{hi} - T_{co}}{T_{ho} - T_{ci}} \right]} \qquad (4.34)$$

We consider a specific case to perform the simulation and obtain the operating point for the heat exchanger.

---

[4]Reader may consult a book on heat transfer dealing with analysis of heat exchangers such as S.P. Venkateshan, Heat Transfer 2nd edition, 2011, Ane Publishers

Example 4.18

Oil flowing at the rate of 2.5 kg/s enters the heat exchanger at 90°C. Water enters the heat exchanger at 25°C. Specific heat of oil is known to be 3 kJ/kg°C while the specific heat of water is known to be 4.19 kJ/kg°C. Control action of the sensor valve combination is given by the relation $m_c = 0.4(T_{ho} - 65)$. Obtain the operating point of the heat exchanger by performing a simulation if $U \cdot A = 6 \, kW/°C$.

Solution :

Using the nomenclature introduced earlier the governing equations may be written down as under:

$$
\begin{aligned}
f_1 = m_c - 0.4(T_{ho} - 65) &= 0 \\
f_2 = 4.19 m_c (T_{co} - 25) - 7.5(90 - T_{ho}) &= 0 \\
f_3 = 7.5(90 - T_{ho}) - \frac{6[(90 - T_{co}) - (T_{ho} - 25)]}{\ln\left[\dfrac{90 - T_{co}}{T_{ho} - 25}\right]} &= 0
\end{aligned}
\tag{4.35}
$$

The control equation is solved for oil outlet temperature as

$$
T_{ho} = 65 + \frac{m_c}{0.4}
$$

Equation relating the heat transfers to the two fluids are solved for the coolant mass flow rate as

$$
m_c = \frac{7.5}{4.19} \cdot \frac{90 - T_{ho}}{T_{co} - 25}
$$

Finally the rate equation is solved for coolant outlet temperature as

$$
T_{co} = 90 - (T_{ho} - 25)\exp\left[\frac{115 - T_{ho} - T_{co}}{1.25 \times (90 - T_{ho})}\right]
$$

The above three equations are used in performing the simulation. Gauss Seidel iteration scheme is employed. With the starting set $T_{ho} =, 60 \; T_{co} = 75$ and $m_c = 1$ a single iteration gives the following:

State of affairs after one iteration

	Old	New	Change	Residuals
m_c	1.0000	1.0740	0.0740	3.0000
T_{ho}	60.0000	67.6850	7.6850	-15.5000
T_{co}	75.0000	74.1791	-0.8209	4.9627

After 20 iterations the results converge to four digits after the decimal point as shown below.

State of affairs after 20 iterations

	Old	New	Change	Residuals
m_c	0.8726	0.8726	0.0000	0.0000
T_{ho}	67.1814	67.1814	0.0000	0.0000
T_{co}	71.8106	71.8105	0.0000	0.0000

The operating point for the heat exchanger is thus given by $m_c = 0.873 \, kg/s$, $T_{ho} = 67.2°C$ and $T_{co} = 71.8°C$. Heat transfer taking place in the heat exchanger between the fluids is given by

$$Q = 7.5(90 - 67.2) = 171.1 \, kW$$

The solution can be obtained using Newton Raphson method also. Newton Raphson method involves determining the Jacobian matrix. Equations 4.35 will be used to determine the operation point of the heat exchanger. The Jacobian matrix would be

$$\mathbf{J} = \begin{pmatrix} \dfrac{\partial f_1}{\partial m_c} & \dfrac{\partial f_1}{\partial T_{ho}} & \dfrac{\partial f_1}{\partial T_{co}} \\[2mm] \dfrac{\partial f_2}{\partial m_c} & \dfrac{\partial f_2}{\partial T_{ho}} & \dfrac{\partial f_2}{\partial m_c} \\[2mm] \dfrac{\partial f_3}{\partial m_c} & \dfrac{\partial f_3}{\partial T_{ho}} & \dfrac{\partial f_3}{\partial T_{co}} \end{pmatrix}$$

$$= \begin{pmatrix} 1 & -0.4 & 0 \\[2mm] 4.19(T_{co} - 25) & 7.5 & 4.19 m_c \\[2mm] 0 & -7.5\left(ln\left(\dfrac{90 - T_{co}}{T_{ho} - 25}\right) + \dfrac{90 - T_{ho}}{T_{ho} - 25}\right) + 6 & -7.5\dfrac{90 - T_{ho}}{90 - T_{co}} + 6 \end{pmatrix}$$

The solution is determined by

$$\begin{pmatrix} \Delta m_c \\ \Delta T_{ho} \\ \Delta T_{co} \end{pmatrix} = \mathbf{J}^{-1} \begin{pmatrix} f_1 \\ f_2 \\ f_3 \end{pmatrix}$$

This has been programmed in MATLAB as given below

```
x(1,1)  = 1;                 % m_c
x(2,1)  = 60;                % T_ho
x(3,1)  = 75;                % T_co
delx = x;                    % initialize Δx
count = 0;                   % initialize iteration count
while( abs(delx) > 1e-6) % loop until convergence
f(1,1)  = x(1)-0.4*(x(2)-65); % f1
f(2,1)  = 4.19*x(1)*(x(3)-25)-7.5*(90-x(2)); % f2
f(3,1)  = 7.5*(90-x(2))*log((90-x(3))/(x(2)-25)) ...
        - 6*(90-x(3))+6*(x(2)-25);  % f3
H(1,1)  = 1;                 % calculate Jacobian matrix
H(1,2)  = -0.4;
H(1,3)  = 0;
H(2,1)  = 4.19*(x(3)-25);
H(2,2)  = 7.5;
H(2,3)  = 4.19*x(1);
H(3,1)  = 0;
H(3,2)  = -7.5*log((90-x(3))/(x(2)-25)) + 6 ...
        -7.5*(90-x(2))/(x(2)-25);
H(3,3)  = -7.5*(90-x(2))/(90-x(3)) + 6;
delx = -inv(H)*f;            % calculate Δx
```

```
x = x+ delx;              % update x
count = count + 1;        % update iteration count
end
```

The solution converges with sixth decimal accuracy in 5 iterations. The output of the program is given below

```
x =   [ 0.8726;       67.1814;        71.8105]
```

4.5.4 Automobile problem

An automobile has an on board engine that develops the required power to haul it. Automobile experiences resistance to motion due to three reasons as shown in Figure 4.18. They are:

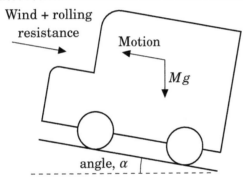

Figure 4.18: *Forces against an automobile climbing a grade*

1. Component of weight in a direction opposite to the direction of motion given by $Mg\sin\alpha$ where M is the mass of the vehicle, g is the acceleration due to gravity and α is the angle of the grade.
2. Wind resistance proportional to square of the velocity of the automobile. It is expressed as $F_d = \dfrac{C_d\,A\rho V^2}{2}$ where F_d is the drag force, ρ is air density, C_d is the drag coefficient, V is the speed of the vehicle and A is the frontal area of the vehicle.
3. Traction force F_t required to move the vehicle against rolling resistance given by $F_r = \mu F_N$ where μ is the coefficient of rolling friction and F_N is the normal reaction at the tyres.

The engine develops torque T_e which varies quadratically with the engine speed, usually represented in revolutions per minute N_e.

$$T_e = A + BN + CN^2$$

where A, B and C are constants specific to an engine. The engine torque is modified at the wheels by the gearbox as well as the differential that turns the wheels. The torque required to propel the vehicle T_w is given by the sum of the opposing forces multiplied by the tyre radius R_t or $\dfrac{D_t}{2}$ where D_t is the tyre diameter.

$$T_w = R_t(F_t + F_d + F_r)$$

This torque is presented usually as a smaller torque T'_e at the engine shaft because of the overall gear ratio N_g which is the product of gear ratio of the gear box and the gear ratio of the differential.

$$T'_e = \frac{R_t(F_t + F_d + F_r)}{N_g}$$

The speed of the car is related to the engine RPM by the relation

$$V = \frac{2\pi R_t N_e}{N_g}$$

Operating point is determined by the condition $T_e = T'_e$. Following example shows such a calculation.

Example 4.19

The torque speed characteristic of an automobile engine is well represented by the relation
$T_e = -104.8 + 0.0664N_e - 4.78 \times 10^{-6}N_e^2$ where T_e represents the torque in N m. The engine
RPM is limited to the range $2700 < N_e < 6500$. The automobile has a mass of $1500\,kg$ and
coefficient of rolling friction of 0.03. The drag coefficient for the automobile is 0.26 based on a
frontal area of $1.9\,m^2$. The automobile runs on $155/70R14$ tyres that have an outer diameter
of $0.573\,m$. The automobile has a 5 speed gearbox with the gear ratios specified below.

Gear Identifier	Gear Ratio
First Gear	3.25
Second Gear	1.8
Third Gear	1.4
Fourth Gear	1
Fifth Gear	0.9
Differential	3.64

The automobile is climbing a steady incline of $5°$ in 2nd gear. Determine the speed of the
automobile.

Solution :

We derive the torque equations based on the data supplied in the problem. Forces on the automobile are calculated first.

1. Force due to grade: $F_r = Mg\sin\alpha = 1500 \times 9.8 \times \sin\left[\dfrac{5\pi}{180}\right] = 1281.19\,N$

2. Force against rolling friction: $F_t = \mu F_N = \mu Mg\cos\alpha = 0.03 \times 1500 \times 9.8 \times \cos\left[\dfrac{5\pi}{180}\right] =$ 439.32 N

3. Wind resistance is calculated assuming density of air to be $\rho = 1.293\,kg/m^3$. Thus
$F_d = \dfrac{0.26 \times 1.9 \times 1.293V^2}{2} = 0.3194V^2$.

It is convenient to represent the vehicle speed in terms of the engine RPM. For this purpose we note that one revolution of the engine shaft corresponds to $\dfrac{1}{N_g} = \dfrac{1}{1.8 \times 3.64} =$ 0.153 revolution of the wheel. N_e RPM of the engine shaft thus corresponds to $V = \dfrac{0.153 \times \pi D_t N_e}{60} = \dfrac{0.153 \times \pi \times 0.573N_e}{60} = 4.58 \times 10^{-3}N_e$ m/s. With this the torque presented by the wheels of the automobile at the engine shaft may be written as

$$T'_e = \frac{R_t}{N_g}(F_t + F_r + F_d)$$

$$= \frac{0.573}{2 \times 1.8 \times 3.64}(1281.19 + 439.32 + 6.7 \times 10^{-6}N_e^2)$$
$$= 75.23 + 2.93 \times 10^{-7}N_e^2$$

The operating point is thus obtained by simultaneous solution of the following two equations:

$$f_1(T, N_e) = T + 104.8 - 0.0664 N_e + 4.78 \times 10^{-6}N_e^2 = 0$$
$$f_2(T, N_e) = T - 75.23 - 2.93 \times 10^{-7}N_e^2 = 0$$

Even though it is possible to eliminate T between these two equations and solve for N we use the Newton Raphson method to solve the problem. The components of the Jacobian are written down as

$$\mathbf{J} = \begin{pmatrix} f_{1T} & f_{1N_e} \\ f_{2T} & f_{2N_e} \end{pmatrix} = \begin{pmatrix} 1 & -0.0664 + 9.56 \times 10^{-6}N_e \\ 1 & -5.86 \times 10^{-7}N_e \end{pmatrix}$$

We start with the initial guess values $T = 50\,N\,m$ and $N_e = 3000\,RPM$ and run through the calculations to obtain the following table of values.

Iteration	0	1	2	3
T	50	79.2	79.5	**79.5**
N	3000	3737	3833	**3835**
f_{1T}	1	1	1	1
f_{1N_e}	-0.038	-0.031	-0.030	-0.030
f_{2T}	1	1	1	1
f_{2N_e}	-0.002	-0.002	-0.002	-0.002
$-f_1$	1.380	-2.593	-0.045	0.000
$-f_2$	27.867	0.159	0.003	0.000
ΔT	29.162	0.370	0.007	0.000
ΔN_e	736.527	96.597	1.721	0.001

The speed of the vehicle corresponding to the engine RPM obtained above is $V = 63.2\,km/h$.

The above example has looked at the steady state solution to the automobile speed problem. However, if the automobile were to start from rest, a dynamic analysis would be required, taking into account the initial accelerating part of the motion. This would require the solution of a differential equation based on Newton's laws of motion.

Concluding remarks

Nonlinear equations occur in various engineering applications and hence are dealt with in books dealing with simulation, optimization, design etc. In the present chapter we have provided an introduction to some of the methods that may be used for solving either one or a set of nonlinear equations. Relative merits of different methods have also been discussed. In general it is necessary to explore different methods and use the best suited one for a particular application. The reader may consult advanced texts to move beyond what has been presented here.

4.A MATLAB routines related to Chapter 4

MATLAB routine	Function
roots(c)	roots of a polynomial function c
fzero(fun)	determines roots of a continuous function 'fun' of one variable
fsolve	solves system of nonlinear equations
fminbnd	determines minimum of single-variable function on fixed interval
fminunc	determines minimum of unconstrained objective function
fminsearch	determines minimum of unconstrained objective function using derivative free method
simulink	a standalone tool for designing and simulating systems
optimtool	optimization toolbox with graphical user interface

4.B Suggested reading

1. **S.S. Rao** *Optimization-Theory and Applications (Second Edition)* Wiley Eastern Ltd.,New Delhi, 1984
2. **Dennis, J.** and **Schnabel, R.** *Numerical methods for unconstrained optimization and nonlinear equations* Society for Industrial and Applied Mathematics, 1996
3. **Fletcher, R.** *Practical methods of optimization*, Wiley, 2000
4. **Nocedal, J.** and **Wright, S.** *Numerical optimization* Springer verlag, 1999
5. **Arora, J.** *Introduction to optimum design* Academic Press, 2004

Exercise I

I.1 Solution of linear equations

Ex I.1: After verifying that the matrix given below is non singular invert it by use of minors and cofactors as explained in the text.

$$\begin{pmatrix} 0.4 & 0.9 & 1.2 \\ 0.6 & 0.7 & 2 \\ 2 & 0.4 & 0.3 \end{pmatrix}$$

Round all matrix elements of the inverse matrix to 3 significant digits after decimals. Calculate the inverse of this matrix and compare with the given matrix. Comment based on your observations.

Obtain, using the above, the solution to the matrix equations.

$$\begin{pmatrix} 0.4 & 0.9 & 1.2 \\ 0.6 & 0.7 & 2 \\ 2 & 0.4 & 0.3 \end{pmatrix} \begin{pmatrix} x_1 \\ x_2 \\ x_3 \end{pmatrix} = \begin{pmatrix} 1.4 \\ 2.3 \\ 4.4 \end{pmatrix}$$

What is the condition number of the coefficient matrix?

Ex I.2: Solve the set of three equations in Exercise I.1 using Kramer's rule. Compare the solution thus obtained with that obtained there.

Ex I.3: Solve the following system of equations by reducing it to upper triangle form using appropriate elementary row operations.

$$\begin{pmatrix} 1 & 0 & -1 & 2 \\ 0 & 2 & 0 & 1 \\ 2 & 1 & 3 & 0 \\ 0 & 2 & 1 & 4 \end{pmatrix} \begin{pmatrix} x_1 \\ x_2 \\ x_3 \\ x_4 \end{pmatrix} = \begin{pmatrix} 2 \\ 3 \\ 4 \\ 1 \end{pmatrix}$$

Ex I.4: Solve the system of equations in Exercise I.3 by LU factorization.

Ex I.5: Determine the inverse of the following matrix by Gauss Jordan method. Also determine the condition number of the matrix.

$$\begin{pmatrix} 0.6 & 0.9 & 1.2 \\ 0.8 & 0.7 & 1.9 \\ 2.2 & 0.4 & 0.3 \end{pmatrix}$$

Ex I.6: Determine the inverse of the following tri-diagonal matrix by any method.

$$\begin{pmatrix} 2.04 & -1 & 0 & 0 & 0 \\ -1 & 2.04 & -1 & 0 & 0 \\ 0 & -1 & 2.04 & -1 & 0 \\ 0 & 0 & -1 & 2.04 & -1 \\ 0 & 0 & 0 & -1 & 1.02 \end{pmatrix}$$

What is the condition number of the above tri-diagonal matrix?

Ex I.7: Solve the system of equations in Exercise I.6 with $\mathbf{b}^T = \begin{pmatrix} 100 & 0 & 0 & 0 & 0 \end{pmatrix}$ by Cholesky decomposition. Solve the equations also by TDMA. Which decomposition is superior and why?

Ex I.8: Figure I.1 shows an electrical circuit consisting of 8 equal resistances of 100 Ω each. Voltages at three nodes are as indicated in the figure. Voltages at nodes 1 through 4 are required. Represent the nodal equations in matrix form. Determine the condition number of the coefficient matrix.

Determine the voltages at nodes 1 through 4 with a precision of two digits after decimals. Use LU decomposition.

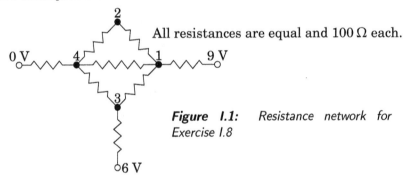

All resistances are equal and 100 Ω each.

Figure I.1: Resistance network for Exercise I.8

Ex I.9: Solve the following set of linear equations by Gauss elimination. Attempt the solution with and without pivoting. Comment on the results. Look for precision to 2 digits after decimals.

$$\begin{pmatrix} 1 & 2 & 4 & 3 \\ 6 & 5 & 8 & 7 \\ 10 & 12 & 11 & 9 \\ 15 & 13 & 14 & 16 \end{pmatrix} \begin{pmatrix} x_1 \\ x_2 \\ x_3 \\ x_4 \end{pmatrix} = \begin{pmatrix} 5 \\ 4 \\ 9 \\ 6 \end{pmatrix}$$

Ex I.10: Solve the set of equations in Exercise I.9 by QR factorization. Make use of Gram Schmidt method.

Ex I.11: Solve the following set of equations by QR transformation of the coefficient matrix using Householder transformations.

$$\begin{pmatrix} 1 & 2 & 4 & 1 \\ 2 & -5 & 3 & -5 \\ 1 & -4 & 4 & 7 \\ 2 & -3 & 2 & -5 \end{pmatrix} \begin{pmatrix} x_1 \\ x_2 \\ x_3 \\ x_4 \end{pmatrix} = \begin{pmatrix} 2 \\ 4 \\ 4 \\ 9 \end{pmatrix}$$

Ex I.12: Reduce the following matrix to QR form by using Givens transformations.

$$\mathbf{A} = \begin{pmatrix} 1 & -0.286 & -0.116 & -0.598 \\ -0.06 & 1 & -0.02 & -0.12 \\ -0.05 & -0.05 & 1 & -0.1 \\ -0.24 & -0.24 & -0.1 & 0.59 \end{pmatrix}$$

If $\mathbf{b}^T = \begin{pmatrix} 1 & 0.029 & 0.59 & 0 \end{pmatrix}$ determine the solution to equation $\mathbf{Ax} = \mathbf{b}$.

Ex I.13: Solve the system of equations in Exercise I.12 by Gauss Seidel iteration technique. Set tolerance as 10^{-3}. Start iteration with $\mathbf{x}^{(0)T} = \begin{pmatrix} 1 & 1 & 1 \end{pmatrix}$ and find the number of iterations required to obtain the solution with 3 digit accuracy.

Ex I.14: Solve for all the nodal voltages in the triangular resistance network shown in Figure I.2.

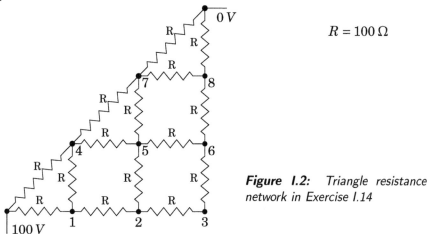

$R = 100 \, \Omega$

Figure I.2: *Triangle resistance network in Exercise I.14*

Ex I.15: Heat transfer problems in steady conduction may sometimes be represented by an electrical analog using a resistance network. Heat transfer from an extended surface[5] may be represented by a resistance network such as that shown in Figure I.3. Voltages replace temperatures and electrical resistances replace thermal resistances in the electrical analog. Formulate the problem as a set of linear equations for the nodal voltages. Solve the set of equations by (a) TDMA and (b) Gauss Seidel iteration. Voltages so determined should have a precision of a mV.

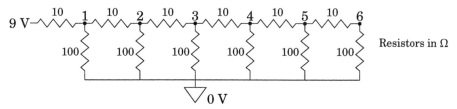

Figure I.3: *Resistance network for Exercise I.15*

---

[5]Refer to a text on heat transfer such as S.P. Venkateshan, "Heat Transfer" 2nd Edition, Ane Books Pvt. Ltd., 2011

Ex I.16: Solve the set of linear equations in Exercise I.15 by the conjugate gradient method.

Ex I.17: Show that the following 3×3 symmetric matrix is positive definite.

$$\mathbf{A} = \begin{pmatrix} 9 & -1 & 2 \\ -1 & 7 & -3 \\ 2 & -3 & 7 \end{pmatrix}$$

Solve the set of three linear equations $\mathbf{Ax} = \mathbf{b}$ where $\mathbf{b}^T = \begin{pmatrix} 4 & 8 & 12 \end{pmatrix}$ by Steepest Descent and Conjugate Gradient methods. Compare the solution with that obtained by Cholesky decomposition.

I.2   Evaluation of eigenvalues

Ex I.18: Consider the set of two coupled first order ordinary differential equations given by

$$\begin{pmatrix} \dfrac{dy_1}{dt} \\[2ex] \dfrac{dy_2}{dt} \end{pmatrix} = \begin{pmatrix} -\dfrac{74}{1113} & \dfrac{1}{21} \\[2ex] \dfrac{1}{65} & -\dfrac{1}{65} \end{pmatrix} \begin{pmatrix} y_1 \\[2ex] y_2 \end{pmatrix}$$

Determine the eigenvalues and thence write down the solution to the problem if $y_1 = y_2 = 1$ at $t = 0$. Note that the above coupled equations model a second order thermal system.

Ex I.19: An example involving two springs and masses arranged in series was given as an example leading to an eigenvalue problem with the eigenvalues representing the natural frequencies of the system. Using the notation introduced in Figure 3.4, the following table gives the masses and spring constants.

$$\begin{array}{lll} m_1 = & 1 & kg \\ m_2 = & 0.5 & kg \\ k_1 = & 100 & N/m \\ k_2 = & 50 & N/m \end{array}$$

Determine both the eigenvalues and the corresponding frequencies for the system. What are the modes that correspond to these two frequencies?

Ex I.20: Consider a system of coupled first order equations given by

$$\begin{pmatrix} \dfrac{dy_1}{dt} & \dfrac{dy_2}{dt} & \dfrac{dy_3}{dt} \end{pmatrix}^T = \begin{pmatrix} 1 & -1 & -1 \\ -1 & 1 & -1 \\ 1 & 1 & -1 \end{pmatrix} \begin{pmatrix} y_1 & y_2 & y_3 \end{pmatrix}^T$$

What are the three eigenvalues of the coefficient matrix? Write the Jordan form of the coefficient matrix. Obtain the general solution to the equations.

Ex I.21: Determine the eigenvalues of the following matrix by solving the characteristic equation.

$$\mathbf{A} = \begin{pmatrix} 3 & -1 & 1 \\ -1 & 2 & 1 \\ 1 & -1 & 3 \end{pmatrix}$$

What are the eigenvectors? Verify whether the eigenvectors are orthogonal to one another. Write down the Schur form of the matrix.

Ex I.22: Determine the largest and smallest eigenvalues, respectively, of the coefficient matrix in Exercise I.17 by the power and inverse power method. What is the third eigenvalue? Are the eigenvectors orthogonal to one another? Is the matrix diagonalizable?

Ex I.23: Determine one eigenpair of the matrix in Exercise I.21 by Rayleigh quotient iteration. How many iterations are required to get convergence to 4 digits after decimals? Determine a second eigenpair by deflation of the matrix followed by Rayleigh quotient iteration.

Ex I.24: Consider the 4×4 matrix

$$\mathbf{A} = \begin{pmatrix} 3 & 0 & 1 & 0 \\ 2 & 4 & 0 & 1 \\ 1 & 0 & 2 & 0 \\ 2 & 1 & 0 & 2 \end{pmatrix}$$

Write down the matrix after swapping rows 1 and 3 and call it **B**. Premultiplication of **A** by what matrix will accomplish this? Will the eigenvalues of matrix **A** and **B** be the same? Explain.

Reduce **A** to Hessenberg form. Then obtain all the eigenvalues of **A** by QR iteration.

Ex I.25: Obtain all the eigenvalues of the matrix

$$\mathbf{A} = \begin{pmatrix} 2 & 1 & 0.5 & 1 \\ 0 & 3 & 0 & 2 \\ 0 & 0 & 2 & 0 \\ 0 & 0 & 0.75 & 1.5 \end{pmatrix}$$

What is the Schur form of this matrix?

Ex I.26: Obtain all the eigenpairs of the following matrix by QR iteration with shifts, as necessary. If the matrix has complex eigenvalues obtain them using Francis shift, as necessary.

$$\mathbf{A} = \begin{pmatrix} 4 & 2 & 1 & 3 \\ 1 & 5 & 7 & 4 \\ 4 & 6 & 1 & 3 \\ 2 & 5 & 6 & 7 \end{pmatrix}$$

I.3 Solution of algebraic equations

Ex I.27: Find a positive root of the function $f(x) = e^{-x^2} - 1.5 - \dfrac{1}{x + 0.2}$ by the bisection method. Minimum four digit accuracy is required in the root.

Ex I.28: Solve Exercise I.27 by fixed point iteration. Comment on the convergence of fixed point iteration as compared to the bisection method. Root is required with 4 figure accuracy.

Ex I.29: Solve Exercise I.27 by Newton Raphson method. Study the effect of varying the initial guess.

Ex I.30: Obtain the root lying between 2π and 4π of the function $f(x) = \sin x - \dfrac{1}{x + 0.5}$.

Ex I.31: Solve the equation $e^{2x} - e^{-2x} = 12x$ by (a) Newton Raphson method, (b) secant method. In the case of Newton Raphson method start iteration with $x > 1$ and $x < 1$ and make comments based on the outcome.

Ex I.32: Solve for one of the roots of the cubic equation $f(x) = x^3 - 2x - 4 = 0$ by any method. Comment on the nature of the other two roots of this cubic equation.

Ex I.33: Obtain first three positive roots of equation $f(x) = x\sin(x) - \cos(x)$ using the Newton Raphson method. Such an equation is encountered in determining the eigenvalues of one dimensional heat equation.

Ex I.34: Solve for all the roots of the cubic $x^3 - 8.5x^2 + 23.75x - 21.875 = 0$ by Newton Raphson method.

Ex I.35: Roots of a polynomial can be determined based on eigenvalue determination. Consider the polynomial equation of the form

$$a_0 + a_1 x + a_2 x^2 + \cdots + a_{n-1}x^{n-1} + a_n x^n = 0$$

with $a_n = 1$. Then the roots of the polynomial are the eigenvalues of the following matrix

$$\mathbf{A} = \begin{pmatrix} 0 & 1 & 0 & \cdots & 0 \\ 0 & 0 & 1 & \cdots & 0 \\ \cdots & \cdots & \cdots & \cdots & \cdots \\ 0 & 0 & 0 & \cdots & 1 \\ -a_0 & -a_1 & -a_2 & \cdots & -a_{n-1} \end{pmatrix}$$

Write a program to determine the roots of the polynomial using the above method.[6] Determine the roots of a 4th degree Legendre polynomial $y(x) = \dfrac{1}{8}\left(35x^4 - 30x^2 + 3\right)$.

Ex I.36: Bessel function represented by $J_\nu(x)$ (Bessel functions appear in the analytical solution of Laplace equation, a partial differential equation that governs many physical problems, in cylindrical coordinates) where ν is a parameter is a function available in spreadsheet programs. Using a spreadsheet program obtain the first positive root of the equation $f(x) = xJ_1(x) - J_0(x)$ using the secant method. What should you do if you want to use the Newton Raphson method instead of the secant method?

Ex I.37: Find the points of intersection of line $y = 1 + x$ with the ellipse $\dfrac{x^2}{4} + \dfrac{y^2}{9} = 1$. Reduce the problem to finding the root of a single equation in one variable by eliminating one of the variables between the two equations.

Ex I.38: Solve Exercise I.37 by treating the two equations simultaneously. Use Newton Raphson method.

Ex I.39: Solve $f(x) = \dfrac{1}{(x-2)^2 + 0.2} - \dfrac{1}{(x-1)^2 + 0.5} = 0$. How many real roots does this equation have? Use different methods such as the bisection method, fixed point iteration method, Newton Raphson method, secant method to solve the equation and comment on their relative performance.

(Hint: Make a plot to study the nature of the function)

Ex I.40: Solve the following set of equations by Gauss Seidel iteration technique. Obtain all the roots.

$$x^2 + y^2 = 4$$

---

[6]MATLAB function `roots(p)` uses the above algorithm for determining the roots of a polynomial.

$$e^x + y = 1$$

Note that this example was solved using Newton Raphson method in the text. Would it be advisable to use SOR in this case?

Ex I.41: Solve the following two non-linear equations:

$$x^2 + y^2 = 4$$
$$x^2 - y^2 = 3$$

Use the Gauss Seidel method.

Ex I.42: Solve the following two non-linear equations:

$$x^2 - y^2 = 3$$
$$xy = 0.5$$

Use SOR method.

Ex I.43: Solve the following two non-linear equations:

$$x^2 - y^2 - y = 0$$
$$x^2 + y^2 - x - \frac{1}{8} = 0$$

Use $x = -0.1$, $y = 0.1$ to start the Newton Raphson iteration.

Ex I.44: A set of nonlinear equations are represented in matrix form as $\mathbf{Ay} + 0.05\mathbf{Iy'} = \mathbf{b}$ where $\mathbf{I}$ is the identity matrix. In a specific case the following data has been specified.

$$
\mathbf{A} = \begin{pmatrix} 2 & -1 & 0 & 0 & 0 & 0 \\ -1 & 2 & -1 & 0 & 0 & 0 \\ 0 & -1 & 2 & -1 & 0 & 0 \\ 0 & 0 & -1 & 2 & -1 & 0 \\ 0 & 0 & 0 & -1 & 2 & -1 \\ 0 & 0 & 0 & 1 & -4 & 3 \end{pmatrix}
\quad
\mathbf{y} = \begin{pmatrix} y_1 \\ y_2 \\ y_3 \\ y_4 \\ y_5 \\ y_6 \end{pmatrix}
\quad
\mathbf{b} = \begin{pmatrix} 1 \\ 0.0648 \\ 0.0648 \\ 0.0648 \\ 0.0648 \\ 0 \end{pmatrix}
\quad
\mathbf{y'} = \begin{pmatrix} y_1^4 \\ y_2^4 \\ y_3^4 \\ y_4^4 \\ y_5^4 \\ 0 \end{pmatrix}
$$

Solve these equations by Gauss Seidel iteration and the SOR method. Comment on convergence rates of the two methods. Is TDMA a better option?

Module **II**

Interpolation, differentiation and integration

Function approximations assume an important space in computational methods. The utility of approximations would arise in two important class of problems. The first deals with representation of an exact known function by a good alternate approximating function which can be easily evaluated. The approximating function would behave the same way as the exact function. Some examples where such function approximation can be used are given below

1. The approximating function values are close to the exact value (within acceptable limits). Examples are polynomials and orthogonal functions such as circular functions, Bessel functions etc.
2. Derivatives and integrals of the exact function.
   Consider the integral

$$\int_a^b e^{x^2} dx$$

   The above integral cannot be evaluated analytically. There are several other functions which do not have analytical closed form integrals. We will have to use function approximations for such functions to evaluate the integrals as well as derivatives.

The second class of problems where function approximations are important is when the exact function is not known but a set of discrete data representing the exact function are available. Examples of the same include experimental data, tabulations of properties in a handbook such as steam tables. It is often desirable to determine the function value at an intermediate point where the function value is not available. Then one must use methods such as interpolation or regression.

Chapter 5 would consider concepts of interpolation using polynomials and splines in one dimensions. Interpolation in multi-dimensions will be discussed in Chapter 6. Chapter 7 will introduce principles of curve fitting which is used when the discrete data contains inherent noise. Concepts of numerical differentiation would be introduced in Chapter 8. Chapter 9 would consider numerical integration.

Chapter 5

Interpolation

In many applications that use tabulated data it is necessary to evaluate the value of a function for a value of the independent variable(s) that is not one of the tabulated values. It is then necessary to use interpolation to evaluate the function at the desired value of the independent variable(s). Interpolated value is generally an approximation to the actual value of the function that is represented by the tabulated data. Interpolating function is used within the range of the tabulated values in constructing the approximating function. Extrapolation involves the use of the approximating function outside the range of tabulated values in constructing the approximating function. Extrapolation is fraught with errors and hence not generally recommended. Normal practice is to use a polynomial of suitable degree to represent the data and use the polynomial to evaluate the function at the desired location.

5.1 Polynomial interpolation

We look at a table of data wherein a function $f(x)$ is available at discrete set of values of x i.e. pairs x_i, f_i are available for $0 \le i \le n$ as shown in Table 5.1 below.

Table 5.1: *Data in tabular form*

x	x_0	x_1	x_2	$\cdots$	$\cdots$	$\cdots$	x_{n-1}	x_n
$f(x)$	f_0	f_1	f_2	$\cdots$	$\cdots$	$\cdots$	f_{n-1}	f_n

We would like to interpolate between the tabulated values using a polynomial of suitable degree. It can be shown that there is a unique polynomial passing through these points which is of the form

$$f(x) = a_0 + a_1 x + a_2 x^2 + \cdots + a_n x^n \tag{5.1}$$

where a_i are coefficients of the interpolating polynomial and are determined by solving the following system of linear equations.

$$\underbrace{\begin{bmatrix} 1 & x_0 & x_0^2 & \cdots & x_0^n \\ 1 & x_1 & x_1^2 & \cdots & x_1^n \\ \vdots & \vdots & \vdots & \ddots & \vdots \\ 1 & x_n & x_n^2 & \cdots & x_n^n \end{bmatrix}}_{\text{Vandermonde matrix}} \underbrace{\begin{Bmatrix} a_0 \\ a_1 \\ \vdots \\ a_n \end{Bmatrix}}_{\text{Vector of coefficients}} = \begin{Bmatrix} f_0 \\ f_1 \\ \vdots \\ f_n \end{Bmatrix} \tag{5.2}$$

Vandermonde matrices are generally ill-conditioned. Hence solving the above system of linear equations may lead to errors. The same polynomial can also be represented in alternate ways such as Lagrange[1] and Newton polynomials.

Lagrange polynomial representation is useful when the degree of the interpolation polynomial is fixed. Lagrange polynomial representations are commonly used in numerical methods like finite element method (FEM) and finite volume method (FVM) for solving partial differential equations. It is a general practice in numerical methods to discretize the domain into a large number of small sub domains known as elements in FEM and volumes in FVM. These methods assume a functional form for the variable in each sub domain and solve for the variables at the nodes (Modules III and IV).

However, when the degree of the polynomial is not fixed, then it would be advantageous to use Newton polynomial representation. The degree of the Newton polynomial can be increased with very little computational effort. Newton polynomials are advantageous when we have equispaced tabulated data.

The following sections would introduce these polynomial representations.

5.2 Lagrange interpolating (or interpolation) polynomial

The interpolating function can be expressed as weighted sum of the function at each node.

$$f(x) = w_0 f_0 + w_1 f_1 + \cdots + w_n f_n \tag{5.3}$$

where w_i are weight functions each of which is a polynomial of form given in Equation 5.1. The interpolating polynomial is to be such that the values obtained by it are the same as the tabulated values for each x_i, the range of x_i depending on the degree of the polynomial. Hence, $w_i = 1$ at i^{th} node and $w_i = 0$ at all other nodes. We shall take a simple example and then generalize.

---

[1]Joseph Louis Lagrange, 1736 - 1813, Italian mathematician

5.2.1 Linear interpolation

Consider data shown in Table 5.1. Linear interpolation is possible between any two entries in the table such as between x_i and x_{i+1}. Let us consider the following interpolating function:

$$f(x) \approx L_1(x) = \frac{(x - x_{i+1})}{(x_i - x_{i+1})} f_i + \frac{(x - x_i)}{(x_{i+1} - x_i)} f_{i+1} \tag{5.4}$$

in the interval $x_i \leq x \leq x_{i+1}$. The chosen polynomial is of first degree since each term is linear in x. The interpolating polynomial (Equation 5.4) agrees with the function at both the nodes x_i and x_{i+1}. The symbol $L_1(x)$ means that the interpolating function is a polynomial of degree 1, referred to as a Lagrange polynomial of degree 1. Excepting at the nodes the Lagrange polynomial represents the function $f(x)$ with *some* error. Hence we refer to $L_1(x)$ as an approximation to the function $f(x)$. Interpolating function given by Equation 5.4 may also be written as a weighted sum as

$$f(x) \approx L_1(x) = w_i(x) f_i + w_{i+1}(x) f_{i+1} \tag{5.5}$$

where the two weights are linear in x. We note that a Lagrange polynomial of degree 1 passes through two neighboring data points. It is easily seen that a Lagrange polynomial of degree 2 would pass through three consecutive data points and a Lagrange polynomial of degree k would pass through $k + 1$ consecutive data points.

5.2.2 Quadratic interpolation

We may easily write a quadratic interpolating polynomial by requiring it to pass through three consecutive data points as

$$f(x) \approx L_2(x) = w_i(x) f_i + w_{i+1}(x) f_{i+1} + w_{i+2}(x) f_{i+2} \tag{5.6}$$

where each weight is a second degree polynomial in x given by

$$w_i(x) = \frac{(x - x_{i+1})(x - x_{i+2})}{(x_i - x_{i+1})(x_i - x_{i+2})} \quad w_{i+1}(x) = \frac{(x - x_i)(x - x_{i+2})}{(x_{i+1} - x_i)(x_{i+1} - x_{i+2})}$$
$$w_{i+2}(x) = \frac{(x - x_i)(x - x_{i+1})}{(x_{i+2} - x_i)(x_{i+2} - x_{i+1})} \tag{5.7}$$

Expression 5.6 is restricted to the range $x_i \leq x \leq x_{i+2}$. Weight w_i is equal to 1 at x_i but vanishes for both x_{i+1} and x_{i+2}. Weight w_{i+1} is equal to 1 at x_{i+1} but vanishes for both x_i and x_{i+2}. Weight w_{i+2} is equal to 1 at x_{i+2} but vanishes for both x_i and x_{i+1}. Thus $L_2(x)$ passes through the three data points (x_i, f_i); (x_{i+1}, f_{i+1}) and (x_{i+2}, f_{i+2}). Hence it is the required interpolating polynomial of degree 2.

5.2.3 Generalization

It is possible to generalize the above to a kth degree Lagrange polynomial where $k \leq n - 1$. We note that the weight of each term in the Lagrange polynomial of degree k is a kth degree polynomial. General term in the Lagrange polynomial is then given by

$$\frac{(x - x_i)(x - x_{i+1})...(x - x_{i+k})}{(x_{i+l} - x_i)(x_{i+l} - x_{i+1})...(x_{i+l} - x_{i+k})} f_{i+l} \tag{5.8}$$

where factor $(x - x_{i+l})$ does not appear in the numerator and correspondingly the factor $(x_{i+l} - x_{i+l})$ does not appear in the denominator. Thus no division by zero! Note also that $0 \le l \le k$. Expression 5.8 may be recast in the form

$$\frac{\prod_{p=0, p \ne l}^{k} (x - x_{i+p})}{\prod_{p=0, p \ne l}^{k} (x_{i+l} - x_{i+p})} \tag{5.9}$$

where

$$\prod_{p=0, p \ne l}^{k} (x - x_{i+p}) = (x - x_i)(x - x_{i+1}) \cdots (x - x_{i+l-1})(x - x_{i+l+1}) \cdots (x - x_{i+k})$$

Finally the interpolating polynomial of degree k may be written as

$$L_k(x) = \sum_{l=0}^{k} \left[\frac{\prod_{p=0, p \ne l}^{k} (x - x_{i+p})}{\prod_{p=0, p \ne l}^{k} (x_{i+l} - x_{i+p})} \right] f_{i+l} \tag{5.10}$$

The polynomial $L_k(x)$ passes through the $k + 1$ points given by (x_i, f_i); $(x_{i+1}, f_{i+1}) \cdots (x_k, f_{i+k})$ and Expression 5.10 is valid for $x_i \le x \le x_{i+k}$.

MATLAB program has been provided below to evaluate Lagrange polynomials at desired points.

Program 5.1: *Evaluate Lagrange polynomials at given points*

```
1  function y = lagrange(xp,yp,x)
2  %          Input   xp :  nodes
3  %                  yp :  function values at points
4  %                   x :  points where polynomial is to evaluated
5  %          Output   y :  polynomial interpolated at x
6  n = length(xp);              % number of interpolating points
7  m = length(x);               % number of points in x
8  w = ones(n,m);               % initialize Lagrange weights
9  for i=1:n                    % loop to calculate weights
10     for j=1:n
11         if(i~=j)             % i≠j
12             w(i,:) = w(i,:).*(x-xp(j))/(xp(i)-xp(j));
13         end
14     end
15 end
16 y = zeros(m,1);              % initialize y
17 for i=1:n                    % evaluate y
18     y = y + w(i,:)'* yp(i);
19 end
```

Example 5.1

Function vs location data is tabulated below. Obtain the value of the function at $x = 0.07$ and $x = 0.35$ using Lagrange polynomials of various degrees and comment on the results.

i	0	1	2	3	4	5	6	7
x_i	0	0.08	0.15	0.21	0.31	0.38	0.47	0.57
$f(x_i)$	1	0.9057	0.8327	0.7767	0.6956	0.6472	0.5942	0.5466

Solution :

We note that the tabulated data is available with unequal spacing and hence we use Lagrange polynomials based on Equation 5.10. As an example, we show how the quadratically interpolated value is calculated. The weights at $x = 0.07$ are obtained first.

$$w_0 = \frac{(x-x_1)(x-x_2)}{(x_0-x_1)(x_0-x_2)} = \frac{(0.07-0.08)(0.07-0.15)}{(0-0.08)(0-0.15)} = 0.0667$$

$$w_1 = \frac{(x-x_0)(x-x_2)}{(x_1-x_0)(x_1-x_2)} = \frac{(0.07-0)(0.07-0.15)}{(0.08-0)(0.08-0.15)} = 1$$

$$w_2 = \frac{(x-x_0)(x-x_1)}{(x_2-x_0)(x_2-x_1)} = \frac{(0.07-0)(0.07-0.08)}{(0.15-0)(0.15-0.08)} = -0.0667$$

Note that the sum of the weights equals one. We then have

$$f(0.07) = 0.0667 \times 1 + 1 \times 0.9057 - 0.0667 \times 0.8327 = 0.9168$$

Linear, quadratic and cubic interpolated values are obtained for $x = 0.07$ and presented in the following table.

i	x_i	$f(x_i)$	Linear	Quadratic	Cubic
0	0	1			
	0.07		0.9174	0.9168	0.9167
1	0.08	0.9057			
2	0.15	0.8327			
3	0.21	0.7766			

Program 5.1 has been used to obtain the cubic interpolated value for $x = 0.35$.

```
x  = [0.31; 0.38; 0.47; 0.57];
y  = [0.6955; 0.6471;  0.5942; 0.5466];
y1 = lagrange(x,y,0.35);
```

The output of the program is

```
y1 = 0.6670
```

Linear, quadratic and cubic interpolated values are obtained for $x = 0.35$ and presented in table below.

i	x_i	$f(x_i)$	Linear	Quadratic	Cubic
4	0.31	0.6955			
	0.35		0.6678	0.6670	0.6670
5	0.38	0.6471			
6	0.47	0.5942			
7	0.57	0.5466			

An alternate way of interpolating for the function value $x = 0.35$ is to choose the four points shown in table below.

i	x_i	$f(x_i)$	Linear	Quadratic	Cubic
2	0.15	0.8327			
3	0.21	0.7766			
4	0.31	0.6955			
	0.35		**0.6678**	**0.6669**	**0.6670**
5	0.38	0.6471			

It is seen that the quadratically interpolated value is sufficient for practical purposes. Also alternate ways of interpolation using the tabulated values yield consistent interpolated values.

Sometimes it would be desirable to store the weight functions of Lagrange polynomials, in order to perform differentiation and integration. MATLAB program 5.2 evaluates the weight functions of Lagrange polynomial and stores the same in the polynomial representation of form

$$w_i(x) = a_{i,0} + a_{i,1}x + a_{i,2}x^2 + \cdots + a_{i,n}x^n \tag{5.11}$$

The output is a matrix **p** that contains the coefficients of the weight function polynomial for each node.

$$\left(\begin{array}{c} w_0 \\ w_1 \\ \vdots \\ w_n \end{array} \right) = \left(\begin{array}{cccc} a_{0,0} & a_{0,1} & \cdots & a_{0,n} \\ a_{1,0} & a_{1,1} & \cdots & a_{1,n} \\ \vdots & \vdots & \ddots & \vdots \\ a_{n,0} & a_{n,1} & \cdots & a_{n,n} \end{array} \right) \left(\begin{array}{c} 1 \\ x \\ \vdots \\ x^n \end{array} \right)$$

Program 5.2: *Evaluate weight functions of Lagrange polynomials*

```
1  function p = lagrangeweightfunction(x)
2  % Input    x: nodes
3  % Output   p: matrix of coefficients of weight function
4  n = length(x);          % number of nodes
5  p = zeros(n);           % initialize p
6  for i = 1:n
7      w = zeros(n,1);
8      w(1) = 1;
9      w1 = zeros(n,1);
10     den = 1;            % initialize denominator
11     for j = 1:n         % loop for calculating coefficients
12         if(i~=j)
13             w1(2:n) = w(1:n-1);
14             w = w1 -x(j)*w;  % updating coefficients
15             den = den*(x(i)-x(j)); % updating denominator
16         end
17     end
18     p(i,:) = w/den;     % update coefficients of node i
19 end
```

The coefficients of the nth degree polynomial (Equation 5.1) passing through the points can be easily estimated from the weight function matrix.

$$f(x) = w_0 f_0 + w_1 f_1 + \cdots + w_n f_n$$

$$= \underbrace{\sum_{i=0}^{n} a_{i,0} f_i}_{a_0} + \underbrace{\sum_{i=0}^{n} a_{i,1} f_i\, x}_{a_1} + \underbrace{\sum_{i=0}^{n} a_{i,2} f_i\, x^2}_{a_2} + \cdots + \underbrace{\sum_{i=0}^{n} a_{i,n} f_i\, x^n}_{a_n}$$

The following MATLAB program evaluates the weights at the desired points.

Program 5.3: *Evaluate weights at specified points*

```
 1 function w = lagrangeweight(p,x)
 2 % Input     p: matrix of coefficients of weight function
 3 %           x: points where weights have to be evaluated
 4 % Output    w: Lagrange polynomial weights at x
 5 m = length(x);         % number of nodes
 6 n = size(p,1);         % n-1 = degree of Lagrange polynomial
 7 v = ones(m,n);         % initialize Vandermonde matrix
 8 for i=2:n              % construct Vandermonde matrix
 9    v(:,i) = x.*v(:,i-1);
10 end
11 w =v*p';               % Lagrange polynomial weights
```

5.2.4 Lagrange polynomials with equi-spaced data

If the tabulated values of the function are available as equi-spaced data Lagrange polynomial take on a simpler form. For equi-spaced data, we have $x_{i+1} - x_i = h$ for $0 \le i \le n$ where h is the constant spacing. We shall also use the definition $x - x_i = rh$ where r is a number such that the point x is within the range of applicability of the Lagrange polynomial. Equation 5.10 may be recast as

$$L_k(r) = \sum_{l=0}^{k} \left[\frac{\prod_{p=0, p \neq l}^{k} (r - p)}{\prod_{p=0, p \neq l}^{k} (l - p)} \right] f_{i+l} \tag{5.12}$$

As a particular case consider $k = 2$. Then we have

$$L_2(r) = \frac{(r-1)(r-2)}{(-1)(-2)} f_i + \frac{r(r-2)}{(1)(-1)} f_{i+1} + \frac{r(r-1)}{(2)(1)} f_{i+2} \tag{5.13}$$

$$= \frac{(r-1)(r-2)}{2} f_i - r(r-2) f_{i+1} + \frac{r(r-1)}{2} f_{i+2} \tag{5.14}$$

We notice that the three weights, in this case of equi-spaced data, given by

$$w_i = w_0 = \frac{(r-1)(r-2)}{2} = \frac{r^2 - 3r + 2}{2}$$

$$w_{i+1} = w_1 = -r(r-2) = 2r - r^2$$

$$w_{i+2} = w_2 = \frac{r(r-1)}{2} = \frac{r^2 - r}{2}$$

are all quadratic in r. These functions are valid in the range $0 \le r \le 2$ and are shown plotted in Figure 5.1.

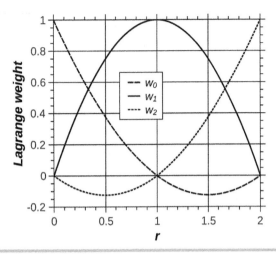

Figure 5.1: Quadratic weights for Lagrange polynomials of degree 2

Example 5.2

Function vs location data is given in the following table with equal spacing of $h = 0.1$. Obtain the value of the function at $x = 0.37$ using Lagrange polynomials of various degrees and comment on the results.

i	0	1	2	3	4	5
x_i	0	0.1	0.2	0.3	0.4	0.5
f_i	0.09091	0.14023	0.18887	0.23659	0.28380	0.32851

Solution :

We use linear, quadratic and cubic interpolations using the data points 2 through 5. The interpolated function value is desired at $x = 0.37$ that corresponds to $r = \dfrac{0.37}{0.1} = 3.7$, where $x - x_0 = rh$ and $h = 0.1$. In the case of linear interpolation the weights are assigned to the data points $i = 3$ and $i = 4$ (these points straddle the point $r = 3.7$) as given below.

$$w_3 = \frac{3.7 - 4}{-1} = 0.3; \; w_4 = \frac{3.7 - 3}{1} = 0.7$$

The linearly interpolated value is then given by

$$f_{0.37} = w_3 f_3 + w_4 f_4 = 0.3 \times 0.23659 + 0.7 \times 0.28380 = 0.26922$$

Data points corresponding to $i = 2$, $i = 3$ and $i = 4$ are used to construct the quadratically interpolated value. The weights are calculated as

$$w_2 = \frac{(3.7 - 3)(3.7 - 4)}{2} = -0.1050$$

$$w_3 = \frac{(3.7 - 2)(3.7 - 4)}{-1} = 0.5100$$

$$w_4 = \frac{(3.7 - 2)(3.7 - 3)}{2} = 0.5950$$

The quadratically interpolated value is then given by

$$f_{0.37} = w_2 f_2 + w_3 f_3 + w_4 f_4$$

$$
\begin{aligned}
&= \quad -0.1050 \times 0.18887 + 0.5100 \times 0.23659 + 0.5950 \times 0.28380 \\
&= \quad 0.26934
\end{aligned}
$$

Data points corresponding to $i = 2$, $i = 3$, $i = 4$ and $i = 5$ are used to construct the cubically interpolated value. The weights are calculated as

$$
w_2 = \frac{(3.7-3)(3.7-4)(3.7-5)}{-6} = -0.0455 \qquad w_3 = \frac{(3.7-2)(3.7-4)(3.7-5)}{2} = 0.3315
$$

$$
w_4 = \frac{(3.7-2)(3.7-3)(3.7-5)}{-2} = 0.7735 \qquad w_5 = \frac{(3.7-2)(3.7-3)(3.7-4)}{6} = -0.0595
$$

The cubically interpolated value is then given by

$$
\begin{aligned}
f_{0.37} &= \quad w_2 f_2 + w_3 f_3 + w_4 f_4 + w_5 f_5 \\
&= \quad -0.0455 \times 0.18887 + 0.3315 \times 0.23659 + 0.7735 \times 0.28380 \\
&\quad -0.0595 \times 0.32851 = 0.26935
\end{aligned}
$$

Representation using local coordinates

It is useful in some applications such as numerical integration, to transform the set of points to local coordinates $-1 < \xi < 1$. If x_1 and x_2 are the boundaries of the domain, the relation between original coordinates and the transformed coordinates can be written as

$$
x = \frac{x_1 + x_2}{2} + \frac{x_2 - x_1}{2}\xi \tag{5.15}
$$

Linear interpolation weights in local coordinates would then be

$$
\begin{aligned}
w_{-1} &= \frac{1-\xi}{2} \\
w_1 &= \frac{1+\xi}{2}
\end{aligned} \tag{5.16}
$$

Quadratic interpolation weights for equi-spaced points in local coordinates would be

$$
\begin{aligned}
w_{-1} &= \frac{\xi(1-\xi)}{2} \\
w_0 &= 1 - \xi^2 \\
w_1 &= \frac{\xi(1+\xi)}{2}
\end{aligned} \tag{5.17}
$$

5.3 Newton Polynomials

Alternate way of expressing an interpolating polynomial is to define it in terms of a Newton polynomial. Newton polynomials use tabulated data and express the polynomials in terms of divided/forward/backward differences generated using the tabulated data. A Newton polynomial may be upgraded to a higher degree by simply adding extra terms without modifying the terms already written down.

5.3.1 Divided differences

Table 5.2: *Data in tabular form*

x	x_0	x_1	x_2	$\cdots$	$\cdots$	$\cdots$	x_{n-1}	x_n
$f(x)$	f_0	f_1	f_2	$\cdots$	$\cdots$	$\cdots$	f_{n-1}	f_n

Consider data shown in Table5.2. We define the first divided difference $f[x_0,x_1]$ at x_0 as follows[2]:

$$f[x_0,x_1] = \frac{f_1 - f_0}{x_1 - x_0} \tag{5.18}$$

Second divided difference $f[x_0,x_1,x_2]$ at x_0 is defined as[3]

$$f[x_0,x_1,x_2] = \frac{f[x_1,x_2] - f[x_0,x_1]}{x_2 - x_0} \tag{5.19}$$

This may be rewritten using Equation 5.18 as

$$f[x_0,x_1,x_2] = \frac{\frac{f_2-f_1}{x_2-x_1} - \frac{f_1-f_0}{x_1-x_0}}{x_2 - x_0} \tag{5.20}$$

Similarly the kth divided difference is defined as[4]

$$f[x_0,x_1,x_2,...,x_k] = \frac{f[x_1,x_2,...,x_k] - f[x_0,x_1,...,x_{k-1}]}{x_k - x_0} \tag{5.21}$$

Above definitions are quite general and valid for data with *equal* or *unequal* spacing. However, with *equi-spaced data* such that $x_{i+1} - x_i = h$ for $0 \le i \le n$ the above may be simplified as

$$f[x_0,x_1,x_2,...,x_k] = \frac{f[x_1,x_2,...,x_k] - f[x_0,x_1,...,x_{k-1}]}{kh} \tag{5.22}$$

Based on the above we write down the first three divided differences for equi-spaced data.

$$f[x_0,x_1] = \frac{f_1 - f_0}{x_1 - x_0} = \frac{f_1 - f_0}{h} \tag{5.23}$$

$$
\begin{aligned}
f[x_0,x_1,x_2] &= \frac{f[x_1,x_2] - f[x_0,x_1]}{x_2 - x_0} \\
&= \frac{\frac{f_2-f_1}{h} - \frac{f_1-f_0}{h}}{2h} = \frac{f_2 - 2f_1 + f_0}{2!h^2}
\end{aligned} \tag{5.24}
$$

$$
\begin{aligned}
f[x_0,x_1,x_2,x_3] &= \frac{f[x_1,x_2,x_3] - f[x_0,x_1,x_2]}{x_3 - x_0} \\
&= \frac{\frac{f_3-2f_2+f_1}{2!h^2} - \frac{f_2-2f_1+f_0}{2!h^2}}{3h} = \frac{f_3 - 3f_2 + 3f_1 - f_0}{3!h^3}
\end{aligned} \tag{5.25}
$$

---

[2]In general, the first divided difference at data point x_i is defined as $f[x_i,x_{i+1}] = \dfrac{f_{i+1} - f_i}{x_{i+1} - x_i}$

[3]In general, the second divided difference at x_i is defined as $f[x_i,x_{i+1},x_{i+2}] = \dfrac{f[x_{i+1},x_{i+2}] - f[x_i,x_{i+1}]}{x_{i+2} - x_i}$

[4]In general, the kth divided difference at data point x_i is defined as $f[x_i,x_{i+1},...,x_{i+k}] = \dfrac{f[x_{i+1},x_{i+2},...,x_{i+k}] - f[x_i,x_{i+1},...x_{i+k-1}]}{x_{i+k} - x_i}$

5.3.2   Forward and backward differences

We see that, for equi-spaced data, divided differences are related to forward differences since the numerators in each of Equations 5.23-5.25 are respectively the first, second and third forward differences given by

$$\Delta f_i = f_{i+1} - f_i; \quad \Delta^2 f_i = f_{i+2} - 2f_{i+1} + f_i; \quad \Delta^3 f_i = f_{i+3} - 3f_{i+2} + 3f_{i+1} - f_i \qquad (5.26)$$

Note that forward difference is defined, in general as,

$$\Delta^k f_i = \Delta^{k-1} f_{i+1} - \Delta^{k-1} f_i \qquad (5.27)$$

Backward differences may be defined, analogous to the forward differences through the following relations.

$$\nabla f_i = f_i - f_{i-1}; \quad \nabla^2 f_i = f_i - 2f_{i-1} + f_{i-2}; \quad \nabla^3 f_i = f_i - 3f_{i-1} + 3f_{i-2} - f_{i-3} \qquad (5.28)$$

In general we also have

$$\nabla^k f_i = \nabla^{k-1} f_i - \nabla^{k-1} f_{i-1} \qquad (5.29)$$

5.3.3   Newton polynomial using divided, forward or backward differences

We would like to derive a polynomial - the Newton polynomial - that is useful for interpolation, based on the divided differences presented above. We define a Newton polynomial $p_{k-1}(x)$ such that it passes through the k points $x_0, x_1, ..., x_{k-1}$. Thus we have

$$p_{k-1}(x_0) = f_0; \ p_{k-1}(x_1) = f_1; \ ....; p_{k-1}(x_{k-1}) = f_{k-1} \qquad (5.30)$$

and, in addition, we define $p_k(x)$ such that

$$p_k(x) = p_{k-1}(x) + g_k(x) \qquad (5.31)$$

satisfies the additional condition $p_k(x_k) = f_k$. Equation 5.31 may be recast as

$$g_k(x) = p_k(x) - p_{k-1}(x) \qquad (5.32)$$

We note that $g_k(x)$ must be zero at $x_0, x_1, ..., x_{k-1}$. Hence it must be of form

$$g_k(x) = a_k(x - x_0)(x - x_1)....(x - x_{k-1}) \qquad (5.33)$$

We also notice that the highest power of x that occurs in $g(x)$ is x^k. Using Equation 5.33 in Equation 5.31 we have

$$p_k(x_k) = p_{k-1}(x_k) + a_k(x_k - x_0)(x_k - x_1)....(x_k - x_{k-1}) = f_k \qquad (5.34)$$

We solve the above for a_k to get

$$a_k = \frac{f_k - p_{k-1}(x_k)}{(x_k - x_0)(x_k - x_1)....(x_k - x_{k-1})} \qquad (5.35)$$

This is in the nature of a recurrence relation for determining the a's as we shall see below.

With $p_0(x) = f_0$, using Equation 5.35 we have

$$a_1 = \frac{f_1 - p_0(x_1)}{x_1 - x_0} = \frac{f_1 - f_0}{x_1 - x_0} = f[x_0, x_1] \tag{5.36}$$

The divided difference has made its appearance now! At this stage we have

$$p_1(x) = p_0(x) + (x - x_0)a_1 = f_0 + (x - x_0)f[x_0, x_1] \tag{5.37}$$

Consider the coefficient a_2 now. This may be written using the previous expressions as

$$\begin{aligned}
a_2 &= \frac{f_2 - p_1(x_2)}{(x_2 - x_0)(x_2 - x_1)} = \frac{f_2 - \{f_0 + (x_2 - x_0)f[x_0, x_1]\}}{(x_2 - x_0)(x_2 - x_1)} \\
&= \frac{f_2 - f_1 + f_1 - f_0 - (x_2 - x_0)f[x_0, x_1]}{(x_2 - x_0)(x_2 - x_1)} \\
&= \frac{f_2 - f_1 + (x_1 - x_0)f[x_0, x_1] - (x_2 - x_0)f[x_0, x_1]}{(x_2 - x_0)(x_2 - x_1)} \\
&= \frac{f_2 - f_1}{(x_2 - x_0)(x_2 - x_1)} + \frac{(x_1 - x_2)f[x_0, x_1]}{(x_2 - x_0)(x_2 - x_1)} \\
&= \frac{f[x_1, x_2] - f[x_0, x_1]}{x_2 - x_0} = f[x_0, x_1, x_2] \tag{5.38}
\end{aligned}$$

Thus (by mathematical induction) the coefficients in the Newton's polynomial are the divided differences of various orders, calculated at the point x_0. We may thus generalize the above and get

$$\begin{aligned}
f(x) \approx p_k(x) &= f_0 + (x - x_0)f[x_0, x_1] + (x - x_0)(x - x_1)f[x_0, x_1, x_2] + \dots \\
&+ (x - x_0)(x - x_1)\cdots(x - x_{k-1})f[x_0, x_1, \dots, x_k] \tag{5.39}
\end{aligned}$$

MATLAB program to calculate the divided difference table and coefficients of Newton polynomial is given below.

Program 5.4: *Coefficients of Newton polynomial using divided differences*

```
1  function pp = newtonpoly(x,f)
2  %    Input     x : interpolation points
3  %              f : function values at interpolation points
4  %    Output    pp: structure related to Newton polynomial
5  %              pp.coef : stores Newton polynomial weights
6  %              pp.x    : stores interpolation points
7  n = size(x,1);        % number of intepolation points
8  dd = zeros(n,n);      % initialize divided difference matrix
9  dd(:,1) = f;
10 for i=2:n             % generate divided difference table
11     for j=1:n-i+1
12         dd(j,i) =  (dd(j+1,i-1)-dd(j,i-1))/(x(j+i-1)-x(j));
13     end
14 end
15 % coefficients = first row of divided difference table
16 %                  (downward diagonal)
17 pp.coef = dd(1,:)';
18 pp.x = x;
```

The output of Program 5.4 is a MATLAB structure storing coefficients of Newton polynomial and corresponding interpolation points. The following MATLAB program evaluates Newton polynomials at desired points.

Program 5.5: *Evaluate Newton polynomial at points input*

```
 1 function y = newtonpolyeval(pp,x)
 2 %  Input   pp: structure related to Newton polynomial
 3 %             pp.coef : stores Newton polynomial weights
 4 %             pp.x    : stores interpolation points
 5 %          x: points where polynomial is to be evaluated
 6 %  Output   y: Newton polynomial evaluated at x
 7 n = size(pp.coef,1);   % number of interpolation points
 8 m = size(x,1);         % number of points in x
 9 x1 = ones(m,1);        % initialize x1 x1=∏_{i=1}^{n}(x-x_{i-1})
10 y = pp.coef(1)*x1;     % initialize y
11 for i=2:n              % loop to calculate y
12     x1 = x1.*(x-pp.x(i-1)); % update x1
13     y = y + pp.coef(i)*x1;  % update y
14 end
```

5.3.4 Newton Gregory formulas with equi-spaced data

Newton polynomials may be simplified further in the case of equi-spaced data. For this purpose let us assume that $r = \dfrac{x - x_0}{h}$. We notice that $\dfrac{x_1 - x_0}{h} = 1$, $\dfrac{x_2 - x_0}{h} = 2,......, \dfrac{x_k - x_0}{h} = k$. Using results from Equations 5.23 - 5.25 in Equation 5.39 we have the following.

$$f(x) \approx p_k(x) = f_0 + r\Delta f_0 + \frac{r(r-1)}{2!}\Delta^2 f_0 + ... + \frac{r(r-1)(r-2)....(r-k+1)}{k!}\Delta^k f_0 \qquad (5.40)$$

The above is referred to as the Newton-Gregory series with forward differences. The forward differences required in Equation 5.40 may be calculated as shown in Table 5.3.

Table 5.3: *Forward difference table for equi-spaced data*

x	$f(x)$	Δf	$\Delta^2 f$	$\Delta^3 f$	$\Delta^4 f$
0	f_0				
		$\underbrace{f_1 - f_0}_{\Delta f_0}$			
h	f_1		$\underbrace{f_2 - 2f_1 + f_0}_{\Delta^2 f_0}$		
		$f_2 - f_1$		$\underbrace{f_3 - 3f_2 + 3f_1 - f_0}_{\Delta^3 f_0}$	
$2h$	f_2		$f_3 - 2f_2 + f_1$		$\underbrace{f_4 - 4f_3 + 6f_2 - 4f_1 + f_0}_{\Delta^4 f_0 \text{ or } \nabla^4 f_4}$
		$f_3 - f_2$		$\underbrace{f_4 - 3f_3 + 3f_2 - f_1}_{\nabla^3 f_4}$	
$3h$	f_3		$\underbrace{f_4 - 2f_3 + f_2}_{\nabla^2 f_4}$		
		$\underbrace{f_4 - f_3}_{\nabla f_4}$			
$4h$	f_4				
x	$f(x)$	∇f	$\nabla^2 f$	$\nabla^3 f$	$\nabla^4 f$

The forward differences associated with the point x_0 are the quantities lying on the downward going diagonal, as indicated in the table. It is also possible to associate backward differences with the entries lying on the upward going diagonal. For example, as indicated in the table, the terms along the upward going diagonal represent backward differences associated with point x_4.

Newton Gregory series may also be written down using backward differences, based on an analysis similar to that presented above. The appropriate formula is given below.

$$f(x) \approx p_k(x) = f_0 + r\nabla f_0 + \frac{r(r+1)}{2!}\nabla^2 f_0 + ... + \frac{r(r+1)(r+2)....(r+k-1)}{k!}\nabla^k f_0 \qquad (5.41)$$

Example 5.3

Construct a table of forward, backward and divided differences using the following data.

i	1	2	3	4	5	6	7
x	0	0.2	0.4	0.6	0.8	1	1.2
f	0	0.0016	0.0256	0.1296	0.4096	1	2.0736

Comment on the nature of the function based on the various differences.

Solution :

Forward and backward differences are easily calculated using a spreadsheet program. An extract from such a spreadsheet is given below. It is observed that 5th forward and backward differences are zero. Hence the data specified in the example is at most a polynomial of degree 4. The reader may note the function is actually $f(x) = x^4$ and hence the above observation is correct.

x	f	Δf	$\Delta^2 f$	$\Delta^3 f$	$\Delta^4 f$	$\Delta^5 f$	$\Delta^6 f$
0	0	0.0016	0.0224	0.0576	0.0384	0.0000	0.0000
0.2	0.0016	0.024	0.08	0.096	0.0384	0.0000	
0.4	0.0256	0.104	0.176	0.1344	0.0384		
0.6	0.1296	0.28	0.3104	0.1728			
0.8	0.4096	0.5904	0.4832				
1	1	1.0736					
1.2	2.0736						
x	f	∇f	$\nabla^2 f$	$\nabla^3 f$	$\nabla^4 f$	$\nabla^5 f$	$\nabla^6 f$

Divided differences may be calculated by noting that the data is equi-spaced with a spacing of $h = 0.2$. Again the calculations are performed using a spreadsheet and given below

x	f	DD_1	DD_2	DD_3	DD_4	DD_5	DD_6
0	0	**0.008**	**0.28**	**1.2**	**1**	**0**	**0**
0.2	0.0016	0.12	1	2	1	0	
0.4	0.0256	0.52	2.2	2.8	1		
0.6	0.1296	1.4	3.88	3.6			
0.8	0.4096	2.952	6.04				
1	1	5.368					
1.2	2.0736						

DD stands for divide difference.

Entries shown in **bold** are the divided differences associated with $x = 0$.

We see again that 5th and higher divided differences are zero. This is as it should be for a polynomial degree 4.

The next example we consider is one where the function form is assumed to be not known. The data is available purely as a table and we would like to use it for interpolation.

Example 5.4

Use the following tabulated data to make a divided differences table. Obtain the value of the function at $x = 0.33$.

i	1	2	3	4	5	6	7
x	0	0.14	0.22	0.28	0.34	0.42	0.51
f	1.2	1.2663	1.3141	1.3551	1.401	1.47	1.5593

Solution :

Note that the data given in this example is not equi-spaced and hence only divided differences make sense. Calculation of divided differences is best done by the use of a spreadsheet, as indicated already in Example 5.3. Such a tabulation is given below.

x	f	DD_1	DD_2	DD_3	DD_4	DD_5	DD_6
0	1.2000	**0.4736**	**0.5633**	**0.1778**	**0.4691**	**-3.3103**	**15.2172**
0.14	1.2663	0.5975	0.6131	0.3373	-0.9212	4.4505	
0.22	1.3141	0.6833	0.6806	0.0794	0.7255		
0.28	1.3551	0.7650	0.6964	0.2898			
0.34	1.4010	0.8625	0.7631				
0.42	1.4700	0.9922					
0.51	1.5593						

Entries shown in **bold** are the divided differences associated with $x = 0$.

By looking at the entries in the table it is not possible to come to any conclusion regarding how well the data is approximated by a polynomial of some degree, say 3 or 4. This can be done only by comparing the interpolated value, for a specified x, by comparing successive approximations to the function value. In the present case we require the interpolated value of the function at $x = 0.33$. Newton polynomials based on the tabulated values of divided differences may be used for this purpose. Approximations may be written down as below:

$$\text{1 term:} \quad f(x) \approx f_0 \qquad \text{2 terms:} \quad f(x) \approx f_0 + (x - x_0)DD_1$$
$$\text{3 terms: } f(x) \approx f_0 + (x - x_0)(DD_1 + (x - x_1)DD_2)$$
$$\text{4 terms: } f(x) \approx f_0 + (x - x_0)(DD_1 + (x - x_1)(DD_2 + (x - x_2)DD_3))$$

and so on, where x_0 is identified with the first data point viz. $x_0 = 0$. The following table indicates the various approximations to the value of $f(0.33)$.

No. of terms	1	2	3	4	5	6	7
$f(0.33)$	1.2	1.3563	1.3916	1.3928	1.3930	1.3930	1.3930

It appears that 5 terms are required for getting an acceptable interpolate for the function at $x = 0.33$.

The last example to be presented involves equi-spaced data of an unknown function. We use Newton-Gregory formula using forward differences to get an interpolated value of the function.

Example 5.5

Obtain an acceptable interpolated value of the function tabulated below at $x = 1.35$ using Newton-Gregory series. Also obtain a quadratic interpolated value using Lagrange polynomial and compare the two.

i	1	2	3	4	5
x	1	1.2	1.4	1.6	1.8
f	-1.0214	-0.7645	-0.4601	-0.1131	0.2727

Solution :

Forward differences are calculated as usual and are tabulated below.

x	f	Δf	$\Delta^2 f$	$\Delta^3 f$	$\Delta^4 f$	$\Delta^5 f$
1	-1.0214	0.2569	0.0475	-0.0049	0.0012	-0.0004
1.2	-0.7645	0.3044	0.0426	-0.0037	0.0008	
1.4	-0.4601	0.3469	0.0389	-0.0029		
1.6	-0.1131	0.3858	0.0360			
1.8	0.2727	0.4218				
2	0.6945					

The required forward differences are shown in *italics*. Newton Gregory series (Equation 5.40) may now be used with $r = \dfrac{1.35 - 1}{1.2 - 1} = 1.75$ and different number of terms to obtain the following:

No. of terms	1	2	3	4	5	6
$f(1.35)$	-1.0214	-0.5718	-0.5407	-0.5404	-0.5404	-0.5404

It is observed that 4 terms are adequate to get an acceptable value for the interpolate. We now calculate the interpolate at $x = 1.35$ by using a second degree Lagrange polynomial. This may be done in two alternate ways, by using the data at x values of 1,1.2 and 1.4 or alternately by using the data at x values of 1.2,1.4 and 1.6. We will show the calculation in the latter case, conveniently arranged in tabular form.

Use data below		$h =$	0.2
x	$f(x)$	$L_2(x = 1.35)=$	-0.5406
1	-1.0214		
1.2	-0.7645		
1.4	-0.4601		

It is seen that $L_2(1.35)$ agrees closely with the three term Newton Gregory value, as it should.

Approximating known functions: We consider another example where the function form is known. We would like to look at the error distribution within the entire range of data when a Newton polynomial of various possible degrees is used for interpolation.

Example 5.6

The following table gives equi-spaced data of $f(x) = e^{0.8x} - 1$. Discuss the error in computation of the function using p_4 that represents Newton polynomial of the largest possible degree.

x	1	1.5	2	2.5	3
$f(x)$	1.225541	2.320117	3.953032	6.389056	10.023176

Comment on what happens when we use a polynomial of lower degree than the maximum possible. What happens to the error when higher degree polynomials are used (such as p_8 and p_{16}).

Solution :

Since the data is equi-spaced we may develop the Newton polynomial using forward differences. The required forward differences are given in the following table.

x	$f(x)$	∇f	$\nabla^2 f$	$\nabla^3 f$	$\nabla^4 f$
1	1.225541				
		1.094576			
1.5	2.320117		0.538340		
		1.632916		0.264769	
2	3.953032		0.803108		0.130220
		2.436024		0.394988	
2.5	6.389056		1.198097		
		3.634120			
3	10.023176				

The required Newton polynomial is given by

$$f(x) \approx p_4(x) = 1.225541 + 1.094576r + \frac{0.538340r(r-1)}{2} +$$
$$\frac{0.264769r(r-1)(r-2)}{6} + \frac{0.130220r(r-1)(r-2)(r-3)}{24}$$

where $r = \dfrac{x-1}{1.5-1} = 2(x-1)$. The error is calculated as the difference $p_4(x) - f(x)$. Table 5.4 gives an extract from a spreadsheet that is used to calculate the functions and the errors with a step size of $\Delta x = 0.04$ or $\Delta r = 0.08$. A plot of the error as a function of x is given in Figure 5.2. Surprisingly the largest errors occur close to the end points i.e. $x = 1$ and $x = 3$. This is referred to as "Runge's" phenomenon[5].

Table 5.4: *Error in Example 5.6*

x	r	$f(x)$	$p_4(x)$	$p_4(x) - f(x)$
1	0	1.2255	1.2255	0.0000
1.04	0.08	1.2979	1.2973	-0.0006
1.08	0.16	1.3726	1.3716	-0.0010
...	...	...	...	...
2.2	2.4	4.8124	4.8119	-0.0006
2.24	2.48	5.0014	5.0008	-0.0006
2.28	2.56	5.1966	5.1960	-0.0006
...	...	...	...	...
2.48	2.96	6.2718	6.2717	-0.0001
2.52	3.04	6.5082	6.5083	0.0001
2.56	3.12	6.7524	6.7527	0.0004
...	...	...	...	...
2.92	3.84	9.3398	9.3411	0.0013
2.96	3.92	9.6760	9.6768	0.0008
3	4	10.0232	10.0232	0.0000

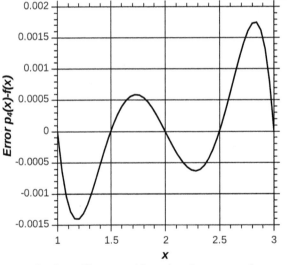

Figure 5.2: *Variation of error in the case of $p_4(x)$ in Example 5.6*

Extrapolation: Now consider what happens when we use $p_2(x)$ or $p_3(x)$ instead of $p_4(x)$ for evaluating the function in the range $1 \leq x \leq 3$. Note that use of $p_2(x)$ will amount to extrapolation for $x > 2$ and use of $p_3(x)$ will amount to extrapolation for $x > 2.5$. Error between the respective interpolating polynomials and the actual function are plotted as shown in Figure 5.3.

This figure shows that large errors may occur if we use the interpolating polynomial to extrapolate the function beyond the range of its applicability. Hence, in general, it is unwise to use an approximating interpolating function to extrapolate the function.

Let us see what happens to the maximum error as the degree of the interpolating polynomial is increased. We use programs 5.4 and 5.5 to estimate the maximum error incurred in interpolation. The following code determines the maximum error for p_8.

---

[5]Named after Carle David Tolmé Runge 1856-1927, German mathematician and physicist. His name is well known for the Runge Kutta method used for numerical solution of ordinary differential equations.

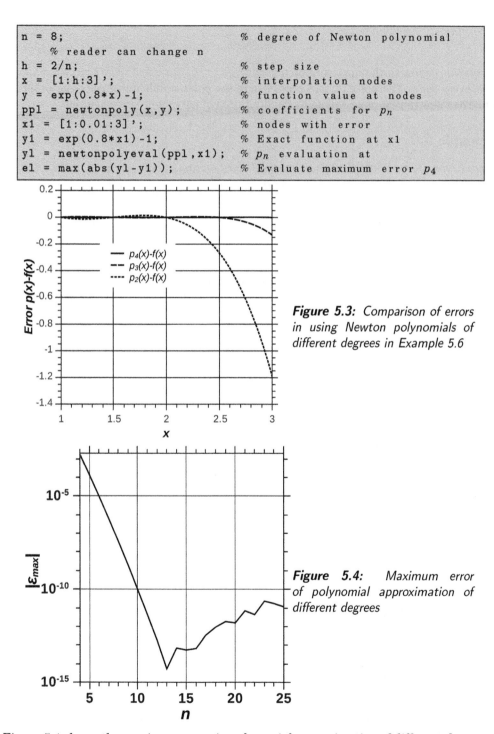

```
n = 8;                          % degree of Newton polynomial
      % reader can change n
h = 2/n;                        % step size
x = [1:h:3]';                   % interpolation nodes
y = exp(0.8*x)-1;               % function value at nodes
pp1 = newtonpoly(x,y);          % coefficients for pₙ
x1 = [1:0.01:3]';               % nodes with error
y1 = exp(0.8*x1)-1;             % Exact function at x1
yl = newtonpolyeval(pp1,x1);    % pₙ evaluation at
el = max(abs(yl-y1));           % Evaluate maximum error p₄
```

Figure 5.3: Comparison of errors in using Newton polynomials of different degrees in Example 5.6

Figure 5.4: Maximum error of polynomial approximation of different degrees

Figure 5.4 shows the maximum error in polynomial approximation of different degrees. It is seen that the error in approximation decreases up to $n = 13$ following which the error increases. This is because of roundoff errors. Round off errors become important while using higher degree polynomial approximations. In general, it would be advisable to avoid higher degree polynomials.

In the previous example, the approximation converges to the function. However, we may not be fortunate every time. There are certain cases where the polynomials never converge to the exact solution. The following example will show the limitation of polynomial interpolation.

Example 5.7

Determine the approximate Newton polynomials for function[6] $f(x) = \dfrac{1}{1+25x^2}$ *in the interval* $[-1, 1]$. *Discuss the error in computation of the function using* p_4, p_8 *and* p_{16}.

Solution :

MATLAB programs have been used to estimate the Newton's polynomials of degree 4, 8 and 16. Figure 5.5 shows the comparison of Newton's polynomial with the exact function. Figure 5.6 shows the error of approximation with different approximating polynomials.

From the figures it is evident that the errors are large for the higher degree polynomials. Moreover, these errors are concentrated close to the end points. This is due to Runge phenomenon already introduced in the previous example. In the previous example, the Runge phenomenon was less serious and hence convergence was achieved. In the present example Runge phenomenon causes divergence of the polynomial approximation.

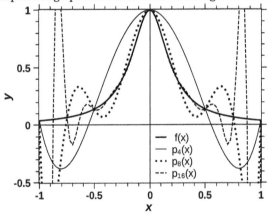

Figure 5.5: *Comparison of different Lagrange polynomial approximations with exact function*

5.4   Error estimates of polynomial approximations

The polynomial interpolation is analogous to the Taylor' series of a function. In fact, the structure of Newton's polynomial (Equation 5.39) resembles a Taylor's series. The Taylor's series of a function is given as follows

$$f(x) = f_0 + f_0^{(1)}\Delta x + \frac{f_0^{(2)}}{2!}\Delta x^2 + \cdots + \frac{f_0^{(n)}}{n!}\Delta x^n + O(\Delta x^{n+1}) \tag{5.42}$$

---

[6]This function is of the same form as used by Runge to demonstrate Runge phenomenon.

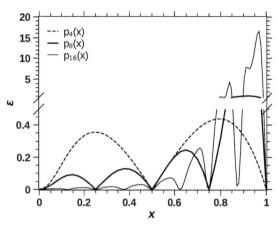

Figure 5.6: Error of approximation of different Lagrange polynomials

A polynomial approximation of degree n will also satisfy Taylor series of degree n and the truncation error of the approximation is of order $n+1$. The magnitude of error of the polynomial approximation in the interpolation range would be

$$\epsilon = |f(x) - p(x)| \approx \left| \frac{f^{(n+1)}}{(n+1)!}(x - x_0)(x - x_1)(x - x_2)\cdots(x - x_n) \right| \tag{5.43}$$

where $f^{(n+1)}$ is maximum value of $(n+1)^{th}$ derivative of the function in the interpolation range. As the derivative used is the maximum value somewhere in the interval, the above estimate represents the upper bound for error. Practically, it may be difficult to determine the derivatives of the function. In such a case, if function value is known at x_{n+1}, one can estimate the error in the polynomial approximation as

$$\epsilon \approx |f[x_0, x_1, x_2, \cdots, x_n, x_{n+1}](x - x_0)(x - x_1)(x - x_2)\cdots(x - x_n)| \tag{5.44}$$

It is evident from the above two estimates for error, that the error is zero at all nodes (property of polynomial interpolation). But what is of interest to us is to minimize the *maximum* error of the approximation in the interval. Ideally, as $n \to \infty$, ϵ_{max} should $\to 0$. In such a case, we say the approximation **converges** to the exact solution. The above requirement has been violated in both the previous examples.

The magnitude of the error depends on two terms namely the $f^{(n+1)}$ and distribution of nodes $\prod_{i=0}^{n} |x - x_i|$. Let us consider the two factors independently to clearly state the importance of each factor. If the bounds of the interpolation domain and the degree of the polynomial is fixed, the error depends on the distribution of nodes within the domain. Let us consider the nodes to be equi-spaced. Then, the truncation error can be written as

$$\epsilon = |f(x) - p(x)| \approx \left| r(r-1)(r-2)\cdots(r-n)h^{n+1} \frac{f^{(n+1)}}{(n+1)!} \right| \tag{5.45}$$

where $r = \dfrac{x - x_0}{h}$. Let us scrutinize the factor $R = |r(r-1)(r-2)\cdots(r-n)|$ more closely. Figure 5.7 indicates the distribution of R as a function of r for $n = 6$. It is seen that the factor assumes large values very close to the end of the interpolation domain. Hence, it can be concluded that interpolation from equi-spaced nodes would promote amplification of *round off errors* close to the ends of the domain.

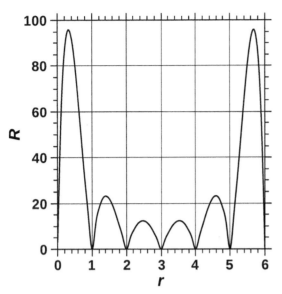

Figure 5.7: *Distribution of factor R with r for n = 6*

However, the nature of approximation of Examples 5.6 and 5.7 have different characteristics. The polynomial approximation in Example 5.6 converges to the exact solution for $n = 13$ and starts diverging for higher degree polynomial approximation. Whereas polynomial approximation for Example 5.7 is poor for small degrees of approximation. Here one has to understand the contribution of $f^{(n+1)}$ also to the error.

Let us approximate e^x by polynomial function. e^x can be represented as an infinite series

$$e^x = 1 + x + \frac{x^2}{2} + \frac{x^3}{3!} + \cdots + \frac{x^i}{i!} + \cdots \tag{5.46}$$

n^{th} degree polynomial curve would satisfy first n terms of the infinite series. The truncation error because of nth degree polynomial approximation would be

$$|\epsilon| = \left| \frac{(x - x_0)(x - x_1)(x - x_2) \cdots (x - x_n)}{(n+1)!} \right| < \left| \frac{(x_n - x_0)^{n+1}}{n+1!} \right| \tag{5.47}$$

As $n \longrightarrow \infty$, $\left| \frac{(x_n - x_0)^{n+1}}{n+1!} \right| \longrightarrow 0$. This means infinite series is convergent for all values of x in the interpolation range. Such a function is also called as *analytic*. Therefore, theoretically as $n \longrightarrow \infty$, the polynomial should approximate the function exactly. However, the accuracy of the approximation is limited by round off errors. The point of divergence is where the round-off error becomes larger than the truncation error.

However, the divergence of polynomial approximation applied to Example 5.7 i.e. to function $\left(\frac{1}{1 + 25x^2} \right)$ is due to Runge phenomenon. We briefly explain the cause of the phenomenon. $f(x) = \frac{1}{1 + 25x^2}$ can also be represented as an infinite series

$$\frac{1}{1 + 25x^2} = 1 - 25x^2 + 5^4 x^4 + \cdots + (-1)^i 5^{2i} x^{2i} + \cdots \tag{5.48}$$

The above equation is nothing but the Taylor series expansion of the function about $x = 0$. Therefore,

$$\left|\frac{f^{(n)}}{n!}\right| = \begin{cases} 5^n & \text{n is even} \\ 0 & \text{n is odd} \end{cases} \tag{5.49}$$

The magnitude of truncation error of nth (n is even) degree polynomial approximation would be

$$|\epsilon| = 5^{n+2}|(x-x_0)(x-x_1)(x-x_2)\cdots(x-x_n)| < 5^{n+2}\left|(x_n-x_0)^{n+2}\right| \tag{5.50}$$

Therefore

$$n \longrightarrow \infty \begin{cases} |x_n - x_0| < 0.2 & 5^{n+2}|(x_n-x_0)^{n+2}| \longrightarrow 0 \\ |x_n - x_0| > 0.2 & 5^{n+2}|(x_n-x_0)^{n+2}| \longrightarrow \infty \end{cases} \tag{5.51}$$

This means the function is convergent only for certain values of x and hence is not analytic. The implication of the above deduction is that the upper bound of error of approximation depends on the interval of interpolation. If the domain of interpolation is less than 0.2 ($-0.2 < x < 0.2$) then polynomial approximation would converge towards the exact solution. However, when the size of the interpolation interval is greater than $x = 0.2$, the convergence may be poor. When the nodes are equi-spaced, both the derivative and $\prod |x - x_i|$ are significant and hence the polynomial approximation would produce large errors.

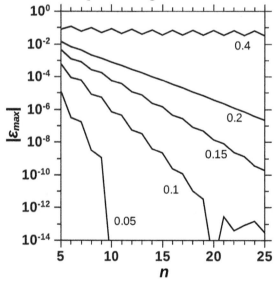

Figure 5.8: Error in polynomial approximations of various degrees for different range of interpolation.

Figure 5.8 indicates the effect of domain size on maximum error in polynomial approximation. Polynomial approximation is excellent when the step size is small and deteriorates as the domain size increases. However, the limitations of Runge phenomenon can be overcome by using the following strategies:

1. Chebyshev nodes
2. Piecewise polynomials

5.5 Polynomial approximation using Chebyshev nodes

When equi-spaced nodes are used, the errors are high close to the edges of the interpolation domain. Although it is easy to analyze equi-spaced data, it is computationally inefficient. If we

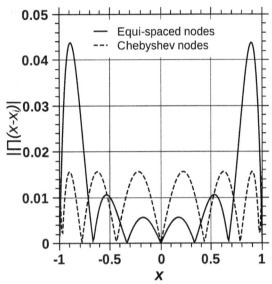

Figure 5.9: *Distribution of $\prod(x - x_i)$ for equi-spaced and Chebyshev nodes $(n = 6)$*

have the freedom to choose $n + 1$ points in the interpolation domain, the nodes have to be chosen such that the maximum of $\prod_{i=0}^{n} (x - x_i)$ in the interpolation range is minimum. As the oscillations occur close to the end points, the errors can be suppressed by having denser distribution of points close to the ends. It can be shown that polynomial function through Chebyshev [7] nodes produces minimum error. For a function defined in the interval $[-1, 1]$, Chebyshev nodes are given by the formula $x(i) = \cos\left(\dfrac{2i - 1}{2(n + 1)}\pi\right)$, where n is the degree of the polynomial and i varies from 1 to $n + 1$. Observe that the end points of the interpolation domain are not included when Chebyshev nodes are made use of.

Figure 5.9 indicating the magnitude of $\prod_{i=0}^{n=6} (x - x_i)$ for equi-spaced and Chebyshev nodes clearly shows the superiority of Chebyshev nodes over equi-spaced nodes. For intervals other that $[-1, 1]$, the interval has to be transformed to $[-1, 1]$. Figure 5.10 shows the maximum error when different degrees of polynomial approximation are applied to Example 5.6 using equi-spaced and Chebyshev nodes. It is evident that both equi-spaced and Chebyshev nodes are equally good for such problems. But the significance of Chebyshev nodes over equi-spaced nodes will become clear when the function is not analytic.

Example 5.8

Revisit function $f(x) = \dfrac{1}{1 + 25x^2}$ in the interval $[-1, 1]$. Estimate the errors for Newton polynomial of degree 4,8 and 16 passing through the Chebyshev nodes. Also estimate the errors for equi-spaced piecewise quadratic interpolation over 2, 4 and 8 segments.

Solution :

Chebyshev nodes:

The Chebyshev nodes are determined using the formula

$$x(i) = \cos\left(\frac{(2i - 1)}{2(n + 1)}\pi\right)$$

---

[7]Pafnuty Lvovich Chebyshev 1821-1894, Russian mathematician

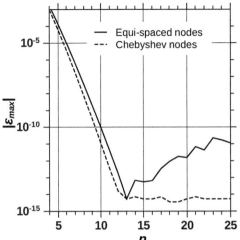

Figure 5.10: *Maximum error of polynomial approximation using equi-spaced nodes and Chebyshev nodes.*

The Chebyshev nodes and the corresponding function evaluations for the polynomial degree of 4, 8 and 16 have been listed in the table below.

L_4		L_8		L_{16}	
x	y	x	y	x	y
±0.9511	0.0424	± 0.9848	0.0396	± 0.9957	0.0388
±0.5878	0.1038	± 0.8660	0.0506	± 0.9618	0.0414
0	1	± 0.6428	0.0883	± 0.8952	0.0475
		± 0.3420	0.2548	± 0.7980	0.0591
		0	1	± 0.6737	0.0810
				± 0.5264	0.1261
				± 0.3612	0.2346
				± 0.1837	0.5423
				0	1

Note that, the end points are not part of Chebyshev nodes.

Programs 5.4 and 5.5 have been used to perform the interpolation. Figure 5.11 indicates the error for Newton polynomial that uses Chebyshev nodes.

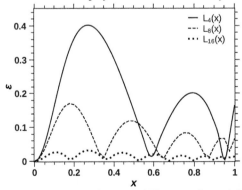

Figure 5.11: *Error of approximation of Newton Polynomials using Chebyshev nodes*

The maximum error between different polynomials is given below.

Polynomial degree	Maximum error	
	Equi-spaced nodes	Chebyshev nodes
L_4	0.4381	0.4018
L_8	1.04264	0.1705
L_{16}	13.8294	0.0325

The improvement in using the Chebyshev nodes for polynomial interpolation is clearly evident.

5.6   Piecewise polynomial interpolation

Instead of considering the entire domain at once, the domain can be divided into several smaller segments and a local polynomial can be defined within each of the segments. Reducing the interval suppresses the oscillatory phenomenon. Examples include piecewise interpolation segments, Hermite interpolating polynomials and splines. Piecewise interpolating segments have zero order continuity i.e. function values alone are continuous at the nodes. Hermite interpolating polynomials and splines have higher order continuity at the nodes.

Example 5.9

Revisit function $f(x) = \dfrac{1}{1+25x^2}$ *in the interval* $[-1,1]$. *Estimate the errors for equi-spaced piecewise quadratic interpolation over 2, 4 and 8 segments.*

Solution :

The number of segments over which quadratic interpolation is to be applied is 2, 4 and 8. Each segment for a quadratic contains 3 nodes, with one node shared between two consecutive segments as shown in the following figure.

The total number of nodes in n segments would be $2n+1$. The nodes and the corresponding function evaluations for piecewise quadratic interpolation have been listed in the following table.

x	y	
	2 segments	4 segments
-1.00	⎰ 0.0385	⎰ 0.0385
-0.75		1⎰ 0.0664
-0.50	1⎰ 0.1379	⎱ 0.1379 ⎱
-0.25		0.3902 ⎰2
0.00	⎱ 1.0000 ⎱	⎰ 1.0000 ⎱
0.25		3⎰ 0.3902
0.50	0.1379 ⎰2	⎱ 0.1379 ⎰
0.75		0.0664 ⎰4
1.00	0.0385 ⎱	0.0385 ⎱

Quadratic interpolation has been applied within each segment using Programs 5.4 and 5.5. Figure 5.12 shows the comparison of the interpolation polynomials with exact function. It is evident from the figure that the Runge phenomenon has indeed been suppressed. As the number of segments is increased the approximation polynomial comes closer to the exact function. This can be confirmed from Figure 5.13. However, it is interesting to note that the error in the region close to $x = 0$ is quite high. This is because the approximation is only continuous in function and not in the derivative. Notice the first derivative of the approximation function is not continuous at $x = 0$. The maximum error in the approximation function has been provided below.

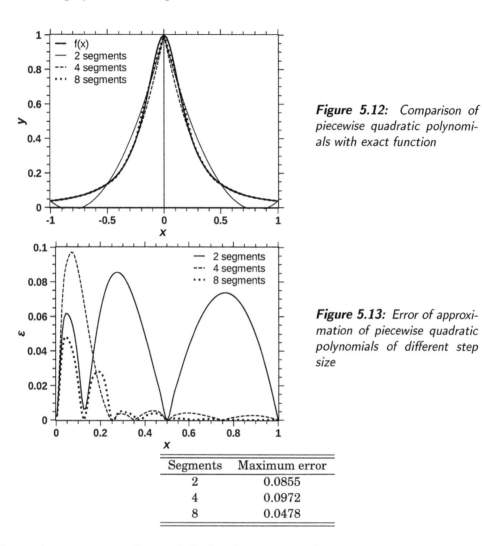

Figure 5.12: *Comparison of piecewise quadratic polynomials with exact function*

Figure 5.13: *Error of approximation of piecewise quadratic polynomials of different step size*

Segments	Maximum error
2	0.0855
4	0.0972
8	0.0478

The reader is encouraged to apply higher degree interpolating piecewise polynomials and see how they affect the errors.

Application in numerical integration Composite quadrature rules for integration use piecewise interpolating polynomial functions. Composite Simpson's $1/3^{rd}$ integration formula uses piecewise quadratic interpolating polynomial where as Composite Simpson's $3/8^{th}$ rule uses piecewise cubic interpolating polynomial(Chapter 9).

Piecewise polynomials considered earlier have zero order continuity (function alone is continuous) at each segment. Sometimes, it may be required or desirable to have higher order continuity i.e. functions as well as derivatives up to some order are continuous across each segment. This can be achieved by using Hermite and cubic spline approximations.

5.7 Hermite interpolation

Hermite[8] interpolation is closely related to Newton polynomials and serves as a precursor to the spline interpolation that will be presented later. We consider a table of data consisting of a function and possibly its derivatives. Hermite interpolation aims to find an interpolating polynomial that agrees with the function as well as its derivatives up to some order, at the nodal points, referred to as *knots*.

5.7.1 Cubic Hermite interpolating polynomial

We shall consider first the case where the function and the first derivatives alone are to be matched at the knots. The given data is represented by the following table.

x	$f(x)$	$f^{(1)}(x)$
x_0	f_0	$f_0^{(1)}$
x_1	f_1	$f_1^{(1)}$

where superscript $f^{(1)}$ represents the first derivative $\dfrac{df}{dx}$. The polynomial that is being sought has four conditions to satisfy (*number of degrees of freedom is four*) and hence it can be a polynomial of up to degree 3 i.e. it is a cubic polynomial. Let the polynomial be given by $H(x) = a_0 + a_1 x + a_2 x^2 + a_3 x^3$ such that its first derivative is $H^{(1)}(x) = a_1 + 2a_2 x + 3a_3 x^2$. Here a_0 through a_3 are constants to be determined by requiring that $H(x)$ and $H^{(1)}(x)$ agree with the tabulated values at the two end points x_0 and x_1. Since the interpolating polynomial passes through (x_0, f_0) and (x_1, f_1) we have

$$a_0 + a_1 x_0 + a_2 x_0^2 + a_3 x_0^3 = f_0$$
$$a_0 + a_1 x_1 + a_2 x_1^2 + a_3 x_1^3 = f_1$$

Since the interpolating polynomial also satisfies the given first derivatives at these two points, we have

$$a_1 + 2a_2 x_0 + 3a_3 x_0^2 = f_0^{(1)}$$
$$a_1 + 2a_2 x_1 + 3a_3 x_1^2 = f_1^{(1)}$$

The four equations may be written in the matrix form as

$$\begin{pmatrix} 1 & x_0 & x_0^2 & x_0^3 \\ 1 & x_1 & x_1^2 & x_1^3 \\ 0 & 1 & 2x_0 & 3x_0^2 \\ 0 & 1 & 2x_1 & 3x_1^2 \end{pmatrix} \begin{pmatrix} a_0 \\ a_1 \\ a_2 \\ a_3 \end{pmatrix} = \begin{pmatrix} f_0 \\ f_1 \\ f_0^{(1)} \\ f_1^{(1)} \end{pmatrix} \tag{5.52}$$

We may solve these easily to get the column vector **a**. Consider a simple case where $x_0 = 0$ and $x_1 = 1$ (this is always possible by a suitable transformation). The above equation set takes the simpler form given below.

$$\begin{pmatrix} 1 & 0 & 0 & 0 \\ 1 & 1 & 1 & 1 \\ 0 & 1 & 0 & 0 \\ 0 & 1 & 2 & 3 \end{pmatrix} \begin{pmatrix} a_0 \\ a_1 \\ a_2 \\ a_3 \end{pmatrix} = \begin{pmatrix} f_0 \\ f_1 \\ f_0^{(1)} \\ f_1^{(1)} \end{pmatrix} \tag{5.53}$$

---

[8]named after Charles Hermite 1822-1901, French mathematician

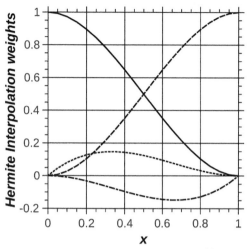

Figure 5.14: *Hermite interpolation weights*

We immediately see that $a_0 = f_0$ and $a_1 = f_0^{(1)}$. Equations governing the other two coefficients are then given by

$$\begin{pmatrix} 1 & 1 \\ 2 & 3 \end{pmatrix} \begin{pmatrix} a_2 \\ a_3 \end{pmatrix} = \begin{pmatrix} f_1 - f_0 - f_0^{(1)} \\ f_1^{(1)} - f_0^{(1)} \end{pmatrix} \tag{5.54}$$

By Gauss elimination we then have $a_2 = 3(f_1 - f_0) - (f_1^{(1)} + 2f_0^{(1)})$ and $a_3 = -2f_1 + 2f_0 + f_1^{(1)} + f_0^{(1)}$. We may then rearrange the Hermite interpolating polynomial as shown below.

$$
\begin{aligned}
H(x) &= f_0 \underbrace{(2x^3 - 3x^2 + 1)}_{w_0(x)} + f_1 \underbrace{(-2x^3 + 3x^2)}_{w_1(x)} \\
&\quad + f_0^{(1)} \underbrace{(x^3 - 2x^2 + x)}_{w_0^1(x)} + f_1^{(1)} \underbrace{(x^3 - x^2)}_{w_1^1(x)} \\
H^{(1)}(x) &= f_0 \underbrace{(6x^2 - 6x)}_{w_0(x)} + f_1 \underbrace{(-6x^2 + 6x)}_{w_1(x)} \\
&\quad + f_0^{(1)} \underbrace{(3x^2 - 4x + 1)}_{w_0^1(x)} + f_1^{(1)} \underbrace{(3x^2 - 2x)}_{w_1^1(x)}
\end{aligned} \tag{5.55}
$$

We notice that $H(x)$ may be interpreted as a weighted sum of nodal values with the weights given by the cubic polynomials (weights are shown as the w's) shown within braces. Similarly $H^{(1)}(x)$ may be interpreted as a weighted sum of nodal values with the weights given by the second degree (quadratic) polynomials shown within braces. The weighting polynomials are shown plotted in Figure 5.14. MATLAB program to calculate coefficients of the cubic Hermite polynomial is given below

Program 5.6: *Coefficients of cubic Hermite polynomial*

```
1 function pp = hermitepoly(x,y,z)
2 %    Inputs      x : interpolation points
3 %                y : function value at interpolating points
4 %                z : derivatives at interpolating points
5 %    Outputs    pp: structure related to Hermite polynomial
6 %                pp.coef : stores Hermite polynomial weights
7 %                pp.x    : stores interpolation points
8 n = size(x,1);                % number of interpolating points
```

```
 9  B = zeros(4,1);                % initialize force vector
10  pp.x = x;                      % store interpolation points
11  pp.coef= zeros(n-1,4);         % initialize coefficients
12  for i=1:n-1                    % loop for calculating coefficients
13  A = hermitemat(x(i:i+1));      % evaluate coefficient matrix
14  B(1:2) = y(i:i+1);             % function values
15  B(3:4) = z(i:i+1);             % derivatives
16  pp.coef(i,:) = A\B;            % evaluate coefficients
17  end
18  %
19  % subfunction hermitemat to evaluate coeficient matrix
20  % called from the main function hermitepoly
21  function A = hermitemat(x)
22  A = zeros(4,4);
23  A(1:2,1) = 1;
24  for i=2:4
25  A(1:2,i) = x.^(i-1);           % function values
26  A(3:4,i) = x.^(i-2)*(i-1);     % derivatives
27  end
```

The output of the program is a structure containing the coefficients and the interpolation points. The following MATLAB program is to be used for evaluating Hermite polynomials at given points.

Program 5.7: *Evaluation Hermite polynomial at input points*

```
 1  function y = hermitepolyeval(pp,x)
 2  % Inputs      pp: structure related to Hermite polynomial
 3  %                 pp.coef : stores Hermite polynomial weights
 4  %                 pp.x    : stores interpolation points
 5  %             x: points where function is to be evaluated
 6  % Output      y: evaluated Hermite polynomial
 7  m = length(x);           % number of points in x
 8  n = length(pp.x);        % number of interpolation points
 9  y = zeros(m,1);          % initialize y
10  for i=1:m                % loop for evaluating polynomial
11      if ( x(i) < pp.x(1)) % check for location of x(i)
12          j1 = 1;
13      elseif ( x(i) > pp.x(n))
14          j1 = n-1;
15      else
16          j=1;
17          while( j < n)
18              if (x(i) >= pp.x(j) && x(i) <= pp.x(j+1) )
19                  j1 = j; break;  % x(i) located break loop
20              else
21                  j = j + 1;
22              end
23          end
24      end        % j1 indicates indices of location of x(i)
25      % evaluate hermite polynomial at x(i)
26      y(i) = pp.coef(j1,1) + pp.coef(j1,2)*x(i) ...
27              +pp.coef(j1,3)*x(i)^2 + pp.coef(j1,4)*x(i)^3;
28  end
```

5.7.2 Hermite interpolating polynomial as Newton polynomial

It is possible to interpret Hermite interpolating polynomials in terms of Newton polynomials considered earlier. For this purpose we need to recognize that the derivative is related to divided difference in the limit of step size tending to zero. Consider the cubic case again. The function and first derivatives are specified at two points $x_0 = 0$ and $x_1 = 1$. Divided difference table may be constructed as under, keeping in mind the comment made above about the first divided difference. Point $x_0 = 0$ is written twice and point $x_1 = 1$ is also written twice to complete the divided difference Table 5.6. The boxed quantities use the derivative data. Other entries in the table are made using

Table 5.6: Divided difference table for cubic Hermite interpolating polynomial

x	$f(x)$	DD_1	DD_2	DD_3
$x_0 = 0$	f_0			
		$\boxed{f_0^{(1)}}$		
$x_0 = 0$	f_0		$f_1 - f_0 - f_0^{(1)}$	
		$f_1 - f_0$		$(f_1^{(1)} - f_1 + f_0) - (f_1 - f_0 - f_0^{(1)})$
$x_1 = 1$	f_1		$f_1^{(1)} - f_1 + f_0$	
		$\boxed{f_1^{(1)}}$		
$x_1 = 1$	f_1			

the formulae for divided differences.

We are now able to write a Newton polynomial (cubic) as

$$
\begin{aligned}
H(x) &= p_3(x) \\
&= f_0 + (x - 0)f_0^1 + (x - 0)(x - 0)(f_1 - f_0 - f_0^{(1)}) + \\
&\quad (x - 0)(x - 0)(x - 1)[(f_1^{(1)} - f_1 + f_0) - (f_1 - f_0 - f_0^{(1)})] \\
&= f_0 + xf_0^{(1)} + x^2(f_1 - f_0 - f_0^{(1)}) \\
&\quad + (x^3 - x^2)(f_1^{(1)} + f_0^{(1)} - 2f_1 - 2f_0)
\end{aligned}
\tag{5.56}
$$

Reader may verify that this reduces, after rearrangement, to the first of Equations 5.55.

Example 5.10

Following table gives a function and its first derivative at four points.

x	0	1	2	3
$f(x)$	0	0.587785	0.951057	0.951057
$f^{(1)}(x)$	0.628319	0.50832	0.194161	-0.194161

Use cubic Hermite interpolating polynomials between successive points and tabulate the results with a constant spacing of 0.2. Compute the error by noting that the function being approximated is $f(x) = \sin\left(\dfrac{\pi x}{5}\right)$. Make a suitable plot. Construct a cubic Newton polynomial with the given data and evaluate the errors.

Solution :

Between two consecutive entries in the tabulated data we may use a Hermite interpolation polynomial of degree 3. Hermite interpolating polynomial uses the consecutive points to be $x' = 0$ and $x' = 1$. Programs 5.6 and 5.7 have been used to evaluate the interpolated values using Hermite polynomial. The following program lists the sequence of operations.

```
x  = [0:3]';                    % interpolation points
y  = sin(pi*x/5);               % function values
z  = pi*cos(pi*x/5)/5;          % derivative values
pp = hermitepoly(x,y,z);        % estimate Hermite
                  % coefficients
x1 = [0:0.2:3]';
y1 = hermitepolyeval(pp,x1);
```

The coefficients of the Hermite polynomial obtained are given in table below:

x_0	x_1	a_0	a_1	a_2	a_3
0	1	0	0.6283	-0.0016	-0.0389
1	2	-0.0175	0.6781	-0.0488	-0.0241
2	3	-0.2139	0.9708	-0.1942	-0.0000

We make use of first of Equations 5.55 replacing x by x' to obtain the interpolated values shown in the last column of the spreadsheet.

x	x'	$f(x)$	$f^{(1)}(x)$	$H(x)$
0	0	0	0.628319	0
0.2	0.2			0.125288
0.4	0.4			0.248580
0.6	0.6			0.368005
0.8	0.8			0.481697
1	1	0.587785	0.508320	0.587785
1	0	0.587785	0.508320	0.587785
1.2	0.2			0.684417
1.4	0.4			0.770215
1.6	0.6			0.844025
1.8	0.8			0.904690
2	1	0.951057	0.194161	0.951057
2	0	0.951057	0.194161	0.951057
2.2	0.2			0.982122
2.4	0.4			0.997655
2.6	0.6			0.997655
2.8	0.8			0.982122
3	1	0.951057	-0.194161	0.951057

A plot of the interpolating function and its derivative is shown in Figure 5.15. Error between the function and the interpolating polynomial may be calculated since the function is known. The calculations are made using a spreadsheet program and presented in table below.

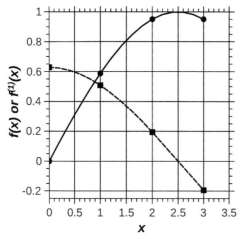

Figure 5.15: Hermite interpolation of data in Example 5.10

x	$f(x)$	$H(x)$	Error $H-f$	x	$f(x)$	$H(x)$	Error $H-f$
0	0	0	0.0000	1.6	0.8443	0.844	-0.0003
0.2	0.1253	0.1253	0.0000	1.8	0.9048	0.9047	-0.0001
0.4	0.2487	0.2486	-0.0001	2	0.9511	0.9511	0.0000
0.6	0.3681	0.368	-0.0001	2.2	0.9823	0.9821	-0.0002
0.8	0.4818	0.4817	-0.0001	2.4	0.998	0.9977	-0.0004
1	0.5878	0.5878	0.0000	2.6	0.998	0.9977	-0.0004
1.2	0.6845	0.6844	-0.0001	2.8	0.9823	0.9821	-0.0002
1.4	0.7705	0.7702	-0.0003	3	0.9511	0.9511	0.0000

We thus see that the maximum error in using the cubic interpolating polynomial is 0.0004. The approximation is very good.

In order now to obtain the cubic Newton polynomial passing through the four data points we construct a divided difference table.

x	$f(x)$	DD_1	DD_2	DD_3
0	0			
		0.5878		
1	0.5878		-0.2245	
		0.3633		-0.1388
2	0.9511		-0.3633	
		0		
3	0.9511			

The cubic Newton polynomial is then given by

$$p_3(x) = 0 + 0.5878x - \frac{0.2245x(x-1)}{2} - \frac{0.1388x(x-1)(x-2)}{6}$$

x	$f(x)$	$p_3(x)$	Error $p_3 - f$	x	$f(x)$	$p_3(x)$	Error $p_3 - f$
0	0	0	0.0000	1.6	0.8443	0.8416	-0.0028
0.2	0.1253	0.1289	0.0035	1.8	0.9048	0.903	-0.0018
0.4	0.2487	0.2532	0.0045	2	0.9511	0.9511	0.0000
0.6	0.3681	0.3718	0.0037	2.2	0.9823	0.9846	0.0023
0.8	0.4818	0.4837	0.0020	2.4	0.998	1.0024	0.0044
1	0.5878	0.5878	0.0000	2.6	0.998	1.0035	0.0055
1.2	0.6845	0.6828	-0.0017	2.8	0.9823	0.9868	0.0045
1.4	0.7705	0.7678	-0.0027	3	0.9511	0.9511	0.0000

Cubic Newton polynomial interpolation has a much higher error (maximum in the range is 0.0055) than the Hermite interpolating polynomial (maximum in the range is 0.0004) .

5.7.3   Generalization

Now consider the case where the interpolating polynomial passes through $N + 1$ points and satisfies derivatives up to order p. We then require the polynomial to satisfy the following:

$$
\begin{aligned}
H(x_i) &= f_i, \ i = 0 \text{ to } N \\
H^{(1)}(x_i) &= f_i^{(1)}, \ i = 0 \text{ to } N \\
\cdots &= \cdots \\
H^{(p)}(x_i) &= f_i^{(p)}, \ i = 0 \text{ to } N
\end{aligned}
\tag{5.57}
$$

where superscript (p) represents p^{th} derivative $\dfrac{d^p H}{dx^p}$. There are $p + 1$ equations in the above set. Each equation in the set has $N + 1$ conditions or constraints to be satisfied. The total number of constraints is thus equal to $(p + 1) \times (N + 1)$. The number of coefficients in the polynomial is the same as this number. The degree of the polynomial is hence given by $d = (p + 1) \times (N + 1) - 1$ with the -1 accounting for a constant in the polynomial. The following table indicates a few possibilities.

No.	p	N	d
1	1	1	3
2	1	2	5
3	2	1	5
4	2	2	8

Case 1 was considered earlier and the equations to be solved was obtained as Equation 5.52. We may write by inspection the equations for case 3 as

$$
\begin{pmatrix}
1 & x_0 & x_0^2 & x_0^3 & x_0^4 & x_0^5 \\
1 & x_1 & x_1^2 & x_1^3 & x_1^4 & x_1^5 \\
0 & 1 & 2x_0 & 3x_0^2 & 4x_0^3 & 5x_0^4 \\
0 & 1 & 2x_1 & 3x_1^2 & 4x_1^3 & 5x_1^4 \\
0 & 0 & 2 & 6x_0 & 12x_0^2 & 20x_0^3 \\
0 & 0 & 2 & 6x_1 & 12x_1^2 & 20x_1^3
\end{pmatrix}
\begin{pmatrix}
a_0 \\ a_1 \\ a_2 \\ a_3 \\ a_4 \\ a_5
\end{pmatrix}
=
\begin{pmatrix}
f_0 \\ f_1 \\ f_0^{(1)} \\ f_1^{(1)} \\ f_0^{(2)} \\ f_1^{(2)}
\end{pmatrix}
\tag{5.58}
$$

In case we take $x_0 = 0$ and $x_1 = 1$ as we did earlier the above equations will simplify to

$$
\begin{bmatrix}
1 & 0 & 0 & 0 & 0 & 0 \\
1 & 1 & 1 & 1 & 1 & 1 \\
0 & 1 & 0 & 0 & 0 & 0 \\
0 & 1 & 2 & 3 & 4 & 5 \\
0 & 0 & 2 & 0 & 0 & 0 \\
0 & 0 & 2 & 6 & 12 & 20
\end{bmatrix}
\begin{Bmatrix}
a_0 \\ a_1 \\ a_2 \\ a_3 \\ a_4 \\ a_5
\end{Bmatrix}
=
\begin{Bmatrix}
f_0 \\ f_1 \\ f_0^{(1)} \\ f_1^{(1)} \\ f_0^{(2)} \\ f_1^{(2)}
\end{Bmatrix}
\tag{5.59}
$$

these equations may be solved easily to get all the constants. It is also possible to construct the Hermite interpolating polynomial by using divided differences as indicated earlier.

5.8 Spline interpolation and the cubic spline

In section 5.7 we have seen how adjacent points in a table of function and its derivative values may be used to fit a Hermite interpolating polynomial. We considered specifically the case of a cubic Hermite interpolating polynomial there. Spline interpolation is a method of passing a polynomial of desired degree, usually cubic, when only the function data is available in the form of a table i.e. pairs (x_i, f_i) are available for $0 \le i \le n$. As opposed to Hermite interpolating polynomial in which function and derivative values agreed with those at the knots, the cubic spline is continuous and has continuous first and second derivatives at the knots. This means that the cubic spline does not need the derivatives to be known at the knots but only that the spline yield the same first and second derivatives when approached from either side of the knot. Figure 5.16

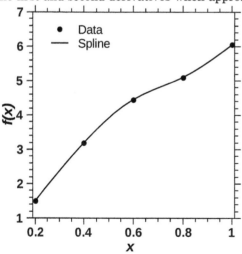

Figure 5.16: *Cubic spline passing through data points having continuous first and second derivatives*

shows an example of a cubic spline that passes through the data points (shown by filled circles) and has continuous first and second derivatives at these points.

5.8.1 General case with non-uniformly spaced data

Consider data shown in the Table 5.7 below.

Table 5.7: *Data in tabular form*

x	x_0	x_1	x_2	$\cdots$	$\cdots$	$\cdots$	x_{n-1}	x_n
$f(x)$	f_0	f_1	f_2	$\cdots$	$\cdots$	$\cdots$	f_{n-1}	f_n

Consider a third degree polynomial to be used to approximate the function between two consecutive points in the table (say between x_i and x_{i+1}). In view of the fact that we are passing a cubic through these two points, the second derivative of the cubic spline should be a linear function of x in the region $x_i \leq x \leq x_{i+1}$. We may thus write the second derivative, using Lagrange polynomial, as

$$f^{(2)}(x) = \frac{x - x_{i+1}}{x_i - x_{i+1}} \cdot f^{(2)}(x_i) + \frac{x - x_i}{x_{i+1} - x_i} \cdot f^{(2)}(x_{i+1}) \qquad (5.60)$$

Let us treat the second derivatives at the nodes as unknowns and use the notation $f^2(x_i) = c_i$. Then we may write Equation 5.60 as

$$f^{(2)}(x) = c(x) = \frac{x - x_{i+1}}{x_i - x_{i+1}} \cdot c_i + \frac{x - x_i}{x_{i+1} - x_i} \cdot c_{i+1} \qquad (5.61)$$

We integrate Equation 5.61 once with respect to x to obtain the first derivative given by

$$f^{(1)}(x) = \frac{(x - x_{i+1})^2}{2(x_i - x_{i+1})} \cdot c_i + \frac{(x - x_i)^2}{2(x_{i+1} - x_i)} \cdot c_{i+1} + b_i \qquad (5.62)$$

where b_i is a constant of integration. Integrate Equation 5.62 once again with respect to x to get

$$f(x) = \frac{(x - x_{i+1})^3}{6(x_i - x_{i+1})} \cdot c_i + \frac{(x - x_i)^3}{6(x_{i+1} - x_i)} \cdot c_{i+1} + b_i x + a_i \qquad (5.63)$$

where a_i is a second constant of integration. The constants a_i and b_i are determined by requiring that the spline pass through x_i, f_i and x_{i+1}, f_{i+1}. Introduce the notation $x_{i+1} - x_i = h_i$. Then we have

$$f_i = \frac{h_i^2}{6} \cdot c_i + b_i x_i + a_i \qquad (5.64)$$

and

$$f_{i+1} = \frac{h_i^2}{6} \cdot c_{i+1} + b_i x_{i+1} + a_i \qquad (5.65)$$

Subtracting Equation 5.64 from Equation 5.65 we get for b_i the relation

$$b_i = \frac{f_{i+1} - f_i}{h_i} - \frac{h_i}{6}(c_{i+1} - c_i) \qquad (5.66)$$

Multiply Equation 5.65 by x_i and Equation 5.64 by x_{i+1} and subtract the latter from the former to get

$$a_i = \frac{h_i}{6}(x_i c_{i+1} - x_{i+1} c_i) - \frac{x_i f_{i+1} - x_{i+1} f_i}{h_i} \qquad (5.67)$$

We may insert these in Equation 5.63 to get

$$f(x) = \frac{(x_{i+1} - x)^3 \cdot c_i + (x - x_i)^3 \cdot c_{i+1}}{6h_i} + \frac{6f_i - h_i^2 c_i}{6h_i} \cdot (x_{i+1} - x) + \frac{6f_{i+1} - h_i^2 c_{i+1}}{6h_i} \cdot (x - x_i) \qquad (5.68)$$

valid for $x_i \leq x \leq x_{i+1}$. From Equation 5.62 we also have

$$f^{(1)}(x) = \frac{1}{6h_i} \left[\{3(x - x_i)^2 - h_i^2\} c_{i+1} - \{3(x_{i+1} - x)^2 - h_i^2\} c_i \right] + \frac{f_{i+1} - f_i}{h_i} \qquad (5.69)$$

Inserting $x = x_i$ in Equation 5.69 the first derivative at $x = x_i$ is obtained.

$$f^{(1)}(x_i) = -\frac{h_i}{3} \cdot c_i - \frac{h_i}{6} \cdot c_{i+1} + \frac{f_{i+1} - f_i}{h_i} \qquad (5.70)$$

In order to satisfy the requirement that the first derivative be continuous at the nodes we should equate the first derivative at node x_i given by Equation 5.70 with that evaluated at the same point by using the spline between x_{i-1} and x_i. Letting $x_i - x_{i-1} = h_{i-1}$, we may easily write for $f^{(1)}(x_i)$ the relation

$$f^{(1)}(x_i) = \frac{h_{i-1}}{3} \cdot c_i + \frac{h_{i-1}}{6} \cdot c_{i-1} + \frac{f_i - f_{i-1}}{h_{i-1}} \tag{5.71}$$

Equating the above two expressions we obtain a relationship between c_{i-1}, c_i and c_{i+1} given by

$$h_{i-1}c_{i-1} + 2(h_i + h_{i-1})c_i + h_i c_{i+1} = 6\left[\frac{f_{i-1} - f_i}{h_{i-1}} - \frac{f_i - f_{i+1}}{h_i}\right] \tag{5.72}$$

Such equations may be written for each of the interior nodes. However at either boundary we need to specify a suitable condition. Usually we use the *natural spline boundary condition* by setting the second derivative as zero at the two boundaries. Other boundary conditions are possible at the boundary nodes. For example, if the first derivative is known at a boundary, it may be imposed in lieu of the natural boundary condition. For example, if the data represents temperature data with adiabatic condition at one of the boundaries, we may specify the first derivative of temperature to be zero there.

The set of equations to be solved may then be written as follows:

$$\begin{aligned}
c_0 &= 0 \\
2(h_1 + h_0)c_1 + h_1 c_2 &= 6\left[\frac{f_0 - f_1}{h_0} - \frac{f_1 - f_2}{h_1}\right] \\
h_1 c_1 + 2(h_2 + h_1)c_2 + h_2 c_3 &= 6\left[\frac{f_1 - f_2}{h_1} - \frac{f_2 - f_3}{h_2}\right] \\
\dots\dots &= \dots\dots \\
h_{n-2}c_{n-2} + 2(h_{n-1} + h_{n-2})c_{n-1} &= 6\left[\frac{f_{n-2} - f_{n-1}}{h_{n-2}} - \frac{f_{n-1} - f_n}{h_{n-1}}\right] \\
c_n &= 0
\end{aligned} \tag{5.73}$$

5.8.2 Special case with equi-spaced data

In this special case all we have to do is to set $h_i = h$, a constant for all i in Equation 5.72 to get

$$c_{i-1} + 4c_i + c_{i+1} = \frac{6(f_{i-1} - 2f_i + f_{i+1})}{h^2} \tag{5.74}$$

The set of equations to be solved may then be written as follows:

$$\begin{aligned}
c_0 &= 0 \\
4c_1 + c_2 &= \frac{6(f_0 - 2f_1 + f_2)}{h^2} \\
c_1 + 4c_2 + c_3 &= \frac{6(f_1 - 2f_2 + f_3)}{h^2} \\
\dots\dots &= \dots\dots \\
c_{n-2} + 4c_{n-1} &= \frac{6(f_{n-2} - 2f_{n-1} + f_n)}{h^2} \\
c_n &= 0
\end{aligned} \tag{5.75}$$

Both set of Equations 5.73 and 5.75 are tri-diagonal in nature and may easily be solved by the tri-diagonal matrix algorithm (TDMA, see Chapter 2).

Once the c's have been determined the spline fit may be calculated using Equation 5.68.

MATLAB program to determine the coefficients of a natural cubic spline is given below.

Program 5.8: *Coefficients of natural cubic spline*

```matlab
 1 function pp = cubicspline(x,y)
 2 %       Inputs    x : interpolation points
 3 %                 y : function value at interpolation points
 4 %       Output    pp: structure related to cubic spline
 5 %                 pp.coef : stores cubic spline weights
 6 %                 pp.x    : stores interpolation points
 7 %                 pp.y    : stores fuction value at pp.x
 8 n = length(y);           % number of interpolating points
 9 pp.x     = x;            % store interpolating points
10 pp.y     = y;            % store function values
11 a = ones(n,1);          % TDMA coefficient a
12 a1 = zeros(n,1);        % TDMA coefficient a'
13 a2 = zeros(n,1);        % TDMA coefficient a''
14 b = zeros(n,1);
15 for i = 2:n-1            % evaluate coefficients of TDMA
16     a(i) = 2*(x(i+1)-x(i-1));
17     a1(i) = -(x(i+1)-x(i));
18     a2(i) = -(x(i)-x(i-1));
19     b(i) = -6*(y(i+1)/b(i)-y(i)*(1/b(i)+1/c(i))+y(i-1)/c(i));
20 end
21 pp.coef  = tdma(a2,a,a1,b);   % evaluate coefficients
```

The output of the above program is a structure containing necessary information to evaluate cubic splines. The program below is used to evaluate cubic splines at input points.

Program 5.9: *Evaluation of cubic spline at input points*

```matlab
 1 function y = cubicsplineval(pp,x)
 2 % Inputs    pp: structure related to cubic spline
 3 %                 pp.coef : stores cubic spline weights
 4 %                 pp.x    : stores interpolation points
 5 %                 pp.y    : stores fuction value at pp.x
 6 %           x: points where function is to be evaluated
 7 % Output    y: evaluated function using cubic spline
 8 m = length(x);           % number of points in x
 9 n = length(pp.x);        % number of interpolating points
10 y = zeros(m,1);          % initialize y
11 for i=1:m                % loop for evaluating y
12     if ( x(i) < pp.x(1)) % check for location of x(i)
13         j1 = 1;
14     elseif ( x(i) > pp.x(n))
15         j1 = n-1;
16     else
17         j=1;
18         while( j < n)
19             if (x(i) >= pp.x(j) && x(i) <= pp.x(j+1) )
```

```
20              j1 = j; break;    % x(i) located break loop
21          else
22              j = j + 1;
23          end
24       end
25    end    % j1 indcates indices of location of x(i)
26    h = pp.x(j1+1)-pp.x(j1);    % step size
27    % evaluate cubic spline at x(i)
28    y(i) = (pp.coef(j1+1)*(x(i)-pp.x(j1))^3 ...
29    +pp.coef(j1)*(pp.x(j1+1)-x(i))^3 ...
30    +(6*pp.y(j1)-pp.coef(j1)*h^2)*(pp.x(j1+1)-x(i)) ...
31    +(6*pp.y(j1+1)-pp.coef(j1+1)*h^2)*(x(i)-pp.x(j1)))/(6*h);
32 end
```

Example 5.11

Pass a cubic spline between all the points given in the following table.

x	1.00	1.20	1.40	1.60	1.80	2.00
$f(x)$	1.9000	2.4380	3.2237	4.3668	6.0374	8.5000

x	2.20	2.40	2.60	2.80	3.00	
$f(x)$	12.1667	17.6858	26.0848	39.0066	59.1	

Solution :

We use Programs 5.8 and 5.9 to perform the cubic spline interpolation.

```
x = [1:0.2:3]';
y = [1.9; 2.438; 3.2237; 4.3668; 6.0374; 8.5; ...
     12.1667; 17.6858; 26.0848; 39.0066; 59.1];
c = cubicspline(x,y);
```

Detailed procedure involved in cubic spline interpolation is given below.

We note that the data is equi-spaced with spacing of $h = 0.2$. We assume that the spline satisfies the natural boundary conditions at the two end points. Thus $c_0 = c_{10} = 0$. It is necessary to determine only c_1 - c_9 in order to get the cubic spline. Equations governing these are obtained by the use of Equations 5.75. The required data may be generated using a spreadsheet program and expressed as an augmented matrix.

	Augmented matrix				TDMA		
i	a_i	a_i'	a_i''	b_i	p_i	q_i	c_i
1	4	-1	0	37.1459	-0.2500	9.2865	7.1938
2	4	-1	-1	53.6017	-0.2667	11.8174	8.3708
3	4	-1	-1	79.1478	-0.2679	18.0349	12.9246
4	4	-1	-1	118.7836	-0.2679	26.9949	19.0784
5	4	-1	-1	180.6241	-0.2679	41.1648	29.5453
6	4	-1	-1	277.8502	-0.2679	63.4197	43.3647
7	4	-1	-1	431.9870	-0.2679	98.7573	74.8462
8	4	-1	-1	678.4228	-0.2679	155.3209	89.2375
9	4	0	-1	1075.7446	0	246.6268	246.6268

5th column is based on right side of Equations 5.75

Note that a, a' and a'' are the coefficients of the c's and b is the non-homogeneous term in Equations 5.75. The augmented matrix may be used to apply the TDMA to solve for the c's. Result of TDMA is given by the last three columns of the table.

With all the c's available now the spline fit may be calculated using Equation (5.68) at a sufficiently large number of intermediate points so that a plot may be made.

```
x1 = [1:0.04:3]';
y1 = cubicsplineval(c,x1);
```

Data generated for $1 \le x \le 1.2$, with a spacing of $\Delta x = 0.04$ is shown in tabular form below. This data is generated using c_0 and c_1.

x	1	1.04	1.08	1.12	1.16	1.2
$f(x)$ (Spline)	1.9	1.9984	2.0991	2.2044	2.3166	2.438
$f(x)$ (Data)	1.9					2.438

Similarly data may be generated for other intervals also (the reader is encouraged do this). Figure 5.17 shows a plot of the given data points with a smooth curve joining them, based on the spline fit developed here.

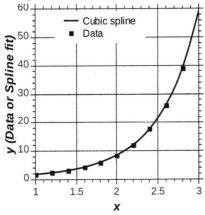

Figure 5.17: *Cubic spline passing through data points of Example 5.11*

Now let us see how the spline and Hermite polynomial approximate a known non-analytic function.

Example 5.12

Consider the function $f(x) = \dfrac{1}{1+25x^2}$ in the interval $[-1,1]$ from Example 5.7. Estimate the errors for Hermite polynomial and cubic spline with the interval divided into 4, 8 and 16 segments.

Solution :

The nodes for Hermite and cubic spline are the same as those used for piecewise quadratic interpolation scheme (Example 5.8). However, the number of points in a segment of a quadratic polynomial is 3 while a Hermite polynomial and cubic spline is estimated between

two nodes.
Cubic Hermite polynomial
Hermite polynomials require function and derivatives at the nodes. The first derivative of
the function is given by

$$f^{(1)}(x) = -\frac{50x}{(1+25x^2)^2}$$

Programs 5.6 and 5.7 have been used to evaluate the approximate functions. Figure 5.18
compares the Hermite polynomial approximations with exact function. Figure 5.19 shows
the error distribution for the Hermite polynomial approximations.   Programs 5.8 and 5.9

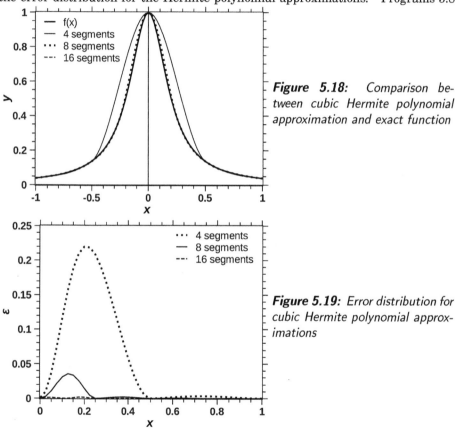

Figure 5.18: Comparison between cubic Hermite polynomial approximation and exact function

Figure 5.19: Error distribution for cubic Hermite polynomial approximations

have been used to evaluate cubic spline polynomials. Natural boundary conditions have
been used. The nature of cubic spline interpolation is not very different from a Hermite
polynomial. A cubic spline ensures higher order continuity at the nodes using only function
values where as Cubic Hermite polynomial requires the derivatives at the node in addition
to function values at nodes. Figure 5.20 indicates the distribution of error for cubic spline
interpolation scheme.

Cubic spline polynomial The following table lists the maximum error for different
piecewise approximation schemes for the function considered in the example.

Nodes	Piecewise quadratic	Cubic Hermite	Cubic spline
5	0.0855	0.2194	0.2791
9	0.0972	0.0355	0.0554
17	0.0478	0.0016	0.0036

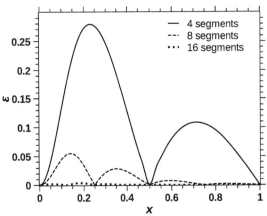

Figure 5.20: *Error distribution for cubic spline approximations*

---

*C*oncluding remarks

> *In this chapter we have looked at in great detail the various ways of interpolating a function of a single variable, available as tabulated data. Lagrange and Newton polynomials of low degree are suitable for this purpose. Hermite interpolating polynomials and cubic splines have some advantages in that they guarantee higher order continuity at the knots. In general they are useful in smoothing functions available in tabular form, as for example, in making a plot.*
>
> *For function approximation high degree polynomials are not suitable since they may suffer from Runge phenomenon accentuated by round off during the calculations. The effect of Runge phenomenon may be mitigated by using Chebyshev nodes.*
>
> *Interpolation will be seen to play a major role in later chapters.*

5.A MATLAB routines related to Chapter 5

MATLAB routine	Function
interp1	performs linear or cubic spline or cubic Hermite polynomial interpolation
spline	cubic spline interpolation

In addition curve fitting tool box provides additional tools for interpolation and splines.

5.B Suggested reading

1. **C. De Boor** *A Practical guide to splines (rev. Ed)* Applied Mathematical Sciences, Vol. 27, 2001

Chapter 6

Interpolation in two and three dimensions

Multidimensional interpolation is commonly encountered in numerical methods such as the <u>F</u>inite <u>E</u>lement <u>M</u>ethod (FEM) and the <u>F</u>inite <u>V</u>olume <u>M</u>ethod (FVM) used for solving partial differential equations. It is a general practice in numerical methods to discretize a two (three) dimensional domain into large number of small areas (volumes) known as elements in FEM and volumes in FVM. These methods assume a functional form for variable within each sub domain based on the nodal values and solve for the variables at the nodes. Development of multidimensional interpolation has also applications in computer graphics where surfaces are represented by Bezier curves and NURBS (Non-uniform rational B-splines). The present chapter intends to introduce the readers to multi-dimensional interpolation, with simple examples, to be made use of in later chapters.

6.1   Interpolation over a rectangle

6.1.1   Linear interpolation

The simplest interpolation technique is to use a linear variation along the two directions, say x and y for a function $f(x,y)$. An engineering application will be in using table of properties of a pure substance such as water (referred to as Steam Tables). A part of such a table would look like this:

	y_1	y_2
x_1	f_1	f_3
x_2	f_2	f_4

where the columns correspond to constant y's, rows correspond to constant x's and the entries correspond to the function $f(x,y)$. We assume that the function varies linearly with both x and y. This means that, for a fixed y, the function varies linearly with x (along a column in the table). Suppose we desire to have the interpolated value of the tabulated function at (x_i, y_i) within the range $x_1 < x_i < x_2, y_1 < y_i < y_2$ defining a rectangular region (Figure 6.1) in the x, y plane.

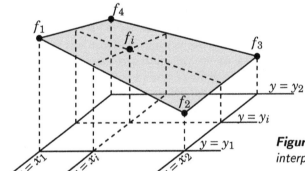

Figure 6.1: *Schematic of linear interpolation in two dimensions*

We hold y fixed at y_1 and obtain the interpolated value of the function at x_i, y_1 by linear interpolation as

$$f(x_i, y_1) = \frac{(x_i - x_2)}{(x_1 - x_2)} f_1 + \frac{(x_i - x_1)}{(x_2 - x_1)} f_2 = w_{1x} f_1 + w_{2x} f_2$$

based on L_1, Lagrange polynomial of degree 1. The weights have been identified with suitable subscripts. Similarly we hold y fixed at y_2 and obtain the interpolated value of the function at x_i, y_2 by linear interpolation as

$$f(x_i, y_2) = \frac{(x_i - x_2)}{(x_1 - x_2)} f_3 + \frac{(x_i - x_1)}{(x_2 - x_1)} f_4 = w_{1x} f_3 + w_{2x} f_4$$

Now we hold x fixed at x_i and obtain the interpolated value of function at x_i, y_i as

$$
\begin{aligned}
f(x_i, y_i) &= \frac{(y_i - y_2)}{(y_1 - y_2)} f(x_i, y_1) + \frac{(y_i - y_1)}{(y_2 - y_1)} f(x_i, y_2) \\
&= w_{1y} f(x_i, y_1) + w_{2y} f(x_i, y_2)
\end{aligned}
\tag{6.1}
$$

where again the weights have been identified by suitable subscripts. Specifically, x in the subscript means the weight is a function of x while y in the subscript means the weight is a function of y. Substituting Equations 6.1 and 6.1 in Equations 6.1 we get

$$
\begin{aligned}
f(x_i, y_i) &= w_{1x}w_{1y}f_1 + w_{2x}w_{1y}f_2 + w_{1x}w_{2y}f_3 + w_{2x}w_{2y}f_4 \\
&= w_1 f_1 + w_2 f_2 + w_3 f_3 + w_4 f_4
\end{aligned} \tag{6.2}
$$

where the final weights w_1 - w_4 are functions of both x and y. For example, we have

$$
w_1 = w_{1x}w_{1y} = \frac{(x_i - x_2)\,(y_i - y_2)}{(x_1 - x_2)\,(y_1 - y_2)} \tag{6.3}
$$

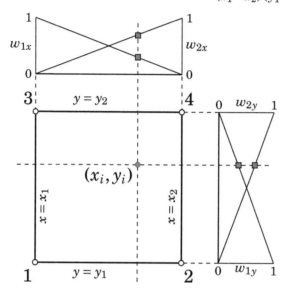

Figure 6.2: Graphical representation of weight functions

The reader may similarly write down the other weights also. Thus the interpolated value is a weighted sum of the function at the 4 nodes. We see that the weights are products of Lagrange weights in one dimension. We also see that the weight is unity at (x_1, y_1) and vanishes for $x_i = x_2$ or $y_i = y_2$ which means that this weight vanishes at the three nodes given by $(x_2, y_1), (x_1, y_2)$ and (x_2, y_2). The reader may verify, in a similar fashion, the properties of the other three weights also. Figure 6.2 illustrates the construction of weight or shape functions over a rectangular domain.

Example 6.1

Following data has been taken from a table of property values of a substance, after suppressing information on the units.

$x \downarrow \ y \rightarrow$	500	600
15	1286	1334.2
25	1285	1333.6

Obtain the value of the function at $x = 20$, $y = 530$ using linear interpolation.

Solution :

The weights are calculated with $x_i = 20$, $y_i = 530$ as given below.

$$
w_1 = \frac{(20 - 25)(530 - 600)}{(15 - 25)(500 - 600)} = 0.35 \qquad w_2 = \frac{(20 - 15)(530 - 600)}{(25 - 20)(500 - 600)} = 0.35
$$

$$w_3 = \frac{(20-25)(530-500)}{(15-25)(600-500)} = 0.15 \qquad w_4 = \frac{(20-15)(530-500)}{(25-15)(600-500)} = 0.15$$

The interpolated value of the function is then obtained as

$$f(20,530) = 0.35 \times 1286 + 0.35 \times 1285 + 0.15 \times 1334.2 + 0.15 \times 1333.6 = 1300.02$$

Linear interpolation; alternate way of representation

Consider the weight w_1 given in Equation 6.3. We may rewrite it in full as

$$w_1 = \frac{x_i y_i - x_2 y_i - y_2 x_i + x_2 y_2}{x_1 y_1 - x_2 y_1 - y_2 x_1 + x_2 y_2} = a_{10} + a_{11} x_i + a_{12} y_i + a_{13} x_i y_i$$

where

$$a_{10} = \frac{x_2 y_2}{x_1 y_1 - x_2 y_1 - y_2 x_1 + x_2 y_2} \qquad a_{11} = -\frac{y_2}{x_1 y_1 - x_2 y_1 - y_2 x_1 + x_2 y_2}$$

$$a_{12} = -\frac{x_2}{x_1 y_1 - x_2 y_1 - y_2 x_1 + x_2 y_2} \qquad a_{13} = \frac{1}{x_1 y_1 - x_2 y_1 - y_2 x_1 + x_2 y_2} \tag{6.4}$$

It is easily shown all the other weights have the same form but with different coefficients (the a's). It is thus seen that the interpolating function may be written in the alternate form

$$f(x_i, y_i) = c_0 + c_1 x_i + c_2 y_i + c_3 x_i y_i \tag{6.5}$$

where the c's are constants determined in terms of the four function values. In fact, we have the following equation set to obtain the c's.

$$\begin{pmatrix} 1 & x_1 & y_1 & x_1 y_1 \\ 1 & x_2 & y_1 & x_2 y_1 \\ 1 & x_1 & y_2 & x_1 y_2 \\ 1 & x_2 & y_2 & x_2 y_2 \end{pmatrix} \begin{pmatrix} c_0 \\ c_1 \\ c_2 \\ c_3 \end{pmatrix} = \begin{pmatrix} f_1 \\ f_2 \\ f_3 \\ f_4 \end{pmatrix} \tag{6.6}$$

These equations may easily be solved to get the vector $\mathbf{c}$ of coefficients in terms of the function values at the corners of the rectangle.

6.1.2 Local coordinate system for a rectangular element

A simple linear transformation may be used to transform the rectangular domain in x, y to a square of sides =2 in ξ, η such that $-1 < \xi, \eta < 1$ with the origin at the center of the square (see Figure 6.3). The required transformation may easily shown to be

$$x = \frac{x_1 + x_2}{2} + \frac{x_2 - x_1}{2}\xi; \quad y = \frac{y_1 + y_2}{2} + \frac{y_2 - y_1}{2}\eta \tag{6.7}$$

Instead of interpolating in (x, y) we may interpolate in (ξ, η).

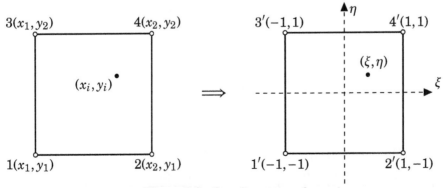

Figure 6.3: *Coordinate transformation*

Such a transformation is useful in engineering applications such as the finite element method. The given data may now be recast as follows:

	$\eta = -1$	$\eta = 1$
$\xi = -1$	f_1	f_3
$\xi = 1$	f_2	f_4

Equation set 6.6 is recast as

$$\begin{pmatrix} 1 & -1 & -1 & 1 \\ 1 & 1 & -1 & -1 \\ 1 & -1 & 1 & -1 \\ 1 & 1 & 1 & 1 \end{pmatrix} \begin{pmatrix} c_0 \\ c_1 \\ c_2 \\ c_3 \end{pmatrix} = \begin{pmatrix} f_1 \\ f_2 \\ f_3 \\ f_4 \end{pmatrix} \tag{6.8}$$

These equations may be solved easily (eg. Gauss elimination method) to get

$$c_0 = \frac{f_1 + f_2 + f_3 + f_4}{4} \qquad c_1 = \frac{-f_1 + f_2 - f_3 + f_4}{4}$$

$$c_2 = \frac{-f_1 - f_2 + f_3 + f_4}{4} \qquad c_3 = \frac{f_1 - f_2 - f_3 + f_4}{4} \tag{6.9}$$

We would now like to write the interpolating function in terms of weighted sum as given earlier. We easily see that $w_1(\xi,\eta)$ will be given by $\dfrac{1 - \xi - \eta + \xi\eta}{4}$ which may be recast in the form

$$w_1(\xi,\eta) = \frac{(1-\xi)}{2}\frac{(1-\eta)}{2} \tag{6.10}$$

It is deliberately written in the form of a product of two Lagrange polynomials of first degree. We may show similarly that the following hold:

$$w_2(\xi,\eta) = \frac{(1+\xi)}{2}\frac{(1-\eta)}{2} \qquad w_3(\xi,\eta) = \frac{(1-\xi)}{2}\frac{(1+\eta)}{2}$$

$$w_4(\xi,\eta) = \frac{(1+\xi)}{2}\frac{(1+\eta)}{2} \tag{6.11}$$

Of course we may revert to (x,y) once the interpolation has been worked out in (ξ,η).

Coordinate transformation The advantage of coordinate transformations is that operations such as differentiation, integration and transforms can be applied on a local coordinate system rather than the actual coordinates. The mapping of any arbitrary quadrilateral to the local coordinate system can be achieved in a number of ways such as affine, projective transformations. The discussion on these methods are beyond the scope of the book and the interested reader should look into books on computational graphics for the same.

Isoparametric transformations are most commonly employed in numerical methods such as FEM and FVM. The transformation maps line to line such that both *variables* (f) and the *real coordinates* (x, y) are evaluated using the *same* shape functions. For an arbitrary quadrilateral considered below, the coordinate transformation can be simply written as

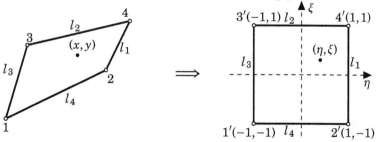

$$x = w_1 x_1 + w_2 x_2 + w_3 x_3 + w_4 x_4; \qquad y = w_1 y_1 + w_2 y_2 + w_3 y_3 + w_4 y_4$$

where w's are the weight functions (evaluated in (ξ, η) coordinates) we are familiar with. If we closely observe, the coordinate transformation equation is of the form

$$x = a_0 + a_1 \xi + a_2 \eta + a_3 \xi \eta \qquad y = b_0 + b_1 \xi + b_2 \eta + b_3 \xi \eta \qquad (6.12)$$

The two dimensional domain need not be represented by straight lines but could be any four arbitrary curves in which case coordinate transformation becomes murkier.

Example 6.2

Redo example 6.1 by transforming the coordinates and using the formulation given above

Solution :

The transformations are easily obtained as follows. With $x_2 = 25$ and $x_1 = 15$, we have $\frac{x_1 + x_2}{2} = \frac{15 + 25}{2} = 20; \frac{x_2 - x_1}{2} = \frac{25 - 15}{2} = 5$. Hence $x = 5\xi + 20$ or $\xi = \frac{x - 20}{5}$. Similarly we may write the second transformation as $y = 50\eta + 550$ or $\eta = \frac{y - 550}{50}$. Thus the interpolation point corresponds to $\xi = \frac{20 - 20}{2} = 0$ and $\eta = \frac{530 - 550}{50} = -0.4$. The weights are then calculated using Equations 6.10 and 6.11.

$$w_1 = \frac{(1 - 0)[1 - (-0.4)]}{4} = 0.35$$

$$w_2 = \frac{(1 + 0)[1 - (-0.4)]}{4} = 0.35$$

$$w_3 = \frac{(1 - 0)[1 + (-0.4)]}{4} = 0.15$$

$$w_4 = \frac{(1 + 0)[1 + (-0.4)]}{4} = 0.15$$

The interpolated value is the same as $f(0, 0.4) = 1300.02$ obtained earlier in Example 6.1.

Program 6.1: *Program to calculate weights of a linear quadrilateral element*

```
1 function w = weightquadlin(x,y)
2 % Input   x,y : local coordinates of the point(s)
3 % Output    w : weights at point x,y
4 n = size(x);           % no of points
5 w = zeros(n,4);        % initialize weight
6 w(:,1) = (1-x).*(1-y)/4;  % calculate weight at (-1,-1)
7 w(:,2) = (1+x).*(1-y)/4;  % calculate weight at (1,-1)
8 w(:,3) = (1-x).*(1+y)/4;  % calculate weight at (-1,1)
9 w(:,4) = (1+x).*(1+y)/4;  % calculate weight at (1,1)
```

6.1.3   Interpolating polynomials as products of 'lines'

Now we are ready for another interpretation of the interpolating function we have obtained above. Consider Equation 6.10 as an example. It contains two factors, the first one containing the equation of a line $l_1(\xi) = \frac{1 - \xi}{2}$ which is a line lying in a plane that is normal to the ξ, η plane that contains the ξ axis. This line passes through the points $l_1(-1) = 1$ and $l_1(1) = 0$. Similarly the second factor represents the equation of a line given by $l_2(\eta) = \dfrac{1-\eta}{2}$ which is a line lying in a plane that is normal to the ξ, η plane that contains the η axis. This line passes through the points $l_2(-1) = 1$ and $l_2(1) = 0$. Thus the weight is the product of equations of lines l_1 and l_2. Similar interpretation is possible for the other weights also. What is important to notice is that the weight is a *product* of two lines. These lines represent Lagrange polynomials of first degree (see Figure 6.4).

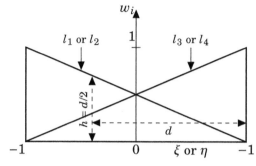

Figure 6.4: *Four 'lines' used in linear interpolation over a standard square*

Using the 'lines' defined above we may recast the weights as

$$w_1 = l_1 l_2; \; w_2 = l_2 l_3; \; w_3 = l_1 l_4 \text{ and } w_4 = l_3 l_4 \tag{6.13}$$

As indicated in the figure the height of the 'line' is equal to half the distance from the farthest point. Hence we may use the side of the square as a proxy for the corresponding line by simply halving d itself (i.e. by dividing the distance by the total distance, which is 2). This is shown schematically in Figure 6.5. The nodes are numbered as indicated and the lines are identified with the sides of the square domain. The weights are shown at the corresponding nodes as product of 'lines'. This way of visualizing the weights in terms of lines is particularly useful in dealing with higher order interpolation and also interpolation over a triangular domain.

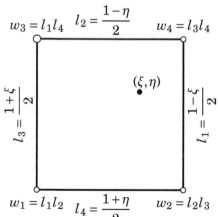

Figure 6.5: *Rectangular element in local coordinates showing the lines*

6.1.4 Lagrange Quadratic rectangular element

Figure 6.6 shows a 9 noded quadratic rectangular element also known as Lagrange quadratic element. The interpolating function for the element will be of the form

$$f(\xi,\eta) = a_0 + a_1\xi + a_2\eta + a_3\xi\eta + a_4\xi^2 + a_5\eta^2 + a_6\xi\eta^2 + a_7\xi^2\eta + a_8\xi^2\eta^2 \tag{6.14}$$

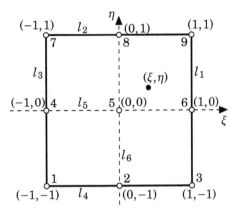

Figure 6.6: *Lagrange quadratic rectangular element*

The constants a_i have to be determined such that $f(\xi,\eta)$ is the same as the given data at all the nine nodes. Alternatively, the interpolation function can be estimated using Lagrange approach such that

$$f(\xi,\eta) = \sum_{i=1}^{9} w_i(\xi,\eta)f_i$$

We can define two independent quadratic Lagrange polynomial functions along η and ξ directions. The total weight functions, following the procedure used in the case of linear interpolation, is the product of the two quadratic polynomial functions. The weight functions are of the form

$$w(\xi,\eta) = (a_0 + a_1\xi + a_2\xi^2)(b_0 + b_1\eta + b_2\eta^2) = F(\xi)G(\eta)$$

Let us decompose the weight function and determine the weights of two separate functions $F(\xi)$ and $G(\eta)$. Then the weight functions at the nine points would be

$$\begin{array}{ccc}
w_1 = F_{-1}G_{-1} & w_2 = F_0G_{-1} & w_3 = F_1G_{-1} \\
w_4 = F_{-1}G_0 & w_5 = F_0G_0 & w_6 = F_1G_0 \\
w_7 = F_{-1}G_1 & w_8 = F_0G_1 & w_9 = F_1G_1
\end{array}$$

Based on Equation 5.17, the Lagrange weights $L_2(\xi)$ and $L_2(\eta)$ are written in vector form as

$$\mathbf{F}^T(\xi) = \left(\frac{\xi(\xi-1)}{2} \quad -(\xi-1)(\xi+1) \quad \frac{\xi(\xi+1)}{2} \right) = \left(\frac{\xi(\xi-1)}{2} \quad (1-\xi^2) \quad \frac{\xi(\xi+1)}{2} \right)$$

$$\mathbf{G}^T(\eta) = \left(\frac{\eta(\eta-1)}{2} \quad -(\eta-1)(\eta+1) \quad \frac{\eta(\eta+1)}{2} \right) = \left(\frac{\eta(\eta-1)}{2} \quad (1-\eta^2) \quad \frac{\eta(\eta+1)}{2} \right)$$

The nodal weights may then be written down as products $\mathbf{GF}^T$ and are given as

$$\left\{ \begin{array}{ccc}
w_1 & w_2 & w_3 \\
w_4 & w_5 & w_6 \\
w_7 & w_8 & w_9
\end{array} \right\} = \mathbf{GF}^T$$

$$= \left\{ \begin{array}{ccc}
\dfrac{\xi\eta(\xi-1)(\eta-1)}{4} & \dfrac{(1-\xi^2)\eta(\eta-1)}{2} & \dfrac{\xi\eta(\xi+1)(\eta-1)}{4} \\[3mm]
\dfrac{\xi(\xi-1)(1-\eta^2)}{2} & (1-\xi^2)(1-\eta^2) & \dfrac{\xi(\xi+1)(1-\eta^2)}{2} \\[3mm]
\dfrac{\xi\eta(\xi-1)(\eta+1)}{4} & \dfrac{(1-\xi^2)\eta(\eta+1)}{2} & \dfrac{\xi\eta(\xi+1)(\eta+1)}{4}
\end{array} \right\} \qquad (6.15)$$

Alternate approach: product of lines

The intersection points of the lines l_1 to l_6 are the nodes of the square domain. The nodes can be represented in terms of these lines. The equation of these lines are

$$l_1 \Longrightarrow 1-\xi = 0; \quad l_2 \Longrightarrow 1-\eta = 0; \quad l_3 \Longrightarrow 1+\xi = 0;$$
$$l_4 \Longrightarrow 1+\eta = 0; \quad l_5 \Longrightarrow \eta = 0; \quad l_6 \Longrightarrow \xi = 0$$

Node 1: The lines passing through node 1 are l_3 and l_4. At node 1, all the weights except w_1 are zero. In order to achieve this, all the weight functions except w_1 should have either l_3 or l_4 as factors. Similarly, node 1 would be a factor of remaining lines l_1, l_2, l_5 and l_6. Therefore the weight function for node 1 can be written down as

$$w_1 = \frac{l_1 l_2 l_5 l_6}{l_1(-1,-1)l_2(-1,-1)l_5(-1,-1)l_6(-1,-1)} = \frac{\xi\eta(\xi-1)(\eta-1)}{4}$$

which is the same as that obtained earlier. Similar approach can be adopted to derive weight functions for the other nodes.

Example 6.3

Obtain value of the tabulated function at $x = 0.14$, $y = 0.07$ by quadratic interpolation.

$x \downarrow \ y \rightarrow$	0	0.1	0.2
0	1.000000	1.010050	1.040811
0.1	1.010050	1.020201	1.051271
0.2	1.040811	1.051271	1.083287

Use a nine noded element to solve the problem.

Solution :

We transform the domain to the standard square domain by the transformation $\xi = 10(x - 0.1)$, $\eta = 10(y - 0.1)$. The elements are numbered as shown in Figure 6.6. In the present example, the function is required at $x = 0.14$ which corresponds to $\xi = 10(0.14 - 0.1) = 0.4$ and $y = 0.07$ which corresponds to $\eta = 10(0.07 - 0.1) = -0.3$. Substituting these in the matrix of weights (Equation 6.15) we get

$$\mathbf{w} = \begin{pmatrix} -0.0234 & 0.1638 & 0.0546 \\ -0.1092 & 0.7644 & 0.2548 \\ 0.0126 & -0.0882 & -0.0294 \end{pmatrix}$$

The interpolated value of the function is then obtained as sum of product of weights and the function values. We thus obtain

$$f(\xi = 0.4, \eta = -0.3) = f(x = 0.14, y = 0.07) = \sum_{i=1}^{9} w_i f_i = 1.024826$$

The reader may verify that linear interpolation using only the corner nodes would give a value for this as 1.043259. The function used for generating the data was $f(x,y) = e^{(x^2+y^2)}$ which has a value of $f(0.14, 0.07) = 1.024803$. The quadratic interpolated value has an error of 0.000024 while the linear interpolated value has an error of $0.018457 \approx 0.019$, with respect to the exact value. There is a considerable improvement in using quadratic interpolation over linear interpolation in the present case.

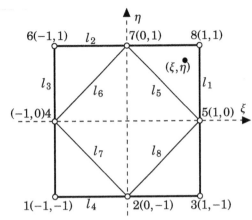

Figure 6.7: *Eight noded rectangular element*

6.1.5 Quadratic eight noded rectangular element

The above element is also known as 'serendipity' quadratic element.[1] The interpolation function in the domain is of the form

$$f(\xi,\eta) = a_0 + a_1\xi + a_2\eta + a_3\xi\eta + a_4\xi^2 + a_5\eta^2 + a_6\xi\eta^2 + a_7\xi^2\eta \qquad (6.16)$$

The present domain does not have a point at the origin. Therefore the present interpolation formula does not support the term involving $\xi^2\eta^2$. Unlike the previous example, the weight function of the present interpolation formula cannot be simplified as product of two univariate functions in ξ and η. Let us adopt the Lagrange interpolation approach based on product of lines. The nodes of the domain are the points of intersection of lines l_1 to l_8. Hence these lines can be used to represent the nodes. Equations of lines indicated in Figure 6.7 are given as a table below.

Equations of lines:

Line	Formula	Line	Formula
l_1	$1-\xi$	l_5	$\xi+\eta-1$
l_2	$1-\eta$	l_6	$-\xi+\eta-1$
l_3	$1+\xi$	l_7	$-\xi-\eta-1$
l_4	$1+\eta$	l_8	$\xi-\eta-1$

Calculation of nodal weights as product of lines:

For node 1 the procedure is as follows:

l_1 and l_2 must be factors of w_1 as these will be zero at nodes 3,5,6,7 and 8. The function should be equal to zero at nodes 2 and 4 as well. This means l_7 is also a factor of the weight function. Therefore the weight function can be written as

$$w_1 = \frac{l_1 l_2 l_7}{l_1(-1,-1)l_2(-1,-1)l_7(-1,-1)} = \frac{(1-\xi)(1-\eta)(-\xi-\eta-1)}{4}$$

Similarly we treat other nodes to arrive at weight functions as given below.

---

[1]For a serendipity element function values are specified only at points on the boundary.

Node Number	Weight formula in terms of product of lines	Weight expressions
1	$(l_1 \times l_2 \times l_7)/4$	$(1-\xi)(1-\eta)(-\xi-\eta-1)/4$
2	$(l_1 \times l_2 \times l_3)/2$	$(1-\xi)(1-\eta)(1+\xi)/2$
3	$(l_2 \times l_3 \times l_8)/4$	$(1-\eta)(1+\xi)(\xi-\eta-1)/4$
4	$(l_1 \times l_2 \times l_4)/2$	$(1-\xi)(1-\eta)(1+\eta)/2$
5	$(l_2 \times l_3 \times l_4)/2$	$(1-\eta)(1+\xi)(1+\eta)/2$
6	$(l_1 \times l_4 \times l_6)/4$	$(1-\xi)(1+\eta)(-\xi+\eta-1)/4$
7	$(l_1 \times l_3 \times l_4)/2$	$(1-\xi)(1+\xi)(1+\eta)/2$
8	$(l_3 \times l_4 \times l_5)/2$	$(1+\xi)(1+\eta)(\xi+\eta-1)4$

Note: One may be tempted to use lines $\eta = 0$ and $\xi = 0$ instead of l_5 to l_8. However, such an interpolation polynomial would not be consistent with Equation 6.16.

Example 6.4

A certain function is available at eight points as given in the following table. Obtain value of the function at $x = 0.14$, $y = 0.07$ by quadratic interpolation.

$x \downarrow$ $y \rightarrow$	0	0.1	0.2
0	1.000000	1.010050	1.040811
0.1	1.010050	$\cdots$	1.051271
0.2	1.040811	1.051271	1.083287

Solution :

Note that this set of data is the same as that used in Example 6.3 obtained by removing the data at the origin (removing data at point 5 in Figure 6.6). Nodes are numbered as shown in Figure 6.7. In the present example interpolated value is desired at $\xi = 0.4$, $\eta = -0.3$. We obtain all the weights using the weight functions and tabulate the results.

Node No.	$f(x,y)$	*Weight*	$Weight \times f(x,y)$
1	1.000000	-0.2145	-0.214500
2	1.010050	0.546	0.551487
3	1.040811	-0.1365	-0.142071
4	1.010050	0.273	0.275744
5	1.051271	0.637	0.669660
6	1.040811	-0.1785	-0.185785
7	1.051271	0.294	0.309074
8	1.083287	-0.2205	-0.238865
Sum =	$\cdots$	1	**1.024744**

Bold entry in the table is the required value of the function at the interpolation point.

6.2   Interpolation over a triangle

Interpolation over a triangle is an important exercise since most numerical methods for solving partial differential equations use triangular elements. Consider a triangular element shown in Figure 6.8.

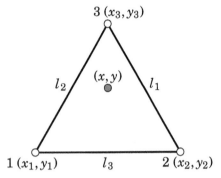

Figure 6.8: *Linear triangular element*

l_1, l_2 and l_3 are the equations of the lines forming the edges of the triangle. As there are three points, the interpolating function is of the form[2]

$$f(x,y) = a_0 + a_1 x + a_2 y \tag{6.17}$$

The constants a_0, a_1 and a_2 can be determined by solving the system of equations at the three nodes of the triangle. Alternatively, we can apply the Lagrange interpolation scheme for the triangle such that $f(x,y) = \sum_{i=1}^{3} w_i f_i$.

Consider node 1. The weight w_1 should be equal to 1 at node 1 and should be equal to zero at other two nodes. l_2 and l_3 are equal to zero at node 1. Therefore the weight function w_1 is proportional to l_1 and is of the form given by Equation 6.17. Meeting all the required conditions, w_1 can be written as

$$w_1 = \frac{l_1(x,y)}{l_1(x_1,y_1)}$$

In general, the weight function at node i can be written as

$$w_i = \frac{l_i(x,y)}{l_i(x_i,y_i)} \tag{6.18}$$

Local coordinate system for a triangular element

Let the fraction $\dfrac{l_i(x,y)}{l_i(x_i,y_i)} = \eta_i$. η_i is nothing but the ratio of normal distance of point (x,y) from line l_i to the normal distance of node i from the line. When the point lies on l_i, $\eta_i = 0$ and when the point is at node i, $\eta_i = 1$. Hence, the following representation also serves as an alternative coordinate system for the triangular element with η_i as the basis. Any point in (x,y) coordinates can be represented in terms of (η_1, η_2, η_3) as indicated in the following figure.

---

[2]Number of nodes less than four would mean that there can be only three terms in the interpolating function. The xy term would be missing.

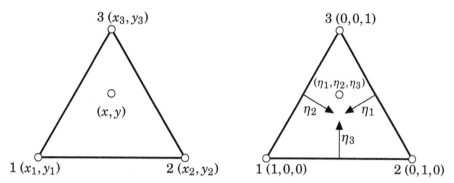

Figure 6.9: *Transformation of coordinates*

Hence the equation of the three lines l_1, l_2 and l_3 in local coordinate system would be $\eta_1 = 0$, $\eta_2 = 0$ and $\eta_3 = 0$.

For a linear triangular element, $w_i = \eta_i$

The area of a triangle is proportional to product of height and base. Hence η_i is also the ratio of the area of the triangles made by points (x, y) and (x_i, y_i) with line l_i. This can be understood from Figure 6.10.

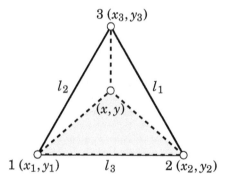

Figure 6.10: *Calculation of weight functions for a triangular element*

η_3 is the ratio of the area of the shaded triangle to the area of the triangle itself. The weight function for a linear triangular element can be generalized as $w_i = \eta_i = A_i/A$. Since $\sum_{i=1}^{n} A_i = A$, sum of weights at any point is equal to 1 i.e. $\sum_{i=1}^{3} w_i(x, y) = 1$. The area of the shaded triangle is given by the following determinant

$$A_3 = \frac{1}{2} \begin{vmatrix} 1 & x_1 & y_1 \\ 1 & x_2 & y_2 \\ 1 & x & y \end{vmatrix} = \frac{1}{2} \{x(y_1 - y_2) + (x_2 - x_1)y + (x_1 y_2 - x_2 y_1)\}$$

When (x, y) lies on the line l_3, $A_3 = 0$ and when $(x, y) = (x_3, y_3)$, $A_3 = A$, where A is the area of the triangular element given by

$$A = \frac{1}{2} \begin{vmatrix} 1 & x_1 & y_1 \\ 1 & x_2 & y_2 \\ 1 & x_3 & y_3 \end{vmatrix}$$

η_i is of the form $(a_i + b_i x + c_i y)$ where a_i, b_i and c_i are constants and are evaluated as follows

$$a_1 = \frac{(x_2 y_3 - y_2 x_3)}{2A} \quad b_1 = \frac{(y_2 - y_3)}{2A} \quad c_1 = \frac{(x_3 - x_2)}{2A}$$
$$a_2 = \frac{(x_3 y_1 - y_3 x_1)}{2A} \quad b_2 = \frac{(y_3 - y_1)}{2A} \quad c_1 = \frac{(x_1 - x_3)}{2A} \tag{6.19}$$
$$a_3 = \frac{(x_1 y_2 - y_1 x_2)}{2A} \quad b_3 = \frac{(y_1 - y_2)}{2A} \quad c_1 = \frac{(x_2 - x_1)}{2A}$$

Interpretation using local coordinates: Any local coordinate can be represented by other two local coordinates i.e. $\eta_3 = 1 - \eta_1 - \eta_2$. The right angled triangle on η_1, η_2 plane in Figure 6.11 represents the map of the actual triangle. In a sense, the transformation exercise can be looked upon as mapping triangle onto a right angled isosceles triangle. MATLAB program has been given

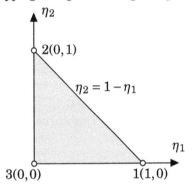

Figure 6.11: Representation of local coordinates of a triangle

below to calculate the coefficients for a triangular element

Program 6.2: Calculate coefficients for local coordinates of a triangular element

```
1  function [a,b,c,A] = trlocalcoeff(p1,p2,p3)
2  %   Input   p1,p2,p3 :   coordinates  of  the  vertices  p1(1) =x1;  p1(2)
                              %  =  y1
3  %                  p2(1) =x2;  p2(2)  =  y2;  p3(1) =x3;  p3(2)  =  y3
4  %   Output   a,b,c   :   coefficients  η=a+bx+cy
5  %             A        :   area  of  triangle
6  a(1)  =  p2(1)*p3(2)-p3(1)*p2(2);   %  calculate  a
7  a(2)  =  p3(1)*p1(2)-p1(1)*p3(2);
8  a(3)  =  p1(1)*p2(2)-p2(1)*p1(2);
9  A  =  (a(1)+a(2)+a(3))/2;                 %  A  =  2*area  of  triangle
10 a  =  a/(2*A);                            %  normailze  a
11 b(1)  =  (p2(2)-p3(2))/(2*A);            %  calculate  b
12 b(2)  =  (p3(2)-p1(2))/(2*A);
13 b(3)  =  (p1(2)-p2(2))/(2*A);
14 c(1)  =  (p3(1)-p2(1))/(2*A);            %  calculate  c
15 c(2)  =  (p1(1)-p3(1))/(2*A);
16 c(3)  =  (p2(1)-p1(1))/(2*A);
```

The following Matlab function can be used to convert from actual coordinates to local coordinates.

Program 6.3: Convert from (x, y) to local coordinates (η_1, η_2, η_3) of a triangle

```
1  function eta = trianglelocalcord(a,b,c,x,y)
2  %  Input    a,b,c:  coefficients  η=a+bx+cy
3  %           x,y :  coordinates  of  points
```

```
4 % Output     eta :  local coordinates of points x,y
5 n =  size(x,1);
6 eta = zeros(n,3);
7 for i=1:3
8    eta(:,i) = a(i) + b(i)*x + c(i)*y;
9 end
```

Quadratic triangular element

Now we derive expressions for weight functions for a quadratic triangular element.

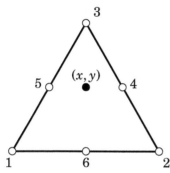

Figure 6.12: Quadratic triangular element

There are six points. Three nodes occupy the vertices of the triangle and three points are located at the midpoints of the three edges of the triangular element. The interpolation function would be of the form[3]

$$f(x,y) = a_0 + a_1 x + a_2 y + a_3 xy + a_4 x^2 + a_5 y^2 \qquad (6.20)$$

l_1, l_2 and l_3 are the equations of the lines forming the edges of the triangle. l_4, l_5 and l_6 are the

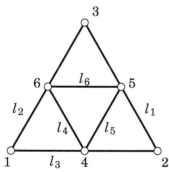

Figure 6.13: Lines representing quadratic triangular element

equation of the lines joining the midpoints of the edges of the triangle. Let us represent the nodes in terms of the triangular coordinate system introduced earlier.

$$1 \Longrightarrow (1,0,0) \qquad 2 \Longrightarrow (0,1,0) \qquad 3 \Longrightarrow (0,0,1)$$
$$4 \Longrightarrow (0.5,0.5,0) \quad 5 \Longrightarrow (0,0.5,0.5) \quad 6 \Longrightarrow (0.5,0,0.5) \qquad (6.21)$$

From Equation 6.21, the equation of the lines can be written down as

$$l_1 \Longrightarrow \eta_1 = 0 \qquad l_2 \Longrightarrow \eta_2 = 0 \qquad l_3 \Longrightarrow \eta_3 = 0$$

---

[3]With only six points available the quadratic is constrained to the form shown here. Recall that nine points gave the more general expression (Equation 6.14)

$$l_4 \Longrightarrow \eta_1 - 0.5 = 0 \quad l_5 \Longrightarrow \eta_2 - 0.5 = 0 \quad l_6 \Longrightarrow \eta_3 - 0.5 = 0$$

Now, we derive the weight function for all the nodes.

Consider node 1: The weight w_1 should be equal to 1 at node 1 and should be equal to zero at the other nodes. Lines l_2 and l_3 are zero where as $l_1 = 1$ at node 1. Therefore, w_1 should be proportional to l_1. This means $w_1 = 0$ at nodes 2, 3 and 5. As w_1 should be zero at nodes 4 and 6, w_1 should be proportional to l_4 also. Hence, meeting all the required conditions w_1 can be written as

$$w_1 = \frac{l_1(\eta_1,\eta_2,\eta_3)l_4(\eta_1,\eta_2,\eta_3)}{l_1(1,0,0)l_4(1,0,0)} = 2\eta_1(\eta_1 - 0.5)$$

Consider Node 4: As node 4 is a point on l_3, $\eta_3 = 0$. However, lines l_1 and l_2 are not equal to zero at node 4. This means w_4 is proportional to $l_1 l_2$. Therefore $w_4 = 0$ at other nodes except node 4. Meeting all the required conditions w_4 can be written as

$$w_4 = \frac{l_1(\eta_1,\eta_2,\eta_3)l_2(\eta_1,\eta_2,\eta_3)}{l_1(0.5,0.5,0)l_2(0.5,0.5,0)} = 4\eta_1\eta_2$$

The weights for all other points may be similarly determined and are summarized below

$$w_1 = 2\eta_1(\eta_1 - 0.5); \quad w_2 = 2\eta_2(\eta_2 - 0.5); \quad w_3 = 2\eta_3(\eta_3 - 0.5);$$
$$w_4 = 4\eta_1\eta_2; \quad\quad\quad w_5 = 4\eta_2\eta_3; \quad\quad\quad w_6 = 4\eta_3\eta_1$$

Example 6.5

Function value is available at six points as shown in the table below.

Point No.	x	y	$f(x,y)$
1	0.2	0.6	2.2255
4	0.25	0.35	1.8221
2	0.3	0.1	1.4918
5	0.325	0.4	2.0647
3	0.35	0.7	2.8577
6	0.275	0.65	2.5219

Obtain the value of the function at $x = 0.28, y = 0.5$ using linear and quadratic interpolation.

Solution :

Figure 6.14 shows the node numbering system used in the analysis.

Linear interpolation

Nodes at the apex of triangle are used for linear interpolation. We evaluate all the weights using the determinants presented in the text. Factor $\frac{1}{2}$ is dropped in all the area calculations since the weights are normalized with the area of the triangle. The area of the triangle is proportional to A given by

$$A = \begin{vmatrix} 1 & 0.2 & 0.6 \\ 1 & 0.3 & 0.1 \\ 1 & 0.35 & 0.7 \end{vmatrix} = 0.085$$

Weights are now calculated as given below. Weight are given by the following.

$$\eta_1 = w_1 = \frac{1}{0.085} \begin{vmatrix} 1 & 0.3 & 0.1 \\ 1 & 0.35 & 0.7 \\ 1 & 0.28 & 0.5 \end{vmatrix} = 0.3765$$

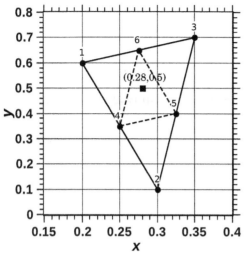

Figure 6.14: *Numbering of nodes for linear and quadratic interpolation over a triangle showing the interpolation point*

$$\eta_2 = w_2 = \frac{1}{0.085} \begin{vmatrix} 1 & 0.35 & 0.7 \\ 1 & 0.2 & 0.6 \\ 1 & 0.28 & 0.5 \end{vmatrix} = 0.2706$$

$$\eta_3 = w_3 = \frac{1}{0.085} \begin{vmatrix} 1 & 0.2 & 0.6 \\ 1 & 0.3 & 0.1 \\ 1 & 0.28 & 0.5 \end{vmatrix} = 0.3529$$

Note that the weights add up to 1. Using function values given in the table, the interpolated value of the function at $x = 0.28$, $y = 0.5$ is obtained as

$$f(0.28, 0.5) = 0.3765 \times 2.2255 + 0.2706 \times 1.4918 + 0.3529 \times 2.8577 = 2.2501$$

Quadratic interpolation

For quadratic interpolation, the data at the midpoints of the sides of the triangle element are also used. The weights already calculated for linear interpolation may be used to obtain the solution in the present case. We interpret the weights calculated there in terms of local coordinate system introduced in the text. Hence the coordinates of the given point $x = 0.28, y = 0.5$ may be identified in terms of the local coordinates as $\eta_1 = 0.3765$, $\eta_2 = 0.2706$, and $\eta_3 = 0.3529$. The following program has been written to perform quadratic interpolation

```
p1 = [0.2  0.6];         % vertices of the triangle (x,y)
p2 = [0.3  0.1];
p3 = [0.35 0.7];
f = [2.2255 1.4918 2.8577 1.8221 2.0647  2.5219]; % f(x,y)
x = 0.28; y = 0.5;
[a,b,c] = trlocalcoeff(p1,p2,p3);      % calculate coefficients
eta = trianglelocalcord(a,b,c,x,y);    % calculate η
w(1)  = 2*eta(1)*(eta(1)-0.5);         % calculate weights
w(2)  = 2*eta(2)*(eta(2)-0.5);
w(3)  = 2*eta(3)*(eta(3)-0.5);
w(4)  = 4*eta(1)*eta(2);
w(5)  = 4*eta(2)*eta(3);
w(6)  = 4*eta(3)*eta(1);
flin = sum(eta*f(1:3));                % linear interpolation
fquad = sum(w.*f);                     % quadratic interpolation
```

The weights for all the points for quadratic interpolation are tabulated below.

Point No.	η	Weight formula	Weight
1	0.3765	$= 2\eta_1(\eta_1 - 0.5)$	-0.0930
4	$\cdots$	$= 4\eta_1\eta_2$	-0.1242
2	0.2706	$= 2\eta_2(\eta_2 - 0.5)$	-0.1038
5	$\cdots$	$= 4\eta_2\eta_3$	0.4075
3	0.3529	$= 2\eta_3(\eta_3 - 0.5)$	0.3820
6	$\cdots$	$= 4\eta_3\eta_1$	0.5315

Using the weights and the corresponding function values, we get

Point No.	w	f	$w \times f$
1	-0.0930	2.2255	-0.2070
4	-0.1242	1.4918	-0.1852
2	-0.1038	2.8577	-0.2966
5	0.4075	1.8221	0.7425
3	0.3820	2.0647	0.7887
6	0.5315	2.5219	1.3403
Sum:	1	$\cdots$	**2.1827**

Bold entry in the last column is the desired value of the function at $x = 0.28, y = 0.5$. This is an improved value as compared to that obtained by linear interpolation. The function is known to be $f(x, y) = e^{(x+y)}$ and has an exact value of 2.1815 at $x = 0.28, y = 0.5$. The error for linear interpolation is 0.0686 while it is only 0.0012 in the case of quadratic. There is a huge improvement!

We have discussed interpolation polynomials for two dimensional geometries commonly used in FEM and FVM. However, there are many more interpolation techniques like Shepard interpolation, Wachspress interpolation of irregular polynomials, Bsplines and NURBS. Interested readers are advised to read the cited literature.

6.3   Interpolation in three dimensions

The concepts from two dimensional interpolation can be easily extended to three and higher dimensions. The most common three dimensional shapes are tetrahedral (extension of triangle) and hexahedral (extension of quadrilateral). Interpolation formulas can be represented as product of '*planes*' in place of *lines* used in the two dimensional cases.

6.3.1   Hexahedral element

Hexahedral element is a solid volume with six faces making it analogous to a quadrilateral in two dimensions. The method of deriving the weight functions is the same, but with product of lines in two dimensions becoming product of planes in the case of three dimensions. Consider the cube in Figure 6.15[4].

---

[4]The cube is the local coordinate representation of an arbitrary hexahedral.

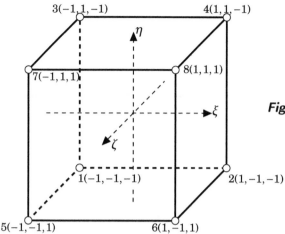

Figure 6.15: Hexahedral element

The weight functions for all the points have been given below

$$
\begin{aligned}
w_1 &= \frac{(1-\xi)(1-\eta)(1-\zeta)}{8} & w_2 &= \frac{(1+\xi)(1-\eta)(1-\zeta)}{8} \\
w_3 &= \frac{(1-\xi)(1+\eta)(1-\zeta)}{8} & w_4 &= \frac{(1+\xi)(1+\eta)(1-\zeta)}{8} \\
w_5 &= \frac{(1-\xi)(1-\eta)(1+\zeta)}{8} & w_6 &= \frac{(1+\xi)(1-\eta)(1+\zeta)}{8} \\
w_7 &= \frac{(1-\xi)(1+\eta)(1+\zeta)}{8} & w_8 &= \frac{(1+\xi)(1+\eta)(1+\zeta)}{8}
\end{aligned}
\tag{6.22}
$$

6.3.2 Tetrahedral element

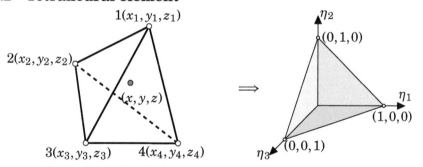

Figure 6.16: Tetrahedral element and its transformation to local coordinates

A tetrahedral element is a volume with four faces and is analogous to a triangle in two dimensions. The derivation of weight functions for the volume element is similar to the one for triangles. Planes forming the volume are analogous to the lines forming the triangle. Similar to the case of a triangle, the tetrahedron can be transformed to local coordinates $(\eta_1, \eta_2, \eta_3, \eta_4)$. The transformation maps the actual tetrahedral unto right angled tetrahedral as shown in Figure 6.16 (similar to transformation of triangular coordinates) η_1 is the ratio of the distance of the point and distance of node 1 from plane opposite to the node i.e. plane containing nodes 2, 3 and 4. It is also the ratio of the volume of the tetrahedron made by the point with the plane containing nodes 2, 3 and 4 (shaded volume in Figure 6.17) to the volume of the tetrahedral element. It can be easily inferred that for a linear tetrahedral element $w_i = \eta_i$.

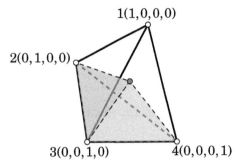

Figure 6.17: *Transformation to local coordinates: ratio of volume of the shaded tetrahedral to volume of tetrahedral element*

Concluding remarks

In this chapter, we have introduced, interpolation in two and three dimensions. These are very important since they are applied in all programs written for solving field problems in engineering. In fact the developments in interpolation techniques have paralleled the developments in FVM, FEM and other commercial software.

Chapter 7

Regression or curve fitting

*Regression analysis consists of expressing data, usually experimental data, in a succinct form as a formula thus giving a **global** interpolating formula. Experimental data is usually error ridden and hence analysis has to identify a possible regression law or model that best fits the data. The regression formula does not attempt to pass through all data points as a piecewise interpolating function is expected to do. Invariably the regression formula does not agree with the tabulated values exactly, but only in an average sense. The regression model is expected to pass close to **all** the data points and yield the smallest overall error, defined in a suitable fashion.*

7.1 Introduction

Data collected in an experiment that involves a cause and an effect gives information in the form of an ordered pairs such as (x_1, y_1), (x_2, y_2), $\cdots\cdots$, (x_i, y_i), $\cdots\cdots$, (x_n, y_n). Here x is the cause and y is the effect. For example, the cause may be the temperature to which a sensor is exposed to and the effect may be the resistance of the sensor. Both x and y are measured and are prone to measurement error. Figure 7.1 shows a plot between x and y with spline interpolation.

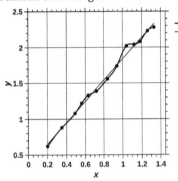

Figure 7.1: *Cubic spline is not appropriate for a set of data with errors*

Interpolation curves such as Lagrange polynomial, Hermite and Cubic spline pass through all the available data points. From the plot we can see that the spline or Lagrange polynomial curves (though being higher degree curves) are not suitable for data laden with errors. The approximation may be worse when derivatives of the curve are desired. In fact, from the plot we can ascertain that the thin curve in the figure is a more proper representation of the relationship between y and x. The approximation curve to the data points should be such that the error is uniformly distributed around the curve.

In practice one looks at such a plot and decides what type of relationship exists between the two variates. The relationship may be of two types.

1. Essentially a linear relationship such as $y_i = ax_i + b$, $y_i = ax_i^2 + bx_i + c$ or any other polynomial of suitable degree; it may involve other elementary functions such as $\sin(x), \cos(x), e^x$ etc. For example $y_i = ae^{bx_i}$ or $y_i = a\ln(x_i) + b$. In these $a, b...$ etc. are constants that are referred to as parameters. Linear refers to the fact that these relations are linear in the *parameters* that are used to represent the relationship while the relationship between the dependent and independent variables may be nonlinear.
2. A non-linear relationship such as $y_i = ae^{bx_i} + cx_i^d + \sin(ex_i)$. The relationship is non-linear in the parameters that characterize the fit.

In either case our job is to determine the best possible values for the parameters that represent the given data to our satisfaction. The degree of satisfaction or *the goodness of fit* as it is referred to, is based on tests which will be described later on.

Sometimes there may be more than one cause which leads to an effect. For example, if humidity of air also affects the resistance of the sensor described previously, we have to look for a relation between the sensor resistance as a function of temperature and humidity. This relation may either be linear or non-linear, in the sense used above. If, for example, this relationship is linear, we refer to it as multiple linear regression. In what follows we will discuss the methods useful for both types of regression models.

7.2   Method of least squares for linear regression

Consider an example of data with one cause and one effect. The data is error prone because of measurement errors and hence appears distributed around a line as shown in Figure 7.2. Note that the line does not pass through any data point. Presumably the line parameters are adjusted such that the line is close to the data in a *statistical* sense.

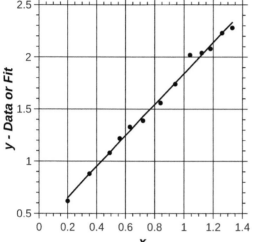

Figure 7.2: Data showing a linear trend

Corresponding to each data point - y_i - the equation of the line gives a point - y_i^l - and the difference between these two may be treated as an error e_i i.e. $e_i = y_i - y_i^l$. Sum of squares of these errors (it will always be positive) is a quantity that certainly tells us how good the line represents the data. A possible way of explaining this is to consider the data and fit as vectors in $\mathbf{R}^n$. The error itself is another vector in $\mathbf{R}^n$. The sum of squares alluded to above will be the square of the *distance* between the two vectors - the data vector and the fit vector. This is also referred to as the ℓ^2 norm of the error vector. Hence it is natural to look for the line parameters that minimize this sum (and thus the variance) and hence we have what is known as the *method of least squares* (also referred to as 'ordinary least squares') for estimating the parameters that describe the line.[1] The method of least squares is justified if the errors are distributed according to the normal distribution.[2]

7.2.1   Linear regression by least squares

Consider the sum of squares of error with respect to regression line given by

$$S = \sum_{i=1}^{n} e_i^2 = \sum_{i=1}^{n} \left[y_i - y_i^l \right]^2 = \sum_{i=1}^{n} [y_i - mx_i - c]^2 \tag{7.1}$$

Here n represents the number of data points and m and c are the parameters that characterize the regression line. Minimization of S requires that $\dfrac{\partial S}{\partial a} = 0$ and $\dfrac{\partial S}{\partial b} = 0$. Hence we have the following two equations.

$$\frac{\partial S}{\partial m} = 2 \sum_{i=1}^{n} [y_i - mx_i - c](-x_i) = 0$$

[1]Alternately one may also use the "method of total least squares". See, for example, P. de Groen, An Introduction to Total Least Squares, Nieuw Archief voor Wiskunde, Vierde serie, deel 14, 1996, pp. 237-253.

[2]See, for example, John E. Freund, Mathematical Statistics with Applications (7th Edition), Prentice Hall 2003. Also this reference is useful for understanding the statistical terms used in the present chapter.

$$\frac{\partial S}{\partial c} = 2\sum_{i=1}^{n}[y_i - mx_i - c](-1) = 0 \tag{7.2}$$

These may be recast as

$$\sum_{i=1}^{n} x_i^2 \, m + \sum_{i=1}^{n} x_i \, c = \sum_{i=1}^{n} x_i y_i \tag{i}$$

$$\sum_{i=1}^{n} x_i \, m + n \, c = \sum_{i=1}^{n} y_i \tag{ii} \tag{7.3}$$

These equations are known as normal equations. Noting that $\sum_{i=1}^{n} x_i = n\bar{x}$ and $\sum_{i=1}^{n} y_i = n\bar{y}$, where $\bar{x}$ and $\bar{y}$ are the mean respectively of the x's and the y's, Equation 7.3(ii) may be rewritten as $m\bar{x} + c = \bar{y}$. This may be rearranged to get

$$c = \bar{y} - m\bar{x} \tag{7.4}$$

Introduce this in Equation 7.3(i) to get $m\sum x_i^2 + \sum x_i[\bar{y} - m\bar{x}] = m\sum x_i^2 + n\bar{x}[\bar{y} - m\bar{x}] = m\left(\sum x_i^2 - n\bar{x}^2\right) + n\bar{x}\bar{y} = \sum x_i y_i$. This may be solved for m to get

$$m = \frac{\dfrac{\sum_{i=1}^{n} x_i y_i}{n} - \bar{x}\bar{y}}{\dfrac{\sum_{i=1}^{n} x_i^2}{n} - \bar{x}^2} \tag{7.5}$$

The numerator of Equation 7.5 is recognized as covariance represented by σ_{xy} while the denominator of Equation 7.5 is recognized as the variance of x represented by σ_x^2. Hence the above equation may be rewritten as

$$m = \frac{\sigma_{xy}}{\sigma_x^2} \tag{7.6}$$

parameter m represents the slope of the regression line while c represents the intercept of the regression line.

7.2.2 Coefficient of correlation and goodness of fit

We have given above the method of estimating the fit parameters of a linear relationship by least squares. y was treated as the effect while x was treated as the cause. It is also possible to treat x as the effect and y as the cause and look for a straight line fit in the form $x = m'y_d + c'$. If we run through the analysis in the previous section, interchanging the roles of x and y we should get the following:

$$c' = \bar{x} - m'\bar{y}$$
$$m' = \frac{\sigma_{xy}}{\sigma_y^2} \tag{7.7}$$

Slope of this regression line is given by $\dfrac{1}{m'}$, which, in general, is not the same as m, the slope of the first regression line. The ratio of the two slopes is given by

$$\frac{m}{\frac{1}{m'}} = mm' = \frac{\sigma_{xy}^2}{\sigma_x^2 \sigma_y^2} = \rho^2 \tag{7.8}$$

where ρ represents the square root of the ratio of the slopes of the two regression lines. It is also known as the correlation coefficient and hence

$$\rho = \pm\frac{\sigma_{xy}}{\sigma_x \sigma_y} \tag{7.9}$$

The sign of ρ is the same as the sign of σ_{xy}. If it is positive, x and y vary alike i.e. y increase when x increases. If it is negative y decreases when x increases.

If $\rho = \pm 1$ the fit and data are in extremely good agreement and we have noise free data! $|\rho| \approx 0$ means the *linear* relationship sought for is extremely unlikely. However, it has to be noted that the correlation coefficient does not indicate if the variables share a non-linear relationship. In practice the linear fit is said to be good if $|\rho|$ is close to one.

Example 7.1

The following data is expected to be represented by a linear relationship. Obtain such a fit by least squares and comment on the goodness of fit.

x	0.6	0.9	1.3	2.1	2.9	3.4	3.7	4.6	5.2
y	1.08	1.11	1.31	2.27	2.9	3	3.35	4.14	4.35

Solution :

Fit parameters are obtained by first calculating all the required statistical parameters. This is best done by the use of a spreadsheet as given below.

Data No.	x	y	x^2	y^2	xy
1	0.6	1.08	0.360	1.166	0.648
2	0.9	1.11	0.810	1.229	0.998
3	1.3	1.31	1.690	1.726	1.708
4	2.1	2.27	4.410	5.172	4.776
5	2.9	2.90	8.410	8.437	8.424
6	3.4	3.00	11.560	8.977	10.187
7	3.7	3.35	13.690	11.190	12.377
8	4.6	4.14	21.160	17.155	19.052
9	5.2	4.35	27.040	18.919	22.618
Column Sum	24.70	23.51	89.130	73.972	80.788
Column Mean	2.74444	2.61268	9.90333	8.21908	8.97640

The column means required for calculating the fit parameters are shown in the last row. The statistical parameters are now calculated:

$$\sigma_x^2 = 9.90333 - 2.74444^2 = 2.37136$$
$$\sigma_y^2 = 8.21908 - 2.61268^2 = 1.39299$$
$$\sigma_{xy} = 8.97640 - 2.74444 \times 2.61268 = 1.80605$$

Using Equation 7.6 the slope of the fit line is given by

$$m = \frac{1.80605}{2.37136} = 0.76161 \approx 0.762$$

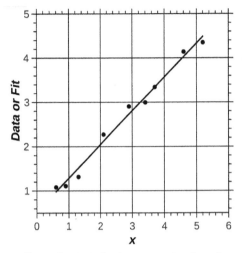

Figure 7.3: *Data and fit line in Example 7.1*

Using Equation 7.4 the intercept is given by

$$c = 2.61268 - 0.76161 \times 2.74444 = 0.52248 \approx 0.522$$

The goodness of fit is indicated by the coefficient of correlation given by (Equation 7.9)

$$\rho = \frac{1.80605}{\sqrt{2.37136 \times 1.39299}} = 0.994$$

The fit is thus seen to be an excellent representation of the data provided in the problem. A plot of the data and the fit line (also referred to as trend line) is shown in Figure 7.3.

7.2.3   Index of correlation and goodness of fit

Alternately we may use the more general 'index of correlation' to comment on the goodness of fit. Index of correlation compares the spread in the y values with respect to $\bar{y}$ with the spread in the error with respect to the regression line or in the more general case with the regression relation which may be non-linear. The index of correlation is defined as

$$\rho = \sqrt{1 - \frac{\sum_{i=1}^{n} e_i^2}{\sum_{i=1}^{n} (y - \bar{y})^2}} \tag{7.10}$$

Note the use of the same symbol ρ to represent the index of correlation. Most software programs use the symbol R^2 to represent the square of the coefficient of correlation or the index of correlation. Unlike correlation coefficient (Equation 7.9), index of correlation (Equation 7.10) can be applied also to nonlinear relationships. Thus we write

$$R^2 = 1 - \frac{\sum_{i=1}^{n} e_i^2}{\sum_{i=1}^{n} (y - \bar{y})^2} \tag{7.11}$$

It may easily be shown (the reader is encouraged to prove this) that in the linear regression case the coefficient of correlation and index of correlation are identical.

Adjusted index of correlation

In practice it is usual to use the "adjusted" index of correlation or the "adjusted R^2" i.e. R^2_{adj} to account for the loss of information in arriving at a fit.[3] Consider as an example the linear regression case presented earlier. We use the least squares method to obtain two parameters - the slope parameter and the intercept parameter - to obtain the regression line. Hence there is a reduction in the number of degrees of freedom given by the number of parameters p. Thus the number of degrees of freedom becomes $n - p = n - 2$ in the linear regression case. Also, in calculating the variance of the y's we lose one degree of freedom in calculating $\bar{y}$. Hence we modify Equation 7.10 as

$$R^2_{adj} = 1 - \frac{(n-1)}{(n-p)} \frac{\sum_{i=1}^{n} e_i^2}{\sum_{i=1}^{n}(y - \bar{y})^2} \tag{7.12}$$

Note that $R^2_{adj} \to R^2$ when n is very large.

7.2.4 Error estimate

Having obtained a fit it is possible to get an estimate of the error in the data. Since we are using the fit as a proxy for the data the error is related to the sum of squares of errors with respect to the fit. We make use of the number of degrees of freedom and define a standard estimate for the error as

$$\sigma_e = \pm \sqrt{\frac{\sum_{i=1}^{n} e_i^2}{n - p}} \tag{7.13}$$

The fit parameters (the slope and intercept) are themselves subject to uncertainties since they are calculated based on error prone data. It is possible to estimate the error in both these parameters but the analysis involves material which is outside the scope of this book. However we present below formulae which may be used to estimate the errors in the slope and intercept parameters.

$$\sigma_m^2 = \frac{\sigma_{y_i}^2}{n\sigma_x^2}; \; \sigma_c^2 = \frac{\sigma_{y_i}^2}{n} + \frac{\sigma_{y_i}^2 \bar{x}^2}{n\sigma_x^2} \tag{7.14}$$

Example 7.2

Use the data of Example 7.1 to calculate R^2_{adj} and comment on the goodness of fit. Also estimate the error in using the fit to represent the data. What are the uncertainties in the slope and intercept parameters?

Solution :

The fit parameters determined in Example 7.1 may be made use of to calculate the 'fit line' and then use the data generated thus to calculate both R^2_{adj} and σ_e. Extract of a spreadsheet shown below helps in arriving at the results.

---

[3]Analysis shows that the adjustment to account for degrees of freedom is related to estimate of the statistical parameters of data drawn from a population. Thus 'adjusted' accounts for the data used being a sample from a large population.

x	y	y^l	$(y - y^l)^2$	$(y - \bar{y})^2$
0.6	1.0798	0.9794	0.010071	2.349714
0.9	1.1087	1.2079	0.009847	2.261949
1.3	1.3139	1.5126	0.039472	1.686824
2.1	2.2743	2.1219	0.023237	0.114500
2.9	2.9047	2.7312	0.030119	0.085277
3.4	2.9962	3.1120	0.013399	0.147089
3.7	3.3451	3.3404	0.000022	0.536442
4.6	4.1418	4.0259	0.013436	2.338215
5.2	4.3496	4.4829	0.017756	3.016899
Sum =	23.5141	23.5141	0.157359	12.536909

With $n = 9$, $p = 2$, R^2_{adj} may be calculated using defining Equation 7.12 as

$$R^2_{adj} = 1 - \frac{(9-1)}{(9-2)} \times \left(\frac{0.157359}{12.536909} \right) = 0.985655 \approx 0.986$$

The value of $R^2_{adj} = 0.986$ represents a good fit. An estimate of the standard error is calculated using defining Equation 7.13 as

$$\sigma_e = \pm \sqrt{\frac{0.157359}{(9-2)}} = \pm 0.149933 \approx \pm 0.150$$

Thus the data is represented by the regression line as

$$y^l = (0.762x + 0.522) \pm 0.150 \tag{7.15}$$

i.e. the line represents the data within an error band of ± 0.150. We shall use this as an estimate for σ in estimating the errors in the parameters. Using Equation 7.14 and $\sigma^2 = 0.149933^2 = 0.0224798$ we get

$$\sigma_m = \pm \sqrt{\frac{0.0224798}{9 \times 2.37136}} = \pm \sqrt{0.0070253117} = \pm 0.0324546273 \approx \pm 0.033$$

$$\sigma_c = \sqrt{\frac{0.0224798}{9} + \frac{0.0224798 \times 2.744^2}{9 \times 2.37136}} = \sqrt{0.0695740031} = 0.102133 \approx \pm 0.102$$

Thus the slope and intercept parameters are given by

$$m = 0.762 \pm 0.033; \quad c = 0.522 \pm 0.102$$

More often in practice, the dependent and independent variables may share a non-linear relationship. However, the fit itself can be transformed into the linear form $Y = A + BX$ so that the linear regression may be used. *Note that the equation of the fit is linear with respect to the coefficients.* The following table lists some of non-linear regression equations which may be linearized

Table 7.1: *Non-linear regression fit and its corresponding linearized form*

S.no	Fit equation	Linearized form, $Y = A + BX$			
		Y	A	B	X
1	$y = ae^{bx}$	$\ln y$	$\ln a$	b	x
2	$xy = ax + b$	y	a	b	$1/x$
3	$y = ax^b$	$\log_{10} y$	$\log_{10} a$	b	$\log_{10} x$
4	$y = a + bx^2$	y	a	b	x^2
5	$y = a + be^x$	y	a	b	e^x
6	$y = ae^{bx} + cx + d$	cannot be linearized			

Not all non-linear relations can be linearized as shown by case 6 above. We will discuss such cases later. First, let us consider an example which can fit a non-linear curve to the data using linear regression.

Example 7.3

It is intended to determine the relation between Nusselt number Nu and Rayleigh number Ra for a vertical heated wall losing heat by natural convection. Heat transfer experiments have been conducted on a heated vertical flat plate and the data has been tabulated below.

	Ra	Nu		Ra	Nu
1	33740	4.550	9	261300	8.144
2	57440	5.174	10	397200	9.119
3	73070	5.606	11	450300	8.994
4	90770	6.027	12	509500	9.639
5	134300	6.524	13	571900	9.740
6	161300	6.935	14	715700	10.196
7	190800	7.334	15	795900	10.874
8	224700	7.745			

The relationship between Nusselt number and Rayleigh number is of the form $Nu = aRa^b$. Fit a least squares curve to the above experimental data and comment on the goodness of the fit.

Background :

Nusselt number and Rayleigh number are non-dimensional numbers used in heat transfer. These represent respectively the non-dimensional convective heat transfer rate and the non-dimensional buoyancy parameter.

Solution :

Nusselt number and Rayleigh number share a non-linear relationship. The equation can be linearized by taking $\log_{10}$ on both the sides. Therefore, we have

$$\log_{10} Nu = \log_{10} a + b \log_{10} Ra$$

which is of the form

$$Y = bX + c$$

where $Y = \log_{10} Nu$, $X = \log_{10} Ra$ and $c = \log_{10} a$ and the above equation is linear with respect to the coefficients b and c. We can now apply linear regression to the data as summarized in the following spreadsheet.

	X	Y	XY	X^2
1	4.5281	0.6580	2.9796	20.5041
2	4.7592	0.7138	3.3973	22.6501
3	4.8637	0.7487	3.6413	23.6560
4	4.9579	0.7801	3.8677	24.5812
5	5.1281	0.8145	4.1769	26.2972
6	5.2076	0.8410	4.3799	27.1195
7	5.2806	0.8653	4.5695	27.8845
8	5.3516	0.8890	4.7577	28.6397
9	5.4171	0.9108	4.9341	29.3454
10	5.5990	0.9599	5.3748	31.3489
11	5.6535	0.9540	5.3932	31.9621
12	5.7071	0.9840	5.6160	32.5715
13	5.7573	0.9886	5.6915	33.1467
14	5.8547	1.0084	5.9041	34.2779
15	5.9009	1.0364	6.1156	34.8201
Sum	79.9666	13.1527	70.7989	428.8048

The fit parameters are obtained as $b = 0.2729$ and $c = -0.5781$. Then $a = 10^c = 0.2642$. Therefore the regression fit is given by

$$Nu = 0.2642 Ra^{0.2729} \approx 0.26 Ra^{0.27}$$

Figure 7.4 shows the comparison between data and fit, plotted on log-log paper.

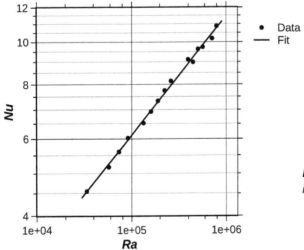

Figure 7.4: *Data and fit line in Example 7.3*

Observe that the curve is a straight line on a log-log graph. Now we shall calculate R^2_{Adj} to ascertain the quality of the fit. Table given below helps in this.

X	Y	Y^l	$(Y - Y^l)^2$	$(Y - \bar{Y})^2$
4.528	0.658	0.658	9.336E-08	4.789E-02
4.759	0.714	0.721	4.823E-05	2.657E-02
4.864	0.749	0.749	4.128E-07	1.643E-02
4.958	0.780	0.775	2.596E-05	9.359E-03
5.128	0.815	0.821	4.790E-05	3.885E-03
5.208	0.841	0.843	4.406E-06	1.281E-03
5.281	0.865	0.863	5.225E-06	1.323E-04
5.352	0.889	0.882	4.333E-05	1.483E-04
5.417	0.911	0.900	1.106E-04	1.156E-03
5.599	0.960	0.950	9.983E-05	6.906E-03
5.654	0.954	0.965	1.183E-04	5.946E-03
5.707	0.984	0.979	2.084E-05	1.149E-02
5.757	0.989	0.993	2.120E-05	1.248E-02
5.855	1.008	1.020	1.281E-04	1.731E-02
5.901	1.036	1.032	1.646E-05	2.545E-02
Sum	13.153	13.153	6.908E-04	0.186

With $n = 15$, $p = 2$, R^2_{adj} may be calculated using defining Equation 7.12 as

$$R^2_{adj} = 1 - \frac{(15-1)}{(15-2)} \times \left(\frac{0.6.908E-04}{0.186} \right) = 0.996$$

The value of $R^2_{adj} = 0.996$ represents a good fit to the above data.

Common misconceptions about R^2

R^2 is large

Larger values of R^2 means the fit is a good representation of the data. The data used to determine the fit often represents only a small sample size. If the sample does not represent the entire distribution properly, the resulting fit will not represent the distribution accurately. Therefore, a good fit to a data may not be a good fit in the entire range. In fact the main challenge faced by one is to obtain proper sample data.

Sampling of data can be poor for following possible reasons

1. Number of data points are small (see Figure 7.5). As the number of data points are small, the uncertainty in the coefficients would be high. Also, the R^2_{adj} term would be usually small. The confidence in the fit increases with large volumes of data points. As the number of data points become large, R^2_{adj} tends to R^2.

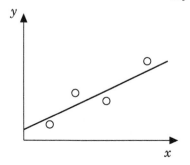

Figure 7.5: A linear fit across four data points may not be accurate fit even though R^2 is close to 1

2. Data points may be concentrated only in some regions of the entire domain. It may be possible that the sampled data corresponds to a small region and does not cover the entire range. In that sense, the regression fit is valid only in the sampled region. Use of the fit beyond may amount to extrapolation.

Example 7.4

Represent the following data by linear fit.

x	0.7	0.75	0.8	0.85	0.9	0.95	1
y	0.49	0.5625	0.64	0.7225	0.81	0.9025	1

The above data corresponds to the exact curve $y = x^2$. Compare the linear fit to the exact curve.

Solution :

All the statistical parameters are determined using the table below.

No.	x	y	x^2	xy
1	0.7	0.49	0.49	0.343
2	0.75	0.5625	0.5625	0.421875
3	0.8	0.64	0.64	0.512
4	0.85	0.7225	0.7225	0.614125
5	0.9	0.81	0.81	0.729
6	0.95	0.9025	0.9025	0.857375
7	1	1	1	1
Column sum	5.95	5.1275	5.1275	4.477375
Column mean	0.85	0.7325	0.7325	0.639625

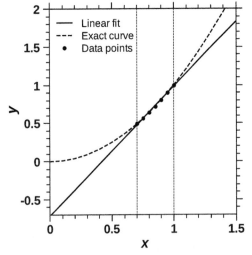

Figure 7.6: Comparison between linear fit and exact curve in the range $0 < x < 1.5$

The linear fit to the data is $y = -0.7125 + 1.7x$. The R^2 for the linear fit is 0.9977. Therefore, the linear fit is a good representation of the data in the region $0.7 < x < 1$. However, we are aware that the exact curve is of quadratic form and the fit is of linear form. Figure 7.6

compares the linear fit curve and the exact curve in the range $0 < x < 1.5$. It is clear from the figure that the linear fit is a good approximation of the curve in the range $0.7 < x < 1$ while the fit represents the curve poorly away from the region.

The example clearly illustrates the importance of sampling. The sample being restricted to a small range can mislead that x and y share linear relationship where as the actual relationship may be different. Therefore R^2 being close to 1 is an indicator that the fit is good in the data range and may not indicate the true relationship at all!

R^2 is small

Small value of R^2 suggests that the fit is not good. This may lead to a conclusion that x and y are unrelated which may be totally untrue. Similarly correlation index used to indicate the goodness of a linear fit indicates the linear relationship between x and y to be poor. It is possible that x and y share a nonlinear relationship! Essentially R^2 only indicates the goodness of the fit to the available data but may not be a true representer of the relationship between dependent (effect) and independent (cause) variables.

7.3 Multi-linear regression

In general a multi-liner fit is of form $y^f = a_0 + a_1 x_1 + a_2 x_2 + \cdots + a_m x_m$ where x_j are m causes and y^f is the effect. We have to determine the parameters $a_0, a_1, \ldots a_m$ by the least squares method.

$$S = \sum_{i=1}^{n} \left[y_i - y^f \right]^2 = \sum_{i=1}^{n} \left[y_i - a_0 - a_1 x_{1,i} - a_2 x_{2,i} - \cdots - a_m x_{m,i} \right]^2 \tag{7.16}$$

The sum of the squares of the error has to be minimized. For minimum of S, the partial derivatives $\dfrac{\partial S}{\partial a_i} = 0$. We then obtain the following normal equations

$$
\begin{aligned}
\sum y_i &= & n a_0 + a_1 \sum x_{1,i} + a_2 \sum x_{2,i} & \quad + \cdots + & a_m \sum x_{m,i} \\
\sum y_i x_{1,i} &= & a_0 \sum x_{1,i} + a_1 \sum x_{1,i}^2 + a_2 \sum x_{2,i} x_{1,i} & \quad + \cdots + & a_m \sum x_{m,i} x_{1,i} \\
&\cdots & \cdots & \cdots & \cdots \\
\sum y_i x_{m,i} &= & a_0 \sum x_{m,i} + a_1 \sum x_{1,i} x_{m,i} + a_2 \sum x_{2,i} x_{m,i} & \quad + \cdots + & a_m \sum x_{m,i}^2
\end{aligned}
\tag{7.17}
$$

Rewriting the above equations in matrix form

$$
\begin{pmatrix}
n & \sum x_{1,i} & \cdot\cdot & \sum x_{j,i} & \cdot\cdot & \sum x_{m,i} \\
\sum x_{1,i} & \sum x_{1,i}^2 & \cdot\cdot & \sum x_{j,i} x_{1,i} & \cdot\cdot & \sum x_{m,i} x_{1,i} \\
\sum x_{2,i} & \sum x_{1,i} x_{2,i} & \cdot\cdot & \sum x_{j,i} x_{2,i} & \cdot\cdot & \sum x_{m,i} x_{2,i} \\
\cdot\cdot & \cdot\cdot & \cdot\cdot \ \cdot\cdot & & \cdot\cdot \ \cdot\cdot \\
\sum x_{m,i} & \sum x_{1,i} x_{m,i} & \cdot\cdot & \sum x_{j,i} x_{m,i} & \cdot\cdot & \sum x_{m,i}^2
\end{pmatrix}
\begin{pmatrix}
a_0 \\ a_1 \\ a_2 \\ \cdots \\ a_m
\end{pmatrix}
=
\begin{pmatrix}
\sum y_i \\ \sum y_i x_{1,i} \\ \sum y_i x_{2,i} \\ \cdots \\ \sum y_i x_{m,i}
\end{pmatrix}
\tag{7.18}
$$

where the summations are from $i = 1$ to $i = n$. The coefficient matrix is *symmetric* and the above set of equations may be solved by matrix manipulations such as conjugate gradient method we are already familiar with. It has to be noted that all least square problems result in *symmetric* matrices. The goodness of fit may be gaged by the R_{adj}^2 value.

Program 7.1: *Multi-linear regression*

```
 1  function [coeff,ss,R2] = multilinearregression(x,y)
 2  %  Input  x: independent variables
 3  %         y: dependent variable
 4  %  Output coeff: coefficients
 5  %         ss  : standard error
 6  %         R2  : index of correlation
 7  [n,m] = size(x);          % size of x
 8  %       m = number of independent variables
 9  %       n = number of data points
10  A = zeros(m+1,m+2);       % initializing Coefficient Matrix
11  r = [ones(n,1) x y];      % initializing r matrix r = {1 X Y}
12  i = 1;
13  while(i<=m+1)             % constructing coefficient matrix
14      for j =1:m+2
15          r1(:,j) = r(:,j).*r(:,i);
16      end
17      A(i,:) = sum(r1);
18      i = i+1;
19  end
20  coeff = choleskydecomposition(A(:,1:m+1),A(:,m+2));
21  yfit = r(:,1:m+1)*coeff;          % calculate yfit
22  ss = sqrt(sum((y-yfit).^2)/(n-m-1)); % calculate ss
23  R2 = 1-ss^2/var(y);       % calculate R2
24  %         var(y) builtin function to calculate variance
```

Example 7.5

Following data has been gathered where the effect y depends on two causes x_1 and x_2. Obtain the fit parameters that characterize a multi-linear fit. Also evaluate the goodness of fit.

No.	1	2	3	4	5	6	7	8	9	10
x_1	1.2	2.15	2.66	2.95	3.42	3.65	3.92	4.22	4.33	4.56
x_2	0.85	0.96	1.12	1.45	1.79	2.06	2.44	2.54	2.76	3.07
y	9.65	12.97	14.98	18.76	20.8	22.4	26.1	27.8	29.1	30.6

Solution :

The normal equations for this case are obtained as

$$
\begin{Bmatrix}
n & \sum_i x_{1,i} & \sum_i x_{2,i} \\
\sum_i x_{1,i} & \sum_i x_{1,i}^2 & \sum_i x_{1,i}x_{2,i} \\
\sum_i x_{2,i} & \sum_i x_{2,i}x_{1,i} & \sum_i x_{2,i}^2
\end{Bmatrix}
\begin{Bmatrix}
a_0 \\ a_1 \\ a_2
\end{Bmatrix}
=
\begin{Bmatrix}
\sum_i y_i \\
\sum_i x_{1,i}y_i \\
\sum_i x_{2,i}y_i
\end{Bmatrix}
$$

where $1 \le i \le 10$. The required sums are easily calculated as given in the following table:

x_1	x_2	y	x_1^2	x_2^2	$x_1 x_2$	$x_1 y$	$x_2 y$
1.2	0.85	9.65	1.44	0.7225	1.02	11.58	8.2025
2.15	0.96	12.97	4.6225	0.9216	2.064	27.8855	12.4512
2.66	1.12	14.98	7.0756	1.2544	2.9792	39.8468	16.7776
2.95	1.45	18.76	8.7025	2.1025	4.2775	55.342	27.202
3.42	1.79	20.8	11.6964	3.2041	6.1218	71.136	37.232
3.65	2.06	22.4	13.3225	4.2436	7.519	81.76	46.144
3.92	2.44	26.1	15.3664	5.9536	9.5648	102.312	63.684
4.22	2.54	27.8	17.8084	6.4516	10.7188	117.316	70.612
4.33	2.76	29.1	18.7489	7.6176	11.9508	126.003	80.316
4.56	3.07	30.6	20.7936	9.4249	13.9992	139.536	93.942
33.06	19.04	213.16	119.5768	41.8964	70.2151	772.7173	456.5633

The normal equations then are

$$\begin{pmatrix} 10 & 33.06 & 19.04 \\ 33.06 & 119.5768 & 70.2151 \\ 19.04 & 70.2151 & 41.8964 \end{pmatrix} \begin{pmatrix} a_0 \\ a_1 \\ a_2 \end{pmatrix} = \begin{pmatrix} 213.16 \\ 772.7173 \\ 456.5633 \end{pmatrix}$$

The above set of equations are solved by Gauss elimination (bring the augmented matrix to upper triangle form) to get $a_1 = 1.70$, $a_2 = 2.95$ and $a_3 = 5.19$. The desired fit to data is given by

$$y^f(x_1, x_2) = 1.70 + 2.95x_1 + 5.19x_2$$

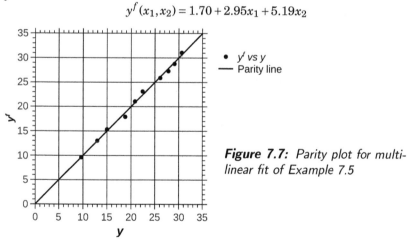

Figure 7.7: *Parity plot for multi-linear fit of Example 7.5*

Since there are two independent variables and one dependent variable a '**parity plot**' helps in assessing the fit. In essence we make a plot of y vs y^f to obtain a parity plot. If the scales along the two axes are chosen properly the data points should lie close to the 45° line passing through the origin, known as the *partiy line* (see Figure 7.7). Goodness of fit may, of course, be gaged from the value of R_{adj}^2 calculated as usual. In the present case $R_{adj}^2 = 0.994$ which indicates a very good fit. The estimated error of the fit is given by $\sigma_e = 0.54$.

7.4 Polynomial regression

Linear regression methodology may easily be applied to a case where the cause effect dependence is a polynomial. The coefficients in the polynomial take on the roles of parameters.

Consider the case where the data set (x_i, y_i), $1 \le i \le n$ is represented by a polynomial of form $y^l(x) = a_1 + a_2 x + a_3 x^2 \ldots + a_m x^{m-1}$ where $m < n$. We have to determine the parameters $a_1, a_2, \ldots a_m$ by the least squares method. It may easily be shown that the parameters are the solution of the following set of normal equations.

$$
\begin{pmatrix}
n & \sum x_i & \cdots & \sum x_i^j & \cdots & \sum x_i^{m-1} \\
\sum x_i & \sum x_i^2 & \cdots & \sum x_i^{j+1} & \cdots & \sum x_i^m \\
\sum x_i^2 & \sum x_i^3 & \cdots & \sum x_i^{j+2} & \cdots & \sum x_i^{m+1} \\
\cdots & \cdots & \cdots & \cdots & \cdots & \cdots \\
\sum x_i^{m-1} & \sum x_i^{m+2} & \cdots & \sum x_i^{j+m-1} & \cdots & \sum x_i^{2m-2}
\end{pmatrix}
\begin{pmatrix}
a_1 \\ a_2 \\ a_3 \\ \cdots \\ a_m
\end{pmatrix}
=
\begin{pmatrix}
\sum y_i \\ \sum y_i x_i \\ \sum y_i x_i^2 \\ \cdots \\ \sum y_i x_i^{m-1}
\end{pmatrix}
\tag{7.19}
$$

where the summations are all from $i = 1$ to $i = n$. The above set of equations may be solved by matrix manipulations we are already familiar with. The goodness of fit may again be gaged by the R^2_{adj} value.

Example 7.6

The location of a particle changes along a line with time. It is desired to look for a quadratic fit to the time vs location data (SI units) globally. Obtain such a fit by least squares.

No.	Time t	Location y	No.	Time t	Location y
1	0.094	1.356	11	2.245	1.918
2	0.204	1.382	12	2.884	2.087
3	0.538	1.462	13	3.112	2.157
4	0.922	1.562	14	3.568	2.312
5	1.088	1.582	15	3.798	2.458
6	1.465	1.725	16	4.154	2.576
7	1.798	1.817	17	4.248	2.662
8	1.442	1.688	18	4.653	2.789
9	1.576	1.756	19	5.123	2.875
10	1.785	1.814	...	...	...

Also discuss the quality of the fit. What is the expected error in using the fit instead of the data?

Solution :

In this case $m = 3$ and there are three parameters a_1, a_2, a_3 to be determined such that the fit is given by $y^f(t) = a_1 + a_2 t + a_3 t^2$. The sums required in formulating the normal equations are easily obtained as $\sum_i t_i = 44.697$, $\sum_i t_i^2 = 148.896$, $\sum_i t_i^3 = 572.973$, $\sum_i t_i^4 = 2377.375$, $\sum_i y_i = 37.798$, $\sum_i t_i y_i = 102.928$, $\sum_i t_i^2 y_i = 368.051$ where i goes from 1 to 19. The normal equations are then given by

$$
\begin{pmatrix}
19 & 44.697 & 148.896 \\
44.697 & 148.896 & 572.973 \\
148.896 & 572.973 & 2377.375
\end{pmatrix}
\begin{pmatrix}
a_1 \\ a_2 \\ a_3
\end{pmatrix}
=
\begin{pmatrix}
37.978 \\ 102.928 \\ 368.051
\end{pmatrix}
$$

In the above and what follows numbers are shown truncated to three significant digits after decimals. However the calculations have been performed with available computer precision and finally rounded as described.

These equations may easily be solved (for example, by Gauss elimination) to get the parameter set as $a_1 = 1.337, a_2 = 0.226, a_3 = 0.017$. The fit and data are compared in the following table.

No.	t	y	y^f	No.	t	y	y^f
1	0.094	1.356	1.358	11	2.245	1.918	1.928
2	0.204	1.382	1.384	12	2.884	2.087	2.127
3	0.538	1.462	1.463	13	3.112	2.157	2.201
4	0.922	1.562	1.560	14	3.568	2.312	2.355
5	1.088	1.582	1.603	15	3.798	2.458	2.435
6	1.465	1.725	1.704	16	4.154	2.576	2.562
7	1.798	1.817	1.797	17	4.248	2.662	2.597
8	1.442	1.688	1.698	18	4.653	2.789	2.748
9	1.576	1.756	1.735	19	5.123	2.875	2.930
10	1.785	1.814	1.793	...	...	...	...

The goodness of fit may be checked by calculating the R^2_{adj}. In the present case $n = 19$ and $p = 3$. Calculation of the sums of squares may easily be made using a spreadsheet program. Leaving these to the reader, the calculated value of $R^2_{adj} = 0.995$. The fit is a very good representation of the data. Further we calculate the error of the fit as $\sigma_e = 0.034$.

Polynomial fit considered in Example 7.6 may be visualized as multi-linear fit if we set $t^2 = u$ (say) such that the fit is of the form $y^f = a_1 + a_2 t + a_3 u$.

Program 7.2: *Polynomial regression*

```
 1 function [coeff,ss,R2] = polyregression(x,y,m)
 2 %   Input    x: independent variable
 3 %            y: dependent variable
 4 %            m: degree of the fit
 5 %   Output: coeff: coefficients
 6 %   convert polynomial regression to multiple linear regression
 7         % [x x^2 ... x^m] = [x1 x2 ... xm]
 8 for i=1:m   % constructing multiple parameter matrix
 9    r(:,i) = x.^(i);
10 end
11 [coeff,ss,R2] = multilinearregression(r,y); % call function
                        %multiregression
```

7.5 Non-linear regression

By far the most involved case is that of non-linear regression. Minimization of sum of squares of the residuals requires the application of an optimization method considered earlier in Chapter 4.

The method of solving for the parameters representing a non-linear fit is perforce iterative. The normal equations are seldom explicitly written down. The iterative scheme actually looks at the minimization of the sum of squares of residuals, within a reasonable tolerance set by the user. Jacobian matrix of a least square method is always symmetric and hence methods such as Conjugate gradient can be applied. The sum of squares of residuals will always be non-zero positive quantity. An example is used to demonstrate the method useful for non-linear regression.

Example 7.7

The following tabulated data is expected to be well represented by the relation $y^f = a_1 e^{a_2 x} + a_3 x$. Determine the three fit parameters $a_1 - a_3$ by non-linear least squares.

x	0	0.2	0.4	0.6	0.8
y	1.196	1.379	1.581	1.79	2.013

x	1	1.2	1.4	1.6	1.8
y	2.279	2.545	2.842	3.173	3.5

Also discuss the quality of the fit. Estimate the error of fit.

Solution :

The sum of squares of residuals is given by

$$S = \sum_{i=1}^{10} \left[y_i - y_i^f \right]^2 = \sum_{i=1}^{10} \left[y_i - a_1 e^{a_2 x_i} - a_3 x_i \right]^2$$

In order to find the best values of the parameters we need to set the three partial derivatives of S to zero i.e.

$$\frac{\partial S}{\partial a_1} = -2 \sum_{i=1}^{10} \left[y_i - a_1 e^{a_2 x_i} - a_3 x_i \right] e^{a_2 x_i} = 0$$

$$\frac{\partial S}{\partial a_2} = -2 \sum_{i=1}^{10} \left[y_i - a_1 e^{a_2 x_i} - a_3 x_i \right] a_1 x_i e^{a_2 x_i} = 0$$

$$\frac{\partial S}{\partial a_3} = -2 \sum_{i=1}^{10} \left[y_i - a_1 e^{a_2 x_i} - a_3 x_i \right] x_i = 0$$

Since it is not feasible to solve this system directly we look at the corresponding optimization problem where we obtain the parameter set that minimizes the sum of squares of residuals. We start the iteration process by assuming an initial set of parameters given by

$$a_1^0 = 1, \ a_2^0 = 0.2, \ a_3^0 = 0.1$$

With these starting set of values the sum of squares of residuals and partial derivatives are calculated using a spreadsheet shown as Table 7.2.

The magnitude of the gradient vector $|g|$ and corresponding unit vector $\mathbf{u}$ along the gradient are obtained as

$$|g| = \sqrt{\sum_{i=1}^{3} \left(\frac{\partial S}{\partial a_i} \right)^2} = \sqrt{24.023^2 + 30.682^2 + 23.003^2} = 45.250968$$

and

$$\mathbf{u}^T = \left(-0.531 \ -0.678 \ -0.508 \right)$$

We now take a step in a direction opposite to $\mathbf{u}$. The step size is determined by quadratic search method. We start with $\alpha = 0.5$ along the descent direction. The minimization step

Table 7.2: Spreadsheet for Example 7.7

x	y	y^f	$(y - y^f)^2$	$\partial S/\partial a_1$	$\partial S/\partial a_2$	$\partial S/\partial a_3$
0	1.196	1.000	0.038	-0.392	0.000	0.000
0.2	1.379	1.061	0.101	-0.662	-0.132	-0.127
0.4	1.581	1.123	0.210	-0.992	-0.397	-0.366
0.6	1.790	1.187	0.363	-1.359	-0.815	-0.723
0.8	2.013	1.254	0.577	-1.783	-1.426	-1.215
1	2.279	1.321	0.917	-2.339	-2.339	-1.915
1.2	2.545	1.391	1.331	-2.933	-3.520	-2.769
1.4	2.842	1.463	1.901	-3.649	-5.108	-3.861
1.6	3.173	1.537	2.676	-4.506	-7.209	-5.235
1.8	3.500	1.613	3.560	-5.408	-9.735	-6.792
Column Sums			11.674	-24.023	-30.682	-23.003

size is then given by $\alpha = 0.4176$. The parameter set gets updated as $a_1^1 = 1.2217$, $a_2^1 = 0.483$, $a_3^1 = 0.312$. Correspondingly $S = 0.0059$ which is a significant reduction after just one descent step. The minimization process using steepest descent converges in 1533 iterations to yield the parameter set $a_1 = 1.194$, $a_2 = 0.512$, $a_3 = 0.282$ with $S = 0.0004$. However, the same results are obtained, using CG method in just 101 iterations. The fit is a good representation of the data as indicated by Figure 7.8. R_{adj}^2 is calculated as 0.9999 which fortifies the previous observation about the quality of the fit. Error of fit is estimated as $\sigma_e = 0.0076$.

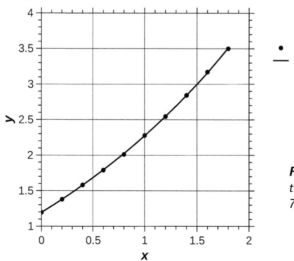

Figure 7.8: Comparison between data and fit in Example 7.7

Concluding remarks

In this chapter we have given a brief introduction to techniques used in data fit. Data fit is an essential part of any experimental research irrespective of the field of study. The goal of experimental research is to look for patterns in the data so that simple and useful relationships may be brought out. Such analysis also exposes any factors that may not have been considered in the experimental data.

7.A MATLAB routines related to Chapter 7

MATLAB provides a curve fitting toolbox, a graphical user interface, which can be used for performing linear and nonlinear regression, interpolation using splines, b-splines and smoothing functions. Here are some of the built in functions provided by MATLAB for regression

MATLAB routine	Function
cftool	opens curve fitting toolbox
fit	performs curve fit for a range of library functions provided by MATLAB
nlinfit	nonlinear regression
nlintool	graphical user interface for `nlinfit`
confint	confidence intervals for fit coefficients

7.B Suggested reading

1. **J.E. Freund** *Mathematical Statistics with Applications (7th Edition)* Prentice Hall, 2003

Chapter 8

Numerical Differentiation

In many engineering applications such as structural mechanics, heat transfer and fluid dynamics numerical methods are employed to solve the governing partial differential equations. The numerical data is available at discrete points in the computational domain. It is then necessary to use numerical differentiation to evaluate or estimate derivatives. Usually finite difference approximations are used to evaluate derivatives. Forward, backward and divided differences dealt with in section 5.3 are in fact related to approximations of derivatives of a given function. We look at numerical differentiation in greater detail in this chapter.

8.1   Introduction

Derivative is a mathematical operation indicating the rate of change of dependent variable f with respect to the independent variable x. Mathematically one can define the derivative at a point x_0 as

$$\frac{df}{dx}\bigg|_{x_0} = f^{(1)}(x_0) = \lim_{h \to 0} \frac{f(x_0+h)-f(x_0)}{h} \qquad (8.1)$$

First derivative is nothing but the slope of the tangent to the curve at the point where as the second derivative is related to the curvature. Recollect, in Chapter 4, we have discussed about Newton Raphson method where the root of the algebraic equation was approximated as the intercept of the tangent with the x axis. In secant method, a secant line joining two points are used instead of the tangent. Slope of secant line is an approximation of the first derivative as shown in Figure 8.1. As the step size h decreases and tends towards zero, the slope of the secant line becomes the slope of the tangent. Therefore, the estimate of the first derivative can be written as

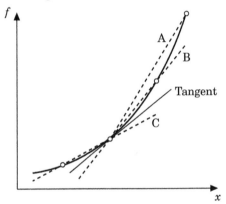

Figure 8.1: *Approximation of a tangent with secants A, B and C*

$$f^{(1)}(x_0) \approx \frac{f(x_0+h)-f(x_0)}{h} \qquad (8.2)$$

Numerical differentiation deals with estimating the derivatives using *function evaluations alone*. We have already dealt with some examples in Chapter 4 where Hessian and Jacobian matrices are evaluated numerically. Differential equations are used to model most physical phenomena and numerical differentiation is employed extensively in solving them. We introduce different aspects of numerical differentiation in the following sections.

8.2   Finite difference formulae using Taylor's series

Forward and backward difference

Consider a function $f(x)$ whose first derivative is required at a point x_0. A Taylor series representation of the function in the neighborhood of x_0 may be written as

$$f(x) = f(x_0) + (x-x_0)f^{(1)}(x_0) + O(x-x_0)^2 \qquad (8.3)$$

where the symbol O is the order symbol which means that the term adjoining the symbol is roughly the size of the term, within a multiplicative constant. If $(x-x_0)$ is sufficiently small we may *truncate* the right hand side by neglecting the third term onwards to get an estimate of the first derivative as

$$f^{(1)}(x_0) \approx \frac{f(x) - f(x_0)}{x - x_0} \tag{8.4}$$

It is seen at once that this is nothing but the first divided difference at x_0. This formula is referred to as the forward difference approximation of the first derivative at x_0. In normal practice $x - x_0$ is specified as the step size h so that the above formula becomes

$$f^{(1)}(x_0) \approx \frac{f(x_0 + h) - f(x_0)}{h} \tag{8.5}$$

It is also seen that the neglected term is of $O(h)$. The above estimate is exact for a linear equation and hence we say that the forward difference formula for the first derivative is first order accurate.

The reader will easily be able to realize that the above is also the backward difference approximation (recall the discussion in section 5.3) of the first derivative at $x_0 + h$. Hence the backward difference formula for first derivative can be written as

$$f^{(1)}(x_0) \approx \frac{f(x_0) - f(x_0 - h)}{h} \tag{8.6}$$

Central difference

First order accurate formulae are seldom used in practice. Higher order finite differences are possible using, for example, central differences that are second order accurate. We consider equi-spaced data first. We choose three consecutive equi-spaced points $x_i - h$, x_i and $x_i + h$ and using Taylor expansion write

$$
\begin{aligned}
f(x_i + h) &= f(x_i) + f^{(1)}(x_i)h + f^{(2)}(x_i)\frac{h^2}{2} + f^{(3)}(x_i)\frac{h^3}{6} + O(h^4) \cdots (a) \\
f(x_i - h) &= f(x_i) - f^{(1)}(x_i)h + f^{(2)}(x_i)\frac{h^2}{2} - f^{(3)}(x_i)\frac{h^3}{6} + O(h^4) \cdots (b)
\end{aligned} \tag{8.7}
$$

Subtracting Equation 8.7(b) from Equation 8.7(a) we get

$$f(x_i + h) - f(x_i - h) = 2f^{(1)}(x_i)h + O(h^3)$$

This may be rewritten in the form

$$f^{(1)}(x_i) = \frac{f(x_i + h) - f(x_i - h)}{2h} + O(h^2) \tag{8.8}$$

Thus we get the central difference approximation for the first derivative and it is second order accurate i.e. the derivative estimate is exact for a quadratic curve. This may be recast as

$$f^{(1)}(x_i) = \frac{\frac{f(x_i + h) - f(x_i)}{h} + \frac{f(x_i) - f(x_i - h)}{h}}{2} + O(h^2) \tag{8.9}$$

showing thereby that the central difference approximation for the first derivative is the average of the forward and backward difference approximations to the first derivative at the central node x_i (Figure 8.2).

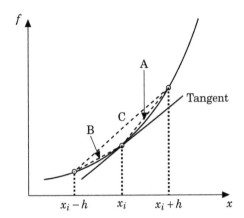

Figure 8.2: *Forward (A), backward (B) and central difference (C) approximations of first derivative*

Second derivative using central differences

Adding Equations 8.7(a) and (b) we get

$$f(x_i + h) + f(x_i - h) = 2f(x_i) + f^{(2)}(x_i)h^2 + O(h^4)$$

This may be rewritten in the form

$$f^{(2)}(x_i) = \frac{f(x_i + h) - 2f(x_i) + f(x_i - h)}{h^2} + O(h^2) \tag{8.10}$$

Thus we get the central difference approximation for the second derivative and it is second order accurate.

8.3 Differentiation of Lagrange and Newton polynomials

It is possible to obtain finite difference approximations to the derivatives by term by term differentiation of the approximating polynomials such as Lagrange or Newton polynomials. Additional advantage of this method is that finite difference formulae may easily be derived for arbitrarily spaced data.

8.3.1 Derivatives of Lagrange polynomials: arbitrarily spaced data

Lagrange polynomials may be differentiated term by term to obtain the derivatives. Usually our interest is limited to the first and second derivatives. Consider first the case with arbitrarily spaced data and second degree Lagrangian polynomial given by Equation 5.6. Term be term differentiation once with respect to x gives

$$
\begin{aligned}
f^{(1)}(x) &\approx \frac{dL_2}{dx} = \frac{dw_i}{dx}f_i + \frac{dw_{i+1}}{dx}f_{i+1} + \frac{dw_{i+2}}{dx}f_{i+2} \\
&= \frac{(2x - x_{i+1} - x_{i+2})f_i}{(x_i - x_{i+1})(x_i - x_{i+2})} + \frac{(2x - x_i - x_{i+2})f_{i+1}}{(x_{i+1} - x_i)(x_{i+1} - x_{i+2})} \\
&\qquad\qquad + \frac{(2x - x_i - x_{i+1})f_{i+2}}{(x_{i+2} - x_i)(x_{i+2} - x_{i+1})}
\end{aligned}
\tag{8.11}
$$

By substituting a value of x in the range $x_i \le x \le x_{i+2}$ we will be able to obtain the first derivative at any point in the interval. Specifically we put $x = x_i$ to get the first derivative at x_i as

$$f^{(1)}(x_i) \approx \left.\frac{dL_2}{dx}\right|_{x_i} = \frac{(2x_i - x_{i+1} - x_{i+2})f_i}{(x_i - x_{i+1})(x_i - x_{i+2})} + \frac{(x_i - x_{i+2})f_{i+1}}{(x_{i+1} - x_i)(x_{i+1} - x_{i+2})}$$
$$+ \frac{(x_i - x_{i+1})f_{i+2}}{(x_{i+2} - x_i)(x_{i+2} - x_{i+1})} \qquad (8.12)$$

Obviously this is second order accurate and is referred to as the *one sided (forward) three point formula* for the first derivative. Another differentiation of Equation 8.11 with respect to x yields

$$f^{(2)}(x_i) \approx \frac{d^2 L_2}{dx^2} = \frac{2f_i}{(x_i - x_{i+1})(x_i - x_{i+2})} + \frac{2f_{i+1}}{(x_{i+1} - x_i)(x_{i+1} - x_{i+2})}$$
$$+ \frac{2f_{i+2}}{(x_{i+2} - x_i)(x_{i+2} - x_{i+1})} \qquad (8.13)$$

The second derivative, as it should be, is seen to be independent of x.

Example 8.1

Function vs location data is given in the following table. Obtain the first derivative with respect to x at $x = 0$ and $x = 0.42$. Compare different estimates.

i	0	1	2	3	4	5	6	7
x_i	0	0.07	0.15	0.23	0.34	0.42	0.47	0.52
f_i	1	0.91681	0.83271	0.75928	0.67403	0.62239	0.59421	0.56900

Solution :

At $x = 0$ the first derivative may be evaluated using 1) forward difference or 2) one sided (forward) three point formula. The former is easily evaluated as

$$f^{(1)}(0) \approx \frac{0.91681 - 1.00000}{0.07 - 0} = -1.18843$$

The latter uses Equation 8.11 at $x = x_0 = 0$ and uses the data at $i = 0$, $i = 1$ and $i = 2$. Equation 8.11 is recast as

$$f^{(1)}(0) \approx \frac{(2x_0 - x_1 - x_2)f_0}{(x_0 - x_1)(x_0 - x_2)} + \frac{(x_0 - x_2)f_1}{(x_1 - x_0)(x_1 - x_2)} + \frac{(x_0 - x_1)f_2}{(x_2 - x_0)(x_2 - x_1)}$$

Introducing numbers, we get

$$f^{(1)}(0) \approx \frac{(0 - 0.07 - 0.15)1.00000}{(0 - 0.07)(0 - 0.15)} + \frac{(0 - 0.15)0.91681}{(0.07 - 0)(0.07 - 0.15)}$$
$$+ \frac{(0 - 0.07)0.83271}{(0.15 - 0)(0.15 - 0.07)} = -1.25245$$

This may be compared with the known exact value of the derivative of -1.25637.

At the point $x = 0.42$ we have several options for calculating the first derivative. They are: 1) Forward difference, 2) Backward difference, 3) One-sided three point - forward, 4)

One-sided three point - backward and 5) Centered three point formula. These calculations are given below.

1) Forward difference:

$$f^{(1)}(x = 0.42) \approx \frac{0.59421 - 0.62239}{0.47 - 0.42} = -0.56360$$

2) Backward difference:

$$f^{(1)}(x = 0.42) \approx \frac{0.62239 - 0.67403}{0.42 - 0.34} = -0.64550$$

3) One-sided three point - forward:
We use Equation 8.11 at $x = x_5 = 0.42$ and use the data at $i = 5$, $i = 6$ and $i = 7$. Equation 8.11 is recast as

$$f^{(1)}(0.42) \approx \frac{(2x_5 - x_6 - x_7)f_5}{(x_5 - x_6)(x_5 - x_7)} + \frac{(2x_5 - x_5 - x_7)f_6}{(x_6 - x_5)(x_6 - x_7)} + \frac{(2x_5 - x_5 - x_6)f_7}{(x_7 - x_5)(x_7 - x_6)}$$

Introducing numbers we get

$$
\begin{aligned}
f^{(1)}(0.42) \approx{} & \frac{(2 \times 0.42 - 0.47 - 0.52)0.62239}{(0.42 - 0.47)(0.42 - 0.52)} + \frac{(0.42 - 0.52)0.59421}{(0.47 - 0.42)(0.47 - 0.52)} \\
& + \frac{(0.42 - 0.47)0.56900}{(0.52 - 0.42)(0.52 - 0.47)} = -0.59330
\end{aligned}
$$

3) One-sided three point - backward:
We use Equation 8.11 at $x = x_5 = 0.42$ and use the data at $i = 5$, $i = 4$ and $i = 3$. The details are left to the reader. The resulting first derivative is equal to -0.59097.

4) Centered three point formula:
We use Equation 8.11 at $x = x_5 = 0.42$ and use the data at $i = 4$, $i = 5$ and $i = 6$. Equation 8.11 is recast as

$$
\begin{aligned}
f^{(1)}(0.42) \approx{} & \frac{(2x_5 - x_5 - x_6)f_4}{(x_4 - x_5)(x_4 - x_6)} + \frac{(2x_5 - x_4 - x_6)f_5}{(x_5 - x_4)(x_5 - x_6)} \\
& + \frac{(2x_5 - x_4 - x_5)f_6}{(x_6 - x_4)(x_6 - x_5)}
\end{aligned}
$$

Introducing numbers we get

$$
\begin{aligned}
f^{(1)}(0.42) \approx{} & \frac{(0.42 - 0.47)0.67403}{(0.34 - 0.42)(0.34 - 0.47)} + \frac{(2 \times 0.42 - 0.34 - 0.47)0.62239}{(0.42 - 0.34)(0.42 - 0.47)} \\
& + \frac{(0.42 - 0.34)0.59421}{(0.47 - 0.34)(0.47 - 0.42)} = -0.59510
\end{aligned}
$$

All of the above may be compared with the exact derivative of -0.59431.

8.3.2 Derivatives of $L_n(x)$

In general, derivatives of $L_n(x)$ may be obtained by term by term differentiation. The first derivative of L_n would be

$$\frac{dL_n}{dx} = \frac{dw_0}{dx}f_0 + \frac{dw_1}{dx}f_1 + \cdots + \frac{dw_n}{dx}f_n = \sum_{i=0}^{n} f_i \frac{dw_i}{dx} \tag{8.14}$$

The mth derivative of L_n ($m \le n$) is

$$\frac{d^m L_n}{dx^m} = \frac{d^m w_0}{dx^m}f_0 + \frac{d^m w_1}{dx^m}f_1 + \cdots + \frac{d^m w_n}{dx^m}f_n = \sum_{i=0}^{n} f_i \frac{d^m w_i}{dx^m} \tag{8.15}$$

The weight function w_i is of form

$$w_i(x) = a_{i0} + a_{i1}x + a_{i2}x^2 + \cdots + a_{in}x^n \tag{8.16}$$

Hence, the first derivative of the above would be

$$\frac{dw_i}{dx} = a_{i1} + 2a_{i2}x + 3a_{i3}x^2 + \cdots + na_{in}x^{n-1} \tag{8.17}$$

The second derivative would be

$$\frac{d^2 w_i}{dx^2} = 2a_{i2} + 6a_{i3}x + 12a_{i4}x^2 + \cdots + n(n-1)a_{in}x^{n-2} \tag{8.18}$$

The mth derivative of the function would be of form

$$\frac{d^m w_i}{dx^m} = \frac{m!}{0!}a_{im} + \frac{m+1!}{1!}a_{im+1}x + \cdots + \frac{r!}{r-m!}a_{ir}x^{r-m} + \cdots + \frac{n!}{n-m!}a_{in}x^{n-m} \tag{8.19}$$

MATLAB program has been given below to differentiate *any* n^{th} degree polynomial. The program can be used to differentiate the weight function polynomials also. In such a case, the input to the program is the coefficient matrix obtained from Program 5.2.

Program 8.1: *Differentiate polynomial functions*

```
1  function pder = polyderivative(p,m)
2  %  Input     p  : matrix of polynomial coefficients of weight
                       %function
3  %            m  : order of differentiation
4  %  Output   pder: matrix of polynomial coefficients of
                       %differentiated polynomial
5  n = size(p,2);          % degree of polynomial
6  pder = zeros(size(p));  % initialize derivative polynomials
7  for i= m+1:n            % evaluate coefficient of derivative
8      pder(:,i-m) = p(:,i); % polynomials
9      for j=1:m
10         pder(:,i-m) = (i-j)*pder(:,i-m);
11     end
12 end
```

8.3.3   Derivatives of Lagrange polynomials: equi-spaced data

Consider the second degree Lagrangian polynomial given by Equation 5.13, valid when the data is equi-spaced. We note that $dx = h\,dr$ and hence the first derivative is obtained by term by term differentiation as

$$
\begin{aligned}
f^{(1)}(x) = \frac{1}{h}\frac{df}{dr} \quad &\approx \quad \frac{1}{h}\frac{dL_2(r)}{dr} \\
&= \quad \frac{(r-1)+(r-2)}{2h}f_i - \frac{r+r-2}{h}f_{i+1} + \frac{r+(r-1)}{2h}f_{i+2} \\
&= \quad \frac{2r-3}{2h}f_i - \frac{2r-2}{h}f_{i+1} + \frac{2r-1}{2h}f_{i+2}
\end{aligned}
\tag{8.20}
$$

$x = x_i$ corresponds to $r = 0$ and the first derivative at x_i is given by putting $r = 0$ in Equation 8.20. Alternately we may substitute $x_{i+1} - x_i = x_{i+2} - x_{i+1} = h$, $x_{i+2} - x_i = 2h$ in Equation 8.12 to get the following.

$$
\begin{aligned}
f^{(1)}(x_i) \quad &\approx \quad \frac{1}{h}\frac{dL_2}{dr}\bigg|_{r=0} = \frac{-3}{2h}f_i - \frac{-2}{h}f_{i+1} + \frac{-1}{2h}f_{i+2} \\
&= \quad \frac{-3f_i + 4f_{i+1} - f_{i+2}}{2h}
\end{aligned}
\tag{8.21}
$$

This is one sided three point formula (forward) with equi-spaced data. We may obtain the first derivative at x_{i+1} by substituting $r = 1$ in Equation 8.20 to get

$$
\begin{aligned}
f^{(1)}(x_{i+1}) \quad &\approx \quad \frac{1}{h}\frac{dL_2(r)}{dr}\bigg|_{r=1} = \frac{-1}{2h}f_i - \frac{0}{h}f_{i+1} + \frac{1}{2h}f_{i+2} \\
&= \quad \frac{f_{i+2} - f_i}{2h}
\end{aligned}
\tag{8.22}
$$

This, as we have seen earlier, is the central difference approximation to first derivative. Lastly we may obtain the first derivative at x_{i+2} by putting $r = 2$ in Equation 8.20.

$$
\begin{aligned}
f^{(1)}(x_{i+2}) \quad &\approx \quad \frac{1}{h}\frac{dL_2}{dr}\bigg|_{r=2} = \frac{1}{2h}f_i - \frac{2}{h}f_{i+1} + \frac{3}{2h}f_{i+2} \\
&= \quad \frac{f_i - 4f_{i+1} + 3f_{i+2}}{2h}
\end{aligned}
\tag{8.23}
$$

This is one sided three point formula (backward). The second derivative will, of course, be independent of x since it is given by a second differentiation of Equation 8.20 as

$$
\begin{aligned}
f^{(2)}(x) = \frac{1}{h^2}\frac{d^2f}{dr^2} \approx \frac{1}{h^2}\frac{d^2L_2(r)}{dr^2} \quad &= \quad \frac{2}{2h^2}f_i - \frac{2}{h^2}f_{i+1} + \frac{2}{2h^2}f_{i+2} \\
&= \quad \frac{f_i - 2f_{i+1} + f_{i+2}}{h^2}
\end{aligned}
\tag{8.24}
$$

This is central difference formula for the second derivative. One may similarly differentiate Newton polynomials to obtain finite difference approximations to the derivatives of a function. This we shall demonstrate by obtaining higher order (>2) formulas by differentiating Newton polynomials. This demonstration is restricted to equi-spaced data.

8.3.4 Higher order formulae using Newton polynomials

Sometimes we require higher order formulae to evaluate numerically the first and higher order derivatives of a function. A fourth degree Newton polynomial provides appropriate formulae for doing this. We make use of five equi-spaced data points given at $i-2, i-1, i, i+1$ and $i+2$ and hence the derived formulae will be referred to as five point formulae. Setting $i = 2$, for convenience, these five points become $0, 1, 2, 3$ and 4. The highest degree Newton polynomial is written down using the Newton Gregory series as

$$f(r) \approx p_4(r) = f_0 + r\Delta f_0 + \frac{r(r-1)}{2!}\Delta^2 f_0 + \frac{r(r-1)(r-2)}{3!}\Delta^3 f_0$$
$$+ \frac{r(r-1)(r-2)(r-3)}{4!}\Delta^4 f_0 \qquad (8.25)$$

where $r = \frac{x - x_0}{h}$, $f(x) = f(x_0 + hr)$ and all forward differences are with respect to x_0. First derivative of the above polynomial with respect to x is written down by term by term differentiation as

$$f^{(1)}(x) \approx \frac{1}{h}\frac{dp_4}{dr} = \Delta f_0 + \frac{(2r-1)}{2!h}\Delta^2 f_0 + \frac{(3r^2 - 6r + 2)}{3!h}\Delta^3 f_0 + \frac{(4r^3 - 18r^2 + 22r - 6)}{4!h}\Delta^4 f_0 \qquad (8.26)$$

Setting $r = 0$ in the above, and making use of forward difference Table 5.3, we get the one sided five point forward difference rule.

$$f^{(1)}(x_0) = \frac{-25f_0 + 48f_1 - 36f_2 + 16f_3 - 3f_4}{12h} + O(h^4) \qquad (8.27)$$

Similarly setting $r = 2$, we get the five point mid point rule given by

$$f^{(1)}(x_2) = \frac{f_0 - 8f_1 + 8f_3 - f_4}{12h} + O(h^4) \qquad (8.28)$$

As indicated both formulae are fourth order accurate.

Second derivative of the polynomial with respect to x is written down by term by term differentiation as

$$f^{(2)}(x) \approx \frac{1}{h^2}\frac{d^2 p_4}{dr^2} = \frac{1}{h^2}\Delta^2 f_0 + \frac{(6r - 6)}{3!h^2}\Delta^3 f_0 + \frac{(12r^2 - 36r + 22)}{4!h}\Delta^4 f_0 \qquad (8.29)$$

Setting $r = 2$ the five point centered formula for second derivative is obtained as

$$f^{(2)}(x_2) = \frac{-f_0 + 16f_1 - 30f_2 + 16f_3 - f_4}{12h^2} + O(h^4) \qquad (8.30)$$

Example 8.2

Function vs location data is given in the following table with equal spacing of $h = 0.1$. Obtain the first derivative with respect to x at $x = 0$ and $x = 0.3$. Compare different estimates.

i	0	1	2	3	4
x_i	0.1	0.2	0.3	0.4	0.5
f_i	0.09983	0.19867	0.29552	0.38942	0.47943

Solution :

At the first point in the table we may use 1) forward difference, 2) one sided three point or 3) one sided five point rule.

1) Forward difference:

$$f^{(1)}(0.1) = \frac{0.19867 - 0.09983}{0.1} = 0.98840$$

2) One sided three point:

$$f^{(1)}(0.1) = \frac{-3 \times 0.09983 + 4 \times 0.19867 - 0.29552}{2 \times 0.1} = 0.99835$$

3) One sided five point:

$$f^{(1)}(0.1) = (-25 \times 0.09983 + 48 \times 0.19867 - 36 \times 0.29552$$
$$+ 16 \times 0.38942 - 3 \times 0.47943) \div (12 \times 0.1) = 0.99510$$

The best estimate is the one given by the use of the five point rule i.e. $f^{(1)}(0.1) = 0.99510$. The exact value of the derivative is known to be 0.99500 in this case.

The point $x = 0.3$ corresponds to the mid point in the table. The options for calculation of first derivative are: 1) Forward difference 2) Backward difference 3) three point forward 4) three point backward 5) central difference and 6) five point centered. These are worked out below.

1) Forward difference: $f^{(1)}(0.3) = \dfrac{0.38942 - 0.29552}{0.1} = 0.93900$

2) Backward difference: $f^{(1)}(0.3) = \dfrac{0.29552 - 0.19867}{0.1} = 0.96850$

3) Three point forward: $f^{(1)}(0.3) = \dfrac{-3 \times 0.29552 + 4 \times 0.38942 - 0.47943}{2 \times 0.1} = 0.95845$

4) Three point backward: $f^{(1)}(0.3) = \dfrac{3 \times 0.29552 - 4 \times 0.19867 - 0.09983}{2 \times 0.1} = 0.95855$ 5)

Central difference: $f^{(1)}(0.3) = \dfrac{0.38942 - 0.19867}{2 \times 0.1} = 0.95375$

6) Five point centered:

$$f^{(1)}(0.3) = \frac{0.09983 - 8 \times 0.19867 + 8 \times 0.38942 - 0.47943}{12 \times 0.1} = 0.95533$$

The last value is the closest to the exact value of 0.95534.

Example 8.3

The function $f(x) = x^x$ is to be differentiated at $x = 0.5$ numerically. Compare the numerical estimates with the exact derivative. Study the influence of step size on the numerically obtained derivative.

Solution :

Before looking at the numerical part we obtain the exact derivative by logarithmic differentiation. Taking logarithms of $f(x) = x^x$ we get $\ln(f) = x \ln x$. On differentiation this yields

$$\frac{d \ln(f)}{dx} = \frac{1}{f} \frac{df}{dx} = \ln x + 1$$

Thus we have

$$f^{(1)}(x) = f(\ln x + 1) = x^x(\ln x + 1)$$

We now consider various approximations to the first derivative with step size of $h = 0.05$. Results are tabulated below.

x	0.4	0.45	0.5	0.55	0.60
$f(x)$	0.69315	0.69815	0.70711	0.71978	0.73602
Backward Difference			0.17924		
Forward Difference			0.25345		
Central Difference			0.21635		
Three point backward			0.21885		
Three point forward			0.21776		
Exact			0.21698		

It is seen that the error is unacceptable in the case of both forward and backward differences. However the three estimates that use central, three point forward or three point backward all have error with respect to the exact of similar order. The absolute deviation is less than 0.002. We now change the step size to $h = 0.025$, redo the calculations and present the results below.

x	0.45	0.475	0.5	0.525	0.55
$f(x)$	0.69815	0.70215	0.70711	0.71299	0.71978
Backward Difference			0.19830		
Forward Difference			0.23534		
Central Difference			0.21682		
Three point backward			0.21736		
Three point forward			0.21723		
Exact			0.21698		

We see that the error has moved to the fourth place after the decimal point in the case of central difference and the three point formulae. The absolute maximum error is limited to 0.0004. Both forward and backward difference estimates are unacceptable. Figure 8.3 shows error of numerical differentiation formulae for different step sizes using double precision (16 digits after decimal place). The figure clearly asserts the importance of order of accuracy. The slope of the error vs step size plot indicates the order of accuracy of the approximation, in the linear portion. The forward difference of $O(h)$ has the least slope where as the central difference of $O(h^4)$ has the highest slope. Also, it must be noted that the error of the approximation reaches a minimum and then starts increasing with smaller step size. Theoretically, as the step size tends towards zero, the approximation should tend towards the exact solution. However, this is not the case due to round off errors i.e. limited numerical precision of the computer.

The following table indicates the role of machine precision on the error in numerical differentiation.

Scheme	Single precision		Double precision	
	h_{min}	ϵ	h_{min}	ϵ
Forward Difference $O(h)$	2.05E-04	1.81E-04	3.44E-09	4.72E-09
Central Difference $O(h^2)$	5.12E-04	6.69E-06	2.10E-06	4.71E-12
Central Difference $O(h^4)$	2.00E-02	9.91E-07	5.12E-04	1.19E-13

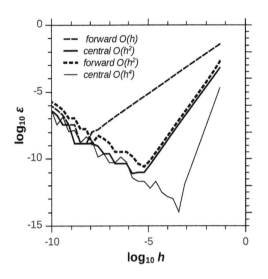

Figure 8.3: Error in derivatives calculated using different difference schemes and step sizes (double precision)

8.4 Numerical partial differentiation

8.4.1 First derivatives in a rectangular domain

Consider the rectangular domain in Figure 8.4. It is intended to calculate first derivatives at any point inside the domain. Linear interpolation over a rectangle has been considered in an earlier chapter where we had

$$f = w_1 f_1 + w_2 f_2 + w_3 f_3 + w_4 f_4 \tag{8.31}$$

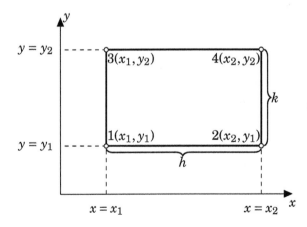

Figure 8.4: Rectangular domain

Hence, the partial first derivatives at a point (x_i, y_i) are given by

$$\frac{\partial f}{\partial x} = \frac{\partial w_1}{\partial x} f_1 + \frac{\partial w_2}{\partial x} f_2 + \frac{\partial w_3}{\partial x} f_3 + \frac{\partial w_4}{\partial x} f_4 \tag{8.32}$$

$$\frac{\partial f}{\partial y} = \frac{\partial w_1}{\partial y} f_1 + \frac{\partial w_2}{\partial y} f_2 + \frac{\partial w_3}{\partial y} f_3 + \frac{\partial w_4}{\partial y} f_4 \tag{8.33}$$

where the indicated partial derivatives of the weights are evaluated at (x_i, y_i). The weight functions being linear functions of x and y, the evaluation of the first derivatives are simple. As the weight functions are product of two functions -one of x and another of y (see Equation 6.2), the derivatives can also be written as

$$\frac{\partial w_1}{\partial x} = \frac{dw_{1x}}{dx} w_{1y} \qquad \frac{\partial w_1}{\partial y} = w_{1x} \frac{dw_{1y}}{dy}$$

$$\frac{\partial w_2}{\partial x} = \frac{dw_{2x}}{dx} w_{1y} \qquad \frac{\partial w_2}{\partial y} = w_{2x} \frac{dw_{1y}}{dy}$$

$$\frac{\partial w_3}{\partial x} = \frac{dw_{1x}}{dx} w_{2y} \qquad \frac{\partial w_3}{\partial y} = w_{1x} \frac{dw_{2y}}{dy} \qquad (8.34)$$

$$\frac{\partial w_4}{\partial x} = \frac{dw_{2x}}{dx} w_{2y} \qquad \frac{\partial w_4}{\partial y} = w_{2x} \frac{dw_{2y}}{dy}$$

The first partial derivatives of the weight functions are given below

$$\frac{\partial w_1}{\partial x} = \frac{(y_i - y_2)}{(x_1 - x_2)(y_1 - y_2)} \qquad \frac{\partial w_1}{\partial y} = \frac{(x_i - x_2)}{(x_1 - x_2)(y_1 - y_2)}$$

$$\frac{\partial w_2}{\partial x} = \frac{(y_i - y_2)}{(x_2 - x_1)(y_1 - y_2)} \qquad \frac{\partial w_2}{\partial y} = \frac{(x_i - x_1)}{(x_2 - x_1)(y_1 - y_2)}$$

$$\frac{\partial w_3}{\partial x} = \frac{(y_i - y_1)}{(x_1 - x_2)(y_2 - y_1)} \qquad \frac{\partial w_3}{\partial y} = \frac{(x_i - x_2)}{(x_1 - x_2)(y_2 - y_1)}$$

$$\frac{\partial w_4}{\partial x} = \frac{(y_i - y_1)}{(x_2 - x_1)(y_2 - y_1)} \qquad \frac{\partial w_4}{\partial y} = \frac{(x_i - x_1)}{(x_2 - x_1)(y_2 - y_1)} \qquad (8.35)$$

Differentiation formulas using local coordinates

The first derivatives can be determined from local coordinate representation also. Consider the local coordinate of a linear rectangular element shown in Figure 6.5. Following previous section, the partial derivatives in the local coordinates would be

$$\frac{\partial w_1}{\partial \xi} = -\frac{1-\eta}{4} \qquad \frac{\partial w_1}{\partial \eta} = -\frac{1-\xi}{4}$$

$$\frac{\partial w_2}{\partial \xi} = \frac{1-\eta}{4} \qquad \frac{\partial w_2}{\partial \eta} = -\frac{1+\xi}{4}$$

$$\frac{\partial w_3}{\partial \xi} = -\frac{1+\eta}{4} \qquad \frac{\partial w_3}{\partial \eta} = \frac{1-\xi}{4} \qquad (8.36)$$

$$\frac{\partial w_4}{\partial \xi} = \frac{1+\eta}{4} \qquad \frac{\partial w_4}{\partial \eta} = \frac{1+\xi}{4}$$

The partial derivatives in local coordinates can be transformed into x, y coordinates. Using chain rule, the partial derivatives in x and y directions would be

$$\frac{\partial w_i}{\partial x} = \frac{\partial w_i}{\partial \xi} \frac{\partial \xi}{\partial x} + \frac{\partial w_i}{\partial \eta} \frac{\partial \eta}{\partial x} \qquad (8.37)$$

$$\frac{\partial w_i}{\partial y} = \frac{\partial w_i}{\partial \xi} \frac{\partial \xi}{\partial y} + \frac{\partial w_i}{\partial \eta} \frac{\partial \eta}{\partial y} \qquad (8.38)$$

where

$$\frac{d\xi}{dx} = \frac{2}{x_2 - x_1} = \frac{2}{h}; \quad \frac{d\xi}{dy} = 0; \quad \frac{d\eta}{dx} = 0; \quad \frac{d\xi}{dy} = \frac{2}{y_2 - y_1} = \frac{2}{k}; \tag{8.39}$$

The partial derivatives in original coordinates can be written as

$$\frac{\partial w_i}{\partial x} = \frac{\partial w_i}{\partial \xi}\frac{d\xi}{dx} = \frac{2}{h}\frac{\partial w_i}{\partial \xi}; \quad \frac{\partial w_i}{\partial y} = \frac{\partial w_i}{\partial \eta}\frac{d\eta}{dy} = \frac{2}{k}\frac{\partial w_i}{\partial \eta} \tag{8.40}$$

Evaluating the first derivatives at the nodes, we have

$$\text{Nodes 1 and 2:} \quad \frac{\partial f}{\partial x} = \frac{f_2 - f_1}{h} \quad \text{Nodes 3 and 4:} \quad \frac{\partial f}{\partial x} = \frac{f_4 - f_3}{h}$$

$$\text{Nodes 1 and 3:} \quad \frac{\partial f}{\partial y} = \frac{f_3 - f_1}{k} \quad \text{Nodes 2 and 4:} \quad \frac{\partial f}{\partial y} = \frac{f_4 - f_2}{k} \tag{8.41}$$

It is not surprising to see that the partial derivatives at the nodes are the first order forward and backward difference formulae over a line.

Second order first derivative formulae

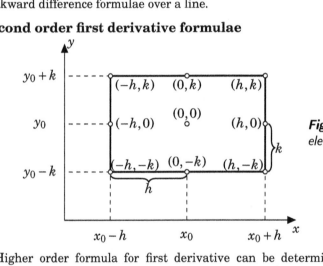

Figure 8.5: *Quadratic rectangular element*

Higher order formula for first derivative can be determined by differentiating the weight functions of rectangular quadratic element. Therefore, we can write the first derivatives as

$$\frac{\partial f}{\partial \xi} = \sum_{i=1}^{9} \frac{\partial w_i}{\partial \xi} f_i; \quad \frac{\partial f}{\partial \eta} = \sum_{i=1}^{9} \frac{\partial w_i}{\partial \eta} f_i \tag{8.42}$$

The weight functions for 9 noded quadratic element is a product of two quadratic functions along ξ and η as given below

$$w_i = \mathbf{G}(\eta)\mathbf{F}(\xi)^T = \begin{pmatrix} \dfrac{\eta(\eta-1)}{2} \\ (1-\eta^2) \\ \dfrac{\eta(\eta+1)}{2} \end{pmatrix} \begin{pmatrix} \dfrac{\xi(\xi-1)}{2} & (1-\xi^2) & \dfrac{\xi(\xi+1)}{2} \end{pmatrix} \tag{8.43}$$

and the derivatives are as given below

$$\frac{\partial w_i}{\partial \xi} = G(\eta)\frac{dF(\xi)}{d\xi} = \begin{pmatrix} \dfrac{\eta(\eta-1)}{2} \\ (1-\eta^2) \\ \dfrac{\eta(\eta+1)}{2} \end{pmatrix} \begin{pmatrix} \dfrac{2\xi-1}{2} & -2\xi & \dfrac{2\xi+1}{2} \end{pmatrix} \tag{8.44}$$

$$\frac{\partial w_i}{\partial \eta} = \frac{dG(\eta)}{d\eta}F(\xi) = \begin{pmatrix} \dfrac{2\eta-1}{2} \\ -2\eta \\ \dfrac{2\eta+1}{2} \end{pmatrix} \begin{pmatrix} \dfrac{\xi(\xi-1)}{2} & (1-\xi^2) & \dfrac{\xi(\xi+1)}{2} \end{pmatrix} \qquad (8.45)$$

The derivatives in the original coordinates can be determined using chain rule. For a rectangular domain as in Figure 8.5 the derivatives would be

$$\frac{d\xi}{dx} = \frac{1}{h}; \quad \frac{d\xi}{dy} = 0; \quad \frac{d\eta}{dx} = 0; \quad \frac{d\xi}{dy} = \frac{1}{k}; \qquad (8.46)$$

Therefore, the second order accurate first derivatives are as given below

$$\frac{\partial w_i}{\partial x} = \frac{1}{h}\frac{dF(\xi)}{d\xi}G(\eta) \qquad \frac{\partial w_i}{\partial y} = \frac{1}{k}F(\xi)\frac{dG(\eta)}{d\eta} \qquad (8.47)$$

The first derivatives at the nodes would be

$$\frac{\partial f}{\partial x} = \frac{-3f(-h,j)+4f(0,j)-f(h,j)}{2h} \quad j=-k,0,k$$

$$\frac{\partial f}{\partial x} = \frac{f(h,j)-f(-h,j)}{2h} \quad j=-k,0,k$$

$$\frac{\partial f}{\partial x} = \frac{f(-h,j)-4f(0,j)+3f(h,j)}{2h} \quad j=-k,0,k$$

$$\frac{\partial f}{\partial y} = \frac{-3f(i,-k)+4f(i,0)-f(i,k)}{2k} \quad i=-h,0,h$$

$$\frac{\partial f}{\partial y} = \frac{f(i,k)-f(i,-k)}{2k} \quad i=-h,0,h$$

$$\frac{\partial f}{\partial y} = \frac{f(i,-k)-f(i,0)+3f(i,k)}{2k} \quad i=-h,0,h \qquad (8.48)$$

which are nothing but the second order forward, backward and central difference formulae applied over a line.

8.4.2 Derivatives for an arbitrary quadrilateral

In the last section, we derived first derivative formulas for a rectangular domain. For an arbitrary quadrilateral, the domain is transformed to local coordinates. Using chain rule, the partial derivatives in x and y directions would be

$$\frac{\partial w_i}{\partial x} = \frac{\partial w_i}{\partial \xi}\frac{\partial \xi}{\partial x} + \frac{\partial w_i}{\partial \eta}\frac{\partial \eta}{\partial x} \qquad (8.49)$$

$$\frac{\partial w_i}{\partial y} = \frac{\partial w_i}{\partial \xi}\frac{\partial \xi}{\partial y} + \frac{\partial w_i}{\partial \eta}\frac{\partial \eta}{\partial y} \qquad (8.50)$$

Writing the above equations in the matrix form

$$\begin{pmatrix} \dfrac{\partial w_i}{\partial x} \\ \dfrac{\partial w_i}{\partial y} \end{pmatrix} = \underbrace{\begin{bmatrix} \dfrac{\partial \xi}{\partial x} & \dfrac{\partial \eta}{\partial x} \\ \dfrac{\partial \xi}{\partial y} & \dfrac{\partial \eta}{\partial y} \end{bmatrix}}_{\mathbf{J}} \begin{pmatrix} \dfrac{\partial w_i}{\partial \xi} \\ \dfrac{\partial w_i}{\partial \eta} \end{pmatrix} \qquad (8.51)$$

where $\mathbf{J}$ is the Jacobian matrix of ξ, η with respect x, y. For a rectangular domain, the Jacobian matrix is

$$\mathbf{J} = \begin{pmatrix} \dfrac{2}{(x_2 - x_1)} & 0 \\ 0 & \dfrac{2}{(y_2 - y_1)} \end{pmatrix} \tag{8.52}$$

However, for arbitrary quadrilaterals it would be difficult to determine the Jacobian matrix. As x and y can be represented as functions of ξ and η, it is possible to determine the inverse of $\mathbf{J}$ as

$$\mathbf{J}^{-1} = \begin{pmatrix} \dfrac{\partial x}{\partial \xi} & \dfrac{\partial x}{\partial \eta} \\ \dfrac{\partial y}{\partial \xi} & \dfrac{\partial y}{d \eta} \end{pmatrix}$$

or

$$\mathbf{J} = \begin{pmatrix} \dfrac{\partial \xi}{\partial x} & \dfrac{\partial \eta}{\partial x} \\ \dfrac{\partial \xi}{\partial y} & \dfrac{\partial \eta}{\partial y} \end{pmatrix} = \dfrac{1}{\frac{\partial x}{\partial \xi}\frac{\partial y}{d \eta} - \frac{\partial x}{\partial \eta}\frac{\partial y}{\partial \xi}} \begin{pmatrix} \dfrac{\partial y}{\partial \eta} & -\dfrac{\partial y}{\partial \xi} \\ -\dfrac{\partial x}{\partial \eta} & \dfrac{\partial x}{d \xi} \end{pmatrix} \tag{8.53}$$

Second partial derivatives

A similar approach as that for first derivatives is to be followed for higher derivatives. We need to determine the second partial derivative of form $\dfrac{\partial^2}{\partial x_2 \partial x_1}$. x_1 and x_2 can be either x or y. First let us write down the expression for $\dfrac{\partial}{\partial x_1}$.

$$\frac{\partial w}{\partial x_1} = \frac{\partial w}{\partial \xi}\frac{\partial \xi}{\partial x_1} + \frac{\partial w}{\partial \eta}\frac{\partial \eta}{\partial x_1} \tag{8.54}$$

Differentiating the above equation with respect to x_2 we obtain

$$\frac{\partial^2 w}{\partial x_1 \partial x_2} = \frac{\partial^2 w}{\partial \xi \partial x_2}\frac{\partial \xi}{\partial x_1} + \frac{\partial w}{\partial \xi}\frac{\partial^2 \xi}{\partial x_1 \partial x_2} + \frac{\partial^2 w}{\partial \eta \partial x_2}\frac{\partial \eta}{\partial x_1} + \frac{\partial w}{\partial \eta}\frac{\partial^2 \eta}{\partial x_1 \partial x_2} \tag{8.55}$$

Let us consider the first term on the right hand side alone. Applying chain rule to the term we get

$$\frac{\partial^2 w}{\partial \xi \partial x_2} = \frac{\partial^2 w}{\partial \xi^2}\frac{\partial \xi}{\partial x_2} + \frac{\partial^2 w}{\partial \xi \partial \eta}\frac{\partial \eta}{\partial x_2} \tag{8.56}$$

Therefore the final expression for the second partial derivative would be

$$\begin{aligned}\frac{\partial^2 w}{\partial x_1 \partial x_2} &= \frac{\partial^2 w}{\partial \xi^2}\frac{\partial \xi}{\partial x_1}\frac{\partial \xi}{\partial x_2} + \frac{\partial^2 w}{\partial \xi \partial \eta}\left(\frac{\partial \xi}{\partial x_1}\frac{\partial \eta}{\partial x_2} + \frac{\partial \xi}{\partial x_2}\frac{\partial \eta}{\partial x_1}\right) \\ &+ \frac{\partial w}{\partial \xi}\frac{\partial^2 \xi}{\partial x_1 \partial x_2} + \frac{\partial w}{\partial \eta}\frac{\partial^2 \eta}{\partial x_1 \partial x_2} + \frac{\partial^2 w}{\partial \eta^2}\frac{\partial \eta}{\partial x_1}\frac{\partial \eta}{\partial x_2}\end{aligned} \tag{8.57}$$

Choosing x_1 and x_2 suitably we get the required three second partial derivatives. The mixed derivative is the same as Equation 8.57 with $x_1 = x$ and $x_2 = y$ or vice versa. The other two derivatives would be given by

$$\frac{\partial^2 w}{\partial x^2} = \frac{\partial^2 w}{\partial \xi^2}\left(\frac{\partial \xi}{\partial x}\right)^2 + 2\frac{\partial^2 w}{\partial \xi \partial \eta}\frac{\partial \xi}{\partial x}\frac{\partial \eta}{\partial x} + \frac{\partial w}{\partial \xi}\frac{\partial^2 \xi}{\partial x^2} + \frac{\partial w}{\partial \eta}\frac{\partial^2 \eta}{\partial x^2} + \frac{\partial^2 w}{\partial \eta^2}\left(\frac{\partial \eta}{\partial x}\right)^2 \tag{8.58}$$

$$\frac{\partial^2 w}{\partial y^2} = \frac{\partial^2 w}{\partial \xi^2}\left(\frac{\partial \xi}{\partial y}\right)^2 + 2\frac{\partial^2 w}{\partial \xi \partial \eta}\frac{\partial \xi}{\partial y}\frac{\partial \eta}{\partial y} + \frac{\partial w}{\partial \xi}\frac{\partial^2 \xi}{\partial y^2} + \frac{\partial w}{\partial \eta}\frac{\partial^2 \eta}{\partial y^2} + \frac{\partial^2 w}{\partial \eta^2}\left(\frac{\partial \eta}{\partial y}\right)^2 \tag{8.59}$$

8.4.3   Second derivative formulas for a rectangle

The second derivative formula can be derived by differentiating the interpolation formula over a nine noded rectangle which would be

$$\frac{\partial^2 f}{\partial \xi^2} = \sum_{i=1}^{9} \frac{\partial^2 w_i}{\partial \xi^2} f_i; \quad \frac{\partial^2 f}{\partial \eta^2} = \sum_{i=1}^{9} \frac{\partial^2 w_i}{\partial \eta^2} f_i \tag{8.60}$$

Hence the second derivative formulas would be

$$\frac{\partial^2 w_i}{\partial \xi^2} = G(\eta)\frac{d^2 F(\xi)}{d\xi^2} = \begin{pmatrix} \dfrac{\eta(\eta-1)}{2} \\[6pt] (1-\eta^2) \\[6pt] \dfrac{\eta(\eta+1)}{2} \end{pmatrix} \begin{pmatrix} 1 & -2 & 1 \end{pmatrix} \tag{8.61}$$

$$\frac{\partial^2 w_i}{\partial \xi \partial \eta} = \frac{dG(\eta)}{d\eta}\frac{\partial F(\xi)}{d\xi} = \begin{pmatrix} \dfrac{2\eta-1}{2} \\[6pt] -2\eta \\[6pt] \dfrac{2\eta+1}{2} \end{pmatrix} \begin{pmatrix} \dfrac{2\xi-1}{2} & -2\xi & \dfrac{2\xi+1}{2} \end{pmatrix} \tag{8.62}$$

$$\frac{\partial^2 w_i}{\partial \eta^2} = \frac{d^2 G(\eta)}{d\eta^2} F(\xi) = \begin{pmatrix} 1 \\[6pt] -2 \\[6pt] 1 \end{pmatrix} \begin{pmatrix} \dfrac{\xi(\xi-1)}{2} & (1-\xi^2) & \dfrac{\xi(\xi+1)}{2} \end{pmatrix} \tag{8.63}$$

For a rectangular domain, the transformation is linear and hence all higher order derivatives of the transformation are zero i.e.

$$\frac{\partial^2 x}{\partial \xi^2} = 0; \quad \frac{\partial^2 y}{\partial \eta^2} = 0 \tag{8.64}$$

Therefore, the second derivatives in original coordinates can be written as

$$\frac{\partial^2 w_i}{\partial x^2} = \frac{\partial^2 w_i}{\partial \xi^2}\left(\frac{dx}{d\xi}\right)^2 = \frac{\partial^2 w_i}{\partial \xi^2}\frac{1}{h^2};$$

$$\frac{\partial^2 w_i}{\partial x \partial y} = \frac{\partial^2 w_i}{\partial \xi \partial \eta}\frac{dx}{d\xi}\frac{dy}{d\eta} = \frac{\partial^2 w_i}{\partial \xi \partial \eta}\frac{1}{hk}$$

$$\frac{\partial^2 w_i}{\partial y^2} = \frac{\partial^2 w_i}{\partial \eta^2}\left(\frac{d\eta}{dy}\right)^2 = \frac{\partial^2 w_i}{\partial \eta^2}\frac{1}{k^2}$$

Evaluating the second derivatives at the nodes we get

$$\frac{\partial^2 f}{\partial x^2} = \frac{f(-h,j)-2f(0,j)+f(h,j)}{h^2} \quad j = -k,0,k$$

$$\frac{\partial^2 f}{\partial y^2} = \frac{f(i,-k)-2f(i,0)+f(i,k)}{k^2} \quad i = -h,0,h \tag{8.65}$$

which are the same as the second order approximate second derivatives over lines.

Example 8.4

A certain function is available as a table of data specified at 9 points on the standard square domain of side 2.

Node i	ξ_i	η_i	$f_i = f(\xi_i, \eta_i)$	Node i	ξ_i	η_i	$f_i = f(\xi_i, \eta_i)$
1	-1	-1	1.221403	6	1	0	1.105171
2	0	-1	1.105171	7	-1	1	1.221403
3	1	-1	1.221403	8	0	1	1.105171
4	-1	0	1.105171	9	1	1	1.221403
5	0	0	1.000000				

Obtain the function, first and second derivatives at $\xi = 0.4$, $\eta = -0.3$.

Solution :

Point at which the function and its derivatives are required is given as $\xi = 0.4$, $\eta = -0.3$. The weights in this example are the same as those that were calculated in Example 6.3. Using the weights and the function values given in this example, the function value at $\xi = 0.4$, $\eta = -0.3$ is calculated using a spreadsheet, as shown below.

Node i	w_i	$w_i \times f_i$	Node i	w_i	$w_i \times f_i$
1	-0.0234	-0.028581	6	0.2548	0.281598
2	0.1638	0.181027	7	0.0126	0.015390
3	0.0546	0.066689	8	-0.0882	-0.097476
4	-0.1092	-0.120685	9	-0.0294	-0.035909
5	0.7644	0.764400	Sum =	1	**1.026452**

The **bold** entry in the above table is the desired function value. First derivatives $\dfrac{\partial f}{\partial \xi}$ and $\dfrac{\partial f}{\partial \eta}$ of the function may be obtained as the sum of product of partial derivatives of the respective weights evaluated at $\xi = 0.4$, $\eta = -0.3$ and the function values at the nodes. The weights have been given in function form in Example 6.3. Partial derivatives of these have been used in arriving at the following spreadsheet shown as Table 8.1.

Table 8.1: *First partial derivatives with data in Example 8.4*

Node i	$\dfrac{\partial w_i}{\partial \xi}$	$\dfrac{\partial w_i}{\partial \xi} \times f_i$	$\dfrac{\partial w_i}{\partial \eta}$	$\dfrac{\partial w_i}{\partial \eta} \times f_i$
1	-0.0195	-0.023817	0.096	0.117255
2	-0.156	-0.172407	-0.672	-0.742675
3	0.1755	0.214356	-0.224	-0.273594
4	-0.091	-0.100571	-0.072	-0.079572
5	-0.728	-0.728000	0.504	0.504000
6	0.819	0.905135	0.168	0.185669
7	0.0105	0.012825	-0.024	-0.029314
8	0.084	0.092834	0.168	0.185669
9	-0.0945	-0.115423	0.056	0.068399
Sum =		**0.084933**		**-0.064164**

Bold entries in Table 8.1 give the desired derivatives as $\dfrac{\partial f}{\partial \xi} = 0.084933$ and $\dfrac{\partial f}{\partial \eta} = -0.064164$ at $\xi = 0.4$, $\eta = -0.3$.

Similarly we may obtain second derivatives (there are three of them) as presented in the spreadsheet Table 8.2.

Table 8.2: Second partial derivatives with data in Example 8.4

Node i	$\dfrac{\partial^2 w_i}{\partial \xi^2}$	$\dfrac{\partial^2 w_i}{\partial \xi^2} \times f_i$	$\dfrac{\partial^2 w_i}{\partial \xi \partial \eta}$	$\dfrac{\partial^2 w_i}{\partial \xi \partial \eta} \times f_i$	$\dfrac{\partial^2 w_i}{\partial \eta^2}$	$\dfrac{\partial^2 w_i}{\partial \eta^2} \times f_i$
1	0.195	0.238174	0.08	0.097712	-0.12	-0.146568
2	-0.39	-0.431017	0.64	0.707309	0.84	0.928344
3	0.195	0.238174	-0.72	-0.879410	0.28	0.341993
4	0.91	1.005706	-0.06	-0.066310	0.24	0.265241
5	-1.82	-1.820000	-0.48	-0.480000	-1.68	-1.680000
6	0.91	1.005706	0.54	0.596792	-0.56	-0.618896
7	-0.105	-0.128247	-0.02	-0.024428	-0.12	-0.146568
8	0.21	0.232086	-0.16	-0.176827	0.84	0.928344
9	-0.105	-0.128247	0.18	0.219852	0.28	0.341993
Sum =		**0.212333**		**-0.005309**		**0.213881**

Bold entries in Table 8.2 give the second derivatives as $\dfrac{\partial^2 f}{\partial \xi^2} = 0.212333$, $\dfrac{\partial^2 f}{\partial \xi \partial \eta} = -0.005309$ and $\dfrac{\partial^2 f}{\partial \eta^2} = 0.213881$.

The function used in this example was $f(\xi, \eta) = e^{\frac{(\xi^2 + \eta^2)}{10}}$. The exact values and the numerically obtained values evaluated above are compared in the following table.

	Numerical	Exact	Difference
f	1.026452	1.025315	0.001137
$\dfrac{\partial f}{\partial \xi}$	0.084933	0.082025	0.002908
$\dfrac{\partial f}{\partial \eta}$	-0.064164	-0.061519	-0.002645
$\dfrac{\partial^2 f}{\partial \xi^2}$	0.212333	0.211625	0.000708
$\dfrac{\partial^2 f}{\partial \xi \partial \eta}$	-0.005309	-0.004922	-0.000388
$\dfrac{\partial^2 f}{\partial \eta^2}$	0.213881	0.208754	0.005127

8.4.4   Linear triangular domain

The expressions for the first derivatives can be written as

$$\frac{\partial f}{\partial x} = \frac{\partial w_1}{\partial x} f_1 + \frac{\partial w_2}{\partial x} f_2 + \frac{\partial w_3}{\partial x} f_3$$

$$\frac{\partial f}{\partial y} = \frac{\partial w_1}{\partial y} f_1 + \frac{\partial w_2}{\partial y} f_2 + \frac{\partial w_3}{\partial y} f_3 \qquad (8.66)$$

The weight functions are of the form $w_i(x,y) = a_i + b_i x + c_i y$ where a_i, b_i and c_i are constants defined in Equation 6.19. It may be inferred that the partial derivatives of the weights are $\dfrac{\partial w_i}{\partial x} = b_i$ and $\dfrac{\partial w_i}{\partial y} = c_i$. The derivatives may be represented in terms of local coordinates also which is left as an exercise to the reader.

Example 8.5

Function value is available at three points as shown in the table below.

Point No.	x	y	$f(x,y)$
1	0	0	1
2	0.1	0	1.007975
3	0	0.1	1.012699

Obtain the first derivatives of the function at $x = 0.03$, $y = 0.04$.

Solution :

The area of the triangle formed by the three points is determined first.

$$A = \frac{1}{2} \begin{vmatrix} 1 & 0 & 0 \\ 1 & 0.1 & 0 \\ 1 & 0 & 0.1 \end{vmatrix} = 0.005$$

Partial derivatives of weights are calculated next.

$$\frac{\partial w_1}{\partial x} = \frac{(y_2 - y_3)}{2A} = \frac{0 - 0.1}{2 \times 0.005} = -10 \; ; \frac{\partial w_1}{\partial y} = \frac{(x_3 - x_2)}{2A} = \frac{0.1 - 0}{2 \times 0.005} = 10$$

$$\frac{\partial w_2}{\partial x} = \frac{(y_3 - y_1)}{2A} = \frac{0.1 - 0}{2 \times 0.005} = 10 \; ; \frac{\partial w_2}{\partial y} = \frac{(x_1 - x_3)}{2A} = \frac{0 - 0}{2 \times 0.005} = 0$$

$$\frac{\partial w_3}{\partial x} = \frac{(y_1 - y_2)}{2A} = \frac{0 - 0}{2 \times 0.005} = 0 \; ; \frac{\partial w_3}{\partial y} = \frac{(x_1 - x_2)}{2A} = \frac{0.1 - 0}{2 \times 0.005} = 10$$

We then have

$$\left.\frac{\partial f}{\partial x}\right|_{(x=0.03,\, y=0.04)} = -10 \times 1 + 10 \times 1.007975 + 0 \times 1.012699 = 0.079749$$

$$\left.\frac{\partial f}{\partial y}\right|_{(x=0.03,\, y=0.04)} = -10 \times 1 + 0 \times 1.007975 + 10 \times 1.012699 = 0.126991$$

Concluding remarks

> *Numerical differentiation is at the back of all finite difference methods used in the solution of ordinary and partial differential equations. The material of the present chapter will be very useful later on in Modules III and IV where we will be considering numerical solution of both ODE and PDE.*

8.A   MATLAB routines related to Chapter 8

MATLAB routine	Function
diff(u)	difference between adjacent elements of array u
diff(u,n)	n^{th} order difference for array u
gradient(u)	calculates first derivatives along x and y directions
del2(u)	Finite difference approximation of Laplacian in two dimensions $\left(\dfrac{\partial^2 u}{\partial x^2} + \dfrac{\partial^2 u}{\partial y^2} \right)$

Chapter 9

Numerical Integration

Integration of a function is best performed, when possible, by analytical methods to get the integral in closed form. However, many a time it is difficult or impossible to obtain result in closed form. In such a case one has to obtain an integral numerically. There are many methods available, of which the following will be dealt with here:

- *Trapezoidal rule*
- *Simpson's rule*
- *Romberg method*
- *Gauss quadrature*

Apart from line integrals we shall also look at numerical evaluation of double integrals.

9.1   Introduction

Integration is one of the most important mathematical operations used by scientists and engineers. Integration over a line is defined as

$$I = \int_a^b f(x)dx \qquad (9.1)$$

The operation simply indicates the area under the curve $f(x)$ and the x axis between $x = a$ and $x = b$. For example, the area under a Gaussian distribution represents the cumulative probability of occurrence of an event between the two limits.

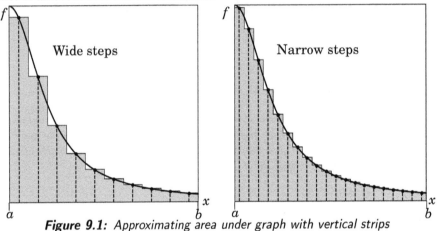

Figure 9.1: *Approximating area under graph with vertical strips*

The integrals have to be obtained numerically when analytical expressions are not available for them. Consider the area under the curve to be approximated by a large number of vertical strips as indicated in Figure 9.1 (Figure shows "wide" strips at the left and "narrow" strips at the right). The sum total of the area under these strips would be close to the actual area under the curve. The area of each vertical strip is the product of function value at the middle of the strip and the width of the strip. As the number of strips increases, the sum total area of these vertical strips would tend towards the exact area. In the limit the number of strips tend to infinity, the sum total area of the strips would be equal to the exact value. However, it would be practically impossible to achieve this.

In general the integral can be approximated as

$$\int_a^b f(x)dx = \sum_i^n w_i f_i + \epsilon = w_0 f_0 + w_1 f_1 + \cdots + w_n f_n + \epsilon \qquad (9.2)$$

where f_i is the function value at a node (points at which function is evaluated) and w_i is the weightage associated with the node, ϵ is the error associated with the approximation method. The nodes are points which lie in the range $a \le x \le b$. The above representation of the integral is known as **quadrature** formula. The aim of any numerical integration procedure is to estimate the area within satisfactory limits and with least number of vertical strips or function evaluations. The following sections will introduce the reader to different strategies to evaluate the integrals accurately.

9.2 Trapezoidal rule

This is the simplest numerical integration scheme and yields exact result for a function that is either a constant are a linear function of x. It may not be bad if the range of integration is divided into small segments over each of which the function may be approximated by a linear function. Let us assume that each segment is of uniform length h such that

$$h = \frac{b-a}{n} \tag{9.3}$$

where the range of integration is covered by n steps of length h each. The integral over the step x_i, x_{i+1} may be approximated as the area of a trapezium with heights of sides of f_i, f_{i+1} and of width h (see Figure 9.2).

$$\int_{x_i}^{x_{i+1}} f(x)dx = \frac{f_i + f_{i+1}}{2} h \tag{9.4}$$

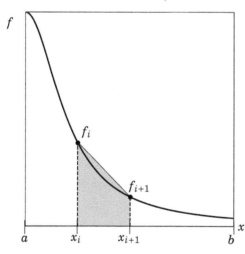

Figure 9.2: Integral over step h as area of trapezium

Integral from $x = a$ to $x = b$ may be obtained by summing integral over each of the n steps. Except the end points all other points will occur twice and hence it is easily seen that the integral must be given by

$$\int_a^b f(x)dx \approx T(h) = \left\{ \frac{f_0 + f_n}{2} + \sum_{i=1}^{n-1} f_i \right\} h \tag{9.5}$$

where $T(h)$ represents the *trapezoidal* estimate of the integral.

The following MATLAB program evaluates the trapezoidal estimate of the integral for equi-spaced nodes.

Program 9.1: *Trapezoidal rule*

```
1 function in = trapezoidal(f,h)
2 % Input   f : function values at nodes
3 %         h : step sizes
4 % Output in : trapezoidal estimate
5 n = length(f);     % number of nodes
6 in = (f(1)+f(n)+2*sum(f(2:n-1)))*h/2;
```

Error of trapezoidal rule

Trapezoidal rule integrates a linear function exactly and produces errors for polynomial functions of degree 2 or higher. Let us estimate the error of trapezoidal rule. Consider the integration applied over a single segment $x_i \leq x \leq x_i + h$.

$$I_i = \int_{x_i}^{x_i+h} f(x)dx \tag{9.6}$$

Express the function as a Taylor's series around point x_i to get

$$f(x) = f(x_i) + f^{(1)}(x_i)(x - x_i) + \frac{1}{2}f^{(2)}(x_i)(x - x_i)^2 + O(x - x_i)^3 \tag{9.7}$$

Integrating the function analytically we get

$$I_i = \int_{x_i}^{x_i+h} f(x)dx = f(x_i)h + f^{(1)}(x_i)\frac{h^2}{2} + f^{(2)}(x_i)\frac{h^3}{6} + O(h^4) \tag{9.8}$$

However, the trapezoidal estimate for the integral is

$$T_i = \frac{f(x_i) + f(x_i + h)}{2}h \tag{9.9}$$

Taylor expand $f(x_i + h)$ in the above to get

$$T_i = f(x_i)h + f^{(1)}(x_i)\frac{h^2}{2} + f^{(2)}(x_i)\frac{h^3}{4} + O(h^4) \tag{9.10}$$

The difference between Equations 9.8 and 9.10 is the error of the trapezoidal estimate. Thus

$$\epsilon_i = I_i - T_i = -\frac{1}{12}f^{(2)}(x_i)h^3 + O(h^4) \tag{9.11}$$

The above error is for one segment. Summing up errors over all segments between a and b we get

$$\epsilon = \sum_{i=1}^{n} \epsilon_i \approx -\frac{n}{12}f^{(2)}_{max}h^3 \approx -\frac{b-a}{12}f^{(2)}_{max}h^2 \tag{9.12}$$

where $f^{(2)}_{max}$ is the maximum second derivative in the interval $[a,b]$. We can see that the error of trapezoidal rule is proportional to h^2 and hence it is a *second order* method.

Example 9.1

Integrate $y = e^{-x^2}$ between 0 and 1 using Trapezoidal rule with a step size of $h = 0.125$. Compare this with the exact value of 0.746824.

Background :

We note that $\int_0^x e^{-x^2}dx = \frac{\sqrt{\pi}\,\text{erf}(x)}{2}$ where erf(x) is the error function, which is a tabulated function. We come across this function in various areas of physics such as transient diffusion.

Solution :

We compute function values with required spacing of $h = 0.125$ and tabulate below.

x	$f(x)$	x	$f(x)$
0	1	0.625	0.676634
0.125	0.984496	0.75	0.569783
0.25	0.939413	0.875	0.465043
0.375	0.868815	1	0.36788
0.5	0.778801		

Trapezoidal estimate is obtained by using Equation 9.5 where $n = 8$. Thus

$$T(0.125) = \left\{ \frac{1 + 0.367880}{2} + 0.984496 + 0.939413 + 0.868815 \right.$$

$$\left. + 0.778801 + 0.676634 + 0.569783 + 0.465043 \right\} \times 0.125$$

$$= 0.745866$$

The error in this estimate with respect to the exact value is given by

$$\text{Error} = 0.745866 - 0.746824 = -0.000958 \approx 0.001$$

9.3 Simpson's rule

In order to improve the accuracy of the integral estimated as above, we consider an improvement that has an order of accuracy up to 3 i.e. integrals up to cubic polynomials are exact. Simpson's rule comes in two variants, the first one - the Simpson's 1/3 rule - uses three points in each segment and the second one - the Simpson's 3/8 rule - uses four points in each segment. However both of these have errors of order five (proportional to h^5), as will be seen below.

9.3.1 Simpson's 1/3 rule

Derivation of Simpson's 1/3 rule - Integration of a second degree polynomial

Consider three points defining a segment of width $2h$ for integration, as shown in Figure 9.3. Using the trick that was employed in defining central difference formulae, we can easily show that a *quadratic* that passes through the three points is given by

$$f(x) = f_i + \frac{f_{i+1} - f_{i-1}}{2h}(x - x_i) + \frac{f_{i-1} - 2f_i + f_{i+1}}{2h^2}(x - x_i)^2 \tag{9.13}$$

which is valid in the region $x_{i-1} \le x \le x_{i+1}$. The integral over the segment may be obtained by integrating term by term Equation 9.13 from $x = x_{i-1}$ to $x = x_{i+1}$. This is equivalent to integrating with respect to $t = x - x_i$ between $-h$ and h. Thus we have

$$\int_{x_{i-1}}^{x_{i+1}} f(x)dx = \int_{-h}^{h} f(t)dt$$

$$S(h) = \left[f_i t + \frac{f_{i+1} - f_{i-1}}{2h} \frac{t^2}{2} + \frac{f_{i-1} - 2f_i + f_{i+1}}{2h^2} \frac{t^3}{3} \right]\Big|_{-h}^{h}$$

$$= \left[2f_i + \frac{f_{i-1} - 2f_i + f_{i+1}}{3} \right] h = (f_{i-1} + 4f_i + f_{i+1})\frac{h}{3} \tag{9.14}$$

Note that the linear term yields zero on integration. Similarly a cubic term, if present, would also integrate out to zero. Hence the error term must be proportional to h^5.

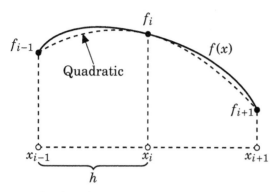

Figure 9.3: *Segment of width 2h for Simpson's 1/3 rule*

Alternate derivation - Method of undetermined coefficients

A second method of arriving at the above is given now. We require that the Simpson estimate of the integral over $t = -h$ to $t = h$ be given by

$$\int_{-h}^{h} f(t)dt = w_0 f_{i-1} + w_1 f_i + w_2 f_{i+1} \tag{9.15}$$

and yield the exact integral for

$$f(t) = 1, \ f(t) = t, \ f(t) = t^2 \tag{9.16}$$

since any quadratic may be represented as a linear combination of these three functions. Noting that $t_{i-1} = -h$, $t_i = 0$, $t_{i+1} = h$, we should then have

$$f(t) = 1; \int_{-h}^{h} f(t)dt \quad = \quad w_0 + w_1 + w_2 = 2h \cdots\cdots\cdots\cdots(a)$$

$$f(t) = t; \int_{-h}^{h} f(t)dt \quad = \quad w_0(-h) + w_1 0 + w_2 h = 0 \cdots\cdots(b)$$

$$f(t) = t^2; \int_{-h}^{h} f(t)dt \quad = \quad w_0(h^2) + w_1 0 + w_2 h^2 = \frac{2h^3}{3} \cdots(c) \tag{9.17}$$

From Equation 9.17(b) $w_0 = w_2$. Substitute this in Equation 9.17(c) to get $2w_0 = 2w_2 = \dfrac{2h}{3}$ or $w_0 = w_2 = \dfrac{h}{3}$. This in Equation 9.17(a) gives $w_1 = 2h - w_0 - w_2 = 2h - \dfrac{h}{3} - \dfrac{h}{3} = \dfrac{4h}{3}$. Immediately we get back Equation 9.14.

Alternate derivation - using Newton's polynomial

A third method of arriving at the Simpson's 1/3 rule is to represent the quadratic passing through the three points by a Newton polynomial given by

$$f(r) \approx p_2(r) = f_{i-1} + \Delta_{i-1} r + \frac{\Delta_{i-1}^2}{2!} r(r-1) \tag{9.18}$$

where $r = \dfrac{x - x_{i-1}}{h}$. We note that the forward differences occurring in Equation 9.18 are given by

$$\Delta_{i-1} = f_i - f_{i-1} \text{ and } \Delta_{i-1}^2 = f_{i-1} - 2f_i + f_{i+1} \tag{9.19}$$

The integral required is given by $\int_{x_{i-1}}^{x_{i+1}} f(x)dx = h\int_0^2 f(r)dr$. Term by term integration of Equation 9.18 gives

$$
\begin{aligned}
h\int_0^2 f(r)dr \approx S(h) = h\int_0^2 p_2(r)dr &= h\int_0^2\left[f_{i-1}+\Delta_{i-1}r+\frac{\Delta_{i-1}^2}{2!}r(r-1)\right]dr \\
&= \left[2f_{i-1}+2\Delta_{i-1}+\frac{2}{3}\frac{\Delta_{i-1}^2}{2!}\right]h \qquad (9.20)
\end{aligned}
$$

Introduce Expressions 9.19 in Equation 9.20 to get the result given in Equation 9.14.

Error of Simpson's rule for one segment

Simpson's rule produces exact integrals up to cubic polynomials. Error in Simpson's estimate arises from term of degree 4 or higher. The error for one segment may be derived as in the case of trapezoidal rule to obtain

$$
\epsilon = I - S = -\frac{1}{90}h^5 f^{(4)} \qquad (9.21)
$$

9.3.2 Simpson's 3/8 rule

Simpson's 3/8 rule uses a segment containing four points and three steps as indicated in Figure 9.4. We are looking for a rule which will yield exact integral when $f(t)=1$, $f(t)=t$, $f(t)=t^2$, $f(t)=t^3$. We look for a rule in the form

$$
S(h) = \int_{t=0}^{3h} f(t)dt = w_0 f_i + w_1 f_{i+1} + w_2 f_{i+2} + w_3 f_{i+3} \qquad (9.22)
$$

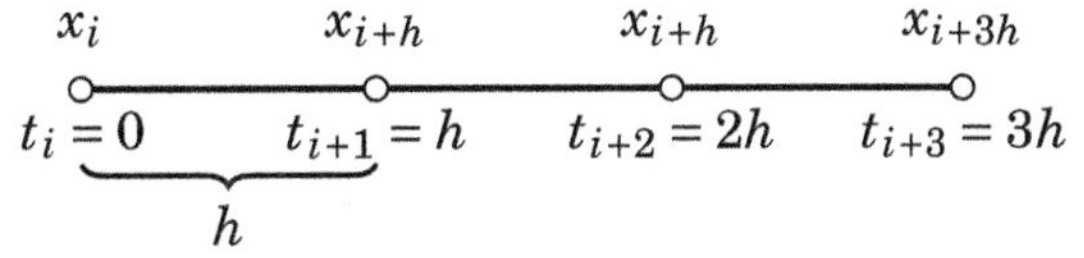

$$
\begin{array}{cccc}
x_i & x_{i+h} & x_{i+h} & x_{i+3h} \\
\circ & \circ & \circ & \circ \\
t_i=0 & t_{i+1}=h & t_{i+2}=2h & t_{i+3}=3h \\
& h & &
\end{array}
$$

Figure 9.4: *Segment of width 3h for Simpson's 3/8 rule*

Following the logic of Equations 9.17 we then have

$$
\begin{aligned}
f(t)=1;\ \int_0^{3h} f(t)dt &= w_0+w_1+w_2+w_3 = 3h & (a)\\
f(t)=t;\ \int_0^{3h} f(t)dt &= w_0 0+w_1 h+w_2 2h+w_3\cdot 3h = \frac{9h^2}{2} & (b)\\
f(t)=t^2;\ \int_0^{3h} f(t)dt &= w_0 0+w_1 h^2+w_2 4h^2+w_3 9h^2 = 9h^3 & (c)\\
f(t)=t^3;\ \int_0^{3h} f(t)dt &= w_0 0+w_1 h^3+w_2 8h^3+w_3 27h^3 = \frac{81h^4}{4} & (d)
\end{aligned}
$$

Eliminating w_1 between Equation 9.23(b) and (c) we get

$$
2w_2+6w_3 = \frac{9h}{2} \qquad (9.23)
$$

Eliminating w_1 between Equation 9.23(c) and (d) we get

$$4w_2 + 18w_3 = \frac{45h}{4} \tag{9.24}$$

From the above two equations we eliminate w_2 to get

$$6w_3 = \left(\frac{45}{4} - 9\right)h \text{ or } w_3 = \frac{3h}{8} \tag{9.25}$$

Use this in Equation 9.23 to get

$$2w_2 = \left(\frac{9h}{2} - 6 \times \frac{3h}{8}\right) \text{ or } w_2 = \frac{9h}{8} \tag{9.26}$$

Using the above two equations in Equation 9.23(b) we then get

$$w_1 = \frac{9h}{2} - 2 \times \frac{9h}{8} - 3 \times \frac{3h}{8} = \frac{9h}{8} \tag{9.27}$$

Finally from Equation 9.23(a) we get

$$w_0 = 3h - \frac{9h}{8} - \frac{9h}{8} - \frac{3h}{8} = \frac{3h}{8} \tag{9.28}$$

Hence we get the Simpson's 3/8 rule given by

$$S(h) = \int_{t=0}^{3h} f(t)dt = [f_i + 3f_{i+1} + 3f_{i+2} + f_{i+3}]\frac{3h}{8} \tag{9.29}$$

The error of 3/8 rule is also proportional to h^5

Composite Simpson's 1/3 rule

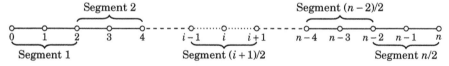

Figure 9.5: *Division of range into segments for applying Simpson's 1/3 rule*

Now consider the range $x = a$, $x = b$ to be divided into a number of segments as shown in Figure 9.5. Obviously n is an even number (with $x = a$ counted as 0). Expression 9.14 holds for the general segment $\frac{i+1}{2}$. The Simpson's 1/3 rule is nothing but integration of piecewise quadratic interpolating polynomial considered in Example 5.8. The desired integral is obtained by summing the contribution to it from each segment. When we do this the end points of each segment (barring the end points 0 and n) contributes to the integral to the two adjoining segments and hence

$$\int_{x_0}^{x_n} f(x)dx \approx S(h) = \left[(f_0 + 4f_1 + f_2) + (f_2 + 4f_3 + f_4) + \cdots \right.$$
$$+(f_{i-1} + 4f_i + f_{i+1}) + \cdots\cdots\cdots + (f_{n-4} + 4f_{n-3} + f_{n-2})$$
$$\left. +(f_{n-2} + 4f_{n-1} + f_n)\right]\frac{h}{3}$$
$$= [f_0 + 4f_1 + 2f_2 + 4f_3 + 2f_4 + \cdots + 2f_{n-2} + 4f_{n-1} + f_n]\frac{h}{3} \tag{9.30}$$

where $S(h)$ stands for Simpson estimate with step h. The above may be rewritten in the form of a weighted sum given by

$$S(h) = h \sum_{i=0}^{i=n} w_i f_i \qquad (9.31)$$

with

$$w_0 = w_n = \frac{1}{3}, \; w_1 = w_3 = \cdots w_{odd} = \frac{4}{3}, \; w_2 = w_4 = \cdots w_{even} = \frac{2}{3} \qquad (9.32)$$

Error of composite Simpson's 1/3 rule

Summing up errors over all the segment we get

$$\epsilon = \sum_{i=1}^{n/2} \epsilon_i \approx -\frac{n}{180} f_{max}^{(4)} h^5 \approx -\frac{b-a}{180} f_{max}^{(4)} h^4 \qquad (9.33)$$

where $f_{max}^{(4)}$ is the maximum fourth derivative in the interval $[a, b]$. We can see that the error of composite Simpson's 1/3 rule is proportional to h^4. Simpson's rule is a *fourth order* rule.

Composite Simpson's 3/8 rule: We may again divide the interval $x = a$ to $x = b$ into a number of segments having four nodes each and obtain the integral by summing over these segments. n should be a multiple of 3 (with node at $x = a$ counted as 0). The result is left to the reader to write down.

When n is neither an even number or a multiple of 3, a combination of Simpson's 1/3 rule and 3/8 rule can be applied. The error of such a scheme is still proportional to h^4. The simplest scheme is to divide the range into segments such that the last segment contains 4 nodes where as all other segments contain 3 nodes. Then 1/3 rule is applied to all segments having 3 nodes where as 3/8 rule is applied to the last segment. A MATLAB program has been provided below to perform the

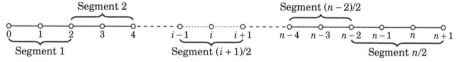

Figure 9.6: *Division of range into segments for applying composite Simpson's rule*

Composite Simpson's rule.

Program 9.2: Composite Simpson rule

```
 1 function in = compsimpson(f,h)
 2 %   Input    f: function values at nodes
 3 %            h: step size
 4 %   Output in: composite Simpson estimate
 5 n = length(f);        % number of nodes
 6 if (mod(n,2) == 1)    % if n is odd, Simpson's 1/3 rule
 7    in = (f(1)+f(n)+4*sum(f(2:2:n))+2*sum(f(3:2:n-2)))*h/3;
 8 else                  % if n is even Simpson's 1/3 + 3/8 rule
 9    in = (f(1)+f(n-3)+4*sum(f(2:2:n-3))+2*sum(f(3:2:n-4)))*h/3;
10    in = in + 3*(f(n-3)+3*f(n-2)+3*f(n-1)+f(n))*h/8;
11 end
```

Example 9.2

Integrate $y = e^{-x^2}$ between 0 and 1 using Simpson rule with a step size of $h = 0.125$. Compare this with the exact value of 0.746824.

Solution :

Function values in Example 9.1 are useful in this case also. Simpson estimate is based on Equation 9.30. Use Program 9.2 to evaluate the integral.

```
x = [0:0.125:1]';
y = exp(-x.^2);
in = compsimpson(y,0.125);
```

The output of the program is as follows

$$
\begin{aligned}
S(0.125) &= \left\{ (1 + 0.367880) + 4 \times (0.984496 + 0.868815 + 0.676634 + \right. \\
&\quad \left. 0.465043) + 2 \times (0.939413 + 0.778801 + 0.569783) \right\} \times \frac{0.125}{3} \\
&= 0.746826
\end{aligned}
$$

The Simpson estimate differs from the exact by +0.000002. This is surprising since the same data as in the Trapezoidal estimate is used in this case also!

9.4   Integration of functions

Examples that have been considered until now use equi-spaced *tabulated* data. We do not rule out the possibility of having different step sizes in different segments. If only data points are available and the functional relationship between the variables is not known, the accuracy of the integration is limited by the number of available data points. We encounter such situations frequently in engineering practice while using property tables, material design handbooks etc.

On the other hand, if the functional relationship between x and y is known, the accuracy would depend on the step size h and the quadrature scheme that has been employed. How do we decide the appropriate step size or quadrature scheme? In practical problems, we do not know the exact value for the integral and hence we are not sure of the error. Errors for trapezoidal and Simpson's rule can be estimated by determining the second and fourth derivatives of the function respectively. But determining the derivatives may not be possible for all functions. There can be two ways by which the error can be reduced namely

1. h refinement: As we have seen in Example 9.3, the error of the trapezoidal and Simpson's estimates depend on h. The error of the quadrature rule can be reduced by decreasing h systematically.
2. The accuracy of the estimate can also be increased by using a higher order accurate quadrature rule.

9.4.1 h refinement: Error estimation

Let us consider trapezoidal rule. We note that the trapezoidal rule uses a linear function between nodes and hence the dominant error is proportional to h^2. If a quadratic function is used, as in the Simpson's rule, the dominant error is proportional to h^4. Hence we see that the error should go as $A_1 h^2 + A_2 h^4 + \ldots$, involving only even powers of h. Thus we may write

$$I = \int_a^b f(x)dx = T(h) + A_1 h^2 + A_2 h^4 + \cdots \cdots \tag{9.34}$$

We may take two step sizes $h_1 = h$ and $h_2 = h/2$ and write

$$\begin{aligned} I &= T(h) + A_1 h^2 + A_2 h^4 + \cdots \cdots \\ I &= T(h/2) + A_1 h^2/4 + A_2 h^4/16 + \cdots \cdots \end{aligned} \tag{9.35}$$

Subtracting Equations 9.35 from 9.34 we get,

$$T(h) - T(h/2) \approx -A_1 h^2 \frac{3}{4} + O(h^4) \tag{9.36}$$

Rearranging the above equation and neglecting the higher order terms we get

$$A_1 \frac{h^2}{4} = \epsilon_T(h/2) = \frac{T(h/2) - T(h)}{3} \tag{9.37}$$

which is nothing but the error estimate for $T(h/2)$. If $\epsilon_T(h/2) = A_1 \dfrac{h^2}{4}$ is less than the desired tolerance, the estimate of the desired output is accepted. A simple algorithm can hence be proposed as follows

Step 1 Start with h

Step 2 Evaluate T(h) and T(h/2)

Step 3 If $\dfrac{|T(h/2) - T(h)|}{3} <$ tolerance, stop iteration

Step 4 else refine grid, $h = h/2$

Grid refinement is done in a similar way for Simpson's 1/3 rule. The error for Simpson's rule can be estimated as

$$\epsilon_S(h/2) = \frac{S(h/2) - S(h)}{15} \tag{9.38}$$

One observes that in both trapezoidal and Simpson's 1/3 rule, the function values are reused after refining the grid. The advantage of refining the grid by halving h is that the calculated functional values at the existing nodes are reused. However, in case of Simpson's 3/8 rule, the already calculated function values are not used on grid refinement. Hence, Simpson's 1/3 rule is preferred over Simpson's 3/8 rule (even though both have same order of accuracy). The same is applicable for Newton Cotes (treated later) rules with odd number of points (5,7 etc).

Let us reexamine Example 9.3. It is intended to obtain the integral within a tolerance of 1×10^{-6}. A MATLAB program has been written to determine the desired estimate.

```
in = 0;              % initialize integral
i = 2;               % initialize i
tol = 1e-6           % tolerance
while 1              % start loop
    h = 1/2^i;       % step size
    h1(i-1) = h;
```

```
x = [0:h:1]';          % x
f = 1./(1+25*x.^2); % evaluate function
in(i) = trapezoidal(f,h);    % trapezoidal
% in(i) = compsimpson(f,h); % (uncomment for Simpson's )
if( abs(in(i)-in(i-1))/3 < tol ) % for trapezoidal rule
%if( abs(in(i)-in(i-1))/15 < tol ) % for Simpson's rule
    break;             % if error < tolerance , break loop
end
i = i+1;               % increment i for grid refinement
end
```

Table 9.1 summarizes the grid refinement procedure for trapezoidal and Simpson's rules. For

Table 9.1: *Effect of grid refinement*

	Trapezoidal		Simpson's	
h	estimate	ϵ_T	estimate	ϵ_S
1/4	0.278449		0.261738	
1/8	0.274611	1.28E-03	0.273332	7.73E-04
1/16	0.274656	1.50E-05	0.274671	8.93E-05
1/32	0.274674	6.01E-06	0.274680	6.05E-07
1/64	0.274679	1.50E-06		
1/128	0.274680	3.76E-07		

achieving the desired tolerance, the number of function evaluations for Trapezoidal and Simpson's rule are respectively 129 and 17.

9.4.2   Closed Newton Cotes quadrature rules

Newton Cotes[1] integration rules can be used when equi-spaced data is available. Closed means the data including those at the end points are made use of by the quadrature formula. Trapezoidal and Simpson rules naturally fall under the category of Newton Cotes formulae. $\int_a^b f(x)dx$ is calculated by integrating the Lagrange polynomial passing through the data points.

Consider $n+1$ points, x_0 to x_n, equally distributed in the range $a \le x \le b$ with a step size of $h = \dfrac{b-a}{n}$. Newton Cotes integration formula can be written as

$$\int_a^b f(x)dx = \int_{x_0}^{x_0+nh} L_n(x)dx = ch\sum_{i=0}^n w_i f_i \qquad (9.39)$$

where c is a multiplication constant. w_i is the weight for each node which is obtained by integrating the Lagrange weight functions

$$w_i = \int_a^b l_i(x)dx \qquad (9.40)$$

where l_i are the Lagrange weight functions[2] . A MATLAB routine has been provided below to integrate the polynomial function. The function can be used to integrate Lagrange polynomial functions.

---

[1]Roger Cotes, 1682 - 1716, English mathematician who worked closely with Newton.

[2]Note that the Lagrange weights were indicated by w_i in Chapter 5 dealing with interpolation.

Program 9.3: *Integration of polynomial functions*

```
1  function in = polyintegral(pp,x1,x2)
2  %  Input   pp: coefficients of polynomials
3  %          x1: lower bound
4  %          x2: upper bound
5  %  Output in: integral
6  n = size(pp,2)+1;    % number of points
7  m = size(pp,1);      % number of polynomial equations
8  pint = zeros(size(pp,1),size(pp,2)+1); % initialize pint
9  in = zeros(m,1);     % initialize in
10 for i= 2:n           % evaluate integrated polynomial
11         pint(:,i) = pp(:,i-1)/(i-1);
12 end
13 for i=1:n            % evalauate integral
14 in = in + pint(:,i).*(x2.^(i-1)-x1.^(i-1));
15 end
```

Table 9.2 lists the weights for various Newton Cotes formulae.

Table 9.2: *Weights for Newton Cotes quadrature rules*

	c	Weights							$\|\epsilon\|^{\dagger}$
n		0	1	2	3	4	5	6	
1	$\frac{1}{2}$	1	1			**Trapezoidal**			$\frac{h^3}{12}f^{(2)}$
2	$\frac{1}{3}$	1	4	1		**Simpson's** $1/3^{rd}$			$\frac{h^5}{90}f^{(4)}$
3	$\frac{3}{8}$	1	3	3	1	**Simpson's** $3/8^{th}$			$\frac{3h^5}{8}f^{(4)}$
4	$\frac{2}{45}$	7	32	12	32	7			$\frac{8h^7}{945}f^{(6)}$
5	$\frac{5}{288}$	19	75	50	50	75	19		$\frac{275h^7}{12096}f^{(6)}$
6	$\frac{1}{140}$	41	216	27	272	27	216	41	$\frac{9h^9}{1400}f^{(8)}$

†Error estimate is over a single segment

Error estimate   The error estimate for i^{th} order accurate Newton quadrature rule can be determined in the same way as in the previous section i.e. based on the difference between estimates with two grids h and $h/2$ which is

$$\epsilon = \frac{I(h/2) - I(h)}{2^i - 1} \tag{9.41}$$

Runge phenomenon and Newton Cotes Quadrature   Use of higher order quadrature rules do not ensure accuracy of the integration. Recollect Runge phenomenon in Chapter 5, where higher order interpolation over equi-spaced data can cause large errors. The same error would reflect in integration while using higher order quadrature schemes. The weights of higher order Newton Cotes formulae can be **negative** making the integration unstable and also unphysical. Consider a rod of length L having a temperature distribution $T(x)$. The mean temperature of the rod is defined as

$$T_{mean} = \frac{1}{L}\int_0^L T(x)dx = w_0 T_0 + w_1 T_1 + \cdots + w_n T_n$$

If a certain node is having a negative weight in a quadrature rule, it means that the node contributes negatively to the mean temperature which is obviously unphysical.

Example 9.3

Evaluate integral $\int_0^1 \frac{1}{1+25x^2} dx$ using the following schemes

1. *Composite trapezoidal*
2. *Composite Simpson's 1/3*
3. *Newton Cotes quadrature rules*

with $h = 1/8$, 1/16 and 1/32. Compare the individual results with the exact value.

Solution :

Integrating the function analytically we get

$$\int_0^1 \frac{1}{1+25x^2} = \frac{1}{5}\tan^{-1}(5x)\Big|_0^1 = \frac{1}{5}\tan^{-1}5 = 0.274680 \tag{9.42}$$

Programs 9.1, 9.2 and 9.3 have been used to apply trapezoidal rule, Simpson's 1/3 rule and Newton Cotes rules. A sample of the program has been given below.

```
h = 1/8;                              % step size
x = [0:h:1]';                         % node positions
f = 1./(1+25*x.^2);                   % function values at nodes
trap = trapezoidal(f,h);              % trapezoidal estimate
simp = compsimpson(f,h);              % Simpson's estimate
pp=lagrangeweightfunction(x);         % Lagrange weight functions
w = polyintegral(pp,0,1);
nc = sum(w.*f);                       % Newton Cotes estimate
```

The following table lists estimates of the integral evaluated by these methods and the corresponding errors with respect to the exact value (difference between exact and estimate).

h	Trapezoidal estimate	ϵ	Simpson's estimate	ϵ	Newton Cotes estimate	ϵ
0.1250	0.2746	6.90E-05	0.2733	1.35E-03	0.2752	5.51E-04
0.0625	0.2747	2.41E-05	0.2747	9.08E-06	0.2746	4.44E-05
0.0312	0.2747	6.02E-06	0.2747	4.55E-09	**2.51E+08**	**2.51E+08**

Accuracy of Simpson's and Trapezoidal estimates increase as the number of nodes increase. However, observe that the Newton Cotes approximation diverges rapidly for higher order approximation. The Runge phenomenon makes the estimate poor. The divergence of higher order Newton Cotes formulae occurs because of the rounding errors of the Lagrange interpolation weights. Direct integration of higher order interpolation formulae must be avoided. Also observe that the convergence of Simpson's estimate to the exact solution is much faster than that of Trapezoidal rule. Figure 9.7 shows the relationship between error and step size for different schemes.

Composite Newton Cotes rule: Like composite Simpson's rule, We can divide the range into smaller segments and apply Newton Cotes quadrature rules over each of these segments. Such a scheme is known as *Composite Newton Cotes rule*. The following example considers composite Newton Cotes quadrature.

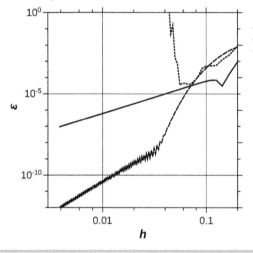

Figure 9.7: *Error vs step size for different integration schemes*

Example 9.4

Integrate $y = e^{-x^2}$ between 0 and 1 using Newton Cotes rules with $n = 4$ and 8 with a step size of $h = 0.125$. Compare this with the exact value of 0.746824.

Solution :

Function values are listed in Example 9.1. The number of nodes is equal to 9. We have already calculated the integral with $n = 1$ (trapezoidal) and $n = 2$ (Simpson's 1/3) in earlier examples.

Let us apply Newton Cotes quadrature rule for $n = 4$. The range is divided into two segments ($0 \leq x \leq 0.5$ and $0.5 \leq x \leq 1$). Each segment has four points with $x = 0.5$ shared between the two segments. The weight functions for all the nodes except $x = 0.5$ are those listed in Table 9.2. As node at $x = 0.5$ is shared between two segments, the total weight at the node is sum of the weights from individual segments. Integration procedure has been summarized in Table 9.3.

The integral is equal to $ch \sum_{i=0}^{8} w_i f_i = \frac{2}{45} \times 0.125 \times 134.428351 = 0.746824$. The error of the estimate is found to be 3.71×10^{-08}

We can use program 9.3 to evaluate the integral with $n = 8$

```
x = [0:1/8:1]';
f = exp(-x.^2);
```

```
pp=lagrangeweightfunction(x);   % evaluate  weight  functions
w = polyintegral(pp,0,1);       % integrate  weights
int = sum(w.*f);                % integral
er = abs(int-sqrt(pi)*erf(1)/2); % error
```

Table 9.3: *Spreadsheet for Example 9.4*

i	x	f_i	w_i	$w_i f_i$
0	0	1.000000	7	7.000000
1	0.125	0.984496	32	31.503886
2	0.25	0.939413	12	11.272957
3	0.375	0.868815	32	27.802082
4	0.5	0.778801	14	10.903211
5	0.625	0.676634	32	21.652283
6	0.75	0.569783	12	6.837394
7	0.875	0.465043	32	14.881382
8	1	0.367879	7	2.575156
Sum				134.428351

The output of the program is

```
int = 0.7468
er = 9.0250e-09
```

We can see that the error has reduced quite significantly!

In practice, higher order Newton Cotes quadrature rules are avoided. It may be advisable to employ composite Newton Cotes rules up to order 7.

9.4.3   Romberg method: Richardson extrapolation

In earlier sections, we were able to estimate the error and used it as an indicator to stop grid refinement. Richardson extrapolation goes one step further by utilizing the calculated error to push the error to higher order. The method does not suffer from instabilities due to Runge phenomenon as seen for higher order Newton quadrature rules, at the same time achieves estimates with higher order accuracy. Also, the procedure determines the trapezoidal estimate (lowest order of accuracy) and then gradually pushes the error to higher and higher orders.

Again we consider trapezoidal estimates of two step sizes $h_1 = h$ and $h_2 = h/2$ and write

$$
\begin{aligned}
I &= T(h) + A_1 h^2 + A_2 h^4 + \cdots \cdots \\
I &= T(h/2) + A_1 h^2/4 + A_2 h^4/16 + \cdots \cdots
\end{aligned}
\tag{9.43}
$$

We may divide the first of these by h^2 and the second by $(h/2)^2$ and subtract the first from the second to get

$$
\frac{4I}{h^2} - \frac{I}{h^2} = \frac{4T(h/2)}{h^2} - \frac{T(h)}{h^2} + A_2(h^2/4 - h^2) + \cdots \cdots
\tag{9.44}
$$

This may be rearranged to get

$$
I = \frac{\frac{4T(h/2)}{h^2} - \frac{T(h)}{h^2}}{\frac{4}{h^2} - \frac{1}{h^2}} + A_2 \frac{(h^2/4 - h^2)}{\frac{4}{h^2} - \frac{1}{h^2}} = \frac{4T\left(\frac{h}{2}\right) - T(h)}{4 - 1} - \frac{A_2}{4} h^4 \cdots \cdots
\tag{9.45}
$$

Rearranging the above equation we obtain

$$I \approx T\left(h, \frac{h}{2}\right) = T\left(\frac{h}{2}\right) + \underbrace{\frac{T(\frac{h}{2}) - T(h)}{3}}_{\epsilon_T(h/2)} \tag{9.46}$$

The above equation is a better estimate for the integral since the error is proportional to h^4. Observe that the above estimate is the same as Simpson's estimate. This extrapolation process may be continued by noting that

$$I = T\left(h, \frac{h}{2}\right) + A_2 h^4 + \cdots\cdots \tag{9.47}$$

$$I = T\left(\frac{h}{2}, \frac{h}{4}\right) + A_2 \frac{h^4}{16} + \cdots\cdots \tag{9.48}$$

We may eliminate A_2 from these two equations to get a better estimate to the integral given by

$$I = T\left(\frac{h}{2}, \frac{h}{4}\right) + \underbrace{\frac{T\left(\frac{h}{2}, \frac{h}{4}\right) - T\left(h, \frac{h}{2}\right)}{2^{2\cdot2} - 1}}_{\epsilon_S(h/2)} + A_3 h^6 \tag{9.49}$$

Thus $T\left(h, \frac{h}{2}, \frac{h}{4}\right) = T\left(\frac{h}{2}, \frac{h}{4}\right) + \frac{T\left(\frac{h}{2}, \frac{h}{4}\right) - T\left(h, \frac{h}{2}\right)}{2^{2\cdot2} - 1}$ is a superior estimate to the integral since it involves error of order $h^6 = h^{2\cdot3}$. This process may be extended to push the error further to higher orders. In general, we may write

$$T\left(h, \frac{h}{2}, \ldots, \frac{h}{2^n}\right) = T\left(\frac{h}{2}, \ldots, \frac{h}{2^n}\right) + \frac{T\left(\frac{h}{2}, \ldots, \frac{h}{2^n}\right) - T\left(h, \frac{h}{2}, \ldots, \frac{h}{2^{n-1}}\right)}{2^{2\cdot n} - 1} + A_n h^{2\cdot(n+1)} \tag{9.50}$$

The above procedure leads to a Romberg table (Table 9.4) as given below. The entries along the upward going diagonal, starting in any row, represent the best estimates for the integral. Usually one stops when two entries along this diagonal differ by less than a user defined tolerance.

Table 9.4: *Romberg table*

Error order				
h^2	h^4	h^6	h^8	h^{10}
$T(h)$	$T\left(h, \frac{h}{2}\right)$	$T\left(h, \frac{h}{2}, \frac{h}{4}\right)$	$T\left(h, \frac{h}{2}, \frac{h}{4}, \frac{h}{8}\right)$	$T\left(h, \frac{h}{2}, \frac{h}{4}, \frac{h}{8}, \frac{h}{16}\right)$
$T\left(\frac{h}{2}\right)$	$T\left(\frac{h}{2}, \frac{h}{4}\right)$	$T\left(\frac{h}{2}, \frac{h}{4}, \frac{h}{8}\right)$	$T\left(\frac{h}{2}, \frac{h}{4}, \frac{h}{8}, \frac{h}{16}\right)$	
$T\left(\frac{h}{4}\right)$	$T\left(\frac{h}{4}, \frac{h}{8}\right)$	$T\left(\frac{h}{4}, \frac{h}{8}, \frac{h}{16}\right)$		
$T\left(\frac{h}{8}\right)$	$T\left(\frac{h}{8}, \frac{h}{16}\right)$			
$T\left(\frac{h}{16}\right)$				

Example 9.5

Use Romberg method and integrate $f(x) = \dfrac{1}{0.3 - x^4}$ *between 1 and 0.8. Compare the value you obtain with the exact value.*

Solution :

Romberg method requires trapezoidal estimates with h, $\dfrac{h}{2}$ etc. In the present case we take a h value of $-\dfrac{1}{30} = -0.033333$. Assuming that at least three Trapezoidal estimates are needed, we calculate the function data with a step size of $\dfrac{h}{4} = \dfrac{-0.033333}{4} = -0.00833333$ as tabulated below.

i	x	$f(x)$	i	x	$f(x)$	i	x	$f(x)$
0	1	-1.42857	9	0.925	-2.31431	17	0.858333	-4.11895
1	0.991667	-1.49907	10	0.916667	-2.46265	18	0.85	-4.50437
2	0.983333	-1.57485	11	0.908333	-2.62647	19	0.841667	-4.95455
3	0.975	-1.65648	12	0.9	-2.80820	20	0.833333	-5.48687
4	0.966667	-1.74463	13	0.891667	-3.010820	21	0.825	-6.12556
5	0.958333	-1.84005	14	0.883333	-3.237992	22	0.816667	-6.90537
6	0.95	-1.94361	15	0.875	-3.494284	23	0.808333	-7.87804
7	0.941667	-2.05634	16	0.866667	-3.785471	24	0.8	-9.12409
8	0.933333	-2.17943						

The required Trapezoidal estimates are obtained using Equation 9.5. With these the Romberg table is constructed and tabulated below

$T(h)$	$T(h, h/2)$	$T(h, h/2, h/4)$
0.70937		
	0.69487	
0.69850		0.69477
	0.69478	
0.69571		

It appears that the results are good to 5 places after the decimal point. We shall evaluate the exact value of the integral now. The integrand is expanded in terms of partial fractions. Setting $a = 0.3^{\frac{1}{4}}$ the integrand is written as

$$\text{Integrand} = \frac{1}{a^4 - x^4} = \frac{1}{4a^3}\left[\frac{1}{a-x} + \frac{1}{a+x}\right] + \frac{1}{2a^2}\frac{1}{a^2 + x^2}$$

On integrating term by term and substituting the limits, the integral in closed form turns out to be

$$\int_1^{0.8} \frac{dx}{a^4 - x^4} = \frac{1}{4a^3}\ln\left[\frac{(a-1)(a+0.8)}{(a+1)(a-0.8)}\right]$$
$$+ \frac{1}{2a^3}\left[\tan^{-1}\left(\frac{0.8}{a}\right) - \tan^{-1}\left(\frac{1}{a}\right)\right] = 0.69477$$

The Romberg estimate is in agreement with the exact value of the integral given above.

Example 9.6

Use Romberg method to integrate $\int_0^1 \dfrac{1}{1+25x^2}\,dx$ with $h = 1/8$. Compare the results with exact value.

Solution :

We have already calculated the trapezoidal estimates for $h = 1/8, 1/16$, and $1/32$ in Example 9.3.

$T(h)$	$T(h/2)$	$T(h/4)$
0.274611	0.274656	0.274674

We shall apply Richardson extrapolation to the trapezoidal estimates.

$$T\left(h, \frac{h}{2}\right) = \frac{4 \times 0.274656 - 0.274611}{3} = 0.274671$$

$$T\left(\frac{h}{2}, \frac{h}{4}\right) = \frac{4 \times 0.274674 - 0.274656}{3} = 0.274680$$

Observe that the Richardson extrapolated values are the same as the Simpson's estimates (see Example 9.3). It turns out that the first extrapolated value of trapezoidal rule is Simpson's 1/3 rule. Let us extrapolate the integral further

$$T\left(h, \frac{h}{2}, \frac{h}{4}\right) = \frac{16 \times 0.274680 - 0.274671}{15} = 0.274681$$

The extrapolation can stop when the desired degree of accuracy is achieved. The following table lists the Romberg estimates and its error.

h	$T(h)$	$T(h, h/2)$	$T(h, h/2, h/4)$	Error		
1/8	0.274611	0.274671	0.274681	6.90E-05	9.08E-06	6.01E-07
1/16	0.274656	0.274680	**0.274680**	2.41E-05	4.55E-09	2.53E-11
1/32	0.274674	0.274680		6.02E-06	2.61E-10	
1/64	0.274679			1.5E-06		

One important point to be noted is that Romberg method being an extrapolation method, round-off errors accumulated at a particular level will continue to accumulate in further extrapolations.

The distribution of nodes in the methods considered so far are uniform. The number of function evaluations required to achieve desired accuracy is considerably high. We shall look at two other important methods which can compute the integral with lesser number of function evaluations.

9.5   Quadrature using Chebyshev nodes

In chapter 5, it was seen that Lagrange polynomial passing through Chebyshev nodes was not affected by Runge phenomenon. Hence if we integrate a polynomial passing through the Chebyshev nodes, we can expect accurate results. Chebyshev nodes are roots of Chebyshev polynomials which are orthogonal polynomials. We take up an example to check out the quality of the integral estimate on using Chebyshev nodes.

Example 9.7

Revisit Example 9.3 and estimate the integral $I = \int_0^1 \dfrac{1}{1+25x^2} dx$ *by using Chebyshev nodes. Compare the estimate with the exact solution for the integral and also obtained with other quadrature rules.*

Solution :

Chebyshev nodes are defined as $\xi(i) = \cos\left(\dfrac{2i-1}{2(n+1)}\pi\right)$ in the range $(-1,1)$ where n is the degree of the polynomial. We transform the coordinates using Equation 5.15 which is given by

$$x = \frac{0+1}{2} + \frac{1-0}{2}\xi = 0.5 + 0.5\xi \tag{9.51}$$

The following program has been used to estimate the integral using Chebyshev nodes

```
n = 8;
i = [1:n+1]';
x = cos((2*i-1)*pi*0.5/(n+1));
f = 1./(1+25*(0.5*x+0.5).^2);
pp=lagrangeweightfunction(x);
w = polyintegral(pp,-1,1);
in = sum(w.*f)/2;
er = abs(int-atan(5)/5);
```

The output of the program is given below

```
in = 0.274680
er = 3.2379e-07
```

Figure 9.8 shows the error in integration using equi-spaced nodes and Chebyshev nodes.

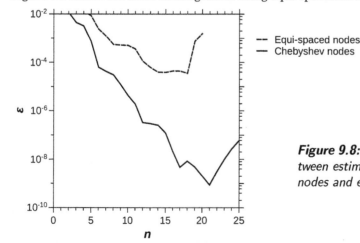

Figure 9.8: *Error comparisons between estimations with Chebyshev nodes and equi-spaced nodes*

Estimate using quadrature rule with 9 Chebyshev nodes meets the required tolerance compared to 32 function evaluations required for composite Simpson's rule! Quadrature using Chebyshev nodes is more accurate compared to equi-spaced nodes. This is because quadrature using equi-spaced nodes correspond to Newton Cotes rules which are sensitive to Runge phenomenon. The divergence of higher order polynomials does not indicate the failure of the quadrature rule. The divergence is due rather to the round off error of Lagrange functions. The intention of considering the above example is to make the reader aware of a powerful quadrature scheme.

Fejer quadrature rules and Clenshaw-Curtis quadrature rules[3] use Chebyshev polynomials to estimate the integral. Chebyshev quadrature rules exactly integrate polynomial up to degree n. In fact, the integration performed in Example 9.7 corresponds to one of the Fejer quadrature rules. However in practice, quadrature rules based on Chebyshev polynomials use **Discrete Cosine Transform** to estimate the integral. More discussion on this quadrature rule is available in advanced texts.

Both Newton Cotes quadrature and rules based on Chebyshev polynomials integrate polynomials exactly up to degree n. Now we shall look at Gauss quadrature rule which can integrate polynomials of degree $2n - 1$ exactly.

9.6 Gauss quadrature

Gauss quadrature deals with integration over a symmetrical range of x from -1 to +1. The important property of Gauss quadrature is that it yields exact values of integrals for polynomials of degree up to $2n - 1$. Gauss quadrature uses the function values evaluated at a number of interior points (hence it is an open quadrature rule) and corresponding weights to approximate the integral by a weighted sum.

$$I = \int_{-1}^{1} f(x)dx = \sum_{i=0}^{n-1} w_i f(x_i) \tag{9.52}$$

A Gauss quadrature rule with 3 points will yield exact value of integral for a polynomial of degree $2 \times 3 - 1 = 5$. Simpson's rule also uses 3 points, but the order of accuracy is 3. Gauss quadrature evaluates integrals with higher order accuracy with smaller number of points and hence smaller number of function evaluations.[4] Moreover Gauss quadrature integration is not affected by Runge phenomenon.

If the limits of integration are $x = a$ and $x = b$, it is possible to use a simple linear transformation to bring the limits to the standard [-1,1] using Equation 5.15. The transformation would be

$$x = \frac{a + b}{2} + \frac{b - a}{2}\xi \tag{9.53}$$

---

[3]C.W. Clenshaw and A.R. Curtis "*A method for numerical integration on an automatic computer*" 1960, Numerische Mathematik, 2(1), 197-205.

[4]Performance of Gauss quadrature may not be better than other methods discussed earlier for certain functions. Please refer to L.N. Trefethen, "Is Gauss quadrature better than Clenshaw-Curtis?" 2008, SIAM review, 50(1), 67-87.

We also have $dx = \dfrac{b-a}{2} d\xi$ and hence the required integral is given by

$$I = \int_a^b f(x)dx = \frac{b-a}{2} \int_{-1}^{1} f\left(\frac{b-a}{2}\xi + \frac{a+b}{2}\right)d\xi \qquad (9.54)$$

With the above proviso we now consider integration of a function over the interval [-1,1]. All the evaluation points are within the interval i.e. $-1 < x_i < 1$.

2 point Gauss Quadrature

A two point Gauss rule produces exact results for polynomials up to degree 3. The points and the weights are to be chosen such that the above condition is met. More or less, we are fitting a straight line to the cubic polynomial. As the integration is performed over symmetric region, contributions due to odd functions such as x, x^3 disappear. As there are even number of points, the points are symmetric about the origin i.e. $x_1 = -x_2$. Also, the weights of both the points are equal i.e. $w_1 = w_2$. The approximation already works for x and x^3. Let us determine the points and weights such that the approximation is exact for $f(x) = 1$ and $f(x) = x^2$. Thus we have

$$f(x) = 1; \int_{-1}^{1} f(x)dx \quad = \quad w_1 + w_2 = 2 \longrightarrow w_1 = w_2 = 1 \cdots (a)$$

$$f(x) = x^2; \int_{-1}^{1} f(x)dx \quad = \quad w_1 x_1^2 + w_2 x_2^2 = \frac{2}{3} \longrightarrow x_1 = \pm\sqrt{\frac{1}{3}} \cdots (b) \qquad (9.55)$$

Hence the 2 point Gauss rule would be

$$I_{G_2} = \int_{-1}^{1} f(x)dx = f\left(\sqrt{\frac{1}{3}}\right) + f\left(-\sqrt{\frac{1}{3}}\right) \qquad (9.56)$$

where the subscript G_2 indicates a two point Gauss quadrature.

3 point Gauss Quadrature

We consider next the three point rule which is explained by Figure 9.9. In case of odd number

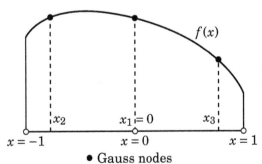

Figure 9.9: Gauss quadrature with three points

• Gauss nodes

of points such as the case shown in the figure origin will automatically be one Gauss point and the other Gauss points are symmetrically arranged with respect to the origin. Hence in the case shown in the figure the Gauss points are $x_0 = 0$, $x_1 = -x_2$ and the weights for the two symmetrically placed points are the same. Thus $w_1 = w_2$. These automatically satisfy the condition that integrals of x,

x^3 and x^5 over the limits [-1,1] vanish, as they should. The points and the weights are determined by requiring that the integral be exact for $f(x) = 1$, $f(x) = x^2$ and $f(x) = x^4$. Thus we have

$$f(x) = 1; \int_{-1}^{1} f(x)dx = w_0 + w_1 + w_2 = w_0 + 2w_1 = 2\cdots(a)$$

$$f(x) = x^2; \int_{-1}^{1} f(x)dx = w_1 \cdot x_1^2 + w_2 \cdot x_2^2 = 2w_1 \cdot x_1^2 = \frac{2}{3}\cdots(b)$$

$$f(x) = x^4; \int_{-1}^{1} f(x)dx = w_1 x_1^4 + w_2 x_2^4 = 2w_1 x_1^4 = \frac{2}{5}\cdots\cdots(c) \tag{9.57}$$

Dividing Equation 9.57(c) by Equation 9.57(b) we get

$$x_1^2 = \frac{3}{5} \text{ or } x_1 = \sqrt{\frac{3}{5}} \tag{9.58}$$

Hence $x_2 = -x_1 = -\sqrt{\frac{3}{5}}$. Then Equation 9.57(b) gives

$$2w_1 x_1^2 = \frac{2}{3} \text{ or } w_1 = \frac{5}{9} \tag{9.59}$$

Hence $w_2 = \frac{5}{9}$ and from Equation 9.57(a) we have

$$w_0 = 2 - 2w_1 = 2 - 2\frac{5}{9} = \frac{8}{9} \tag{9.60}$$

Thus the Gauss three point rule is

$$I_{G_3} = \int_{-1}^{1} f(x)dx = \frac{8}{9}f(0) + \frac{5}{9}\left[f\left(\sqrt{\frac{3}{5}}\right) + f\left(-\sqrt{\frac{3}{5}}\right)\right] \tag{9.61}$$

Legendre polynomials

The three Gauss points (points at which the function values are evaluated in Gauss quadrature) in the above indeed represent the three zeros of the Legendre polynomial $P_3(x) = \frac{x(5x^2 - 3)}{2}$. This connection is not fortuitous but *may* be shown rigorously to hold for any number of Gauss points i.e. for a n point Gauss rule the points are the zeroes of $P_n(x)$. Legendre polynomials defined as

$$P_n(x) = \frac{1}{2^n n!} \frac{d^n}{dx^n}\left[(x^2 - 1)^n\right] \tag{9.62}$$

satisfy the Legendre differential equation. Alternatively, P_n may be determined by

$$P_n(x) = 2^n \sum_{i=0}^{n} \binom{n}{i}\binom{(n+i-1)/2}{n} x^i \tag{9.63}$$

where $\binom{n}{i} = {}^nC_i = \frac{n!}{(n-i)!i!}$. For this reason Gauss quadrature is also referred to as Gauss Legendre quadrature. The weights of Gauss Legendre quadrature can be determined using method of undetermined coefficients. Table 9.5 is a short listing of Gauss points and weights. For a complete listing of Gauss nodes and weights, reader may refer to the book by Abramovitz and Stegun[5].

Table 9.5: *Gauss points and weights*

n	x	w	$2n-1$	n	x	w	$2n-1$
1	0	2	1		0	0.568889	
2	±0.577350	1	3	5	±0.538469	0.478629	9
3	0	0.888889	5		±0.906180	0.236927	
	±0.774597	0.555556			±0.238619	0.467914	
4	±0.339981	0.652145	7	6	±0.661209	0.360762	11
	±0.861136	0.347855			±0.932470	0.171324	

MATLAB program has been given below to compute the integral using 5 Gauss points.

Program 9.4: *5 point Gauss quadrature*

```
1 function in = gauss5pt(f,a,b)
2 % Input    f :    function to be integrated
3 %          a :    lower bound
4 %          b :    upper bound
5 % Output   in:    Gauss 5 pt estimate
6 x = [-0.906179846;      -0.538469310;      0.000000000;
7        0.538469310;      0.906179846];   % gauss nodes
8 w = [0.236926885;       0.478628670;    0.568888889;
9        0.478628670; 0.236926885];         % weights
10 x = (a+b)/2 + (b-a)*x/2;        % transform coordinates
11 in = (b-a)*sum(w.*f(x))/2;      % compute integral
```

Similar programs can be written for other order Gauss quadratures also. It may be better to store all the weights and nodes in a single file and use them as required. Algorithms are available to determine the nodes and weights efficiently for higher order Legendre polynomials. A simple algorithm to determine the weights of Gauss quadrature is to integrate Lagrange polynomial through the roots of Legendre polynomial between the interval $(-1, 1)$. However, as the order of polynomial increases, round off errors would make it difficult to determine the weights accurately. Alternatively, the nodes and weights can be determined using eigenvalue computations.[6]

Example 9.8

Use Gauss quadrature and integrate $f(x) = \dfrac{1}{0.3 - x^4}$ *between 1 and 0.8. Compare the value you obtain with the exact value.*

Solution :

In applying Gauss quadrature the limits of integration have to be -1 and $+1$. In this example the lower limit is $a = 1$ and the upper limit is $b = 0.8$. Transformation required

[5]**Abramowitz** and **Stegun** *Handbook of Mathematical Functions* Available online at http://people.math.sfu.ca/~cbm/aands/. To best of our knowledge the book is not copyrighted.

[6]G.H. Golub and J.H. Welsch, "Calculation of Gauss quadrature rules" 1969, Math. Comp, 23(106), 221-230.

to convert the limits to [-1, +1] is $x = \dfrac{a+b}{2} + t\dfrac{b-a}{2} = \dfrac{1+0.8}{2} + t\dfrac{0.8-1}{2} = 0.9 - 0.1t$. We also have $dx = -0.1dt$. Hence the desired integral is written as

$$\int_1^{0.8} \frac{dx}{a^4 - x^4} = -0.1 \int_{-1}^1 \frac{dt}{a^4 - (0.9 - 0.1t)^4}$$

One may use different Gauss quadrature rules to solve this problem. We show the use of Gauss quadrature with $n = 5$ as an example. Only 5 function evaluations are required to get an estimate for the integral. We make use of the Gauss points and weights given in Table 9.5 for this purpose. The computation is tabulated as shown below.

t	$f(t)$	$w(t)$	I_{G_5}
0.000000	-2.808200	0.568889	0.69473
0.538469	-4.703220	0.478629	
-0.538469	-1.894730	0.478629	
0.906180	-7.742635	0.236927	
-0.906180	-1.508303	0.236927	

It is interesting to note that the Gauss estimate of the integral has an error of only -0.00004 when compared to the exact value.

Composite Gauss quadrature Like other quadrature methods discussed before, composite rules can be applied using Gauss quadratures also. The entire integration range is divided into small segments and Gauss quadrature is applied to each segment. The following MATLAB program performs composite Gauss quadrature.

Program 9.5: *Composite 5 point Gauss quadrature*

```
 1 function in = compositegauss5pt(f,a,b,n)
 2 %  Input    f :   function to be integrated
 3 %           a :   lower bound
 4 %           b :   upper bound
 5 %           n :   number of segments
 6 % Outpur  in:   composite Gauss 5 pt estimate
 7 in = 0;                 % initialize integral
 8 h = (b-a)/n;            % step size
 9 for i=1:n               % Gauss 5 pt over each segment
10    in = in + gauss5pt(f,a+(i-1)*h,a+i*h);
11 end
```

Example 9.9

Integrate $\displaystyle\int_0^1 \frac{1}{1 + 25x^2} dx$ using Gauss quadratures. Refine the grid and apply composite Gauss quadrature rules. Comment on the errors.

Solution :

Program 9.5 is used to estimate the integral. Following table summarizes the estimate and error of Gauss quadrature.

h	estimate	actual error	estimated error
1	0.274321	3.59E-04	
0.5	0.274667	1.28E-05	3.46E-04
0.25	0.274680	2.71E-07	1.25E-05
0.125	0.274680	9.51E-10	2.70E-07

Estimated error in the above table indicates the difference between two consecutive quadrature estimates. One can obtain good convergence using small number of segments. Alternatively, one can use higher order Gauss quadrature formulae over the entire interval. But one must be careful not to round off the weights and nodes of the Gauss quadrature rules so that accuracy is not compromised.

Other Gauss quadrature rules   While Gauss Legendre quadrature is most commonly used, there are other rules which can be used under special conditions. All Gauss quadrature rules are exact up to polynomial degree of $2n - 1$. Consider the integral of the form

$$I = \int_{-1}^{1} f(x)dx = \int_{-1}^{1} W(x)g(x)dx \tag{9.64}$$

where $f(x)$ is a convolution (product) of two functions $W(x)$ and $g(x)$. Then for certain known functions $W(x)$ (orthogonal functions), the integral can be approximated as

$$I = \int_{-1}^{1} W(x)g(x)dx \approx \sum_{i=1}^{n} w_i g(x_i) \tag{9.65}$$

When $W(x) = 1$, we get the standard Gauss Legendre quadrature. Table 9.6 indicates some of the other quadrature rules.

Table 9.6: *Variants of Gauss quadrature rules*

Quadrature	Integral	$W(x)$	Orthogonal polynomial
Gauss-Legendre	$\int_{-1}^{1} g(x)dx$	1	Legendre
Chebyshev-Gauss	$\int_{-1}^{1} \dfrac{1}{\sqrt{1-x^2}} g(x)dx$	$\dfrac{1}{\sqrt{1-x^2}}$	Chebyshev
Gauss-Laguerre	$\int_{0}^{\infty} e^{-x} g(x)dx$	e^{-x}	Laguerre
Gauss-Hermite	$\int_{-\infty}^{\infty} e^{-x^2} g(x)dx$	e^{-x^2}	Hermite

The nodes and weight functions for rules given in Table 9.6 are available in handbooks. Alternatively, the nodes and weights of these orthogonal polynomials can be determined by solving eigenvalue problems.

9.7 Singular integrals

Consider the integral

$$\int_0^1 \frac{1}{\sqrt{x}}dx = 2\sqrt{x}\Big|_0^1 = 2 \tag{9.66}$$

$x = 0$ is a singular point since the integrand becomes infinite as $x \to 0$. Such integrals are known as singular integrals and are also classified as improper integrals. There are abundance of practical examples which involve singular integrals. Closed Newton Cotes quadrature rules cannot be used for a singular integral as they include the end points in their function evaluation.

9.7.1 Open Newton Cotes quadrature

Open Newton Cotes rules are similar to the closed quadrature rules discussed earlier. The nodes of open quadrature rules are distributed within the integration domain and the end points are avoided. The quadrature rule involves extrapolation of the Lagrange polynomial outside the range of the quadrature points. Like the closed formulae, the Lagrange polynomial are integrated over the entire domain. Hence open quadrature rules are useful when integrand is singular at one/both of the end points.

Mid point rule

The simplest open quadrature rule is the mid-point rule. As the name suggests, the mid-point of the integration segment is used as the node. Let the integration range be divided into n segments. Then the step size of each segment is given by $h = (b - a)/n$. The mid-point rule applied over a segment is given by

$$\int_{x_i}^{x_i+h} f(x)dx \approx hf(x_i + h/2) \tag{9.67}$$

The mid-point rule over the entire integration range can be written as

$$\int_a^b f(x)dx \approx h\sum_{i=0}^{n-1} f(x_i + h/2) \tag{9.68}$$

Mid-point rule has been illustrated in Figure 9.1. The quadrature rule is equivalent to trapezoidal rule as it also integrates a linear function exactly. The error of mid-point rule is proportional to h^2. However near a singularity the error of the integral is going to be large and very small step sizes have to be chosen.

Example 9.10

Integrate $\int_0^1 \frac{\exp\left(-\frac{1}{t}\right)}{t}dt$. *Use step sizes of* $h = 0.25$. *Refine the grid and discuss the improvement in the estimate.*

Background :

Integral $E_1(x) = \int_0^1 \frac{\exp\left(-\frac{x}{t}\right)}{t}dt$ is known as exponential integral of order 1. An nth order exponential function is defined as

$$E_n(x) = \int_0^1 t^{n-2}\exp\left(-\frac{x}{t}\right)dt$$

These integrals occur in solution of neutron transport equation and radiative transfer equation.

Solution :

At $t = 0$, we have $\lim\limits_{t \to 0} \dfrac{\exp\left(-\frac{1}{t}\right)}{t} \to \dfrac{0}{0}$. We cannot use Simpson's rule or trapezoidal rule for this example. We use the midpoint rule to evaluate the integral. The following table summarizes the results for various step sizes using the mid point rule

h	Integral	Error
1/4	0.218868	...
1/8	0.219416	5.48E-04
1/16	0.219384	3.19E-05
1/32	0.219384	3.49E-07

Error refers to difference between the estimates for two consecutive step sizes. One can apply Richardson extrapolation to the above data and further improve the accuracy.

Example 9.11

Integrate $\int_0^1 \dfrac{1}{\sqrt{x}} dx$ using mid point rule and compare the same with exact value for $h = 0.25$. Also refine the grid and discuss the improvement in the estimate.

Solution :

The following table summarizes the mid-point rule calculation with $h = 0.25$

x_i	0.125	0.375	0.625	0.875
$f(x_i)$	0.707107	0.408248	0.316228	0.267261
			Integral =	**1.698844**

The exact value of the integral is 2. The estimate of the integral does not improve much with grid refinement as seen in the following table.

h	estimate
1/4	1.6988
1/8	1.7865
1/16	1.8489
1/128	1.9465
1/1028	1.9811

Even after using very fine grids, the estimate is poor!

The derivation of higher order open quadrature rules are similar to closed quadrature rules and can be obtained by integrating Lagrange polynomials passing through the nodes.

Open Trapezoidal rule		
a $\quad$ 1 $\quad$ 2 $\quad$ b h	$I = \dfrac{b-a}{2}(f_1 + f_2)$	Degree of accuracy = 1
Milne's rule		
a $\quad$ 1 $\quad$ 2 $\quad$ 3 $\quad$ b h	$I = \dfrac{b-a}{3}(2f_1 - f_2 + 2f_3)$	Degree of accuracy = 3

The reader is encouraged to apply the higher order formulae to Example 9.11. Observe that the weights of the lower order quadrature formulae contain negative coefficients. Going by the arguments for closed Newton Cotes formulae, such methods are bound to be unstable. Alternatively, one can use a combination of open and closed Newton quadratures. For example apply open quadrature rule such as midpoint rule in a segment containing the singular point with rest of the region estimated by closed quadrature rule.

Alternative to Newton Cotes formulae, Gauss Legendre quadrature can be used as they naturally do not include the end points.

Example 9.12

Integrate $\int_0^1 \dfrac{1}{\sqrt{x}} dx$ using 5 point Gauss quadrature rule. Use step sizes of $h = 0.25$. Refine the grid and discuss the improvement in the estimate.

Solution :

We use program 9.5 to evaluate the integrals using composite Gauss quadrature. The results are summarized in the following table

h	estimate
1/4	1.920800
1/8	1.943997
1/16	1.960400
1/128	1.985999
1/1028	1.993000

Gauss quadrature performs better than mid-point rule but is not good enough. The reader can try higher order Gauss rules and check the results.

9.8 Integrals with infinite range

In all the integrals we have considered, the bounds a and b were finite. Sometimes, we encounter *finite* integrals which have a or b tending towards ∞. Such integrals are also known as improper integrals and special attention needs to be paid in evaluating these integrals. Examples are:

1. Cumulative probability of normal distribution is given by

$$p(x) = \frac{1}{\sqrt{2\pi}} \int_{-\infty}^{x} e^{-\frac{1}{2}x^2} dx$$

2. $\int_1^\infty \dfrac{1}{x} dx = \ln(x) \Big|_1^\infty$ is also an improper integral, but the integral of this function is divergent and quadrature rules are bound to fail for such an integral.

9.8.1   Coordinate transformation

Quadrature rules such as Newton Cotes and Gauss Legendre cannot be applied directly to the integral and would require some special operations. It always helps to reorganize the integral such that the resultant integral(s) have finite limits making it possible for the quadrature rules to be applied. This can be done in several ways. Consider the integral which is continuous in the interval and the integral converges to a finite value.

$$I = \int_0^\infty f(x)dx \tag{9.69}$$

Method 1 The integral can be broken into two parts as

$$I = \int_0^1 f(x)dx + \int_1^\infty f(x)dx \tag{9.70}$$

The first integral can be determined by any of the methods discussed earlier. A coordinate transformation can be performed on the second integral so that the limits are finite. In the above integral, the coordinate transformation would be $x = \dfrac{1}{t}$ and $dx = -\dfrac{1}{t^2}dt$ so that $x = 1$ becomes $t = 1$ and $x \to \infty$ becomes $t = 0$. Then the integral can be rewritten as

$$I = \int_0^1 f(x)dx + \int_0^1 \frac{1}{t^2} f\left(\frac{1}{t}\right) dt \tag{9.71}$$

Method 2 Alternatively, one can use the transformation $x = \dfrac{1-t}{t}$ and $dx = -\dfrac{dt}{t^2}$. The transformation is equivalent to $t = \dfrac{1}{1+x}$ so that $x = 0$ becomes $t = 1$ and $x \to \infty$ becomes $t = 0$. Then the integral becomes

$$I = \int_0^1 \frac{1}{t^2} f\left(\frac{1-t}{t}\right) dt \tag{9.72}$$

Therefore by suitable coordinate transformations one can convert infinite limit to finite limits.

Example 9.13

Integrate $I = \displaystyle\int_0^\infty \frac{1}{1+25x^2} dx$.

Solution :
We have already handled this function earlier.

Step 1 The exact solution for the integral is $I = \dfrac{1}{5} \tan^{-1} \infty = \dfrac{\pi}{10} = 0.314159$.

Step 2 We will split the interval of integration into two parts as follows

$$I = \int_0^1 \frac{1}{1+25x^2} dx + \int_1^\infty \frac{1}{1+25x^2} dx$$

Step 3 Numerically solve first integral (Example 9.6) to get $I_1 = 0.274680$.

Step 4 Transform coordinate for second integral with $x = 1/t$ to get

$$I_2 = \int_0^1 \frac{1}{t^2} \frac{t^2}{t^2 + 25} dt = \int_0^1 \frac{1}{t^2 + 25} dt$$

Step 5 Numerically integrating I_2 (up to six digits accuracy) we get $I_2 = 0.039479$.

Finally the desired integral is $I = I_1 + I_2 = 0.314159$.

Example 9.14

Integrate $I = \int_0^\infty e^{-x} dx$.

Solution :

Step 1 The exact solution for the integral is $I = -e^{-x}\big|_0^\infty = 1$.

Step 2 Transform the integral using $x = \dfrac{1-t}{t}$ as

$$I = \int_0^1 \frac{1}{t^2} \exp\left(\frac{1-t}{t}\right) dt$$

This integral is similar to the integral in Example 9.10

Step 3 Apply 5 point composite Gauss quadrature rule with n segments. The results have been tabulated below

n	estimate	error
1	1.002635	2.64E-03
2	0.999888	-1.12E-04
3	1.000188	1.88E-04
4	0.999954	-4.58E-05
5	0.999988	-1.15E-05
6	1.000003	3.01E-06

Alternatively, as the integral is of the form $\int_0^\infty \underbrace{e^{-x}}_{W} f(x) dx$ with $f(x) = 1$, Gauss-Laguerre quadrature can be used to estimate the integral.

9.9 Adaptive quadrature

For a given function, there could be regions where there is a steep variation in the integrand compared to others. Large number of integration nodes would be required in such regions where as only a few nodes would be sufficient in regions of gentle variations. In the methods discussed thus far, there was complete disregard for the nature of variation of the function. This means, even in regions where the function variation is gentle, fine grids were used. Adaptive quadrature becomes useful when the integral is of the following forms:

1. The integrand is singular at one or both the boundaries.
2. The limits of integration are $\pm\infty$
3. Integrand has singular derivatives and hence cannot be represented by a polynomial. Even Gauss quadrature may converge slowly in such cases.

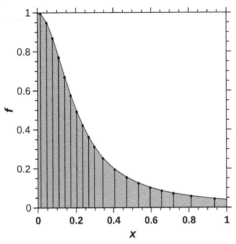

Figure 9.10: *Distribution of integrating segments for adaptive quadrature*

An adaptive quadrature method can be formulated such that it automatically decides the step size. Instead of applying the refinement uniformly in the entire region, the refinement is applied in required regions as shown in Figure 9.10. The procedure for one segment can be written as

Step 1 Determine estimate for grid size h and $h/2$ for the segment

Step 2 If $|T(h) - T(h/2)| < 3\epsilon$ for trapezoidal or $|S(h) - S(h/2)| < 15\epsilon$ for Simpson's, stop refinement

Step 3 Else, equally divide the segment into two segments and repeat the steps again for each segment. The tolerance for the new segments is half the tolerance for the big segment.

Therefore adaptive Simpson's rule becomes a recursive program[7] where the function calls itself until the desired conditions are met.

Warnings: While programming, one must be careful about two aspects. As the program is recursive in nature, one must specify a maximum level of recursion of the program i.e. number of times the program calls itself. Also, one must terminate the recursion if the step size reached is smaller than the precision of the computer.

A MATLAB program has been written to perform adaptive Simpson's rule. We have fixed the maximum recursion level at 15 and fixed minimum step size at 10^{-12}

Program 9.6: *Adaptive Simpson's rule*

```
 1  function in = adaptsimpsons(f,a,c,b,f1,f3,f2,tol,i)
 2  %   Input    f   :   function to be integrated
 3  %            a   :   lower bound
 4  %            c   :   (a+b)/2
 5  %            b   :   upper bound
 6  %            f1  :   f(a)
 7  %            f3  :   f(c)
 8  %            f2  :   f(b)
 9  %            tol:   tolerance
10  %            i   :   recursion level (i=0 for level 1)
11  %  Output    in  :   integral
12  i = i+1;                        % increment recursion level
13  d = (a+c)/2; e = (c+b)/2; % mid points of two sub segments
14  f4 = f(d);    f5 = f(e);   % function at two subsegments
```

---

[7]A program which calls itself is called as recursive and recursion would continue as long as specified conditions are not met. For example to determine $n!$ we can write a recursive program to multiply n with $n - 1!$. The program continues to call itself until $n = 1$.

```
15  in1 = (f1 + 4*f3+ f2)*(b-a)/6;              % S(h)
16  in2 = (f1 + 4*f4 + 2*f3 + 4*f5 + f2)*(b-a)/12;    % S(h/2)
17  if( abs(in1-in2) < 15*tol || i == 15 || b-c < 10^-12)
18  % if diff < tol,   or i = max recursion level
19  %    or step size < 10^-12
20     in = in2;     % in = S(h/2)
21  else
22     % if diff > tol, determine integrals for two segments
23     in1 = adaptsimpsons(f,a,d,c,f1,f4,f3,tol/2,i);
24     in2 = adaptsimpsons(f,c,e,b,f3,f5,f2,tol/2,i);
25     in = in1 + in2;   % total integral over segment
26  end
```

Applying adaptive Simpson's rule to Example 9.3 with tolerance 1×10^{-6}.

```
f = @(x) 1./(1+25*x.^2);
f1 = f(0);
f2 = f(1);
f3 = f(0.5);
[in,er] = adaptsimpsons(f,0,0.5,1,f1,f3,f2,1e-6,0);
```

The output of the program is

```
in = 0.274680
er = 8.2176e-07
```

The number of segments required for the above estimate is 66 with about 132 function evaluations. The error of the estimate compared to the exact value is 3.37×10^{-8}. It seems that the adaptive Simpson's rule has no added advantage over Composite Simpson's rule. The real advantage of adaptive quadrature comes when the integral does not converge easily. We consider an example to illustrate the point.

Example 9.15

Integrate $I = \int_0^1 \sqrt{x}\,dx$ up to six digits accuracy using composite Simpson's rule and adaptive Simpson's rule.

Solution :

Programs 9.2 and 9.6 have been used to estimate the integral.

```
inc = 0;
i = 2;
f = @(x) sqrt(x);
while 1
    h = 1/2^i;
    h1(i-1) = h;
    x = [0:h:1]';
    y = f(x);
    in(i) = compsimpson(y,h);
    if( abs(in(i)-in(i-1))/15 < 1e-6 )
        break;
    end
    i = i+1;
end
ina = adaptsimpsons(f,0,0.5,1,f(0),(0.5),f(1),1e-6,0);
```

The integral estimates using composite Simpson's rule, the actual error (difference between integral estimate and exact value) and estimated errors for various sizes are tabulated below.

h	Integral	Error	
		ϵ_S	actual
1/4	0.656526		1.01E-02
1/8	0.663079	4.37E-04	3.59E-03
1/16	0.665398	1.55E-04	1.27E-03
1/32	0.666218	5.47E-05	4.48E-04
1/64	0.666508	1.93E-05	1.59E-04
1/128	0.666611	6.83E-06	5.61E-05
1/256	0.666647	2.42E-06	1.98E-05
1/512	0.666660	8.54E-07	7.01E-06

The adaptive Simpson's rule converged to the correct solution with actual error of 10^{-9} and required only 67 segments. The number of function evaluations of adaptive Simpson's rule is significantly lower than composite rule.

Adaptive Gauss quadrature

Adaptive quadrature can be used with Gauss Legendre rules. The procedure for adaptive Gauss quadrature is similar to those of trapezoidal and Simpson's rules. The Gauss quadrature points of two different orders (apart from $x = 0$) never coincide. To estimate error of a Gauss quadrature, one must compute the function values for each quadrature afresh. Thus, Gauss quadrature loses the advantage of nesting as in Simpson's rule. Rather nested Gauss quadratures (Gauss Kronrod quadrature) can be used in which the nodes of the lower order quadrature (Gauss) are also nodes of the higher quadrature formula (Kronrod)[8]. The error estimate of the integral is the difference between Gauss and Kronrod rule. The weights and nodes of 7 point Gauss-15 point Kronrod quadrature rule has been provided below.

Nodes	Weights	
	Gauss	Kronrod
± 0.991455		0.022935
± 0.949108	0.129485	0.063092
± 0.864864		0.104790
± 0.741531	0.279705	0.140653
± 0.586087		0.169005
± 0.405845	0.381830	0.190351
± 0.207785		0.204433
0	0.417959	0.209482

Adaptive Gauss Kronrod quadrature can be invoked using the command **quadgk** in MATLAB. The usage of the command is as follows

---

[8]A brief history of Gauss Kronrod rules is given by W. Gautschi "A historical note on Gauss-Kronrod quadrature" 2005 Numerische Mathematik, 100(3), 483-484. For more mathematical insight on nested quadrature, refer to T.N.L. Patterson "The optimum addition of points to quadrature formulae" 1968, Math. Comput, 22(104), 847-856.

```
f = @(x) sqrt(x);
in = quadgk(f,0,1)
```

The output of the program is

```
in = 0.6667
```

In general for all practical non-singular integrals, adaptive Simpson's rule would be sufficient. However, for an improper integral, adaptive Gauss quadrature may be necessary. Quadrature rules based on Chebyshev polynomials (Clenshaw-Curtis) are also extensively used in such cases. Gauss-Kronrod adaptive quadratures have been used extensively in most software packages including MATLAB.

9.10 Multiple integrals

Multiple integrals refer to integrals of a function of more than one variable along more than one direction. A double integral integrates the function along two directions where as a triple integral involves integration along three directions. The concepts that have been presented until now for line integrals can easily be adopted for multiple integrals. Geometrically, a double integral of a function with respect to x and y is equal to the volume between the surface represented by the function and the $x - y$ plane. Like we represented the line integral as summation of areas of vertical strips, the double integral can be approximated as the total sum of volumes of vertical boxes. Geometrical representation of higher dimensional integrals become difficult to visualize. A multiple integral can also be represented by a quadrature rule which is a weighted sum of function values at a few suitably chosen points in the integration range.

$$I = \int \int \cdots \int f(x_1, x_2 \cdots x_m) dx_1 dx_2 \cdots dx_m \approx \sum_i^n w_i f_i \qquad (9.73)$$

where m refers to the dimension of the integral and n refers to number of points used in the quadrature rule. The following section introduces methods to construct multi-dimensional quadrature rules from the concepts used in evaluating line integrals.

9.10.1 Double integral with fixed limits for both x and y

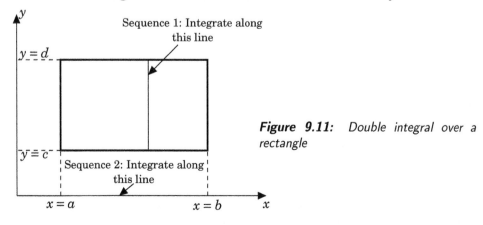

Figure 9.11: *Double integral over a rectangle*

Now consider a double integral over a rectangular domain as shown in Figure 9.11. The required integral has constant upper and lower limits along x and y directions and is defined by

$$I = \int_{x=a}^{b} \int_{y=c}^{d} f(x,y)\,dx\,dy \tag{9.74}$$

Sequential integration is possible and hence one may perform first an integration along the vertical line at any x followed by a second integration along the horizontal line as indicated below.

$$I = \int_{x=a}^{b} \left[\int_{y=c}^{d} f(x,y)\,dy \right] dx = \int_{x=a}^{b} g(x)\,dx \tag{9.75}$$

where $g(x) = \displaystyle\int_{y=c}^{d} f(x,y)\,dy$. Thus the first integration yields a function of x that may be integrated with respect to x in the second sequence. This is indeed the method that is employed in analytical integration. Of course it is possible to integrate first with respect to x along a horizontal line at any y and subsequently with respect to y along the vertical line. The numerical integration scheme also realizes the double integral as a sequential one as we shall see below.

9.10.2　Double integrals using Newton Cotes quadrature

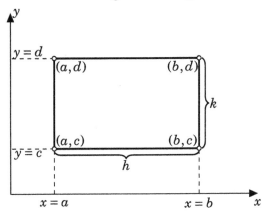

Figure 9.12: *Trapezoidal rule for integration in two dimensions*

Consider the rectangle shown in Figure 9.12. Let us apply sequential integration over the rectangle. The integral in each direction is approximated by trapezoidal rule. Therefore the integral along y can be written as

$$g(x) = \int_{y=c}^{d} f(x,y)\,dy \approx \frac{f(x,c)+f(x,d)}{2}(d-c) \tag{9.76}$$

Now integrating the above approximation along x direction we get

$$\begin{aligned}
I = \int_{x=a}^{b} g(x)\,dx &\approx \frac{g(a)+g(b)}{2}(b-a) \\
&\approx \frac{f(a,c)+f(a,d)+f(b,c)+f(b,d)}{4}hk \tag{9.77}
\end{aligned}$$

where $h = b-a$ and $k = d-c$ are the step sizes along x and y directions respectively. This represents the trapezoidal rule in two dimensions and the estimate is exact for functions linear in x and y.

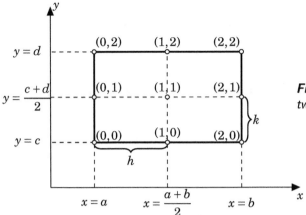

Figure 9.13: *Simpson's rule for two dimensions*

Similarly, Simpson's rule for two dimensions can be derived by performing sequential Simpson's rule along x and y directions. Consider the rectangular domain of width $2h$ and height $2k$ as shown in Figure 9.13. The domain of integration has been covered by two steps of length h each along the x direction and two steps of length k each along the y direction. The nodes are identified by i, j where $0 \le i \le 2$ and $0 \le j \le 2$. Consider sequential integration as discussed previously. Using Simpson's 1/3 rule, the integrals are given by

$$
\begin{aligned}
g(x=a) &= \frac{k}{3}\left[f_{0,0}+4f_{0,1}+f_{0,2}\right] \\
g\left(x=\frac{a+b}{2}\right) &= \frac{k}{3}\left[f_{1,0}+4f_{1,1}+f_{1,2}\right] \\
g(x=b) &= \frac{k}{3}\left[f_{2,0}+4f_{2,1}+f_{2,2}\right]
\end{aligned}
\tag{9.78}
$$

We now use Simpson's rule to integrate $g(x)$ over the range $0, 2h$ to get

$$
\begin{aligned}
I &= \frac{h}{3}\left[g(a)+4g\left(\frac{a+b}{2}\right)+g(b)\right] \\
&= \frac{hk}{9}\left[\begin{array}{ccc} f(0,0)+ & 4f(1,0)+ & f(2,0)+ \\ 4f(0,1)+ & 16f(1,1)+ & 4f(2,1)+ \\ f(0,2)+ & 4f(1,2)+ & f(2,2) \end{array}\right]
\end{aligned}
\tag{9.79}
$$

Simpson's quadrature for a double integral integrates exactly a function containing x^3 and y^3. The errors in the estimate would occur due to terms involving x^4 and y^4. In general the quadrature rules arising out of sequential integration can be written as

$$
I = \int_{x=a}^{b}\int_{y=c}^{d} f(x,y)\,dx\,dy = \sum_{i=0}^{n}\sum_{j=0}^{m} w_{ix}w_{jy}f(x_i,y_j) = \sum_{i=0}^{n}\sum_{j=0}^{m} w_{i,j}f(x_i,y_j)
\tag{9.80}
$$

where w_x and w_y refer to the weights respectively along x and y directions. Table 9.7 summarizes the weights for trapezoidal, Simpson's 1/3 and Simpson's 3/8 rule over two dimensions.

Table 9.7: *Weights for trapezoidal, Simpsons 1/3 and 3/8 two dimensional quadrature rules*

Trapezoidal

$$\frac{1}{4}\begin{array}{|c|c|}\hline 1 & 1 \\\hline 1 & 1 \\\hline\end{array}$$

Simpson's 1/3

$$\frac{1}{9}\begin{array}{|c|c|c|}\hline 1 & 4 & 1 \\\hline 4 & 16 & 4 \\\hline 1 & 4 & 1 \\\hline\end{array}$$

Simpson's 3/8

$$\frac{9}{64}\begin{array}{|c|c|c|c|}\hline 1 & 3 & 3 & 1 \\\hline 3 & 9 & 9 & 3 \\\hline 3 & 9 & 9 & 3 \\\hline 1 & 3 & 3 & 1 \\\hline\end{array}$$

The process of sequential integration can be applied to other quadrature rules such as Newton cotes, Chebyshev and Gauss quadratures. Also, a combination of quadrature rules can be applied over different directions. For example if the function is linear with respect to x and quadratic with respect to y, one can apply trapezoidal rule along x direction and Simpson's rule along y direction.

Alternative view of integration over a rectangle

Let us transform the coordinates to local coordinate system (Figure 9.14) for a rectangular element introduced in earlier chapter.

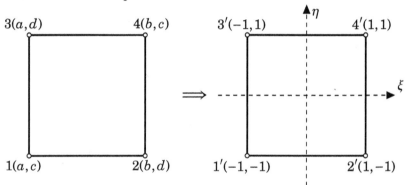

Figure 9.14: *Transforming coordinates to local coordinate system*

The coordinate transformation can be written as

$$x = \frac{b+a}{2} + \frac{b-a}{2}\xi; \quad y = \frac{c+d}{2} + \frac{d-c}{2}\eta; \tag{9.81}$$

$$dx = \frac{b-a}{2}d\xi; \quad dy = \frac{d-c}{2}d\eta; \tag{9.82}$$

Therefore the integral becomes

$$I = \int_a^b \int_c^d f(x,y)dy dx = \frac{(b-a)}{2}\frac{(d-c)}{2}\int_{-1}^1\int_{-1}^1 f(\xi,\eta)d\eta d\xi \tag{9.83}$$

Recollect two dimensional interpolation over a rectangle from Chapter 5. The function within the rectangle is represented in Lagrange form and the integral becomes

$$I = \frac{(b-a)}{2}\frac{(d-c)}{2}\int_{-1}^1\int_{-1}^1 f(\xi,\eta)d\eta d\xi = \frac{(b-a)}{2}\frac{(d-c)}{2}\int_{-1}^1\int_{-1}^1 \sum_{i=1}^n l_i f_i d\eta d\xi$$

$$= \frac{(b-a)}{2}\frac{(d-c)}{2}\sum_{i=1}^n f_i \int_{-1}^1\int_{-1}^1 l_i d\eta d\xi \tag{9.84}$$

where l_i is the Lagrangian weight functions of the two dimensional element. Going by the definition of quadrature rules, the weights of the quadrature rule can be written as

$$w_i = \frac{(b-a)}{2}\frac{(d-c)}{2}\int_{-1}^{1}\int_{-1}^{1} l_i d\xi d\eta \tag{9.85}$$

Integration over a linear rectangular element

The function within the linear element is written as

$$f(\xi,\eta) = l_1 f_1 + l_2 f_2 + l_3 f_3 + l_4 f_4 \tag{9.86}$$

and the Lagrange weight function is given by Equation 6.11

$$l_1(\xi,\eta) = \frac{(1-\xi)}{2}\frac{(1-\eta)}{2} \quad l_2(\xi,\eta) = \frac{(1+\xi)}{2}\frac{(1-\eta)}{2}$$
$$l_3(\xi,\eta) = \frac{(1-\xi)}{2}\frac{(1+\eta)}{2} \quad l_4(\xi,\eta) = \frac{(1+\xi)}{2}\frac{(1+\eta)}{2} \tag{9.87}$$

Integrating the first Lagrange weight function we get

$$\begin{aligned} w_1 &= \frac{(b-a)}{2}\frac{(d-c)}{2}\int_{-1}^{1}\int_{-1}^{1}\frac{(1-\xi)}{2}\frac{(1-\eta)}{2} d\xi d\eta \\ &= \frac{hk}{16}\left(\xi - \frac{\xi^2}{2}\right)\Big|_{-1}^{1}\left(\eta - \frac{\eta^2}{2}\right)\Big|_{-1}^{1} = \frac{hk}{4} \end{aligned} \tag{9.88}$$

Similarly, other weights may be derived to get

$$w_1 = w_2 = w_3 = w_4 = \frac{hk}{4} \tag{9.89}$$

These are the same as for the trapezoidal rule.

Integration over a quadratic rectangular element

The weight functions for a Lagrange quadratic element are

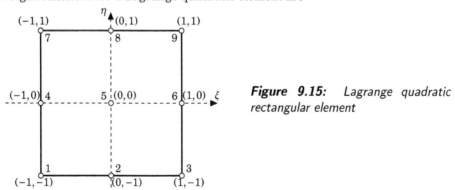

Figure 9.15: *Lagrange quadratic rectangular element*

$$l_1 = \frac{\xi\eta(\xi-1)(\eta-1)}{4} \quad l_2 = \frac{(1-\xi^2)\eta(\eta-1)}{2} \quad l_3 = \frac{\xi\eta(\xi+1)(\eta-1)}{4}$$
$$l_4 = \frac{\xi(\xi-1)(1-\eta^2)}{2} \quad l_5 = (1-\xi^2)(1-\eta^2) \quad l_6 = \frac{\xi(\xi+1)(1-\eta^2)}{2}$$
$$l_7 = \frac{\xi\eta(\xi-1)(\eta+1)}{4} \quad l_8 = \frac{(1-\xi^2)\eta(\eta+1)}{2} \quad l_9 = \frac{\xi\eta(\xi+1)(\eta+1)}{4} \tag{9.90}$$

Integrating the first Lagrange weight function we get

$$
\begin{aligned}
w_1 &= \frac{(b-a)}{2}\frac{(d-c)}{2}\int_{-1}^{1}\int_{-1}^{1}\frac{\xi\eta(\xi-1)(\eta-1)}{4}d\xi d\eta \\
&= \frac{4hk}{4}\frac{1}{4}\left(\frac{\xi^3}{3}-\frac{\xi^2}{2}\right)_{-1}^{1}\left(\frac{\eta^3}{3}-\frac{\eta^2}{2}\right)_{-1}^{1} = \frac{hk}{9}
\end{aligned}
\tag{9.91}
$$

Similarly all the other weights of the quadrature rule are determined and we end up with Simpson's 1/3 rule applied along the two directions.

A similar process can be used to derive higher order Newton Cotes quadrature rules. But as stated for line integral, it is advisable to use composite lower order quadrature rules.

Double integral using composite trapezoidal rule

Let us consider the use of composite trapezoidal rule in determining double integral over a rectangle as shown in Figure 9.13. The rectangle is of width $2h$ and height $2k$ as shown. The domain of integration has been covered by two steps of length h each along the x direction and two steps of length k each along the y direction. Let us perform sequential integration. The integrals are given by

$$
\begin{aligned}
g(x=0) &= k\left[\frac{f_{0,0}}{2}+f_{0,1}+\frac{f_{0,2}}{2}\right] \\
g(x=h) &= k\left[\frac{f_{1,0}}{2}+f_{1,1}+\frac{f_{1,2}}{2}\right] \\
g(x=2h) &= k\left[\frac{f_{2,0}}{2}+f_{2,1}+\frac{f_{2,2}}{2}\right]
\end{aligned}
\tag{9.92}
$$

We now use composite trapezoidal rule to integrate $g(x)$ over the range $0,2h$ to get

$$
\begin{aligned}
I &= k\left[\frac{g(x=0)}{2}+g(x=h)+\frac{g(x=2h)}{2}\right] \\
&= hk\left[\frac{f_{0,0}+f_{0,2}+f_{2,0}+f_{2,2}}{4}+\frac{f_{1,0}+f_{0,1}+f_{2,1}+f_{1,2}}{2}+f_{1,1}\right]
\end{aligned}
\tag{9.93}
$$

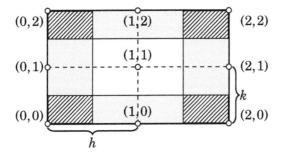

Figure 9.16: *Use of composite trapezoidal rule for numerical integration over a rectangle*

It is clear that the expression given above may be interpreted as a weighted sum of function values at the nodes, the weights being fractions of the elemental area hk. The corner nodes have weights each of $\frac{1}{4}$, nodes at the centers of sides have weights each of $\frac{1}{2}$ and the center node has a

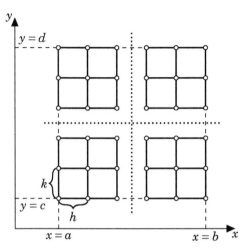

Figure 9.17: Distribution of nodes and segments for composite trapezoidal rule

weight of 1. The nodes and weights are shown by the areas associated with each of the nodes in Figure 9.16. The area in white is associated with the central node. The hatched areas represent those associated with the corner nodes. The areas filled with gray indicate the areas associated with the nodes lying on the middle of the sides of the rectangle.

We may extend the above to an arbitrary rectangle lying between $x = a$, $x = b$ and $y = c$, $y = d$ as shown in Figure 9.17. The width of the rectangle is divided into uniform segments of width $h = \dfrac{b-a}{m}$ where m is an integer. Similarly the height of the rectangle is divided into uniform segments of height $k = \dfrac{d-c}{n}$ where n is an integer. The integral is replaced by a double sum given by

$$I = \int_{x=a}^{x=b} \int_{y=c}^{y=d} f(x,y)dxdy = hk \sum_{i=0}^{m} \sum_{j=0}^{n} w_{i,j} f(x_i, y_j) \tag{9.94}$$

where the weights are as indicated in Array 9.95.

$$w_{i,j} = \frac{1}{4}$$

	0	1	2	3	$\cdots\cdots$	$n-1$	n
0	1	2	2	2	$\cdots\cdots$	2	1
1	2	4	4	4	$\cdots\cdots$	4	2
2	2	4	4	4	$\cdots\cdots$	4	2
3	2	4	4	4	$\cdots\cdots$	4	2
$\cdots$	$\cdots$	$\cdots$	$\cdots$	$\cdots$	$\cdots\cdots$	$\cdots$	$\cdots$
$\cdots$	$\cdots$	$\cdots$	$\cdots$	$\cdots$	$\cdots\cdots$	$\cdots$	$\cdots$
$m-1$	2	4	4	4	$\cdots\cdots$	4	2
m	1	2	2	2	$\cdots\cdots$	2	1

$$\tag{9.95}$$

Example 9.16

Obtain the double integral $I = \displaystyle\int_{x=2}^{3} \int_{y=2}^{3} (x^2 + y^2)dxdy$ by trapezoidal rule.

Solution :

We use step size of $h = 0.25$ and $k = 0.25$ to obtain the trapezoidal estimate of the double integral. The function values required are calculated and tabulated below.

$x \downarrow y \rightarrow$	2	2.25	2.5	2.75	3
2	8	9.0625	10.25	11.5625	13
2.25	9.0625	10.125	11.3125	12.625	14.0625
2.5	10.25	11.3125	12.5	13.8125	15.25
2.75	11.5625	12.625	13.8125	15.125	16.5625
3	13	14.0625	15.25	16.5625	18

Entries in the table are function values

Multiplying the entries in the table by the appropriate weights given by Array 9.95 the following table may be constructed.

2	4.53125	5.125	5.78125	3.25
4.53125	10.125	11.3125	12.625	7.03125
5.125	11.3125	12.5	13.8125	7.625
5.78125	12.625	13.8125	15.125	8.28125
3.25	7.03125	7.625	8.28125	4.5

Entries in the table are product of function and weights

The required double integral is then given by the sum of all entries in the above table multiplied by $hk = 0.25 \times 0.25 = 0.0625$. Thus we have $I_T = 203 \times 0.0625 = 12.6875$. The exact value of the integral may easily be obtained by sequential integration as $I_E = 12.6667$. Trapezoidal estimate has an error of $12.6875 - 12.6667 = 0.0208$.

Double integral using composite Simpson's rule

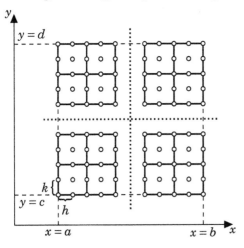

Figure 9.18: *Distribution of nodes and segments for composite Simpson's rule*

We shall now see what happens in the case of a rectangle bound by $x = a$, $x = b$ and $y = c$, $y = d$ as shown in Figure 9.18. The rectangle is divided into m steps of size h along x and n steps of size

k along y where both $m \geq 2$ and $n \geq 2$ are even integers. The reader may easily show that Equation 9.94 is valid with weights alone being replaced by those shown by Array 9.96.

$$w_{i,j} = \frac{1}{9}$$

	0	1	2	3	......	$n-1$	n
0	1	4	2	4	......	4	1
1	4	16	8	16	......	16	4
2	2	8	4	8	......	8	2
3	4	16	8	16	......	16	4
...	...	...	...	...	......	...	...
...	...	...	...	...	......	...	...
$m-1$	4	16	8	16	......	16	4
m	1	4	2	4	......	4	1

(9.96)

Example 9.17

Obtain the double integral $I = \int_{x=0}^{1} \int_{y=0}^{1} e^{-(x^2+y^2)} dx dy$ by Simpson rule. Use equal step sizes of 0.25 in the two directions.

Solution :

We use step size of $h = 0.25$ and $k = 0.25$ to obtain the Simpson estimate of the double integral. The function values required are calculated and tabulated below.

$x \downarrow y \rightarrow$	0	0.25	0.5	0.75	1
0	1	0.939413	0.778801	0.569783	0.367879
0.25	0.939413	0.882497	0.731616	0.535261	0.345591
0.5	0.778801	0.731616	0.606531	0.443747	0.286505
0.75	0.569783	0.5352613	0.4437473	0.3246523	0.2096113
1	0.3678793	0.345591	0.286505	0.209611	0.135335

Entries in the table are function values

Multiplying the entries in the table by the appropriate weights given by Array 9.96 the following table may be constructed.

0.111111	0.417517	0.173067	0.253237	0.040875
0.417517	1.568883	0.650325	0.951576	0.153596
0.173067	0.650325	1.078277	0.394442	0.063668
0.253237	0.951576	0.394442	0.577160	0.093161
0.040875	0.153596	0.063668	0.093161	0.015037

Entries in the table are product of function and weights

The required double integral is then given by the sum of all entries in the above table multiplied by $hk = 0.25 \times 0.25 = 0.0625$. Thus we have $I_S = 9.733395 \times 0.0625 = 0.608337$. This may be compared with the exact value of $I_E = 0.557747$. We see that the error is an unacceptably high value of $0.608337 - 0.557747 = 0.050590$.

9.10.3   Double integrals using Gauss quadrature

The method essentially consists in obtaining the double integral by sequential integration. First we may integrate with respect to y using a Gauss rule with N_y points. The limits of integration are changed to -1,1 by the transformation $y = \dfrac{c+d}{2} + \dfrac{d-c}{2}\eta$ such that integration with respect to y is written down as

$$g(x) = \frac{d-c}{2} \int_{-1}^{1} f\left(x, \frac{c+d}{2} + \frac{d-c}{2}\eta\right) d\eta \qquad (9.97)$$

We make use of Gauss quadrature with N_y Gauss points to write the above integral as

$$g(x) = \frac{d-c}{2} \sum_{i=0}^{N_y} w_{yi} f\left(x, \frac{c+d}{2} + \frac{d-c}{2}\eta_i\right) \qquad (9.98)$$

where w_{yi}'s are the Gauss weights. We now integrate $g(x)$ with respect to x by using Gauss quadrature with N_x points. Again the limits of integration are changed to -1,1 by the transformation $x = \dfrac{a+b}{2} + \dfrac{b-a}{2}\xi$. The double integral is then written down as

$$I = \frac{(b-a)}{2} \frac{(d-c)}{2} \sum_{j=0}^{N_x} \sum_{i=0}^{N_y} w_{xj} w_{yi} f\left(\frac{a+b}{2} + \frac{b-a}{2}\xi_j, \frac{c+d}{2} + \frac{d-c}{2}\eta_i\right) \qquad (9.99)$$

where w_{xj}'s are the Gauss weights.

Specific example

A specific example will make the above procedure clear. Consider, for simplicity, a square domain of side 2 each. Assume that we want to use a three point Gauss quadrature formula for integration with respect to both x and y. The Gauss points are then as indicated in Figure 9.19(a), based on data from Table 9.5. The weights are now calculated as products of weights, as required by Equation 9.99 and indicated in Figure 9.19(b).

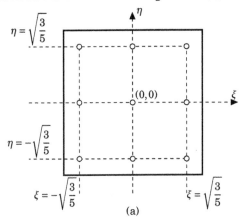

	25	40	25
$\frac{1}{81}$	40	64	40
	25	40	25

(b)

Figure 9.19: *Gauss points and weights for double integration over the standard square*

(a)

We may thus write the integral as

$$I = \frac{64}{81} f(0,0)$$

$$+ \quad \frac{40}{81}\left\{f\left(\sqrt{\frac{3}{5}},0\right)+f\left(-\sqrt{\frac{3}{5}},0\right)+f\left(0,-\sqrt{\frac{3}{5}}\right)+f\left(0,\sqrt{\frac{3}{5}}\right)\right\}$$

$$+ \quad \frac{25}{81}\left\{f\left(\sqrt{\frac{3}{5}},\sqrt{\frac{3}{5}}\right)+f\left(-\sqrt{\frac{3}{5}},\sqrt{\frac{3}{5}}\right)+f\left(-\sqrt{\frac{3}{5}},-\sqrt{\frac{3}{5}}\right)+f\left(\sqrt{\frac{3}{5}},-\sqrt{\frac{3}{5}}\right)\right\}$$

Example 9.18

Obtain the double integral $I = \int_{x=0}^{1}\int_{y=0}^{1}e^{-(x^2+y^2)}dxdy$ *by Gauss quadrature. Use three Gauss points along the two directions.*

Solution :

In the first step we transform the unit square (domain of integration) to a standard square with side 2 with $-1 \leq \xi$ or $\eta \leq 1$. We then have $dxdy = 0.25d\xi d\eta$. The desired transformations are easily shown to be $x = 0.5(1+\xi)$ and $y = 0.5(1+\eta)$. The Gauss points and weights are as in Section 9.10.3. The function values required are calculated and tabulated below.

$\xi\downarrow\ \eta\rightarrow$	$-\sqrt{\dfrac{3}{5}}$	0	$\sqrt{\dfrac{3}{5}}$
$-\sqrt{\dfrac{3}{5}}$	0.974917	0.768971	0.449329
0	0.768971	0.606531	0.354411
$\sqrt{\dfrac{3}{5}}$	0.449329	0.354411	0.207091

Entries in the table are function values

Multiplying the entries in the table by the appropriate weights given in Figure 9.19 the following table may be constructed.

0.300900	0.379739	0.138682
0.379739	0.479234	0.175018
0.138682	0.175018	0.063917

Entries in the table are product
of function and weights

The required double integral is then given by the sum of all entries in the above table multiplied by 0.25 to account for the transformation. Thus we have $I_G = 2.230928 \times 0.25 = 0.557732$. This may be compared with the exact value of $I_E = 0.557747$. We see that the error is a mere $0.557732 - 0.557747 = -0.000015$.

9.10.4 Double integral with variable limits on x or y

We now consider a double integral in the form

$$I = \int_{x=a}^{b}\int_{y=c(x)}^{d(x)}f(x,y)dydx \tag{9.100}$$

The integration is over an area bounded by the two curves $y = c(x)$ and $y = d(x)$ and the lines $x = a$ and $x = b$, as shown in Figure 9.20. Integration again proceeds sequentially to get

$$I = \int_{x=a}^{b} \left[\int_{y=c(x)}^{d(x)} f(x,y) dy \right] dx = \int_{x=a}^{b} g(x) dx \qquad (9.101)$$

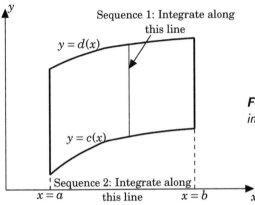

Sequence 1: Integrate along this line

$y = d(x)$

$y = c(x)$

Sequence 2: Integrate along this line

$x = a$ $x = b$

Figure 9.20: *General double integration evaluation method*

Composite Newton Cotes

Consider the application of a composite trapezoidal rule to evaluate this integral. The step size along y used in integration at a fixed $x_i = a + ih$ will be given by $k_i = \dfrac{d(x_i) - c(x_i)}{m}$ assuming that the number of steps in the y direction is a constant equal to m. The weights are now changed as shown in Array 9.102.

$$hkw_{i,j} = \frac{h}{4} \begin{pmatrix} & 0 & 1 & 2 & 3 & \cdots\cdots & m-1 & m \\ \hline 0 & k_0 & 2k_0 & 2k_0 & 2k_0 & \cdots\cdots & 2k_0 & k_0 \\ 1 & 2k_1 & 4k_1 & 4k_1 & 4k_1 & \cdots\cdots & 4k_1 & 2k_1 \\ 2 & 2k_2 & 4k_2 & 4k_2 & 4k_2 & \cdots\cdots & 4k_2 & 2k_2 \\ 3 & 2k_3 & 4k_3 & 4k_3 & 4k_3 & \cdots\cdots & 4k_3 & 2k_3 \\ \cdots & \cdots & \cdots & \cdots & \cdots & \cdots\cdots & \cdots & \cdots \\ \cdots & \cdots & \cdots & \cdots & \cdots & \cdots\cdots & \cdots & \cdots \\ n-1 & 2k_{n-1} & 4k_{n-1} & 4k_{n-1} & 4k_{n-1} & \cdots\cdots & 4k_{n-1} & 2k_{n-1} \\ n & k_n & 2k_n & 2k_n & 2k_n & \cdots\cdots & 2k_n & k_n \end{pmatrix} \qquad (9.102)$$

Figure 9.21 indicates the geometrical representation of trapezoidal rule applied to a variable

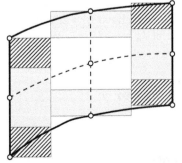

Figure 9.21: *Composite trapezoidal rule over a two dimensional domain with variable limits*

domain. Reader may similarly work out the weights for composite Simpson's rule also.

Gauss quadrature

Once we decide the number of Gauss points, the procedure of obtaining the integral is quite straightforward. Corresponding to the Gauss point with the x component x_i the limits of integration are given by $y = c(x_i)$ and $x = d(x_i)$. This region is easily transformed to the limits $\eta = -1$ to $\eta = 1$ by the transformation $y = \dfrac{c(x_i) + d(x_i)}{2} + \dfrac{d(x_i) - c(x_i)}{2}\eta$. The integral $g(x_i)$ is thus obtained by using Gauss quadrature with N_y Gauss points after the transformation. The second integration with respect to x is as before in the case of double integral with fixed limits of integration on x and y.

Example 9.19

Obtain the double integral $I = \displaystyle\int_{x=1}^{2}\int_{y=0.5}^{(0.5+0.5x)} (x^2 y - xy^2)\,dx\,dy$ *by using 3 Gauss points along each direction of integration.*

Solution :

We see that the double integral is required over the quadrilateral ABCD shown in Figure 9.22. The width of the domain is constant and is from $x = 1$ to $x = 2$. We transform this to

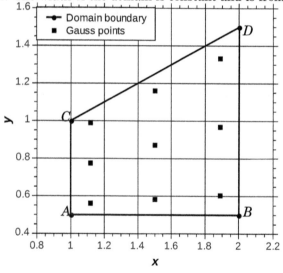

Figure 9.22: *Domain of integration in Example 9.19 showing also the Gauss points*

standard $-1 \le \xi \le 1$ by the transformation $x = 1.5 + 0.5\xi$. We also have $dx = 0.5d\xi$. The three Gauss points along ξ situated at $0, \pm\sqrt{\dfrac{3}{5}}$ translate to x values as shown below:

ξ	x
-0.774597	1.112702
0	1.5
0.774597	1.887298

Corresponding to these x values the ordinates on the top boundary are given by the following table:

x	y
1.112702	1.056351
1.5	1.25
1.887298	1.443649

We choose to transform the ordinates at each of these points such that the interval between a point on the bottom boundary and a corresponding one on the top boundary is the standard interval $-1, 1$. This is done by transformations shown in the following tabulation:

y_l	y_u	$\dfrac{y_l + y_u}{2}$	$\dfrac{y_u - y_l}{2}$	Formula
0.5	1.056351	0.778175	0.278175	$y = 0.778175 + 0.278175\eta$
0.5	1.25	0.875	0.375	$y = 0.875 + 0.375\eta$
0.5	1.443649	0.971825	0.471825	$y = 0.978175 + 0.478175\eta$

In the table above y_l and y_u represent respectively the y values on lower and upper boundary at the chosen x values that correspond to Gauss points along the x direction. It is now possible to get the ordinates of all the Gauss points by substituting suitable η's in the formulae shown in the last column of the table above. The x, y coordinates of all nine Gauss points are thus obtained and tabulated below and also are shown in Figure 9.22.

x_G	y_G		
1.112702	0.562702	0.778175	0.993649
1.5	0.584526	0.875	1.165474
1.887298	0.606351	0.971825	1.337298

The table shows the three y_G values corresponding to each x_G value. Function values are now calculated and tabulated below.

0.344365	0.289659	0.131629
0.802678	0.820313	0.584822
1.465871	1.679091	1.388135

The weights have to be evaluated as follows. Transformation from x to ξ involves a scale factor of 0.5. For each fixed value of Gauss point x there corresponds an additional scale factor given by the coefficients of η in the formulae presented in an earlier table. Thus the factors due to scaling will be given by $0.5 \times 0.278175 = 0.139088$ corresponding to $x = 1.056351$, $0.5 \times 0.375 = 0.1875$ corresponding to $x = 1.25$ and $0.5 \times 0.278175 = 0.139088$ corresponding to $x = 1.443649$. These row weights will multiply the Gauss weights given earlier in Figure 9.19 to get the weights shown below.

0.042928	0.068685	0.042928
0.092593	0.148148	0.092593
0.072812	0.116500	0.072812

Finally each function value is multiplied by the corresponding weight and summed to get the integral as $I_G = 0.69375$. Incidentally this coincides with the exact value (the reader is encouraged to work it out). This is as expected since Gauss quadrature with three points will be exact for a power of x or y up to 5.

Integration over an arbitrary quadrilateral

Consider the integration over a quadrilateral as shown in Figure 9.23.

$$I = \int \int f(x, y) dx dy \tag{9.103}$$

The arbitrary quadrilateral can be transformed to local coordinates and the integration may be

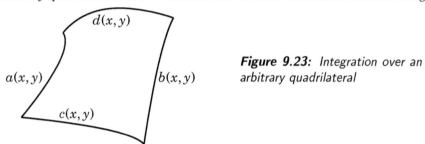

Figure 9.23: *Integration over an arbitrary quadrilateral*

performed using the local coordinates. If the quadrilateral is represented by straight lines, we can use isoparametric transformation discussed earlier

$$x = l_1 x_1 + l_2 x_2 + l_3 x_3 + l_4 x_4; \quad y = l_1 y_1 + l_2 y_2 + l_3 y_3 + l_4 y_4 \tag{9.104}$$

where l_i are the shape functions for a rectangular element. The total derivatives of x and y can be written as

$$\left(\begin{array}{c} dx \\ dy \end{array} \right) = \underbrace{\left(\begin{array}{cc} \dfrac{\partial x}{\partial \xi} & \dfrac{\partial x}{\partial \eta} \\ \dfrac{\partial y}{\partial \xi} & \dfrac{\partial y}{\partial \eta} \end{array} \right)}_{\mathbf{J}} \left(\begin{array}{c} d\xi \\ d\eta \end{array} \right) \tag{9.105}$$

With the transformation, the integral becomes

$$I = \int_{-1}^{1} \int_{-1}^{1} \underbrace{|\mathbf{J}(\xi, \eta)| f(x, y)}_{g(\xi, \eta)} \, d\xi d\eta \tag{9.106}$$

where $|\mathbf{J}|$ is the determinant of the Jacobian. Now, any quadrature rule can be applied over the local coordinates. The advantage of using local coordinates for integration is that the distribution of nodes within the domain is simpler.

Example 9.20

Obtain the double integral in Example 9.19 using trapezoidal, Simpson's and Gauss quadrature rules using local coordinates

Solution :

The coordinates of the quadrilateral are $(x_1, y_1) = (1, 0.5)$, $(x_2, y_2) = (2, 0.5)$, $(x_3, y_3) = (1, 1)$ and $(x_4, y_4) = (2, 1.5)$.

Step 1 Let us first transform the geometry to local coordinates using isoparametric transformation and determine the Jacobian of the transformation.

$$x = \frac{(1-\xi)(1-\eta)}{4} 1 + \frac{(1+\xi)(1-\eta)}{4} 2 + \frac{(1-\xi)(1+\eta)}{4} 1 + \frac{(1+\xi)(1+\eta)}{4} 2$$

$$
\begin{aligned}
&= 1.5 + 0.5\xi \\
y &= \frac{(1-\xi)(1-\eta)}{4}0.5 + \frac{(1+\xi)(1-\eta)}{4}0.5 + \frac{(1-\xi)(1+\eta)}{4}1 + \frac{(1+\xi)(1+\eta)}{4}1.5 \\
&= \frac{1-\eta}{4} + \frac{(1+\eta)(5+\xi)}{8}
\end{aligned}
$$

The partial derivatives would be

$$
\frac{\partial x}{\partial \xi} = \frac{1}{2}; \quad \frac{\partial x}{\partial \eta} = 0; \quad \frac{\partial y}{\partial \xi} = \frac{1+\eta}{8}; \quad \frac{\partial y}{\partial \eta} = \frac{3+\xi}{8}
$$

Jacobian of the transformation is

$$
|\mathbf{J}| = \begin{vmatrix} \dfrac{1}{2} & 0 \\ \dfrac{1+\eta}{8} & \dfrac{3+\xi}{8} \end{vmatrix} = \frac{3+\xi}{16} \tag{9.107}
$$

Step 2 Integration in the transformed coordinates is

$$
I = \int_{-1}^{1}\int_{-1}^{1} J(\xi,\eta)f(x,y)d\eta d\xi = \int_{-1}^{1}\int_{-1}^{1} g(\xi,\eta)d\eta d\xi \tag{9.108}
$$

Trapezoidal and Simpson's rules are applied to the above integral and have been summarized in following table.

ξ	η	x	y	$J(\xi,\eta)$	$f(x,y)$	$w_i g_i$ Trapezoidal	$w_i g_i$ Simpson's 1/3
-1	-1	1	0.5	0.125	0.25	0.007812	0.003472
-1	0	1	0.75	0.125	0.1875	0.011719	0.010417
-1	1	1	1	0.125	0	0.000000	0.000000
0	-1	1.5	0.5	0.1875	0.75	0.070312	0.062500
0	0	1.5	0.875	0.1875	0.8203125	0.153809	0.273438
0	1	1.5	1.25	0.1875	0.46875	0.043945	0.039062
1	-1	2	0.5	0.25	1.5	0.093750	0.041667
1	0	2	1	0.25	2	0.250000	0.222222
1	1	2	1.5	0.25	1.5	0.093750	0.041667
					Integral estimate	0.725098	0.694444

Step 3 Gauss quadrature rule is applied to the above integral and has been summarized in following table.

ξ	η	x	y	$J(\xi,\eta)$	$f(x,y)$	$w_i g_i$
-0.774597	-0.774597	1.112702	0.562702	0.139088	0.344365	0.014783
-0.774597	0.000000	1.112702	0.778175	0.139088	0.289659	0.019895
-0.774597	0.774597	1.112702	0.993649	0.139088	0.131629	0.005651
0.000000	-0.774597	1.500000	0.584526	0.187500	0.802678	0.074322
0.000000	0.000000	1.500000	0.875000	0.187500	0.820312	0.121528
0.000000	0.774597	1.500000	1.165474	0.187500	0.584822	0.054150
0.774597	-0.774597	1.887298	0.606351	0.235912	1.465871	0.106734
0.774597	0.000000	1.887298	0.971825	0.235912	1.679091	0.195614
0.774597	0.774597	1.887298	1.337298	0.235912	1.388135	0.101073
					Integral estimate	0.693750

Notice that the above procedure is no different from Example 9.19. The integral estimated by the use of Gauss quadrature agrees with the exact value while the estimates using trapezoidal and Simpson's 1/3 rule do not. It has to be noted that Simpson's 1/3 rule is accurate up to first 9 terms of the binomial series. Although Simpson's rule should exactly integrate the original function $x^2 y - xy^2$, the coordinate transformation makes the number of terms of the series more than 9 and hence the observed error in the estimate. For the present case, Simpson's 3/8 rule will integrate exactly.

9.10.5 Quadrature rules for triangle

Integration over a triangle is important in numerical methods such as FEM and FVM and we will briefly introduce some ideas on the quadrature rules for triangular elements. Consider integration over an arbitrary triangle

$$I = \iint_\triangle f(x,y)dydx \qquad (9.109)$$

In an earlier chapter, conversion of a triangle to its local coordinate system was considered. Also, it was shown that the transformation maps the actual triangle onto an isosceles right angled triangle (Figure 9.24). Thus the integral after transformation becomes

$$I = 2A \int_0^1 \int_0^{1-\eta_1} f d\eta_2 d\eta_1 \qquad (9.110)$$

where A is the area of the triangle.

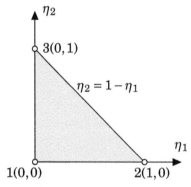

Figure 9.24: *Integration over a triangle*

Integration over a linear element

Lagrange weight functions for the linear triangular element are

$$l_1 = 1 - \eta_1 - \eta_2; \quad l_2 = \eta_1; \quad l_3 = \eta_2 \qquad (9.111)$$

The integral over the triangle can be written as

$$I = 2A \sum_{i=1}^3 f_i \int_0^1 \int_0^{1-\eta_1} l_i d\eta_2 d\eta_1 \qquad (9.112)$$

Integrating first Lagrange weight function l_1

$$I = 2A \int_0^1 \int_0^{1-\eta_1} (1 - \eta_1 - \eta_2)d\eta_2 d\eta_1 = 2A \int_0^1 \left[\eta_2(1-\eta_1) - \frac{\eta_2^2}{2} \right]\Big|_0^{1-\eta_1} d\eta_1$$

$$= \qquad 2A \int_0^1 \frac{(1-\eta_1)^2}{2} d\eta_1 = \qquad A\left[\eta_1 - \eta_1^2 + \frac{\eta_1^3}{3}\right]\Big|_0^1 = \frac{A}{3} \tag{9.113}$$

Performing integration of other two Lagrange weight functions, we finally get

$$I = \frac{A}{3}(f_1 + f_2 + f_3) \tag{9.114}$$

Integration over a quadratic element

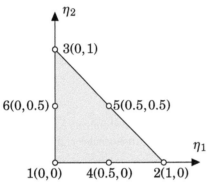

Figure 9.25: Integration over a quadratic triangular element

Lagrange weight functions for quadratic triangular element are

$$l_1 = 2(1-\eta_1-\eta_2)(0.5-\eta_1-\eta_2) \quad l_2 = 2\eta_1(\eta_1-0.5) \quad l_3 = 2\eta_2(\eta_1-0.5)$$
$$l_4 = 4(1-\eta_1-\eta_2)\eta_1 \qquad\qquad l_5 = 4\eta_1\eta_2 \qquad l_6 = 4(1-\eta_1-\eta_2)\eta_2 \tag{9.115}$$

The integral over the triangle can be written as

$$I = 2A \sum_{i=1}^{6} f_i \int_0^1 \int_0^{1-\eta_1} l_i d\eta_2 d\eta_1 \tag{9.116}$$

Integrating the Lagrange weight functions, we get

$$I = \frac{A}{3}(f_4 + f_5 + f_6) \tag{9.117}$$

It is interesting that the weights for the corner nodes are zero and therefore the method requires only three function evaluations. Quadrature rules for higher order elements can be derived similarly. In fact, the above rules are also referred to as closed Newton Cotes quadrature rules for a triangle.

Gauss Quadrature rules for triangles

Similar to Gauss quadrature rules for integration over a line, we can construct quadrature rules for a triangle. Gauss quadrature rule for a triangle are exact up to $2n - 1$ terms of the binomial equation. The quadrature rule assumes the form

$$I = 2A \int_0^1 \int_0^{1-\eta_1} f d\eta_2 d\eta_1 \approx A \sum_{i=1}^{n} w_i f_i \tag{9.118}$$

The quadrature rules can be constructed using the method of undetermined coefficients already familiar to us.

One point formula

First let us consider quadrature rule based on a single point. Quadrature rule based on one point has to give exact integral with a function that is linear in η_1 and η_2. The error arises due to the higher order terms involving η_1 and η_2. Therefore the integral is exact for $f = 1$, η_1 and η_2. As there is only one point, $w_1 = 1$ and the quadrature rule automatically satisfies the requirement for $f = 1$.

$$f = \eta_1; \ I = 2A \int_0^1 \int_0^{1-\eta_1} \eta_1 d\eta_2 d\eta_1 = \frac{A}{3} = A\eta_1 \tag{9.119}$$

$$f = \eta_2; \ I = 2A \int_0^1 \int_0^{1-\eta_1} \eta_2 d\eta_2 d\eta_1 = \frac{A}{3} = A\eta_2 \tag{9.120}$$

following which we get the quadrature point as $\left(\frac{1}{3}, \frac{1}{3}\right)$ which is nothing but the centroid of the triangle.

Three point formula

A three point formula (η_{11}, η_{21}), (η_{12}, η_{22}) and (η_{13}, η_{23}) would give exact integral for $f = 1$, η_1, η_2, $\eta_1 \eta_2$, η_1^2 and η_2^2. Conveniently we choose the points to be symmetric with respect to the centroid of the triangle. This means the three weights of the quadrature rule are equal to $\frac{1}{3}$. This condition automatically satisfies the required condition when $f = 1$. Integrating $f = \eta_1$, we get

$$\begin{aligned} f = \eta_1; \ I &= 2A \int_0^1 \int_0^{1-\eta_1} \eta_1 d\eta_2 d\eta_1 = \frac{A}{3} \\ I &= A\left(w_1 \eta_{11} + w_2 \eta_{12} + w_3 \eta_{13}\right) \\ &\longrightarrow \ \eta_{11} + \eta_{12} + \eta_{13} = 1 \end{aligned}$$

The coordinates of the points should be chosen such that all other conditions are also satisfied. Thus we have

f	Condition		
η_1	$\eta_{11} + \eta_{12} + \eta_{13}$	$=$	1
η_2	$\eta_{21} + \eta_{22} + \eta_{23}$	$=$	1
$\eta_1 \eta_2$	$\eta_{11}\eta_{21} + \eta_{12}\eta_{22} + \eta_{13}\eta_{23}$	$=$	0.25
η_1^2	$\eta_{11}^2 + \eta_{12}^2 + \eta_{13}^2$	$=$	0.5
η_2^2	$\eta_{21}^2 + \eta_{22}^2 + \eta_{23}^2$	$=$	0.5

Due to symmetry, $\eta_{11} = \eta_{22}$ and $\eta_{12} = \eta_{21}$. Therefore $\eta_{13} = \eta_{23} = 1 - \eta_{11} - \eta_{12}$. On further simplification we find that, there are two sets of points that satisfy the conditions, namely

1. $(0.5, 0)$, $(0, 0.5)$ and $(0.5, 0.5)$
2. $\left(\frac{2}{3}, \frac{1}{6}\right)$, $\left(\frac{1}{6}, \frac{2}{3}\right)$ and $\left(\frac{1}{6}, \frac{1}{6}\right)$

Higher order quadrature rules can be derived similarly. Table 9.8 indicates some of the Gauss quadrature rules for triangle (reproduced from Cowper[9]). Multiplicity refers to number of possible combinations with the specified coordinates.

---

[9]G. R. Cowper "Gaussian quadrature formulae for triangles" International Journal for Numerical Methods in Engineering Vol. 7(3), pages 405-408, 1973

Table 9.8: *Gauss quadrature rules for a triangle*

No. of points	Degree of accuracy	w_i	η_1	η_2	η_3	Multiplicity
1	1	1	0.333333	0.333333	0.333333	1
3	2	0.333333	0.666667	0.166667	0.166667	3
3	2	0.333333	0.500000	0.500000	0.000000	3
6	3	0.166667	0.659028	0.231933	0.109039	6
6	4	0.109952	0.816848	0.091576	0.091576	3
		0.223382	0.108103	0.445948	0.445948	3

Example 9.21

Obtain the double integral $I = \int_{x=1}^{2} \int_{y=0.5}^{(0.5+0.5x)} (x^2 y - xy^2)dxdy$ by splitting the domain into a rectangle and triangle.

Solution :

The domain of integration considered here is shown in Figure 9.26. The domain is split into

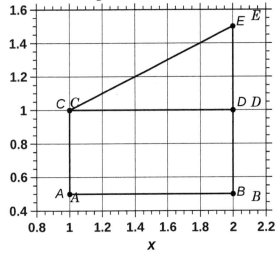

Figure 9.26: *Domain of integration in Example 9.21*

two *sub-domains*, rectangle ABCD and triangle CDE. The integral would be

$$I = \underbrace{\int_{x=1}^{2} \int_{y=0.5}^{1} (x^2 y - xy^2)dxdy}_{Rectangle} + \underbrace{\int_{x=1}^{2} \int_{y=1}^{(0.5+0.5x)} (x^2 y - xy^2)dxdy}_{Triangle}$$

The integrand is of degree three and we must use quadrature rules that would produce exact integral for the function.

Step 1 First, let us integrate in the rectangular domain. Simpson's 1/3 rule will integrate the function exactly in the domain. The results are summarized in the following table

x	y	w_i	$f(x,y)$	$w_i f_i$
1	0.5	0.111111	0.25	0.027778
1.5	0.5	0.444444	0.75	0.333333
2	0.5	0.111111	1.5	0.166667
1	0.75	0.444444	0.1875	0.083333
1.5	0.75	1.777778	0.84375	1.5
2	0.75	0.444444	1.875	0.833333
1	1	0.111111	0	0
1.5	1	0.444444	0.75	0.333333
2	1	0.111111	2	0.222222
			$I_1 = hk \sum w_i f_i =$	0.4375

Step 2 Now integration is performed over the triangular domain. The triangle is first transformed into local coordinates such that $\eta_i = a_i + b_i x + c_i y$. The area of the triangle is

$$A = \frac{1}{2} \times 0.5 \times 1 = 0.25$$

The constants (Equations 6.19) corresponding to the coordinate transformation are given below.

x	y	a_i	b_i	c_i
1	1	2	-1	0
2	1	1	1	-2
2	1.5	-2	0	2

We choose a third order accurate quadrature rule from Table 9.8. The quadrature nodes are specified in the local coordinates and corresponding location in actual coordinates are given by

$$x = \frac{(\eta_1 - a_1)c_2 - (\eta_1 - a_2)c_1}{b_1 c_2 - b_2 c_1}; \qquad y = \frac{(\eta_1 - a_1)b_2 - (\eta_1 - a_2)b_1}{b_2 c_1 - b_1 c_2}$$

The quadrature rule over the triangular region is summarized in the table below.

η_1	η_2	x	y	f(x,y)
0.659028	0.231933	1.340972	1.054520	0.405067
0.659028	0.109039	1.340972	1.115967	0.336716
0.231933	0.659028	1.768067	1.054520	1.330382
0.109039	0.659028	1.890961	1.115966	1.635431
0.231933	0.109039	1.768067	1.329514	1.030893
0.109039	0.231933	1.890961	1.329514	1.411511
			$I_2 = A \sum w_i f_i =$	0.256250

Step 3 The estimate for the integral is

$$I = I_1 + I_2 = 0.4375 + 0.256250 = 0.693750$$

which is the same as the exact value.

Transforming triangle to a rectangle

Before ending this topic, we will draw the attention of the reader to an important transformation that converts a triangular domain to a rectangular domain. The first transformation is from (x,y) to (η_1, η_2). Subsequently, we introduce a transformation given below and as illustrated in Figure 9.27

$$\eta_1 = \frac{1+u}{2}; \qquad \eta_2 = \frac{(1-u)(1+v)}{4} \tag{9.121}$$

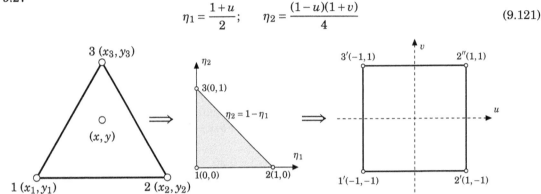

Figure 9.27: *Transformation of a triangular domain to a rectangular domain*

The Jacobian of the transformation is given by

$$\mathbf{J} = \begin{pmatrix} \dfrac{1}{2} & 0 \\ \dfrac{1+v}{4} & \dfrac{1-u}{4} \end{pmatrix} \tag{9.122}$$

The integral thus becomes

$$I = 2A \int_{-1}^{1} \int_{-1}^{1} \frac{1-u}{8} f \, du \, dv \tag{9.123}$$

The above integral can be evaluated using quadrature rules over a rectangle given in earlier sections.

Let us evaluate the integral over the triangular region in Example 9.21 by transforming the domain into a rectangular domain. 3 point Gauss quadrature rule is applied over the transformed rectangular region along the two directions and have been summarized in the following table.

u	v	η_1	η_2	x	y	$\dfrac{1-u_i}{8}f_i$
0.774597	0.774597	0.887298	0.100000	1.112702	1.006351	0.003355
0.000000	0.774597	0.500000	0.443649	1.500000	1.028175	0.090960
-0.774597	0.774597	0.112702	0.787298	1.887298	1.050000	0.368061
0.774597	0.000000	0.887298	0.056351	1.112702	1.028175	0.002725
0.000000	0.000000	0.500000	0.250000	1.500000	1.125000	0.079102
-0.774597	0.000000	0.112702	0.443649	1.887298	1.221825	0.340400
0.774597	-0.774597	0.887298	0.012702	1.112702	1.050000	0.002064
0.000000	-0.774597	0.500000	0.056351	1.500000	1.221825	0.063728
-0.774597	-0.774597	0.112702	0.100000	1.887298	1.393649	0.288020
$I = 2A \sum w_i \dfrac{1-u_i}{8}f_i$						0.256250

Concluding remarks

> *Numerical integration is central to most analysis of engineering problems. Specifically, integration over a line, area or a volume is routinely met with in analysis. Interpolation of functions, introduced in Chapters 5 and 6 are very useful in understanding numerical integration.*
>
> *We have presented integration techniques which are traditionally used by engineers in routine evaluations as well as special techniques useful for improper integrals, integral over rectangles, quadrilaterals and triangles which are used in advanced applications such as FEM and FVM.*

9.A MATLAB routines related to Chapter 9

MATLAB routine	Function
trapz(x,y)	integration using trapezoidal rule
quad(fun,a,b)	integrates function fun between the limits a and b using adaptive Simpson's quadrature
quadv	integration using adaptive Simpson's quadrature (vectorized quad)
quadl(fun,a,b)	integrates function fun between the limits a and b using adaptive Lobatto quadrature
quadgk(fun,a,b)	integrates function fun between the limits a and b using adaptive Gauss Kronrod quadrature
dblquad(fun,a,b,c,d)	evaluates double integral of a function fun over a rectangle $a \leq x \leq b$ and $c \leq y \leq d$
quad2d(fun,a,b,c,d)	evaluates double integral of a function fun over a 2d region $a \leq x \leq b$ and $c(x) \leq y \leq d(x)$
triplequad	evaluates triple integral over a cuboid

9.B Suggested reading

1. **P.J. Davis** and **P. Rabinowitz, P** *Methods of numerical integration* Dover Publications, 2007
2. **P.K. Kythe** and **M.R. Schaferkotter** *Handbook of computational methods for integration* Chapman and Hall, 2004
3. **Abramowitz** and **Stegun** *Handbook of Mathematical Functions* Available online at http://people.math.sfu.ca/ cbm/aands/. To best of our knowledge the book is not copyrighted.

Exercise II

II.1 Interpolation

Ex II.1: Variation of density of air at atmospheric pressure with temperature is provided in the following table. Obtain the value of density at $50°C$ by interpolation. Both Lagrange and Newton interpolations may be tested out for this purpose.

Temperature: $°C$	0	20	40	60	80	100
Density: kg/m^3	1.2811	1.1934	1.1169	1.0496	0.9899	0.9367

The answer must be accurate to four digits after decimals.

Ex II.2: The following table gives equi-spaced data of a certain function. By interpolation obtain the value of the function at $x = 0.5$ by the following three methods and compare them with the value $f(0.5) = 1.12763$.
a) Linear interpolation, b) Lagrangian interpolation using a cubic and
c) Newton interpolation based on forward differences.

x	0	0.2	0.4	0.6	0.8	1.0
$f(x)$	1.00000	1.08107	1.02007	1.18547	1.33744	1.54308

Ex II.3: Obtain the value of the tabulated function $f(x)$ at $x = 0.3$ by linear and quadratic interpolations. Obtain the second derivative at $x = 0.4$ by central difference formulae. Obtain the first derivative at $x = 0.4$ by central difference formula and the three point one sided formula.

x	0	0.2	0.4	0.6	0.8
$f(x)$	-0.200	0.152	0.936	2.344	4.568

Ex II.4: Function and its first derivatives have been provided in the following table with uniform spacing of $h = 0.2$. Fit a Hermite interpolating polynomial through the data by writing a computer program.

x	1	1.2	1.4	1.6	1.8	2
$f(x)$	0.582	1.072	1.760	2.653	3.742	5.009
$f^{(1)}(x)$	1.989	2.934	3.950	4.966	5.911	6.729

Use the Hermite interpolating polynomial to obtain the values of the function at $x = 1.5$ and $x = 1.9$. Compare these with the values obtained by using Newton polynomial of highest possible degree.

Ex II.5: Obtain a cubic spline fit using the function data given in Exercise II.4. Use the cubic spline to obtain $f(1.5)$ and $f(1.9)$. Compare these with the values obtained in Exercise II.4. Evaluate the first derivative of the cubic spline fit at $x = 1.4$ and compare it with the first derivative data supplied in Example II.4. Make suitable comment(s) based on the comparison.

Ex II.6: Use the first derivative data in Exercise II.4 to obtain the second derivative at $x = 1.4$ using central differences. Compare this with second derivative obtained at the same point by the cubic spline. Compare these two and make suitable comments.

Ex II.7: A certain function has been provided in the form of a table of values as shown below. Obtain a cubic spline fit to the data over the entire range from $x = 0.6$ to $x = 1.5$. Estimate the second derivative at $x = 1.2$ using the spline fit. Compare this with the values obtained by using centered three and five point rules.

x	0.6	0.7	0.8	0.9	1
$f(x)$	1.04932	1.14100	1.23646	1.33732	1.44535
x	1.1	1.2	1.3	1.4	1.5
$f(x)$	1.56245	1.69064	1.83213	1.98929	2.16465

Obtain interpolated value of the function at $x = 1.45$ using Newton Gregory series and compare this with that given by the cubic spline.

Ex II.8: It is desired to represent the function $f(x) = \dfrac{1 + 2x - x^2}{x^4 - 2x^3 + 3}$ in the range $0 \le x \le 2$ by an approximating polynomial. Use a single polynomial of highest possible degree and also piecewise polynomials to study the utility of each in representing the function. Use data generated with $\Delta x = 0.1$ for the purpose of this exercise.

Try also a cubic spline for the purpose of approximating the function. Compare the performance of the cubic spline with those obtained above. Use of computer is essential for this exercise.

II.2 Interpolation in two dimensions

Ex II.9: Tabulated data below is of a function $f(x, y)$. Obtain the value of the function at $(0.25, 0.35)$ by linear and quadratic interpolations.

$x \downarrow \ y \rightarrow$	0.15	0.3	0.45
0.2	1.3221	1.9596	2.8201
0.3	1.9596	2.8201	3.9817
0.4	2.8201	3.9817	5.5496

Compare the results obtained with nine noded and eight noded (serendipity) elements.

Ex II.10: A function $f(x, y)$ is shown as a table of values as indicated below. Obtain by linear and quadratic interpolation the value of the function at $x = 0.35$, $y = 0.22$ and compare it with the exact value of 0.64380.

x	y	$f(x,y)$
0.20	0.30	0.59400
0.40	0.00	0.66000
0.60	0.60	1.29200
0.30	0.15	0.59675
0.50	0.30	0.80400
0.40	0.45	0.84225

Note that the data is provided over
a six noded triangular (serendipity) element

Ex II.11: The coordinates of four corners of a quadrilateral are given by the entries in the table.

Point No.	x	y
1	0	0
2	1	0
3	0	1
4	1.5	2.5

Transform the quadrilateral to the standard square of side equal to 2? What happens to the point $x = 0.5$, $y = 0.5$ under this transformation?

Ex II.12: Data below is given for an eight noded serendipity element. Determine the value of the function at $x = 0.75$, $y = 0.95$.

$x \downarrow y \rightarrow$	0.6	0.9	1.2
0.6	-1.208	-0.182	1.816
0.8	-0.928		2.096
1	-0.568	0.458	2.456

Ex II.13: Three vertices of a triangular element are given below along with the values of a certain function. Convert the triangle to local coordinates and find the value of the function at $x = 0.4$, $y = 0.3$.

x	y	$f(x,y)$
0.20	0.30	-0.019
0.40	0.00	0.064
0.60	0.60	0.000

II.3 Regression

Ex II.14: y and x in the following table are expected to follow the relation $y = e^{\left(A + \frac{B}{x} + \frac{C}{x^2}\right)}$. Estimate the values of the parameters $A - C$ using least squares.

x	0.27	0.29	0.31	0.33	0.35	0.37	0.39	0.41	0.43	0.45
y	0.01	0.02	0.07	0.20	0.47	1.01	1.99	3.61	6.78	10.03

Discuss the quality of the fit. Estimate the value of y at $x = 4$ using the fit. Compare this value with those obtained by linear and quadratic interpolations.

Ex II.15: The following data is expected to follow the relation $y = c_1 x^{c_2}$. Obtain the best estimates for the parameters c_1 and c_2. Discuss the quality of the fit. What is the error with respect to fit?

x	1.0	1.2	1.8	2.2	3.0	4.2	5.1	6.3	7.0
y	37.0	39.5	60.6	67.8	93.8	122.1	127.3	155.8	179.0

Ex II.16: The data shown in the table below is expected to be well represented by a relation of form $y = \dfrac{1}{c_1 + c_2 x^2}$. Obtain best estimates for the fit parameters using least squares. Discuss the quality of the fit.

x	1.2	1.4	1.6	1.8	2	2.2	2.4	2.6	2.8
y	0.434	0.394	0.373	0.370	0.285	0.261	0.258	0.209	0.187

Ex II.17: Data shown in the table below follows the relation $y = c_1 e^{c_2 x}$. Determine the two fit parameters by least squares. Comment on the goodness of fit. What are the uncertainties in the parameters?

x	1	1.2	1.4	1.6	1.8	2	2.2	2.4	2.6
y	0.661	0.568	0.479	0.412	0.338	0.283	0.243	0.207	0.170

Ex II.18: Data tabulated below is expected to follow the relation $y = \dfrac{x^{c_0}}{c_1 + c_2 x}$. Estimate the fit parameters using least squares. Comment on the quality of the fit based on the index of correlation.

x	0.2	0.4	0.6	0.8	1
y	0.6493	0.9039	1.1774	1.3019	1.4682
x	1.2	1.4	1.6	1.8	2
y	1.5075	1.5980	1.6496	1.7441	1.7322

Ex II.19: Fit a relation of form $y = c_0 + c_1 x_1 + c_2 x_2$ to the data given in the following table. Make a parity plot and discuss the quality of the fit. Estimate also the standard error of the fit.

x_1	x_2	y
0.15	0.23	0.593
0.22	0.28	0.729
0.27	0.34	0.818
0.31	0.39	0.903
0.36	0.45	1.020
0.44	0.52	1.174
0.49	0.56	1.256
0.56	0.64	1.405
0.61	0.69	1.506
0.69	0.77	1.659
0.73	0.83	1.758
0.80	0.95	1.846

II.4 Numerical differentiation

Ex II.20: Consider the function $y = e^{\left[1 - \frac{5}{x} + \frac{1}{x^2}\right]}$ in the range $1 \le x \le 3$. Obtain numerically, using suitable finite difference formulae, the derivative of this function at $x = 2$ that agrees with the exact value of the derivative to 4 digits after decimals.

Ex II.21: Estimate numerically the first and second derivatives of the function $y = 2\cos(x) - 4\sin(5x)$ at $x = 0.6$ that agree with the exact values to 5 digits after decimals.

Ex II.22: Estimate numerically the first and second derivatives of the function $y = \sinh(x) - \cosh(3x)$ at $x = 0.5$ that agree with the exact values to 5 digits after decimals.

Ex II.23: Table below gives an extract from data pertaining to the distribution of a function of $f(x, y)$ with respect to x, y. Estimate the following:

- Second order accurate first derivative with respect to x at $x = 0.4$, $y = 0.3$
- Second derivatives with respect to x and y at $x = 0.2$, $y = 0.2$
- Second order accurate first derivative with respect to y at $x = 0.1$, $y = 0.1$
- Second order accurate mixed derivative $\left(\text{i.e. } \dfrac{\partial^2 f}{\partial x \partial y}\right)$ at $x = 0.2$, $y = 0.2$

$y \downarrow x \rightarrow$	0.1	0.2	0.3	0.4
0.1	100.00	100.00	100.00	100.00
0.2	48.26	66.10	73.05	74.93
0.3	26.93	43.11	51.18	53.61
0.4	16.36	28.20	34.97	37.14

Ex II.24: Table below gives an extract from data pertaining to the distribution of a function of $f(t, x)$ with respect to t, x. Estimate the following:

- Second order accurate first derivative with respect to x at $t = 0.016$, $x = 0.0250$
- Second derivatives with respect to x and t at $t = 0.016$, $x = 0.025$
- Second order accurate first derivative with respect to t at $t = 0.048$, $x = 0$
- Second order accurate mixed derivative $\left(\text{i.e. } \dfrac{\partial^2 f}{\partial t \partial x}\right)$ at $t = 0.032$ $x = 0.0250$

$x \downarrow t \rightarrow$	0.000	0.016	0.032	0.048	0.064
0.0000	1.0000	0.0000	0.0000	0.0000	0.0000
0.0125	1.0000	0.4639	0.3797	0.3077	0.2627
0.0250	1.0000	0.8557	0.6633	0.5675	0.4832
0.0375	1.0000	0.9588	0.8508	0.7317	0.6308
0.0500	1.0000	0.9794	0.9048	0.7912	0.6812

Ex II.25: Consider the function $f(x, y) = e^x \sin(y) - y\cos(x)$. Obtain all first and second partial derivatives at the point $x = 0.5$, $y = 0.5$ using suitable finite difference approximations. The results are required with four digit accuracy.

What is the numerical value of $\dfrac{\partial^2 f}{\partial x^2} + \dfrac{\partial^2 f}{\partial y^2}$ at the point $x = 0.5$, $y = 0.5$? Use the partial derivatives obtained previously based on finite differences.

Ex II.26: Consider the data given in Exercise II.10. Use finite differences and estimate all first and second partial derivatives at the centroid of the triangle.

Ex II.27: Consider the data given in Exercise II.12. Use finite differences and estimate all first and second partial derivatives at the point $x = 0.8$, $y = 0.9$.

II.5   Numerical integration

Ex II.28: Integrate using trapezoidal rule the function $f(x) = \dfrac{1}{x + 0.1}$ between $x = 0$ and $x = 1$. Four figure accuracy is desired for the integral.

Ex II.29: Obtain numerically the line integral $\displaystyle\int_0^1 \ln(x)dx$ accurate to four digits after the decimal point. Make use of (a) trapezoidal rule coupled with Romberg and (b) Gauss quadrature with a suitable number of Gauss points, for this purpose. Comment on which is a better method in the present case and why.

Ex II.30: (a) Find the mean value of the function $y = 3x^4 - 2x^2 + 2$ in the interval $0 \le x \le 1$ accurate to six digits after decimals. Use a suitable quadrature and verify by comparison with the result obtained by analysis.

(b) Find the weighted mean of the same function as in part (a) with the weight $w(x) = x + 0.5$. The weighted mean is again required with six digit accuracy.

Hint: Weighted mean is defined as $\dfrac{\int_{x_1}^{x_2} w(x)f(x)dx}{\int_{x_1}^{x_2} w(x)dx}$ where x_1 and x_2 are the lower and upper values defining the range of x.

Ex II.31: Find the area between the curve $y = \sin(x)$ and the line $y = 0.5x$ between $x = 0$ and $x = \dfrac{\pi}{2}$. Use suitable quadrature and obtain the area correct to at least four digits after decimals. Try to simplify the calculations so that the variable limits do not pose much of a problem.

Ex II.32: Obtain the integral $\displaystyle\int_2^3 \left[x^3 + \ln(x + 3)\right] dx$ by Gauss quadrature using 4 Gauss points. Compare this with the exact value of the integral.

Ex II.33: Determine the definite integral $I = \displaystyle\int_0^1 x^3 e^{-x} dx$ by (a) Simpson rule with $\Delta x = h = 0.125$ and (b) Gauss quadrature with 3 Gauss points. Show that the two results agree to four digits after decimals.

Ex II.34: Complete the Romberg table by supplying the missing entries. What is the best estimate of the line integral based on this table?

$T(h)$	$T\left(h, \dfrac{h}{2}\right)$	$T\left(h, \dfrac{h}{2}, \dfrac{h}{4}\right)$	$T\left(h, \dfrac{h}{2}, \dfrac{h}{4}, \dfrac{h}{8}\right)$
0.113448			
	0.125099		
0.124146		0.124801	
	0.124801		
0.124637			

Ex II.35: Obtain, by a suitable integration scheme, the integral $I = \displaystyle\int_0^\infty \dfrac{x^3}{e^x - 1} dx$, accurate to at least 4 digits after decimals. Analytically derived value of this integral is $\dfrac{\pi^4}{15}$.

Ex II.36: Obtain $I = \int_{x=0}^{1} \int_{y=0}^{\frac{1}{2}} e^{-(x+2y)} \sin(\pi xy) \, dx dy$ by a suitable numerical integration scheme.

Ex II.37: Obtain the double integral $I = \int_{0}^{1} \int_{0}^{1} (x + y^2) e^{x^2 + 2y} dx dy$ by the following methods and make comparisons:

- Trapezoidal rule with $h = k = 0.25$
- Simpson rule with $h = k = 0.25$
- Gauss quadrature with four Gauss points along each direction

Ex II.38: Consider the integral in Exercise II.37 again. Change step size in the Simpson scheme and obtain the integral convergent to 5 digits after decimals.

Ex II.39: Consider the integral in Exercise II.37 again. Divide the domain into sub domains and obtain the integral by using Gauss quadrature with four Gauss points over each of the sub domains. Goal is to obtain the value of the integral good to 5 digits after decimals.

Ex II.40: Determine the mean value of the function $f(x, y) = e^{(x-y)} \cos[\pi(x+y)]$ over the rectangle $-0.5 \le x \le 0.5$ and $-1 \le y \le 1$.

Ex II.41: Determine the mean value of the function in Exercise II.40 over the triangle given by the vertices $A(0.0), B(0, 1)$ and $C(0.5, 0)$.

Ex II.42: We come across moment of inertia I in rotational motion that is analogous to mass in linear motion. From Newton's law (of mechanics) we have $T = I\alpha$ where T is the torque and α is the angular acceleration. Moment of inertia of an arbitrary shaped rigid body about an axis of rotation is given by

$$\rho \int_{v} r^2 dv$$

where ρ is the density and integration is performed over entire three dimensional volume and r is the distance from the axis of rotation. Moment of inertia for a plane region such as disks or plates is known as area moment of inertia or second moment and is defined as

$$I = \int_{A} r^2 dA = \int_{A} r^2 dx dy$$

Determine the area moment of inertia of an ellipse with major axis 3 and minor axis 2 for the following cases

- axis of rotation is aligned with major axis
- axis of rotation is aligned with minor axis
- axis of rotation is perpendicular to the plane of the ellipse and passing through the origin.
- axis of rotation is perpendicular to the plane of the ellipse passing through coordinates $(1, 0)$

$$\frac{x^2}{9} + \frac{y^2}{4} = 0$$

Module **III**

Ordinary differential equations

Ordinary differential equations naturally occur when we try to describe the behavior of physical systems with respect to spatial (one dimension) or temporal variations. The former occur as a class of problems in one-dimension in areas such as electricity, magnetism, fluid dynamics, structural dynamics etc. The latter describe dynamical systems such as structural, thermal electrical etc. Solution of ODEs are hence of great importance in most applications. In this module we consider both types of ODEs namely initial value problems corresponding to dynamical systems as well as boundary value problems corresponding to one dimensional field problems. Various numerical methods of solving both initial and boundary value problems form the content of this module.

Chapter 10

Initial value problems

Ordinary differential equations (ODE) occur very commonly in analysis of problems of engineering interest. Analytical solution of such equations is many times difficult or the evaluation using closed form solution itself may be as laborious as a numerical solution. Hence numerical solution of ODE is a topic of much practical utility.

We develop numerical methods to solve first order ODEs and extend these to solve higher order ODEs, as long as they are initial value problems (IVP). The following methods of solution of IVPs will be discussed here:

- *Euler method*
- *Modified Euler or Heun method or the second order Runge Kutta (RK2) method*
- *Runge Kutta methods*
- *Predictor corrector methods*
- *Backward difference formulae based methods (BDF methods)*

10.1   Introduction

A first order ordinary differential equation is the starting point for our discussion here. A first order system is governed, in general, by the first order ODE given by

$$\frac{dy}{dt} = f(t, y) \text{ or } y^{(1)} = f(t, y) \tag{10.1}$$

subject to the initial condition

$$t = t_0, \; y = y_0 \tag{10.2}$$

Consider the ODE

$$y^{(1)} = t \tag{10.3}$$

Integrating the above equation we get,

$$y = 0.5t^2 + C \tag{10.4}$$

where C is a constant of integration that depends on the initial value. Therefore we have a family of curves satisfying the same differential equation, satisfying different initial values, as shown in Figure 10.1

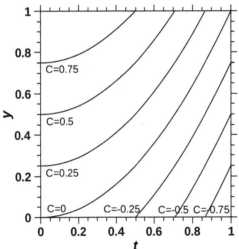

Figure 10.1: *Family of curves satisfying* $y^{(1)} = t$

Equation 10.1 is classified as a linear first order ODE if $f(t, y)$ is linear in y. Equation 10.1 is said to be non-linear if $f(t, y)$ is non-linear in y. Examples of these two types are:

$$
\begin{array}{lll}
\text{Linear ODE:} & y^{(1)}(t, y) & = \quad 3t^2 y + \cos(t) \\
\text{Non-linear ODE:} & y^{(1)}(t, y) & = \quad 3\sin(y) + 4t^3 y + 7t^2 + 10
\end{array} \tag{10.5}
$$

Note that $f(t, y)$ is **not** non-linear if it involves non-linear functions of t alone.

Geometric interpretation of a first order ODE is possible if we recognize that the given ODE is a relation between the slope and the function since $f(t, y)$ represents the slope of the tangent drawn at any point P on the solution curve, as shown in Figure 10.2. For a general first order ODE, the derivative could be a function of t and y and direct integration may not be possible. However for certain systems of differential equations exact solutions are available. When it is difficult to obtain

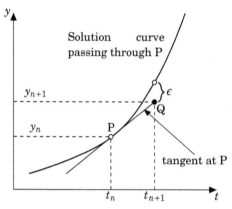

Figure 10.2: *Geometric interpretation of a first order ODE*

exact solution to the differential equation, one must resort to numerical methods. The derivative at a point t_n is defined as

$$y^{(1)}(t_n) = f(t_n, y_n) = \lim_{h \to 0} \frac{y(t_n + h) - y(t_n)}{h} \tag{10.6}$$

Numerical solution of the ODE consists in taking a finite sized step $h = t_{n+1} - t_n$, starting at t_n, where the solution is known to be y_n. Then, the solution to the differential equation at t_{n+1} can be evaluated as $y_{n+1} \approx y_n + hf(t_n, y_n)$. As $h \to 0$, the approximate solution would tend to the exact value. It is not practical to have very small step sizes as the number of computations increase and also round-off errors would become important as $h \to 0$. Thus point P on the solution curve is known and the numerical method aims to determine a point on the solution curve with abscissa t_{n+1}, possibly within a small but acceptable error. At every step of the numerical method there would be a local error and these errors would accumulate to produce cumulative errors. If the step size is too large the numerical solution may show unwanted oscillations or even may diverge due to accumulation of these errors thus making the solution *unstable*. For any numerical integration scheme for ODE, the step size will have to be chosen such that the solution is *stable* and *accurate*. In the following sections, concepts of stability and accuracy of different numerical methods for solving initial value problems will be introduced.

10.2   Euler method

The simplest method of solving a first order ODE is by the Euler[1] method. This method is explicit and uses only the information known at t_n to obtain the solution at t_{n+1}. By Taylor expansion of y around t_n, we have

$$y_{n+1} = y_n + y_n^{(1)}(t_{n+1} - t_n) + O(t_{n+1} - t_n)^2 = y_n + f(t_n, y_n)h + O(h^2) \tag{10.7}$$

If we ignore the second order term (term proportional to h^2) in Equation 10.7 we get the Euler scheme given by

$$y_{n+1} \approx y_n + hf(t_n, y_n) \tag{10.8}$$

The second order term that was ignored in writing the above is the local truncation error (LTE) and is proportional to h^2. In practice we will be integrating the ODE from the starting point $t = a$ to an end point $t = b$ after taking n steps of size h. At each step there is an LTE of order h^2.

---

[1]Leonhard Euler, 1707-1783, was a pioneering Swiss mathematician and physicist

Hence the total or global truncation error (GTE) will be proportional to $nh^2 = \dfrac{a-b}{h}h^2 = (a-b)h$, which is proportional to h and is of first order. Hence the Euler method is first order accurate. In general a kth order accurate method would drop terms of order $k+1$ in each integration step i.e. dropped term would be proportional to h^{k+1} or LTE is proportional to h^{k+1}. GTE over the range of the solution would however be proportional to h^k. Figure 10.2 shows that the error is the departure of the actual solution (shown by the dot on the solution curve) from Q. The Euler method approximates the solution by a straight line segment lying along the tangent at the point P.

Example 10.1

Use step size of $h = 0.1$ to obtain the solution of the first order ordinary differential equation $y^{(1)} + y = 0$ up to $t = 0.5$ using the Euler method. The initial value is specified to be $fy(0) = 1$.

Background :

The above first order differential equation is known as exponential decay equation. General exponential decay ODE is of the form $y^{(1)} = \lambda y$. λ is known as the decay constant and controls the rate at which the quantity y decreases. We come across these equations in modeling radioactive decay, a first order thermal system, and charging-discharging of a capacitor.

Temperature history of a hot object losing heat to surroundings at T_∞.

| Rate of change of internal energy $mc_p T_w$ of system | $=$ | Heat transfer q from system to the surroundings |

$$q = h_c A(T_w - T_\infty)$$
$$m, c_p, T_w$$

$$mc_p \frac{d(T_w - T_\infty)}{dt} = -h_c A(T_w - T_\infty)$$

$$\text{or } \frac{d\theta}{dt} = -\frac{h_c A}{mc_p}\theta$$

where m is the mass, c_p is the specific heat capacity and A is the surface area of the object, h_c is the heat transfer coefficient. The above is known as a **Lumped** model.

Voltage E across a capacitor during charging or discharging

$$RC\frac{dE}{dt} = E_0 - E$$

$$\text{or } \frac{d\delta E}{dt} = -\frac{1}{RC}\delta E$$

where $\delta E = E_0 - E$, R is the resistance and C is the capacitance.

Solution :

For the given ODE we have $f(t, y) = -y$, which does not contain t explicitly. The ODE is said to be autonomous. At $t = 0$, $y = 1$ and hence $f(t, y) = f(0, 1) = -1$. With $h = 0.1$, first Euler step yields $y(t = 0 + 0.1) = y(0.1) = y(0) + h \times f(0, 1) = 1 + 0.1 \times (-1) = 0.9$. This process is repeated till we reach $t = 0.5$. The results are shown conveniently in tabular form. We also show the exact solution given by $y_E = e^{-t}$.

t	$f(t,y)$	Euler	Exact	Error	
		y_{Eu}	y_E	ϵ_y	$\%\epsilon_y$
0	-1	1	1	0	0
0.1	-0.9	0.9	0.9048	-0.0048	-0.53
0.2	-0.81	0.81	0.8187	-0.0087	-1.06
0.3	-0.729	0.729	0.7408	-0.0118	-1.59
0.4	-0.6561	0.6561	0.6703	-0.0142	-2.12
0.5	-0.5905	0.5905	0.6065	-0.016	-2.64

The last two columns indicate the absolute error ϵ_y - difference between the Euler solution and the Exact solution and relative error given by $\%\epsilon_y = \dfrac{\epsilon_y \times 100}{y_E}$. Note that the error increases progressively with t indicating accumulation of error at each step. The Euler solution becomes practically useless since the error is in the second place after decimals. Accuracy may be improved by using a smaller step size.

> **Convergence and stability** Stability and convergence are two characteristics of a numerical scheme. Stability means the error caused during numerical solution does not grow with time. Convergence means the numerical solution tends to the exact solution as the step size tends to zero.

10.2.1 Stability of Euler method

Consider now the first order ODE $y^{(1)} = \lambda y$ subject to the initial condition $y(0) = y_0$ and λ a constant. Analytical solution to this problem is obtained as an exponential given by $y(t) = e^{\lambda t}$. If $\lambda > 0$, the function value grows exponentially with t and if $\lambda < 0$, we have an exponential decay. Numerical solution to the ODE will have a truncation error at every step. It is desired to have the error under control and the approximate solution to be close to the exact solution. However, the truncation error at every step accumulates.

Figure 10.3 shows the solution to the above differential equation using Euler method for $\lambda = -1$

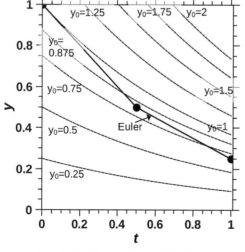

Figure 10.3: Migration of Euler solution away from the exact solution due to error buildup

and step size $h = 0.5$. Also indicated is the exact solution to the differential equation (by dashed lines) for various initial values of y at $t = 0$ (y_0). All these curves are part of the same family as they satisfy the same differential equation but different initial values. When performing integration of the ODE at the starting point using Euler method, a small error has occurred. The new point would actually fall on a different solution curve (y_0 close to 0.875) but from the same family as the original solution. While integrating the ODE about the new point, the derivatives are evaluated on this new curve and not the original curve. Therefore, at every step of integration, we move away from the original solution i.e. all points lie on different curves of the same family. If the accumulated error at a given step is large the system can diverge away from the exact solution. Therefore, in numerical solution of ODE, it would be important to control the error to ensure stability as well as accuracy.

Consider the first order ODE $y^{(1)} = \lambda y$ subject to the initial condition $y(0) = 1$. The exact solution to the above equation is $y = e^{\lambda t}$. When applying explicit numerical methods to solve ODE, the function value at the beginning of a given step is taken for all calculations. Then the solution to ODE becomes

$$y_n^{(1)} = \lambda y_n \tag{10.9}$$

The error in the numerical approximation at a given step would thus be

$$
\begin{aligned}
E_n &= y_n - e^{\lambda t_n} \\
&\quad \text{or} \\
y_n &= e^{\lambda t_n} + E_n
\end{aligned}
\tag{10.10}
$$

Equation 10.9 can hence be written as

$$
(e^{\lambda t_n} + E_n)^{(1)} = \lambda(e^{\lambda t_n} + E_n)
$$
$$
\implies E_n^{(1)} = \lambda E_n \tag{10.11}
$$

The above differential equation represents the error accumulation after n time steps. If the error at the starting step is known to be E_0, the analytical solution for the above equation would also be an exponential function $E_n = E_0 e^{\lambda t_n}$. Following this we can conclude that for positive values of λ the error would grow and for negative values of λ, the error would decay. Therefore, λ has to be always negative for a stable solution.

However, numerical methods involve truncation errors and $\lambda < 0$ does not ensure stability. Now we look at the Euler solution applied to this equation with a step size $\Delta t = h$. The error at next step would be

$$E_1 = E(h) = E_0(1 + \lambda h) \tag{10.12}$$

After n Euler steps we have

$$E_n = E(nh) = E_{n-1}(1 + \lambda h) = E_0(1 + \lambda h)^n \tag{10.13}$$

We define the amplification factor as $\left|\dfrac{E_n}{E_{n-1}}\right| = |1 + \lambda h|$ which is the ratio of two consecutive error evaluations. For the analytical solution showing exponential decay to 0, the amplification factor of the numerical scheme should be less than 1. Otherwise, for amplification factors greater than 1, the solution would diverge away from the exact solution. A numerical scheme is termed as *stable*, if the approximate solution closely follows the exact solution. Therefore, the Euler scheme will be stable only if the amplification factor $\left|\dfrac{E_n}{E_{n-1}}\right| = |1 + \lambda h| < 1$ or if $\lambda h < -2$ or $h < -\dfrac{2}{\lambda}$.

For $|\lambda h| < 1$ the solution is bounded and monotonic. For $|\lambda h| > 1$ the solution is oscillatory. The amplitude diverges with h for $|\lambda h| \geq 2$. Hence the Euler solution is conditionally stable i.e. the step size has to be chosen carefully such that the solution is stable. Accuracy of the solution, in fact, requires much smaller h since the method is only first order accurate i.e. the error is proportional to h itself.

More generally if λ is a complex number, $\lambda h = z$ is also complex[2] and the stability criterion reads as $|1 + z| < 1$ where z is a complex number. This represents the region inside a unit circle centered at $z = -1$ in the z plane (shaded region in Figure 10.4). In order to clarify these ideas a simulation

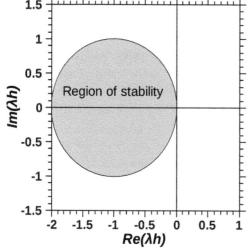

Figure 10.4: *Region of stability of Euler method*

was performed by taking various step sizes and obtaining the value of the function at $t = 0.5$ for the above simple first order ODE. Table 10.1 shows the result of such a simulation.

Table 10.1: *Error at $t = 0.5$ with different step sizes*

h	$y_{Euler}(0.5)$	$y_E(0.5)$	ϵ_y	$\%\epsilon_y$
0.025	0.6027	0.6065	-0.0038	-0.62
0.05	0.5987	0.6065	-0.0078	-1.28
0.1	0.5905	0.6065	-0.016	-2.64
0.25	0.5625	0.6065	-0.044	-7.25
0.5	0.5	0.6065	-0.1065	-17.56

We observe that the step size is to be chosen such that accuracy is guaranteed. Even $h = 0.025$ has an error of ≈ 4 in the third decimal place. Now we look at what happens if $h > 0.5$. We do another simulation and show the results in Table 10.2.

The oscillatory nature of the numerical solution for $h > 1$ is seen from Figure 10.5 where we have used Euler scheme with a step size of $h = 1.5$. The exact solution is also shown for comparison. The numerical solution still follows the exact solution (on an average) which is not the case when $h > 2$.

> To summarize, step size used in a numerical method of solving an ODE has to be chosen such that the solution is stable and also accurate. In most applications the latter is the deciding factor.

---

[2]A complex number is defined as $z = x + jy$ where x and y are real numbers and $j = \sqrt{-1}$ is a pure imaginary number. z represents a point in the (x, y) plane. x is referred to as the real part of z (written as $Re(z)$) while y is referred to as the imaginary part of z (written as $Im(z)$).

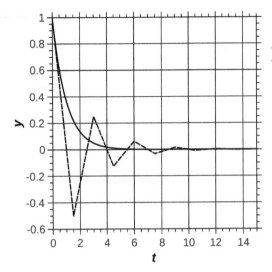

Figure 10.5: Numerical oscillations while using Euler scheme
with $h = 1.5$

Table 10.2: Error at $t = 5$ with different step sizes

h	$y_{Euler}(5)$	$y_E(5)$	ϵ_y	$\%\epsilon_y$
0.5	0.001	0.0067	-0.0058	-86.57
0.5556	0.0007	0.0067	-0.0061	-91.04
0.625	-0.0007	0.0067	-0.0074	-110.45
0.8333	0	0.0067	-0.0067	-100.00
1	0	0.0067	-0.0067	-100.00
1.25	0.0039	0.0067	-0.0028	-41.79
1.6667	-0.2963	0.0067	-0.3031	-4523.88

However there may be specific cases where stability of the solution is also an important criterion
for determining the step size. Such cases will be treated later on, while dealing with *stiff* ODEs.

10.3   Modified Euler method or Heun method

The modified Euler method is also known as the Heun[3] method and accounts for the change in
the slope of the tangent to the solution curve as we move from t_n to t_{n+1}. The first step in the
calculation is the Euler step that gives an estimate for y_{n+1} as y_{n+1}^P (Equation 10.8), a predictor
for the nodal value at node $n + 1$. The slope at the target point is calculated using this predicted
value so that the average slope over the step h is given by

$$\text{Average Slope} = \frac{f(t_n, y_n) + f(t_{n+1}, y_{n+1})}{2} \approx \frac{f(t_n, y_n) + f(t_{n+1}, y_{n+1}^P)}{2} \tag{10.14}$$

We then get a better estimate for y_{n+1} as

$$y_{n+1} = y_n + h \frac{f(t_n, y_n) + f(t_{n+1}, y_{n+1}^P)}{2} \tag{10.15}$$

The Heun method is second order accurate and hence is superior to the Euler method. We note
that the method is equivalent to using a Taylor expansion that includes the second derivative at t_n,
using however, only the predictor value of the function f at t_{n+1}. Figure 10.6 shows the geometric

---

[3]after Karl Heun,1859-1929, a German mathematician

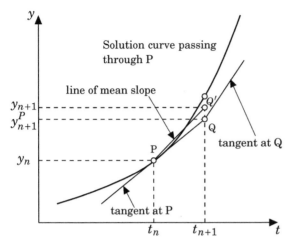

Figure 10.6: *Geometric construction for the Heun method*

equivalent of the Heun scheme. Line PQ is along the tangent at the known point P on the solution curve. y at point Q is the Euler estimate y_{n+1}^P for y_{n+1}. Function $f(t,y)$ calculated at Q represents the slope of the tangent at Q. The line of mean slope is the direction that is chosen for estimating y_{n+1} by the Heun method (point Q'). Geometric construction in the figure shows that the error is considerably smaller in this method. That the error is of second order will be shown later while discussing the Runge Kutta methods.

10.4 Runge Kutta (RK) methods

Runge Kutta[4] methods are very popular because they are *single* step methods and hence *self starting*. These methods, in general, estimate y_{n+1} by making use of the function $f(t,y)$ evaluated at several points within the step y_n - y_{n+1}, updating the y's at these points by suitable approximations. We shall consider second order RK method in detail in what follows since the algebra is tractable and it yields the core idea used in the RK methods.

10.4.1 Second order Runge Kutta method (RK2)

A second order method requires that it be equivalent to using Taylor expansion that includes terms proportional to h^2. Hence we have

$$y_{n+1} = y_n + y_n^{(1)}h + \frac{1}{2!}y_n^{(2)}h^2 + O(h^3) \qquad (10.16)$$

where the subscript n indicates that the calculation is made at t_n. LTE is proportional to h^3 (this is the dominant term that is ignored in the Taylor expansion). However GTE is proportional to h^2 (reader may use a procedure that was used in discussing the order of Euler scheme). Hence the method is second order accurate. Noting that $y^{(1)} = f(t,y)$, we may write the second derivative as

$$y^{(2)} = f^{(1)}(t) = \frac{\partial f}{\partial t} + \frac{\partial f}{\partial y}y^{(1)} = \frac{\partial f}{\partial t} + f\frac{\partial f}{\partial y} \qquad (10.17)$$

Introduce Equation 10.17 in 10.16 to get

$$y_{n+1} = y_n + f_n h + [f_{t,n} + f_{y,n}f_n]\frac{h^2}{2} + O(h^3) \qquad (10.18)$$

---

[4]after Martin Wilhelm Kutta, 1867-1944, a German mathematician

where $f_t \equiv \dfrac{\partial f}{\partial t}$ and $f_y \equiv \dfrac{\partial f}{\partial y}$. The RK2 method aims to achieve equivalence with the above expression by using the following algorithm

$$
\begin{aligned}
k_1 &= hf(t_n, y_n) \\
k_2 &= hf(t_n + \alpha h, y_n + \beta k_1) \\
y_{n+1} &= y_n + w_1 k_1 + w_2 k_2
\end{aligned}
\tag{10.19}
$$

where weights w_1, w_2 and fractions α and β have to be chosen suitably. The last step in Equation 10.19 may be rewritten as

$$
y_{n+1} = y_n + w_1 f_n h + w_2 h f(t_n + \alpha h, y_n + \beta f_n h) + O(h^3)
\tag{10.20}
$$

We use Taylor expansion to get

$$
f(t_n + \alpha h, y_n + \beta f(t_n, y_n)h) = f_n + f_{t,n}\alpha h + f_{y,n} f_n \beta h
\tag{10.21}
$$

Using Equation 10.21 in Equation 10.20 we get

$$
y_{n+1} = y_n + (w_1 + w_2)f_n h + w_2(f_{t,n}\alpha + f_y f_n \beta)h^2 + O(h^3)
\tag{10.22}
$$

Term by term comparison of Equations 10.18 and 10.22 shows that we must have

$$
w_1 + w_2 = 1; \quad w_2\alpha = \frac{1}{2}; \quad w_2\beta = \frac{1}{2}
\tag{10.23}
$$

Since the number of equations is less than the number of unknowns multiple solutions are possible for Equation 10.23. A choice that is commonly made is $\alpha = 1$. It is then seen that $w_2 = \dfrac{1}{2}$, $w_1 = \dfrac{1}{2}$ and $\beta = 1$. The RK2 algorithm given by Equation 10.19 may then be written as

$$
\begin{aligned}
k_1 &= hf(t_n, y_n) \\
k_2 &= hf(t_n + h, y_n + k_1) = hf(t_{n+1}, y_n + k_1) \\
y_{n+1} &= y_n + h\left(\frac{k_1 + k_2}{2}\right)
\end{aligned}
\tag{10.24}
$$

The reader may note that the above algorithm is essentially the same as the Heun method. Hence both Heun and RK2 are second order accurate methods.

Consider the special case where $f(t, y) = f(t)$ only. Then last of Equation 10.24 takes the form

$$
y_{n+1} = y_n + h\left[\frac{k_1 + k_2}{2}\right] = y_n + h\left[\frac{f(t_n) + f(t_{n+1})}{2}\right]
$$

This is nothing but the trapezoidal rule introduced earlier while dealing with numerical integration over a line.

Example 10.2

Use step size of $h = 0.1$ to obtain the solution of the first order ordinary differential equation $y^{(1)} + y = 0$ up to $t = 0.5$ using RK2. The initial value is specified to be $y(0) = 1$.

Solution :

The ODE is the same as that considered in Example 10.1. We would like to improve the solution by using a higher order scheme viz. RK2 which is second order accurate. Starting with the initial solution, using a step size of $h = 0.1$, we have $k_1 = hf(0,1) = 0.1 \times (-1) = -0.1$. We update t_0 to $t_1 = t_0 + h = 0 + 0.1 = 0.1$ and y_0 to $y_0 + k_1 = 1 + (-0.1) = 0.9$ and evaluate k_2 as $k_2 = hf(t_0 + h, y_0 + k_1) = 0.1 \times (-0.9) = -0.09$. Finally RK2 yields $y_1 = y_0 + \dfrac{k_1 + k_2}{2} = 1 + \dfrac{-0.1 - 0.09}{2} = 0.9050$. Compare this with the Euler estimate of 0.9 obtained in Example 10.1. The exact value is given by $e^{-0.1} = 0.9048$. The error is thus given by $0.9050 - 0.9048 = 0.0002$ which is much less than what we saw in the Euler case. We show the results up to $t = 0.5$ as a table below.

x	y_{RK2}	k_1	k_2	y_E	$\epsilon_{RK2,abs}$	$\%\epsilon_{RK2,rel}$
0	1	-0.1	-0.09	1	0	0
0.1	0.905	-0.0905	-0.0815	0.9048	0.0002	0.022
0.2	0.819	-0.0819	-0.0737	0.8187	0.0003	0.036
0.3	0.7412	-0.0741	-0.0667	0.7408	0.0004	0.053
0.4	0.6708	-0.0671	-0.0604	0.6703	0.0005	0.074
0.5	0.6071	-0.0607	-0.0546	0.6065	0.0005	0.082

Subscript abs is absolute while subscript rel is relative

We notice that there has been a good improvement in the accuracy of the numerical solution.

Accuracy and stability of RK2

In order to look at the stability and accuracy of the second order RK method (the same as the Heun method) we present below the effect of step size on the solution. Table 10.3 shows the effect of h on the solution at $t = 0.5$. The error with respect to the exact solution is also shown in the last column of the table.

Table 10.3: Error at $t = 0.5$ with different step sizes

h	$y_{RK2}(0.5)$	$y_E(0.5)$	$\epsilon_{RK2,abs}$	$\%\epsilon_{RK2,rel}$
0.025	0.6066	0.6065	0.00003	0.005
0.05	0.6067	0.6065	0.00013	0.021
0.1	0.6071	0.6065	0.00055	0.091
0.125	0.6074	0.6065	0.00087	0.143
0.25	0.6104	0.6065	0.00382	0.630

For practical purposes a step size as large as 0.1 may be acceptable since the solution at $t = 0.5$ is accurate to three digits after decimals. Table 10.4 shows what happens when the step size is bigger than 0.25. The function value at $t = 5$ is being looked at.

Table 10.4: Error at $t = 5$ with different step sizes

h	$y_{RK2}(0.5)$	$y_E(0.5)$	$\epsilon_{RK2,abs}$	$\%\epsilon_{RK2,rel}$
0.25	0.0072	0.0067	0.0004	5.97
0.5	0.0091	0.0067	0.0024	35.82
1	0.0313	0.0067	0.0245	365.67
1.25	0.0797	0.0067	0.0729	1088.06
1.6667	0.3767	0.0067	0.37	5522.39
2.5	2.6406	0.0067	2.6339	39311.94

Simulation shows that the numerical solution is stable up to $h = 2$. For example the RK2 solution with $h = 1.6667$ shows a monotonic decay (hence stable) but with significant deviation from the exact solution as shown in Figure 10.7.

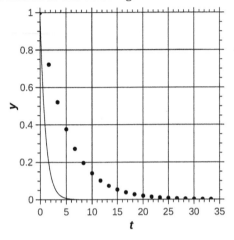

Figure 10.7: *Comparison of RK2 solution with the exact solution using a step size of $h = 1.6667$*

The RK2 method relates two neighboring values of the dependent variable by an expression that includes the second derivative term. Thus the Taylor expansion for this case is given by

$$y_{n+1} = y_n + h f_n + \frac{h^2}{2} f_n^{(1)} \tag{10.25}$$

For the standard problem $y^{(1)} - \lambda y = 0$ where λ is complex the above becomes

$$y_{n+1} = y_n + h \lambda y_n + \frac{\lambda^2 h^2}{2} y_n = y_n \left(1 + \lambda h + \frac{\lambda^2 h^2}{2} \right) \tag{10.26}$$

In general, stability of the RK2 method is governed by the magnification factor $\left| \dfrac{y_{n+1}}{y_n} \right| < 1$ or $\left| 1 + z + \dfrac{z^2}{2} \right| < 1$ where z stands for λh. We at once see that $z = 0$ and $z = -2$ bracket the stability region along the real axis. Thus λ must be negative and $|\lambda h| < 2$. The region of stability in the complex plane is now the slightly oblong region shown in Figure 10.8. It is also noticed that the region of stability of RK2 is bigger than that for the Euler scheme.

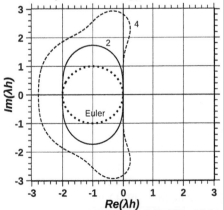

Figure 10.8: *Stability regions for various methods 2 - RK2, 4 - RK4*

10.4.2 Fourth order Runge Kutta method (RK4)

RK4 is by far the most popular Runge Kutta method and is given below without proof.[5] It uses f calculated at the end points as well at a point within the step.

$$
\begin{aligned}
k_1 &= hf(t_n, y_n) \\
k_2 &= hf\left(t_n + \frac{h}{2}, y_n + \frac{k_1}{2}\right) \\
k_3 &= hf\left(t_n + \frac{h}{2}, y_n + \frac{k_2}{2}\right) \\
k_4 &= hf(t_n + h, y_n + k_3) \\
y_{n+1} &= y_n + \frac{k_1 + 2k_2 + 2k_3 + k_4}{6}
\end{aligned}
\tag{10.27}
$$

We see that the solution at the end of the step is obtained as a weighted sum of the f values calculated as k_1 to k_4. Note that the sum of the weights (=6) is the same as the denominator, in the final summation step.

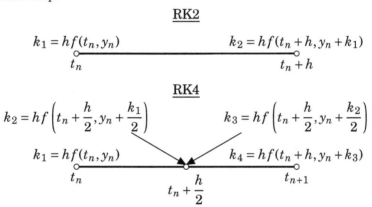

Figure 10.9: *Function calculations in RK2 and RK4*

The function calculations involved in RK2 and RK4 are shown in Figure 10.9. RK2 involves two function calculations per step while RK4 involves four function calculations per step. RK4 is fourth order accurate which means that the LTE is proportional to h^5 and GTE is proportional to h^4.

Again, in the special case when $f(t, y) = f(t)$ RK4 reduces to the Simpson 1/3 rule introduced earlier while dealing with numerical integration over a line. Note that h in the present context is $2h$ in the context of Simpson rule.

Stability of the RK4 scheme is governed by the inequality

$$
\left| 1 + z + \frac{z^2}{2} + \frac{z^3}{6} + \frac{z^4}{24} \right| < 1
$$

In the complex plane the region of stability is as shown in Figure 10.8. It is seen that the region of stability of RK4 is even bigger than that of RK2. A MATLAB program has been provided below to solve ODE using RK4. The program can also solve a system of linear ODEs (will be dealt in detail in a later section).

---

[5]See Runge - Kutta methods for ordinary differential equations, J Butcher - 2005, http://www.math.auckland.ac.nz

Program 10.1: Initial value problem using Runge Kutta 4

```
 1 function  [X,Y]  =  ODERK4(x1,y1,h,n,F,varargin)
 2 % Input  :   x1:  starting  point
 3 %             y1:  initial  values  of  y's  at  x=x1
 4 %             h :  step  size
 5 %             n :  number  of  intervals
 6 %             F :  external  function  containing  derivatives
 7 %                 y' = F(x,y)
 8 %             varargin : optional  arguments  to  be  passed  to  F
 9 % Output:
10 %             X :  abcissca
11 %             Y :  ordinates
12 x2  =  x1+n*h;              %  end  point  of  interval
13 m  =  length(y1);          %  number  of  first  order  ODEs
14 Y  =  zeros(n+1,m);        %  initialize  Y
15 X  =  x1:h:x2;             %  define  grid
16 Y(1,:)  =  y1;             %  initial  condition  at  t=0
17 for  i  =  1:n             %  loop  for  RK4  calculation
18     k1  =  h*F(X(i),Y(i,:),varargin{:});
19     k2  =  h*F(X(i)+h/2,Y(i,:)+k1/2,varargin{:});
20     k3  =  h*F(X(i)+h/2,Y(i,:)+k2/2,varargin{:});
21     k4  =  h*F(X(i)+h,Y(i,:)+k3,varargin{:});
22     Y(i+1,:)  =  Y(i,:)  +  (k1+2*k2+2*k3+k4)/6;   %y_{n+1}
23 end
```

varargin is useful to perform parametric study, which will be demonstrated in some of the examples that follow.

Example 10.3

Solve the first order ordinary differential equation $y^{(1)} = y - 3t^2$ subject to initial condition $y(0) = 1$. Use RK4 with a step size of $h = 0.1$ and obtain the solution till $t = 0.5$. Discuss the error by comparing the numerical solution with the exact solution.

Solution :

The exact solution to the given non-homogeneous ODE may be obtained by summing complementary function and the particular integral and making sure that the initial condition is satisfied. The reader may verify that the exact solution is given by

$$y_E = 3t^2 + 6t + 6 - 5e^t$$

In order to apply the RK4 we note that the function is given by $f(t, y) = y - 3t^2$. We start with $t = 0$, $y = 1$ and use a step size of $h = 0.1$ as specified in the problem. We go through one RK4 cycle in detail below.

$$k_1 = hf(t_0, y_0) = 0.1(1 - 3 \times 0^2) = 0.1$$

$$k_2 = hf\left(t_0 + \frac{h}{2}, y_0 + \frac{k_1}{2}\right) = 0.1\left[1 + \frac{0.1}{2} - 3 \times \left(\frac{0.1}{2}\right)^2\right] = 0.1043$$

$$k_3 = hf\left(t_0 + \frac{h}{2}, y_0 + \frac{k_2}{2}\right) = 0.1\left[1 + \frac{0.1043}{2} - 3 \times \left(\frac{0.1}{2}\right)^2\right] = 0.1045$$

$$k_4 = hf(t_0 + h, y_0 + k_3) = 0.1\left[1 + 0.1045 - 3 \times 0.1^2\right] = 0.1074$$

$$y_1 = y_0 + \frac{k_1 + 2k_2 + 2k_3 + k_4}{6} = 1 + \frac{0.1 + 2(0.1043 + 0.1045) + 0.1074}{6} = 1.1041$$

The exact solution is the same as above, when rounded to four significant digits. It may be ascertained that the error is equal to -2×10^{-7} when more digits are retained in the calculations. The calculations continue till $t = 0.5$ as given in Table 10.5. The numerical solution, for all practical purposes, is as good as the exact solution.

Table 10.5: RK4 solution till $t = 0.5$ to ODE in Example 10.3

t	y_{RK4}	k_1	k_2	k_3	k_4	y_{Ex}	$\epsilon_{RK4,abs}$
0	1.0000	0.1000	0.1043	0.1045	0.1074	1.0000	0
0.1	1.1041	0.1074	0.1090	0.1091	0.1093	1.1041	-2.01E-07
0.2	1.2130	0.1093	0.1080	0.1079	0.1051	1.2130	-3.79E-07
0.3	1.3207	0.1051	0.1006	0.1003	0.0941	1.3207	-5.27E-07
0.4	1.4209	0.0941	0.0860	0.0856	0.0757	1.4209	-6.35E-07
0.5	1.5064	0.0756	0.0637	0.0631	0.0489	1.5064	-6.95E-07

Error estimation

In practice, the exact solution of ODE will not be available and there is no information on error. However, an estimate for error can be determined (similar to numerical integration) using the following methods

1. Difference between estimates using two grid sizes h and $h/2$
2. Difference between two estimates using ODE solvers of order n and $n + 1$

The former procedure is demonstrated in the next example. The latter option is considered later on.

Example 10.4

Solve the first order ordinary differential equation $y^{(1)} = \sin(t) - y\cos(t)$ subject to initial condition $y(0) = 1$. Use RK4 and obtain the solution till $t = 0.5$.

Solution :

We use a step size of $h = 0.1$, use RK4 and obtain the solution up to $t = 0.5$. The results are tabulated below.

t	y_{RK4}	k_1	k_2	k_3	k_4
0	1.000000	-0.100000	-0.089883	-0.090389	-0.080523
0.1	0.909822	-0.080544	-0.071035	-0.071505	-0.062294
0.2	0.838503	-0.062312	-0.053484	-0.053912	-0.045403
0.3	0.784751	-0.045418	-0.037294	-0.037676	-0.029868
0.4	0.747213	-0.029881	-0.022441	-0.022776	-0.015633
0.5	0.724556	-0.015643	-0.008835	-0.009125	-0.002583

We want to find out how good is the solution, realizing that we are unable to obtain the exact solution. In order to do this we recalculate the solution using a smaller step size, say $h = \dfrac{0.1}{2} = 0.05$. The computed results are shown in the table below.

t	y_{RK4}	k_1	k_2	k_3	k_4
0	1.000000	-0.050000	-0.047485	-0.047548	-0.045064
0.05	0.952478	-0.045065	-0.042620	-0.042681	-0.040271
0.1	0.909822	-0.040272	-0.037903	-0.037962	-0.035632
0.15	0.871883	-0.035633	-0.033346	-0.033402	-0.031155
0.2	0.838502	-0.031156	-0.028954	-0.029007	-0.026846
0.25	0.809515	-0.026847	-0.024732	-0.024782	-0.022708
0.3	0.784751	-0.022709	-0.020680	-0.020728	-0.018740
0.35	0.764040	-0.018741	-0.016798	-0.016843	-0.014940
0.4	0.747213	-0.014941	-0.013081	-0.013123	-0.011302
0.45	0.734105	-0.011303	-0.009523	-0.009563	-0.007821
0.5	0.724555	-0.007822	-0.006119	-0.006156	-0.004488

We have two estimates for the nodal values of y at $t = 0.5$ viz. 0.724556 obtained by RK4 with step 0.1 and 0.724555 obtained by RK4 with step 0.05. The change is noted to be -4.16×10^{-7}. We may accept the solution with $h = 0.1$ as being accurate to at least six significant digits after decimals.

10.4.3   Embedded Runge Kutta methods

An alternate way of estimating the error of RK methods is to determine the difference between two schemes having order of accuracy n and $n+1$. For example, difference between Euler estimate and RK2 estimate is an indicator for error. Similar procedure can be applied for higher order Runge Kutta methods also. Two consecutive RK methods do not share any common points (except the first and last point of the segment) between them and hence both the estimates have to be evaluated independently. Following such conditions, this approach for error estimation is not very advantageous. Instead embedded RK methods can be applied where the points of function evaluation are common for the two numerical schemes. These methods are suited for automatic control of step size.

Runge Kutta Fehlberg method

The method is another member of Runge Kutta family proposed by German mathematician Erwin Fehlberg, where the ODE is solved using Runge Kutta method of orders 4 and 5 and the error is estimated from the difference between these two estimates.[6] The main advantage of this method over classical RK methods is that the points of function evaluations are shared between 4th and 5th order solvers. This is economical compared to using classical RK methods as well as using two steps of h and $h/2$. Functions are evaluated at six points within the interval. Four of these points are used for evaluating fourth order estimate and five points are used for estimating

---

[6]Erwin Fehlberg (1969), Low-order classical Runge-Kutta formulas with step size control and their application to some heat transfer problems, NASA Technical Report 315.

fifth order estimate and the difference between the two solutions is the estimate for error. The six function evaluations are given below:

$$k_1 = hf(t_n, y_n) \qquad k_2 = hf\left(t_n + \frac{h}{4}, y_n + \frac{k_1}{4}\right)$$

$$k_3 = hf\left(t_n + \frac{3h}{8}, y_n + \frac{3k_1}{32} + \frac{9k_2}{32}\right)$$

$$k_4 = hf\left(t_n + \frac{12h}{13}, y_n + \frac{1932k_1}{2197} - \frac{7200k_2}{2197} + \frac{7296k_3}{2197}\right) \qquad (10.28)$$

$$k_5 = hf\left(t_n + h, y_n + \frac{439k_1}{216} - 8k_2 + \frac{3680k_3}{513} - \frac{845k_4}{4104}\right)$$

$$k_6 = hf\left(t_n + \frac{h}{2}, y_n - \frac{8k_1}{27} + 2k_2 - \frac{3544k_3}{2565} + \frac{1859k_4}{4104} - \frac{11k_5}{40}\right)$$

The estimate of y_{n+1} and error are provided below.

$$\begin{array}{|c|c|}
\hline
y_{n+1,4} = & y_n + \dfrac{25}{216}k_1 + \dfrac{1408}{2565}k_3 + \dfrac{2197}{4101}k_4 - \dfrac{1}{5}k_5 \\
\hline
\epsilon = & \dfrac{1}{360}k_1 - \dfrac{128}{4275}k_3 - \dfrac{2197}{75240}k_4 + \dfrac{1}{50}k_5 + \dfrac{2}{55}k_6 \\
\hline
\end{array} \qquad (10.29)$$

The following MATLAB program performs one step of Runge Kutta Fehlberg procedure

Program 10.2: *Runge Kutta Fehlberg 45*

```
1  function [Y,er] = ODERKF45(x1,y1,h,F,varargin)
2  % Input  :  x1: starting point
3  %            y1: initial values of y's at x=x1
4  %            h : step size
5  %            F : external function containing derivatives
6  %                 y' = F(x,y)
7  %            varargin : optional arguments to be passed to F
8  % Output:
9  %            Y : ordinates
10 %            er: error estimate
11 a1 = 1/4;        b11=1/4;     % RKF constants
12 a2 = 3/8;        b21=3/32;        b22=9/32;
13 a3 = 12/13; b31=1932/2197; b32=-7200/2197; b33 = 7296/2197;
14 a4 = 1;      b41= 439/216 ; b42=-8;
15 b43 = 3680/513; b44= -845/4104;
16 a5 = 1/2;        b51 = -8/27; b52 = 2; b53 =      -3544/2565;
17 b54 = 1859/4104; b55 = -11/40;
18 k1 = h*F(x1,y1,varargin{:});
19 k2 = h*F(x1+a1*h,y1+b11*k1,varargin{:});
20 k3 = h*F(x1+a2*h,y1+b21*k1+b22*k2,varargin{:});
21 k4 = h*F(x1+a3*h,y1+b31*k1+b32*k2 ...
22     + b33*k3,varargin{:});
23 k5 = h*F(x1+a4*h,y1+b41*k1+b42*k2 ...
24     + b43*k3 + b44*k4,varargin{:});
25 k6 = h*F(x1+a5*h,y1+b51*k1+b52*k2 ...
26     + b53*k3 + b54*k4+ b55*k5,varargin{:});
27 er = abs(k1/360 - 128*k3/4275 - ...
28     2197*k4/75240 +k5/50 +2*k6/55);  % error
29 Y = y1 + 25*k1/216 + 1408*k3/2565 ...
30     + 2197*k4/4104  - 1*k5/5;       % estimate
```

Since the method estimates the error, it can be used to automatically to select the step size.

10.4.4   Adaptive Runge Kutta methods

Similar to numerical integration, adaptive methods automatically decide on the step size required to achieve sufficient accuracy. The error estimates calculated using the two methods discussed earlier can be used to decide the step size. Two sets of tolerance values have to be provided (tol_{max} and tol_{min}) to control the step size. The following simple strategy can be used to apply adaptive techniques with RK methods

Step 1 Determine the error estimate using two grids h and $h/2$ or RKF method.

Step 2 If the error estimate is greater than specified tolerance tol_{max}, reduce the step size by half ($h = h/2$) and recalculate the error.

Step 3 Else if the error estimate is less than a specified maximum tolerance tol_{max}, stop iteration and accept the present value of t_n and y_n.

Step 4 If the error estimate is smaller than tol_{min}, double the step size for next point.

MATLAB program has been provided below to solve an ODE adaptively. The following program can also take into account a system of linear ODEs (will be discussed later).

Program 10.3: *Adaptive RK4 using h and h/2 segments*

```
 1 function [X,Y] = ODEadaptRK4(x1,x2,y1,h,F,tol,varargin)
 2 % Input :   x1:  starting point
 3 %           x2:  end point
 4 %           y1:  initial values of y's at x=x1
 5 %           h :  step size
 6 %           F :  external function containing derivatives
 7 %                y' = F(x,y)
 8 %           tol: tolerance
 9 %           varargin : optional arguments to be passed to F
10 % Output :
11 %           X :  abcissca
12 %           Y :  ordinates
13 X(1) = x1;              % initialize x(1)
14 Y(1,:) = y1;           % y(x=0)
15 count = 1;             % no. of points in interval
16 while (abs(X(count)-x2) >= 1e-15)  % outer loop
17     if( x2 - X(count) < h )
18         h = x2-X(count);
19     end
20     [xh,yh] = ODERK4(X(count),Y(count,:),h,1,F,varargin{:});
21     while 1                        % inner loop
22         [xh2,yh2] = ODERK4(X(count),Y(count,:), ...
23                     h/2,2,F,varargin{:});
24         if(max(abs(yh2(3,:)-yh(2,:))) > tol & h > 1e-6)
25             h = h/2;               % halve the interval
26             yh = yh2;
27         else
28             X(count+1) = X(count) + h;  % update X
29             Y(count+1,:) = yh2(3,:);    % update Y
30             count = count + 1;          % update point count
31             max(abs(yh2(3,:)-yh(2,:)))  % maximum error
32             if(max(abs(yh2(3,:)-yh(2,:))) < tol/10 )
```

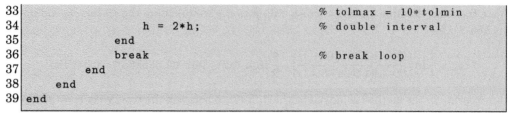

```
33 |                                              % tolmax  =  10*tolmin
34 |                    h  =  2*h;                 % double  interval
35 |              end
36 |              break                            % break  loop
37 |         end
38 |    end
39 | end
```

Note: `tolmax = 10tolmin`

Example 10.5

Use RK2 and RK4 to solve the following initial value problem.

$$y^{(1)} = -(y^4 - 1), \quad y(0) = y_0 = 2$$

Choose suitable step size for obtaining the solution by comparison with the exact solution. Also apply adaptive method and comment on the solution.

Background :

Radiative heat transfer is the only mode of heat transfer available to cool components in a spacecraft. Heat is rejected to space by radiation and hence maintains the temperature of components at desired level. If a component of heat capacity mc_p and temperature T is exposed to space at temperature T_∞, the temperature of the component will be governed by

$$mc_p \frac{dT}{dt} = -\varepsilon\sigma(T^4 - T_\infty^4)$$

where ε is surface emissivity and σ is the Stefan Boltzmann constant. Equation being considered in this example is a specific case based on this equation.

Solution :

Before we look at the numerical solution we obtain the exact solution by noting that the given first order ODE is in variable separable form. The exact solution must therefore be given by

$$-\int_{y_0}^{y} \frac{dy}{(y^4 - 1)} = \int_0^t dt$$

The integrand in the integral on the left hand side may be simplified by using partial fractions as

$$\frac{1}{y^4 - 1} = \frac{1}{2}\left(\frac{1}{y^2 - 1} - \frac{1}{y^2 + 1}\right) = \frac{1}{4}\left(\frac{1}{y - 1} - \frac{1}{y + 1}\right) - \frac{1}{2}\frac{1}{y^2 + 1}$$

Term by term integration yields the solution given below.

$$t = \frac{1}{4}\ln\left(\frac{(y_0 - 1)(y + 1)}{(y_0 + 1)(y - 1)}\right) - \frac{1}{2}\left(\tan^{-1}(y_0) - \tan^{-1}(y)\right)$$

Now we look at the numerical solution. Starting at $t = 0$, $y_0 = 2$ we get the solution using a step $h = 0.05$ and the RK2. One pass of the calculation is shown below.

$$
\begin{aligned}
t_1 &= t_0 + h = 0 + 0.05 = 0.05 \\
k_1 &= hf(t_0, y_0) = -0.05(2^4 - 1) = -0.75 \\
k_2 &= hf(t_1, y_0 + k_1) = -0.05((2 - 0.75)^4 - 1) = -0.0720703 \\
y_1 &= y_0 + \frac{k_1 + k_2}{2} = 2 + \frac{-0.75 - 0.0720703}{2} = 1.588965
\end{aligned}
$$

Since the exact solution yields t corresponding to a y we calculate the t_E corresponding to the above y as

$$t_{E,1} = \frac{1}{4}\ln\left(\frac{(2-1)(1.588965+1)}{(2+1)(1.588965-1)}\right) - \frac{1}{2}\left(\tan^{-1}(2) - \tan^{-1}(1.588965)\right) = 0.046475$$

The accuracy of the method is judged by the difference between the two t's i.e. $\epsilon_t = t_1 - t_{E,1} = 0.05 - 0.046475 = 0.003525$ which is quite poor!

We may improve the solution by taking a smaller h or by improving the method. If we use RK4 and keep the step size the same, we get the following results in tabular form.

t	y	k_1	k_2	k_3	k_4
0	2.00000	-0.75000	-0.29865	-0.53653	-0.17935
0.05	1.56672				

Corresponding to $y_1 = 1.56672$ we have $t_1 = 0.0507567$. The error now is given by $\epsilon_t = 0.05 - 0.050757 = 0.00076$ which is somewhat better. The error has migrated to the fourth place after decimals. Finally, if we take a step size of $h = 0.02$ (50 steps) the error gets reduced to approximately 3×10^{-6}, for all t. This last one may be taken as an acceptable solution. We may use Program 10.3 to solve the ODE using adaptive method.

```
[X,Y] = ODEadaptRK4(0,1,2,0.05,@odeexample,1e-6);
```

The adaptive method requires 29 steps. A comparison of RK4 with exact solution is shown in Figure 10.10. The two solutions appear to agree closely with each other.

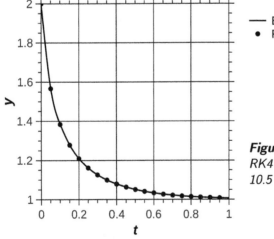

Figure 10.10: *Comparison of adaptive RK4 with the exact solution in Example 10.5*

RK methods evaluates the function several times within one interval. Sometimes, evaluation of function may be computationally expensive. Now we shall look at predictor corrector methods which can be used under such situations.

10.5 Predictor corrector methods

We have already introduced Heun method as a predictor corrector method. Calculation of y_{n+1}^P was a predictor, corrected by function calculation at the target point. Adams Bashforth and Adams Moulton methods are multi-step predictor corrector methods that are very useful in practice. Both of these methods can be applied independently to solve an ODE. AB being an explicit method would be prone to numerical errors. However AM being an implicit method would become difficult to treat non-linear ODE (will be treated in detail later). Most frequently AB and AM methods are applied together as predictor corrector pairs overcoming the limitations of each. Multi-step methods use the already calculated function values to extend the solution as opposed to the Runge Kutta methods that calculate function values at a number of points within a step.

Explicit and implicit methods: Methods to treat Initial value problems can be classified as explicit or implicit solvers. Consider a first order ODE

$$\frac{dy}{dt} = f(t,y)$$

Let us replace the derivative with a first order forward difference formula.

$$\frac{y_{n+1} - y_n}{\Delta t} = f(t,y)$$

An important question remains is at what t should $f(t,y)$ be evaluated. From calculus, it is known that the point lies between t_n and t_{n+1}. However, as the function itself is not known such evaluation is not straightforward. Hence in practice, $f(t,y)$ is evaluated as a weighted sum of function evaluations at a number of points. In an explicit method, functions are evaluated at previous nodes i.e. function evaluations are already available for future calculations. Euler and RK methods are both explicit. On the other hand when $f(t,y)$ is expressed as a function of the current point t_{n+1}, $f(t_{n+1}, y_{n+1})$ is not known and becomes part of the solution procedure, the method is implicit. Implicit methods become computationally intensive for nonlinear differential equations. Explicit methods are computationally simple and can be applied to a wide range of problems. However, implicit methods become necessary when stability issues arise. Implicit methods will be treated in more detail later.

10.5.1 Adams Bashforth Moulton (ABM2) second order method

The second order predictor corrector method uses function evaluations at two points in the explicit Adams Bashforth step ($n-1$ and n) and function evaluations at two points in the implicit Adams Moulton step (n and $n+1$) as illustrated in Figure 10.11.

Adams Bashforth predictor step (explicit)

Adams Bashforth method (AB2) is the predictor step of the method. This step is explicit i.e. it is assumed that y_{n-1} and y_n are available and we intend to obtain y_{n+1}. We may use a Newton polynomial of degree one to extrapolate the function $f(t_i, y_i)$ outside the interval $t_n - t_{n-1}$ as

$$f(t_i, y_i) \approx p_1(r) = f_n + r \nabla_n \tag{10.30}$$

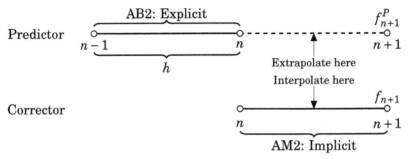

Figure 10.11: *Steps in 2nd order predictor corrector method*

where $r = \dfrac{t_i - t_n}{h}$ and $\nabla_n = f_n - f_{n-1}$. This step is *explicit* since the function f at the target point is not involved in the calculation. The predictor y_{n+1}^P may then be written down as

$$
\begin{aligned}
y_{n+1}^P &= y_n + h \int_0^1 [f_n + r\nabla f_n]\,dr = y_n + h \left[f_n r + \nabla f_n \frac{r^2}{2} \right]\Big|_0^1 \\
&= y_n + h \left[f_n + \frac{\nabla f_n}{2} \right] = y_n + h \left[\frac{3}{2} f_n - \frac{1}{2} f_{n-1} \right]
\end{aligned}
\tag{10.31}
$$

Adams Moulton corrector step (implicit)

Second order accurate Adams Moulton (AM2) step is implicit and uses the function values at n and $n+1$ to construct an approximation given by

$$
f(t_i, y_i) \approx p_1(r) = f_{n+1} + r\nabla_{n+1}
\tag{10.32}
$$

where $r = \dfrac{t_i - t_{n+1}}{h}$ and $\nabla_{n+1} = f_{n+1} - f_n$. The backward difference is usually written as $\nabla_{n+1} = f_{n+1}^P - f_n$ where the Adams Bashforth predictor y_{n+1}^P is used to start the calculation. It is possible to then use the Adams Moulton step as a fixed point iteration scheme to determine y_{n+1}. We thus have

$$
\begin{aligned}
y_{n+1} &= y_n + h \int_{-1}^0 \left[f_{n+1}^P + r\nabla_{n+1} \right] dr = y_n + h \left[f_{n+1}^P r + \nabla_{n+1} \frac{r^2}{2} \right]\Big|_{-1}^0 \\
&= y_n + h \left[f_{n+1}^P - \frac{\nabla_{n+1}}{2} \right] = y_n + h \left[\frac{f_n + f_{n+1}^P}{2} \right]
\end{aligned}
\tag{10.33}
$$

This last expression is nothing but the familiar trapezoidal scheme.

Note that the Adams-Bashforth scheme requires starting values at two nodes n and $n-1$. Hence the multi-step method is not self starting. If we are starting the solution at $t = 0$ we have one initial condition specified as $y(t = 0) = y_0$. Hence $y(t = h) = y_1$ may be generated by a single Euler step which is second order accurate. The solution may then be continued using the multi-step method.

Example 10.6

Solve the first order ordinary differential equation $y^{(1)} = y - 3t^2$ subject to the initial condition $y(0) = 1$. Use AB2-AM2 combination (i.e. ABM2) after the first Euler step. Use $h = 0.05$.

Solution :

Starting from the initial condition $t = t_0 = 0$, $y = y_0 = 1$ Euler step gives

$$y_1 = y_0 + h(y_0 - 3t_0^2) = 1 + 0.05(1 - 3 \times 0^2) = 1.05$$

at $t = t_1 = t_0 + h = 0 + 0.05 = 0.05$. Using AB2 we get the predictor y_2^P as

$$
\begin{aligned}
y_2^P &= y_1 + \frac{h}{2}\left[3(y_1 - 3t_1^2) - (y_0 - 3t_0^2)\right] \\
&= 1.05 + \frac{0.05}{2}[3(1.05 - 3 \times 0.05^2) - (1 - 3 \times 0^2)] = 1.1032
\end{aligned}
$$

Using AM2 we get the corrector y_2 as

$$
\begin{aligned}
y_2 &= y_1 + \frac{h}{2}[(y_2^P - 3t_2^2) + (y_1 - 3t_1^2)] \\
&= 1.05 + \frac{0.05}{2}[(1.1032 - 3 \times 0.1^2) + (1.05 - 3 \times 0.05^2)] = 1.1029
\end{aligned}
$$

This completes one AB2 AM2 pass. The calculations may be continued as required. The exact solution may easily be shown to be given by $y_E = 3t^2 + 6t + 6 - 5e^t$. The exact values of y are given by $y_E(0) = 1$, $y_E(0.05) = 1.0511$ and $y_E(0.1) = 1.1041$. Results are tabulated below up to $t = 0.5$. The last column shows the error of the numerical solution with respect to the exact solution.

The error appears to be severe since it is in the third place after decimals. The solution may be improved by taking a smaller step size.

t	$y(AM2)$	$y^P(AB2)$	y_E	ϵ_y
0	1.0000	Initial Condition	1.0000	0.0000
0.05	1.0500	Euler	1.0511	-0.0011
0.1	1.1029	1.1032	1.1041	-0.0013
0.15	1.1570	1.1573	1.1583	-0.0014
0.2	1.2115	1.2118	1.2130	-0.0015
0.25	1.2657	1.2661	1.2674	-0.0016
0.3	1.3189	1.3193	1.3207	-0.0018
0.35	1.3702	1.3706	1.3722	-0.0019
0.4	1.4188	1.4192	1.4209	-0.0021
0.45	1.4637	1.4641	1.4659	-0.0023
0.5	1.5039	1.5044	1.5064	-0.0025

10.5.2 Fourth order method

This is a three step method that uses f_n, f_{n-1}, f_{n-2}, f_{n-3} in the Adams-Bashforth explicit step and f_{n+1}^P, f_n, f_{n-1}, f_{n-2} in the implicit Adams-Moulton step (see Figure 10.12). Here f_{n+1}^P is the value based on predictor y_{n+1}^P.

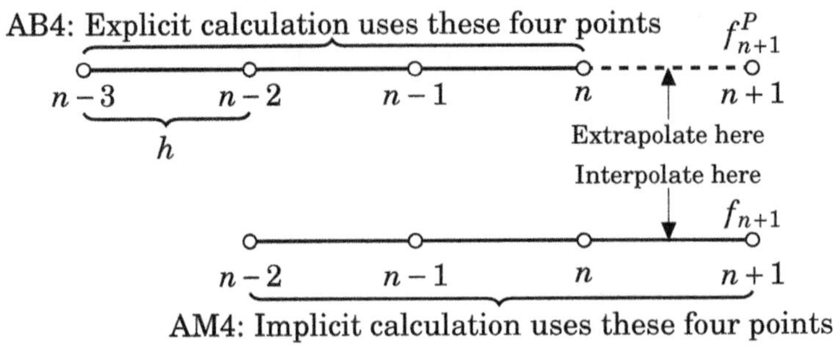

AB4: Explicit calculation uses these four points f^P_{n+1}

Figure 10.12: Steps in 4th order predictor corrector method

Adams-Bashforth explicit step

We shall use backward differences and define a polynomial approximation to f that passes through the points n, $n-1$, $n-2$ and $n-3$. This polynomial may be written down, using Newton Gregory series with $(t-t_n) = hr$ as

$$f(r) = f_n + r\nabla_n + \frac{r(r+1)}{2}\nabla_n^2 + \frac{r(r+1)(r+2)}{6}\nabla_n^3 \tag{10.34}$$

This step is explicit since the function f at the target point is not involved in the calculation. The predictor y^P_{n+1} may then be written down as

$$
\begin{aligned}
y^P_{n+1} &= y_n + h\int_0^1\left[f_n + r\nabla_n + \frac{r(r+1)}{2}\nabla_n^2 + \frac{r(r+1)(r+2)}{6}\nabla_n^3\right]dr \\
&= y_n + h\left[f_n r + \nabla_n\frac{r^2}{2} + \nabla_n^2\left(\frac{r^3}{6} + \frac{r^2}{4}\right) + \nabla_n^3\left(\frac{r^4}{24} + \frac{r^3}{6} + \frac{r^2}{6}\right)\right]\Big|_0^1 \\
&= y_n + h\left[f_n + \frac{\nabla_n}{2} + \frac{5}{12}\nabla_n^2 + \frac{9}{24}\nabla_n^3\right]
\end{aligned}
\tag{10.35}
$$

The last expression is simplified after writing the backward differences in terms of function values. The reader may verify that we should get

$$y^P_{n+1} = y_n + \frac{h}{24}[55f_n - 59f_{n-1} + 37f_{n-2} - 9f_{n-3}] \tag{10.36}$$

Adams-Moulton implicit step

In the Adams-Moulton implicit step we use a polynomial representation of f based on nodes $n+1$, n, $n-1$, $n-2$ but using f^P_{n+1} calculated as $f(t_{n+1}, y^P_{n+1})$. Again we make use of Newton Gregory series using backward differences to get

$$f(r) = f^P_{n+1} + r\nabla^P_{n+1} + \frac{r(r+1)}{2}\nabla^{2P}_{n+1} + \frac{r(r+1)(r+2)}{6}\nabla^{3P}_{n+1} \tag{10.37}$$

where $r = \dfrac{t-t_{n+1}}{h}$. This step is implicit since the function value at the target point $n+1$ is used in the calculation. However this step has become explicit here since it is based on approximate predictor value. We have to integrate the above between t_n and t_{n+1} or $r = -1$ to $r = 0$ to get the corrector as

$$
\begin{aligned}
y_{n+1}^C &= y_n + h \int_{-1}^0 \left[f_{n+1}^P + r \nabla_{n+1}^P + \frac{r(r+1)}{2} \nabla_{n+1}^{2P} + \frac{r(r+1)(r+2)}{6} \nabla_{n+1}^{3P} \right] dr \\
&= y_n + h \left[f_{n+1}^P r + \nabla_{n+1}^P \frac{r^2}{2} + \nabla_{n+1}^{2P} \left(\frac{r^3}{6} + \frac{r^2}{4} \right) + \nabla_{n+1}^{3P} \left(\frac{r^4}{24} + \frac{r^3}{6} + \frac{r^2}{6} \right) \right]\Big|_{-1}^0 \\
&= y_n + h \left[f_{n+1}^P - \frac{\nabla_{n+1}^P}{2} - \frac{1}{12} \nabla_{n+1}^{2P} - \frac{1}{24} \nabla_{n+1}^{3P} \right]
\end{aligned}
\tag{10.38}
$$

The last expression is simplified after writing the backward differences in terms of function values. The reader may verify that we should get

$$
y_{n+1}^C = y_n + \frac{h}{24} \left[9 f_{n+1}^P + 19 f_n - 5 f_{n-1} + f_{n-2} \right]
\tag{10.39}
$$

In principle the Adams-Moulton step may be used iteratively, as the reader will appreciate. We notice that the 4th order predictor corrector method requires four nodal values to be known before we can use the Adams-Bashforth step. In an initial value problem we have only one known nodal value. We may use a method such as the RK4 to evaluate the required number of nodal values and use Adams-Bashforth-Moulton scheme thereafter. A MATLAB program has been provided below to solve ODE using ABM fourth order method.

Program 10.4: *Fourth order Adams Bashforth Moulton method*

```
 1 function [X,Y] = ODEABM4(x1,y1,h,n,F,varargin)
 2 % Input  :   x1:  starting  point
 3 %             y1:  initial  values  of  y's  at  x=x1
 4 %             h  :  step  size
 5 %             n  :  number  of  intervals
 6 %             F  :  external  function  containing  derivatives
 7 %                 y' = F(x,y)
 8 %             varargin : optional  arguments  to  be  passed  to  F
 9 % Output:   X :  abcissca
10 %           Y :  ordinates
11 x2 = x1+n*h;            %
12 m = length(y1);        % no  of  ODEs
13 Y = zeros(n+1,m);      % initialize  Y
14 X = [x1:h:x2]';        % initialize  grid
15 Y(1,:) = y1;           % initial  condition
16 Z = zeros(n+1,m);      % initialize  function  values
17 [X(1:4),Y(1:4,:)] = ODERK4(x1,y1,h,3,F,varargin{:});
18                        % RK4  for  three  steps
19 for i=1:4              % function  evaluation  at  first
20    Z(i,:) = F(X(i),Y(i,:),varargin{:}); % four  points
21 end
22 for i=4:n              % loop  for  ABM
23    yp = Y(i,:) + h*(55*Z(i,:) -59*Z(i-1,:) ...
24                 +37*Z(i-2,:)-9*Z(i-3,:))/24; % AB step
25    zp = F(X(i+1),yp);  % function  at  predictor
26    Y(i+1,:) = Y(i,:) + h*(9*zp+19*Z(i,:)  ...
27                 -5*Z(i-1,:)+Z(i-2,:))/24;    % AM step
28    Z(i+1,:) =  F(X(i+1),Y(i+1,:),varargin{:});
```

```
29                              % function  at  X( i +1)
30 end
```

Stability of Adams Bashforth and Adams Moulton methods

Now we look at the stability aspect of AB2. We consider the standard problem $y^{(1)} = f = \lambda y$. AB2 step will then read as

$$y_{n+1} = y_n + h\left[\frac{3}{2}f_n - \frac{1}{2}f_{n-1}\right] = y_n + h\lambda\left[\frac{3}{2}y_n - \frac{1}{2}y_{n-1}\right] \tag{10.40}$$

Noting that the solution is of form $y_n = y_0 z^n$ where z is the gain parameter, possibly complex, the above equation is recast as

$$y_0 z^{n+1} = y_0 z^n + h\lambda\left[\frac{3}{2}y_0 z^n - \frac{1}{2}y_0 z^{n-1}\right] \tag{10.41}$$

or solving for λh in terms of z we get

$$h\lambda = \frac{2z(z-1)}{(3z-1)} \tag{10.42}$$

Limit of stability is governed by $z = e^{j\theta} = \cos\theta + j\sin\theta$. This represents $|z| = 1$ in polar form and hence we have the stability condition given by

$$h\lambda = \frac{2e^{j\theta}(e^{j\theta}-1)}{(3e^{j\theta}-1)} \tag{10.43}$$

AB2 is conditionally stable within the region of stability given by the closed contour shown in Figure 10.13 (the closed contour is a plot of Expression 10.43).

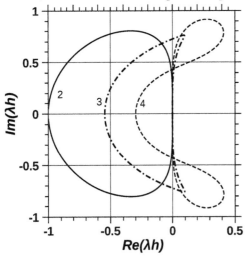

Figure 10.13: *Stability regions Adams Bashforth methods: 2 - AB2, 3 - AB3 and 4 - AB4*

Consider now the AM2. We may apply AM2 to the model problem and recast it as

$$z = \frac{y_{n+1}}{y_n} = \frac{1+\frac{\lambda h}{2}}{1-\frac{\lambda h}{2}} \tag{10.44}$$

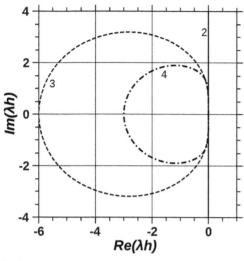

Figure 10.14: *Stability regions Adams Moulton 2 - AM2, 3 - AM3 and 4 - AM4*

The gain factor in this case satisfies $|z| < 1$ as long as λh is in the left half of the z plane.[7] The method is said to be A-stable (Absolute stable). Stability regions of three Adams Moulton methods are shown in Figure 10.14. Thus when we use AB2 as the predictor and AM2 as the corrector stability is affected by the conditionally stable AB2 step. In order to see this we consider AB2 followed by AM2 such that, for the standard problem, we have

$$y_{n+1} = y_n + \frac{\lambda h}{2}\left[y_n + \lambda h\left(\frac{3}{2}y_n - \frac{1}{2}y_{n-1}\right)\right]$$

Inserting $y_n = y_0 z^n$, we get

$$z^2 - z\left[1 + a + \frac{3}{4}a^2\right] + \frac{a^2}{4} = 0 \tag{10.45}$$

where $a = \lambda h$. We may recast this equation as

$$\frac{(1-3z)}{4}a^2 - az + (z^2 - z) = 0 \tag{10.46}$$

We may now substitute $z = e^{j\theta}$ and plot the locus of $a = \lambda h = \lambda_r h + j\lambda_i h$ to get the region of stability for the AB2 AM2 combination. This is shown in Figure 10.15 by the region enclosed by the closed contour labeled as 2. Without dwelling on the details of stability analysis we show the stability region for AB4 and AM4 in Figures 10.13 and 10.14 respectively. We see that the stability regions decrease as we increase the order of the Adams Moulton method. We also show in Figure 10.15 the stability region for AB4 - AM4 combination. These are obtained by following the stability analysis procedure given in the case of AB2, AM2 and AB2 -AM2 combination. Combination schemes have stability regions in between those of AB and AM methods.

Example 10.7

Consider again the ODE of Example 10.6. Use AB4-AM4 combination after the required number of RK4 steps using $h = 0.1$. Comment on the accuracy of the solution by comparing it with the exact solution.

Solution :

[7]i.e. $\lambda = \lambda_r + j\lambda_i$ with λ_r, the real part being negative. Note that h is always real and positive.

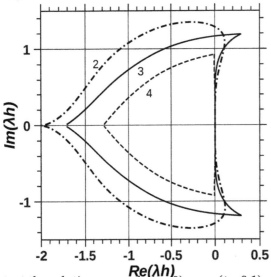

Figure 10.15: Stability regions Adams Bashforth Moulton combinations 2 - AB2-AM2, 3 - AB3-AM3 and 4 - AB4-AM4

To start the solution we need $y(t=0) = y_0, y(t=0.1) = y_1, y(t=0.2) = y_2, y(t=0.3) = y_3$. These are generated using RK4 with step size of 0.1. RK4 generated values are given by the first four entries in the second column of the table given below. The function values are calculated as shown in the third column of the table. Using these values we obtain the predictor and corrector values at $t = 0.4$ as under.

$$y_4^P = 1.320705 + \frac{1}{24}[55 \times 1.050705 - 59 \times 1.092986 + 37 \times 1.074145 - 9 \times 1] = 1.420897$$

Using this the corrector value is obtained as

$$y_4^C = 1.320705 + \frac{1}{24}[9 \times 0.940897 + 19 \times 1.050705 - 5 \times 1.092986 + 1.074145] = 1.420875$$

The solution is continued till $t = 0.6$ and the results are shown in the table below.

t	y	$f(t,y)$	y_{n+1}^P	$f(t,y^P)$	y_E	ϵ_y
0	1.000000	1.000000	Initial Value		1.000000	0
0.1	1.104145	1.074145	RK4 generated value		1.104145	-2.01E-07
0.2	1.212986	1.092986	RK4 generated value		1.212986	-3.79E-07
0.3	1.320705	1.050705	RK4 generated value		1.320706	-5.27E-07
0.4	1.420875	0.940875	1.420897	0.940897	1.420877	-1.52E-06
0.5	1.506391	0.756391	1.506415	0.756415	1.506394	-2.73E-06
0.6	1.569402	0.489402	1.569429	0.489429	1.569406	-4.18E-06

We see from the last column that the error is small and no more than 10^{-5}, that too with $h = 0.1$ as compared to the AB2-AM2 which had an error of $\approx 10^{-3}$ with $h = 0.05$.

Error estimates of ABM

Adams Bashforth step We may now make an estimate for the error by calculating the next term that involves the next higher derivative of the function at t_n. This term is equal to $\frac{r(r+1)(r+2)(r+3)}{12}\nabla_n^4$ which integrates to $\frac{251}{720}h^5\nabla_n^4 = \frac{251}{720}h^5 f_n^{(4)}$. Thus we have

$$\text{Error of predictor: } E_{n+1}^P = y_{n+1} - y_{n+1}^P = \frac{251}{720}h^5 f_n^{(4)} \qquad (10.47)$$

where y_{n+1} is the exact value.

Adams Moulton step We may now make an estimate for the error by calculating the next term that involves the next higher derivative of the function at $t = t_{n+1}$. This term is equal to $\dfrac{r(r+1)(r+2)(r+3)}{12}\nabla_{n+1}^4$ which integrates to $-\dfrac{19}{720}h^5\nabla_{n+1}^4 \approx -\dfrac{19}{720}h^5 f_n^{(4)}$. The last approximation assumes that the fourth derivative varies very little in the interval t_n, t_{n+1}. Thus we have

$$\text{Error of corrector: } E_{n+1}^C = y_{n+1} - y_{n+1}^C = -\frac{19}{720}h^5 f_n^{(4)} \tag{10.48}$$

10.5.3 Improving accuracy of ABM methods

There are several variants of Adams Bashforth Moulton methods that can improve the accuracy of the solution. A simple variation is to repeat the corrector step till convergence. This resembles a fixed point iteration of implicit Adams Moulton method. This entails increased computational effort and hence is not a very clever thing to do.

Another method of improving accuracy of ABM method is to eliminate the error from the solution estimates (similar to Richardson extrapolation in integration). Equations 10.47 and 10.48 are the errors in the estimation for predictor and corrector step respectively. With a simple manipulation, we can eliminate truncation errors from the estimate. Subtract Equation 10.47 from Equation 10.48 to get

$$-y_{n+1}^C + y_{n+1}^P = -\frac{270}{720}h^5 f_n^4 \text{ or } h^5 f_n^4 = \frac{720}{270}(y_{n+1}^C - y_{n+1}^P) \tag{10.49}$$

Using this estimate for $h^5 f_n^4$ we may modify the predictor as

$$y_{n+1}^M = y_{n+1}^P + \frac{251}{270}(y_{n+1}^C - y_{n+1}^P) \tag{10.50}$$

As we do not have the values of y_{n+1}^C, in this equation we may use the already available predictor (y_n^P) and corrector (y_n^C) estimates to get

$$y_{n+1}^M = y_{n+1}^P + \frac{251}{270}(y_n^C - y_n^P) \tag{10.51}$$

The corrector itself may be modified as

$$y_{n+1} = y_{n+1}^C - \frac{19}{270}(y_{n+1}^C - y_{n+1}^P) \tag{10.52}$$

Thus we may summarize the predictor corrector algorithm as follows:

$$\begin{aligned}
\text{Predictor:} \quad & y_{n+1}^P &=& \quad y_n + \frac{h}{24}[55f_n - 59f_{n-1} + 37f_{n-2} - 9f_{n-3}] \\[4pt]
\text{Modifier:} \quad & y_{n+1}^M &=& \quad y_{n+1}^P + \frac{251}{270}(y_n^C - y_n^P) \quad \text{Skip modifier for first step.} \\[4pt]
\text{Corrector:} \quad & y_{n+1}^C &=& \quad y_n + \frac{h}{24}\left[9f_{n+1}^M + 19f_n - 5f_{n-1} + f_{n-2}\right] \\[4pt]
\text{Target value:} \quad & y_{n+1} &=& \quad y_{n+1}^C - \frac{19}{270}(y_{n+1}^C - y_{n+1}^P)
\end{aligned} \tag{10.53}$$

where $f_{n+1}^M = f(t_{n+1}, y_{n+1}^M)$.

Example 10.8

Solve the first order ordinary differential equation $y^{(1)} = \sin(t) - y\cos(t)$ subject to initial condition $y(0) = 1$. Use RK4 and obtain the solution till $t = 0.3$ and extend it beyond by Adams-Bashforth-Moulton predictor corrector method

Solution :

The required RK4 solution with $h = 0.1$ is available as a part of solution worked out in Example 10.4. The y values and f values are as given in the following table.

t	y	$f(t,y)$
0	1.000000	-1.000000
0.1	0.909822	-0.805443
0.2	0.838503	-0.623119
0.3	0.784751	-0.454181

Adam-Bashforth-Moulton without correction:

Using the four tabulated values of $f(t,y)$ we use the Adams-Bashforth predictor to get

$$y_4^P = y_3 + \frac{h(55f_3 - 59f_2 + 37f_1 - 9f_0)}{24}$$

$$= 0.784751 + \left(\frac{0.1}{24}\right)[55 \times (-0.454181) - 59 \times (-0.623119) + 37 \times (-0.805443)$$

$$-9 \times (-1)] = 0.747179$$

Using the predictor value the function is updated as $f_4^P = f(0.4, y_4^P) = \sin(0.4) - 0.747179 \times \cos(0.4) = -0.298779$. Use this in the Adams-Moulton corrector to get

$$y_4^C = y_3 + \frac{h(9f_4^P + 19f_3 - 5f_2 + f_1)}{24}$$

$$= 0.784751 + \left(\frac{0.1}{24}\right)[9 \times (-0.298779) + 19 \times (-0.4541819) - 5 \times (-0.623119)$$

$$+1 \times (-0.805443)] = 0.747217$$

The process may be continued to get solution for higher values of t.

Adams-Bashforth-Moulton with correction:

Using the four tabulated values of $f(t,y)$ we use the Adams-Bashforth predictor to get y_4^P as in the case of Adams-Bashforth-Moulton without correction. The function is updated as done there. Use this in the Adams-Moulton corrector to get

$$y_4^C = y_3 + \frac{h(9f_4^P + 19f_3 - 5f_2 + f_1)}{24}$$

$$= 0.784751 + \left(\frac{0.1}{24}\right)[9 \times (-0.298779) + 19 \times (-0.4541819) - 5 \times (-0.62311)$$

$$+1 \times (-0.805443)] = 0.747217$$

The target value may then be calculated as

$$y_4 = c_4 - \frac{19}{270}(y_4^C - y_4^P) = 0.747217 - \frac{19}{270}(0.747217 - 0.747179) = 0.747214$$

This process is repeated to extend the solution to next higher value of $t = 0.5$ now. Using the four tabulated values of $f(t, y)$ we use the Adams-Bashforth predictor to get

$$
\begin{aligned}
y_5^P &= y_4 + \frac{h(55f_4 - 59f_3 + 37f_2 - 9f_1)}{24} \\
&= 0.747214 + \left(\frac{0.1}{24}\right)[55 \times (-0.298811) - 59 \times (-0.454181) + 37 \times (-0.623119) \\
&\quad -9 \times (-0.805443)] = 0.724529
\end{aligned}
$$

The modifier is now calculated as

$$
y_5^M = p_5 + \frac{251}{270}(y_4^C - y_4^P) = 0.724531 + \frac{251}{270}(0.747217 - 0.747179) = 0.724564
$$

The function value based on y_5^M is calculated as $f_5^M = \sin(0.5) - 0.724566 \times \cos(0.5) = -0.156439$. The corrector is then calculated as

$$
\begin{aligned}
y_5^C &= y_4 + \frac{h(9f_5^M + 19f_4 - 5f_3 + f_2)}{24} \\
&= 0.747214 + \left(\frac{0.1}{24}\right)[9 \times (-0.156439) + 19 \times (-0.298814) - 5 \times (-0.454181) \\
&\quad +1 \times (-0.623119)] = 0.724557
\end{aligned}
$$

Finally the target value is obtained as

$$
y_5 = y_5^C - \frac{19}{270}(y_5^C - y_5^P) = 0.724557 - \frac{19}{270}(0.724557 - 0.724531) = 0.724555
$$

We could have continued the solution using RK4. In the following table we compare the results obtained by Adams-Bashforth-Moulton (ABM) no error correction, Adams-Bashforth-Moulton (ABM) with error correction and the RK4. Difference with respect to the RK4 solution (it is valid with at least six decimal accuracy as shown in Example 10.4) indicates the accuracy of the Predictor Corrector method.

t	ABM No error correction	ABM With error correction	RK4	Difference RK4-ABM (No error correction)	Difference RK4-ABM (With error correction)
0.4	0.747217	0.747214	0.747213	-3.26E-06	-6.01E-07
0.5	0.724561	0.724555	0.724556	-5.36E-06	1.67E-07
0.6	0.715538	0.715531	0.715531	-6.60E-06	6.94E-07

The ABM with error corrections is preferred over the ABM without error corrections. The solution differs from RK4 by at the most 1 digit in the sixth place after decimals. This example shows that, in general, RK4 is better than ABM in spite of error correction.

10.5.4 Adaptive ABM method: change of step size

We have seen that estimates of Adams Bashforth Moulton method can be improved by eliminating error in the predictor and corrector steps. However, there was no automatic control of step size. With the knowledge of the error for corrector step, the solution to ODE can be made adaptive i.e. the step size is doubled or halved depending on the error estimated. This exercise also ensures the error is below a threshold tolerance throughout the solution range. Being a multi-step method, the step size of ABM has to be changed with proper care.

In Example 10.8 we solved a particular ODE with a step size of $h = 0.1$. However one may want to either reduce or increase the step size depending on the error estimate, similar to RK methods. In case the step size is to be reduced, say, to $\dfrac{h}{2}$ we need starting values with this step size. One may easily obtain the desired values using the previously calculated values with step size h and an interpolating polynomial such as $p_3(t)$. In case we would like to increase the step size, say to $2h$ all we have to do is to use the already available data skipping alternate values (Figure 10.16). An example is worked out below to demonstrate this.

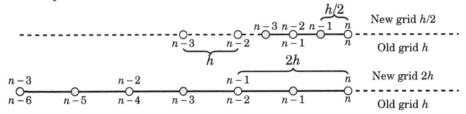

Figure 10.16: Modified grid for $h/2$ and $2h$

Example 10.9

Extend the RK4 solution beyond $t = 0.3$ in Example 10.8 by ABM method using a step size of 0.05 i.e. halving the step size used in RK4.

Solution :

The available RK4 solution with $h = 0.1$ is made use of to obtain the following forward difference table.

t	y	Δy	$\Delta^2 y$	$\Delta^3 y$
0	1	-0.0901779	0.0188582	-0.00128999
0.1	0.909822	-0.0713196	0.0175683	
0.2	0.838503	-0.0537513		
0.3	0.784751			

Defining $\dfrac{x-0}{0.1} = r$ a cubic Newton polynomial may be written down as

$$
\begin{aligned}
p_3(r) &= y_0 + r\Delta y_0 + \frac{r(r-1)}{2}\Delta^2 y_0 + \frac{r(r-1)(r-2)}{6}\Delta^3 y_0 \\
&= 1 - 0.0901779r + 0.0094291r(r-1) - 0.000214998r(r-1)(r-2)
\end{aligned}
$$

The function values required at $t = 0.15$ or $r = 1.5$ and $t = 0.25$ or $r = 2.5$ are obtained and shown below.

t	r	y_{int}	
0.15	1.5	0.871886	Value Needed
0.2	2	0.838503	Already Available
0.25	2.5	0.809512	Value Needed
0.3	3	0.784751	Already Available

Calculation proceeds now as in Example 10.8 but with a step size of $h = 0.05$. Suppressing details we show below the results.

t	y	$f(t,y)$	p	$f(t,y^P)$	y^M	$f(t,y^M)$
0.15	0.871886	-0.712657		$\cdots$Interpolation$\cdots$		
0.2	0.838503	-0.623119		$\cdots$Solution by $RK4\cdots$		
0.25	0.809512	-0.536942		$\cdots$Interpolation$\cdots$		
0.3	0.784751	-0.454181		$\cdots$Solution by $RK4\cdots$		
0.35	0.764040	-0.374821	0.764039	-0.374820	0.764039	-0.374820
0.4	0.747213	-0.298811	0.747213	-0.298810	0.747214	-0.298811
0.45	0.734105	-0.226057	0.734104	-0.226056	0.734105	-0.226057
0.5	0.724555	-0.156432	0.724555	-0.156431	0.724556	-0.156432

We also show the solution based on $RK4$ below with a step size of $h = 0.05$.

t	y_{RK4}	k_1	k_2	k_3	k_4
0	1.000000	-0.050000	-0.047485	-0.047548	-0.045064
0.05	0.952478	-0.045065	-0.042620	-0.042681	-0.040271
0.1	0.909822	-0.040272	-0.037903	-0.037962	-0.035632
0.15	0.871883	-0.035633	-0.033346	-0.033402	-0.031155
0.2	0.838502	-0.031156	-0.028954	-0.029007	-0.026846
0.25	0.809515	-0.026847	-0.024732	-0.024782	-0.022708
0.3	0.784751	-0.022709	-0.020680	-0.020728	-0.018740
0.35	0.764040	-0.018741	-0.016798	-0.016843	-0.014940
0.4	0.747213	-0.014941	-0.013081	-0.013123	-0.011302
0.45	0.734105	-0.011303	-0.009523	-0.009563	-0.007821
0.5	0.724555	-0.007822	-0.006119	-0.006156	-0.004488

It is seen that the two solutions agree to at least six digits after decimals.

10.6 Set of first order ODEs

In many engineering applications we need to solve a set of *coupled* first order ordinary differential equations. Two first order systems that interact with each other will be governed by two coupled first order differential equations that need to be solved simultaneously. Thus we need to solve simultaneously two equations such as

$$y_1^{(1)} = f_1(t, y_1, y_2)$$
$$y_2^{(1)} = f_2(t, y_1, y_2) \tag{10.54}$$

subject to initial conditions $y_1(0) = y_{1,0}$ and $y_2(0) = y_{2,0}$. Coupling between the two equations is indicated by the appearance of y_1 and y_2 in both the equations. In general the functions f_1 and f_2 may be non-linear i.e. involve non-linear functions of y_1 and y_2. However there are some special cases that are of interest in applications. The two equations may be linear in y_1 and y_2 and be of form

$$y_1^{(1)} = a(t)y_1 + b(t)y_2 + s_1(t)$$

$$y_2^{(1)} \quad = \quad c(t)y_1 + d(t)y_2 + s_2(t) \tag{10.55}$$

These equations are referred to as non-homogeneous first order ordinary differential equations with variable coefficients. If, in addition, $a - d$ are constant we refer to the equations as non-homogeneous first ordinary differential equations with constant coefficients. If both s_1 and s_2 are zero we have homogeneous first order ordinary differential equations with variable or constant coefficients, respectively.

Exact solution to a set of homogeneous first order ordinary differential equations with constant coefficients

Consider a set of first order differential equations with constant coefficients written in matrix form below.

$$\frac{d\mathbf{y}}{dt} = \mathbf{A}\mathbf{y} \tag{10.56}$$

Here $\mathbf{y}$ is a vector and $\mathbf{A}$ is a square matrix of the coefficients. Recall that the solution consists of exponentials and hence $\mathbf{y} = \mathbf{y}_0 e^{\lambda t}$. Introduce this in Equation 10.56 to get

$$\mathbf{y}_0 \lambda e^{\lambda t} = \mathbf{A}\mathbf{y}_0 e^{\lambda t} \tag{10.57}$$

This may be recast as

$$\lambda\mathbf{y} = \mathbf{A}\mathbf{y} \tag{10.58}$$

Thus the λ's are the eigenvalues of the coefficient matrix $\mathbf{A}$. We have seen earlier that we can find a similarity transformation such that the coefficient matrix is brought to Schur form. This means that we can find a matrix $\mathbf{P}$ such that $\mathbf{P}^{-1}\mathbf{A}\mathbf{P} = \Lambda$ where Λ is a diagonal matrix, if the eigenvalues are distinct. Let $\mathbf{z} = \mathbf{P}\mathbf{y}$. Then we have

$$\frac{d\mathbf{P}\mathbf{y}}{dt} = \frac{d\mathbf{z}}{dt} = \mathbf{P}\mathbf{A}\mathbf{y} = \mathbf{P}\mathbf{A}\mathbf{P}^{-1}\mathbf{P}\mathbf{y} = \Lambda\mathbf{P}\mathbf{y} = \Lambda\mathbf{z} \tag{10.59}$$

Thus we end up with a set of uncoupled equations for $\mathbf{z}$. Easily the solution is written down as

$$z_i = c_i e^{\lambda_i t} \tag{10.60}$$

In case the given set of ordinary differential equations are non-homogeneous we may use superposition principle to write down the solution. In terms of z the equations will be of form

$$\frac{d\mathbf{z}}{dt} = \Lambda\mathbf{z} + \mathbf{P}\mathbf{s} \tag{10.61}$$

If we put $\mathbf{P}\mathbf{s} = \mathbf{s}'$ the particular integral corresponding to eigenvalue λ_i may be written down as $e^{-\lambda_i t} \int_0^t s_i' e^{\lambda_i t} dt$. This may be added to the complementary function to get the complete solution as

$$z_i = c_i e^{\lambda_i t} + e^{-\lambda_i t} \int_0^t s_i' e^{\lambda_i t} dt \tag{10.62}$$

Once $\mathbf{z}$ has been obtained we get the desired solution as

$$\mathbf{y} = \mathbf{P}^{-1}\mathbf{z} \tag{10.63}$$

Example 10.10

Obtain the exact solution for the set of two ODEs

$$y_1^{(1)} = -3y_1 + y_2$$
$$y_2^{(1)} = y_1 - y_2$$

subject to the initial conditions $y_1(0) = y_2(0) = 1$.

Solution :

Writing the two equations in matrix form, the coefficient matrix is identified as

$$\mathbf{A} = \begin{pmatrix} -3 & 1 \\ 1 & -1 \end{pmatrix}$$

The characteristic equation is given by $(-3 - \lambda)(-1 - \lambda) - 1 = \lambda^2 + 4\lambda + 2 = 0$. The two eigenvalues are given by $\lambda = \dfrac{-4 \pm \sqrt{8}}{2}$ or

$$\lambda_1 = \frac{-4 + \sqrt{8}}{2} = -0.5858$$

$$\lambda_2 = \frac{-4 - \sqrt{8}}{2} = -3.4142$$

The eigenvalues are thus real and distinct. We may obtain easily the corresponding eigenvectors and construct the matrix $\mathbf{P}$ as

$$\mathbf{P} = \begin{pmatrix} 0.3827 & 0.9239 \\ -0.9239 & 0.3827 \end{pmatrix}$$

The solution is given by

$$y_1 = 0.3827 c_1 e^{-0.5858t} - 0.9239 c_2 e^{-3.4142t}$$
$$y_2 = 0.9239 c_1 e^{-0.5858t} + 0.3827 c_2 e^{-3.4142t}$$

Constants c_1 and c_2 are obtained by requiring satisfaction of the initial conditions. Thus we have $c_1 = 1.3066$, $c_2 = -0.5412$ and finally

$$y_1 = 0.5 e^{-0.5858t} + 0.5 e^{-3.4142t}$$
$$y_2 = 1.2071 e^{-0.5858t} - 0.2071 e^{-3.4142t}$$

10.6.1   Euler and RK2 applied to a set of first order ordinary differential equations

The simplest algorithm to apply to a set of first order ODEs is the Euler method. The scheme is first order accurate and is given by the following.

$$t_{n+1} = t_n + h$$
$$y_{1,n+1} = y_{1,n} + h f_1(t_n, y_{1,n}, y_{2,n}) \tag{10.64}$$

$$y_{2,n+1} \quad = \quad y_{2,n} + h f_2(t_n, y_{1,n}, y_{2,n})$$

Here we are considering a set of two coupled first order ODEs. In the case of RK2 the above Euler step is used as a predictor to get

$$
\begin{aligned}
t_{n+1} &= t_n + h \\
y_{1,n+1}^P &= y_{1,n} + h f_1(t_n, y_{1,n}, y_{2,n}) \\
y_{2,n+1}^P &= y_{2,n} + h f_2(t_n, y_{1,n}, y_{2,n})
\end{aligned}
\tag{10.65}
$$

This is followed by the corrector step as follows.

$$
\begin{aligned}
t_{n+1} &= t_n + h \\
y_{1,n+1} &= y_{1,n} + \frac{h}{2}[f_1(t_n, y_{1,n}, y_{2,n}) + f_1(t_{n+1}, y_{1,n+1}^P, y_{2,n+1}^P)] \\
y_{2,n+1} &= y_{2,n} + \frac{h}{2}[f_2(t_n, y_{1,n}, y_{2,n}) + f_2(t_{n+1}, y_{1,n+1}^P, y_{2,n+1}^P)]
\end{aligned}
\tag{10.66}
$$

These may be modified to suit the special linear cases, alluded to earlier, without much effort. In fact we work out a simple case of a set of two homogeneous first order ordinary differential equations with constant coefficients below by the use of RK2 method.

Example 10.11

Solve the following set of two homogeneous first order ordinary differential equations with constant coefficients

$$
\begin{aligned}
y_1^{(1)} &= -3y_1 + y_2 \\
y_2^{(1)} &= y_1 - y_2
\end{aligned}
$$

subject to the initial conditions $y_1(0) = y_2(0) = 1$. Use a step size of $\Delta t = h = 0.05$. Comment on how good is the solution obtained numerically by the RK2 method.

Solution :

We start the calculations from $t = 0$, $y_1(0) = y_2(0) = 1$ and follow through one pass of the RK2 method.

Predictor step:

$$
\begin{aligned}
t_1 &= t_0 + 0.05 = 0 + 0.05 = 0.05 \\
y_{1,1}^P &= y_{1,0} + h[-3y_{1,0} + y_{2,0}] = 1 + 0.05(-3 \times 1 + 1) = 0.9000 \\
y_{2,1}^P &= y_{2,0} + h[y_{1,0} - y_{2,0}] = 1 + 0.05(1 - 1) = 1.0000
\end{aligned}
$$

Corrector step:

$$
\begin{aligned}
t_1 &= t_0 + 0.05 = 0 + 0.05 = 0.05 \\
y_{1,1} &= y_{1,0} + \frac{h}{2}[-3y_{1,0} + y_{2,0} - 3y_{1,0}^P + y_{2,0}^P] \\
&= 1 + \frac{0.05}{2}(-3 \times 1 + 1 - 3 \times 0.9 + 1) = 0.9075 \\
y_{2,1} &= y_{2,0} + \frac{h}{2}[-3y_{1,0} + y_{2,0} - 3y_{1,0}^P + y_{2,0}^P] \\
&= 1 + \frac{0.05}{2}(1 - 1 + 0.9000 - 1) = 0.9975
\end{aligned}
$$

This process may be continued and the results tabulated as under:

t	y_1	y_2	y_{1E}	y_{2E}	ϵ_{y_1}	ϵ_{y_2}
0	1.0000	1.0000	1.0000	1.0000	0.0000	0.0000
0.05	0.9075	0.9975	0.9071	0.9977	0.0004	-0.0002
0.1	0.8276	0.9910	0.8269	0.9912	0.0007	-0.0003
0.15	0.7584	0.9811	0.7575	0.9815	0.0009	-0.0003
0.2	0.6983	0.9687	0.6973	0.9690	0.0010	-0.0004
0.25	0.6459	0.9541	0.6448	0.9545	0.0010	-0.0004
0.3	0.6000	0.9378	0.5990	0.9382	0.0010	-0.0004
0.35	0.5597	0.9203	0.5587	0.9206	0.0010	-0.0004
0.4	0.5241	0.9017	0.5232	0.9021	0.0010	-0.0004
0.45	0.4926	0.8825	0.4917	0.8828	0.0009	-0.0003
0.5	0.4646	0.8627	0.4637	0.8631	0.0009	-0.0003

At $x = 0.5$ the RK2 solution is given by $y_1 = 0.4646$ and $y_2 = 0.8627$ as compared to exact solution $y_{1E} = 0.4637$ and $y_{2E} = 0.8631$. The error is limited, respectively, to 9 units and 4 units in the fourth place after decimals. A full table showing both RK2 and exact values are given above. The error seems to be larger (in the third place after decimals) for short times. It may be necessary to reduce the step size if better results are required.

Stability issues of a system of ODEs

In the examples considered above, the solution is made of two exponentials, the first one with an eigenvalue of $\lambda_1 = -0.5858$ and the second one $\lambda_2 = -3.4142$. Both are real and are negative. From the earlier stability analysis we have seen that the condition for stability is that $|\lambda h| < 2$. Since there are two λ's in the present problem the stability requirement imposes an upper limit on the h given by $h < \dfrac{2}{3.4142} = 0.5858 \approx 0.6$. Thus the bigger eigenvalue imposes a restriction on the step size. We show in Figure 10.17 the results of varying the step size. When $h = 0.2$ the RK2 solution follows closely the exact solution. When we choose $h = 0.6$ the RK2 solution diverges and is entirely worthless.

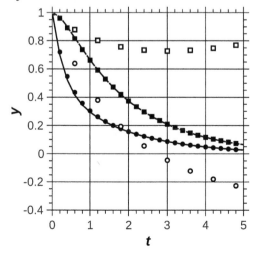

Figure 10.17: Effect of step size h on the solution by RK2

Differential equations where eigenvalues differ considerably in magnitude with a large gap between the smallest and the biggest eigenvalues are termed as *stiff* equations. Such equations show propensity for instabilities for the reason given above. Hence special methods are needed to solve such problems. These will be taken up later on, after dealing with higher order differential equations, that may also exhibit stiff nature.

10.6.2 Application of RK4 to two coupled first order ODEs

Consider first the general case of two coupled first order ODEs. Consider two equations given by

$$\frac{dy_1}{dt} = f_1(t, y_1, y_2)$$

$$\frac{dy_2}{dt} = f_2(t, y_1, y_2) \tag{10.67}$$

with the initial conditions specified for both the dependent variables y_1 and y_2. Let $k's$ and $l's$ represent function values associated respectively with the two first order ODEs. The RK4 will then be written down as follows.

$$
\begin{aligned}
t_{n+1} &= t_n + h \\
k_1 &= h \cdot f_1(t_n, y_{1,n}, y_{2,n}) \qquad\qquad l_1 = h \cdot f_2(t_n, y_{1n}, y_{2n}) \\
k_2 &= h \cdot f_1\left(t_n + \frac{h}{2}, y_{1,n} + \frac{k_1}{2}, y_{2,n} + \frac{l_1}{2}\right) \\
l_2 &= h \cdot f_2\left(t_n + \frac{h}{2}, y_{1,n} + \frac{k_1}{2}, y_{2,n} + \frac{l_1}{2}\right) \\
k_3 &= h \cdot f_1\left(t_n + \frac{h}{2}, y_{1,n} + \frac{k_2}{2}, y_{2,n} + \frac{l_2}{2}\right) \\
l_3 &= h \cdot f_2\left(t_n + \frac{h}{2}, y_{1,n} + \frac{k_2}{2}, y_{2,n} + \frac{l_2}{2}\right) \\
k_4 &= h \cdot f_1(t_n + h, y_{1,n} + k_3, y_{2,n} + l_3) \\
l_4 &= h \cdot f_2(t_n + h, y_{1,n} + k_3, y_{2,n} + l_3) \\
y_{1,n+1} &= y_{1,n} + \frac{k_1 + 2k_2 + 2k_3 + k_4}{6} \qquad y_{2,n+1} = y_{2,n} + \frac{l_1 + 2l_2 + 2l_3 + l_4}{6}
\end{aligned}
\tag{10.68}
$$

Example 10.12

Solve the two first order ODEs of Example 10.11 by the RK4 method. Discuss the effect of change in step size.

Solution :

As usual we will show detailed calculations for one pass of the RK4 starting from the initial value. The initial values are taken as $t_0 = 0$, $y_{1,0} = y_{2,0} = 1$. We take a step of $h = 0.2$ and get the following by the use of Algorithm 10.68.

$$
\begin{aligned}
t_1 &= 0 + 0.2 = 0.2 \\
k_1 &= 0.2(-3 + 1) = -0.4 \\
l_1 &= 0.2(1 - 1) = 0 \\
k_2 &= 0.2\left[-3\left(1 + \frac{-0.4}{2}\right) + 1\right] = -0.28 \\
l_2 &= 0.2\left[\left(1 + \frac{-0.4}{2} - 1\right)\right] = -0.04
\end{aligned}
$$

$$k_3 = 0.2\left[-3\left(1+\frac{-0.28}{2}\right)+\left(1+\frac{-0.04}{2}\right)\right] = -0.32$$

$$l_3 = 0.2\left[\left(1+\frac{-0.28}{2}\right)-\left(1+\frac{-0.04}{2}\right)\right] = -0.024$$

$$k_4 = 0.2[-3(1-0.32)+(1-0.024)] = -0.2128$$

$$l_4 = 0.2[(1-0.32-(1-0.024))] = -0.0592$$

$$y_{1,1} = 1+\frac{-0.4+2(-0.28-0.32)-0.2128}{6} = 0.6979$$

$$y_{2,1} = 1+\frac{0+2(-0.04+-0.024)-0.0592}{6} = 0.9688$$

The corresponding exact values are $y_{1E} = 0.6973$ and $y_{2E} = 0.9690$. The RK4 values have errors respectively of 0.0006 and -0.0002. These seem to be somewhat excessive. However the errors migrate to the 7th digit after decimals for $t = 4$ as shown by the first table below.

We may improve the solution for small t by using a smaller step size in the first few steps of the RK4 algorithm. For example, if we use $h = 0.05$ we get the results in the second table below.

t	y_1	y_2	y_{1E}	y_{2E}	ϵ_{y_1}	ϵ_{y_2}
0	1.000000	1.000000	1.000000	1.000000	0.000000	0.000000
0.2	0.697867	0.968800	0.697312	0.969029	0.000555	-0.000229
0.4	0.523720	0.901871	0.523159	0.902103	0.000561	-0.000232
0.6	0.416714	0.822506	0.416288	0.822682	0.000426	-0.000176
0.8	0.345782	0.741872	0.345495	0.741991	0.000287	-0.000118
...	...	...	...	...	...	...
2.4	0.122717	0.295863	0.122713	0.295864	0.000004	-0.000001
2.6	0.109096	0.263177	0.109093	0.263177	0.000002	0.000000
2.8	0.097007	0.234093	0.097006	0.234093	0.000001	0.000000
...	...	...	...	...	...	...
3.6	0.060692	0.146518	0.060692	0.146517	0.000000	0.000000
3.8	0.053982	0.130320	0.053981	0.130320	0.000000	0.000000
4	0.048013	0.115913	0.048013	0.115912	0.000000	0.000000

t	y_1	y_2	y_{1E}	y_{2E}	ϵ_{y_1}	ϵ_{y_2}
0	1.000000	1.000000	1.000000	1.000000	0.000000	0.000000
0.05	0.907101	0.997659	0.907100	0.997660	0.000001	0.000000
0.1	0.826933	0.991224	0.826932	0.991224	0.000001	0.000000
0.15	0.757550	0.981465	0.757549	0.981465	0.000001	-0.000001
0.2	0.697313	0.969029	0.697312	0.969029	0.000001	-0.000001
0.25	0.644837	0.954458	0.644835	0.954458	0.000001	-0.000001
0.3	0.598952	0.938204	0.598950	0.938205	0.000002	-0.000001
0.35	0.558671	0.920647	0.558669	0.920648	0.000001	-0.000001
0.4	0.523161	0.902102	0.523159	0.902103	0.000001	-0.000001

We see that the errors have migrated to the 6th digit after decimals. Since RK4 is a fourth order method and also has a larger stability region (see Figure 10.8) the solution does not get affected as severely by the stiffness of the equations as in the case of RK2.

Application of RK4 to a set of first order equations

The methods that have been discussed thus far can be easily extended to a set of m differential equations considered below

$$
\begin{aligned}
y_1^{(1)} &= f_1(t, y_1, y_2, \cdots, y_m) \\
y_2^{(1)} &= f_2(t, y_1, y_2, \cdots, y_m) \\
&\cdots \\
y_m^{(1)} &= f_m(t, y_1, y_2, \cdots, y_m)
\end{aligned}
\tag{10.69}
$$

Applying RK4 to the above set of equations we get

$$
\begin{aligned}
k_{i,1} &= h f_i(t, y_{1,n}, y_{2,n}, \cdots, y_{m,n}) \\
k_{i,2} &= h f_i\left(t + \frac{h}{2}, y_{1,n} + \frac{k_{1,1}}{2}, y_{2,n} + \frac{k_{2,1}}{2}, \cdots, y_{m,n} + \frac{k_{m,1}}{2}\right) \\
k_{i,3} &= h f_i\left(t + \frac{h}{2}, y_{1,n} + \frac{k_{1,2}}{2}, y_{2,n} + \frac{k_{2,2}}{2}, \cdots, y_{m,n} + \frac{k_{m,2}}{2}\right) \\
k_{i,4} &= h f_i\left(t + h, y_{1,n} + k_{1,3}, y_{2,n} + k_{2,3}, \cdots, y_{m,n} + k_{m,3}\right) \\
y_{i,n+1} &= y_{i,n} + \frac{k_{i,1} + 2k_{i,2} + 2k_{i,3} + k_{i,4}}{6}
\end{aligned}
\tag{10.70}
$$

Similarly, other methods including adaptive strategies discussed so far can be extended to a set of differential equations.

Example 10.13

Solve the following system of differential equations using adaptive RK4 method

$$
\frac{dx}{dt} = \sigma(y - x); \quad \frac{dy}{dt} = x(\rho - z) - y; \quad \frac{dz}{dt} = xy - \beta z
$$

where $\sigma = 10$, $\beta = 8/3$ and $\rho = 28$. Initial conditions are $x = 1$, $y = 0$ and $z = 0$.

Background :

The above set of differential equations is a simplified model for fluid circulation in a fluid layer which is heated from below and cooled from above. These equations were derived by Edward Lorenz and are also known as Lorenz attractor. The equations are known to exhibit chaotic solutions for certain choice of parameters σ, β and ρ. The model is highly sensitive to initial conditions, where a small change in initial conditions brings a huge change in the final solution. This makes the solution to the problem highly unpredictable. This means even small errors from the numerical model can render the solution inaccurate. This is the main reason why numerical weather forecasting remains a challenge![8]

Solution :

We have a set of three coupled first order differential equations. A MATLAB function has been written to define the above system of governing equations.

```
function z = odelorenzprob(t,y,p)
% Input t :independent variable
%       y : dependent variables
```

[8]R. C. Hilborn *"Sea gulls, butterflies, and grasshoppers: A brief history of the butterfly effect in nonlinear dynamics"* 2004, American Journal of Physics, 72, 425.

```
%              y(1) = x,  y(2) = y  and  y(3) = z
%              p : parameters
%              p(1) = σ,  p(2) = β  and  p(3) = ρ
% Output z :  function  evaluation
z(1) =  p(1)*(y(2)-y(1));
z(2) =  y(1)*(p(3)-y(3))-y(2);
z(3) =  y(1)*y(2)-p(2)*y(3);
end
```

In the above function variable p has to be input by the user in the main function as given below

```
[t,y] = ODEadaptRK4(0,100,[1;0;0],0.1,@odelorenzprob, ...
                    1e-6,[10;8/3;28]);
```

Figure 10.18 shows the trajectory plot[9] of the three variables. Readers are encouraged

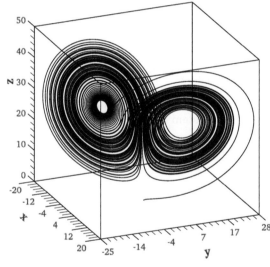

Figure 10.18: Trajectory of Lorenz attractor. Note that this figure is used on the cover of the present book.

to change the initial conditions and check the sensitivity of the trajectory to the initial conditions.

10.7 Higher order ODEs

Higher order ODEs are essentially converted to a set of simultaneous first order ODEs and solved by the many schemes that are available to solve first order ODEs. Consider a second order ODE as a typical example. A general second order ODE is

$$\frac{d^2y}{dt^2} = f\left(t, y, \frac{dy}{dt}\right) \text{ or } y^{(2)}(t) = f(t, y, y^{(1)}) \tag{10.71}$$

---

[9]A trajectory is the path of the dependent variables with respect to the independent variable (usually time). Time appears implicitly in the plot.

Restricting our attention to initial value problems here (boundary value problems will be considered separately later), initial conditions are specified as

$$t = 0, \ y = y_0; \qquad \left.\frac{dy}{dt}\right|_{t=0} = y^{(1)}(0) = y_0^{(1)} \tag{10.72}$$

Noting that the first derivative itself may be a function of t and y, we introduce the following transformation

$$u_0(t) = y; \qquad u_1(t,y) = y^{(1)}(t,y) \tag{10.73}$$

The above transformation helps in converting a second order ODE to a set of two first order ODEs given by

$$u_0^{(1)} = u_1; \qquad u_1^{(1)} = f(t, u_0, u_1) \tag{10.74}$$

The initial conditions may then be recast as

$$t = 0, \ u_0(0) = y_0, \ u_1(0) = y_0^{(1)} \ \text{(say)} \tag{10.75}$$

Thus the original second order ODE given by Equation 10.71 is equivalent to two first order ODEs given by Equations 10.74. Both of these are amenable to solution by any of the numerical methods that have been discussed earlier.

Consider m^{th} order differential equation (initial value problem)

$$y^{(m)} = f(t, y, y^{(1)}, y^{(2)}, \cdots, y^{(m-1)}) \tag{10.76}$$

with m initial conditions. $y, y^{(1)}, \ldots y^{(m)}$ can be transformed as $u_0, u_1, \ldots, u_m$ respectively. Then, the above higher order differential equation reduces to a system of first order ODEs as follows

$$\begin{aligned}
u_m^{(1)} &= f(t, u_0, u_1, u_2, \cdots, u_m) \\
u_{m-1}^{(1)} &= u_m \\
&\cdots \\
u_0^{(1)} &= u_1
\end{aligned} \tag{10.77}$$

The above can also be written in matrix form

$$\begin{Bmatrix} u_m^{(1)} \\ u_{m-1}^{(1)} \\ \cdots \\ u_0^{(1)} \end{Bmatrix} = \begin{bmatrix} 0 & 0 & \cdots & 0 & 0 & f(t, u_0, u_1, u_2, \cdots, u_m) \\ 1 & 0 & \cdots & 0 & 0 & 0 \\ \cdots & \cdots & \cdots & \cdots & \cdots & \cdots \\ 0 & 0 & \cdots & 1 & 0 & 0 \end{bmatrix} \begin{Bmatrix} u_m \\ \cdots \\ u_1 \\ u_0 \\ 1 \end{Bmatrix} \tag{10.78}$$

10.7.1 Euler method applied to second order ODE

Having recast the second order ODE as two first order ODEs the Euler method may be used to obtain the solution. The algorithm is easily written down as follows.

$$t_{n+1} = t_n + h; \ u_{1,n+1} = u_{1,n} + h f(t_n, u_{0,n}, u_{1,n}); \ u_{0,n+1} = u_{0,n} + h u_{1,n} \tag{10.79}$$

An example follows.

Example 10.14

Solve equation $y^{(2)} = \dfrac{t + y + y^{(1)} + 2}{2}$ *with* $y(0) = 0$, $y^{(1)}(0) = 0$. *Use a step size of* $h = 0.1$ *and obtain the solution for* $0 \le x \le 0.5$ *using the Euler method.*

Solution :

First we transform the variables as follows

$$u_0 = y; \qquad u_1 = y^{(1)};$$

Then the differential system reduces to

$$u_1^{(1)} = \frac{t + u_0 + u_1 + 2}{2}$$
$$u_0^{(1)} = u_1$$

with initial conditions $u_0(0) = 0$ and $u_1(0) = 0$ We see that the algorithm is very simple and given by

$$t_{i+1} = t_i + h$$
$$u_{1,i+1} = u_{1,i} + hu_{1,i}^{(1)} = u_{1,i} + h\frac{t_i + u_{0,i} + u_{1,i} + 2}{2}$$
$$u_{0,i+1} = u_{0,i} + 0.1u_{1,i}$$

We start with $u_{0,0} = u_{1,0} = 0$ to get $u_{1,0}^{(1)} = \dfrac{0 + 0 + 0 + 2}{2} = 1$ and hence

$$t_1 = t_0 + 0.1 = 0 + 0.2 = 0.1$$
$$u_{1,1} = u_{1,0} + 0.1u_{1,0}^{(1)} = 0 + 0.2 \times 1 = 0.1$$
$$u_{0,1} = u_{0,0} + 0.2u_{1,0} = 0 + 0.1 \times 0 = 0$$

The algorithm is used again and again to get the following tabulated values.

i	t	y	$y^{(1)}$	$y^{(2)}$	y_E	ϵ_y
0	0	0.0000	0.0000	1.0000	0.0000	0.0000
1	0.1	0.0000	0.1000	1.1000	0.0052	-0.0052
2	0.2	0.0100	0.2100	1.2100	0.0214	-0.0114
3	0.3	0.0310	0.3310	1.3310	0.0499	-0.0189
4	0.4	0.0641	0.4641	1.4641	0.0918	-0.0277
5	0.5	0.1105	0.6105	1.6105	0.1487	-0.0382

In the above table y_E is the exact solution which may easily be shown to be given by $y_E = e^t - t - 1$. The error with respect to the exact is excessive. Also it increases with t and the solution quickly becomes useless! One may need to use a very small h to get an acceptable solution.

10.7.2   RK2 method applied to second order ODE

With a little more effort we get a second order accurate solution by the use of the RK2 method (or the Heun method). We easily see that the algorithm is as follows.

$$t_{i+1} \quad = \quad t_i + h$$

Predictor step:

$$u_{0,i+1}^P \quad = \quad u_{0,i} + h u_{1,i}$$

$$u_{1,i+1}^P \quad = \quad u_{1,i} + h f(t_i, u_{0,i}, u_{1,i})$$

Corrector step:

$$u_{0,i+1} \quad = \quad u_{0,i} + \frac{h(u_{1,i} + u_{1,i+1}^P)}{2}$$

$$u_{1,i+1} \quad = \quad u_{1,i} + h \frac{\left[f(t_i, u_{0,i}, u_{1,i}) + f(t_{i+1}, u_{0,i+1}^P, u_{1,i+1}^P) \right]}{2} \qquad (10.80)$$

Example 10.15

Solve the second order ODE of Example 10.14 by the Heun method with a step size of $h = 0.1$. Obtain the solution for $0 \le x \le 0.5$.

Solution :

We make use of the algorithm given by Equation 10.80 and give below the calculations for a single pass starting at $t = 0$.

$$t_1 \quad = \quad t_0 + h = 0 + 0.1 = 0.1$$

Predictor step:

$$u_{0,1}^P \quad = \quad u_{0,0} + h u_{1,0} = 0 + 0.1 \times 0 = 0$$

$$u_{1,1}^P \quad = \quad u_{1,0} + h f(t_0, u_{0,0}, u_{1,0}) = 0 + 0.1 \frac{(0+0+0+2)}{2} = 0.1$$

Corrector step:

$$u_{0,1} \quad = \quad u_{0,0} + \frac{h(u_{1,0} + u_{1,1}^P)}{2} = 0 + 0.1 \left(\frac{0 + 0.1}{2} \right) = 0.0050$$

$$u_{1,1} \quad = \quad u_{1,0} + h \frac{\left[f(t_0, u_{0,0}, u_{1,0}) + f(t_1, u_{0,1}^P, u_{1,1}^P) \right]}{2}$$

$$= \quad 0 + 0.1 \left(\frac{0 + 0 + 0 + 2 + 0.1 + 0 + 0.005 + 2}{4} \right) = 0.1050$$

The algorithm is used repeatedly to get the following tabulated values.

i	x	y	u_1	y^P	u_1^P	y_E	ϵ_y
0	0	0.0000	0.0000	0.0000	0.1000	0.0000	0.0000
1	0.1	0.0050	0.1050	0.0155	0.2155	0.0052	-0.0002
2	0.2	0.0210	0.2210	0.0431	0.3431	0.0214	-0.0004
3	0.3	0.0492	0.3492	0.0842	0.4842	0.0499	-0.0006
4	0.4	0.0909	0.4909	0.1400	0.6400	0.0918	-0.0009
5	0.5	0.1474	0.6474	0.2122	0.8122	0.1487	-0.0013

In the above table y_E is the exact solution given by $y_E = e^t - t - 1$. The error with respect to the exact is moderate. However it increases with t and the solution is probably adequate with an error in the third decimal place. It is possible to improve the solution by reducing the step size. Indeed improved solution shown below with $h = 0.05$ is better with error in the fourth place after the decimal point.

i	x	y	u_1	y^P	u_1^P	y_E	ϵ_y
0	0	0.0000	0.0000	0.0000	0.0500	0.0000	0.0000
1	0.05	0.0013	0.0513	0.0038	0.1038	0.0013	0.0000
2	0.1	0.0051	0.1051	0.0104	0.1604	0.0052	0.0000
3	0.15	0.0118	0.1618	0.0199	0.2199	0.0118	-0.0001
4	0.2	0.0213	0.2213	0.0324	0.2824	0.0214	-0.0001
5	0.25	0.0339	0.2839	0.0481	0.3481	0.0340	-0.0001
6	0.3	0.0497	0.3497	0.0672	0.4172	0.0499	-0.0002
7	0.35	0.0689	0.4189	0.0898	0.4898	0.0691	-0.0002
8	0.4	0.0916	0.4916	0.1162	0.5662	0.0918	-0.0002
9	0.45	0.1180	0.5680	0.1464	0.6464	0.1183	-0.0003
10	0.5	0.1484	0.6484	0.1808	0.7308	0.1487	-0.0003

Application of RK4 to a second order ODE

Now consider a second order ODE converted to two first order ODEs as given by Equations 10.73 and 10.74. At any stage of the calculation we have $u_{0,n}$ and $u_{1,n}$. We may identify the k's with Equation 10.74 and the l's with Equation 10.73. The RK4 algorithm will then be written down as follows:

$$t_{n+1} = t_n + h$$

$$k_1 = h \cdot f(t_n, u_{0,n}, u_{1,n}) \qquad l_1 = h \cdot u_{1,n}$$

$$k_2 = h \cdot f\left(t_n + \frac{h}{2}, u_{0,n} + \frac{l_1}{2}, u_{1,n} + \frac{k_1}{2}\right) \qquad l_2 = h \cdot \left(u_{1,n} + \frac{k_1}{2}\right)$$

$$k_3 = h \cdot f\left(t_n + \frac{h}{2}, u_{0,n} + \frac{l_2}{2}, u_{1,n} + \frac{k_2}{2}\right) \qquad l_3 = h \cdot \left(u_{1,n} + \frac{k_2}{2}\right) \qquad (10.81)$$

$$k_4 = h \cdot f(t_n + h, u_{0,n} + l_3, u_{1,n} + k_3) \qquad l_4 = h \cdot \left(u_{1,n} + k_3\right)$$

$$u_{0,n+1} = u_{0,n} + \frac{l_1 + 2l_2 + 2l_3 + l_4}{6}$$

$$u_{1,n+1} = u_{1,n} + \frac{k_1 + 2k_2 + 2k_3 + k_4}{6}$$

Example 10.16

Solve the second order ODE of Example 10.14 by the RK4 method. Choose a step size of $h = 0.2$ and obtain the solution up to $t = 1$.

Solution :

Using the starting values provided (at $n = 0$) earlier, one pass of calculations using Equations 10.81 gives the following:

$$t_1 = 0 + 0.2 = 0.2$$

$$k_1 = 0.2 \times \frac{0+0+0+2}{2} = 0.2$$

$$l_1 = 0.2 \times 0 = 0$$

$$k_2 = 0.2 \times \frac{0+0.2/2+0+0/2+0+0.2/2+2}{2} = 0.22$$

$$l_2 = 0.2(0+0.2/2) = 0.02$$

$$k_3 = 0.2 \times \frac{0+0.2/2+0+0.02/2+0+0.22/2+2}{2} = 0.222$$

$$l_3 = 0.2(0+0.22/2) = 0.022$$

$$k_4 = 0.2 \times \frac{0+0.2+0+0.02+0+0.222+2}{2} = 0.2444$$

$$l_4 = 0.2(0+0.222) = 0.0444$$

$$y_1 = 0 + \frac{0+2(0.02+0.022)+0.0444}{6} = 0.0214$$

$$y_{1,1} = 0 + \frac{0.2+2(0.22+0.222)+0.2444}{6} = 0.2214$$

The exact solution at $t = 0.2$ is $y_E(0.2) = e^{0.2} - 0.2 - 1 = 0.2214$. Actually the RK4 value has an error of only -2.76×10^{-6} with respect to the exact value. The above calculation process is repeated till $t = 1$ is reached and the results are tabulated below.

x	0.000000	0.200000	0.400000	0.600000	0.800000	1.000000
y_{RK4}	0.000000	0.021400	0.091818	0.222106	0.425521	0.718251
$u_{1,RK4}$	0.000000	0.221400	0.491818	0.822106	1.225521	1.718251
l_1	0.000000	0.044280	0.098364	0.164421	0.245104	0.343650
l_2	0.020000	0.068708	0.128200	0.200863	0.289615	0.398015
l_3	0.022000	0.071151	0.131184	0.204508	0.294066	0.403452
l_4	0.044400	0.098510	0.164600	0.245323	0.343917	0.464341
k_1	0.200000	0.244280	0.298364	0.364421	0.445104	0.543650
k_2	0.220000	0.268708	0.328200	0.400863	0.489615	0.598015
k_3	0.222000	0.271151	0.331184	0.404508	0.494066	0.603452
k_4	0.244400	0.298510	0.364600	0.445323	0.543917	0.664341
y_E	0.000000	0.021403	0.091825	0.222119	0.425541	0.718282
ϵ_y	0.000000	-0.000003	-0.000007	-0.000012	-0.000020	-0.000031
$u_{1,E}$	0.000000	0.221403	0.491825	0.822119	1.225541	1.718282
ϵ_{u_1}	0.000000	-0.000003	-0.000007	-0.000012	-0.000020	-0.000031

The maximum error between the RK4 and exact solutions is -3.1×10^{-5} and occurs at $t = 1$. In this case also the error increases with t but is not excessive as in the case of the Euler method (note that $h = 0.2$ in the present case while it was 0.1 in both the Euler and Heun solutions). RK4 is superior to the Heun method since the error is smaller and has moved from the 4th (with $h = 0.1$) place to the 5th place after decimals with $h = 0.2$. If the error is seen to be excessive it is easy to reduce the step size and improve the accuracy.

There can be practical examples where we have a set of coupled higher order differential equations. We will consider one such example here.

Example 10.17

A projectile of mass 1 (kg) is fired from ground (0,0) with a velocity of $v_0 = 20$ (m/s) at an angle of 30° to the horizontal. The magnitude of wind resistance faced by the projectile is equal to kv^2 (N) where $k = 0.02$. Determine the time and distance the projectile travels before striking the ground.

Background :

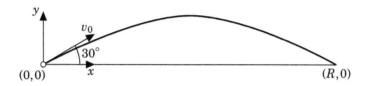

The forces acting on the projectile are acceleration due to gravity acting along $-y$ and wind resistance acting opposite to the direction of motion. Resolving the forces along x and y directions we have

$$F_x = ma_x = -kv^2\cos\theta; \quad F_y = ma_y = -mg - kv^2\sin\theta;$$

where F and a represent respectively force and acceleration. We have $a_x = \dfrac{d^2x}{dt^2}$, $a_y = \dfrac{d^2y}{dt^2}$, $v_x = \dfrac{dx}{dt}$ and $v_y = \dfrac{dy}{dt}$. The governing equations of motion can hence be written as

$$\frac{dv_x}{dt} = -\frac{k}{m}v_x\sqrt{v_x^2 + v_y^2}; \quad \frac{dv_y}{dt} = -g - \frac{k}{m}v_y\sqrt{v_x^2 + v_y^2};$$
$$\frac{dx}{dt} = v_x \qquad\qquad\qquad \frac{dy}{dt} = v_y$$

Solution :

A MATLAB function to calculate acceleration and velocity is created as given below.

```
function U = projectile(t,y,k)
% Inputs :   t : time
%            y(1) : x; y(2) = y; y(3) = vx; y(4) = vy
%            k : parameter related to wind resistance
g = 9.81;
U(1) = y(3);
U(2) = y(4);
U(3) = -k*(y(3)^2+y(4)^2)^0.5*y(3);
U(4) = -g-k*(y(3)^2+y(4)^2)^0.5*y(4);
```

MATLAB program has been written to determine the trajectory of the projectile.

```
y = [0 0 20*cosd(30) 20*sind(30)];    % initial conditions
dt = 0.1;                             % step size = 0.1
t = 0;
k = 0.02;         % parameter related to wind resistance
count = 1;        % initialize number of points
```

```
while 1           % loop to determine  trajectory
    [y1,er]=ODERKF45(t(count),y(count,:),dt,@projectile,k);
    t(count+1) = t(count) + dt;
    y(count+1,:) = y1;
    count = count + 1;
    if( y(count,2) < 0 )
        break      % break loop if projectile reaches ground
    end
end
```

A step size of $dt = 0.2\,(s)$ is used. The results have been summarized in the table below.

$t\,(s)$	$x\,(m)$	$y\,(m)$	$v_x\,(m/s)$	$v_y\,(m/s)$
0	0.00	0.00	17.32	10.00
0.2	3.33	1.73	16.07	7.38
0.4	6.44	2.97	15.03	5.01
0.6	9.36	3.75	14.15	2.81
0.8	12.11	4.10	13.38	0.75
1	14.71	4.06	12.70	-1.20
1.2	17.19	3.63	12.08	-3.06
1.4	19.55	2.84	11.49	-4.82
1.6	21.79	1.70	10.93	-6.50
1.8	23.92	0.25	10.38	-8.08
2	25.94	-1.52	9.83	-9.57

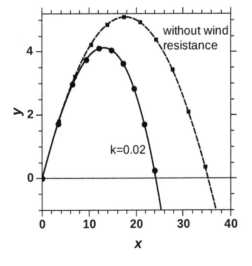

Figure 10.19: *Path of the projectile with and without wind resistance*

The exact location where the projectile hits the ground may be obtained by interpolation as $24.2\,m$. The governing equations have also been solved with zero wind resistance. The projectile travels higher as well as farther without wind resistance. Further the flight time of the projectile is about 22 seconds without wind resistance and 20 seconds for $k = 0.02$. The paths traversed by the projectile with and without wind resistance have been shown in Figure 10.19.

We consider now a second order ODE with constant coefficients that will indicate the performance of RK4 when applied to the case of a stiff ODE.

Example 10.18

Solve the second order ODE $y^{(2)} + 25y^{(1)} + y = 0$ by the RK4 method. The initial conditions are specified as $y(0) = 1$, $y^{(1)}(0) = 0$. Comment on the effect of step size on the numerically computed solution.

Background :

Practical examples of the above type of differential equation are spring mass damper system (vehicle suspension), Resistance-capacitance -inductance (RLC) circuits and second order thermal systems.

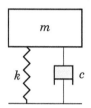

The oscillations of a spring mass damper system is governed by the following equation

$$\frac{d^2x}{dt^2} + \frac{c}{m}\frac{dx}{dt} + \frac{k}{m}x = 0$$

where c is the damping parameter

The current i in the circuit is governed by the following equation

$$\frac{d^2i}{dt^2} + \frac{R}{L}\frac{di}{dt} + \frac{1}{LC}i = 0$$

where L is the inductance

In general the above equations can be rewritten as

$$\frac{d^2x}{dt^2} + 2\zeta\omega_0\frac{dx}{dt} + \omega_0^2 x = 0$$

where ζ is known as damping ratio and ω_0 is known as undamped natural frequency. For a spring mass system $\omega_0 = \sqrt{k/m}$ and for a RLC circuit, $\omega_0 = 1/\sqrt{LC}$. Similarly, the damping ratio for spring mass system and RLC circuit are respectively given by $\dfrac{c}{2\sqrt{mk}}$ and $\dfrac{R}{2}\sqrt{\dfrac{C}{L}}$. The value of ζ controls the response of the system to any external disturbance. For $\zeta = 0$, the equation reduces to a spring mass system with zero damping (Section 3.1.5). $\zeta < 1$ represents an under damped system where the system oscillates about its mean position with gradual reduction in amplitude with time. When $\zeta \geq 1$, the system does not oscillate about its mean position but returns to the equilibrium position monotonically but more slowly. $\zeta > 1$ is referred to as an over damped system and $\zeta = 1$ is referred to as a critically damped system.

Solution :

The present example corresponds to a over damped system. We look at the exact solution first. The characteristic equation is $m^2 + 25m + 1 = 0$ which has the two roots $m_1 = \dfrac{-25 + \sqrt{25^2 - 4}}{2} = -0.040064$ and $m_2 = \dfrac{-25 - \sqrt{25^2 - 4}}{2} = -24.959936$. Thus the given second order ODE is stiff. The solution consists of two exponentials and is given by $y_E = 1.0016077e^{-0.040064t} - 0.0016077e^{-24.959936t}$.

Stability using RK4 requires that $|\lambda_{max}h| < 24.959936h < 2.784$ or $25h \lesssim 2.75$ or $h \lesssim 0.11$ (refer Figure 10.8). If indeed we choose this value of h the solution is quite useless. We have to choose a step size much smaller than this to obtain a useful solution. We indicate below what happens when we vary h and look at the results at the t value shown in column 2.

h	t	y(RK4)	y_E	ϵ_y	Comment
0.02	0.4	0.985684	0.985684	-5.84E-10	Stable
0.04	0.4	0.985684	0.985684	-1.55E-08	Stable
0.05	0.4	0.985684	0.985684	-5.56E-08	Stable
0.08	0.4	0.985678	0.985684	-6.44E-06	Stable
0.1	0.4	0.985407	0.985684	-2.77E-04	Stable
0.104	0.416	0.984544	0.985053	-5.09E-04	Stable
0.11	0.44	0.982841	0.984106	-1.26E-03	Stable
0.12	0.48	0.976942	0.982530	-5.59E-03	Unstable

We see that the error steadily increases with h and the solution is hardly useful beyond $h = 0.05$. When $h = 0.12$ the solution is unstable and diverges for large t. For example the value of y(RK4) at $t = 4.8$ is -412.961485!

10.8 Stiff equations and backward difference formulae (BDF) based methods

As seen in the previous section stability considerations impose stringent conditions when we intend to solve stiff differential equations. All the methods we have discussed till now are explicit methods (either single or multi-step) that have restrictions on step size. In the case of stiff equations we have seen the errors to be excessive even when relatively small values of h are used. If we are able to devise methods that have larger stability regions we should expect the numerical methods to perform better in solving stiff equations. We look at some of these here.

10.8.1 Implicit Euler or Backward Euler scheme

Consider the solution of a general first order ODE given by Equation 10.1. We approximate the derivative at t_{n+1} by the backward difference formula given by $\dfrac{y_{n+1} - y_n}{h}$ so that the following scheme emerges for the solution of the first order ODE.

$$
\begin{aligned}
t_{n+1} &= t_n + h & (a) \\
y_{n+1} &= y_n + hf(t_{n+1}, y_{n+1}) & (b)
\end{aligned}
\qquad (10.82)
$$

The scheme is implicit since y_{n+1} occurs on both sides of Equation 10.82(b). In case the function $f(t, y)$ is nonlinear, this equation needs to be solved using an iterative method such as the Newton Raphson method. We shall look at the stability of this method. For this purpose we apply the scheme to the standard equation $y^{(1)} = -\lambda y$. We then have

$$
y_{n+1} = y_n - h\lambda y_{n+1} \quad \text{or} \quad y_{n+1} = \frac{y_n}{1 + \lambda h} \qquad (10.83)
$$

The magnification factor is given by $z = \dfrac{y_{n+1}}{y_n} = \dfrac{1}{1 + \lambda h}$. The stability requirement is obtained by replacing z by $e^{j\theta}$ and looking for the locus of λh such that $|z| < 1$ i.e. by looking at $\lambda h = e^{-j\theta} - 1$.

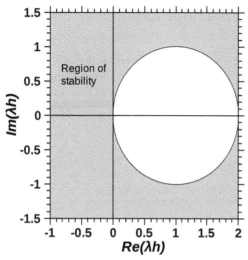

Figure 10.20: *Stability region of implicit Euler method*

This is a circle centered at $\lambda h = 1$ and radius 1 in the z plane. The method is stable *outside* this circle and specifically, it is stable in the entire left half of the z plane as shown in Figure 10.20. Hence the method is A-stable. However the method is first order accurate and it is advisable to look for a higher order method. For this purpose we look at a second order accurate method in what follows.

10.8.2 Second order implicit scheme

Consider a first order ODE whose solution is available as $y = y_n$ at $t = t_n$. Now take a step h to define a point $t_{n+1} = t_n + h$. We consider a quadratic function of form

$$p_2(t) = a + b(t - t_n) + c(t - t_n)^2 \tag{10.84}$$

We require that this quadratic satisfy the following three conditions viz. $p_2(t_n) = y_n$, $p_2^{(1)}(t_n) = f(t_n, y_n)$ and $p_2^{(1)}(t_{n+1}) = f(t_{n+1}, y_{n+1})$. We at once see that $a = y_n$. Taking the first derivative with respect to t of the quadratic and setting $t = t_n$ we get $b = f(t_n, y_n)$, and by setting $t = t_{n+1}$ we get $c = \dfrac{f(t_{n+1}, y_{n+1}) - f(t_n, y_n)}{2h}$. Thus the desired quadratic is obtained as

$$p_2(t) = y_n + (t - t_n)f(t_n, y_n) + \frac{f(t_{n+1}, y_{n+1}) - f(t_n, y_n)}{2h}(t - t_n)^2 \tag{10.85}$$

Hence we have

$$p_2^{(1)}(t) = f(t_n, y_n) + \frac{f(t_{n+1}, y_{n+1}) - f(t_n, y_n)}{h}(t - t_n) \tag{10.86}$$

We integrate this with respect to t between $t = t_n$ and $t = t_{n+1}$ to get

$$
\begin{aligned}
y_{n+1} &= y_n + hf(t_n, y_n) + h\frac{f(t_{n+1}, y_{n+1}) - f(t_n, y_n)}{2} \tag{10.87}\\
&= y_n + h\frac{f(t_n, y_n) + f(t_{n+1}, y_{n+1})}{2} \tag{10.88}
\end{aligned}
$$

This scheme is second order accurate and is also implicit. We refer to this scheme as implicit trapezoidal scheme. It is the same as the AM2 scheme that has been presented earlier. The reader may note that this is also an implicit version of the Heun or the modified Euler or the RK2 method. This scheme is A-stable.

The above procedure, when applied to a second order ODE with constant coefficients, requires the simultaneous solution of two linear equations which may easily be accomplished by a method such as Cramer's rule. When solving a set of linear ODEs we may use any of the methods described in the chapter on linear equations. Consider a system of linear ODE.

$$\mathbf{Y}^{(1)} = \mathbf{A}(t)\mathbf{Y} + \mathbf{b}(t) \tag{10.89}$$

Solving the above ODE using trapezoidal rule we get

$$\mathbf{Y}_{n+1} = \mathbf{Y}_n + h\frac{\mathbf{A}(t_n)\mathbf{Y_n} + \mathbf{A}(t_{n+1})\mathbf{Y_{n+1}}}{2} + h\frac{\mathbf{b}(t_n) + \mathbf{b}(t_{n+1})}{2} \tag{10.90}$$

On further simplification we get

$$\left(\mathbf{I} - \frac{h}{2}\mathbf{A}(t_{n+1})\right)\mathbf{Y}_{n+1} = \left(\mathbf{I} + \frac{h}{2}\mathbf{A}(t_n)\right)\mathbf{Y_n} + h\frac{\mathbf{b}(t_n) + \mathbf{b}(t_{n+1})}{2} \tag{10.91}$$

where $\mathbf{I}$ is identity matrix. A MATLAB program has been provided below to solve a system of linear ODEs using trapezoidal method.

Program 10.5: *Solution of system of linear ODE using implicit trapezoidal method*

```
1  function [X,Y] = ODEtrapezoidal(x1,y1,h,n,F)
2  % Input  :   x1: starting point
3  %             y1: initial values of y's at x=x1
4  %             h : step size
5  %             n : number of intervals
6  %             F : external function containing derivatives
7  %                 [A,b]  = F(x)
8  % Output :
9  %             X : abcissca
10 %             Y : ordinates
11 X = x1:h:x1+n*h;          % initialize grid
12 m = length(y1);           % no of ODEs
13 Y = zeros(n+1,m);         % initialize Y
14 Y(1,:) = y1;              % initial conditions
15 [A1,B1] = F(X(1));
16 for i= 1:n                % loop for trapezoidal method
17     [A2,B2] = F(X(i));    %
18     A = eye(m) - h*A2*0.5;       % calculate coefficients
19     B = h*(B1+B2)*0.5 + Y(i,:)' +0.5*h*A1*Y(i,:)';
20     y1 = ludecomposition(A,B);  % solve for Y
21     Y(i+1,:) =y1';        % update Y
22     A1 = A2; B1 = B2;     %
23 end
```

Example 10.19

Obtain the solution of the second order ODE with constant coefficients of Example 10.18 by second order implicit method. Comment on the effect of step size on the solution.

Solution :

The given ODE is written down, as usual, as two first order ODEs. We identify the functions involved as

$$u_0^{(1)} = u_1$$
$$u_1^{(1)} = -25u_1 - u_0$$

With a time step of h, the algorithm given by Equation 10.87 translates to the following two equations:

$$u_{1,n+1} = u_{1,n} + h\frac{(-25u_{1,n} - u_{0,n}) + (-25u_{1,n+1} - u_{0,n+1})}{2}$$

$$u_{0,n+1} = u_{0,n} + h\frac{u_{1,n} + u_{1,n+1}}{2}$$

These may be rewritten as

$$\frac{h}{2}u_{0,n+1} + \left(1 + \frac{25h}{2}\right)u_{1,n+1} = u_{1,n} + h\frac{(-25u_{1,n} - u_{0,n})}{2}$$

$$u_{0,n+1} - \frac{h}{2}u_{1,n+1} = u_{0,n} + h\frac{u_{1,n}}{2}$$

The above may be recast in the matrix form as

$$\begin{pmatrix} \frac{h}{2} & \left(1 + \frac{25h}{2}\right) \\ 1 & -\frac{h}{2} \end{pmatrix} \begin{pmatrix} u_{0,n+1} \\ u_{1,n+1} \end{pmatrix} = \begin{pmatrix} u_{1,n} + h\frac{(-25u_{1,n} - u_{0,n})}{2} \\ u_{0,n} + h\frac{u_{1,n}}{2} \end{pmatrix}$$

The determinant of the coefficient matrix is given by

$$\Delta = -\left[1 + \frac{25h}{2} + \frac{h^2}{4}\right]$$

Using Cramer's rule the solution is obtained as

$$u_{0,n+1} = \frac{-u_{1,n}\left(\frac{29}{4}h - \frac{h^2}{4}\right) - u_{0,n}\left(1 + \frac{25h}{2} - \frac{h^2}{4}\right)}{-\left(1 + \frac{25h}{2} + \frac{h^2}{4}\right)}$$

$$u_{1,n+1} = \frac{u_{0,n}h - u_{1,n}\left(1 - \frac{25h}{2} - \frac{h^2}{4}\right)}{-\left(1 + \frac{25h}{2} + \frac{h^2}{4}\right)}$$

Before we look at the effect of h on the solution we show below the results obtained by taking a step size of $h = 0.1$. The solution starts with the initial values provided in the problem and is carried forward up to $x = 1$.

t	u_0	u_1	y_E	ϵ_y
0	1.00000	0.00000	1.00000	0.00E+00
0.1	0.99778	-0.04440	0.99747	3.10E-04
0.2	0.99359	-0.03932	0.99360	-8.66E-06
0.3	0.98964	-0.03970	0.98964	3.04E-06
0.4	0.98568	-0.03948	0.98568	-1.85E-07
0.5	0.98174	-0.03933	0.98174	6.08E-09
0.6	0.97782	-0.03918	0.97782	-3.38E-08
0.7	0.97391	-0.03902	0.97391	-3.62E-08
0.8	0.97001	-0.03886	0.97001	-4.16E-08
0.9	0.96614	-0.03871	0.96614	-4.66E-08
1	0.96227	-0.03855	0.96227	-5.16E-08

The numerical solution is compared with the exact solution (see Example 10.18). The biggest error occurs after the very first step and is equal to 3.10×10^{-4}. At $t = 0.4$ the error has reduced to -1.85E-07. However, with the RK4 method and $h = 0.1$ the error at $t = 0.4$ was much larger and equal to -2.77E-04, even though RK4 is a fourth order accurate method. We also saw that $h > 0.11$ with RK4 leads to an unstable solution. However in the case of implicit RK2 which is only second order accurate the solution is much better. With $h = 0.2$ we still have an error of only -2.94×10^{-4} at $x = 0.4$. It is clear that the implicit RK2 is performing much better than RK4 in this case of a stiff ODE.

In the following table we show the error in the numerical solution with respect to the exact at $t = 1$ using different step sizes.

h	t	y	y_E	ϵ_y
0.05	1	0.96227	0.96227	-1.29E-08
0.1	1	0.96227	0.96227	-5.16E-08
0.2	1	0.96230	0.96227	2.29E-05
0.25	1	0.96216	0.96227	-1.13E-04
0.5	1	0.96143	0.96227	-8.43E-04

10.8.3   Higher order implicit schemes based on BDF

Higher order implicit schemes may also be devised by the use of backward difference formulae of various orders. Consider a two step backward difference formula (referred to earlier as three point one sided rule) given by $y_n^{(1)} = \dfrac{3y_n - 4y_{n-1} + y_{n-2}}{2h}$. We may apply this to the case of a first order ODE to get

$$f(t_{n+1}, y_{n+1}) = \frac{3y_{n+1} - 4y_n + y_{n-1}}{2h} \tag{10.92}$$

which is in the implicit form. Clearly the method is second order accurate since the BDF is second order accurate. We may apply this to the case of the standard problem $y^{(1)} = \lambda y$ to bring out the stability region for this method. Denoting the magnification of the solution in each step as z, we obtain from Equation 10.92 the following expression for λh.

$$\lambda h = \frac{3z^2 - 4z + 1}{2z^2} \tag{10.93}$$

The stability region is obtained by setting $z = e^{j\theta}$. The region outside the closed curve indicated as 2 in Figure 10.21 is the region of stability. Note that the method is A-stable in this case also. A third order BDF scheme uses 3 steps and calculates the derivative as $y_n^{(1)} =$

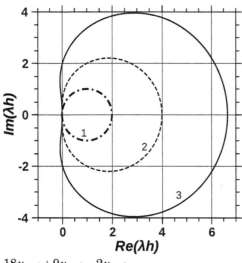

Figure 10.21: *Stability diagram for the BDF methods (region outside the closed curves are stable)*

$\dfrac{11y_n - 18y_{n-1} + 9y_{n-2} - 2y_{n-3}}{6h}$ to write the solution to a first order ODE as

$$f(t_{n+1}, y_{n+1}) = \frac{11y_{n+1} - 18y_n + 9y_{n-1} - 2y_{n-2}}{6h} \tag{10.94}$$

The region of stability for this scheme may be easily shown to be governed by the equation

$$\lambda h = \frac{11z^3 - 18z^2 + 9z - 2}{6z^3} \tag{10.95}$$

The stability region is the region outside the curve shown with label 3 in Figure 10.21. The stability region now excludes a small region in the left half of z plane and hence the scheme is not A-stable.

Example 10.20

Solve the equation $y^{(1)} = y - 3t^2$ with $y_0 = 1$. Use implicit RK2 for the first step and 2nd order BDF thereafter (two step method).

Solution :
Since both implicit RK2 and two step BDF are second order accurate it is possible to start the solution and use implicit RK2 for the first step and continue with the two step BDF thereafter. Advantage of BDF is that f value need not be calculated again and again as in the case of implicit RK2.

Using implicit RK2 for the first step we have

$$y_1 = \left(y_0 \left[1 + \frac{h}{2} \right] \right) + \frac{h}{2} \frac{(-3t_0^2 - 3t_1^2)}{\left(1 - \frac{h}{2} \right)}$$

$$= \left(1\left[1+\frac{0.05}{2}\right]\right) + \frac{0.05}{2}\frac{(-3\times 0^2 - 3\times 0.05^2)}{\left(1-\frac{0.05}{2}\right)} = 1.051090$$

Two step BDF approximates the ODE as $y_{n+1}^{(1)} = \frac{3y_{n+1}-4y_n+y_{n-1}}{2h} = y_{n+1} - 3t_{n+1}^2$ which

may be solved for y_{n+1} to get $y_{n+1} = \frac{4y_n - y_{n-1} - 6ht_{n+1}^2}{3-2h}$. The next step onwards we use the two step BDF.

$$
\begin{aligned}
y_2 &= \frac{(4y_1 - y_0 - 6ht_2^2)}{(3-2h)} \\
&= \frac{(4\times 1.051090 - 1 - 6\times 0.05 \times 0.1^2)}{(3-2\times 0.05)} = 1.103917
\end{aligned}
$$

We may obtain the solution for larger t and tabulate the results as below.

t	y	y_E	ϵ_y
0	1.0000	1.0000	0.00E+00
0.05	1.0511	1.0511	-5.48E-05
0.1	1.1039	1.1041	-2.29E-04
0.15	1.1579	1.1583	-4.57E-04
0.2	1.2123	1.2130	-7.21E-04
0.25	1.2664	1.2674	-1.01E-03
0.3	1.3194	1.3207	-1.34E-03
0.35	1.3705	1.3722	-1.69E-03
0.4	1.4188	1.4209	-2.08E-03
0.45	1.4634	1.4659	-2.50E-03
0.5	1.5034	1.5064	-2.96E-03

The error has been calculated by comparing the numerical solution with the exact solution given by $y_E = 3t^2 + 6t + 6 - 5e^t$. Since the error is excessive we need to use a smaller step size and redo the calculations.

Higher order implicit schemes are not self starting. Therefore, the solution procedure is started with lower order implicit schemes and higher order schemes are employed when sufficient number of function evaluations are available. The error can be estimated by comparing the estimates of order n and $n+1$. Hence, one can automatically control the error using the strategies discussed earlier. Hence, an implicit scheme is more computationally intensive than an explicit scheme especially when a large number of equations have to be solved.

10.8.4 Non-linear ODEs

While using explicit solvers the nature of the ODE is not important. However stability of explicit solvers may be poor especially for nonlinear ODEs. For an implicit solver iterative methods are needed. Methods such as Newton Raphson is needed to linearize the nonlinear terms in each iteration step. When dealing with a large number of ODEs, we need to evaluate a large Jacobian matrix at each iteration step. Let us consider a non-linear first order ODE

$$y^{(1)} = f(t,y) \tag{10.96}$$

where $f(t,y)$ is non linear with respect to y. Solving the equations using Backward Euler method we get

$$y_{n+1} - hf(t_{n+1}, y_{n+1}) = y_n \qquad (10.97)$$

The above nonlinear algebraic equation can be solved using any of the methods discussed in Chapter 3. Applying Newton Raphson method to the above equation we get

$$\left(1 - h\frac{df(t_{n+1}, y_{n+1,i})}{dy_{n+1}}\right)\Delta y_{n+1,i+1} = -\left[y_{n+1,i} - hf(t_{n+1}, y_{n+1,i}) - y_n\right] \qquad (10.98)$$

where i is the iteration number.[10]

Example 10.21

Solve the following initial value problem

$$y^{(1)} = -(y^4 - 1), \quad y(0) = y_0 = 2$$

using implicit Backward Euler ODE with $h = 0.1$.

Solution :

Exact solution for the above ODE has been given in Example 10.5.
We have

$$f(x,y) = -y^4 + 1; \qquad \frac{df(x,y)}{dy} = -4y^3$$

Applying Newton Raphson method to the present problem

$$\underbrace{\left(1 + 4hy_{n+1,i}^3\right)}_{A}\Delta y_{n+1,i+1} = \underbrace{-\left[y_{n+1,i} + h(y_{n+1,i}^4 - 1) - y_n\right]}_{B} \qquad (10.99)$$

where i is the iteration number. The following table summarizes the iteration procedure for first step at $t = 0.1$

i	y_1	A	B	Δy_1
0	2.000000	4.200000	-1.500000	-0.357143
1	1.642857	2.773615	-0.271306	-0.097817
2	1.545040	2.475297	-0.014889	-0.006015
3	1.539025	2.458134	-0.000052	-0.000021
4	1.539004	2.458074	-6.28E-10	-2.56E-10

A MATLAB program has been written to solve the ODE using implicit backward Euler method

```
n = 10;              % no. of nodes
h = 1/n;             % step size
x = 0:h:1;           % nodes
y = zeros(n+1,1);    % initialize y
y(1) = 2;            % initial condition
```

---

[10]Equation 10.98 is similar to $f^{(1)}(x)\Delta x = -f(x)$ which is Newton Raphson formulation for solving $f(x) = 0$. In the present case y_{n+1} takes on the role of x

```
for i =1:n              % loop for solving ODE using Backward
                           %Euler
    while 1             % loop for Newton Raphson
        A = 1 + 4*h*y(i)^3;
        B = -(y(i+1)+h*(y(i+1)^4-1)-y(i));
        dy = B/A;
        y(i+1) = y(i+1)+dy;
        if( abs(dy) < 1e-6)
            break   % if dy<tolerance break Newton Raphson
                           %loop
        end
    end
end
```

The comparison of Backward Euler estimate and exact solution is shown in Figure 10.22. Backward Euler method is only first order accurate. The accuracy of the estimate can be improved by refining the grid. Alternatively, more accurate estimates can be obtained by using higher order implicit methods. Of course the present problem is not stiff and explicit methods themselves produce accurate results and implicit methods are not required.

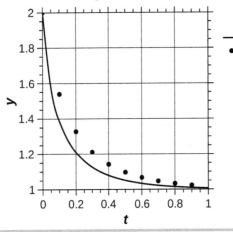

Figure 10.22: *Comparison of solution of implicit Euler method with exact solution for non-linear ODE of Example 10.21*

For a system of non-linear ODEs

$$\mathbf{Y}^{(1)} = f(t,\mathbf{Y}) \tag{10.100}$$

Applying Newton's method to the above equation we get

$$\left(\mathbf{I} - h\mathbf{J}(t_{n+1},\mathbf{Y}_{n+1,i})\right)\Delta\mathbf{Y}_{n+1,i+1} = -\left(\mathbf{Y}_{n+1,i} - hf(t_{n+1},\mathbf{Y}_{n+1,i}) - \mathbf{Y}_n\right) \tag{10.101}$$

$\mathbf{J}$ is the Jacobian matrix of $f(t,\mathbf{Y})$ and $\mathbf{I}$ is the identity matrix. Similar expressions can be derived for higher order implicit schemes. For trapezoidal rule we obtain the following

$$\left(\mathbf{I} - \frac{h}{2}\mathbf{J}(t_{n+1},\mathbf{Y}_{n+1,i})\right)\Delta\mathbf{Y}_{n+1,i+1} = -\left(\mathbf{Y}_{n+1,i} - h\frac{f(t_{n+1},\mathbf{Y}_{n+1,i}) + f(t_n,\mathbf{Y}_n)}{2} - \mathbf{Y}_n\right)$$

Program 10.6: *Solution of system of nonlinear ODE using implicit trapezoidal method*

```
1  function y2 = ODEtrapezoidalnonlin(x1,y1,h,F,J)
2  % Input : x1: starting point
```

```
 3 %              y1:  initial  values  of  y's  at  x=x1
 4 %              h :  step  size
 5 %              F :  external  function  derivatives  F(x,y)
 6 %              J :  external  function  Jacobian  of  derivatives
 7 % Output:
 8 %              X :  abcissca
 9 %              Y :  ordinates
10 x2 = x1+h;                    %
11 y2 = y1;                      % guess  value
12 m = length(y1);              % no.  of  ODEs
13 f1 = F(x1,y1)';              % derivatives  at  tn
14 while 1                       % loop  for  Newton's  iteration
15     A = eye(m) - 0.5*h*J(x2,y2);          % coefficients
16     B = -(y2-0.5*h*(F(x2,y2)'+f1)-y1);
17     dy = ludecomposition(A,B);            % Δy
18     y2 = y2+dy;              % update  yn
19     if( abs(dy) < 1e-6) % if  dy  <  tolerance
20         break              % break  Newton  Raphson  loop
21     end
22 end
```

Example 10.22

Solve the following second order non-linear equation

$$\frac{d^2y}{dt^2} - \lambda\left(1-y^2\right)\frac{dy}{dt} + y = 0$$

with initial conditions $y(t=0) = 2$ and $\dfrac{dy}{dt}(t=0) = 0$ using Explicit RK4 scheme and implicit trapezoidal scheme for $\lambda = 1$ and $\lambda = 10$. Use a step size of $h = 0.1$. Also apply adaptive RK4 scheme and comment on the results.

Background :

The above nonlinear differential equation corresponds to Van der Pol Oscillator and finds application in a number of electronic circuits, laser systems etc. The main feature of the oscillator is nonlinear damping of oscillations.

Solution :

Firstly We will transform the above second order differential equation into two first order couped differential equations. We have $u_0 = y$ and $u_1 = y^{(1)}$. Therefore the two first order differential equations are

$$\begin{aligned} u_0^{(1)} &= u_1 \\ u_1^{(1)} &= \lambda\left(1-u_0^2\right)u_1 - u_0 \end{aligned} \tag{10.102}$$

with initial conditions $u_0(t=0) = 2$ and $u_1(t=0) = 0$. MATLAB programs will be used to solve the above system of differential equations. A MATLAB function to calculate the derivative functions is created.

```
function A = vanderpoloscill(x,y,varargin)
%   Input : x,y    independent  and  dependent  variables
%           y(1) = u0;  y(2) = u1;
%       varargin :  variable  arguments  varargin{1} = λ
%   Output: f(x,y)
```

```
A(1) = y(2);
A(2) = varargin{1}*(1-y(1)^2)*y(2)-y(1);
end
```

For applying implicit trapezoidal method a MATLAB function to calculate the Jacobian of derivative matrix is constructed. The Jacobian matrix of the derivative function is given by

$$\mathbf{J} = \begin{pmatrix} 0 & 1 \\ -2\lambda u_0 u_1 - 1 & \lambda(1 - u_0^2) \end{pmatrix}$$

The MATLAB function is provided below

```
function J = vanderpoloscilljacob(x,y,varargin)
%   Input : x,y     independent and dependent variables
%              y(1) = u_0; y(2) = u_1;
%          varargin : variable arguments varargin{1} = λ
%   Output: J Jacobian
J(1,1) = 0;
J(1,2) = 1;
J(2,1) = -2*varargin{1}*y(1)*y(2)-y(1);
J(2,2) = varargin{1}*(1-y(1)^2);
end
```

MATLAB program has been provided below to solve the system of differential equations with above said methods

```
h = 0.1;
b = 15;
n = b/h;
lambda = 1;
[XaRK4,YaRK4] = ODEadaptRK4(0,b,[2 0],0.05, ...
                @vanderpoloscill,1e-6,lambda);
[XRK4,YRK4] = ODERK4(0,[2 0],0.1,n,@vanderpoloscill,lambda);
Xtr = [0:h:b]';
Ytr = zeros(n+1,2);
Ytr(1,:) = [2 ; 0];
for i=2:n+1
    Ytr(i,:) = ODEtrapezoidalnonlin(Xtr(i-1),Ytr(i-1,:)',h,...
        @vanderpoloscill,@vanderpoloscilljacob,lambda)';
end
```

Figure 10.23a indicates the function values for $\lambda = 1$ estimated by the three methods. All the three numerical schemes seem to be equally adept in capturing the features of the function. The trapezoidal scheme is of lower order accuracy and as expected found to be the least accurate. The explicit schemes are enough for this set of differential equations. Figure 10.23b indicates the function values for $\lambda = 10$. Adaptive RK4 and trapezoidal method perform well where as RK4 diverges. The present set of differential equations represents a stiff system. Hence, RK4 being an explicit scheme without any proper control becomes unstable. However, adaptive RK4 scheme is able to capture the trend. Hence adaptive schemes are very important for stiff equations. Similarly, the implicit trapezoidal rule, though being a lower order scheme does capture the trend correctly. The number of steps taken by trapezoidal scheme is 101 where as for the adaptive RK4 scheme it is 320. However, implicit schemes are computationally intensive and it may be still advantageous

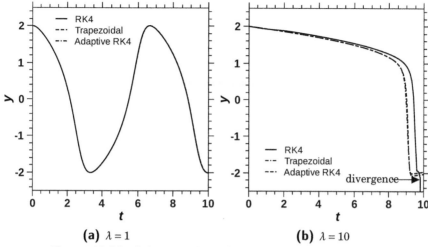

Figure 10.23: *Solution to Van der Pol oscillator for low and high* λ

to use explicit schemes. However, for a stiff system of equations implicit scheme may be necessary. Also, one can incorporate adaptive strategies in an implicit method.

Concluding remarks

Most dynamical systems are governed by initial value ODEs. Hence we have presented all the important techniques of numerically solving such ODEs like single step (RK) and multi-step (ABM) predictor corrector methods. Detailed analysis of stability of various schemes has been presented to highlight the limitation of the numerical schemes as well as to bring out need for strategies to overcome these. Particularly, stiff ODEs pose challenges, which need special methods such as adaptive and implicit methods (BDF). Error estimation (and correction) requires special methods such as those that use sensitivity of the solution to the step size. Alternatively one may combine numerical schemes of lower and higher order to get accurate solutions.

Higher order equations are written as a set of first order ODEs and solved by any of the techniques useful for the solution of first order ODEs.

Nonlinear equations require linearization followed by iteration when implicit methods are made use of.

10.A   MATLAB routines related to Chapter 10

MATLAB offers several inbuilt functions to solve initial value problems.

MATLAB routine	Function
ode45	Solution of ODE using explicit RK45 (non stiff)
ode23	Solution of ODE using explicit RK23 (non stiff)
ode113	Solution of ODE using explicit ABM (non stiff)
ode15s	Solution of ODE based on BDFs (stiff)
ode23s	Solution of ODE based on Rosenbrock formula (stiff)
ode23t	Solution of ODE based on trapezoidal rule (stiff)
ode23tb	Solution of ODE based on implicit RK and trapezoidal rule
ode15i	Solution of ODE based on fully implicit BDF method (stiff)

Reader may refer the MATLAB reference to understand the above functions.

10.B   Suggested reading

1. **E. Kreyszig** *Advanced engineering mathematics* Wiley-India, 2007
2. **L.F. Shampine**, **I. Gladwell** and **S. Thompson** *Solving ODEs with MATLAB* Cambridge University Press, 2003
3. **J.C. Butcher** *Numerical methods for ordinary differential equations* Wiley, 2008

Chapter 11

Boundary value problems (ODE)

Second and higher order ordinary differential equations may also be formulated as problems defined over finite or semi-infinite domains. In such cases the function and its derivatives may be specified at either ends of the interval defining the domain. The ODE becomes a boundary value problem (BVP).

BVPs occur in applications such as heat transfer, electromagnetism, fluid mechanics and, in general, in many field problems in one space dimension. Hence the solution of such equations is important. The following numerical methods are considered in this chapter:

- *Shooting method*
- *Finite difference method (FDM)*
- *Method of weighted residuals (MWR)*
- *Collocation method*
- *Finite element and finite volume methods*

11.1   Introduction

Consider a second order ODE given by

$$\frac{d^2y}{dx^2} + 3\frac{dy}{dx} + 2y = 0 \tag{11.1}$$

valid in the range $0 \le x \le 1$. This equation is similar to the ODE considered in Example 10.18. The exact solution is determined by finding the roots of the characteristic equation $m^2 + 3m + 2 = 0$. The two roots of the characteristic equation are real and distinct, given by $m_1 = -1$ and $m_2 = -2$. Hence the solution to the above ODE can be written as

$$y = Ae^{m_1x} + Be^{m_2x} \tag{11.2}$$

where A and B are two constants. Being a second order equation, two conditions are required to determine the constants A and B. In an initial value problem two conditions are specified at a point. If the initial conditions are specified as $y(x = 0) = y_0$ and $y^{(1)}(x = 0) = y_0^{(1)}$. Then we have

$$\begin{array}{ccccc} A & + & B & = & y_0 \\ m_1A & + & m_2B & = & y_0^{(1)} \end{array} \tag{11.3}$$

Therefore the constants are given by

$$A = \frac{y_0^{(1)} - m_2y_0}{m_1 - m_2}; \quad B = \frac{y_0^{(1)} - m_1y_0}{m_2 - m_1}; \tag{11.4}$$

Figure 11.1 shows a family of curves satisfying the above differential equation and passing through the origin ($y_0 = 0$). The difference between these curves is due to the initial slope $y_0^{(1)}$.

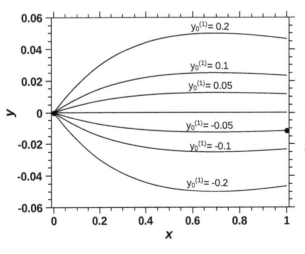

Figure 11.1: Family of curves satisfying a second order ODE passing through the origin

Consider the curve having $y_0^{(1)} = -0.05$. The same solution curve may be obtained by specifying two conditions at different values of x such as those shown in the following table

$x = 0$	$x = 1$
$y = 0$	$y = -0.0116$
$y = 0$	$y^{(1)} = 0.0049$
$y^{(1)} = -0.05$	$y = -0.0116$

Such problems are classified as *boundary value problems*. The difference between IVP and BVP lies in the nature of flow of information. All IVP are marching problems i.e. the information flows along one direction. Where as the information for a BVP flows from both directions i.e. from both *boundaries* of the domain. The boundary conditions may be of any of the following three types:

Boundary condition of first kind or Dirichlet* boundary condition	:	y is specified, in above example $y(x = 1) = 0.0116$
Boundary condition of second kind or Neumann† boundary condition	:	$y^{(1)}$ is specified, in above example $y^{(1)}(x = 1) = 0.0049$
Boundary condition of third kind or Robin+ boundary condition	:	Relation between y and $y^{(1)}$ is specified

* after Johann Peter Gustav Lejeune Dirichlet, 1805-1859, a German mathematician; † after Carl Gottfried Neumann, 1832-1925, a German mathematician; + after Victor Gustave Robin, 1855-1897, a French mathematician

There is a fourth kind of boundary condition called the Cauchy[1] boundary condition. In this case both the function and the derivative at a boundary are specified (as in initial value problems).

The present chapter will discuss different methods to solve boundary value problems.

11.2 The 'shooting method'

Consider a second order ODE given by

$$\frac{d^2y}{dx^2} = f\left(x, y, \frac{dy}{dx}\right) \quad \text{or} \quad y^{(2)} = f(x, y, y^{(1)}) \tag{11.5}$$

valid in the range $0 \le x \le L$. First kind boundary conditions are specified at the two boundaries as

$$y = y_0 \text{ at } x = 0 \text{ and } y = y_L \text{ at } x = L \tag{11.6}$$

The solution can be determined if both the conditions are available at one point. Then, the solution to this problem can be obtained numerically using the methods discussed in Chapter 10 to solve initial value problems.

All the methods that have been presented in Chapter 10 yield numerical solution of a second order ODE if initial conditions are specified. Hence Equation 11.5 may be solved only if both the function and the derivative are known, say at $x = 0$. Since the derivative is not known at $x = 0$ we assume a value for the derivative as $y^{(1)} = s_1$ at $x = 0$ and solve Equation 11.5 with *Cauchy boundary conditions* at $x = 0$. We may use any of the numerical methods discussed earlier and obtain the value of the function at $x = L$ as y_{1L}. In general $y_{1L} \ne y_L$, the specified Dirichlet condition at $x = L$. The way forward depends on the nature of the ODE. If the ODE is linear the following procedure is used.

[1] after Augustin-Louis Cauchy, 1789-1857, a French mathematician

11.2.1   Linear ODE case

In this case we obtain a second solution by choosing $y^{(1)} = s_2$ at $x = 0$ and solve Equation 11.5 by any of the numerical methods and obtain the value of the function at $x = L$ as y_{2L}. Since the ODE is linear, a linear combination of the two solutions is also another solution to the problem (principle of superposition). Thus we may seek the solution to Equation 11.5 in the form of a linear combination of the two solutions that are already available, in the form

$$y = \alpha y_1 + (1 - \alpha)y_2 \qquad (11.7)$$

where α is an as yet undetermined fraction. We require that this solution satisfy the Dirichlet condition at $x = L$. Hence

$$y_L = \alpha y_{1L} + (1 - \alpha)y_{2L}$$

and α is obtained as

$$\alpha = \frac{y_L - y_{2L}}{y_{1L} - y_{2L}} \qquad (11.8)$$

This value of α in Equation 11.7 yields the required solution that satisfies the two boundary conditions.

Since the given ODE is solved by obtaining two trial solutions by using assumed slopes (related to tangent of angle with the x axis) at $x = 0$, the method is referred to as the *shooting method*. Analogy is with a gun which is fired at a suitable angle with respect to the ground so that the projectile hits the target.

Example 11.1

Solve the homogeneous second order ODE with constant coefficients $y^{(2)} = 2y$ subject to the boundary conditions $y(x = 0) = 1$ and $y(x = 1) = 0.6$. Make use of the shooting method. Compare the solution with the exact solution.

Solution :
Exact solution: We obtain the exact solution before taking up the numerical solution. The characteristic equation is $m^2 - 2 = 0$ which has the roots $m = \pm\sqrt{2}$. The general solution may hence be written in terms of hyperbolic functions as

$$y_E = A\cosh(\sqrt{2}x) + B\sinh(\sqrt{2}x)$$

where A and B are constants to be determined by the use of specified boundary conditions. The boundary condition at $x = 0$ is satisfied if we take $A = 1$. The boundary condition at $x = 1$ is satisfied if $B = \dfrac{0.6 - \cosh(\sqrt{2})}{\sinh(\sqrt{2})}$. Thus the constants of integration are given by $A = 1$ and $B = -0.815571$.

Alternately, the governing equation may be written in matrix form $\mathbf{y}^{(1)} = \mathbf{A}\mathbf{y}$ where the coefficient matrix is given by

$$\mathbf{A} = \begin{pmatrix} 0 & 1 \\ 2 & 0 \end{pmatrix}$$

and $\mathbf{y}^T = \begin{pmatrix} y & y^{(1)} \end{pmatrix}$. The eigenvalues of the coefficient matrix are the roots of the characteristic equation that has already been obtained as a part of the exact solution. They are $\lambda_1 = \sqrt{2}$ and $\lambda_2 = -\sqrt{2}$. The roots are real and distinct. We may find eigenvectors and use these to complete the solution (we leave it to the reader).

We are now ready to obtain the numerical solution by the shooting method.

Numerical solution: We convert the second order ODE in to two first order ODEs and write them as

$$y^{(1)} \quad = \quad u_0^{(1)} = u_1(x, u_0)$$
$$y^{(2)} \quad = \quad u_1^{(1)} = 2y = 2u_0$$

The system of ODEs is solved using RK4 method. Two numerical solutions are obtained using $y(0) = u_0(0) = 1$, $u_1(0) = y^{(1)}(0) = -0.75$ and $y(0) = u_0(0) = 1$, $u_1(0) = y^{(1)}(0) = -1.25$ starting at $x = 0$. The calculations in both cases terminate at $x = 1$. The results are shown as a table below.

	First Trial		Second Trial	
x	$y = u_0$	$y^{(1)} = u_1$	$y = u_0$	$y^{(1)} = u_1$
0	1.0	**-0.75**	1.0	**-1.25**
0.1	0.934767	-0.556846	0.884600	-1.061854
0.2	0.888260	-0.374847	0.786921	-0.894981
...	...	...	...	...
0.8	0.974799	0.680416	0.483837	-0.175205
0.9	1.052832	0.882842	0.471105	-0.079870
1	**1.151956**	1.102953	**0.467810**	0.013865

In order to satisfy the Dirichlet condition at $x = 1$ we choose α as

$$\alpha = \frac{0.6 - 0.467810}{1.151956 - 0.467810} = 0.193219$$

We may now combine the two solutions using the above α value to get the solution to the problem. A MATLAB program has been written to carry out the calculations.

```
[X,Y1] = ODERK4(0,[1;-0.75],0.1,10,@odeexshooting);% trial 1
[X,Y2] = ODERK4(0,[1;-1.25],0.1,10,@odeexshooting);% trial 2
alpha  = (0.6-Y2(11,1))./(Y1(11,1)-Y2(11,1));   % calculate α
Y = Y1*alpha + Y2*(1-alpha);    % correct solution
```

where odeexshooting is a MATLAB function defining the ODE given by

```
function z = odeexshooting(x,y)
z(1) =    y(2);     % u_0^(1)
z(2) =    2*y(1);   % u_1^(1)
end
```

Table shows the solution thus obtained and also compares the solution with the exact solution obtained earlier.

x	y	$y^{(1)}$	y_E	ϵ_y	$y_E^{(1)}$	$\epsilon_{y^{(1)}}$
0	1.0	-1.1534	1.0	0.0	-1.1534	9.1E-07
0.1	0.8943	-0.9643	0.8943	4.6E-07	-0.9643	2.7E-07
0.2	0.8065	-0.7945	0.8065	8.1E-07	-0.7945	2.2E-07
0.3	0.7349	-0.6406	0.7349	1.0E-06	-0.6406	5.9E-07
0.4	0.6780	-0.4996	0.678	1.2E-06	-0.4996	8.7E-07
0.5	0.6346	-0.3685	0.6346	1.3E-06	-0.3685	1.1E-06
0.6	0.6040	-0.2449	0.604	1.3E-06	-0.2449	1.3E-06
0.7	0.5855	-0.1261	0.5855	1.3E-06	-0.1261	1.4E-06
0.8	0.5787	-0.0099	0.5787	1.2E-06	-0.0099	1.6E-06
0.9	0.5835	0.1061	0.5835	1.0E-06	0.1061	1.7E-06
1	0.6000	0.2243	0.6	0	0.2243	1.9E-06

The magnitude of the largest error in y is limited to $1e-6$. This is satisfactory for our purpose in modeling a practical problem. If necessary one may improve the solution by reducing h. The following figure illustrates the trial and actual solutions.

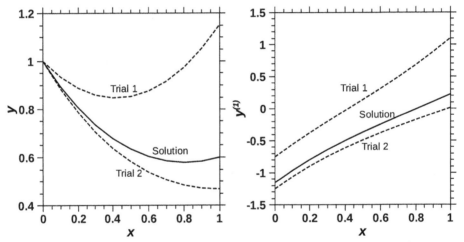

Figure 11.2: Trial and actual solutions using the "Shooting method"

We consider a second example where Neumann boundary condition is specified at $x = 1$ and the second order ODE is non-homogeneous.

Example 11.2

Solve numerically the second order ODE $y^{(2)} = 2y + \cos\left(\dfrac{\pi x}{2}\right)$. The boundary conditions have been specified as $y(x = 0) = 1$ and $y^{(1)}(x = 1) = 0$. Compare the numerical solution with the exact analytical solution.

Solution :

Exact solution: The exact solution to the ODE may easily be obtained by combining the complementary function (solution to the homogeneous equation) and a particular integral. The result is given by

$$y = C_1 \cosh(\sqrt{2}x) + C_2 \sinh(\sqrt{2}x) - \frac{4}{\pi^2 + 8} \cos\left(\frac{\pi x}{2}\right)$$

where the constants are given by

$$C_1 = \frac{\pi^2 + 12}{\pi^2 + 8}$$

$$C_2 = \frac{-\sqrt{2}C_1 \sinh(\sqrt{2}) - \frac{4}{\pi^2+8}}{\sqrt{2}\cosh(\sqrt{2})}$$

Numerical solution: The numerical solution is obtained by applying implicit RK2[2] with $h = 0.1$. Equations to be solved are given by

$$\begin{pmatrix} 1 & -\frac{h}{2} \\ -h & 1 \end{pmatrix} \begin{pmatrix} y_{n+1} \\ y_{n+1}^{(1)} \end{pmatrix} = \begin{pmatrix} y_n + \frac{h y^{(1)} n + 1}{2} \\ y_n^{(1)} + h y_n + \frac{h}{2}\left[\cos\left(\frac{\pi x_n}{2}\right) + \cos\left(\frac{\pi x_{n+1}}{2}\right)\right] \end{pmatrix}$$

We obtain the numerical solution with two different starting values for the slope at $x = 0$ and tabulate the results as given below.

	First trial			Second trial	
x	y	$y^{(1)}$	x	y	$y^{(1)}$
0	1	-1	0	1	-2
...	...	...	...	...	...
1	1.2983	1.5255	1	-0.0736	-0.6573

We linearly interpolate requiring that the slope be zero at $x = 1$. We then have

$$\alpha = \frac{0 - (-0.6573)}{1.5255 - (-0.6573)} = 0.3011$$

The numerical solution satisfying the given boundary conditions are then obtained by linear combination of the two solutions already obtained with weights of 0.3011 and 0.6989. We show the results along with the exact solution as a table below.

x	y	$y^{(1)}$	y_E	ϵ_y	$y_E^{(1)}$	$\epsilon_{y^{(1)}}$
0	1	-1.6989	1.0000	0.0000	-1.6990	0.0002
0.1	0.8443	-1.4151	0.8445	-0.0002	-1.4155	0.0004
0.2	0.7154	-1.1621	0.7159	-0.0004	-1.1627	0.0005
...	...	...	...	...	...	...
...	...	...	...	...	...	...
0.8	0.3556	-0.1693	0.3568	-0.0012	-0.1696	0.0003
0.9	0.3433	-0.0761	0.3447	-0.0014	-0.0763	0.0002
1	0.3395	0.0000	0.3410	-0.0015	0.0000	0.0000

It is seen that the maximum error in y is around 0.0015 or approximately 0.15%. This may be acceptable in a practical application. However it is always possible to improve the solution by taking a smaller h.

---

[2]Implicit RK2 may not be necessary in this example as the given ODE is not stiff. Here we have used it for demonstration purpose.

11.2.2   Non-linear ODE case

In the case of a non-linear ODE the relationship between the initial slope and the final slope follows a non-linear trend, assuming that we are interested in satisfying Neumann condition $y^{(1)} = 0$ at $x = L$. It is clear that it is not possible to use a linear combination in this case. Secant method is used to determine the correct initial slope such that the desired zero slope is obtained at $x = L$. The algorithm can be summarized as follows

- Start with two guess values of initial slopes and solve the ODE as initial value problems
- Determine the new value of initial slope using secant method and solve the ODE.
- When the boundary condition at the second boundary is satisfied within desired tolerance, stop iteration.

Example 11.3

Solve the equation $y^{(2)} - (y^4 - 0.5) = 0$ with $y(x = 0) = 1$ and $y^{(1)}(x = 1) = 0$. Use RK4 with $h = 0.1$ and the secant method to determine the initial slope.

Solution :

As usual the non-linear second order ODE is written as a set of two first order ODEs.

$$
\begin{aligned}
y^{(1)} &= u_0^{(1)} = u_1 \\
y^{(2)} &= u_1^{(1)} = (u_0^4 - 0.5)
\end{aligned}
$$

RK4 may now be used starting at $x = 0$ using the boundary condition $y(0) = 1$ and a guess value for the slope given by $y^{(1)}(x = 0) = s_{11}$. In the present application we have used $s_{11} = -0.2$ as the first guess value. The RK4 solution is obtained such that when $x = 1$ we have a slope $y^{(1)}(x = 1) = s_{21}$. In the present case we obtain a value of $s_{21} = 0.1674$.

We obtain a second solution using RK4 starting at $x = 0$ using the boundary condition $y(0) = 1$ and a guess value for the slope given by $y^{(1)}(x = 0) = s_{12}$. In the present application we have used $s_{12} = -0.3$ as the second guess value. The RK4 solution is obtained such that when $x = 1$ we have a slope $y^{(1)}(x = 1) = s_{22}$. In the present case we obtain a value of $s_{22} = -0.1338$. The fact that the two values of the terminal slope are of opposite sign is encouraging.

Using the secant method we determine the new approximating starting slope s_1 as

$$
s_{13} = s_{12} - \frac{s_{22}(s_{11} - s_{12})}{(s_{21} - s_{22})} = -0.3 - \frac{-0.1338(-0.2 + 0.3)}{(0.1674 + 0.1338)} = -0.2556
$$

We now set $s_{11} = -0.3$, $s_{12} = -0.2556$ and continue the process outlined above. The initial slope converges in a few secant method based iterations to the value -0.2527 as indicated in the following table.

Result of secant method

$y_1^{(1)}(0)$	$y_2^{(1)}(0)$	$y_3^{(1)}(0)$
-0.200000	-0.300000	-0.255579
-0.300000	-0.255579	-0.252511
-0.255579	-0.252511	-0.252672
-0.252511	-0.252672	-0.252671
-0.252672	-0.252671	-0.252671

In fact a plot of target slope as a function of the guess slope is non-linear as shown in Figure 11.3.

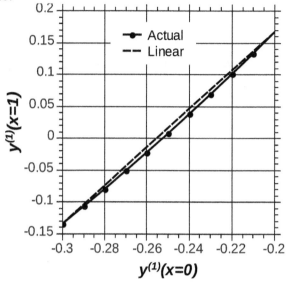

Figure 11.3: Relation between target slope and the guess slope

The line labeled as "Linear" is simply a line joining the two end points in the figure. This line is drawn to indicate that the relationship is indeed non-linear and the secant method is *required* to solve the boundary value problem. The solution is finally obtained with the initial slope determined by the secant method and tabulated below.

x	y	$y^{(1)}$	x	y	$y^{(1)}$
0	1.0000	-0.2527	0.6	0.9128	-0.0673
0.1	0.9771	-0.2073	0.7	0.9070	-0.0488
0.2	0.9583	-0.1697	0.8	0.9030	-0.0318
0.3	0.9429	-0.1381	0.9	0.9007	-0.0157
0.4	0.9305	-0.1112	1	0.8999	0.0000
0.5	0.9206	-0.0879			

We also show the above data in the form of a plot in Figure 11.4.

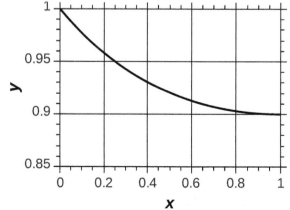

Figure 11.4: Plot of solution in Example 11.3

MATLAB program to solve the above ODE using shooting method has been given below

```
s11 = -0.2;        % initial slope guess 1
s12 = -0.3;        % initial slope guess 2
[X,Y1] = ODERK4(0,[1;s11],0.1,10,@odeexshooting2); % solution
                            %1
ds = abs(s11-s12); % change in initial slope
count =1;          % iteration count
while ds > 1e-6 % loop for secant method
    [X,Y2] = ODERK4(0,[1;s12],0.1,10,@odeexshooting2); %
                            %solution 2
    s21 = Y1(11,2); s22 = Y2(11,2); % slope at x=1
    s13 = s12 - s22*(s11-s12)/(s21-s22); % new initial slope
    ds = abs(s13-s12);  % change in initial slope
    s11 = s12;          % update inital slopes
    s12 = s13;
    Y1 = Y2;
    count = count + 1;  % increment iteration count
end
```

where odeexshooting is a MATLAB function defining the ODE given by

```
function z = odeexshooting2(x,y)
z(1) =   y(2);       % u_0^(1)
z(2) =   y(1)^4-0.5;;  % u_1^(1)
end
```

The non-linear equation of the present example does not have an analytical exact solution. Hence we do not know how good the above solution is. In order to verify the quality of the solution we have obtained a solution to the boundary value problem by halving the step size. The starting slope turns out to be -0.252671 which is different from the previously obtained value by approximately 6×10^{-7}. Hence the solution with $h = 0.1$ is adequate and it is not necessary to reduce the step size.

11.2.3 Boundary value problem over semi-infinite domain

In many engineering applications we encounter boundary value problems where one of the boundaries is at infinity. Examples may be cited from various area such as electromagnetism, fluid mechanics and heat transfer. In order to make the development easy for the reader to follow we take a simple second order ODE. Consider the general second order ODE given by

$$y^{(2)} = f(x,y,y^{(1)}) \tag{11.9}$$

with Dirichlet boundary conditions at both boundaries viz.

$$y(x = 0) = 0, \quad \text{and} \quad y(x \to \infty) = 1 \tag{11.10}$$

Numerically speaking boundary at ∞ is something which is reachable in an asymptotic sense. Any ODE solver can solve the governing equation once a finite but large numerical value is assigned to the boundary at ∞ which we identify as x_∞. The question arises as to how large should the chosen value of x_∞ be. One way of approaching the problem is to repeat the numerical solution of the ODE with different values for x_∞ and find out a value of x_∞ beyond which the solution is insensitive to the chosen value of x_∞. Such a procedure would be an inefficient cut and try type of procedure. Also the solution should be obtained over a variable range depending on the chosen value of x_∞. Hence such a method is avoided in preference to the method that will be described below.

The procedure that we would like to follow uses the transformation $\xi = \dfrac{x}{x_\infty}$ to transform the domain from $x = 0$ to $x = \infty$ to $\xi = 0$ to $\xi = 1$. The governing differential equation then takes the form

$$\frac{d^2y}{d\xi^2} = f\left(\xi, y, \frac{dy}{d\xi}; x_\infty\right) \tag{11.11}$$

where x_∞ appears as a parameter and is shown to the right of the semicolon. The quantities to the left of the semicolon represent the variables on which the second derivative of y depends. The boundary conditions now are written down as

$$y(\xi = 0) = 0, \quad \text{and} \quad y(\xi = 1) = 1 \tag{11.12}$$

Equation 11.11 subject to boundary conditions 11.12 may be easily solved by a method such as the shooting method described earlier, for a chosen value of the parameter x_∞. Two solutions required to perform the shooting method may be evaluated using an accurate algorithm such as the adaptive RK4.

We now turn our attention to the determination of x_∞. After the solution has been obtained with a particular value of x_∞ we determine the trend of slope of y at the outer boundary (the boundary at ∞ is referred to as the outer boundary) after reverting to the original coordinate. The slope will in fact be given by $\dfrac{1}{x_\infty}\dfrac{dy}{d\xi}$ which may be non-zero. We *know* that if the chosen value of x_∞ is satisfactory $y^{(1)}$ should be 0 at the outer boundary. If not, we linearly extrapolate $y^{(1)}$ using two consecutive values, one at the boundary and the other at the previous node, to hit the x axis at a new value of x_∞ which is taken as the next trial values for x_∞.

$$x_\infty^{new} = x_\infty^{old}\left(\xi_n - y_n^{(1)}\frac{\xi_{n-1} - \xi_n}{y_{n-1}^{(1)} - y_n^{(1)}}\right) = x_\infty^{old}\left(1 - y_n^{(1)}\frac{\xi_{n-1} - 1}{y_{n-1}^{(1)} - y_n^{(1)}}\right)$$

since $\xi_n = 1$.

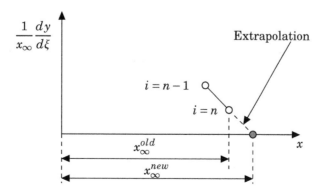

In addition to the above condition, the slope at $x = 0$ should also become insensitive to x_∞. Iteration continues till convergence i.e. till the slope becomes zero at x_∞ within a specified tolerance or change in slope with x_∞ is less than specified tolerance.

Example 11.4

Solve the boundary value problem $y^{(2)} + xy^{(1)} = 0$ subject to the boundary conditions $y(0) = 0$ and $y \to 1$ as $x \to \infty$. Compare the numerical solution with the exact solution and make suitable comments.

Solution :

Exact solution: We obtain first the exact solution. The given equation is in variable separable form and may be written as

$$\frac{y^{(2)}}{y^{(1)}} + x = 0$$

We may integrate once with respect to x to get

$$\ln y^{(1)} = -\frac{x^2}{2} + C_1'$$

where C_1' is a constant of integration. This may be rewritten as

$$y^{(1)} = e^{-\frac{x^2}{2} + C_1'} = C_1 e^{-\frac{x^2}{2}}$$

where we have set $e^{C_1'} = C_1$, another constant. A second integration yields

$$y = C_1 \int_0^x e^{-\frac{x^2}{2}} dx + C_2$$

where C_2 is a second constant of integration. The boundary condition at $x = 0$ requires that C_2 be set to zero. The boundary condition at $x \to \infty$ requires that $1 = C_1 \int_0^\infty e^{-\frac{x^2}{2}} dx$. We may eliminate C_1 and get the following expression for y.

$$y = \frac{\int_0^x e^{-\frac{x^2}{2}} dx}{\int_0^\infty e^{-\frac{x^2}{2}} dx}$$

The quantity on the right hand side is related to the error function and given by

$$y = \operatorname{erf}\left(\frac{x}{\sqrt{2}}\right)$$

where "erf" stands for the error function that is tabulated in handbooks of mathematics.

Numerical solution: Now we look at the numerical solution of the problem. Make a transformation as indicated previously to convert the second order ODE to the following form.

$$\frac{d^2 y}{d\xi^2} + x_\infty^2 \xi \frac{dy}{d\xi} = 0$$

with the boundary conditions $y(0) = 0$, $y(1) = 1$. The above differential equation is a linear equation.

   Step 1 We shall solve the differential equation using adaptive RK4 with tolerance of 10^{-6}. Assuming $x_\infty = 2$ and applying shooting method we get

ξ	x	y	$\dfrac{dy}{d\xi}$	$\dfrac{dy}{dx}$
0	0	0.000000	1.671838	0.835919
0.05	0.1	0.083453	1.663500	0.831750
...	...	...	...	...
0.95	1.9	0.987498	0.274975	**0.137487**
1	2	1.000000	0.226259	**0.113129**

Step 2 The slope $\dfrac{dy}{dx}$ at x_∞ is not equal to zero. Hence, x_∞ is updated using extrapolation.

$$
\begin{aligned}
x_\infty^{new} &= x_\infty^{old}\left(\xi_n - y_n^{(1)}\,\frac{\xi_{n-1}-\xi_n}{y_{n-1}^{(1)}-y_n^{(1)}}\right) \\
&= 2\left(1 - 0.113129 \times \frac{0.95-1}{0.137487-0.113129}\right) = 2.4644
\end{aligned}
$$

Step 3 Applying shooting method with the new x_∞ we get

ξ	x	y	$\dfrac{dy}{d\xi}$	$\dfrac{dy}{dx}$
0	0	0.000000	1.993701	0.815086
0.05	0.1223	0.099433	1.978623	0.808922
...	...	...	...	...
0.975	2.38485	0.997419	0.111157	0.045444
1	2.446	1.000000	0.095680	**0.039117**

The slope at the outer boundary has reduced considerably.However it is still non-zero and x_∞ is updated using extrapolation. The procedure is repeated until a suitable x_∞ is obtained.

A MATLAB program has been written to perform the above operations.

```
xinf = 2;           % x∞
count = 1;          % iteration count
Sold = 0;           % initialize old slope at x = 0 dy/dx
while 1             % loop for solving BVP
    % Applying shooting method
    [X1,Y1] = ODEadaptRK4(0,1,[0 1.8],0.05, ...    % trial 1
                    @seminfiniteexamplefunc,1e-6,xinf);
    [X2,Y2] = ODEadaptRK4(0,1,[0 0.2],0.05, ...    % trial 2
                    @seminfiniteexamplefunc,1e-6,xinf);
    n1 = length(X1);    % no. of nodes for trial 1
    n2 = length(X2);    % no. of nodes for trial 2
    alpha=(1-Y1(n1,1))/(Y2(n2,1)-Y1(n1,1));  % α
            % new slope at x = 0 new dy/dx
    Snew = (Y2(1,2)*alpha + (1-alpha)*Y1(1,2))/xinf;
    [X,Y] = ODEadaptRK4(0,1,[0 Snew*xinf],0.05, ... % solution
                    @seminfiniteexamplefunc,1e-6,xinf);
    % end of shooting method
    count = count + 1;  % increment iteration count
    n = length(X);      % no of nodes for solution
    if ( Y(n,2)/xinf < 1e-4 || abs(Snew-Sold) < 1e-4)
    % check for convergence
```

```
              break % if converged break loop
      end
        % perform  extrapolation  to  update  x∞
      y1 = Y(n,2); y2 = Y(n-1,2); x1 = X(n); x2 = X(n-1);
      xinf = xinf*(x1 - y1*(x2-x1)/(y2-y1));
      Snew = Sold;           % update old slope at x = 0
  end
```

where `seminfiniteexamplefunc` is MATLAB function defining the differential equation. The input to the function varargin passes the parameter x_∞ from main program.

```
function z = seminfiniteexamplefunc(x,y,varargin)
z(2)  =  -varargin{1}^2*x*y(2);      % y(1) = y, z(2) = y^(2)
z(1)  =  y(2);                       % y(2) = y^(1)
end
```

Iteration	x_∞	ϵ_y	$y^{(1)}(x=0)$	$\epsilon_{y^{(1)}}$	$y^{(1)}(x_\infty)$
1	2.0000	4.6E-02	0.8359		0.1131
2	2.4644	1.4E-02	0.8090	2.7E-02	0.0388
3	2.8453	4.4E-03	0.8014	7.5E-03	0.0140
4	3.1668	1.5E-03	0.7991	2.3E-03	0.0053
5	3.4487	5.6E-04	0.7983	7.8E-04	0.0021
6	3.7013	2.1E-04	0.7981	2.8E-04	0.0008
7	3.9312	8.4E-05	0.7980	1.0E-04	0.0004
8	4.1428	3.4E-05	0.7979	4.0E-05	0.0001

ϵ_y is the maximum difference between exact and numerical solution

$\epsilon_{y^{(1)}}$ is the change in the slope at $x=0$ with x_∞

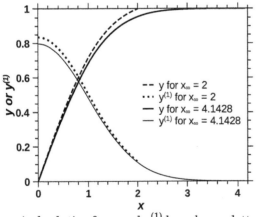

Figure 11.5: Solution to Example 11.4 using $x_\infty = 2$ (initial guess for x_∞) and 4.1428 (final value of x_∞)

Numerical solution for y and $y^{(1)}$ have been plotted for $x_\infty = 2$ and 4.1228 in Figure 11.5. It is clear from the present exercise the size of the domain has to be chosen appropriately for obtaining accurate solutions.

11.2.4 Generalization of shooting method for higher order ODEs

Consider a fourth order boundary value problem

$$y^{(4)} = f\left(x, y, y^1, y^{(2)}, y^{(3)}\right) \tag{11.13}$$

with boundary conditions $y = y_0$ and $y^{(1)} = y_0^{(1)}$ at $x = 0$ and $y = y_1$ and $y^{(1)} = y_1^{(1)}$ at $x = 1$. The problem is solved as an initial value problem with two given conditions (y_0 and $y_0^{(1)}$) and two assumed conditions ($y_0^{(2)}$ and $y_0^{(3)}$). The solution would probably not match the boundary conditions at $x = 1$. The shooting method is a root-finding method where initial conditions (independent parameters) are selected such that all boundary conditions at $x = 1$ (dependent parameters) are determined. The solution procedure is different based on the nature of the ODE.

Linear ODE

Since the ODE is linear, principle of superposition is applicable. Therefore, one has to solve the ODE a certain number of times and the desired solution will be weighted sum of these solutions. For the present case there are two dependent parameters. The ODE is solved with three trial values of initial conditions and the final solution would be a linear combination of the three solutions. Solution $\mathbf{u}_1$ is obtained using initial condition $[y_0 \ y_0^{(1)} \ u_{12} \ u_{13}]^T$, $\mathbf{u}_2$ using initial condition $[y_0 \ y_0^{(1)} \ u_{22} \ u_{23}]^T$ and $\mathbf{u}_3$ using initial condition $[y_0 \ y_0^{(1)} \ u_{32} \ u_{33}]^T$. Then the solution to the above BVP is given by $\mathbf{y} = \alpha_1\mathbf{u}_1 + \alpha_2\mathbf{u}_2 + \alpha_3\mathbf{u}_3$ ($\mathbf{y} = [y \ y^{(1)} \ y^{(2)} \ y^{(3)}]^T$). The constants α_1, α_2 and α_3 have to be determined from boundary conditions at $x = 1$. We have

$$\alpha_1 u_1(x = 1) + \alpha_2 u_2(x = 1) + \alpha_3 u_3(x = 1) = y_1$$
$$\alpha_1 u_1^{(1)}(x = 1) + \alpha_2 u_2^{(1)}(x = 1) + \alpha_3 u_3^{(1)}(x = 1) = y_1^{(1)}$$

In addition to this, the sum of the constants has to be equal to 1 i.e. $\alpha_1 + \alpha_2 + \alpha_3 = 1$, as otherwise, the boundary condition at $x = 0$ will not be satisfied. Writing these in matrix notation we have

$$\begin{pmatrix} u_1(x=1) & u_2(x=1) & u_3(x=1) \\ u_1^{(1)}(x=1) & u_2^{(1)}(x=1) & u_3^{(1)}(x=1) \\ 1 & 1 & 1 \end{pmatrix} \begin{pmatrix} \alpha_1 \\ \alpha_2 \\ \alpha_3 \end{pmatrix} = \begin{pmatrix} y_1 \\ y_1^{(1)} \\ 1 \end{pmatrix}$$

The constants α_i may hence be determined by solving the system of linear equations. We shall consider an example to demonstrate the shooting method for a linear fourth order ODE.

Example 11.5

Solve the fourth order differential equation $y^{(4)} = -1$ with boundary conditions $y(x = 0) = 0$; $y^{(2)}(x = 0) = 0$; $y(x = 1) = 0$ and $y^{(2)}(x = 1) = 0$ and compare with exact solution.

Background :

Deflection of Beams Estimating the deflection of loaded beams is important for designing structures. The governing equations for a beam is

$$EI\frac{d^2y}{dx^2} = M(x)$$

where EI is known as the flexural rigidity of the beam (E is Young's modulus and I is the moment of inertia of the beam) and M is the bending moment acting on the beam. Further the equation can be rewritten as

$$\frac{d}{dx}\left(EI\frac{d^2y}{dx^2}\right) = \frac{dM(x)}{dx} = V(x)$$

$$\frac{d^2}{dx^2}\left(EI\frac{d^2y}{dx^2}\right) = \frac{dV(x)}{dx} = q(x)$$

where V is the shear force and q is the load distribution on the beam. Being a fourth order ODE, four conditions need to be specified for solving the ODE. The following figure shows two configuration of beams subjected to loading.

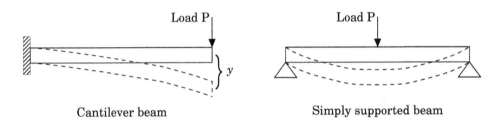

Cantilever beam   Simply supported beam

For a cantilever beam, the deflection y and slope of defection $y^{(1)}$ are zero at $x = 0$. Other boundary conditions are $M = y^{(2)} = 0$ at $x = L$ and $V = y^{(3)} = P$. For a simply supported beam, the deflection y and bending moment M are zero at the two supporting ends.

The present problem corresponds to a simply supported beam having constant flexural rigidity and subjected to uniform loading (self weight of the beam) throughout the span of the beam as indicated below.

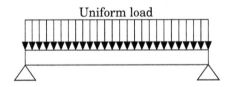

Uniform load

Solution :

Exact solution: The solution to the differential equation can be determined by integrating the equation four times

$$y^{(3)} = -x + C_1$$

$$y^{(2)} = -\frac{x^2}{2} + C_1x + C_2$$

$$y^{(1)} = -\frac{x^3}{6} + C_1\frac{x^2}{2} + C_2x + C_3$$

$$y = -\frac{x^4}{24} + C_1\frac{x^3}{6} + C_2\frac{x^2}{2} + C_3x + C_4$$

where C_1, C_2, C_3 and C_4 are constants. Applying the boundary conditions, we find $C_1 = \dfrac{1}{2}$, $C_2 = 0$, $C_3 = -\dfrac{1}{24}$ and $C_4 = 0$. Therefore $y = \dfrac{1}{24}\left(-x^4 + 2x^3 - x\right)$ and $y^{(2)} = \dfrac{1}{2}\left(-x^2 + x\right)$.

Numerical solution: The ODE is to be solved as an initial value problem using three trial sets of initial conditions given by

$$\mathbf{u}_1^T = [0\ \ 0\ \ 0\ \ 0] \qquad \mathbf{u}_2^T = [0\ \ 1\ \ 0\ \ 0] \qquad \mathbf{u}_3^T = [0\ \ 0\ \ 0\ \ 1]$$

The three trial solutions are obtained using RK4 with step size of 0.1. The values of the function and second derivative at $x = 1$ of the three trial solution are given below

	$i = 1$	$i = 2$	$i = 3$
$u_i(x = 1)$	-0.041667	0.958333	0.125
$u_i^{(1)}(x = 1)$	-0.5	-0.5	0.5

The correct solution is of the form $\mathbf{y} = \alpha_1\mathbf{u}_1 + \alpha_2\mathbf{u}_2 + \alpha_3\mathbf{u}_3$ where the constants have to be determined from given boundary conditions at $x = 1$. Expressing in matrix notation

$$\underbrace{\begin{pmatrix} -0.041667 & 0.958333 & 0.125 \\ -0.5 & -0.5 & 0.5 \\ 1 & 1 & 1 \end{pmatrix}}_{\mathbf{J}} \begin{pmatrix} \alpha_1 \\ \alpha_2 \\ \alpha_3 \end{pmatrix} = \begin{pmatrix} 0 \\ 0 \\ 1 \end{pmatrix}$$

On solving, we get $\alpha_1 = 0.541667$ and $\alpha_2 = -0.041667$ and $\alpha_3 = 0.5$. The procedure has been programmed in MATLAB.

```
ug1 = [0 0 0 0];      % trial 1 initial condition
ug2 = [0 1 0 0];      % trial 2 initial condition
ug3 = [0 0 0 1];      % trial 3 initial condition
   % solve ODE using RK4 for the 3 trials
[X,u1] = ODERK4(0,ug1,0.1,10,@bendingfunc);
[X,u2] = ODERK4(0,ug2,0.1,10,@bendingfunc);
[X,u3] = ODERK4(0,ug3,0.1,10,@bendingfunc);
   % defining coefficient matrix from boundary condition
j(1,1) = u1(11,1); j(1,2) = u2(11,1); j(1,3) = u3(11,1);
j(2,1) = u1(11,3); j(2,2) = u2(11,3); j(2,3) = u3(11,3);
j(3,:) = 1;
alpha =
inv(j)*[0; 0; 1];  % determining α
Y = alpha(1)*u1 + alpha(2)*u2+alpha(3)*u3; % solution
```

where `bendingfunc` is a MATLAB function defining the ODE.

```
function z = bendingfunc(x,y,varargin)
z(1) = y(2);    % y(1) = y; z(1) = y^(1)
z(2) = y(3);    % z(2) = y^(2)
z(3) = y(4);    % z(3) = y^(3)
z(4) = -1;      % z(4) = y^(4)
end
```

The results have been summarized in the following table.

x	y	y_{Exact}	$y^{(1)}$	$y^{(2)}$	$y^{(3)}$
0	0.0000	0.0000	0.0000	0.000	0.5
0.1	-0.0041	-0.0041	-0.0393	0.045	0.4
0.2	-0.0077	-0.0077	-0.0330	0.080	0.3
0.3	-0.0106	-0.0106	-0.0237	0.105	0.2
0.4	-0.0124	-0.0124	-0.0123	0.120	0.1
0.5	-0.0130	-0.0130	0.0000	0.125	0.0
0.6	-0.0124	-0.0124	0.0123	0.120	-0.1
0.7	-0.0106	-0.0106	0.0237	0.105	-0.2
0.8	-0.0077	-0.0077	0.0330	0.080	-0.3
0.9	-0.0041	-0.0041	0.0393	0.045	-0.4
1	0.0000	0.0000	0.0417	0.000	-0.5

As the function y is a polynomial of degree 4 the numerical solution obtained using RK4 is exact.

In general, if we have a nth order ODE with m conditions specified at $x = 0$ and $n - m$ conditions at the other boundary, then the number of trial solutions that have to be solved is $n - m + 1$.

Non-linear ODE

Let us denote the independent parameters $y_0^{(2)} = u_1$ and $y_0^{(3)} = u_2$ and dependent parameters $y_1 = v_1$ and $y_1^{(1)} = v_2$. The relationship between the dependent and independent variables is nonlinear and assumes the form

$$v_1 = g_1(u_1, u_2)$$
$$v_2 = g_2(u_1, u_2)$$

The solution to the differential equation can be determined by solving the above system of non-linear equations using the methods discussed earlier. We start with a guess value of the independent variables and update the values using Newton's method as below

$$\underbrace{\begin{pmatrix} \dfrac{\partial g_1}{\partial u_1} & \dfrac{\partial g_1}{\partial u_2} \\ \dfrac{\partial g_2}{\partial u_1} & \dfrac{\partial g_2}{\partial u_2} \end{pmatrix}}_{\mathbf{J}} \begin{pmatrix} \Delta u_1 \\ \Delta u_2 \end{pmatrix} = \begin{pmatrix} -g_1 \\ -g_2 \end{pmatrix}$$

$$\begin{pmatrix} \Delta u_1 \\ \Delta u_2 \end{pmatrix} = \mathbf{J}^{-1} \begin{pmatrix} -g_1 \\ -g_2 \end{pmatrix}$$

Quasi-Newton methods have to be applied to determine the inverse of the Jacobian matrix. For a nth order differential equation with m boundary conditions specified at $x = 0$ and $n - m$ boundary conditions, the problem reduces to determination of $n - m$ roots. The algorithm can be summarized as follows

- Start with a guess value for independent parameters
- Estimate the Jacobian matrix and the new estimates of independent parameters.
- Iterate until convergence.

Choosing an appropriate initial guess is important for convergence of shooting method. Shooting method becomes cumbersome as the number of dependent variables increases. Also, application of the method to stiff equations is computationally intensive. Shooting methods cannot be extended to higher dimensional problems. We shall look into alternative methods which can accommodate one or more limitations of shooting method.

11.3 Finite difference method

The finite difference method (FDM) is a powerful technique that may be used for the solution of boundary value problems. It is possible to solve both linear and non-linear differential equations by the FDM. It is possible to solve ODEs involving constant as well as variable coefficients. FDM can also be extended to multidimensional problems governed by partial differential equations (See Module IV)

11.3.1 Second order ODE with constant coefficients: a simple example

As an example consider the problem in Example 11.1. The given second order ODE is directly converted to the finite difference form by dividing the domain $0 \le x \le 1$ into segments of width h such that we have n segments. Nodes are placed at $x = 0$, $x = h$, ... $x = 1$. The nodes are numbered as 0, 1, 2, ..., $n-1$, n and we associate $y = y_i$ (i is node number) with each node.

Figure 11.6: *Discretization of the problem domain*

Consider a node i lying between 1 and $n-1$. Using central differences, we have

$$\frac{y_{i-1} - 2y_i + y_{i+1}}{h^2} - 2y_i = 0 \tag{11.14}$$

This may be recast in the form

$$y_{i-1} - 2(1+h^2)y_i + y_{i+1} = 0 \tag{11.15}$$

In case Dirichlet conditions are imposed at both boundaries, we have for example

$$y_0 = 1 \qquad y_n = 0.6 \tag{11.16}$$

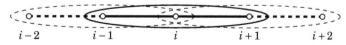

Figure 11.7: *Finite difference and connectivity of nodal function values*

Referring to Figure 11.7 we see that function at node i will appear in the equations for nodes i, $i-1$ and $i+1$. Thus each nodal function value appears in three equations of form 11.14. Thus all the nodal function values are related, end to end, through a set of linear simultaneous equations as given below:

$$
\begin{pmatrix}
1 & 0 & 0 & \cdots & \cdots & 0 \\
1 & -2(1+h^2) & 1 & \cdots & \cdots & 0 \\
0 & 1 & -2(1+h^2) & 1 & \cdots & 0 \\
\cdots & \cdots & \cdots & \cdots & \cdots & \cdots \\
0 & 0 & 0 & 1 & -2(1+h^2) & 1 \\
0 & 0 & 0 & 0 & 0 & 1
\end{pmatrix}
\begin{pmatrix}
y_0 \\ y_1 \\ y_2 \\ \cdots \\ y_{n-1} \\ y_n
\end{pmatrix}
=
\begin{pmatrix}
1 \\ 0 \\ 0 \\ \cdots \\ 0 \\ 0.6
\end{pmatrix}
\qquad (11.17)
$$

Since the coefficient matrix is tri-diagonal we may use TDMA (see Chapter 2) to obtain the solution.

Example 11.6

Based on the formulation given above obtain the solution to the equation in Example 11.1 by finite differences with $h = 0.1$. Comment on the results by comparing the solution with the exact solution. Comment also based on comparison with the solution obtained by the shooting method.

Background :

The above equation represents heat transfer from a pin fin of uniform cross section. Fins are extended surfaces that enhance the convective heat transfer between the object and the environment. Convective heat transfer from any surface is given by

$$
q = h_c A(T_w - T_\infty)
$$

where h_c is the convective heat transfer coefficient, A is the exposed area available for heat transfer and T_w and T_∞ are respectively the temperatures of the surface and environment. In order to increase the heat transfer from the surface $(T_w - T_\infty)$ is constrained for most applications), either h_c or A of the surface can be increased. Increasing the available area of heat transfer is a simple option of improving the heat transfer. Thermal conductivity of fin should be high for good heat transfer enhancement.

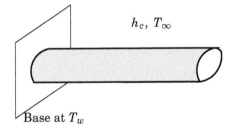

$h_c,\ T_\infty$

Base at T_w

Consider a fin of length L (m), area of cross section A_s (m^2) and perimeter p (m) extending from the base surface maintained at T_w (K). The thermal conductivity of the fin is k (W/m^2K).

The temperature is lumped across the cross section (the temperature variation across the cross section is negligible) and hence is a function of x alone. Then the equation governing heat transfer from the fin to the environment is given by

$$
kA_S \frac{d^2T}{dx^2} - h_c p(T_w - t_\infty) = 0
$$

$$
\text{or} \quad \frac{d^2\theta}{dx^2} - \underbrace{\frac{h_c p}{kA_s}}_{m^2}\theta = 0
$$

where $\theta = \dfrac{T - T_\infty}{T_w - T_\infty}$ is the dimensionless temperature and m is known as the fin parameter. In the absence of fin, heat transfer from the surface would be $q = h_c A_s (T_w - T_\infty)$. The heat transfer in the presence of the fin is given by $q = -k A_s \dfrac{dT}{dx}\Big|_{x=0}$.

Solution :

With $h = 0.1$ there are 11 nodes starting with $i = 0$ and ending with $i = 10$. With $y_0 = 1$ and $y_{10} = 0.6$ we need to essentially solve for the nine nodal values viz. y_1 to y_9. The coefficient matrix is written down as

$$
A = \begin{pmatrix}
-2.02 & 1 & 0 & 0 & 0 & 0 & 0 & 0 & 0 \\
1 & -2.02 & 1 & 0 & 0 & 0 & 0 & 0 & 0 \\
0 & 1 & -2.02 & 1 & 0 & 0 & 0 & 0 & 0 \\
0 & 0 & 1 & -2.02 & 1 & 0 & 0 & 0 & 0 \\
0 & 0 & 0 & 1 & -2.02 & 1 & 0 & 0 & 0 \\
0 & 0 & 0 & 0 & 1 & -2.02 & 1 & 0 & 0 \\
0 & 0 & 0 & 0 & 0 & 1 & -2.02 & 1 & 0 \\
0 & 0 & 0 & 0 & 0 & 0 & 1 & -2.02 & 1 \\
0 & 0 & 0 & 0 & 0 & 0 & 0 & 1 & -2.02
\end{pmatrix}
$$

The solution vector required is of form $\mathbf{y}^T = \begin{pmatrix} y_1 & y_2 & \cdots y_8 & y_9 \end{pmatrix}$ and the right hand vector is $\mathbf{b}^T = \begin{pmatrix} -1 & 0 & 0 & \cdots & 0 & -0.6 \end{pmatrix}$. We use TDMA and tabulate the results as shown below.

Node	a	a'	a''	b	P	Q	y	$y^{(1)}$
0							1.0000	-1.1456
1	2.02	1	0	1	0.4950	0.4950	0.8944	-0.9667
2	2.02	1	1	0	0.6558	0.3246	0.8067	-0.7966
3	2.02	1	1	0	0.7330	0.2380	0.7351	-0.6424
4	2.02	1	1	0	0.7770	0.1849	0.6782	-0.5011
5	2.02	1	1	0	0.8045	0.1487	0.6348	-0.3698
6	2.02	1	1	0	0.8227	0.1224	0.6042	-0.2459
7	2.02	1	1	0	0.8352	0.1022	0.5857	-0.1269
8	2.02	1	1	0	0.8440	0.0863	0.5788	-0.0104
9	2.02	0	1	0.6	0.0000	0.5836	0.5836	0.1058
10							0.6000	0.2225

Note that the entries in column 2 are $-a_i$ and entries in column 5 are $-b_i$. If the derivative is also required at the nodal points we may obtain them by using second order accurate formulae. At either end we may use three point one sided rules while at the interior nodes we may use central differences. These are shown as entries in the last column of the table.

We compare below the finite difference solution with the exact solution as well as that obtained using the shooting method (second order RK).

x	FDM y	SM y	Exact y_E	Error FD ϵ_y	Error SM ϵ_y
0	1.0000	1.0000	1.0000	0.0000	0.0000
0.1	0.8944	0.8941	0.8943	0.0001	-0.0002
0.2	0.8067	0.8062	0.8065	0.0002	-0.0003
0.3	0.7351	0.7345	0.7349	0.0002	-0.0004
0.4	0.6782	0.6775	0.6780	0.0002	-0.0004
0.5	0.6348	0.6342	0.6346	0.0002	-0.0005
0.6	0.6042	0.6036	0.6040	0.0002	-0.0004
0.7	0.5857	0.5851	0.5855	0.0002	-0.0004
0.8	0.5788	0.5784	0.5787	0.0001	-0.0003
0.9	0.5836	0.5833	0.5835	0.0001	-0.0002
1	0.6000	0.6000	0.6000	0.0000	0.0000

FD - Finite Difference, SM - Shooting Method

The errors in the above table are the difference between the respective numerical and exact solutions. Since we have used second order methods in both FD and SM cases the errors are of the same order. The numerical solutions are correct to at least three digits after decimals. The FD solution seems to be slightly superior to the SM solution. Fourth order RK4 method was used in Example 11.1 to solve the same ODE and the errors are found to be of the order of $1e-6$. Similarly, the FD estimate can be improved by using higher order finite difference formulae. Also, the accuracy of the estimate can be improved further by refining the grid.

However as exact solutions are not known for most practical problems, we have to check for the convergence of the solution with grid refinement. The change in solution parameter such as function value or gradient of the function at a specified point with grid refinement can be used as a measure of convergence of numerical solution. In the present problem, we choose the gradient at $x = 0$ as the parameter to check convergence. This is because the above parameter is directly related to the performance of the fin. The following table indicates the first derivative of the function value at $x = 0$ obtained with different step sizes h.

no of nodes	h	$y^{(1)}(x = 0)$	ϵ	ϵ_E
11	0.1	-1.1456		0.00023
21	0.05	-1.1513	0.005	5.7E-05
41	0.025	-1.1529	0.0013	1.4E-05
81	0.0125	-1.1533	0.00035	3.6E-06
161	0.00625	-1.1534	8.8E-05	8.9E-07

In the above table $\epsilon = \left| \dfrac{y_h^{(1)} - y_{h/2}^{(1)}}{y_{h/2}^{(1)}} \right|$ and $\epsilon_E = max|y - y_E|$. The first derivative has been calculated using three point one sided formula.

$$y^{(1)} = \frac{-3y_0 + 4y_1 - y_2}{2h}$$

We see that a grid of 161 is sufficient as the gradient at $x = 0$ does not change significantly.

11.3.2 Second order ODE with constant coefficients: a variant

Now consider the same example as that considered in the previous section but with Neumann boundary condition at $x = 1$, $y^{(1)}(x = 1) = y_1^{(1)}$. The FDM formulation at interior nodes do not change. The boundary at which the Neumann condition is specified needs special attention. The obvious approach of approximating a first derivative is to use forward (backward) difference formula

$$y^{(1)} \approx \frac{y_n - y_{n-1}}{h} = y_1^{(1)}$$

However, as the above is first order accurate, whereas the function values at the interior nodes are approximated using second order accurate formulae it is desirable to use second order accurate formula at the boundary as well.

One approach is to use a second order accurate one sided three point formula for the first derivative.

$$y^{(1)} \approx \frac{y_{n-2} - 4y_{n-1} + 3y_n}{2h} = y_1^{(1)}$$

Representing the set of linear equations in matrix form

$$
\begin{pmatrix}
1 & 0 & 0 & \cdots & \cdots & 0 \\
1 & -2(1+h^2) & 1 & \cdots & \cdots & 0 \\
0 & 1 & -2(1+h^2) & 1 & \cdots & 0 \\
\cdots & \cdots & \cdots & \cdots & \cdots & \cdots \\
0 & 0 & \cdots & 1 & -2(1+h^2) & 1 \\
0 & 0 & \cdots & 1 & -4 & 3
\end{pmatrix}
\begin{pmatrix}
y_0 \\ y_1 \\ y_2 \\ \cdots \\ y_{n-2} \\ y_{n-1} \\ y_n
\end{pmatrix}
=
\begin{pmatrix}
1 \\ 0 \\ 0 \\ \cdots \\ 0 \\ 2hy_1^{(1)}
\end{pmatrix}
\tag{11.18}
$$

The above equations can be solved using the methods discussed earlier. However, one cannot use TDMA directly. To apply TDMA, y_n is eliminated from last two equations to reduce the matrix to a tridiagonal form.

$$
\begin{pmatrix}
1 & 0 & 0 & \cdots & \cdots & 0 \\
1 & -2(1+h^2) & 1 & \cdots & \cdots & 0 \\
\cdots & \cdots & \cdots & \cdots & \cdots & \cdots \\
0 & 0 & \cdots & 1 & -2(1+h^2) & 1 \\
0 & 0 & \cdots & 0 & 2 & -2-6h^2
\end{pmatrix}
\begin{pmatrix}
y_0 \\ y_1 \\ y_2 \\ \cdots \\ y_{n-3} \\ y_{n-2} \\ y_{n-1}
\end{pmatrix}
=
\begin{pmatrix}
1 \\ 0 \\ \cdots \\ 0 \\ -2hy_1^{(1)}
\end{pmatrix}
\tag{11.19}
$$

Alternatively, we may use a ghost point to the right of $x = 1$ or $i = n$ and at a distance of h and use it to evaluate the first derivative at $x = 1$. This means that

$$y^{(1)} \approx \frac{y_{n+1} - y_{n-1}}{2h} = y_1^{(1)}$$

$$y_{n+1} = y_{n-1} + 2hy_1^{(1)}$$

which is nothing but the central difference approximation to the first derivative at $x = 1$.

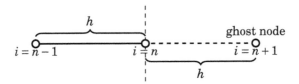

Thus we will be able to satisfy the Neumann condition at $x = 1$. Now we treat the node n as an interior node and write

$$y_{n-1} - 2(1+h^2)y_n + y_{n+1} = 2y_{n-1} - 2(1+h^2)y_n + 2hy_1^{(1)} = 0$$

$$\text{or} \quad y_{n-1} - (1+h^2)y_n = -hy_1^{(1)} \tag{11.20}$$

Equation 11.17 will then be recast as

$$\begin{pmatrix} 1 & 0 & 0 & \cdots & \cdots & 0 \\ 1 & -2(1+h^2) & 1 & \cdots & \cdots & 0 \\ 0 & 1 & -2(1+h^2) & 1 & \cdots & 0 \\ \cdots & \cdots & \cdots & \cdots & \cdots & \cdots \\ 0 & 0 & 0 & 1 & -2(1+h^2) & 1 \\ 0 & 0 & 0 & 0 & 1 & -(1+h^2) \end{pmatrix} \begin{pmatrix} y_0 \\ y_1 \\ y_2 \\ \cdots \\ y_{n-1} \\ y_n \end{pmatrix} = \begin{pmatrix} 1 \\ 0 \\ 0 \\ \cdots \\ 0 \\ -hy_1^{(1)} \end{pmatrix} \tag{11.21}$$

This set of equations may again be solved by TDMA.

The numerical solution produced from the two approaches discussed above would be of identical accuracy but not the same. Both the approaches are second order accurate and the error is proportional to h^2.

Example 11.7

Solve the boundary value problem $y^{(2)} = 2y$ subject to the conditions $y(x = 0) = 1$ and $y^{(1)}(x = 1) = 0$ by FDM. Comment on the results by comparing the solution with the exact solution.

Solution :

With $h = 0.1$ there are 11 nodes starting with 0 and ending with 10. With $y_0 = 1$ and $y_{10}^{(1)} = 0$ we need to essentially solve for the ten nodal values viz. y_1 to y_{10}. The coefficient matrix is written down based on the discussion just preceding this example as

$$\mathbf{A} = \begin{pmatrix} -2.02 & 1 & 0 & 0 & 0 & 0 & 0 & 0 & 0 & 0 \\ 1 & -2.02 & 1 & 0 & 0 & 0 & 0 & 0 & 0 & 0 \\ 0 & 1 & -2.02 & 1 & 0 & 0 & 0 & 0 & 0 & 0 \\ 0 & 0 & 1 & -2.02 & 1 & 0 & 0 & 0 & 0 & 0 \\ 0 & 0 & 0 & 1 & -2.02 & 1 & 0 & 0 & 0 & 0 \\ 0 & 0 & 0 & 0 & 1 & -2.02 & 1 & 0 & 0 & 0 \\ 0 & 0 & 0 & 0 & 0 & 1 & -2.02 & 1 & 0 & 0 \\ 0 & 0 & 0 & 0 & 0 & 0 & 1 & -2.02 & 1 & 0 \\ 0 & 0 & 0 & 0 & 0 & 0 & 0 & 1 & -2.02 & 1 \\ 0 & 0 & 0 & 0 & 0 & 0 & 0 & 0 & 1 & -1.01 \end{pmatrix}$$

The solution vector required is of form $\mathbf{y}^T = \begin{pmatrix} y_1 & y_2 & \cdots y_9 & y_{10} \end{pmatrix}$ and the right hand vector is $\mathbf{b}^T = \begin{pmatrix} 1 & 0 & 0 & \cdots & 0 & 0 \end{pmatrix}$. We use the TDMA and tabulate the results as shown in the first table below.

The exact solution to the problem may easily be shown to be $y_E = \dfrac{\cosh\left[\sqrt{2}(1-x)\right]}{\cosh\sqrt{2}}$.
Comparison of the FDM solution with the exact solution is shown in the second table below.

Node	a	a'	a''	b	P	Q	y
0							1.0000
1	2.02	1	0	1	0.4950	0.4950	0.8841
2	2.02	1	1	0	0.6558	0.3246	0.7859
3	2.02	1	1	0	0.7330	0.2380	0.7033
4	2.02	1	1	0	0.7770	0.1849	0.6349
5	2.02	1	1	0	0.8045	0.1487	0.5791
6	2.02	1	1	0	0.8227	0.1224	0.5350
7	2.02	1	1	0	0.8352	0.1022	0.5015
8	2.02	1	1	0	0.8440	0.0863	0.4781
9	2.02	1	1	0	0.8504	0.0734	0.4642
10	1.01	0	1	0	0.0000	0.4596	0.4596

x	y	y_E	ϵ_y	$y^{(1)}$	$y_E^{(1)}$	$\epsilon_{y^{(1)}}$
0	1.0000	1.0000	0.0000	-1.2476	-1.2564	0.0088
0.1	0.8841	0.8840	0.0001	-1.0707	-1.0683	-0.0025
0.2	0.7859	0.7856	0.0002	-0.9038	-0.9016	-0.0021
0.3	0.7033	0.7030	0.0003	-0.7548	-0.7530	-0.0018
0.4	0.6349	0.6345	0.0004	-0.6210	-0.6194	-0.0016
0.5	0.5791	0.5787	0.0004	-0.4996	-0.4983	-0.0013
0.6	0.5350	0.5345	0.0004	-0.3882	-0.3872	-0.0010
0.7	0.5015	0.5010	0.0005	-0.2846	-0.2838	-0.0008
0.8	0.4781	0.4776	0.0005	-0.1866	-0.1861	-0.0005
0.9	0.4642	0.4637	0.0005	-0.0924	-0.0921	-0.0002
1	0.4596	0.4591	0.0005	0.0000	0.0000	0.0000

The derivatives shown in the last column of the above table have been obtained using second order accurate finite difference formulae using the computed nodal values of y. The procedure has also been programmed in MATLAB.

```
n = 11;                  % no. of nodes
x = [0:1/(n-1):1]';      % nodes
h = 1/(n-1);             % step size
a = zeros(n,1);          % TDMA coefficient node i
a1 = zeros(n,1);             % TDMA coefficient node i+1
a2 = zeros(n,1);             % TDMA coefficient node i-1
b = zeros(n,1);          % TDMA constant
a(1) = 1;                % first node
b(1) = 1;                % Dirichlet condition
for i=2:n-1              % coefficients for
    a(i) = 2+2*h^2;      % interior nodes
    a1(i) = 1;
    a2(i) = 1;
end
a(n) =   1+h^2;          % last node
```

```
a2(n) = 1;                  % Neumann  condition
b(n) = 0;
y = tdma(a2,a,a1,b);        % solve  using  TDMA
yr = cosh(sqrt(2)*(1-x))/cosh(sqrt(2)); % exact  solution
er = abs(yr-y);             % error
```

The reader can check for the convergence of the solution with grid refinement.

11.3.3   Application of FDM using non-uniform grids

Just like adaptive methods used to solve an IVP, non-uniform grids may be useful for BVP. A fine grid may be used in regions of steep function variation and a coarse grid may be used in regions of gentle function variations. Let us consider the second order differential equation considered in Example 11.7. Dirichlet condition $y = 1$ is specified at $x = 0$ and Neumann condition $y^{(1)} = 0$ at the other boundary. The second derivative at the interior nodes are approximated by second derivatives of Lagrange polynomials.

$$
\begin{aligned}
y_i^{(2)} &= \frac{2}{x_{i+1}-x_{i-1}}\left(\frac{y_{i+1}-y_i}{x_{i+1}-x_i} - \frac{y_i-y_{i-1}}{x_i-x_{i-1}}\right) \\
&= \frac{2}{x_{i+1}-x_{i-1}}\left[\frac{y_{i+1}}{x_{i+1}-x_i} - y_i\left(\frac{1}{x_i-x_{i-1}} + \frac{1}{x_{i+1}-x_i}\right) + \frac{y_{i-1}}{x_i-x_{i-1}}\right]
\end{aligned}
$$

Therefore the algebraic equation for node i is given by

$$
a_i'' y_{i-1} - a y_i + a_i' y_{i+1} = 0
$$

where $a_i' = \dfrac{2}{(x_{i+1}-x_{i-1})(x_{i+1}-x_i)}$, $a_i'' = \dfrac{2}{(x_{i+1}-x_{i-1})(x_i-x_{i-1})}$ and $a_i = a_i' + a_i'' + 2$.

Neumann boundary condition at $x = 1$ can be achieved by either of the two approaches discussed earlier. Let us apply the ghost node approach where a ghost node is assumed at a distance $x_n - x_{n-1}$ from x_n.

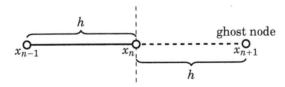

On simplifications we get

$$
a_n'' y_{n-1} - a_n y_n = 0
$$

where $a''_n = \dfrac{2}{(x_n - x_{n-1})^2}$ and $a = (a''_n + 2)$. The system of linear equation is represented in matrix notation of the form $\mathbf{A}\mathbf{y} = \mathbf{b}$ where

$$
\mathbf{A} = \begin{pmatrix}
1 & 0 & 0 & \cdots & \cdots & 0 \\
a''_1 & -(a''_1 + a'_1 + 2) & a'_1 & \cdots & \cdots & 0 \\
0 & a''_2 & -(a''_2 + a'_2 + 2) & a'_2 & \cdots & 0 \\
\cdots & \cdots & \cdots & \cdots & \cdots & \cdots \\
0 & 0 & 0 & a''_{n-1} & -(a''_{n-1} + a'_{n-1} + 2) & a'_{n-1} \\
0 & 0 & 0 & 0 & a''_n & -(a''_n + 2)
\end{pmatrix}
$$

(11.22)

$$
\mathbf{y} = \begin{pmatrix} y_0 \\ y_1 \\ y_2 \\ \cdots \\ y_{n-1} \\ y_n \end{pmatrix} \quad \text{and} \quad \mathbf{b} = \begin{pmatrix} 1 \\ 0 \\ 0 \\ \cdots \\ 0 \\ -h y_1^{(1)} \end{pmatrix}
$$

A MATLAB program is presented below to solve the differential equation of Example 11.7 using nonuniform (Chebyshev) nodes. To treat the second boundary condition ghost node approach has been used in the present case.

```
n = 11;                    % no of nodes
x = zeros(n,1);            % nodes
x(1) = 0;
x(n) = 1;
for i=2:n-1                % Chebyshev nodes
    x(i) = 0.5-0.5*cos((2*(i-1)-1)*pi/(2*(n-2)));
end
a = zeros(n,1);            % TDMA coefficients a
a1 = zeros(n,1);           % TDMA coefficients a'
a2 = zeros(n,1);           % TDMA coefficients a''
b = zeros(n,1);            % TDMA coefficients b
a(1) = 1;                  % first node
b(1) = 1;                  % Dirichlet condition
for i=2:n-1                % internal nodes
    a1(i) = 2/((x(i+1)-x(i))*(x(i+1)-x(i-1)));
    a2(i) = 2/((x(i)-x(i-1))*(x(i+1)-x(i-1)));
    a(i) = a1(i) + a2(i) + 2;
end
a2(n) = 2/(x(n)-x(n-1))^2; % last node
a(n) = c(n)+2;             % Neumann condition
y = tdma(a2,a,a1,b);       % solve using TDMA
```

11.3.4 Solution of non-linear case by FDM

It is possible to apply the FDM to the case of a non-linear second order ODE also. However the solution method is perforce an iterative one. Essentially it consists in the use of a method such as Newton Raphson to linearize the equations, at every step.

For example we consider the second order ODE of Example 11.3. Assume that a trial solution is available as y^k. The fourth degree term may be linearized such that

$$
y^4 = (y^{k+1})^4 = [y^k + (y^{k+1} - y^k)]^4 \approx (y^k)^4 + 4(y^k)^3(y^{k+1} - y^k) = 4(y^k)^3 y^{k+1} - 3(y^k)^4 \qquad (11.23)
$$

y^{k+1} is the new value of the function we seek at the end of the iteration. The finite difference analog of the governing differential equation, applied to an interior node, then takes the form

$$y_{i+1}^{k+1} - 2y_i^{k+1} + y_{i-1}^{k+1} - h^2[4(y_i^k)^3 y_i^{k+1} - 3(y_i^k)^4 - 0.5] = 0 \qquad (11.24)$$

This may be rewritten as.

$$y_n^{k+1}\underbrace{[2 + 4h^2(y_i^k)^3]}_{a_i^k} = y_{i-1}^{k+1} + y_{i+1}^{k+1} + \underbrace{h^2[3(y^k)^4 + 0.5]}_{b_i^k} \qquad (11.25)$$

Reader may verify that the same linearization scheme will result on application of the Newton Raphson method. Observe that coefficients a_i and b_i change from iteration to iteration.

> In general a nonlinear term S can be linearized as
>
> $$S^{k+1} = S^k + \left.\frac{dS}{dy}\right|_k (y_{k+1} - y_k)$$

We thus end up with a set of linear equations that may be solved by TDMA. Iterations stop when the change in any nodal value is less than or equal to a preassigned small number. We present the FDM solution of this nonlinear case in the next example.

Example 11.8

Solve the boundary value problem $y^{(2)} = y^4 - 0.5$ subject to the conditions $y(x = 0) = 1$ and $y^{(1)}(x = 1) = 0$ by FDM.

Solution :

Since the solution is an iterative one based on the linearized set of nodal equations, we start the solution by assuming trial values for all the nodal values of y. In the present case we have taken the nodal values to follow the relation $y(x) = 1 - 0.2x$ for $0 \le x \le 1$. We may use Equation 11.25 to calculate the coefficients in the matrix and tabulate them as shown below.

y^0	a	a'	a''	b
1.00				
0.98	2.0376	1	0	1.0327
0.96	2.0354	1	1	0.0305
0.94	2.0332	1	1	0.0284
0.92	2.0311	1	1	0.0265
0.90	2.0292	1	1	0.0247
0.88	2.0273	1	1	0.0230
0.86	2.0254	1	1	0.0214
0.84	2.0237	1	1	0.0199
0.82	2.0221	1	1	0.0186
0.80	2.0205	0	2	0.0173

The entries in the last row are obtained by imposing the zero slope condition at $x = 1$ using essentially the arguments given earlier. Entries shown as a and b only will change from iteration to iteration. Using the coefficients in the above table and TDMA we get the following update for the nodal values of y.

	P	Q	y^1	Change
			1	0
	0.4908	0.5068	0.9775	-2.54E-03
	0.6474	0.3478	0.9590	-9.62E-04
	0.7216	0.2715	0.9441	4.08E-03
	0.7636	0.2276	0.9321	1.21E-02
	0.7902	0.1993	0.9226	2.26E-02
	0.8084	0.1797	0.9153	3.53E-02
	0.8216	0.1652	0.9100	5.00E-02
	0.8319	0.1540	0.9065	6.65E-02
	0.8402	0.1450	0.9045	8.45E-02
	0.0000	0.9039	0.9039	1.04E-01

This completes one iteration. We use y^1 as the starting set to improve the solution. Iterations stop when the values have converged. In the present case convergence is obtained after three iterations as indicated by the following table.

x	Initial Set	Iter 1	Iter 2	Iter 3
0	1.000000	1.000000	1.000000	1.000000
0.1	0.980000	0.977455	0.977120	0.977119
0.2	0.960000	0.959038	0.958356	0.958354
0.3	0.940000	0.944081	0.943028	0.943025
0.4	0.920000	0.932066	0.930608	0.930603
0.5	0.900000	0.922591	0.920688	0.920682
0.6	0.880000	0.915336	0.912953	0.912946
0.7	0.860000	0.910042	0.907165	0.907157
0.8	0.840000	0.906490	0.903148	0.903140
0.9	0.820000	0.904494	0.900785	0.900776
1	0.800000	0.903882	0.900005	0.899995

The procedure has been programmed in MATLAB and given below

```
n = 11;                   % node number
x = [0:1/(n-1):1]';       % nodes
h = 1/(n-1);              % step size
a = zeros(n,1);           % TDMA coefficient a
a1 = zeros(n,1);          % TDMA coefficient a'
a2 = zeros(n,1);          % TDMA coefficient a''
b = zeros(n,1);           % TDMA constant
yo = 1-0.2*x;             % guess value first set
a(1) = 1;                 % first node
b(1) = 1;                 % Dirichlet condition
a1(2:n-1) = 1;            % coefficients for
a2(2:n-1) = 1;            % internal nodes
a2(n) = 2;                % coefficient last node
tol = 1e-4;               % tolerance for convergence
res = 0;                  % initialize residual
count = 0;                % iteration count
while ( res > tol || count == 0) % loop for convergence
```

```
a(2:n)  =  2+4*yo(2:n).^3*dx^2; % update a and d
b(2:n)  =  3*yo(2:n).^4*dx^2+0.5*dx^2;
y  =  tdma(a2,a,a1,b);            % solve y
res  =   sum(abs(y-yo)); % calculate residual
yo  =  y;                    % update y
count  =  count + 1;  % increment iteration count
end
```

11.3.5 Application of FDM to second order ODE with variable coefficients

We consider next the application of finite difference method to the solution of a boundary value problem involving an ODE with variable coefficients. Consider the second order ODE given by Equation 11.5. Consider it in the specific linear form[3] given by

$$\frac{d^2y}{dx^2} + A(x)\frac{dy}{dx} + B(x)y = f(x) \tag{11.26}$$

The finite difference method uses central differences to rewrite the above in the form

$$\frac{y_{i-1} - 2y_i + y_{i+1}}{h^2} + A(x_i)\frac{y_{i+1} - y_{i-1}}{2h} + B(x_i)y_i = f(x_i)$$

where $h = x_{i+1} - x_i$, $1 \leq i \leq n-1$ corresponding to an internal node. Node 0 represents the boundary at $x = 0$ and node n represents the boundary at $x = L$. The above equation may be recast in the form of an algebraic equation

$$y_{i-1}\left(1 - \frac{A(x_i)h}{2}\right) - y_i(2 - B(x_i)h^2) + y_{i+1}\left(1 + \frac{A(x_i)h}{2}\right) = f(x_i)h^2 \tag{11.27}$$

At the boundary nodes we simply have the specified boundary values. For example, if both boundaries have first kind of boundary conditions, then we simply have

$$y(0) = y_0 \text{ and } y(L) = y_n = y_L \tag{11.28}$$

Thus the nodal equations take the form

$$
\begin{aligned}
y_0 &= y_0 \\
y_0\left(1 - \frac{A(x_1)h}{2}\right) - y_1(2 - B(x_1)h^2) + y_2\left(1 + \frac{A(x_1)h}{2}\right) &= f(x_1)h^2 \\
y_1\left(1 - \frac{A(x_2)h}{2}\right) - y_2(2 - B(x_2)h^2) + y_3\left(1 + \frac{A(x_2)h}{2}\right) &= f(x_2)h^2 \\
\cdots &= \cdots \\
y_{i-1}\left(1 - \frac{A(x_i)h}{2}\right) - y_i(2 - B(x_i)h^2) + y_{i+1}\left(1 + \frac{A(x_i)h}{2}\right) &= f(x_i)h^2 \\
\cdots &= \cdots \\
y_{n-3}\left(1 - \frac{A(x_{n-2})h}{2}\right) - y_{n-2}(2 - B(x_{n-2})h^2) + y_n\left(1 + \frac{A(x_{n-2})h}{2}\right) &= f(x_{n-2})h^2
\end{aligned}
$$

---

[3]There is no difficulty if the equation is non-linear

$$y_{n-2}\left(1 - \frac{A(x_{n-1})h}{2}\right) - y_{n-1}(2 - B(x_{n-1})h^2) + y_n\left(1 + \frac{A(x_{n-1})h}{2}\right) = f(x_{n-1})h^2$$

$$y_n = y_L \qquad (11.29)$$

Letting $a_i = 2 - B(x_i)h^2$, $a_i' = 1 + \frac{A(x_i)h}{2}$, $a_i'' = 1 - \frac{A(x_i)h}{2}$ and $b_i = -f(x_i)h^2$ (in TDMA form), the above equations are rewritten as

$$\mathbf{Ay = b} \qquad (11.30)$$

where

$$[A] = \begin{bmatrix}
1 & 0 & 0 & \cdots & \cdots & \cdots & \cdots & 0 \\
a_1'' & -a_1 & a_1' & 0 & \cdots & \cdots & \cdots & 0 \\
0 & a_2'' & -a_2 & a_2' & 0 & \cdots & \cdots & 0 \\
\cdots & \cdots & \ddots & \ddots & \ddots & \cdots & \cdots & \cdots \\
0 & 0 & 0 & a_i'' & -a_i & a_i' & 0 & 0 \\
\cdots & \cdots & \cdots & \cdots & \ddots & \ddots & \ddots & \cdots \\
0 & 0 & 0 & 0 & \cdots & a_{n-1}'' & -a_{n-1} & a_{n-1}' \\
0 & 0 & 0 & 0 & 0 & 0 & 0 & 1
\end{bmatrix}$$

$$\{y\} = \begin{Bmatrix} y_0 \\ y_1 \\ y_2 \\ \cdots \\ y_i \\ \cdots \\ y_{n-1} \\ y_n \end{Bmatrix} \quad \text{and} \quad \{b\} = \begin{Bmatrix} y_0 \\ f(x_1)h^2 \\ f(x_2)h^2 \\ \cdots \\ f(x_i)h^2 \\ \cdots \\ f(x_{n-1})h^2 \\ y_n \end{Bmatrix}$$

Again the set of equations may be solved by TDMA. When Neumann condition is specified at the second boundary we require a different method as shown in the following example.

Example 11.9

Solve the second order ODE with variable coefficients $y^{(2)} + \frac{1}{x}y^{(1)} - y = 0$ subject to the boundary conditions $y(x = 1) = 1$ and $y^{(1)}(x = 2) = 0$ by the finite difference method.

Solution :

The formulation follows the method described above. We choose a step size of $h = 0.1$ to discretize the equations. Note that $i = 0$ corresponds to $x = 1$ and $i = 10$ corresponds to $x = 2$. At the second boundary i.e. at $x = 2$ Neumann condition has been specified. We can use the ghost node approach for the present problem. Instead we will approximate the first derivative using three point one sided rule. Specifically this requires that $3y_{10} - 4y_9 + y_8 = 0$. In the formulation we would like to preserve the tri-diagonal form and hence some preconditioning is required. We know that $a_9 y_9 = a_9' y_{10} + a_9'' y_8$ as applicable to node 9. From the latter equation we have $y_8 = \frac{a_9}{a_9''}y_9 - \frac{a_9'}{a_9''}y_{10}$. Introducing this in the former equation we get $3y_{10} - 4y_9 + \frac{a_9}{a_9''}y_9 - \frac{a_9'}{a_9''}y_{10} = 0$ or, on simplification $\left(3 - \frac{a_9'}{a_9''}\right)y_{10} = \left(4 - \frac{a_9}{a_9''}\right)y_9$.

We note that the $a - a''$ are given by the following:

$$a_i = 2 + h^2, \quad a_i' = 1 + \frac{h}{2x_i}, \quad a_i'' = 1 - \frac{h}{2x_i}$$

Excepting b_1 all other b's are zero. b_1 alone is given by $1 - \dfrac{h}{2x_1}$. Using these, the required input for performing the TDMA is tabulated as given below:

x	a	a'	a''	b	x	a	a'	a''	b
1					1.6	2.0100	1.0313	0.9688	0
1.1	2.0100	1.0455	0.0000	0.9545	1.7	2.0100	1.0294	0.9706	0
1.2	2.0100	1.0417	0.9583	0	1.8	2.0100	1.0278	0.9722	0
1.3	2.0100	1.0385	0.9615	0	1.9	2.0100	1.0263	0.9737	0
1.4	2.0100	1.0357	0.9643	0	2	1.9459	0.0000	1.9357	0
1.5	2.0100	1.0333	0.9667	0					

Results are rounded to 4 digits after decimals. However, all computations have been made with available machine precision. On applying TDMA we get the solution given in the following table.

x	P	Q	y	$y^{(1)}$	x	P	Q	y	$y^{(1)}$
1			1.0000	-1.0264	1.6	0.8671	0.1056	0.6417	-0.2753
1.1	0.5201	0.4749	0.9058	-0.8579	1.7	0.8811	0.0877	0.6182	-0.1979
1.2	0.6891	0.3011	0.8284	-0.7029	1.8	0.8911	0.0739	0.6021	-0.1275
1.3	0.7707	0.2149	0.7652	-0.5719	1.9	0.8984	0.0630	0.5927	-0.0626
1.4	0.8176	0.1636	0.7140	-0.4595	2	0.0000	0.5896	0.5896	0.0000
1.5	0.8472	0.1296	0.6733	-0.3617					

The derivatives shown in the table have of course been obtained by using the appropriate second order accurate finite difference formulae.

FDM can be applied to higher order BVPs. Also, higher order finite difference formulae can be used. Now we shall look at the Collocation method. A polynomial collocation method is a generalized form of FDM.

11.4   Collocation method

The domain is discretized into a number of points also known as collocation points. The philosophy of collocation method is to choose parameters and basis functions such that the residual is zero at the collocation points. If the basis functions are sinusoidal functions, we refer to the method as Fourier collocation method. If the basis functions are polynomial, we refer to the method as polynomial collocation method. Again the polynomials could be orthogonal polynomials such as Chebyshev, Legendre, Jacobi etc.

We shall consider polynomial collocation method where the function y is assumed to be a Lagrange polynomial passing through all the collocation nodes given by

$$y = y_0 l_0 + y_1 l_1 + y_2 l_2 + \cdots + y_n l_n \qquad (11.31)$$

where l_i are the Lagrange polynomial weights (basis function) and y_i are the unknown function values that have to be determined. The derivatives in the differential equation at each of the nodes is obtained by differentiating the Lagrange polynomial weights.

Reader should to refer to the concepts on numerical differentiation of Lagrange weights presented in Section 8.3.2.

Example 11.10

Solve the following second order ODE using Collocation method.

$$y^{(2)} - 2y = 0 \qquad (11.32)$$

The boundary conditions are $y(x = 0) = 1$ and $y^{(1)}(x = 1) = 0$. Study the grid sensitivity of Collocation method

Solution :

Discretization: The domain is divided into n (say 4) equal segments i.e the number of points would be equal to $n + 1$ as indicated in the following figure.

Collocation method: y is approximated as a Lagrange polynomial passing through all the nodes represented as

$$y = y_0 l_0 + y_1 l_1 + \cdots + y_{n-1} l_{n-1} + y_n l_n$$

where l_i are the Lagrange weights and y_i are the function values at the nodes. y_i are unknown and have to be determined by satisfying the ODE and boundary conditions. At the first node, $y_0 = 1$. At interior nodes i.e. $i = 1, 2$ to $n - 1$, the ODE has to be satisfied and hence we have

$$y_0 \left[\frac{d^2 l_0}{dx^2} \bigg|_{x_i} - 2l_0(x_i) \right] + \cdots y_i \left[\frac{d^2 l_i}{dx^2} \bigg|_{x_i} - 2l_i(x_i) \right] + \cdots + y_n \left[\frac{d^2 l_n}{dx^2} \bigg|_{x_i} - 2l_n(x_i) \right] = 0$$

$$\text{or} \quad y_0 \frac{d^2 l_0}{dx^2} \bigg|_{x_i} + \cdots + y_i \left[\frac{d^2 l_i}{dx^2} \bigg|_{x_i} - 2 \right] + \cdots + y_n \frac{d^2 l_n}{dx^2} \bigg|_{x_i} = 0$$

At $i = n$, Neumann boundary condition has to be satisfied as given below

$$\frac{dy}{dx} \bigg|_{x_n} = \frac{dl_0}{dx} \bigg|_{x_n} y_0 + \frac{dl_1}{dx} \bigg|_{x_n} y_1 + \cdots + \frac{dl_{n-1}}{dx} \bigg|_{x_n} y_{n-1} + \frac{dl_n}{dx} \bigg|_{x_n} y_n = 0$$

Hence we have a system of linear equations of the form $\mathbf{Ay} = \mathbf{b}$ where

$$\mathbf{A} = \begin{pmatrix} 1 & \cdots & 0 & \cdots & 0 \\ \cdots & \cdots & \cdots & \cdots & \cdots \\ \frac{d^2 l_0}{dx^2} \bigg|_{x_i} & \cdots & \left[\frac{d^2 l_i}{dx^2} \bigg|_{x_i} - 2 \right] & \cdots & \frac{d^2 l_n}{dx^2} \bigg|_{x_i} \\ \cdots & \cdots & \cdots & \cdots & \cdots \\ \frac{dl_0}{dx} \bigg|_{x_n} & \cdots & \frac{dl_i}{dx} \bigg|_{x_n} & \cdots & \frac{dl_n}{dx} \bigg|_{x_n} \end{pmatrix}$$

$$\mathbf{y}^T = \begin{pmatrix} y_0 & y_1 & \cdots & y_{n-1} & y_n \end{pmatrix} \quad \text{and} \quad \mathbf{b}^T = \begin{pmatrix} 1 & \cdots & 0 & \cdots & 0 \end{pmatrix}$$

We can solve these equations using any of the methods discussed earlier. A MATLAB program has been written to apply collocation method to the present problem.

```
n = 5;                      % no. of nodes
x = [0:1/(n-1):1]';         % nodes
A = zeros(n);               % initialize coefficient matrix
B = zeros(n,1);             % initialize force vector
A(1,1) = 1;                 % first node
B(1) = 1;                   %
p = lagrangeweightfunction(x);  % generate Li
p1 = polyderivative(p,1);       % first derivative
p2 = polyderivative(p,2);       % second derivative
w  = lagrangeweight(p,x(2:n-1));    % evaluate L at nodes
w2 = lagrangeweight(p2,x(2:n-1));   % evaluate d²L/dx² at nodes
A(2:n-1,:) = w2-2*w;        % coefficient for interior nodes
w1 = lagrangeweight(p1,x(n));   % evaluate dL/dx at last
                            %node
A(n,:) = w1;                % coefficient for last node
y = A\B;                    % solve for y
yr = cosh(sqrt(2)*(1-x))/cosh(sqrt(2)); % exact solution
er = max(abs(yr-y));
```

The results for $n = 4$ are summarized in the following table

x	y	y_E	ε_y
0	1.0000	1.0000	0
0.25	0.7424	0.7425	9.3E-05
0.5	0.5783	0.5787	4.1E-04
0.75	0.4873	0.4881	7.7E-04
1	0.4580	0.4591	1.1E-03

Figure 11.8 shows the results of the grid sensitivity study that has been conducted. Collocation method converges much faster than the finite difference method. However, roundoff error becomes important around $n = 10$.

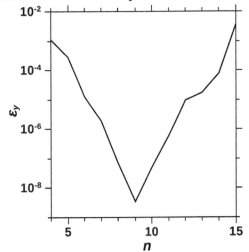

Figure 11.8: *Grid sensitivity study for Example 11.10*

Collocation method can be easily extended to a nonlinear ODE. All nonlinear terms can be linearized using Newton Raphson method. We will consider an example of nonlinear ODE now.

Example 11.11

Solve BVP from Example 11.8 $y^{(2)} = y^4 - 0.5$ *subject to the conditions* $y(x = 0) = 1$ *and* $y^{(1)}(x = 1) = 0$ *using collocation method.*

Solution :

The above equation has a non-linear term y^4 and has to be linearized. Let us denote the nonlinear term as S. The domain is discretized into n equal parts. The function value for node 1 has been specified. The function y in the domain is defined as Lagrange polynomial passing through all the nodes.

$$y = y_0 l_0 + y_1 l_1 + y_2 l_2 + \cdots + y_n l_n$$

Then the equation for internal node i is given by

$$y_0 \left.\frac{d^2 l_0}{dx^2}\right|_{x_i} + \cdots + y_i \left[\left.\frac{d^2 l_i}{dx^2}\right|_{x_i} - S_i\right] + \cdots + y_n \left.\frac{d^2 l_n}{dx^2}\right|_{x_i} = -0.5$$

We apply Newton Raphson method to linearize S

$$S_{new} = S_{old} + 4y_{old}^3(y_{new} - y_{old}) = 4y_{old}^3 y_{new} - 3S_{old}$$

Then the nodal equations are given by

$$y_0 \left.\frac{d^2 l_0}{dx^2}\right|_{x_i} + \cdots + y_i \left[\left.\frac{d^2 l_i}{dx^2}\right|_{x_i} - 4y_{i,old}^3 y_{i,new}\right] + \cdots + y_n \left.\frac{d^2 l_n}{dx^2}\right|_{x_i} = -0.5 - 3S_{i,old}$$

At $i = n$, Neumann boundary condition is satisfied by differentiating the Lagrange polynomial.

$$\left.\frac{dy}{dx}\right|_{x_n} = \left.\frac{dl_0}{dx}\right|_{x_n} y_0 + \left.\frac{dl_1}{dx}\right|_{x_n} y_1 + \left.\frac{dl_2}{dx}\right|_{x_n} y_2 + \cdots + \left.\frac{dl_n}{dx}\right|_{x_n} y_n = 0$$

The system of linear equations is solved iteratively until convergence. A MATLAB program has been written to solve the BVP using collocation method.

```
n = 6;                              % no of nodes
x = [0:1/(n-1):1]';                 % nodes
A = zeros(n);                       % initialize coefficient matrix
B = zeros(n,1);                     % initialize force vector
A(1,1) = 1;                         % first node
B(1) = 1;                           % Dirichlet condition
p = lagrangeweightfunction(x);      % generate L_i
p1 = polyderivative(p,1);           % first derivative
p2 = polyderivative(p,2);           % second derivative
w2 = lagrangeweight(p2,x(2:n-1));   % evaluate d^2L/dx^2 at nodes
w  = lagrangeweight(p,x(2:n-1));    % evaluate L at nodes
w1 = lagrangeweight(p1,x(n));       % evaluate dL/dx at last node
```

```
count = 0;               % initialize iteration count
tol =1e-6;               % tolerance
res = 0;                 % residual
yo = ones(n,1);          % initialize y
while(count == 0 || res > tol) % loop for iteration
    A(2:n-1,:) = w2;     % internal nodes
    for i=2:n-1
        A(i,i)=A(i,i)-4*yo(i)^3;
    end
    B(2:n-1) = -3*yo(2:n-1).^4-0.5;
    A(n,:) = w1;         % last node
    y = A\B;             % solve for y
    res = sum(abs(yo-y));
    yo = y;
    count = count +1;    % update iteration count
end
```

The following table compares the result obtained using FDM and collocation method

x	y_{FDM}	y_c	$|y_{FDM} - y_c|$
0	1.000000	1.000000	0.0E+00
0.1	0.977119	0.977075	4.4E-05
0.2	0.958354	0.958282	7.3E-05
0.3	0.943025	0.942934	9.1E-05
0.4	0.930603	0.930501	1.0E-04
0.5	0.920682	0.920572	1.1E-04
0.6	0.912946	0.912832	1.1E-04
0.7	0.907157	0.907040	1.2E-04
0.8	0.903140	0.903022	1.2E-04
0.9	0.900776	0.900657	1.2E-04
1	0.899995	0.899877	1.2E-04

For the present example, there is not much difference between the two methods.

Recollecting from earlier chapters on function approximation, higher order polynomial approximations are prone to round off errors. Also for equi-spaced nodes, the errors can be larger close to the boundaries. As a remedy one can use Chebyshev nodes. Orthogonal polynomials can also be used as basis functions leading to the Pseudo-spectral method. If the boundary conditions are symmetric, Fourier series can be used to approximate the function y. Now we shall look at method of weighted residuals, which would help us to understand different numerical methods used to solve BVPs.

11.5   Method of weighted residuals

Consider the second order BVP

$$\frac{d^2 y}{dx^2} = f(x, y, y^{(1)}) \tag{11.33}$$

over the interval $0 \le x \le 1$. Another approach of treating the problem is to approximate y as

$$y = a_0\phi_0 + a_1\phi_1 + a_2\phi_2 + \cdots + a_n\phi_n \tag{11.34}$$

where ϕ_i are the basis functions (polynomial, Fourier series etc.) used to approximate the functions. The approximation should be such that the *residual* is minimum. Residual is defined as

$$\mathbf{R} = \frac{d^2y}{dx^2} - f(x, y, y^{(1)}) \tag{11.35}$$

The reader must make a distinction between *error* and *residual*. Error arises from approximating function y whereas residual comes from approximating the differential equation. For most practical problems exact solutions do not exist and it would not be possible to directly calculate the error. However as residual can be calculated with more ease, it can be used as an estimate for accuracy. It is desirable that both residuals and errors are within tolerable limits. Solution to the differential equation can be looked upon as an optimization problem where we look to minimize the residual globally as well as locally through out the domain. The unknown coefficients a_i of the basis functions have to be determined such that the residual is a minimum. The method resembles estimation of parameters using regression. The minimization procedure is achieved using the following form

$$\int_0^1 W_i(x)R(x)dx = 0; \quad i = 0 \text{ to } n \tag{11.36}$$

where $W_i(x)$ is a weighting function which is used to control the distribution of residual along x. The number of weight functions is equal to the number of unknowns. The above operation converts a differential equation into a system of linear equations which can be solved using methods familiar to us. Some of the common weight functions are given below.

Collocation The weight function used in collocation method is Kronecker δ, a mathematical function defined by

$$\delta(x - x_i) = \begin{cases} 1 & x = x_i \\ 0 & \text{otherwise} \end{cases} \tag{11.37}$$

If Kronecker δ is used as the weight function, the residual is equal to zero only at the nodes or *collocation* points. We are not worried about the residuals at other points in the domain. Hence, the equation to be solved becomes

$$\delta(x - x_i)R(x) = 0; \quad i = 1 \text{ to } n$$

$$\text{or}$$

$$R(x_i) = 0; \quad i = 1 \text{ to } n \tag{11.38}$$

FDM as MWR: Finite difference method can also be classified as Method of Weighted residuals. We have already seen that FDM solves the governing equation at each of the nodes. Equation 11.38 is another way of expressing FDM.

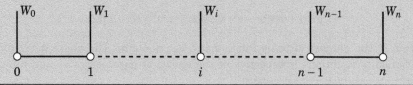

Galerkin methods: In Galerkin[4] methods, the basis functions used to approximate the actual function is used as weight functions. Hence Equation 11.36 becomes

$$\int_0^1 \phi_i R\, dx \quad i = 0 \text{ to } n \tag{11.39}$$

Galerkin methods form the basis of FEM which will be discussed later.

Method of moments: In this method, the weight functions are $1, x, x^2, \cdots, x^n$. Hence Equation 11.36 becomes

$$\int_0^1 x^i R\, dx \quad i = 0 \text{ to } n \tag{11.40}$$

If a polynomial is used as weight function, the method of moments and Galerkin approach are identical.

Least squares method: We have already encountered least squares method in Chapter 5 (regression). In the least squares method we minimize the square of the residual.

$$\text{Minimize:} \quad I = \int_0^1 R^2 dx \tag{11.41}$$

For a function to be minimum, the derivative of the function with respect to the variables should be equal to zero.

$$\frac{\partial I}{\partial a_i} = 2\int_0^1 \frac{\partial R}{\partial a_i} R\, dx = 0; \quad i = 1, \text{no of variables} \tag{11.42}$$

Hence, the weight function for a least square method is the derivative of the residual with respect to the parameter a_i i.e. $\dfrac{\partial R}{\partial a_i}$.

Strong and weak forms of ODE: Strong form of an ODE is nothing but the differential equation itself. FDM and collocation methods consider the strong form of the equation directly i.e. the residual is forced to zero at nodes. On the other hand, methods using Galerkin, least squares and moments as the weight functions are weak form representation of the ODE. As the name suggests, the weak form of ODE weakly enforces the residual equation. The weak form tries to satisfy the ODE on an average in the entire domain instead of satisfying the ODE at a few points.

Example 11.12

Solve $y^{(2)} = 2y$ with boundary conditions $y = 1$ at $x = 0$ and $y^{(1)} = 0$ at $x = 1$ using method of weighted residuals. Choose appropriate basis functions. Compare results obtained using different weight functions.

Solution :

Let us select basis functions such that they satisfy the boundary conditions automatically. On investigation we see that y can assume the form

$$y = 1 + a_1(2x - x^2) + a_2(2x - x^2)^2 + \cdots + a_n(2x - x^2)^n \tag{11.43}$$

where n is the number of basis functions in the domain. The above description satisfies the boundary conditions irrespective of the choice of parameters a_i and number of basis

---

[4]Boris Grigoryevich Galerkin, 1871-1945, a Russian/Soviet mathematician and engineer

functions.

Kronecker δ as weight: Let us consider a single basis function for all our examples. Then $y = 1 + a_1(2x - x^2)$. The residual is

$$\delta(x - x_1)R_1 \quad = y^{(2)} - 2y = -2a_1 - 2(1 + a_1(2x_1 - x_1^2)) \quad = 0$$
$$\longrightarrow \qquad\qquad a_1(2 + 4x_1 - 2x_1^2) \qquad\qquad = -2$$

The method requires us to choose collocation point x_1 where the weight is equal to 1. If the point were to be chosen as $x_1 = 0.5$ we get $a_1 = -0.571428$ and for $x_1 = 1$ we get $a_1 = -0.5$. Therefore we get different solutions based on the choice of the collocation point as indicated in the Figure 11.9. From the figure it is evident that the residual is zero at the collocation points. The reader must make a distinction between residual and error. Though the residual is zero at the collocation points, error is non zero at these points. The solution may be further improved by adding more number of collocation points.

Consider two collocation points $x_1 = 0.5$ and $x_2 = 1$. Then $y = 1 + a_1(2x - x^2) + a_2(2x - x^2)^2$.

$$R_2 = -2 - \{2 + 2(2x - x^2)\}\, a_1 + \{12x^2 - 24x + 8 - 2(2x - x^2)^2\}\, a_2 = 0$$

The residuals have to be zero at these two collocation points and hence we obtain a system of two linear equations.

$$\begin{pmatrix} -3.5 & -2.125 \\ -4 & -6 \end{pmatrix} \begin{pmatrix} a_1 \\ a_2 \end{pmatrix} = \begin{pmatrix} 2 \\ 2 \end{pmatrix}$$

Solving these equations we get $a_1 = -0.62$ and $a_2 = 0.08$. The solution and the residual have been plotted in Figure 11.9. We see that the solution has improved to a large extent (maximum error of 0.0016, Table 11.2) and the residuals are also significantly smaller. The choice of collocation points are important for the accuracy of the method. We have

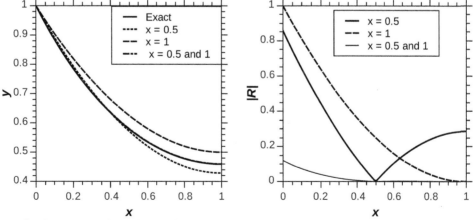

Figure 11.9: Solution and residuals of weighted residuals method with weighing function taken as Kronecker δ

seen approximating a function with Chebyshev nodes produces minimum errors. If the collocation nodes are chosen as Chebyshev nodes, we get $a_1 = -0.627078$ and $a_2 = 0.085511$.

Method of moments: The weight function of moments is x^i. For two basis functions, Equation 11.36 reduces to

$$\int_0^1 R_2 dx \quad \longrightarrow \quad 2 + 3.33333a_1 + 1.06667a_2 = 0$$

$$\int_0^1 x R_2 dx \quad \longrightarrow \quad 1 + 1.83333a_1 + 1.73333a - 2 = 0$$

On solving we get $a_1 = -0.627906$ and $a_2 = 0.0872093$.

Galerkin method: The weighing function for Galerkin method is the basis function itself. For $n = 2$, we have

$$\int_0^1 (2x - x^2) R_2 dx \quad \longrightarrow \quad 1.3333 + 2.4a_1 + 1.9809a_2 = 0$$

$$\int_0^1 (2x - x^2)^2 R_2 dx \quad \longrightarrow \quad 1.0667 + 1.9810a_1 + 2.0317a_2 = 0$$

On solving we get $a_1 = -0.626016$ and $a_2 = 0.085366$.

Least squares: The weighing function for least squares is derivative of the residual. Let us first consider $n = 2$. The derivative of the residual is given by

$$\frac{\partial R_2}{\partial a_1} = -2 - 2(2x - x^2); \qquad \frac{\partial R_2}{\partial a_2} = 12x^2 - 24x + 8 - 2(2x - x^2)^2;$$

Hence the least square weighted residual method reduces to

$$\int_0^1 \frac{\partial R_2}{\partial a_1} R_2 dx \quad \longrightarrow \quad 6.6667 + 11.4667a_1 + 6.0952a_2 = 0$$

$$\int_0^1 \frac{\partial R_2}{\partial a_2} R_2 dx \quad \longrightarrow \quad 2.1333 + 6.0952a_1 + 19.3016a_2 = 0$$

Solving for the parameters we get $a_1 = -0.6280$ and $a_2 = 0.0878$.

The results of all the four methods for two basis functions have been summarized in the table below

Table 11.2: *Comparison of different weighting methods*

Method	a_1	a_2	$\epsilon_{y,max}$	$|R_{max}|$
Collocation (x=0.5 and 1)	-0.620000	0.080000	1.60E-03	1.20E-01
Collocation (Chebyshev)	-0.627078	0.085511	9.55E-04	6.18E-02
Moments	-0.627907	0.087209	6.71E-04	4.65E-02
Galerkin	-0.626016	0.085366	2.51E-04	6.50E-02
LS	-0.628073	0.087813	6.41E-04	4.14E-02

Figure 11.10 shows the distribution of residuals for the different methods considered. From these results the following conclusions emerge

- There is little to choose between different methods applied to the differential equation considered in this example. There may be differential equations where choice of weightage function is crucial.
- Accuracy of collocation method depends on the choice of points.
- Collocation method is the easiest to apply. Galerkin and Least square methods involve elaborate algebra and may be difficult when n is large. Nevertheless, they are powerful methods and are often used in engineering problems

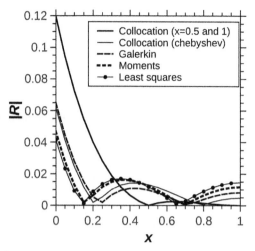

Figure 11.10: *Comparison of residuals for different weight schemes*

The above exercise is only of academic interest to demonstrate different aspects of MWR. In the example considered, basis functions were chosen such that the boundary conditions are automatically satisfied. Formulating such basis functions may be difficult for general problems. Similar to collocation method discussed in Section 11.4, Galerkin and least squares approach can also be applied. The algebra involving these methods are much more demanding. Interested readers are encouraged to refer to more advanced books.

Sub domain methods

As discussed in chapters on interpolation and integration, methods applied over the entire range have limitations. On the other hand, it would be advantageous to divide the domain into a large number of smaller domains or *sub domains* and solve the equations over them individually. This allows the use of adaptive methods (distribution of nodes according to function variation). The methods discussed until now i.e. collocation, Galerkin, moments and least squares can also be applied over smaller domains. Collocation method applied over smaller domain reduces to finite difference method which we have already discussed. Other common tools for solving differential equations are finite volume method and finite element method. We will briefly introduce Galerkin FEM and FVM.

11.6 Finite element method

FEM is also a weighted residual method. FEM was initially developed for solving problems in structural mechanics, but has slowly found applications in other areas of engineering. In this section we will discuss application of Galerkin FEM to a BVP.

11.6.1 Elements

In FEM, the entire domain is discretized into large number of sub domains called elements. The function y is defined within each element by basis functions. In FEM literature, basis functions are referred to as shape functions denoted as N_i. We will use this notation here. If the shape function is linear, the element contains two nodes where as a quadratic element contain three nodes.

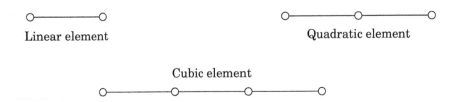

Linear element   Quadratic element

Cubic element

Composite quadrature rules also divide the domain into several segments. Each segment of trapezoidal rule contains two nodes (linear element), Simpson's 1/3 rule has 3 nodes (a quadratic element). Similar to composite quadrature rules, the nodes on the boundaries of the element are shared between two or more elements.

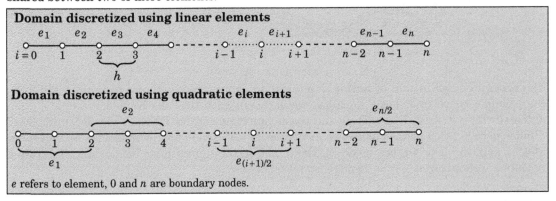

Domain discretized using linear elements

Domain discretized using quadratic elements

e refers to element, 0 and n are boundary nodes.

The shape function for a linear element is same as Lagrange polynomial discussed in interpolation chapter

$$y = N_i y_i + N_{i+1} y_{i+1} \quad \text{for} \quad x_i \le x \le x_{i+1}$$
$$N_i = \frac{x - x_{i+1}}{x_i - x_{i+1}} \qquad N_{i+1} = \frac{x - x_i}{x_{i+1} - x_i}$$

Similarly shape function for a quadratic element would be

$$y = N_i y_i + N_{i+1} y_{i+1} + N_{i+2} y_{i+2} \quad \text{for} \quad x_i \le x \le x_{i+2}$$
$$N_i(x) = \frac{(x - x_{i+1})(x - x_{i+2})}{(x_i - x_{i+1})(x_i - x_{i+2})}$$
$$N_{i+1}(x) = \frac{(x - x_i)(x - x_{i+2})}{(x_{i+1} - x_i)(x_{i+1} - x_{i+2})}$$
$$N_{i+2}(x) = \frac{(x - x_i)(x - x_{i+1})}{(x_{i+2} - x_i)(x_{i+2} - x_{i+1})}$$

Elements with other basis functions such as cubic spline, Hermite polynomial etc. can also be used.

11.6.2 Weightage function

The weightage function is defined locally for each element. Weightage function used in Galerkin method is same as the shape function for the element. For a domain discretized by linear elements, the weightage function has been illustrated in the following figure.

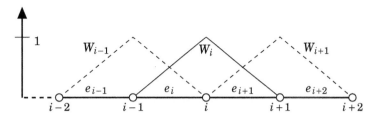

The weightage function for i^{th} node is given by

$$W_i = \begin{cases} \dfrac{x - x_{i-1}}{x_i - x_{i-1}} = N_i^i & \text{for } i^{th} \text{ element} \\[3mm] \dfrac{x - x_{i+1}}{x_i - x_{i+1}} = N_i^{i+1} & \text{for } i+1^{th} \text{ element} \end{cases}$$

The notation followed is N_i^j, i refers the node number where as j refers to element number. The residual equation for i^{th} node would be

$$\int_{i-1}^{i+1} W_i R\, dx = \underbrace{\int_{i-1}^{i} W_i R\, dx}_{e_i} + \underbrace{\int_{i}^{i+1} W_i R\, dx}_{e_{i+1}} = 0 \tag{11.44}$$

Node i is shared between two elements e_i and e_{i+1}. Hence, the residual equation contains contributions from each element shared by the node. Effectively, the residual equations can be applied over each element independently and later assembled together. This is same as what is done for composite quadrature rules, only that function values at the nodes are unknown. Hence, we obtain a system of linear equations which have to be solved for function values at the nodes. First, let us apply FEM to a second order BVP.

11.6.3 Second order ODE with linear element

Let us consider the differential equation $y^{(2)} - 2y = 0$ with boundary conditions $y(x = 0) = 1$ and $y^{(1)}(x = 1) = 0$. The domain is discretized into n equispaced linear elements.

e refers to element, 0 and n are boundary nodes.

The shape function for a linear element would be

$$y = \frac{x - x_{i+1}}{x_i - x_{i+1}} y_i + \frac{x - x_i}{x_{i+1} - x_i} y_{i+1} \quad \text{for} \quad x_i \le x \le x_{i+1}$$

$$y = N_i^{i+1} y_i + N_{i+1}^{i+1} y_{i+1}$$

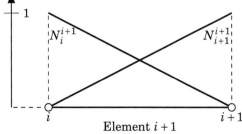

Element $i + 1$

where N_j^i is the shape function for j^{th} node for i^{th} element. The shape function can be represented in terms of local coordinates (Equation 5.15) as

$$y = \frac{1-\xi}{2}y_i + \frac{1+\xi}{2}y_{i+1} \quad \text{for} \quad -1 \le \xi \le 1$$

The first derivative of the shape function can be written as

$$\frac{dN_i^j}{dx} = -\frac{1}{x_{i+1}-x_i}; \quad \frac{dN_{i+1}^j}{dx} = \frac{1}{x_{i+1}-x_i};$$

Let us apply residual equations to an internal node i

$$\int_{x_{i-1}}^{x_{i+1}} W_i R dx = \int_{x_{i-1}}^{x_i} N_i^i(y^{(2)} - 2y)dx + \int_{x_i}^{x_{i+1}} N_i^{i+1}(y^{(2)} - 2y)dx = 0$$

$$\underbrace{\int_{x_{i-1}}^{x_i} N_i^i y^{(2)} dx + \int_{x_i}^{x_{i+1}} N_i^{i+1} y^{(2)} dx}_{I_1} - \underbrace{2\int_{x_{i-1}}^{x_i} N_i^i y dx - 2\int_{x_i}^{x_{i+1}} N_i^{i+1} y dx}_{I_2} = 0$$

The residual equations have two integrals corresponding to $y^{(2)}$ and y. Product of two functions N_i and $y^{(2)}$ can be rewritten as

$$\frac{d}{dx}\left(N_i y^{(1)}\right) = N_i y^{(2)} + \frac{dN_i}{dx} y^{(1)}$$

$$\text{or} \quad N_i y^{(2)} = \frac{d}{dx}\left(N_i y^{(1)}\right) - \frac{dN_i}{dx} y^{(1)}$$

Substituting the above into I_1 gives

$$
\begin{aligned}
I_1 &= \int_{x_{i-1}}^{x_i}\left\{\frac{d}{dx}\left(N_i^i y^{(1)}\right) - \frac{dN_i^i}{dx} y^{(1)}\right\}dx + \int_{x_i}^{x_{i+1}}\left\{\frac{d}{dx}\left(N_i^{i+1} y^{(1)}\right) - \frac{dN_i^{i+1}}{dx} y^{(1)}\right\}dx \\
&= N_i^i y^{(1)}\Big|_{x_{i-1}}^{x_i} + N_i^{i+1} y^{(1)}\Big|_{x_i}^{x_{i+1}} - \frac{y_i - y_{i-1}}{(x_i - x_{i-1})^2}(x_i - x_{i-1}) + \frac{y_{i+1} - y_i}{(x_{i+1} - x_i)^2}(x_{i+1} - x_i) \\
&= y_{e_i}^{(1)}(x_i) - y_{e_{i+1}}^{(1)}(x_i) - \frac{y_i - y_{i-1}}{x_i - x_{i-1}} + \frac{y_{i+1} - y_i}{x_{i+1} - x_i}
\end{aligned}
$$

The term $y_{e_i}^{(1)}(x_i) - y_{e_{i+1}}^{(1)}(x_i)$ is the difference between the left hand side and right hand side first derivatives at x_i. As we have used piecewise linear elements which have zero order continuity (continuous only in y) this term is non-zero (unless the solution itself is linear). Hence, the above term corresponds to the residual and is neglected in numerical computation.[5] The magnitude of the residual decreases as the grid spacing reduces. As the nodes are uniformly spaced we get

$$I_1 = \frac{y_{i+1} - 2y_i + y_{i-1}}{h}$$

---

[5]The term $y_{e_i}^{(1)} - y_{e_{i+1}}^{(1)}$ can be reduced to zero by using a basis function having first order continuity i.e. continuous in $y^{(1)}$ such as cubic spline or Hermite interpolating polynomial

The above equation has a semblance to central difference formula. Now we shall determine the integral I_2.

$$
\begin{aligned}
I_2 &= 2\int_{x_i}^{x_{i+1}} N_i^{i+1}(N_i^{i+1}y_i + N_{i+1}^{i+1}y_{i+1})dx + 2\int_{x_{i-1}}^{x_i} N_i^i(N_i^i y_i + N_{i+1}^i y_{i+1})dx \\
&= (x_{i+1} - x_i)\left(\frac{2y_i}{3} + \frac{y_{i+1}}{3}\right) + (x_i - x_{i-1})\left(\frac{2y_i}{3} + \frac{y_{i-1}}{3}\right) \\
&= \frac{h}{3}(y_{i-1} + 4y_i + y_{i+1})
\end{aligned}
$$

The above expression resembles Simpson's 1/3 rule. Assembling integrals I_1 and I_2 we have

$$
\frac{y_{i+1} - 2y_i + y_{i-1}}{h} - \frac{h}{3}(y_{i-1} + 4y_i + y_{i+1}) = 0
$$

$$
\left(\frac{1}{h} - \frac{h}{3}\right)y_{i+1} - \left(\frac{2}{h} + \frac{4h}{3}\right)y_i + \left(\frac{1}{h} - \frac{h}{3}\right)y_{i-1}
$$

Boundary conditions: Let us apply the above procedure for boundary node n. It has to be noted that the boundary node is associated with a single element.

$$
\int_{x_{n-1}}^{x_n} W_n R dx = \int_{x_{n-1}}^{x_n} N_n^n(y^{(2)} - 2y)dx = 0
$$

$$
y_n^{(1)} - \frac{y_n - y_{n-1}}{h} - \frac{h}{3}(y_{n-1} + 2y_n) = 0
$$

One must concentrate on the term $y_n^{(1)}$ which indicates the derivative at the boundary. Hence, both Neumann and Robin boundary conditions specified at the boundary are represented. Hence, FEM naturally takes into account the boundary conditions. In the present case $y_n^{(1)} = 0$. Hence we have the discretized equation for node n as

$$
\left(\frac{1}{h} - \frac{h}{3}\right)y_{n-1} - \left(\frac{1}{h} + \frac{2h}{3}\right)y_n = 0
$$

The system of equations may be written in matrix form as

$$
\begin{bmatrix}
1 & 0 & 0 & \cdots & \cdots & 0 \\
a' & -2a & a' & \cdots & \cdots & 0 \\
0 & a' & -2a & a' & \cdots & 0 \\
\cdots & \cdots & \cdots & \cdots & \cdots & \cdots \\
0 & 0 & \cdots & a' & -2a & a' \\
0 & 0 & \cdots & 0 & a' & -a
\end{bmatrix}
\begin{bmatrix}
y_0 \\
y_1 \\
y_2 \\
\cdots \\
y_{n-2} \\
y_{n-1} \\
y_n
\end{bmatrix}
=
\begin{bmatrix}
1 \\
0 \\
0 \\
\cdots \\
0 \\
0
\end{bmatrix}
\qquad (11.45)
$$

where $a' = \left(\frac{1}{h} - \frac{h}{3}\right)$ and $a = \left(\frac{1}{h} + \frac{2h}{3}\right)$. A MATLAB program has been provided below to carry out the solution using TDMA.

```
n = 160;                    % number  of  nodes
x = [0:1/(n-1):1]';         % nodes
a = zeros(n,1);             % TDMA  coefficient  node i
a1 = zeros(n,1);                   % TDMA  coefficient  node  i+1
a2 = zeros(n,1);                   % TDMA  coefficient  node  i-1
b = zeros(n,1);             % TDMA  constant
a(1) = 1;                   % first  node
b(1) = 1;                   % Dirichlet  condition
for i=2:n-1                 % coefficients  for
     h1 = x(i)-x(i-1);      % interior  nodes
     h2 = x(i+1)-x(i);
     a1(i) = 1/h2-h2/3;
     a2(i) = 1/h1-h1/3;
     a(i) = 1/h2+2*h2/3 + 1/h1+2*h1/3;
end
h1 = x(n)-x(n-1);           % last  node
a2(n) = 1/h1-h1/3;          % Neumann  boundary  condition
a(n) = 1/h1+2*h1/3;
y = tdma(a2,a,a1,b);        % solve  using  TDMA
```

It is a good practice to apply the residual equations to every element and assemble the
matrix equations together. This becomes important for multidimensional problems when
a node is shared by more than two elements. The residual equations applied to the
element $i+1$ can be represented as

$$\int_{x_i}^{x_{i+1}} \left\{ \begin{array}{c} N_i^{i+1}(y^{(2)}-2y) \\ N_{i+1}^{i+1}(y^{(2)}-2y) \end{array} \right\} dx = 0$$

which on integration yields

$$\left(\begin{array}{cc} -y^{(1)}(x_i) & 0 \\ 0 & y^{(1)}(x_{i+1}) \end{array} \right) + \frac{1}{h} \left(\begin{array}{cc} -1 & 1 \\ 1 & -1 \end{array} \right) \left(\begin{array}{c} y_i \\ y_{i+1} \end{array} \right) - \frac{h}{3} \left(\begin{array}{cc} 2 & 1 \\ 1 & 2 \end{array} \right) \left(\begin{array}{c} y_i \\ y_{i+1} \end{array} \right)$$

Node i is shared between elements i and $i+1$. We would get local elemental matrices of
size 2 of the form

$$\left(\begin{array}{cc} -a_i & a_i' \\ a_{i+1}'' & -a_{i+1} \end{array} \right) \left(\begin{array}{c} y_i \\ y_{i+1} \end{array} \right) = \left(\begin{array}{c} b_i \\ b_{i+1} \end{array} \right)$$

All the local matrices are then to be assembled together to form a global matrix as
illustrated below

		$i-1$	i	$i+1$	
node $i-1$	$\cdots +$	a_{i-1}^i	$a_{i-1}'^i$		
node i		$a_i''^i$	a_i^i	$+$	a_i^{i+1}	$a_i'^{i+1}$
node $i+1$				$a_{i+1}''^{i+1}$	a_{i+1}^{i+1}	$+\cdots$

$$\underbrace{\qquad\qquad}_{\text{Element } i} \quad \underbrace{\qquad\qquad}_{\text{Element } i+1}$$

Example 11.13

Solve the differential equation in Example 11.9 using FEM approach.

Solution :

The differential equation to be solved is

$$y^{(2)} + \frac{1}{x} y^{(1)} - y = 0$$

with boundary conditions $y(x = 1) = 1$ and $y^{(1)}(x = 2) = 0$. The domain is discretized into n equal linear elements, the number of nodes being $n + 1$ (0 to n). Let us consider an element $i + 1$ containing nodes i and $i + 1$, residual equations for which can be written as

$$\int_{x_i}^{x_{i+1}} \begin{pmatrix} N_i^{i+1}(& \overbrace{\boxed{y^{(2)}}}^{I_1} & + & \overbrace{\boxed{\frac{1}{x} y^{(1)}}}^{I_2} & - & \overbrace{\boxed{y}}^{I_3} &) \\ N_{i+1}^{i+1}(& \boxed{y^{(2)}} & + & \boxed{\frac{1}{x} y^{(1)}} & - & \boxed{y} &) \end{pmatrix} dx = 0$$

The residual equation is composed of three integrals, of which two integrals has already been evaluated.

$$I_1 = \begin{bmatrix} -y^{(1)}(x_i) & 0 \\ 0 & y^{(1)}(x_{i+1}) \end{bmatrix} + \frac{1}{h} \begin{bmatrix} -1 & 1 \\ 1 & -1 \end{bmatrix} \begin{pmatrix} y_i \\ y_{i+1} \end{pmatrix}$$

$$I_3 = \frac{h}{6} \begin{bmatrix} 2 & 1 \\ 1 & 2 \end{bmatrix} \begin{pmatrix} y_i \\ y_{i+1} \end{pmatrix}$$

Again, for all the interior nodes we neglect the term related to $y^{(1)}$. At the boundary at $x = 2$, we take into account of the Neumann boundary condition. In the present case, $y^{(1)}(x = 2) = 0$. Now let us concentrate on I_2.

$$I_2 = \int_{x_i}^{x_{i+1}} \frac{1}{x} \begin{pmatrix} N_i^{i+1} \\ N_{i+1}^{i+1} \end{pmatrix} \begin{pmatrix} dN_i^{i+1}/dx & dN_{i+1}^{i+1}/dx \\ dN_i^{i+1}/dx & dN_{i+1}^{i+1}/dx \end{pmatrix} \begin{pmatrix} y_i \\ y_{i+1} \end{pmatrix} dx$$

We will represent the above equation in local coordinates

$$x = \frac{1-\xi}{2} x_i + \frac{1+\xi}{2} x_{i+1} = \frac{x_{i+1} + x_i}{2} + h\frac{\xi}{2}$$

The above integral reduces to

$$I_2 = \int_{-1}^{1} \frac{1}{x_{i+1} + x_i + h\xi} \begin{pmatrix} 1-\xi \\ 1+\xi \end{pmatrix} \begin{pmatrix} -0.5 & 0.5 \\ -0.5 & 0.5 \end{pmatrix} \begin{pmatrix} y_i \\ y_{i+1} \end{pmatrix} d\xi$$

Essentially we have to evaluate two integrals which can be evaluated analytically (in general one may use numerical quadrature)

$$I_4 = \int_{-1}^{1} \frac{1-\xi}{x_{i+1} + x_i + h\xi} d\xi = \frac{2x_{i+1} \log[x_{i+1} + x_i + h\xi] - h\xi}{h^2} \Big|_{-1}^{1} = \log\left(\frac{x_{i+1}}{x_i}\right) \frac{2x_{i+1}}{h^2} - \frac{2}{h}$$

$$I_5 = \int_{-1}^{1} \frac{1+\xi}{x_{i+1}+x_i+h\xi} d\xi = -\left. \frac{2x_i \log(x_{i+1}+x_i+h\xi)-h\xi}{h^2} \right|_{-1}^{1} = -\log\left(\frac{x_{i+1}}{x_i}\right)\frac{2x_i}{h^2}+\frac{2}{h}$$

Therefore the integral I_2 becomes

$$I_2 = \frac{1}{2}\begin{pmatrix} -I_4 & I_4 \\ -I_5 & I_5 \end{pmatrix}\begin{pmatrix} y_i \\ y_{i+1} \end{pmatrix}$$

All elemental equations have to be assembled together to get a global set of linear equations.

$$\begin{pmatrix} 1 & 0 & 0 & \cdots & \cdots & \cdots & 0 \\ a_1'' & -a_1 & a_1' & 0 & \cdots & \cdots & 0 \\ 0 & a_2'' & -a_2 & a_2' & 0 & \cdots & 0 \\ \cdots & \cdots & \ddots & \ddots & \ddots & \cdots & \cdots \\ 0 & 0 & 0 & a_i'' & -a_i & a_i' & \cdots \\ \cdots & \cdots & \cdots & \cdots & \ddots & \ddots & \cdots \\ 0 & 0 & 0 & 0 & \cdots & a_n'' & -a_n \end{pmatrix}\begin{pmatrix} y_0 \\ y_1 \\ y_2 \\ \cdots \\ y_i \\ \cdots \\ y_{n-1} \\ y_n \end{pmatrix} = \begin{pmatrix} 1 \\ 0 \\ 0 \\ \cdots \\ 0 \\ \cdots \\ 0 \end{pmatrix}$$

where $a_i'' = \frac{1}{h}-\frac{h}{6}-\frac{I_5^i}{2}$, $a_i' = \frac{1}{h}-\frac{h}{6}+\frac{I_4^{i+1}}{2}$ and $a_i = \frac{2}{h}+\frac{h}{3}-\frac{I_5^i}{2}+\frac{I_4^{i+1}}{2}$. I_5^i and I_4^{i+1} are integrals calculated for i^{th} and $i+1^{th}$ elements respectively. The system of linear equation is tridiagonal and we can solve the system of linear equations using TDMA. A MATLAB program has been written below to solve the ODE.

```
n = 11;                    % no of nodes
x = [1:1/(n-1):2]';        % nodes
a = zeros(n,1);            % TDMA coefficient node i
a1 = zeros(n,1);           % TDMA coefficient node i+1
a2 = zeros(n,1);           % TDMA coefficient node i-1
b = zeros(n,1);            % TDMA constant
a(1) = 1;                  % first node
b(1) = 1;
for i=1:n-1                % determine elemental equations
    h = x(i+1)-x(i);       % length of element
    % I4 and I5 and integrals as in solution
    I4 =  log(x(i+1)/x(i))*2*x(i+1)/h^2 - 2/h;
    I5 = -log(x(i+1)/x(i))*2*x(i)/h^2 + 2/h;
    % Elements are stored in TDMA format
    if(i~=1)               % function is known at node 1
    a(i) = a(i)+1/h +h/3 + I4/2;
    a1(i) = 1/h -h/6 + I4/2;
    end
    a(i+1) = a(i+1) +1/h +h/3 - I5/2;
    a2(i+1) = 1/h -h/6 - I5/2;
end
y = tdma(a2,a,a1,b);       % solution using TDMA
% note indices of nodes in program and text are not same but
            %identical
```

The following table compares the result obtained using FDM (Example 11.9) and FEM. The solutions from FEM and FDM closely agree with each other! As the exact solution for the above equation is not written down (it involves special functions), we have performed grid sensitivity study to ascertain the convergence of the solution to the exact solution. The derivative of the function value at $x = 1$ has been taken as the marker. The derivative itself is calculated using three point forward difference formula.

x	y_{FDM}	y_{FEM}	$|y_{FDM} - y_{FEM}|$
1.0	1	1	0
1.1	0.9058	0.9056	1.5E-04
1.2	0.8284	0.8282	2.2E-04
1.3	0.7652	0.7650	2.5E-04
1.4	0.7140	0.7138	2.5E-04
1.5	0.6733	0.6731	2.1E-04
1.6	0.6417	0.6415	1.6E-04
1.7	0.6182	0.6181	9.7E-05
1.8	0.6021	0.6021	1.9E-05
1.9	0.5927	0.5928	6.9E-05
2.0	0.5896	0.5898	1.7E-04

Grid sensitivity study has been summarized below as a table. From the above analysis, there is no significant difference between FDM and FEM. One can improve the order of accuracy of FEM using higher order elements such as quadratic and cubic elements. This is equivalent to using higher order finite difference formulae in FDM.

No of	$y^{(1)}(x = 1)$	
elements	FDM	FEM
10	-2.0529	-2.0566
20	-2.0685	-2.0695
40	-2.0730	-2.0733
80	-2.0742	-2.0743
160	-2.0745	-2.0745
320	-2.0746	-2.0746

11.6.4 Finite element method applied to structural problems

Our discussion on FEM would be incomplete without discussing application of FEM to structural problems. Like before, the domain is discretized into elements but the definition of the residual equations is based on principle of minimum potential energy. We shall consider a simple case where strain and stress are linearly related.

Background

Consider the following spring mass system subject to a load F as shown. We are interested in describing the new equilibrium position in the presence of the load.

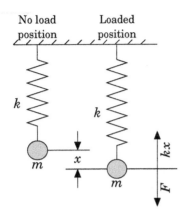

The change in the equilibrium position changes the potential energy of the system.

$$PE = \frac{1}{2}kx^2 - Fx$$

The new equilibrium position is such that the potential energy of the system is minimum.

$$\frac{dPE}{dx} = kx - F = 0$$

The same principle known as the principle of minimum potential energy can be applied to the analysis of structures. At equilibrium position the potential energy of the system would be minimum.

Now let us consider a simple block having uniform cross section area A and length l subjected to axial load.

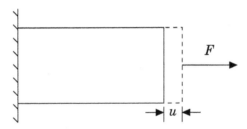

The block elongates by an amount u (displacement) due to the load. For an elastic homogeneous material, stress σ and strain ϵ are linearly related, $\sigma = E\epsilon$. E is known as the Young's modulus. Strain ϵ is the ratio of elongation u to the length of the block L i.e. $\epsilon = u/L$. The stress acting on the member is $\sigma = F/A$. Then

$$\frac{F}{A} = E\frac{u}{L} \rightarrow \underbrace{\frac{AE}{L}}_{k}u - F = 0$$

Hence the above configuration can be treated as an elongated spring subjected to an axial load. The potential energy for a structural system is given by

$$PE = \text{strain energy} - Fu$$

The strain energy is given by $\int_V \frac{\sigma\epsilon}{2}dV$ where V refers to the volume of the block. Then the potential energy of the block would be

$$PE = \frac{1}{2}\sigma\epsilon AL - Fu = \frac{1}{2}E\epsilon^2 AL - Fu$$

The strain can be represented in the differential form as

$$\epsilon = \frac{\partial u}{\partial x}$$

and hence the strain energy for an axially loaded block is given by

$$SE = \frac{1}{2} \int A(x)E \left(\frac{\partial u}{\partial x}\right)^2 dx$$

FEM based on principle of minimum potential energy

Now we shall try to understand the application of FEM to an axially loaded homogeneous block with uniform area of cross section. We shall discretize the block into n linear elements.

Then the total potential energy for the entire domain can be written as

$$PE = \frac{1}{2} \int_0^1 A(x)E \left(\frac{\partial u}{\partial x}\right)^2 dx + \sum_{j}^{\text{no of nodes}} F_j u_j$$

where F_j and u_j are the loads and displacement at node j. As the elements are linear, the total potential energy can be written as

$$PE = \sum_{i}^{\text{no of elements}} \int_e \frac{AE}{2} \left(\frac{u_2 - u_1}{x_2 - x_1}\right)^2 dx + \sum_{j}^{\text{no of nodes}} F_j u_j$$

(the indices u_1 and u_2 are local indices for an element). Principle of minimum potential energy states that at equilibrium, the potential energy is minimum. Hence we have an optimization problem on hand. Hence, the minimum value can be determined by differentiating the potential energy with respect to the unknown displacements u. For an internal node j, on simplification, we have

$$\frac{\partial PE}{\partial u_j} = \frac{AE}{2} \left(\frac{u_j - u_{j-1}}{x_j - x_{j-1}} + \frac{u_{j+1} - u_j}{x_{j+1} - x_j}\right) - F_j = 0$$

Now, we have system of equations which can be solved by methods familiar to us.

As we did earlier, it would be easier to handle element by element. Let us consider an element i.

The index of the elements are local

u_1 and u_2 are the displacements at the two nodes and F_1 and F_2 are the two forces acting on the nodes. The strain energy corresponding to the linear element

$$SE = \int_{x_1}^{x_2} \frac{AE}{2} \left(\frac{u_2 - u_1}{x_2 - x_1} \right)^2 dx = \frac{AE}{2} \left(\frac{u_2 - u_1}{x_2 - x_1} \right)^2 (x_2 - x_1) = \frac{AE}{2} \frac{u_2^2 - 2u_1 u_2 + u_1^2}{x_2 - x_1}$$

Similarly the work done by force F_1 and F_2 are equal to $F_1 u_1$ and $F_2 u_2$. Thus the potential energy of element i is given by

$$PE = \frac{AE}{2} \frac{u_2^2 - 2u_1 u_2 + u_1^2}{x_2 - x_1} - F_1 u_1 - F_2 u_2$$

Note the PE corresponds to a single element and the total potential energy. To minimize PE, we shall consider term by term i.e. SE and F and differentiate these terms with respect to the unknowns u_1 and u_2.

$$\frac{\partial SE}{\partial u_1} = AE \frac{u_2 - u_1}{x_2 - x_1} \qquad \frac{\partial w}{\partial u_1} = -F_1$$

$$\frac{\partial SE}{\partial u_2} = AE \frac{u_1 - u_2}{x_2 - x_1} \qquad \frac{\partial w}{\partial u_2} = -F_2$$

Representing in the matrix form we have

$$\frac{\partial SE}{\partial u} = \underbrace{\frac{AE}{x_2 - x_1} \begin{pmatrix} 1 & -1 \\ -1 & 1 \end{pmatrix}}_{K} \begin{pmatrix} u_1 \\ u_2 \end{pmatrix} ; \qquad \frac{\partial w}{\partial u} = \underbrace{\begin{pmatrix} F_1 \\ F_2 \end{pmatrix}}_{F}$$

The matrix K is referred to as the stiffness matrix and the vector F is referred as force vector in FEM (the name is derived from the analogy with springs). The above equations are applicable for an element. We should assemble all the local elemental equations as discussed earlier to obtain a set of linear equations of the form $\mathbf{Ku} - \mathbf{F} = 0$. We will take up a simple example now.

Example 11.14
Determine the displacements at node 2 and 3 for the stepped block shown below.

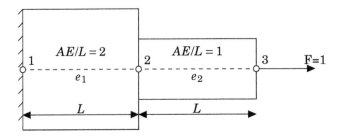

Solution :
Element 1
Displacement at node 1 is zero. Then the nodal equations for node 2 is given by

$$2(u_2 - u_1)$$

Element 2
The nodal equations are

$$
\begin{aligned}
\text{node 2} &\quad u_2 - u_3 \\
\text{node 3} &\quad u_3 - u_2 \quad -1
\end{aligned}
$$

Assembling the elemental equations we have

$$
\begin{aligned}
u_1 &= 0 \\
-2u_1 + 3u_2 - u_3 &= 0 \\
0u_1 - u_2 + u_3 &= 1
\end{aligned}
$$

On solving the above set of equations we get $u_2 = 0.5$ and $u_3 = 1.5$.

Another example involving a tapered block subject to axial load is considered now.

Example 11.15

A tapered block is subjected to an axial load. Determine the elongation of the tapered block.

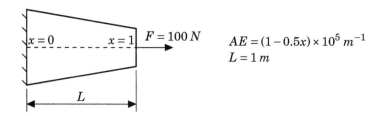

$x = 0$ $x = 1$ $F = 100\,N$ $AE = (1 - 0.5x) \times 10^5 \; m^{-1}$
$L = 1\,m$

L

Solution :
Exact solution It can be shown that the total elongation of the tapered block is given by

$$u = \int_0^1 \frac{F}{AE}\,dx = \int_0^1 \frac{10^{-3}}{1 - 0.5x}\,dx = -0.002\log(1 - 0.5x)|_0^1 = 0.00138629$$

FEM solution We shall discretize the tapered block into n elements. Consider i^{th} element

u_1 F_1 u_2 F_2 The index of the elements are local

Then the elemental equation would be

$$PE = \int_{x_1}^{x_2} \frac{AE}{2}\left(\frac{u_2-u_1}{x_2-x_1}\right)^2 dx - F_1 u_1 - F_2 u_2$$

$$= 10^5[1-0.25(x_2+x_1)]\frac{(u_2-u_1)^2}{x_2-x_1} - F_1 u_1 - F_2 u_2$$

Differentiating the potential energy with respect to the unknown displacement we get

$$K\left(\begin{array}{cc} 1 & -1 \\ -1 & 1 \end{array}\right)\left(\begin{array}{c} u_1 \\ u_2 \end{array}\right) - \left(\begin{array}{c} F_1 \\ F_2 \end{array}\right)$$

where $K = \dfrac{10^5[1-0.25(x_2+x_1)]}{x_2-x_1}$. The displacement at $x=0$ is 0. On assembling all the elemental equations we get system of linear equation of the form

$$\left\{\begin{array}{cccccccc} 1 & 0 & 0 & \cdots & 0 & 0 & 0 \\ -K_1 & K_1+K_2 & -K_2 & \cdots & 0 & 0 & 0 \\ \cdots & \cdots & \cdots & \cdots & \cdots & \cdots & \cdots \\ 0 & 0 & 0 & \cdots & -K_{n-1} & K_{n-1}+K_n & -K_n \\ 0 & 0 & 0 & \cdots & 0 & -K_n & K_n \end{array}\right\}\left\{\begin{array}{c} u_0 \\ u_1 \\ u_2 \\ \cdots \\ u_{n-2} \\ u_{n-1} \\ u_n \end{array}\right\} = \left\{\begin{array}{c} 0 \\ 0 \\ \cdots \\ 0 \\ 100 \end{array}\right\}$$

The above equation can be solved using TDMA and a MATLAB program has been written to solve the same.

```
n=11;                % no of nodes
x=0:1/(n-1):1;       % nodes
a = zeros(n,1);      % TDMA coefficients a
a1 = zeros(n,1);     % TDMA coefficients a'
a2 = zeros(n,1);     % TDMA coefficients a''
b = zeros(n,1);      % TDMA coefficients b
a(1) = 1;            % first node
for i=1:n-1          % element by element calculations
    K = 10^5*(1-0.25*(x(i)+x(i+1)))/(x(i+1)-x(i));
    if(i~=1)         %
        a(i) = a(i) + K; a1(i) = K;
    end
        a(i+1) = K; a2(i+1) = K;
end
d(n) = 100;          % axial load last node
u = tdma(a2,a,a1,d); % solve for deflection using TDMA
```

Using 11 nodes, the total elongation of the block is 0.00138567 which is close to the exact solution.

Now we shall look at a third example related to bending of beams.

Example 11.16

Determine the deflection of the beam at the free end of a cantilever beam of length $L=1\,m$ for *the following configurations.*

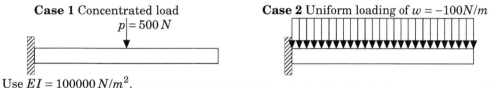

Case 1 Concentrated load
$p = 500\,N$

Case 2 Uniform loading of $w = -100N/m$

Use $EI = 100000\,N/m^2$.

Background :

The principle of minimum potential energy can be applied to beams as well. We have already introduced some concepts on beam deflection earlier. Now, consider a small section of the beam subjected to bending moments and shear forces.

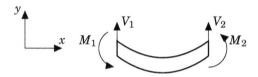

V_1 and V_2 are the shear forces acting at nodes and M_1 and M_2 are the bending moments at the nodes.

The strain for a beam element is given by

$$\epsilon = -y\frac{d^2u}{dx^2}$$

Then the strain energy for a beam element would be

$$SE = \frac{1}{2}\int_V E\left(y\frac{d^2u}{dx^2}\right)^2 dAdx = \frac{1}{2}\int_{x_1}^{x_2} E\left(\frac{d^2u}{dx^2}\right)^2 \underbrace{\left[\int_A y^2 dA\right]}_{I} dx = \frac{1}{2}\int_{x_1}^{x_2} EI\left(\frac{d^2u}{dx^2}\right)^2 dx$$

I is the moment of inertia of the beam cross-section. EI represents the flexural rigidity of the beam. The work done by the external forces V_1 and V_2 are V_1u_1 and V_2u_2 respectively. Similarly, the work done by the bending moments M_1 and M_2 are given by $M_1\theta_1$ and $M_2\theta_2$ respectively. θ is equal to $\dfrac{du}{dx}$. Hence the potential energy of the small section is given by

$$PE = \frac{1}{2}\int_{x_1}^{x_2} EI\left(\frac{d^2u}{dx^2}\right)^2 dx - V_1u_1 - V_2u_2 - M_1\theta_1 - M_2\theta_2$$

To apply FEM, we cannot discretize the domain using linear elements. Also, as the strain energy is proportional to the second derivative, we should have at least first order continuity (first derivative is continuous) at the nodes. In interpolation chapter, we have considered two curves which have first order continuity at the nodes viz. Hermite polynomial and cubic spline. We shall use Hermite polynomial as the interpolation polynomial.

The Hermite polynomial equation for a domain between $0 \leq \xi \leq 1$ (Equation 5.55) is given by

$$u(\xi) \quad = \quad u_1 \underbrace{(2\xi^3 - 3\xi^2 + 1)}_{N_{u,1}(\xi)} + u_2 \underbrace{(-2\xi^3 + 3\xi^2)}_{N_{u,2}(\xi)} + \theta_1 \underbrace{h(\xi^3 - 2\xi^2 + \xi)}_{N_{\theta,1}(\xi)} + \theta_2 \underbrace{h(\xi^3 - \xi^2)}_{N_{\theta,2}(\xi)}$$

where $h = x_2 - x_1$. The second derivative $\dfrac{d^2 u}{d\xi^2}$ can be written as

$$\frac{d^2 u}{d\xi^2} \quad = \quad u_1 N_{u,1}^{(2)} + u_2 N_{u,2}^{(2)} + \theta_1 N_{\theta,1}^{(2)} + \theta_2 N_{\theta,2}^{(2)}$$

$$= \quad u_1(12\xi - 6) + u_2(-12\xi + 6) + \theta_1 h(6\xi - 4) + \theta_2 h(6\xi - 2)$$

The strain energy in the local coordinates is

$$SE = \frac{EI}{2h^3} \int_0^1 \left(\frac{d^2 u}{d\xi^2} \right)^2 d\xi$$

As the potential energy of the beam element has to be minimum, we differentiate the potential energy with respect to the unknowns u_1, u_2, θ_1 and θ_2 to get

$$\frac{EI}{h^3} \int_0^1 \begin{bmatrix} N_{u,1}^{(2)} N_{u,1}^{(2)} & N_{u,1}^{(2)} N_{u,2}^{(2)} & N_{u,1}^{(2)} N_{\theta,1}^{(2)} & N_{u,1}^{(2)} N_{\theta,2}^{(2)} \\ N_{u,2}^{(2)} N_{u,1}^{(2)} & N_{u,2}^{(2)} N_{u,2}^{(2)} & N_{u,2}^{(2)} N_{\theta,1}^{(2)} & N_{u,2}^{(2)} N_{\theta,2}^{(2)} \\ N_{\theta,1}^{(2)} N_{u,1}^{(2)} & N_{\theta,1}^{(2)} N_{u,2}^{(2)} & N_{\theta,1}^{(2)} N_{\theta,1}^{(2)} & N_{\theta,1}^{(2)} N_{\theta,2}^{(2)} \\ N_{\theta,2}^{(2)} N_{u,1}^{(2)} & N_{\theta,2}^{(2)} N_{u,2}^{(2)} & N_{\theta,2}^{(2)} N_{\theta,2}^{(2)} & N_{\theta,2}^{(2)} N_{\theta,2}^{(2)} \end{bmatrix} \begin{Bmatrix} u_1 \\ u_2 \\ \theta_1 \\ \theta_2 \end{Bmatrix} dx - \begin{Bmatrix} V_1 \\ V_2 \\ M_1 \\ M_2 \end{Bmatrix}$$

This can be further simplified to

$$\frac{EI}{h^3} \begin{bmatrix} 12 & -12 & 6h & 6h \\ -12 & 12 & -6h & -6h \\ 6h & -6h & 4h^2 & 2h^2 \\ 6h & -6h & 2h^2 & 4h^2 \end{bmatrix} \begin{Bmatrix} u_1 \\ u_2 \\ \theta_1 \\ \theta_2 \end{Bmatrix} - \begin{Bmatrix} V_1 \\ V_2 \\ M_1 \\ M_2 \end{Bmatrix}$$

All the local elemental equations are assembled together to form a set of linear equations and are solved for the unknowns u and θ.

Solution :

Case 1 Exact solution for the present case is

$$u = \begin{cases} \dfrac{px^2}{6EI}(1.5 - x) & \text{for} \quad 0 < x < 0.5 \\[2mm] \dfrac{p}{48EI}(3x - 0.5) & \text{for} \quad 0.5 < x < 1 \end{cases}$$

Hence, we should expect the FEM solution to be exact. We shall discretize the domain into two elements, $h = 0.5$.

The boundary conditions are $u_1 = 0$ and $\theta_1 = 0$. Now let us write down the elemental equations.

Element 1

$$\frac{EI}{h^3} \begin{pmatrix} -12 & 12 & -6h & -6h \\ 6h & -6h & 2h^2 & 4h^2 \end{pmatrix} \begin{pmatrix} u_1 \\ u_2 \\ \theta_1 \\ \theta_2 \end{pmatrix}$$

Element 2

$$\frac{EI}{h^3} \begin{pmatrix} 12 & -12 & 6h & 6h \\ -12 & 12 & -6h & -6h \\ 6h & -6h & 4h^2 & 2h^2 \\ 6h & -6h & 2h^2 & 4h^2 \end{pmatrix} \begin{pmatrix} u_2 \\ u_3 \\ \theta_2 \\ \theta_3 \end{pmatrix}$$

Combining the two elemental equations

$$\frac{EI}{h^3} \begin{pmatrix} 1 & 0 & 0 & 0 & 0 & 0 \\ -12 & 24 & -12 & -6h & 0 & 6h \\ 0 & -12 & 12 & 0 & -6h & -6h \\ 0 & 0 & 0 & 1 & 0 & 0 \\ 6h & 0 & -6h & 2h^2 & 8h^2 & 2h^2 \\ 0 & 6h & -6h & 0 & 2h^2 & 4h^2 \end{pmatrix} \begin{pmatrix} u_1 \\ u_2 \\ u_3 \\ \theta_1 \\ \theta_2 \\ \theta_3 \end{pmatrix} - \begin{pmatrix} 0 \\ -500 \\ 0 \\ 0 \\ 0 \\ 0 \end{pmatrix} = 0$$

On solving we get $u_2 = -2.083 \times 10^{-4}$ m, $u_3 = -5.208 \times 10^{-4}$ m, $\theta_2 = -6.25 \times 10^{-4}$ rad and $\theta_3 = -6.25 \times 10^{-4}$ rad.

It can be shown that the numerical solution is the same as the exact solution.
Case 2 The exact solution for the present configuration is given below.

$$u = \frac{wx^2}{24EI}\left(x^2 + 6L^2 - 4Lx\right)$$

The strain energy for the present case remains the same as presented earlier. However, we have to consider the work done by the uniform loading on the beam. Consider an element subjected to uniform loading

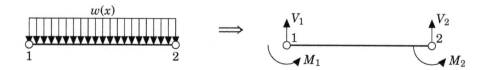

Then the work done by uniform loading is

$$W = h \int_0^1 w(\xi)\left(N_{u,1}u_1 + N_{u,2}u_2 + N_{\theta,1}\theta_1 + N_{\theta,2}\theta_2\right)d\xi$$

$$= h\left(\frac{w}{2}u_1 + \frac{w}{2}u_2 + \frac{wh}{12}\theta_1 - \frac{wh}{12}\theta_2\right)$$

The above equation is equivalent to shear force and bending moments acting at the two nodes. Differentiating the above to minimize the potential energy we get the elemental equations as

$$\frac{EI}{h^3}\begin{bmatrix} 12 & -12 & 6h & 6h \\ -12 & 12 & -6h & -6h \\ 6h & -6h & 4h^2 & 2h^2 \\ 6h & -6h & 2h^2 & 4h^2 \end{bmatrix}\begin{Bmatrix} u_1 \\ u_2 \\ \theta_1 \\ \theta_2 \end{Bmatrix} - \begin{Bmatrix} \dfrac{wh}{2} \\ \dfrac{wh}{2} \\ \dfrac{wh^2}{12} \\ -\dfrac{wh^2}{12} \end{Bmatrix}$$

First let us discretize the beam into two equal elements. The stiffness matrix for the beam would remain the same and only the force vector would change as below.

$$\frac{EI}{h^3}\begin{bmatrix} 1 & 0 & 0 & 0 & 0 & 0 \\ -12 & 24 & -12 & -6h & 0 & 6h \\ 0 & -12 & 12 & 0 & -6h & -6h \\ 0 & 0 & 0 & 1 & 0 & 0 \\ 6h & 0 & -6h & 2h^2 & 8h^2 & 2h^2 \\ 0 & 6h & -6h & 0 & 2h^2 & 4h^2 \end{bmatrix}\begin{Bmatrix} u_1 \\ u_2 \\ u_3 \\ \theta_1 \\ \theta_2 \\ \theta_3 \end{Bmatrix} - \begin{Bmatrix} 0 \\ -50 \\ -25 \\ 0 \\ 0 \\ 2.08333 \end{Bmatrix} = 0$$

Solving the above we get $u_1 = -4.43 \times 10^{-5}\ m$, $u_2 = -1.25 \times 10^{-4}\ m$, $\theta_1 = -1.458 \times 10^{-4}\ rad$ and $\theta_2 = -1.667 \times 10^{-4}\ rad$. We find that the numerical solution matches with the exact solution at the nodes.

11.7   Finite volume method

FVM has been frequently applied to numerical solution of problems in fluid dynamics and heat transfer. The method considers discretizing the domain into large number of sub domains or *volumes* with each volume represented by a node (usually placed at the center of the volume).

11.7.1   Background

Fluid flow and heat transfer are governed by conservation equations involving mass, momentum and energy. Conservation laws occur in also electromagnetism, neutron transfer, chemical reaction engineering etc. The basis of finite volume method is conservation of the variable y over each discretized volume. y could be mass, momentum or any other conserved variable depending on the problem that is being considered.

Let us consider conservation of energy in one dimension.

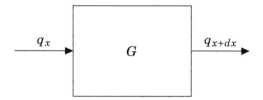

The above figure shows a small one dimensional element of length dx and area of cross section A. G is the amount of energy that is generated per unit volume in the system and q is the amount of energy per unit area entering/leaving the system. Applying energy conservation we get

$$\underbrace{GAdx + q_x A -}_{\text{energy added}} \underbrace{q_{x+dx} A}_{\text{energy removed}} = 0 \tag{11.46}$$

When the length of the element tends to 0, we get the differential equation

$$\frac{dq}{dx} - G = 0 \tag{11.47}$$

For a three dimensional element, we can generalize the conservation equation as

$$\frac{\partial q_x}{\partial x} + \frac{\partial q_y}{\partial y} + \frac{\partial q_z}{\partial z} - G = 0$$

$$\nabla.\mathbf{q} - G = 0$$

where $\mathbf{q} = [q_x \quad q_y \quad q_z]^T$ is the energy flux vector, $\nabla = \dfrac{\partial}{\partial x}\hat{\mathbf{i}} + \dfrac{\partial}{\partial y}\hat{\mathbf{j}} + \dfrac{\partial}{\partial z}\hat{\mathbf{k}}$ is the divergence operator where $\hat{\mathbf{i}}, \hat{\mathbf{j}}$ and $\hat{\mathbf{k}}$ are the unit vectors along the three directions.

Now let us consider an arbitrary system of volume δv. A_i is the area of the sides of the system and q_i is the energy per unit area entering normally the respective sides of the system as shown in the figure below.

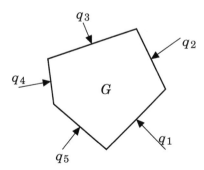

Now, the net amount of energy entering the system should be equal to the volumetric energy dissipated within the system

$$G\Delta v + q_1 A_1 + q_2 A_2 + q_3 A_3 + q_4 A_4 + q_5 A_5 = G\Delta v + \sum_i q_i A_i = 0 \tag{11.48}$$

As the number of sides of the system increase and the area of each side of the system becomes infinitesimally small, the summation operation can be replaced by integration of energy over the boundary. Hence, the above equation reduces to

$$G\Delta v + \oint \mathbf{q}.\mathbf{n}dA = 0 \qquad (11.49)$$

Applying Gauss divergence theorem[6] to the above equation we obtain

$$\oint (\nabla.\mathbf{q} - G)dv = 0 \qquad (11.50)$$

Hence, the above equation is the integral form of differential equation 11.47. Finite volume method considers Equation 11.49 as the governing equation in place of the differential equation. After discretization the integral equation reduces to the form of Equation 11.48. The advantage of finite volume method is that it is naturally *conservative* globally as well as locally.

11.7.2 Discretization

Discretization of the domain can be carried out in several ways.
Discretization 1 The nodes are distributed throughout the domain (uniform or nonuniform). The boundaries of the control volume are placed at the midpoint of two consecutive nodes. The volumes at the boundary are represented by half volumes as indicated in the following figure.

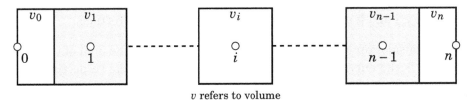

v refers to volume

It would be later shown that above discretization reduces to finite difference formulation under certain conditions.
Discretization 2 The entire domain is discretized into n volumes and two boundary nodes. Following figure indicates domain discretized into uniform volumes.

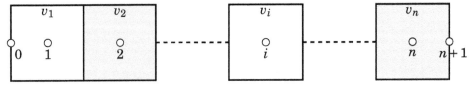

v refers to volume, 0 and $n+1$ are boundary nodes.

---

[6]Refer to a book on vector calculus such as Murray R. Spiegel, "An Introduction to Tensor Analysis", Schaum's outline series, McGraw Hill, 2009.

Governing differential equations are applied over the n volumes and boundary conditions are applied at the respective boundary nodes. This discretization scheme is more frequently used in the finite volume method.

Again the placement of nodes with respect to the volume can be done in two ways viz. cell centered and vertex centered. In cell centered discretization, the internal nodes are placed at the center of each volume. In vertex centered discretization scheme, the boundary of the volumes are placed at the midpoint between two consecutive nodes.

Cell centered discretization

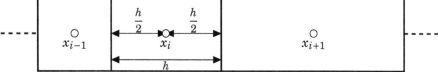

Vertex centered discretization

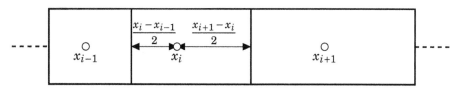

For a grid system with uniform volume elements, there is no distinction between cell-centered and vertex-centered discretization schemes. Hence in FVM, one has to specify the location of the nodes as well as the boundaries of the control volume.

11.7.3 Simple example with discretization scheme 1

Let us consider the differential equation $y^{(2)} - 2y = 0$ with boundary conditions $y(x = 0) = 1$ and $y^{(1)}(x = 1) = 0$. Without delving into the physical attributes of the differential equation, let us understand the concept of FVM. FVM is a weighted residuals method where the weight function is 1 over corresponding sub domain and zero otherwise.

$$W_i = \begin{array}{ll} 1 & \text{for } i^{th} \text{ volume} \\ 0 & \text{for other volumes} \end{array} \tag{11.51}$$

The domain is discretized uniformly using discretization scheme 1. This means, that all internal nodes are placed at the center of the control volume and the vertex of the control volume is equi distant from the two adjacent nodes. Therefore the residual equation for i^{th} *internal* volume is

$$\int_{x_i-0.5h}^{x_i+0.5h} (y^{(2)} - 2y)dx = \left[\frac{dy}{dx}\bigg|_{x_i+0.5h} - \frac{dy}{dx}\bigg|_{x_i-0.5h} \right] - 2\int_{x_i-0.5h}^{x_i+0.5h} ydx \tag{11.52}$$

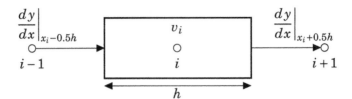

Notice that the derivatives are evaluated at the boundaries of the control volume (just as we did in the derivation of the equations of conservation of energy). The evaluation of the first derivatives and integral depends on the choice of basis functions. A simple choice of basis function is to assume y is constant inside every volume. The first derivatives at the faces can be approximated by central difference formulae. Then Equation 11.52 reduces to

$$\frac{y_{i+1} - y_i}{h} - \frac{y_i - y_{i-1}}{h} - 2y_i h = 0$$

$$y_{i+1} - 2y_i + y_{i-1} - 2h^2 y_i = 0$$

Note the similarities between FVM and FEM. As both treat integral equations of the governing equation, a term related to $y^{(1)}$ arises in the discretized equation. In FEM discussed earlier, this term was conveniently neglected, which leads to non-zero residuals. In FVM, this term is approximated using finite difference approximations. As all the terms in the discretized equation are accounted for, the net residual for a control volume is zero. This makes FVM naturally conservative and hence is favored for treating conservative laws! Note: In the present case, second order central difference was used for calculating the derivatives. Higher order difference formulas can also be used. There is thus a similarity between FVM and FDM.

Boundary conditions: The function value is specified at x_0. Let us consider the other boundary node x_n where Neumann boundary condition has been specified. Applying FVM to the boundary control volume.

From the above figure it is clear that the boundary conditions are included by the numerical method automatically. Therefore the residual equation for n^{th} volume is

$$\int_{x_n-0.5h}^{x_n} (y^{(2)} - 2y)dx = \left[\frac{dy}{dx}\bigg|_{x_n} - \frac{dy}{dx}\bigg|_{x_n-0.5h} \right] - 2\int_{x_n-0.5h}^{x_n} y dx = 0$$

$$0 - \frac{y_n - y_{n-1}}{h} - 2y_n \frac{h}{2} = y_{n-1} - y_n - h^2 y_n = 0$$

The above set of linear equations obtained using FVM is the same as that obtained using FDM with ghost node approach at the boundaries.

11.7.4 Simple example with discretization scheme 2

Let us consider the same differential equation as earlier. Now, the domain is discretized uniformly using scheme 2. The reader may verify that the equations for internal nodes $2 \leq i \leq n-1$ are the same as derived earlier. The difference between the two schemes only arises in the treatment of boundary conditions. First let us apply FVM to node at x_n.

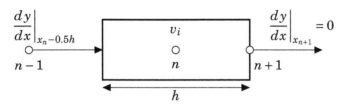

The residual equation for n^{th} volume is

$$\int_{x_n-0.5h}^{x_{n+1}} (y^{(2)} - 2y)dx = \left[\frac{dy}{dx}\Big|_{x_{n+1}} - \frac{dy}{dx}\Big|_{x_n-0.5h}\right] - 2\int_{x_n-0.5h}^{x_{n+1}} ydx = 0$$

$$0 - \frac{y_n - y_{n-1}}{h} - 2y_nh = y_{n-1} - y_n - 2h^2 y_n = 0$$

The above equation has directly accounted for the boundary condition. Notice that function value at $n+1$ does not occur in the above equation, decoupling it from the rest of the nodes. (The same would not be the case if Robin condition was specified at x_{n+1}) The function value at x_{n+1} can be determined by using finite difference approximation at x_{n+1} such as first order formula $\dfrac{y_{n+1} - y_n}{0.5h} = 0$ or three point backward difference formula. The three nodes are not equally spaced and hence we shall differentiate the quadratic polynomial passing through the three nodes and obtain

$$\frac{y_{n-1} - 9y_n + 8y_{n+1}}{3h} = 0$$

Now let us consider node at x_1.

$$\frac{dy}{dx}\Big|_{x_0} \qquad v_i \qquad \frac{dy}{dx}\Big|_{x_1+0.5h}$$

$$0 \qquad 1 \qquad 2$$

$$h$$

The residual equation for 1st volume is

$$\int_{x_0}^{x_1+0.5h} (y^{(2)} - 2y)dx = \left[\frac{dy}{dx}\Big|_{x_1+0.5h} - \frac{dy}{dx}\Big|_{x_0}\right] - 2\int_{x_0}^{x_1+0.5h} ydx = 0$$

We shall use the central difference formula for $\dfrac{dy}{dx}\Big|_{x_1+0.5h}$ and three point forward

difference formula for $\dfrac{dy}{dx}\Big|_{x_0}$ which is

$$\frac{dy}{dx}\Big|_{x_1+0.5h} = \frac{y_2 - y_1}{h}; \qquad \frac{dy}{dx}\Big|_{x_0} = \frac{-8y_0 + 9y_1 - y_2}{3h}$$

Then we obtain,

$$\frac{8}{3}y_0 - \left(4 + 2h^2\right)y_1 + \frac{4}{3}y_2 = 0$$

Hence, the set of linear equations is

$$
\begin{bmatrix}
1 & 0 & 0 & \cdots & \cdots & \cdots & 0 & 0 \\
8/3 & -4-2h^2 & 4/3 & \cdots & \cdots & \cdots & 0 & 0 \\
0 & 1 & -2-2h^2 & 1 & \cdots & \cdots & 0 & 0 \\
\cdots & \cdots & \cdots & \cdots & \cdots & \cdots & \cdots & \cdots \\
0 & 0 & \cdots & \cdots & 1 & -2-2h^2 & 1 & 0 \\
0 & 0 & \cdots & \cdots & 0 & 1 & -1-2h^2 & 0 \\
0 & 0 & \cdots & \cdots & 0 & 1 & -9 & 8
\end{bmatrix}
\begin{pmatrix}
y_0 \\ y_1 \\ y_2 \\ y_3 \\ \cdots \\ y_{n-2} \\ y_{n-1} \\ y_n \\ y_{n+1}
\end{pmatrix}
=
\begin{pmatrix}
1 \\ 0 \\ 0 \\ 0 \\ \cdots \\ 0 \\ 0 \\ 0
\end{pmatrix}
\qquad (11.53)
$$

11.7.5   Example using piecewise linear function

Previously we have assumed that the dependent variable is constant within each volume element. However it is possible to assume a better variation such as a linear variation within each volume element. We then have

$$y = \frac{x - x_{i+1}}{x_i - x_{i+1}}y_i + \frac{x - x_i}{x_{i+1} - x_i}y_{i+1} \quad \text{for } x_i \le x \le x_{i+1}$$

The entire domain is discretized using scheme 1. The first derivatives at the boundaries of a control volume remain the same as FDM. However, the term proportional to y is subjected to change. Evaluating the integral of y over the volume element we get

$$
\begin{aligned}
\int_{x_i-0.5h}^{x_i+0.5h} y\,dx &= \int_{x_i-0.5h}^{x_i}\left(\frac{x - x_i}{x_{i-1} - x_i}y_{i-1} + \frac{x - x_{i-1}}{x_i - x_{i-1}}y_i\right)dx \\
&\quad + \int_{x_i}^{x_i+0.5h}\left(\frac{x - x_{i+1}}{x_i - x_{i+1}}y_i + \frac{x - x_i}{x_{i+1} - x_i}y_{i+1}\right)dx \\
&= 0.125h\,y_{i-1} + 0.75h\,y_i + 0.125h\,y_{i+1}
\end{aligned}
$$

Hence, the equations at the interior nodes can be written as

$$(1 - 0.25h^2)y_{i+1} - (2 + 1.5h^2)y_i + (1 - 0.25h^2)y_{i-1} = 0$$

Similarly, one can derive the equations at the boundary as

$$(1 - 0.25h^2)y_{n-1} - (1 + 0.75h^2)y_n = 0$$

Let $a = 1+0.75h^2$ and $a' = 1-0.25h^2$, we represent the system of linear equation in matrix notation.

$$
\begin{pmatrix}
1 & 0 & 0 & \cdots & \cdots & 0 \\
a' & -2a & a' & \cdots & \cdots & 0 \\
0 & a' & -2a & a' & \cdots & 0 \\
\cdots & \cdots & \cdots & \cdots & \cdots & \cdots \\
0 & 0 & \cdots & a' & -2a & a' \\
0 & 0 & \cdots & 0 & a' & -a
\end{pmatrix}
\begin{pmatrix}
y_0 \\ y_1 \\ y_2 \\ \cdots \\ y_{n-2} \\ y_{n-1} \\ y_n
\end{pmatrix}
=
\begin{pmatrix}
1 \\ 0 \\ 0 \\ \cdots \\ 0 \\ 0
\end{pmatrix}
\tag{11.54}
$$

MATLAB program has been written to carry out the above procedure.

```
n = 11;                        % no of nodes
x = [0:1/(n-1):1]';            % nodes
h = 1/(n-1);                   % step size
a = zeros(n,1);                % TDMA coefficient a
a1 = zeros(n,1);               % TDMA coefficient a'
a2 = zeros(n,1);               % TDMA coefficient a''
b = zeros(n,1);                % TDMA b
a(1) = 1;                      % first node
b(1) = 1;                      % Dirichlet condition
for i=2:n-1                    % internal nodes
    a1(i) = 1-0.25*h^2;
    a2(i) = 1-0.25*h^2;
    a(i) = 2+1.5*h^2;
end
a2(n) = 1-0.25*h^2;            % last node
a(n) = 1+0.75*h^2;             % Neumann condition
y = tdma(a2,a,a1,d);           % solve using TDMA
```

Table 11.3 compares the solution obtained using three methods viz. FDM, FEM and FVM with the exact solution. The reader may note that the exact solution is given by

$$
y_E = \frac{\cosh(\sqrt{2}(1-x))}{\cosh(\sqrt{2})}
$$

It is interesting to notice that FDM, FEM and FVM produce similar solutions. The reader is encouraged to perform the grid sensitivity study and check for the convergence of the three methods.

Table 11.3: *Comparison of FDM, FEM and FVM solutions*

x	y_{FDM}	y_{FEM}	y_{FVM}	Exact
0	1	1	1	1
0.1	0.88408	0.88384	0.88390	0.88396
0.2	0.78585	0.78541	0.78552	0.78563
0.3	0.70333	0.70274	0.70289	0.70304
0.4	0.63488	0.63418	0.63435	0.63453
0.5	0.57913	0.57834	0.57854	0.57874
0.6	0.53496	0.5341	0.53432	0.53453
0.7	0.50149	0.50059	0.50081	0.50104
0.8	0.47805	0.47711	0.47735	0.47758
0.9	0.46417	0.46322	0.46346	0.46370
1	0.45958	0.45862	0.45886	0.45910

Concluding remarks

BVPs involving second or higher order differential equations occur in many areas of science and engineering. As opposed to IVPs the conditions to be satisfied by the function and/or its derivatives are specified at more than one location. Hence, in general, BVPs are more difficult to solve and may involve iterative procedures in the case of nonlinear BVPs. Several methods including FDM have been presented in great detail to afford the reader a choice in the solution of such equations. Basic introduction to collocation, FEM and FVM have also been given. The reader should refer to advanced books to carry forward from here.

We will see later that many of the concepts introduced in the solution of ODEs are also useful in solving PDEs.

11.A MATLAB routines related to Chapter 11

MATLAB routine	Function
bvp4c	solution of BVPs
bvp5c	solution of BVPs

The difference between the above two program is in the specification of error.

11.B Suggested reading

1. **L.F. Shampine**, **I. Gladwell** and **S. Thompson** *Solving ODEs with MATLAB* Cambridge University Press, 2003
2. **B.A. Finlayson** *The method of weighted residuals and variational principles* Academic Press, 1972
3. **L.J. Segerlind** *Applied finite element analysis* Wiley, 1976
4. **J.P. Boyd** *Chebyshev and Fourier spectral methods* Dover publications, 2001

Exercise III

III.1 Initial value problems

Ex III.1: Solve the first order ODE $y^{(1)} = -e^{-2t}y + (t^2 - 1)$ with $y(0) = 1$. Obtain the value of $y(1)$ numerically using (a) Euler method and (b) Heun method. What step size will guarantee 6 digit accuracy in the two methods.

Ex III.2: Solve the ODE of Exercise III.1 using RK4 method. What step size is suitable for obtaining solution good to six digits after decimals? Compare the computational effort required in using RK4 with that required in the case of Euler and Heun schemes, for achieving the same accuracy.

Ex III.3: Solve the ODE of Exercise III.1 using RK4 method to obtain enough number of function values to continue the solution by ABM4 method with corrections. Compare ABM4 with RK4 in terms of the computational effort needed to get the same accuracy of 5 digits after decimals.

Ex III.4: Solve the first order ODE $\dfrac{dy}{dt} + my = n\cos(\omega t)$ where $m = 0.1$, $n = 1.5$ and $\omega = 0.2$. All quantities are in compatible units and hence the reader need not worry about these. At zero time y is given to be 1. Use ABM2 to solve the problem. Choose a suitable step size after giving reason(s) for the choice. Obtain the solution up to a t beyond which the solution shows a steady state behavior (what is it?). Compare the numerical result with the exact solution.[7]

Ex III.5: A first order system is governed by the ODE $y^{(1)} = -y^{0.25}$. The initial condition is specified as $y(0) = 5$. Obtain the solution up to $t = 2$ by RK4. Study the effect of step size on the solution. Starting step size may be taken as $\Delta t = 0.25$.
Use the exact solution for comparison purposes.

Ex III.6: Solve the following set of two ODEs by a numerical scheme of your choice.

$$y_1^{(1)} = y_2 - y_1 \qquad y_2^{(1)} = y_1 - 3y_2$$

Initial conditions are specified as $y_1(0) = 1$, $y_2(0) = 0.5$. Results are required till $t = 5$ and accuracy to five digits after decimals. Solution is to be compared with the exact solution.

---

[7]Based on an example in the book: S.P. Venkateshan, "Heat Transfer", 2nd Edition, Ane Books, 2011. Governing equation models the transient of a lumped system under convection and incident time varying heat flux.

Ex III.7: Numerically solve the following two coupled first order ODEs.

$$y_1^{(1)} = y_2 + 1 \qquad y_2^{(1)} = y_1 + 1$$

The initial conditions are $y_1(0) = 1$, $y_2(0) = 0$. Compare the numerical solution with the exact solution.

Ex III.8: Convert the two coupled ODEs of Exercise III.6 to a single second order ODE by eliminating one of the dependent variables between the two equations. What will be the two initial conditions to the second order ODE? Solve the resulting second order ODE by an initial value solver of your choice. Obtain therefrom the solution to the dependent variable that was eliminated while arriving at the single second order ODE.

Ex III.9: Solve the set of two coupled first order differential equations given below with the initial conditions $y_1(0) = 1$ and $y_2(0) = 0.5$.[8]

$$4y_1^{(1)} - y_2^{(1)} + 3y_1 = \sin(t); \ y_1^{(1)} + y_2 = \cos(t)$$

Ex III.10: Transient behavior of a second order system subject to step input is governed by the ODE $y^{(2)} + 0.25y^{(1)} + y = 1$. The initial conditions have been specified as $y(0) = y^{(1)}(0) = 0$. Solve this equation numerically using adaptive RK4.

Ex III.11: Solve the following coupled second order ordinary differential equations by a numerical scheme of your choice. Solution with five digit precision is required.

$$y_1^{(2)} + y_2^{(1)} = 0; \ y_2^{(2)} - y_1^{(1)} = 0$$

Initial conditions have been specified as $y_1(0) = y_2(0) = 0$; $y_1^{(1)}(0) = 1$, $y_2^{(1)}(0) = 0$. Plot the solution as a trajectory with y_1 along the abscissa and y_2 along the ordinate. Compare your solution with the exact solution.[9]

Ex III.12: Certain object is moving under the influence of gravity and experiences wind resistance proportional to the square of its velocity. The governing differential equation has been derived and is given by $\dfrac{d^2y}{dt^2} + 0.1\left(\dfrac{dy}{dt}\right)^2 = 10$. At $t = 0$ both y and $\dfrac{dy}{dt}$ are specified to be zero. Obtain the solution numerically and determine the value of y at $t = 10$.[10] (Hint: Terminal velocity of the object should be 10 for large t such as $t = 10$.)

Ex III.13: A boat of mass $M = 500 \ kg$ experiences water resistance of $F_D = 50v$ where F_D is the drag force in N and v is its velocity in m/s. Boat is initially moving with a velocity of $v = 15 \ m/s$. How long does it take the boat to reduce the speed to $7.5 \ m/s$? How far will it be from the starting point. Obtain the solution by solving the equation of motion of the boat numerically using a second order accurate method and compare the results with the exact results.

---

[8] Based on an example in the book: H.B. Phillips, "Differential Equations", John Wiley and Chapman and Hall, 1922

[9] Ibid: Motion of an electrically charged particle in a magnetic field is modeled by these equations

[10] Ibid

Ex III.14: Solve the initial value problem $t\dfrac{d^2y}{dt^2} - \dfrac{dy}{dt} + 4t^3y = 0$ subject to the initial conditions $y = 1$ and $\dfrac{dy}{dt} = -1$ at the initial time of $t = \sqrt{\dfrac{\pi}{2}}$. Carry forward the calculations ten steps from the starting value using a step size of $\Delta t = 0.05$. Make use of the RK4 method. Compare the numerical solution with the exact solution given by $y = \sin(t^2) + \dfrac{1}{\sqrt{2\pi}}\cos(t^2)$.

Ex III.15: Dynamics of a linearly damped pendulum is governed by the second order ODE $y^{(2)} + 0.5y^{(1)} + 10\sin(y) = 0$. The initial conditions have been specified as $y(0) = 0.5$ $y^{(1)}(0) = 0$. Obtain the solution to $t = 2.5$ with five digit accuracy. Comment on the results based on the observed behavior of the solution. Can you identify periodicity in the solution?

Ex III.16: Consider the damped pendulum of Exercise III.15 again but with a forcing function of $f(t) = 5\sin(t)$. Solve the resulting non-homogeneous equation by a suitable initial value solver. Make comments on the observed behavior of the pendulum by obtaining the solution to large enough t.

Ex III.17: Solve the first order ODE $y^{(1)} = 1 - y^2$ from $t = 0$ to $t = 5$. Initial value is specified as $y(0) = 0.2$. Use Euler and Backward Euler methods. Make suitable comments based on the solution of the given ODE by these two methods. Write a program to solve the equation by second order BDF method. Compare the numerical solution with analytically obtained solution.

Ex III.18: Solve the initial value problem $y^{(1)} = 3(1 - y^4)$ with $y(0) = 0$ from $t = 0$ to $t = 1.5$. Use any scheme as long as you achieve 4 digit accuracy.[11]

Ex III.19: Obtain the solution of the initial value problem $y^{(2)} = x^2 - \dfrac{2}{x^2} - xy$ in the range $1 \le t \le 2$. The initial conditions are specified as $y(1) = 1$ and $y^{(1)}(0) = -1$. The solution should be accurate to 5 digits after decimals. Use a solver of your choice.

Ex III.20: Solve the IVP $\dfrac{d^2y}{dt^2} + 2\zeta\dfrac{dy}{dt} + 4y = \sin(2t)$ subject to initial conditions $y(0) = 0.2$ and $\dfrac{dy}{dt}\Big|_{t=0} = 0$. Make use of RK4 combined with ABM4 to solve the problem. Compare the results with the exact solution. Consider three cases: (i) $\zeta = 1$ (ii) $\zeta = 2$ and (iii) $\zeta = 2.5$. Comment on the nature of the solution in each case.

Ex III.21: Consider a simple two stage chemical reaction

$$y_1 \xrightarrow{k_1} y_2; \qquad y_2 \xrightarrow{k_2} y_3$$

where k_1 and k_2 are kinetic rate constants of the two reactions. ODEs governing the chemical reactions are given by

$$\frac{dy_1}{dt} = -k_1y_1; \quad \frac{dy_2}{dt} = k_1y_1 - k_2y_2; \quad \frac{dy_3}{dt} = -k_2y_2;$$

Comment on the stiffness of the ODEs for the following combinations of k_1 and k_2. (i) $k_1 = 0.01$ and $k_2 = 0.05$, (ii) $k_1 = 10$ and $k_2 = 0.05$ and (iii) $k_1 = 0.01$ and $k_2 = 10$. Solve the system of ODEs for the above combinations of k_1 and k_2 using a suitable numerical scheme with proper justification. Use the following initial conditions: $y_1(0) = 1$, $y_2(0) = 0$ and $y_3(0) = 0$.

---

[11]It is clear that as $t \to \infty$ the dependent variable approaches unity. When this happens the derivative of the right hand side viz. $f^{(1)}(y) \to -12$ which is large. Hence the given ODE is stiff.

Ex III.22: A simple chemical reaction usually involves a large number of intermediate reaction steps and are nonlinear and stiff making numerical solution very difficult. An interesting example is that proposed by Robertson.[12]

$$y_1 \xrightarrow{k_1} y_2; \qquad y_2 + y_2 \xrightarrow{k_2} y_3 + y_2; \qquad y_2 + y_3 \xrightarrow{k_3} y_1 + y_3;$$

leading to the following set of ODEs

$$\begin{pmatrix} y_1^{(1)} \\ y_2^{(1)} \\ y_3^{(1)} \end{pmatrix} = \begin{pmatrix} -k_1 y_1 + k_3 y_2 y_3 \\ k_1 y_1 - k_2 y_2^2 - k_3 y_2 y_3 \\ k_2 y_2^2 \end{pmatrix}$$

where the rate constants are $k_1 = 0.04$, $k_2 = 3 \times 10^7$ and $k_3 = 10^4$. Solve the above system of ODEs using a suitable method. Reader is encouraged to use the inbuilt MATLAB functions for obtaining the numerical solution. Assume initial conditions as $y_1(0) = 1$, $y_2(0) = 0$ and $y_3(0) = 0$.

III.2 Boundary value problems

Ex III.23: Solve the boundary value problem $\dfrac{d^2 y}{dx^2} + \dfrac{1}{x}\dfrac{dy}{dx} - 3y = 0$ subject to the boundary conditions $y(x = 1) = 1$, $\dfrac{dy}{dx}\Big|_{x=2} = 0$ by the use of shooting method. Fourth order Runge Kutta method may be used for starting the solution followed by the ABM4 method with corrections for continuing the solution to the other boundary. Use a uniform step size of $\Delta x = 0.05$.

Ex III.24: Solve the boundary value problem $\dfrac{d^2 y}{dx^2} = 2y + \left[1 - \sin\left(\dfrac{\pi x}{2}\right)\right]$ subject to the boundary conditions $y(0) = 1$ and $\dfrac{dy}{dx}\Big|_{x=1} = 0$. Use the shooting method. Choose a step size of $\Delta x = 0.1$ and RK4 - ABM4 combination to solve the problem. Compare the numerical solution with the exact solution and comment on the accuracy of the numerical scheme used.

Ex III.25: Solve the boundary value problem $(1 + y)\dfrac{d^2 y}{dx^2} + \left(\dfrac{dy}{dx}\right)^2 = 0$ subject to the boundary conditions $y(0) = 0$; $y(1) = 1$. Use a second order accurate scheme for your calculations. For example, you may make use of the Heun method. Compare the numerical solution with the exact solution.[13]

Ex III.26: Solve the two point non-linear boundary value problem $\dfrac{d^2 y}{dx^2} = y + y^4$ by shooting method. The boundary conditions are specified as $y(0) = 1$ and $\dfrac{dy}{dx}\Big|_{x=1} = 0$.

---

[12]H.H. Robertson *The solution of a set of reaction rate equations* In J. Walsh, editor, "Numerical Analysis: An Introduction", pages 178-182. Academic Press, London, 1966.

[13]A simple one dimensional heat conduction problem with variable thermal conductivity leads to this equation

Ex III.27: Consider the boundary value problem $\dfrac{d^2y}{dx^2} = c\dfrac{dy}{dx}$ where c is a parameter that can take different values. Consider four different values viz. $c = 0.1$, $c = 1$, $c = 10$ and $c = 100$ and solve the equation by a numerical method of your choice. The boundary conditions remain the same in all the cases with $y(0) = 1$ and $\left.\dfrac{dy}{dx}\right|_{x=1} = 0$. Comment on the step size required in the four cases and also the nature of the solution. Compare the solution with the exact solution, in each of the cases.[14]

Ex III.28: Solve the boundary value problem $\dfrac{d^3y}{dx^3} - 4\dfrac{dy}{dx} + 3 = 0$ with $y(0) = 0.1$, $y(1) = 0.5$ and $\left.\dfrac{dy}{dx}\right|_{x=1} = -0.25$. Use shooting method for solving the problem. Six digit accuracy is required.

Ex III.29: Velocity of viscous fluid flow in the boundary layer past a flat plate is governed by the boundary value problem $y^{(3)} + 0.5yy^{(2)} = 0$ where $y^{(1)}$ is the velocity. The applicable boundary conditions are $y(0) = y^{(1)}(0) = 0$ and $y^{(1)} \to 1$ as $y \to \infty$. Obtain the solution to this equation using the shooting method.
Hint: Outer boundary condition needs an iteration scheme.

Ex III.30: Solve the boundary value problem of Exercise III.25 by the finite difference method. Compare the solution obtained by FDM with the exact solution.

Ex III.31: Solve the boundary value problem of Exercise III.26 by the finite difference method. Compare the solution obtained by FDM with that obtained earlier by the shooting method.

Ex III.32: Consider the boundary value problem given in Exercise III.24. Solve it by the finite difference method. Use a suitable step size such that the solution agrees with that obtained there by the shooting method.

Ex III.33: Consider the boundary value problem given in Exercise III.24. Solve it by the collocation method. Compare the solution with that obtained earlier by the shooting method as well as the exact solution.

Ex III.34: Solve the boundary value problem given in Exercise III.23 by the finite difference method. In addition to the boundary conditions specified there, consider also a second case where the boundary conditions are $y(x = 1) = 1$ and $y(x = 2) = 0.5$.

Ex III.35: Inhomogeneous ODE in the form $\dfrac{1}{r}\dfrac{d}{dr}\left(r\dfrac{du}{dr}\right) = \cos\left(\dfrac{\pi(r-1)}{2}\right)$ is satisfied by a function $u(r)$ in an annulus defined by inner boundary $r = 1$ and outer boundary $r = 2$. Neumann condition $\dfrac{du}{dr} = 0$ is specified at the inner boundary and Robin condition $\dfrac{du}{dr} = 0.1(u^4 - 0.4)$ is specified at the outer boundary. Obtain the solution to this problem by the finite difference method.

Ex III.36: The temperature field in a circular disk of radius 1 is governed by the equation $\dfrac{d^2u}{dr^2} + \dfrac{1}{r}\dfrac{du}{dr} - 2u = 0$. The periphery of the disk is specified with the Dirichlet condition $u(r = 1) = 1$. Obtain the solution to this boundary value problem by finite difference method.

---

[14]The ODE becomes stiffer as c is increased

Ex III.37: The governing ODE for steady conduction in a material with temperature dependent thermal conductivity is given by $\dfrac{d}{dx}\left((1+u)\dfrac{du}{dx}\right) - 2u = 0$. Boundary conditions are specified as $u(x = 0) = 1$ and $u(x = 1) = 0.5$. Obtain the solution to this nonlinear BVP by finite differences.

Module **IV**

Partial differential equations

Many problems in science and engineering involve functions of more than one variable. For example, the temperature in a solid depends, in general, on time and space coordinate(s). Equation governing temperature in the solid is the heat equation which is a partial differential equation (PDE) involving derivatives of temperature with respect to each of the independent variables. They may be first or higher partial derivatives. A material under the action of external loads is subject to stresses that vary with more than one independent variable again leading to one or more PDEs. Motion of a fluid in more than one dimension is governed by a set of PDEs involving the pressure and the fluid velocity vector. Problems in electricity and magnetism are governed by appropriate PDEs involving electrical and magnetic field vectors. Thus the solution of PDEs is an important part of analysis in science and engineering. Numerical solution of PDEs are built upon the ideas discussed while dealing with ODEs. Yet, special methods are required in treating PDEs. In this module we look at the fundamental ideas involved in the numerical solution of PDEs.

Chapter 12

Introduction to PDEs

12.1 Preliminaries

In most applications in science and engineering we come across second order partial differential equations. Hence these are of particular interest to us. However it is instructive to look at first order PDEs before turning our attention to PDEs of second order. For most part we will be considering two independent variables such as t- time and x - space or x- space and y- space. Hence the PDEs will involve partial derivatives with respect to t, x or x,y.

To start the discussion consider a first order homogeneous partial differential equation, typically of the form

$$a\frac{\partial u}{\partial x} + b\frac{\partial u}{\partial y} = 0 \qquad (12.1)$$

where a and b are non-zero constants. Using short hand notation $\dfrac{\partial u}{\partial x} = u_x$ and $\dfrac{\partial u}{\partial y} = u_y$ we rewrite the above in the form

$$a u_x + b u_y = 0 \qquad (12.2)$$

The above equation may also be written in the form of a matrix equation given by

$$\begin{pmatrix} a & b \end{pmatrix} \begin{pmatrix} u_x \\ u_y \end{pmatrix} = 0 \qquad (12.3)$$

or in vector form as

$$\mathbf{a} \cdot \nabla u = 0 \qquad (12.4)$$

Thus, at every point the gradient of u is perpendicular to the vector $\mathbf{a}$ whose components are a and b.

We seek solution to the above in the form $u = f(\alpha x + \beta y) = f(\xi)$ where ξ is short hand notation for $\alpha x + \beta y$. We shall also denote the total derivative of u with respect to ξ as u_ξ. Substituting the above in Equation 12.2 we get

$$a\alpha u_\xi + b\beta u_\xi = 0 \quad \text{or} \quad a\alpha = -b\beta$$

With this we have $\alpha x + \beta y = \beta\left(\dfrac{b}{a}x - y\right)$. Note now that for $u = \text{constant} = C$ we have $\dfrac{b}{a}x - y = \text{constant} = c$ and hence it is a straight line with slope equal to $\dfrac{b}{a}$ and intercept $-c$ on the y axis. The line is referred to as a "characteristic" of the first order PDE. For different values of C we get a set of parallel characteristic lines but with different intercepts as shown in Figure 12.1.

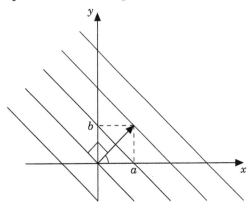

Figure 12.1: *Characteristic lines*

12.2   Second order PDE with constant coefficients

Consider now the second order homogeneous PDE with constant coefficients given by

$$au_{xx} + bu_{xy} + cu_{yy} = 0 \tag{12.5}$$

We would like to test whether the solution can be represented by a function of the form $u = u(\alpha x + \beta y) = u(\xi)$. The derivatives appearing in Equation 12.5 may all be written down as under:

$$u_{xx} = \alpha^2 u_{\xi\xi}, \quad u_{xy} = \alpha\beta u_{\xi\xi}, \quad u_{yy} = \beta^2 u_{\xi\xi} \tag{12.6}$$

Substituting these in Equation 12.5, removing the common term $u_{\xi\xi}$ and denoting the ratio $\dfrac{\alpha}{\beta}$ as γ (note that this will be the same as the slope of the characteristic line) we get the quadratic equation

$$a\gamma^2 + b\gamma + c = 0 \tag{12.7}$$

γ can have two values corresponding to the two roots of the above quadratic given by

$$\gamma = \frac{-b \pm \sqrt{b^2 - 4ac}}{2a} \tag{12.8}$$

The type of PDE is determined by the discriminant (quantity under the square root sign).

Hyperbolic: If it is positive we have two real roots i.e two real and distinct slopes for the characteristic lines. The PDE is said to be "hyperbolic". An example of an PDE of this type is the wave equation in one dimension given by

$$u_{tt} - s^2 u_{xx} = 0 \tag{12.9}$$

where t is time variable and x is the space variable and s is the wave speed. Note that $a = 1$, $b = 0$ and $c = -s^2$ and hence the slope of the two characteristics are $\pm s$. The two characteristics are given by $\xi_1 = st + x$ and $\xi_2 = -st + x$. Since both time and space derivatives are second derivatives, this equation can support two initial conditions and two boundary conditions.

Parabolic: In case the discriminant is zero the two roots are real, identical and given by $-\dfrac{b}{2a}$. Essentially there is only one characteristic. The PDE is said to be "parabolic". A common example of a parabolic type of equation is the heat equation given by

$$u_{xx} - u_t = 0 \tag{12.10}$$

In this case $a = 1$, $b = 0$ and $c = 0$. Hence $\gamma = 0$ is the repeated root. The characteristic simply becomes $\xi = t$. This equation will support an initial condition and two boundary conditions.

Elliptic: In case the discriminant is negative there are no real roots and hence there is no characteristic with real slope. This type of ODE is referred to be "elliptic". A common example of an elliptic equation is the Laplace equation given by

$$u_{xx} + u_{yy} = 0 \tag{12.11}$$

In this case $a = 1$, $b = 0$ and $c = 1$. Hence $b^2 - 4ac = -4$ and hence the two roots are $\pm j$ (pure imaginary). In this case both coordinates are space coordinates and hence the equation supports only boundary conditions. Unlike the other two types of equations which are initial boundary value problems the elliptic PDE is always a boundary value problem.

In practice we may come across PDEs which cannot be classified as parabolic, hyperbolic or elliptic i.e. these equations have terms having different characteristics. Some of the PDEs occurring in practice are given below.

1. One dimensional advection equation

$$\frac{\partial u}{\partial t} + a \frac{\partial u}{\partial x} = 0$$

where a is the advection velocity. Advection equation is important in fluid dynamics and heat transfer.

2. Poisson equation has applications in electrodynamics, heat transfer, mass diffusion, seeping flows, stress in continuum etc. Poisson equation[1] is given by

$$\underbrace{\frac{\partial^2 u}{\partial x^2} + \frac{\partial^2 u}{\partial y^2} + \frac{\partial^2 u}{\partial z^2}}_{\text{Laplacian of } u} = q$$

The above equation is elliptic. Laplacian term can also be written as $\nabla^2 u$. If $q = 0$, we get the Laplace equation.

3. Diffusion equation which occurs in heat transfer and fluid dynamics is elliptic with respect to space and parabolic with respect to time.

$$\frac{\partial u}{\partial t} = \alpha \nabla^2 u$$

α is known as diffusivity.

4. Wave equation occurs in electromagnetism, mechanical vibrations, acoustics and is given by

$$\frac{\partial^2 u}{\partial t^2} + a^2 \nabla^2 u = 0$$

where a is the wave velocity.

5. The following equation is the x momentum equation of the Navier Stokes equations used in fluid dynamics

$$\frac{\partial u}{\partial t} + \underbrace{u \frac{\partial u}{\partial x} + v \frac{\partial u}{\partial y}}_{\text{advection}} = -\frac{1}{\rho}\frac{\partial p}{\partial x} + v \underbrace{\left(\frac{\partial^2 u}{\partial x^2} + \frac{\partial^2 u}{\partial y^2} \right)}_{\text{Laplacian}}$$

where u and v are x and y velocities, p is pressure, ρ is density of fluid and v is kinematic viscosity of fluid.

12.3   Numerical solution methods for PDEs

Numerical methods for ODE can also be extended to solution of PDE. Methods discussed for treating initial value problems can be adopted for parabolic as well as hyperbolic equations. Similarly, methods that have been discussed for treating BVPs can be adopted for solution of elliptic PDEs which are also boundary value problems. However, the extension of the methods to solve PDE is not straightforward.

Methods such as finite difference method (FDM), finite volume method (FVM), finite element method (FEM), boundary element method (BEM) etc are commonly used for treating PDE numerically. All numerical methods used to solve PDEs should have *consistency*, *stability* and *convergence*.

---

[1] after Simeon Denis Poisson, 1781-1840, a French mathematician and physicist

A numerical method is said to be consistent if all the approximations (finite difference, finite element, finite volume etc) of the derivatives tend to the exact value as the step size (Δt, Δx etc) tends to zero. A numerical method is said to be stable (like IVPs) if the error does not grow with time (or iteration). Convergence of a numerical method can be ensured if the method is consistent and stable.

We shall look at different aspects of numerical treatment of different types of PDE in the forthcoming chapters.

12.4 MATLAB functions related to PDE

MATLAB provides an extensive set of tools to solve elliptic, parabolic and hyperbolic PDEs. MATLAB provides a graphical user interface (GUI) toolbox "pdetool" to solve PDE in two dimensional space domain. FEM has been adopted for the solution of PDEs. The interface allows user to construct the geometry, discretize the geometry using triangular meshes, apply boundary conditions, solve the equations and post-process the results. Alternatively for users not intending to use the GUI, the following inbuilt functions may be used

MATLAB routine	Function
pdepe	solves initial value problem in elliptic and parabolic PDEs
assempde	constructs (assemble in FEM terminology) the coefficient matrix and right hand side vector for an elliptic PDE (Poisson equation). The above function can also be used to solve the same.
pdenonlin	solves nonlinear elliptic PDEs
parabolic	solves parabolic PDE using FEM
hyperbolic	solves hyperbolic PDE using FEM
pdeeig	solves eigenvalue PDE problem using FEM

The reader should look at MATLAB help and reference for more details on the above functions.

12.A Suggested reading

1. **S.J. Farlow** *Partial differential equations for scientists and engineers* Dover Publications, 1993

Chapter 13

Laplace and Poisson equations

Laplace and Poisson equations occur as field problems in many areas of engineering and hence have received much attention. Relaxation methods[a] were developed originally in order to solve such field problems. In the linear case the governing equations (Laplace or Poisson) are transformed to a set of linear equations for the nodal values and are solved by the various techniques given in Chapter 2. However when nonlinearities are involved such as when the properties of the medium depends on the dependent variable, discretized equations are linearized and then solved iteratively. Even though analytical solutions are sometimes possible numerical techniques have become the norm in solving elliptic PDEs.

---

[a]Developed by Richard Vynne Southwell, 1888 - 1970, British mathematician. Refer "Relaxation methods in engineering science : a treatise on approximate computation", Oxford Univ. Press - 1940)

13.1 Introduction

Laplace and Poisson equations are elliptic partial differential equations and occur in many practical situations such as inviscid flow (also called potential flow or ideal fluid flow), heat conduction, mass diffusion, electrostatics etc. These equations are boundary value problems applied over multi-dimensions. Poisson[1] equation is given by

$$\underbrace{\frac{\partial^2 u}{\partial x^2} + \frac{\partial^2 u}{\partial y^2} + \frac{\partial^2 u}{\partial z^2}}_{\nabla^2 u} = -q(x, y, z) \tag{13.1}$$

∇^2 (or laplacian) is the divergence operator. In the above equation the variable u could be temperature in which case q represents a heat generation term. When solving an electrostatic problem, u represents electric potential and q represents charge density distribution. If the source term $q(x, y, z)$ is zero we get the Laplace equation.

Such equations can be treated analytically under special conditions and hence numerical treatment of these equations become important. In this chapter we shall discuss the solution of these equations using FDM and also briefly explain the application of FEM and FVM.

13.2 Finite difference solution

Laplace equation in two dimensions has been presented earlier as an example of an elliptic PDE of second order. We have introduced the standard problem with Dirichlet boundary conditions along the four edges of a rectangular domain. We consider this problem to introduce the FDM as applicable to elliptic problems (see Figure 13.1). The domain is discretized by dividing the edges parallel to the x axis into n segments of width h each and the edges parallel to the y axis into m segments of width k each. The number of nodes in the domain will then equal $(n + 1) \times (m + 1)$ of which $2(n + m)$ are nodes on the boundary. In effect the rectangular domain has been divided into many small rectangles of area $h \times k$ each. The boundary conditions are applicable at each boundary node and hence the unknown u's are associated with "internal" nodes which are $(n - 1) \times (m - 1)$ in number. Using subscript i to represent node number along x and subscript j to represent node number along y, the coordinates corresponding to node i, j will be $x = ih$, $y = jk$.

13.2.1 Discretization of computational domain

Nodal equations

Nodal equations for internal nodes are obtained by writing the finite difference analog of the governing equation viz. the Laplace equation in two dimensions given by Equation 12.11. We make use of central differences and write this equation as

$$\frac{u_{i-1,j} - 2u_{i,j} + u_{i+1,j}}{h^2} + \frac{u_{i,j-1} - 2u_{i,j} + u_{i,j+1}}{k^2} = 0 \tag{13.2}$$

---

[1]named after French mathematician Simeon Denis Poisson 1781 - 1840, French mathematician

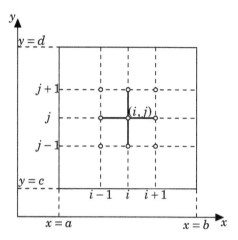

Figure 13.1: Discretization of the computational domain

This equation may be rewritten as

$$u_{i,j} = \frac{k^2(u_{i-1,j} + u_{i+1,j}) + h^2(u_{i,j-1} + u_{i,j+1})}{2(h^2 + k^2)} \tag{13.3}$$

showing that the nodal value at i,j depends on the nodal values at the four nearest neighboring nodes. In case $h = k$ the above equation may be simplified to get

$$u_{i,j} = \frac{(u_{i-1,j} + u_{i+1,j} + u_{i,j-1} + u_{i,j+1})}{4} \tag{13.4}$$

In this case the function at node P is the arithmetic mean of the nodal values at the four nearest neighbors.

Nodal equation such as 13.4 may be written for all the interior nodes i.e. for $1 \le i \le (n-1)$ and $1 \le j \le (m-1)$. All boundary nodes are assigned the specified boundary values. The interior nodal equations relate nodal values to adjoining nodes which may happen to be boundary nodes. This means that the function at all interior nodes get affected by the function values at all boundary nodes. In particular, if any one of the boundary nodal value is perturbed it will affect all the interior nodal values. This is a consequence of the PDE being elliptic.

Solution of nodal equations

As seen above the nodal equations form a set of simultaneous linear equations for the nodal values of the function u. These equations are conveniently solved by methods that have been discussed in great detail in Chapter 2. We take a simple example amenable to solution by Gauss elimination.

Example 13.1

Figure 13.2 shows the temperatures on the boundary of a solid material in which the temperature distribution is governed by the Laplace equation in 2 dimensions. The nodal spacing is the same along the two directions. Determine the temperatures at the four nodes identified by numbers 1 - 4.

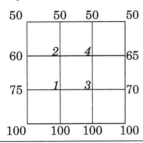

$h = k$

Figure 13.2: *Nodal scheme used in Example 13.1*

Solution :

The nodal equations may easily be written down based on Equation 13.4.

$$T_1 = \frac{175 + T_2 + T_3}{4}; \quad T_2 = \frac{110 + T_1 + T_4}{4}$$

$$T_3 = \frac{170 + T_1 + T_4}{4}; \quad T_4 = \frac{115 + T_2 + T_3}{4}$$

In this form the equations are naturally suited for solution by Jacobi iteration scheme or the variants viz. Gauss Seidel and SOR (see Chapter 2). Alternately we may recast the equations in matrix form given below.

$$\begin{pmatrix} 4 & -1 & -1 & 0 \\ -1 & 4 & 0 & -1 \\ -1 & 0 & 4 & -1 \\ 0 & -1 & -1 & 4 \end{pmatrix} \begin{pmatrix} T_1 \\ T_2 \\ T_3 \\ T_4 \end{pmatrix} = \begin{pmatrix} 175 \\ 110 \\ 170 \\ 115 \end{pmatrix}$$

We can convert the coefficient matrix to lower triangle form by elementary row operations to get the following augmented matrix.

$$\begin{pmatrix} 1.0000 & -0.2500 & -0.2500 & 0.0000 & \vdots & 43.75 \\ 0.0000 & 1.0000 & -0.0667 & -0.2667 & \vdots & 41.00 \\ 0.0000 & 0.0000 & 1.0000 & -0.2857 & \vdots & 60.00 \\ 0.0000 & 0.0000 & 0.0000 & 3.4286 & \vdots & 220.00 \end{pmatrix}$$

Back-substitution yields the desired solution as $\mathbf{T}^T = \begin{pmatrix} 79.17 & 63.33 & 78.33 & 64.17 \end{pmatrix}$.

Next example considers solution of Poisson equation.

Example 13.2

Poisson equation is applicable in a square domain $0 \le x, y \le 1$ with source term given by $q = 100\sin(\pi x)\sin(\pi y)$. The boundaries of the domain are maintained at $u = 0$. Determine the solution using finite difference method.

Solution :

We shall discretize the domain uniformly using step sizes of h and k along x and y direction respectively. The equations at the nodes are given by

$$\frac{u_{i+1,j} - 2u_{i,j} + u_{i-1,j}}{h^2} + \frac{u_{i,j+1} - 2u_{i,j} + u_{i,j-1}}{k^2} = -q_{i,j}$$

or

$$u_{i,j} = \frac{1}{2/h^2 + 2/k^2}\left(\frac{u_{i+1,j} + u_{i-1,j}}{h^2} + \frac{u_{i,j+1} + u_{i,j-1}}{k^2} + q_{i,j}\right)$$

where $q_{i,j} = 100\sin(\pi x_i)\sin(\pi y_j)$. The system of equations can be solved iteratively by Gauss Seidel method. MATLAB program has been written to carry out the above procedure. We have chosen $h = k = 11$ with uniform grid spacing of $\Delta x = \Delta y = 0.1$.

```
n = 11; m = 11;                      % no of nodes
x = 0:1/(n-1):1; y = 0:1/(m-1):1;    % nodes
h = 1/(n-1); k = 1/(m-1);            % step size
[Y,X] = meshgrid(y,x);               %
q = 100*sin(pi*X).*sin(pi*Y);        % source term
u = zeros(n); uo = u;                % initialize function
residual = 1;
while residual > 1e-6                 % Gauss Siedel iteration
  for i=2:n-1
    for j=2:n-1                       % inner nodes
      u(i,j) =    ((u(i+1,j) + u(i-1,j))/h^2 +  ...
           (u(i,j+1) + u(i,j-1))/k^2 + q(i,j)) /(2/h^2+2/k^2);
    end
  end
  residual = max(max(abs(uo-u)));  % residual
  uo = u;                          % update u
end
```

The maximum change in the function value after every iteration has been considered for applying the convergence criterion. Alternatively, one can define the residual as

$$R = \left|\frac{u_{i+1,j} - 2u_{i,j} + u_{i-1,j}}{h^2} + \frac{u_{i,j+1} - 2u_{i,j} + u_{i,j-1}}{k^2} + q_{i,j}\right|$$

and use the same for applying the convergence criterion. Residual indicates how well the solution satisfies the governing equations (in the present case, the discretized equations). Although the residual is reduced to a small value at the end of an iteration process, the solution may not be close to the exact solution. Hence, one must perform a grid sensitivity study and check for convergence of the solution. For carrying out the study, we can choose function or the derivative of the function at a given point as a criterion. For the present problem, let us choose the function value at $x = y = 0.5$ as the parameter. The function value at the above point also represents the maximum value of u inside the domain. The following table shows the convergence of the solution with grid refinement

n	h	$u(0.5, 0.5)$	$\%\epsilon$	no. of iterations
11	0.1	5.10792		133
21	0.05	5.07645	0.62	476
41	0.025	5.06850	0.16	1679
81	0.0125	5.06606	0.05	5814

$\%\epsilon$ refers to percentage difference between $u(0.5, 0.5)$ for two consecutive grid size

Figure 13.3 shows the contour plot of u for two different grid size. It is evident from the plot, that as we refine the grid we are likely to move towards the exact solution.

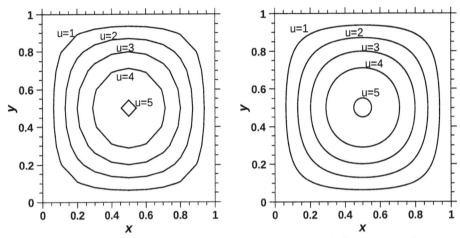

Figure 13.3: *Contour plot of u using a)* 11×11 *and b)* 81×81 *grids*

The convergence of Gauss Seidel is slow. Now we shall look at the relative performance of Jacobi, Gauss Seidel and SOR.

Example 13.3

Obtain the nodal values of a function that satisfies the Laplace equation in two dimensions within a square domain. The domain has been discretized such that there are 5 nodes along each of the directions. The boundaries are specified Dirichlet boundary conditions as indicated in the following figure

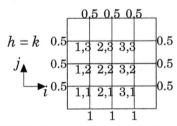

Use Jacobi, Gauss Seidel and SOR (with $\omega = 1.25$) iteration schemes. Compare the three methods from the point of view of number of iterations required for convergence to five significant digits after decimals.

Solution :

From the figure it is clear that we need to solve for the nodal values $u_{i,j}$ at the 9 interior nodes for $1 \le i \le 3$, $1 \le j \le 3$. The three schemes are written down as under:

$$\text{Jacobi: } u_{i,j}^{k+1} = \frac{u_{i-1,j}^{k} + u_{i,j-1}^{k} + u_{i+1,j}^{k} + u_{i,j+1}^{k}}{4}$$

$$\text{Gauss Seidel: } u_{i,j}^{k+1} = \frac{u_{i-1,j}^{k+1} + u_{i,j-1}^{k+1} + u_{i+1,j}^{k} + u_{i,j+1}^{k}}{4}$$

$$\text{SOR: } u_{i,j}^{k+1} = (1-\omega)u_{i,j}^{k} + \omega\frac{\left(u_{i-1,j}^{k+1} + u_{i,j-1}^{k+1} + u_{i+1,j}^{k} + u_{i,j+1}^{k}\right)}{4}$$

where index k stands for the iteration number and ω is taken as 1.25. We initialize all nodal values to 0.5 to start the iterations. After the very first iteration the nodal values according to the three schemes are as indicated below (All values are shown rounded to 4 digits after decimals. Boundary values are not shown in the tables).

After one iteration

Jacobi iteration

	$i=1$	$i=2$	$i=3$
$j=1$	0.6250	0.6250	0.6250
$j=2$	0.5000	0.5000	0.5000
$j=3$	0.5000	0.5000	0.5000

Gauss Seidel Iteration

	$i=1$	$i=2$	$i=3$
$j=1$	0.6250	0.6563	0.6641
$j=2$	0.5313	0.5469	0.5527
$j=3$	0.5078	0.5137	0.5166

SOR with $\omega = 1.25$

	$i=1$	$i=2$	$i=3$
$j=1$	0.6563	0.7051	0.7203
$j=2$	0.5488	0.5793	0.5937
$j=3$	0.5153	0.5296	0.5385

Each of the schemes converges to the same set of final values given below.

After convergence

All three schemes

	$i=1$	$i=2$	$i=3$
$j=1$	0.7143	0.7634	0.7143
$j=2$	0.5938	0.6250	0.5938
$j=3$	0.5357	0.5491	0.5357

Jacobi iteration requires 28 iterations, Gauss Seidel scheme requires 15 iterations whereas the SOR method requires only 10 iterations. In this case SOR is the best bet for solving the problem. The following MATLAB program performs the calculations discussed above. Variable sor in the program is used to choose the iteration method.

```
n = 5;              % number of nodes along x and y
h = 1/(n-1);        % grid spacing
u = zeros(n);       % initialize u
%          Dirichlet boundary conditions
u(n,:) = 0.5; u(:,1) = 0.5; u(:,n) = 0.5; u(1,:) = 1;
uo = u;             % uold (before iteration)
un = u;             % unew (without applying relaxation)
res = 1;            % residual
count = 0;          % iteration count
sor = 1;            % relaxation factor
%    sor = 0 Jacobi;  sor = 1 Gauss Siedel
%    sor < 1 underrelaxation; sor > 1 overrelaxation
while res > 1e-5            % iteration loop
    for i=2:n-1
        for j=2:n-1
            un(i,j) = (u(i-1,j) + u(i+1,j) + ...
                       u(i,j-1) + u(i,j+1))/4;      % unew
        % applying relaxation
```

```
        u(i,j)   = sor*un(i,j)  +  (1-sor)*uo(i,j);
    end
end
res = max(max(abs(un-uo)));    % residual
uo = un;  u = un;              % update  uold  and  u
count   =  count +1;          % update  iteration  count
end
```

13.2.2 Different types of boundary conditions

Example 13.1 has considered the case of Laplace equation with Dirichlet conditions specified along all the boundaries of the rectangular domain. In this section we look at the other types of boundary conditions (see Chapter11) that may be specified at the boundaries.

Neumann or second kind boundary condition

The Neumann boundary condition specifies the normal derivative at a boundary to be zero or a constant. When the boundary is a plane normal to an axis, say the x axis, zero normal derivative represents an adiabatic boundary, in the case of a heat diffusion problem. Conduction heat flux is zero at the boundary. It may also represent a plane of symmetry.

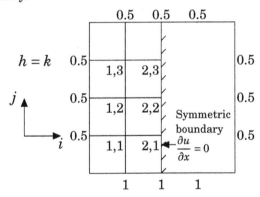

Figure 13.4: *Symmetric boundary condition as applied to Example 13.3*

In Example 13.1 the solution is symmetric with respect to the mid plane parallel to the y axis. We may economize on arithmetic calculations by limiting the domain to the left half of the mid plane referred to above. Instead of 9 nodal equations we have to solve for only 6 nodal equations for $i = 1$ and 2, $j = 1$ to 3. At the mid plane we use the symmetry conditions viz. $u_{1,j} = u_{3,j}$ so that the nodal equations on the mid plane are of form $u_{2,j} = \dfrac{2u_{1,j} + u_{1,j-1} + u_{1,j+1}}{4}$. The previous example is now redone making use of the symmetry condition.

Example 13.4

Redo Example 13.3 by using symmetry boundary condition along mid plane parallel to the y axis.Use SOR with $\omega = 1.25$. *Calculate nodal values after halving the grid spacing and comment on how good the solution is. Use a tolerance of* 10^{-5} *for stopping the iterations.*

Solution :

We repeat the calculations made in Example 13.3 but for only the 6 nodal points alluded to above using SOR with $\omega = 1.25$. The result is the same as that presented there.

Now we reduce the grid size to half its earlier value by introducing extra nodes. The total number of interior nodes at which iterations have to be performed is 28 with $1 \le i \le 4$ and $1 \le j \le 7$. We initialize all nodal function values to 0.5 to start the iterations. We make use of SOR and get the following nodal values after convergence.

	$i = 1$	$i = 2$	$i = 3$	$i = 4$
$j = 7$	0.5087	0.5160	0.5208	*0.5224*
$j = 6$	0.5188	0.5345	0.5446	*0.5481*
$j = 5$	0.5322	0.5584	0.5752	*0.5809*
$j = 4$	0.5515	0.5919	0.6167	*0.6250*
$j = 3$	0.5818	0.6410	0.6748	*0.6857*
$j = 2$	0.6346	0.7155	0.7559	*0.7680*
$j = 1$	0.7413	0.8305	0.8653	*0.8746*

Nodal function values
on and to the left of
the plane of symmetry

In the table the nodal values on the plane of symmetry are shown in *italics*. Every alternate value in this table also is an entry in the table presented earlier with bigger grid spacing. We tabulate the nodal values for the two grid spacing after convergence and also show the change in the values on halving the grid spacing.

Grid spacing h

	$i = 1$	$i = 2$
$j = 3$	0.5357	0.5491
$j = 2$	0.5938	0.6250
$j = 1$	0.7143	0.7634

Grid spacing h/2*

	$i = 2$	$i = 4$
$j = 6$	0.5345	0.5481
$j = 4$	0.5919	0.6250
$j = 2$	0.7155	0.7680

Change due to change
in grid spacing[#]

	$i = 1$	$i = 2$
$j = 1$	0.0013	0.0010
$j = 2$	0.0019	0.0000
$j = 3$	-0.0012	-0.0046

* Note change in node numbers
[#] Node number with h

Since the changes of nodal values are small when the grid spacing is halved the values obtained with the original grid spacing may be considered acceptable.

Robin or third kind boundary condition

This is the most general possible boundary condition that may be imposed at a boundary. Robin condition specifies a relation between the normal derivative and the function at the surface.[2] For example, at a boundary parallel to the y axis, the Robin condition may be specified as

$$f\left(u, \frac{\partial u}{\partial x}\right) = 0 \qquad (13.5)$$

which may, in general, be non-linear. Two examples are given below, the first one linear and the second non-linear.

$$\frac{\partial u}{\partial x} + c_1 u \;\; = \;\; c_2 \qquad (i)$$

$$\frac{\partial u}{\partial x} + c_1 u^4 \;\; = \;\; c_2 \qquad (ii) \qquad (13.6)$$

where c_1 and c_2 are constants. In the context of a problem in heat conduction the former corresponds to convection boundary condition while the latter corresponds to radiation boundary condition.

Consider now a boundary node i,j. The Robin boundary condition in the linear version is written as

$$\frac{3u_{i,j} - 4u_{i-1,j} + u_{i-2,j}}{2h} + c_1 u_{i,j} \;\; = \;\; c_2$$

$$\text{or} \;\; (3 + 2c_1 h)u_{i,j} - 4u_{i-1,j} + u_{i-2,j} - 2c_2 h \;\; = \;\; 0 \qquad (13.7)$$

where h is the grid spacing along the x direction (see Figure 13.5). Note the use of second order accurate three point backward formula for the first partial derivative at the boundary.

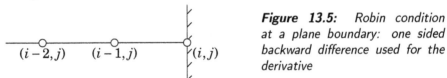

Figure 13.5: *Robin condition at a plane boundary: one sided backward difference used for the derivative*

Alternately, we may use the concept of ghost node and write the boundary condition as

$$\frac{u_{i+1,j} - u_{i-1,j}}{2h} + c_1 u_{i,j} \;\; = \;\; c_2$$

$$\text{or} \;\; u_{i+1,j} \;\; = \;\; u_{i-1,j} - 2c_1 h u_{i,j} + 2c_2 h \qquad (13.8)$$

This is substituted in Laplace equation written for node i,j to get

$$u_{i,j} = \frac{2u_{i-1,j} + 2c_2 h + u_{i,j-1} + u_{i,j+1}}{4 + 2c_1 h} \qquad (13.9)$$

---

[2]Named after Victor Gustave Robin (1855-1897), a French mathematical analyst and applied mathematician

Application of boundary condition to a corner node

Consider a boundary node which is located at a corner as shown in Figure 13.6.

The corner node may be associated partly to the plane parallel to the y axis and partly to the plane parallel to the x axis. We may assume that the Robin condition is the mean of the two[3]. Hence we may write the nodal equation for the corner node i, j as below.

The above formulation is also applicable to the case when we have Robin condition along one plane and Neumann condition along the other. For example, if Neumann condition (zero derivative normal to the plane) is specified along the plane parallel to the x axis, we put $c_3 = c_4 = 0$ to get

$$(3 + c_1 h)u_{i,j} - 2(u_{i-1,j} + u_{i,j-1}) + \frac{u_{i-2,j} + u_{i,j-2}}{2} - c_2 h = 0 \qquad (13.10)$$

$$(3 + [c_1 + c_3]h)u_{i,j} - 2(u_{i-1,j} + u_{i,j-1}) + \frac{u_{i-2,j} + u_{i,j-2}}{2} - (c_2 + c_4)h = 0 \qquad (13.11)$$

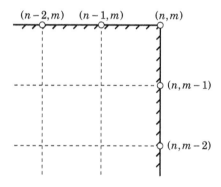

Figure 13.6: *Robin condition at a corner: first method*

Alternately we may use two ghost nodes as shown in Figure 13.7. We introduce one node to the right of the corner and another above the corner , as shown. We shall assume $\Delta x = \Delta y = h$, for simplicity.

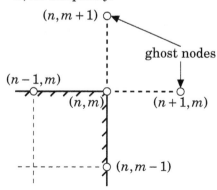

Figure 13.7: *Robin condition at a corner: alternate method using ghost nodes*

---

[3]When h and k are not equal the Robin condition is a weighted sum with respective weights of $\dfrac{h}{h+k}$ and $\dfrac{k}{h+k}$.

The Robin conditions on the two sides forming the corner lead to the following two equations.

$$u_{n+1,m} = u_{n-1,m} - 2c_1 h u_{n,m} + 2hc_2$$
$$u_{n,m+1} = u_{n,m-1} - 2c_3 h u_{n,m} + 2hc_4 \tag{13.12}$$

We now treat the corner node as an internal node to get

$$2u_{n-1,m} - 2[2 + (c_1 + c_3)h]u_{n,m} + 2u_{n,m-1} = -2h(c_2 + c_4) \tag{13.13}$$

Example 13.5

Consider a rectangular domain in which Laplace equation is satisfied. Dirichlet boundary condition is specified along the lower boundary - $u = 1$; Neumann condition along the right boundary - $\dfrac{\partial u}{\partial x} = 0$ and Robin condition is specified along the other two boundaries with $2c_1 h = 2c_3 h = 0.2$ and $2c_2 h = 2c_4 h = 0.1$. Obtain all the nodal function values using SOR with $\omega = 1.25$.

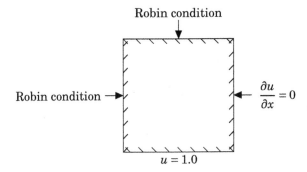

Solution :

We assume that the grid spacing is the same along x and y directions, with 5 nodes along each direction. Function values need to be obtained at all the 20 nodes with $0 \le i \le 4$ and $1 \le i \le 4$. Interior nodal equations are written down as in Example 13.3. Neumann condition along the right boundary is represented by one sided three point formula (left) excepting the corner node which uses the relation given by expression 13.10. The left boundary is treated using an expression following 13.11. The other corner uses expression 13.11 with $c_1 = c_3$ and $c_2 = c_4$. To summarize we have the following, suitable for the application of SOR.

Internal Nodes with $1 \le i \le 4,\ 1 \le j \le 3$:

$$u_{i,j}^{k+1} = (1-\omega)u_{i,j}^{k} + \omega\left(\frac{u_{i-1,j}^{k+1} + u_{i,j-1}^{k+1} + u_{i+1,j}^{k} + u_{i,j+1}^{k}}{4}\right)$$

Nodes on right boundary $i = 4,\ 1 \le j \le 3$:

$$u_{4,j}^{k+1} = (1-\omega)u_{4,j}^{k} + \omega\left(\frac{4u_{3,j}^{k+1} - u_{2,j}^{k+1}}{3}\right)$$

Nodes on left boundary $i = 0$, $1 \le j \le 3$:

$$u_{0,j}^{k+1} = (1-\omega)u_{0,j}^{k} + \omega\left(\frac{4u_{1,j}^{k} - u_{2,j}^{k} + 2c_2h}{3 + 2c_1h}\right)$$

Left corner node $i = 0$, $j = 4$:

$$u_{0,4}^{k+1} = (1-\omega)u_{0,4}^{k} + \omega\left(\frac{2(u_{1,4}^{k} + u_{0,3}^{k+1}) + \frac{u_{2,4}^{k} + u_{0,2}^{k+1} + (c_2 + c_4)h}{2}}{3 + (c_1 + c_3)h}\right)$$

Right corner node $i = 4$, $j = 4$:

$$u_{4,4}^{k+1} = (1-\omega)u_{4,4}^{k} + \omega\left(\frac{2(u_{3,4}^{k+1} + u_{4,3}^{k+1}) + \frac{u_{2,4}^{k+1} + u_{4,2}^{k+1} + c_4h}{2}}{3 + c_3h}\right)$$

We initialize all the nodal function values suitably and use SOR with $\omega = 1.25$ till convergence. Tolerance chosen is $|u_{i,j}^{k+1} - u_{i,j}^{k}| \le 10^{-5}$. The converged nodal values are shown in the following table.

	$i = 0$	$i = 1$	$i = 2$	$i = 3$	$i = 4$
$j = 4$	0.7104	0.7205	0.7410	0.7585	0.7697
$j = 3$	0.7547	0.7778	0.7960	0.8084	0.8125
$j = 2$	0.8134	0.8399	0.8568	0.8665	0.8698
$j = 1$	0.8818	0.9116	0.9249	0.9312	0.9333
$j = 0$	1.0000	1.0000	1.0000	1.0000	1.0000

Robin condition - non-linear case

The non-linear Robin condition given by equation 13.6(ii) is considered now. The solution is sought by the SOR method. Consider the nodal function value $u_{i,j}^{k}$ at a boundary node is to be updated. We may assume that the change $u_{i,j}^{k+1} - u_{i,j}^{k}$ is small and hence we may linearize the fourth power term as

$$(u_{i,j}^{k+1})^4 = [u_{i,j}^{k} + (u_{i,j}^{k+1} - u_{i,j}^{k})]^4 \approx 4(u_{i,j}^{k})^3 u_{i,j}^{k+1} - 3(u_{i,j}^{k})^4$$

The Robin condition applied to the boundary node then takes the form

$$\frac{3u_{i,j}^{k+1} - 4u_{i-1,j}^{k+1} + u_{i-2,j}^{k+1}}{2h} + c_1\left[4(u_{i,j}^{k})^3 u_{i,j}^{k+1} - 3(u_{i,j}^{k})^4\right] = c_2$$

This may be simplified to the following expression suitable for SOR iterations.

$$u_{i,j}^{k+1} = (1-\omega)u_{i,j}^{k} + \omega\left(\frac{4u_{i-1,j}^{k+1} - u_{i-2,j}^{k+1} + 6c_1h(u_{i,j}^{k})^4 + 2c_2h}{3 + 8c_1h(u_{i,j}^{k})^3}\right) \tag{13.14}$$

Example 13.6

Consider a rectangular domain in which Laplace equation is satisfied. Dirichlet boundary condition is specified along the bottom, left and the top boundaries. The right boundary is subject to the non-linear Robin boundary condition with $c_1 h = 0.1$ and $c_2 h = 0.05$ (Equation 13.6(ii)). Obtain all the nodal function values using SOR with $\omega = 1.25$.

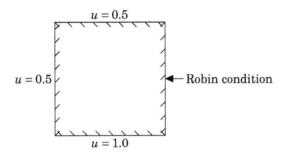

Solution :

We initialize all internal and right boundary nodes to 0.5 to get the initial set given below.

	$i = 0$	$i = 1$	$i = 2$	$i = 3$
$j = 4$		0.5	0.5	0.5
$j = 3$	0.5	0.5	0.5	0.5
$j = 2$	0.5	0.5	0.5	0.5
$j = 1$	0.5	0.5	0.5	0.5
$j = 0$		1	1	1

Entries shown in italics in the table change from iteration to iteration. Nodes on the right boundary satisfy Equation 13.14 (with $i = 1$, $1 \le j \le 3$) while the internal nodes satisfy the Laplace equation. After one iteration the nodal values are given by the following.

	$i = 0$	$i = 1$	$i = 2$	$i = 3$
$j = 4$		0.5000	0.5000	0.5000
$j = 3$	0.5000	0.5153	0.5296	0.5768
$j = 2$	0.5000	0.5488	0.5793	0.6436
$j = 1$	0.5000	0.6562	0.7051	0.8030
$j = 0$		1.0000	1.0000	1.0000

After 11 iterations convergence to 5 digits after decimals is obtained and the result is tabulated below.

	$i = 0$	$i = 1$	$i = 2$	$i = 3$
$j = 4$		0.5000	0.5000	0.5000
$j = 3$	0.5000	0.5528	0.5925	0.6287
$j = 2$	0.5000	0.6188	0.6886	0.7267
$j = 1$	0.5000	0.7338	0.8165	0.8437
$j = 0$		1.0000	1.0000	1.0000

13.2.3 Alternate direction implicit or ADI method

Methods of solving the nodal equations presented above are referred to as point by point iteration methods. An alternate method consists of line by line iteration using the ADI method. In this method we convert the sparse coefficient matrix to tri-diagonal form so that TDMA may be used. This may be accomplished as follows.

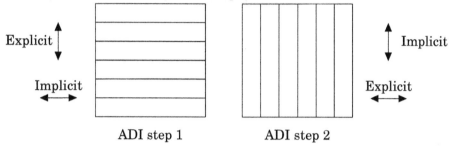

Figure 13.8: *ADI scheme*

Consider an internal node where the Laplace equation is satisfied (with grid spacing the same along the two directions). Then we have

$$4u_{i,j}^{k+1} = u_{i-1,j}^{k+1} + u_{i,j-1}^{k+1} + u_{i+1,j}^{k} + u_{i,j+1}^{k} \tag{13.15}$$

This has been written with point by point (Gauss Seidel) iteration scheme in mind. It may also be recast with SOR in mind. In the first ADI step we recast the above as

$$\underbrace{u_{i-1,j}^{k+1} - 4u_{i,j}^{k+1} + u_{i+1,j}^{k+1}}_{\text{unknown}} = \underbrace{-u_{i,j-1}^{k+1} - u_{i,j+1}^{k}}_{\text{known}} \tag{13.16}$$

Here we hold j fixed and vary i from end to end to get a set of simultaneous equations in the tri-diagonal form (see Figure 13.8(i)). Nodal equations are implicit along the x direction. Essentially the equations are in *one dimensional form*, with the function values along y direction being taken as knowns i.e. explicit. Hence this step considers the rectangular domain to be covered by lines parallel to the x axis and iteration is for all the values along each of these lines. Of course, appropriate expressions will have to be written for the boundary nodes. We may solve these set of equations by TDMA. In the second ADI step we write nodal equations in the form

$$\underbrace{u_{i,j-1}^{k+2} - 4u_{i,j}^{k+2} + u_{i,j+1}^{k+2}}_{\text{unknown}} = \underbrace{-u_{i-1,j}^{k+2} - u_{i+1,j}^{k+1}}_{\text{known}} \tag{13.17}$$

Here we hold i fixed and vary j from end to end to get a simultaneous equations in the tri-diagonal form (see Figure 13.8(ii)). Nodal equations are implicit along the y direction. Hence this step considers the rectangular domain to be covered by lines parallel to the y axis and iteration is for all the values along each of these lines. We may again solve these set of equations by TDMA.

The two step ADI process has to be repeated till convergence. Note that k will be 0 or an even integer.

Example 13.7

Consider a rectangular domain in which Laplace equation is satisfied. Dirichlet boundary condition is specified along all the boundaries (see Figure). Obtain all the nodal function values using ADI method.

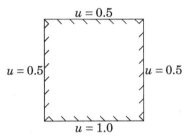

$$u = 0.5$$

$$u = 0.5 \qquad u = 0.5$$

$$u = 1.0$$

Solution :

Let us discretize the domain into 4 equal segments in x and y direction respectively.

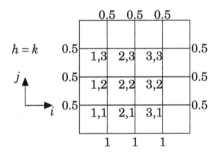

Then the discretized equation would be

$$4u_{i,j} = u_{i+1,j} + u_{i-1,j} + u_{i,j+1} + u_{i,j-1}$$

with Dirichlet conditions specified at the boundaries. Let us initialize the function values at the nodes at 0.5. The following MATLAB program discretizes the domain, calculates the coefficients of nodal equations.

```
n = 5;              % number of nodes along x and y
x = 0:1/(n-1):1;% nodes along x
y = 0:1/(n-1):1;% nodes along y
h = 1/(n-1);        % grid spacing
u = 0.5*ones(n); % initialize u
a = zeros(n);       % coefficient of i,j
bi = zeros(n);      % coefficient of i+1,j
ci = zeros(n);      % coefficient of i-1,j
bj = zeros(n);      % coefficient of i,j+1
cj = zeros(n);      % coefficient of i,j-1
d = zeros(n);       % constant term
a(1,:) = 1; a(n,:) = 1; a(:,1) = 1; a(:,n) =1; % boundary
                 %nodes
```

```
d(1,:) = 1; d(n,:) = 0.5; d(:,1) = 0.5; d(:,n) =0.5;
a(2:n-1,2:n-1) = 4; % interior nodes
bi(2:n-1,2:n-1) = 1;
ci(2:n-1,2:n-1) = 1;
bj(2:n-1,2:n-1) = 1;
cj(2:n-1,2:n-1) = 1;
```

ADI step 1: The equations will be solved implicitly along i and explicitly along j. We start with $j = 1$, then the nodal equations for $1 \le i \le 3$ can be written as

$$4u_{i,1}^1 = \underbrace{u_{i+1,1}^1 + u_{i-1,1}^1}_{\text{implicit}} + \underbrace{u_{i,2}^0 + u_{i,0}^0}_{\text{explicit}}$$

$$4u_{i,1}^1 = u_{i+1,1}^1 + u_{i-1,1}^1 + 1$$

On applying TDMA to above set of linear equations we have

i	a	a'	a''	d	P	Q	$u_{i,1}$
0	1	0	0	1	0	1	1
1	4	1	1	1	0.25	0.5	0.63393
2	4	1	1	1	0.26667	0.4	0.53571
3	4	1	1	1	0.26786	0.375	0.50893
4	1	0	0	0.5	0	0.5	0.5

Similarly, the procedure is repeated for other lines $j = 2$ and 3. At the end of first ADI step, we get

$i\backslash j$	0	1	2	3	4
0	1	1	1	1	1
1	0.5	0.63393	0.67251	0.68444	0.5
2	0.5	0.53571	0.55612	0.56523	0.5
3	0.5	0.50893	0.51626	0.52037	0.5
4	0.5	0.5	0.5	0.5	0.5

ADI step 2 The procedure is now repeated by keeping i fixed i.e. explicit along i and implicit along j. At the end of the procedure we get

$i\backslash j$	0	1	2	3	4
0	1	1	1	1	1
1	0.5	0.69331	0.73753	0.70069	0.5
2	0.5	0.57625	0.60275	0.58095	0.5
3	0.5	0.52921	0.54059	0.53038	0.5
4	0.5	0.5	0.5	0.5	0.5

This completes one iteration of ADI method. The procedure is repeated until the solution does not change in its fifth digit after decimals. The above procedure has been programmed in MATLAB as given below.

```
uo = u;                  % initialize uold
res = 1;                 % residual
count = 0;               % iteration count
while res > 1e-5         % iteration loop
        % ADI step 1, implicit along i, explicit along j
    for j=2:n-1
        d1 = bj(:,j).*u(:,j+1)+cj(:,j).*u(:,j-1) + d(:,j);
        u(:,j) = tdma(ci(:,j),a(:,j),bi(:,j),d1);
    end
        % ADI step 2, implicit along j, explicit along i
    for i=2:n-1
        d1 = bi(i,:).*u(i+1,:)+ci(i,:).*u(i-1,:) + d(i,:);
        u(i,:) = tdma(cj(i,:),a(i,:),bj(i,:),d1);
    end
    res = max(max(abs(u-uo)));   % update residual
    uo = u;                 % update uold
    count  = count +1;   % update iteration count
end
```

The solution converges in 7 iterations where as it took 15 iterations (Example 13.3) using point by point scheme! Further the method can be coupled with relaxation schemes as well. The following MATLAB commands plots the contour map of u.

```
[Y,X] = meshgrid(y,x);
contour(X,Y,u);
```

The following figure shows the contour map of u.

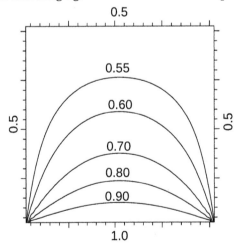

Figure 13.9: Contour map of u

13.3   Elliptic equations in other coordinate systems

We have considered elliptic equations in Cartesian coordinates till now. Many applications in engineering involve other coordinate systems such as the cylindrical and spherical coordinates. The Laplace equation will have to be modified as given below.

Laplace equation in cylindrical coordinates

$$\frac{1}{r}\frac{\partial}{\partial r}\left(r\frac{\partial u}{\partial r}\right) + \frac{1}{r^2}\frac{\partial^2 u}{\partial\theta^2} + \frac{\partial^2 u}{\partial z^2} = 0 \qquad (13.18)$$

where u is a function of r, θ and z. There are interesting applications where u may be a function of either r, z or r, θ - representing two dimensional problems.

Laplace equation in spherical coordinates

$$\frac{1}{r^2}\frac{\partial}{\partial r}\left(r^2\frac{\partial u}{\partial r}\right) + \frac{1}{r^2\sin\theta}\frac{\partial}{\partial\theta}\left(\sin\theta\frac{\partial u}{\partial\theta}\right) + \frac{1}{r^2\sin^2\theta}\frac{\partial^2 u}{\partial\phi^2} = 0 \qquad (13.19)$$

where u is a function of r, θ and ϕ. There are interesting applications where u may be a function of either r, θ or θ, ϕ - representing two dimensional problems.

Example 13.8

Consider a cylindrical annulus of inner radius 1 and outer radius 2 as shown in Figure 13.8. The inner boundary of the annulus has Dirichlet boundary condition specified as $u(r=1) = 1$ for $0 \le \theta \le 2\pi$. One half of the outer boundary is specified with no flux condition i.e. $\dfrac{\partial u}{\partial r} = 0$ for $r = 2, \pi \le \theta \le 2\pi$ while the rest of the outer boundary is specified with Dirichlet boundary condition in the form $u(r = 2) = 0.0$ for $0 \le \theta \le \pi$.

$u(2,\theta) = 0.0$ for $0 \le \theta \le \pi$

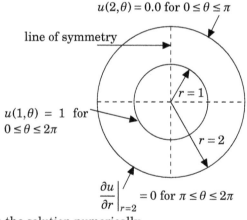

line of symmetry

$r = 1$

$u(1,\theta) = 1$ for
$0 \le \theta \le 2\pi$

$r = 2$

$\dfrac{\partial u}{\partial r}\bigg|_{r=2} = 0$ for $\pi \le \theta \le 2\pi$

Domain showing boundary conditions in Example 13.8

Obtain the solution numerically.

Solution :

PDE governing the problem is given by

$$\frac{1}{r}\frac{\partial}{\partial r}\left(r\frac{\partial u}{\partial r}\right) + \frac{1}{r^2}\frac{\partial^2 u}{\partial\theta^2} = 0$$

$$\text{or}\quad \frac{1}{r}\frac{\partial u}{\partial r} + \frac{\partial^2 u}{\partial r^2} + \frac{1}{r^2}\frac{\partial^2 u}{\partial\theta^2} = 0$$

The boundary conditions are

$$u(r=1) = 1;\quad u(r=2, 0 \le \theta \le \pi) = 0;\quad \frac{\partial u}{\partial r}\bigg|_{r=2,\pi\le\theta\le2\pi} = 0$$

Owing to the symmetry of the problem about $\theta = \pi/2$ and $\theta = 3\pi/2$, only half of the domain may be used for numerical calculation. Then, along the line of symmetry, the boundary condition $\dfrac{\partial u}{\partial \theta} = 0$ is imposed. Let us discretize r coordinate into n equal divisions (say 4, $\Delta r = 0.25$) and θ coordinate into $2m$ equal divisions (say m=3, $\Delta \theta = \pi/3$). Then, there would be $n+1$ nodes along r numbered $i = 0$ to n and $2m+1$ nodes along θ numbered $j = -m$ to m as indicated in the figure below.

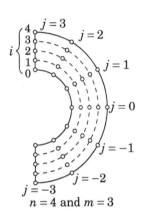

$n = 4$ and $m = 3$

Interior nodes:
$1 \le i \le 3$ and $-2 \le j \le 2$
Boundary nodes:

Condition	Nodes
$u(r = 1) = 1$	$i = 0$ and $-m \le j \le m$
$u(r = 2, 0 \le \theta \le \pi) = 0$	$i = n$ and $0 \le j \le m$
$\dfrac{\partial u}{\partial r}\bigg|_{r=2,\pi \le \theta \le 2\pi} = 0$	$i = n$ and $-m \le j \le 0$
$\dfrac{\partial u}{\partial \theta} = 0$	$j = m$ and $j = -m$ for $0 \le i \le n$

Nodal equations for interior nodes

$$\frac{1}{r_i}\frac{u_{i+1,j} - u_{i-1,j}}{2\Delta r} + \frac{u_{i+1,j} - 2u_{i,j} + u_{i-1,j}}{\Delta r^2} + \frac{1}{r_i^2}\frac{u_{i,j+1} - 2u_{i,j} + u_{i,j-1}}{\Delta \theta^2} = 0$$

$$\text{or} \quad \left(\frac{2}{\Delta r^2} + \frac{2}{r_i^2 \Delta \theta^2}\right) u_{i,j} = \left(\frac{1}{\Delta r^2} + \frac{1}{2r_i \Delta r}\right) u_{i+1,j} + \left(\frac{1}{\Delta r^2} - \frac{1}{2r_i \Delta r}\right) u_{i-1,j}$$

$$+ \frac{1}{r_i^2 \Delta \theta^2}(u_{i,j+1} + u_{i,j-1})$$

Boundary nodes $\dfrac{\partial u}{\partial r} = 0$ ($i = n(4)$ **and** $j = -1, -2$)
We shall use the ghost node approach and applying the boundary condition we have $u_{n+1,j} = u_{n-1,j}$. Hence, the discretized equations reduce to

$$\left(\frac{2}{\Delta r^2} + \frac{2}{r_n^2 \Delta \theta^2}\right) u_{n,j} = \left(\frac{2}{\Delta r^2}\right) u_{n-1,j} + \frac{1}{r_n^2 \Delta \theta^2}(u_{n,j+1} + u_{n,j-1})$$

Boundary nodes $\dfrac{\partial u}{\partial \theta} = 0$ ($i = 1$ **to** 3, $j = -3$)
Again ghost node approach shall be used and we have $u_{i,j+1} = u_{i,j-1}$. The discretized equation reduces to

$$\left(\frac{2}{\Delta r^2} + \frac{2}{r_i^2 \Delta \theta^2}\right) u_{i,j} = \left(\frac{1}{\Delta r^2} + \frac{1}{2r_i \Delta r}\right) u_{i+1,j} + \left(\frac{1}{\Delta r^2} - \frac{1}{2r_i \Delta r}\right) u_{i-1,j} + \frac{2}{r_i^2 \Delta \theta^2} u_{i,j+1}$$

Boundary nodes $\dfrac{\partial u}{\partial \theta} = 0$ ($i = 1$ **to** 3, $j = 3$)

Again ghost node approach shall be used and we have $u_{i,j+1} = u_{i,j-1}$. The discretized equation reduces to

$$\left(\frac{2}{\Delta r^2} + \frac{2}{r_i^2 \Delta \theta^2}\right) u_{i,j} = \left(\frac{1}{\Delta r^2} + \frac{1}{2r_i \Delta r}\right) u_{i+1,j} + \left(\frac{1}{\Delta r^2} - \frac{1}{2r_i \Delta r}\right) u_{i-1,j} + \frac{2}{r_i^2 \Delta \theta^2} u_{i,j-1}$$

Boundary node $i = 4$ **and** $j = -3$, $\dfrac{\partial u}{\partial \theta} = 0$ **and** $\dfrac{\partial u}{\partial r} = 0$

This node is a corner node. We will apply ghost node method along both the directions. Hence, the discretized equations will be

$$\left(\frac{2}{\Delta r^2} + \frac{2}{r_n^2 \Delta \theta^2}\right) u_{n,-m} = \left(\frac{2}{\Delta r^2}\right) u_{n-1,-m} + \frac{2}{r_n^2 \Delta \theta^2} u_{n,-m+1}$$

All the above equations may be written in the form

$$a u_{i,j} = b_i u_{i+1,j} + c_i u_{i-1,j} + b_j u_{i,j+1} + c_j u_{i,j-1} + d$$

The following MATLAB program discretizes the domain and calculates the coefficients corresponding to (i,j) and its neighboring nodes and the right hand side term d.

```
n = 5;              % number of nodes along r
m = 7;              % number of nodes along θ
r = 1:1/(n-1):2;    % nodes along r
theta = -pi/2:pi/(m-1):pi/2;   % nodes along θ
% Note: node numbering in program is different
dr = 1/(n-1);       % Δr
dtheta = pi/(m-1);% Δθ
u = zeros(n,m);     % initialize function
a = zeros(n,m);     % coefficient of i,j
bi = zeros(n,m);    % coefficient of i+1,j
ci = zeros(n,m);    % coefficient of i-1,j
bj = zeros(n,m);    % coefficient of i,j+1
cj = zeros(n,m);    % coefficient of i,j-1
d =  zeros(n,m);    % constant
a(1,:) = 1;         % Dirichlet boundary condition
d(1,:) = 1;         % u=1
u(1,:) = 1;
for i=2:n-1         % coefficients for i=1 to n-1
        bi(i,:) = 1/dr^2+1/(2*dr*r(i)); % interior nodes
        ci(i,:) = 1/dr^2-1/(2*dr*r(i)); % interior nodes
        bj(i,2:m-1) = 1/(r(i)*dtheta)^2;% interior nodes
        cj(i,2:m-1) = 1/(r(i)*dtheta)^2;% interior nodes
        bj(i,1) = 2/(r(i)*dtheta)^2;    % ∂u/∂θ = 0  j=-m
        cj(i,m) = 2/(r(i)*dtheta)^2;    % ∂u/∂θ = 0  j=m
        a(i,:) =  bi(i,:) + ci(i,:) + bj(i,:) + cj(i,:);
end
        a(n,1:(m+1)/2) = 1; % Dirichlet boundary condition
        d(n,1:(m+1)/2) = 0; % u=0,  i=n,  j=0 to m
        % Neumann boundary  condition  i=n,  j=-1  to  -m
        u(n,1:(m+1)/2) = 0;
        ci(n,(m+1)/2+1:m) = 2/dr^2;
```

```
      bj(n,(m+1)/2+1:m-1) = 1/(r(n)*dtheta)^2; % j=1 to m-1
      cj(n,(m+1)/2+1:m-1) = 1/(r(n)*dtheta)^2; % j=1 to m-1
      cj(n,m) = 2/(r(n)*dtheta)^2;    %∂u/∂θ=0 j=m
      a(n,(m+1)/2+1:m) =  bi(n,(m+1)/2+1:m) ...
           + ci(n,(m+1)/2+1:m) ...
      + bj(n,(m+1)/2+1:m) + cj(n,(m+1)/2+1:m);
```

The solution can be obtained from the discretized equations using point by point iterative scheme. Alternatively, one can apply ADI scheme to solve the system of linear equations. The following MATLAB program performs ADI iterations. The iterations stop when maximum change in the values of the variable is less than 1×10^{-6}.

```
uo = u;                    % old value of u
res = 1;                   % initialize residual
count = 0;                 % iteration count
while res > 1e-6           % iteration loop
    % ADI 1st step implicit along r, explicit along θ
    d1 = bj(:,1).*u(:,2)+ d(:,1);  % constant term for TDMA
    u(:,1) = tdma(ci(:,1),a(:,1),bi(:,1),d1);
    for j=2:m-1
        d1 = bj(:,j).*u(:,j+1)+cj(:,j).*u(:,j-1) + d(:,j);
        u(:,j) = tdma(ci(:,j),a(:,j),bi(:,j),d1);
    end
    d1 = cj(:,m).*u(:,m-1) + d(:,m);
    u(:,m) = tdma(ci(:,m),a(:,m),bi(:,m),d1);

    % ADI 2nd step implicit along θ, explicit along r
    for i=2:n-1
        d1 = bi(i,:).*u(i+1,:)+ci(i,:).*u(i-1,:) + d(i,:);
        u(i,:) = tdma(cj(i,:),a(i,:),bj(i,:),d1);
    end
    res = max(max(abs(u-uo)));  % update residual
    uo = u;                % update old value of u
    count  = count +1;     % update iteration count
end
```

The following MATLAB program constructs contour plot of the solution u.

```
[Theta,R] = meshgrid(theta,r);
X = R.*cos(Theta);
Y = R.*sin(Theta);
contour(X,Y,u)
```

Figure 13.10 shows the contour map of u in the annular region (right half). Also shown is the grid sensitivity study. As the number of nodes are increased, the iteration count also increases. In practice, the sensitivity of a desired quantity such as u or gradient at a location is used for grid sensitivity study.

13.4 Elliptic equation over irregular domain

FDM can be applied to irregular geometries as well. Let us consider applying FDM in Cartesian coordinates to the cylindrical domain considered in Example 13.8. A simple

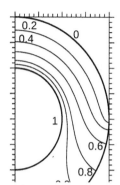

no. of nodes		$u(1.5,0)$	$\%\epsilon$	no. of iterations
r	θ			
4	6	0.473		12
8	12	0.516	8.37	33
16	24	0.545	5.32	110
32	48	0.558	2.27	382
64	96	0.563	0.98	1371
128	192	0.566	0.43	4733

$\%\epsilon$ refers to percentage difference between $u(1.5,0)$ for two consecutive grid size

(b)

(a)

Figure 13.10: (a) Contour map of u (grid size 128 × 192) (b) Grid sensitivity study with r = 1.5, θ = 0 used as the marker node

discretization of the domain in Cartesian coordinates has been shown in the figure below. As the domain is non uniform, the distribution of nodes close to the boundary also are non-uniformly distributed.

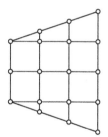

Under such cases, application of Neumann boundary condition is not straightforward. The gradients can be determined using multidimensional interpolation

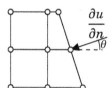

$$\frac{\partial u}{\partial n} = \frac{\partial u}{\partial x}\cos\theta + \frac{\partial u}{\partial y}\sin\theta$$

where θ is the angle made by the normal vector to the horizontal.

Alternatively, one can use boundary fitted coordinate system where the irregular domain is mapped onto a regular domain, say a square. In such a case, the governing equations would transform according to the Jacobian of the transformation. For the case of a cylindrical annulus, the Laplace equation in Cartesian coordinates would transform into that in cylindrical coordinates.

Example 13.9

Poisson equation with a constant source term q = 1 is applicable over a two dimensional trapezoidal domain as shown in the figure below. Discretize the domain using finite difference method and determine the distribution of function inside the domain.

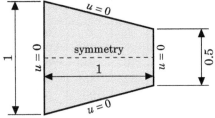

Solution :

As symmetry is applicable, we shall consider only one half of the domain, in which case the condition at the line of symmetry is $\dfrac{\partial u}{\partial y} = 0$. Poisson equation in Cartesian coordinates is given by

$$\frac{\partial^2 u}{\partial x^2} + \frac{\partial^2 u}{\partial y^2} = -1$$

Coordinate transformation Now we shall transform the trapezium to a square and hence from Cartesian coordinates (x, y) to new coordinates (ξ, η).

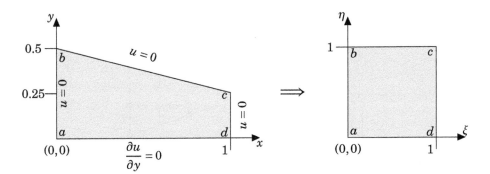

The trapezoidal domain can be transformed to the square domain using isoparametric transformation introduced in Chapter 6. The weight function for the four points are

$$w_a = (1-\eta)(1-\xi); \qquad w_b = (1-\xi)\eta; \qquad w_c = \eta\xi; \qquad w_d = \xi(1-\eta);$$

Then the transformation can be written as

$$x = \quad x_a w_a + x_b w_b + x_c w_c + x_d w_d = \quad \xi$$
$$y = \quad y_a w_a + y_b w_b + y_c w_c + y_d w_d = \quad 0.5\eta - 0.25\xi\eta$$

ξ and x are same and the transformation occurs with respect to η. Jacobian of the transformation can be determined using Equation 8.53.

$$
\mathbf{J} =
\begin{pmatrix}
\dfrac{\partial \xi}{\partial x} & \dfrac{\partial \eta}{\partial x} \\[2mm]
\dfrac{\partial \xi}{\partial y} & \dfrac{\partial \eta}{\partial y}
\end{pmatrix}
=
\frac{1}{\dfrac{\partial x}{\partial \xi}\dfrac{\partial y}{\partial \eta} - \dfrac{\partial x}{\partial \eta}\dfrac{\partial y}{\partial \xi}}
\begin{pmatrix}
\dfrac{\partial y}{\partial \eta} & -\dfrac{\partial y}{\partial \xi} \\[2mm]
-\dfrac{\partial x}{\partial \eta} & \dfrac{\partial x}{\partial \xi}
\end{pmatrix}
$$

We have

$$\frac{\partial x}{\partial \xi} = 1; \quad \frac{\partial x}{\partial \eta} = 0; \quad \frac{\partial y}{\partial \xi} = -0.25\eta; \quad \frac{\partial y}{\partial \eta} = 0.5 - 0.25\xi$$

Then the Jacobian of the transformation becomes

$$\mathbf{J} = \begin{pmatrix} \dfrac{\partial \xi}{\partial x} & \dfrac{\partial \eta}{\partial x} \\[2mm] \dfrac{\partial \xi}{\partial y} & \dfrac{\partial \eta}{\partial y} \end{pmatrix} = \frac{1}{0.5 - 0.25\xi} \begin{pmatrix} 0.5 - 0.25\xi & 0.25\eta \\ 0 & 1 \end{pmatrix}$$

Then applying equations 8.58 and 8.59, second derivatives would become

$$\frac{\partial^2 u}{\partial x^2} = \frac{\partial^2 u}{\partial \xi^2} + 2\frac{0.25\eta}{0.5 - 0.25\xi}\frac{\partial^2 u}{\partial \xi \partial \eta} + \left(\frac{0.25\eta}{0.5 - 0.25\xi}\right)^2 \frac{\partial^2 u}{\partial \eta^2} \qquad \frac{\partial^2 u}{\partial y^2} = \frac{1}{(0.5 - 0.25\xi)^2}\frac{\partial^2 u}{\partial \eta^2}$$

Thus, the Poisson equation in the transformed coordinates is

$$\underbrace{\left[0.5 - 0.25\xi\right]^2 \frac{\partial^2 u}{\partial \xi^2}}_{a} + \underbrace{\left[0.5\eta(0.5 - 0.25\xi)\right] \frac{\partial^2 u}{\partial \xi \partial \eta}}_{b} + \underbrace{\left(1 + 0.0625\eta^2\right)\frac{\partial^2 u}{\partial \eta^2}}_{c} = \underbrace{(0.5 - 0.25\xi)^2}_{d}$$

Poisson equation in the new coordinate system contains two second derivative terms and also an additional mixed derivative term. The extra factor in the source term d accounts for the change in area due to transformation.

One has to transform the boundary conditions as well to the new coordinate system. The Neumann boundary condition at the bottom boundary becomes $\dfrac{\partial u}{\partial \eta} = 0$

Discretization

We shall discretize the square domain uniformly with $n = 11$ segments along ξ direction and $m = 11$ segments along η direction. The step size along the two directions are $\Delta \xi$ and $\Delta \eta$ respectively.

```
n = 11;m = 11;                          % no of nodes
xi = 0:1/(n-1):1; eta = 0:1/(m-1):1;    % nodes
[Eta,Xi] = meshgrid(eta,xi);            % mesh
dxi = 1/(n-1); deta = 1/(m-1);          % step size
u = zeros(n,m);                         % initialize function
uo = u;
a = (0.5-0.25*Xi).^2/dxi^2;             % transformation
b = 0.5*Eta.*(0.5-0.25*Xi)/(4*dxi*deta);% constants
c = (1+0.0625*Eta.^2)/deta^2;           %
```

Discretization at internal node:

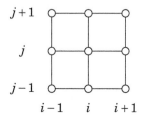

Poisson equation contains three terms of which $\dfrac{\partial^2 u}{\partial \xi^2}$ and $\dfrac{\partial^2 u}{\partial \eta^2}$ are discretized using central difference formula. The third term is a mixed derivative and has not been treated yet. We shall apply central differences to the mixed derivative term

$$\frac{\partial}{\partial \xi}\left(\frac{\partial u}{\partial \eta}\right) = \frac{\left.\frac{\partial u}{\partial \eta}\right|_{i+1} - \left.\frac{\partial u}{\partial \eta}\right|_{i-1}}{2\Delta \xi} = \frac{u_{i+1,j+1} - u_{i+1,j-1} - u_{i-1,j+1} + u_{i-1,j-1}}{4\Delta \eta \Delta \xi}$$

Then the difference equation for an internal node is

$$2\left(\frac{a_{i,j}}{\Delta \xi^2} + \frac{c_{i,j}}{\Delta \eta^2}\right) u_{i,j} = a_{i,j}\frac{u_{i+1,j} + u_{i-1,j}}{\Delta \xi^2} + c_{i,j}\frac{u_{i,j+1} + u_{i,j-1}}{\Delta \eta^2}$$
$$+ b_{i,j}\frac{u_{i+1,j+1} - u_{i+1,j-1} - u_{i-1,j+1} + u_{i-1,j-1}}{4\Delta \eta \Delta \xi} + d_{i,j}$$

Discretization at bottom boundary: We use ghost node approach to get $u_{i,-1} = u_{i,1}$. The mixed derivative term automatically disappears. The difference equation for this boundary would be

$$2\left(\frac{a_{i,0}}{\Delta \xi^2} + \frac{c_{i,0}}{\Delta \eta^2}\right) u_{i,0} = a_{i,0}\frac{u_{i+1,0} + u_{i-1,0}}{\Delta \xi^2} + 2c_{i,0}\frac{u_{i,1}}{\Delta \eta^2} + d_{i,0}$$

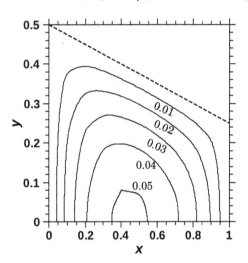

Figure 13.11: *Contour plot of function in the trapezoidal domain*

The above procedure has been programmed in MATLAB. Gauss Seidel iteration has been adopted. The iteration is stopped when the maximum change in the function values is less than the specified tolerance.

```
residual = 1;                      % initialize residual
while residual > 1e-12             % iteration loop
    for i=2:n-1                    % internal nodes
        for j=2:m-1
            u(i,j) = (a(i,j)*(u(i+1,j)+u(i-1,j)) +  ...
            c(i,j)*(u(i,j+1)+u(i,j-1)) + ...
            b(i,j)*(u(i+1,j+1)+u(i-1,j-1) - ...
            u(i-1,j+1)-u(i+1,j-1)) + d(i,j)) ...
```

```
                  /(2*a(i,j)+2*c(i,j));
         end
    end
    j = 1;                          % bottom  boundary  nodes
    for i=2:n-1
         u(i,j) = (d(i,j)+a(i,j)*(u(i+1,j)+u(i-1,j)) +  ...
              2*c(i,j)*u(i,j+1))/(2*a(i,j)+2*c(i,j));
    end
    residual = max(max(abs(uo-u))); % residual
    uo = u;                         % update u
end
X = Xi;                             % transform  coordinates
Y = 0.5*Eta-0.25*Xi.*Eta;
contour(X,Y,u);                     % contour  plot
```

A contour plot of the function within the trapezoidal domain is given in Figure 13.11.

Now we shall look at application of Robin and Neumann boundary condition for an irregular geometry.

Example 13.10

Laplace equation is applicable over a two dimensional trapezoidal domain as shown in figure. Discretize the domain using finite difference method and determine the distribution of function inside the domain.

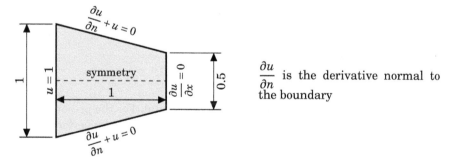

Solution :

As symmetry is applicable, we shall consider only one half of the domain, in which case the condition at the line of symmetry is $\dfrac{\partial u}{\partial y} = 0$. Laplace equation in Cartesian coordinates is given by

$$\frac{\partial^2 u}{\partial x^2} + \frac{\partial^2 u}{\partial y^2} = 0$$

Coordinate transformation Now we shall transform the trapezium to a square and hence from Cartesian coordinates (x, y) to new coordinates (ξ, η).

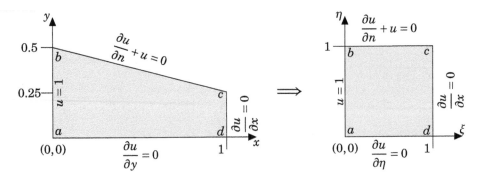

The Laplace equation in the transformed coordinates is (see previous example)

$$\underbrace{\left(0.5 - 0.25\xi\right)^2 \frac{\partial^2 u}{\partial \xi^2}}_{a} + \underbrace{\left(0.5\eta(0.5 - 0.25\xi)\right) \frac{\partial^2 u}{\partial \xi \partial \eta}}_{b} + \underbrace{\left(1 + 0.0625\eta^2\right) \frac{\partial^2 u}{\partial \eta^2}}_{c} = 0$$

Boundary conditions
Bottom boundary

$$\frac{\partial u}{\partial y} = \frac{\partial u}{\partial \xi}\frac{\partial \xi}{\partial y} + \frac{\partial u}{\partial \eta}\frac{\partial \eta}{\partial y} = \frac{1}{0.5 - 0.25\xi}\frac{\partial u}{\partial \eta} = 0$$

Right boundary

$$\frac{\partial u}{\partial x} = \frac{\partial u}{\partial \xi}\frac{\partial \xi}{\partial x} + \frac{\partial u}{\partial \eta}\frac{\partial \eta}{\partial x} = \frac{\partial u}{\partial \xi} + \frac{0.25\eta}{0.5 - 0.25\xi}\frac{\partial u}{\partial \eta} = \frac{\partial u}{\partial \xi} + \eta\frac{\partial u}{\partial \eta} = 0$$

Top boundary
The normal gradient at the top boundary is given by

$$\frac{\partial u}{\partial n} = \frac{\partial u}{\partial x}\cos\theta + \frac{\partial u}{\partial y}\sin\theta$$

θ is the angle made by the normal with respect to horizontal. For the present problem, we have a straight line. For a general curve, slope can be determined by performing gradient operation on the curve. The equation of the top boundary is given by $0.25x + y = 0.5$, gradient of which is $0.25\hat{\mathbf{i}} + \hat{\mathbf{j}}$, with the magnitude of the vector being $\sqrt{1 + 0.25^2} = 1.0308$. Hence $\cos\theta = \dfrac{0.25}{1.0308}$ and $\sin\theta = \dfrac{1}{1.0308}$. Expanding other terms we get

$$\frac{\partial u}{\partial n} + u = \cos\theta\frac{\partial u}{\partial \xi} + \frac{0.25\cos\theta + \sin\theta}{0.5 - 0.25\xi}\frac{\partial u}{\partial \eta} + u = 0$$

Discretization
We shall discretize the square domain uniformly with $n = 11$ segments along ξ direction and $m = 11$ segments along η direction. The discretization can be achieved using the program given in the previous example.
Equations for internal nodes: Then the difference equation for an internal node is

$$2\left(\frac{a_{i,j}}{\Delta\xi^2} + \frac{c_{i,j}}{\Delta\eta^2}\right)u_{i,j} = a_{i,j}\frac{u_{i+1,j} + u_{i-1,j}}{\Delta\xi^2} + c_{i,j}\frac{u_{i,j+1} + u_{i,j-1}}{\Delta\eta^2}$$

$$+b_{i,j}\frac{u_{i+1,j+1}-u_{i+1,j-1}-u_{i-1,j+1}+u_{i-1,j-1}}{4\Delta\eta\Delta\xi}$$

Equations for nodes on the bottom boundary: Again using ghost node approach, the difference equation for the boundary would be

$$2\left(\frac{a_{i,0}}{\Delta\xi^2}+\frac{c_{i,0}}{\Delta\eta^2}\right)u_{i,0}=a_{i,0}\frac{u_{i+1,0}+u_{i-1,0}}{\Delta\xi^2}+2c_{i,0}\frac{u_{i,1}}{\Delta\eta^2}$$

Equations for nodes on the right boundary: We shall use three point one sided difference for ξ coordinate and central difference formula for η coordinate.

$$\frac{u_{i-2,j}-4u_{i-1,j}+3u_{i,j}}{2\Delta\xi}+\eta_{i,j}\frac{u_{i,j+1}-u_{i,j-1}}{2\Delta\eta}=0$$

$$u_{i,j}=\frac{1}{3}\left(4u_{i-1,j}-u_{i-2,j}-\eta_{i,j}\Delta\xi\frac{u_{i,j+1}-u_{i,j-1}}{\Delta\eta}\right)$$

Equations for nodes on the top boundary: We will use three point one sided formula along η and central differences along ξ coordinate.

$$\cos\theta\frac{u_{i+1,m}-u_{i-1,m}}{2\Delta\xi}+K\frac{u_{i,m-2}-4u_{i,m-1}+3u_{i,m}}{2\Delta\eta}+u_{i,j}=0$$

$$u_{i,j}=\frac{1}{\frac{3K}{2\Delta\eta}+1}\left(K\frac{4u_{i,m-1}-u_{i,m-2}}{2\Delta\eta}-\cos\theta\frac{u_{i+1,m}-u_{i-1,m}}{2\Delta\xi}\right)$$

where $K=\dfrac{0.25\cos\theta+\sin\theta}{0.5-0.25\xi_{i,j}}$.

Equation at bottom right corner At bottom right corner we have $\dfrac{\partial u}{\partial\eta}=0$. We use the ghost node approach.

$$2\left(\frac{a_{n,0}}{\Delta\xi^2}+\frac{c_{n,0}}{\Delta\eta^2}\right)u_{n,0}=2a_{n,0}\frac{u_{n-1,0}}{\Delta\xi^2}+2c_{n,0}\frac{u_{n,1}}{\Delta\eta^2}$$

Equation at top right corner At top right corner, we use the Neumann boundary condition $\dfrac{\partial u}{\partial x}=0$. We use three point formula in both directions to get

$$u_{i,j}=\left(\frac{3}{\Delta\xi}+\frac{3\eta_{i,j}}{\Delta\eta}\right)^{-1}\left(\frac{4u_{i-1,j}-u_{i-2,j}}{\Delta\xi}+\eta_{i,j}\frac{4u_{i,j-1}-u_{i,j-2}}{\Delta\eta}\right)$$

The above procedure has been programmed in MATLAB. Gauss Seidel iteration has been adopted. The iteration is stopped when the maximum change in the function values is less than the specified tolerance.

```
residual = 1;                 % residual
cost = 0.25/sqrt(1+0.25^2); % cosθ
sint = 1/sqrt(1+0.25^2);      % sinθ
while residual > 1e-12   % iteration loop
    for i=2:n-1          % internal nodes
        for j=2:m-1
            u(i,j) = (a(i,j)*(u(i+1,j)+u(i-1,j)) +  ...
```

```
                  c(i,j)*(u(i,j+1)+u(i,j-1)) + ...
                  b(i,j)*(u(i+1,j+1)+u(i-1,j-1) - ...
                  u(i-1,j+1)-u(i+1,j-1))) ...
                  /(2*a(i,j)+2*c(i,j));
        end
    end
    j = 1;                    % bottom boundary
    for i=2:n-1
        u(i,j) = (a(i,j)*(u(i+1,j)+u(i-1,j)) +  ...
            2*c(i,j)*u(i,j+1))/(2*a(i,j)+2*c(i,j));
    end
    j = m;                    % top boundary
    for i=2:n-1
        K = (0.25*cost+sint)/(0.5-0.25*Xi(i,j));
        u(i,j) = (K/(2*deta)*(4*u(i,m-1)-u(i,m-2)) - ...
            cost*(u(i+1,m)-u(i-1,m))/(2*dxi))/ ...
            (3*K/(2*deta)+1);
    end
    i = n;                    % right boundary
    for j=2:m-1
        u(i,j) = (4*u(i-1,j) - u(i-2,j)   ...
            -dxi*Eta(i,j)*(u(i,j+1)-u(i,j-1))/deta)/3;
    end
    i=n;j=m;                  % top right node
    u(i,j) = ((4*u(i-1,j) - u(i-2,j))/dxi   ...
        +Eta(i,j)*(4*u(i,j-1)-u(i,j-2))/deta)/ ...
        (3*(1/dxi+Eta(i,j)/deta));
    i=n;j=1;                  % bottom right node
    u(i,j) = (2*a(i,j)*u(i-1,j) +  ...
        2*c(i,j)*u(i,j+1)) ...
        /(2*a(i,j)+2*c(i,j));
    residual = max(max(abs(uo-u))); % residual
    uo = u;                   % update u
end                           % end loop
```

A contour plot of the solution within the trapezoidal domain has given in Figure 13.12.

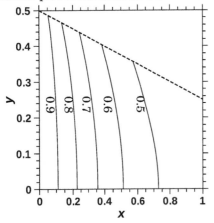

Figure 13.12: *Contour plot of function in the trapezoidal domain*

Another alternative is to divide the geometry into sub domains. Finite difference discretization is applied to each sub domain individually.

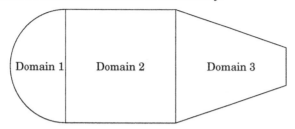

FDM using boundary fitted transformation can be used to solve a wide range of irregular domains. They have been extensively employed in engineering practice. However, they become tedious and involve extensive algebraic manipulations. Solving PDEs over irregular domains is a challenge and there have been many strategies that have been used to treat them. Now we shall introduce briefly application of FEM and FVM techniques to multidimensional problems. FEM and FVM have been prominently used to treat nonuniform domains.

13.5 FEM and FVM applied to elliptic problems

We have already introduced FEM and FVM in an earlier chapter. The advantage of FEM and FVM lies in their application to irregular domains. We shall briefly introduce these concepts towards a multidimensional elliptic equation.

Galerkin FEM

Let us consider the two dimensional Poisson equation over a square

$$\frac{\partial^2 u}{\partial x^2} + \frac{\partial^2 u}{\partial y^2} = -q(x,y) \tag{13.20}$$

The square domain is discretized into a number of linear elements (rectangular or triangular) as shown in the figure below

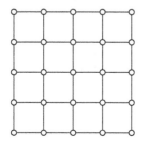
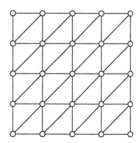

The function within each rectangular element is defined using bilinear interpolation discussed in Chapter 6. The weighted residual (Galerkin formulation) equations for an element would be

$$\int_{A_e} N_i(\nabla^2 u + q) dA = \int_{A_e} N_i \left(\frac{\partial^2 u}{\partial x^2} + \frac{\partial^2 u}{\partial y^2} + q \right) dA = 0$$

where A_e is the area of the two dimensional element and N_i is the shape function associated with node i.   The residual consists of two integrals integrated over the elemental area. The integral $\int_{A_e} N_i q dA$ can be evaluated using any of the quadrature rules we are familiar with. The order of the quadrature rule to be used for evaluation of the integral would affect the accuracy of the final numerical solution.

The second integral $\int_{A_e} N_i \nabla^2 u dA$, just as one dimensional problems, can be rewritten as

$$\int_{A_e} N_i \nabla^2 u dA \;=\; \underbrace{\int_{A_e} \nabla(N_i . \nabla u) dA}_{I_1} - \underbrace{\int_{A_e} \nabla N_i . \nabla u dA}_{I_2}$$

Applying Gauss divergence theorem to I_1, the area integral is converted to line integral

$$I_1 = \int_S N_i . \nabla u ds$$

where s is the boundary of the element. This term is nothing but normal first derivatives to the boundary integrated over the boundary of the element as shown in the following figure.

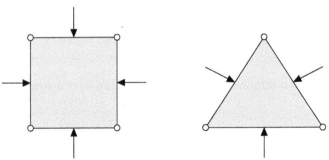

For internal nodes the derivative terms are not available and hence these terms are neglected in the residual equation. Hence this term (similar to one dimensional case) accounts for residual. The residual becomes small as the grids are refined. As this term is related to normal derivatives, both Neumann and Robin condition are accounted for i.e. when such boundary conditions are specified, integral I_1 would be considered in the residual.

I_2 is an area integral given by

$$I_2 = -\int_{A_e} \left(\frac{\partial N_i}{\partial x} \frac{\partial u}{\partial x} + \frac{\partial N_i}{\partial y} \frac{\partial u}{\partial y} \right) dA$$

which can be rewritten as

$$I_2 = -\int_{A_e} \left(\frac{\partial N_i}{\partial x} \sum_j \frac{\partial N_j}{\partial x} u_j + \frac{\partial N_i}{\partial y} \sum_j \frac{\partial N_j}{\partial y} u_j \right) dA$$

The residual equations are solved for each element and are assembled together to form a set of linear equations. We shall take a simple example to illustrate the method.

Example 13.11

Laplace equation is valid in the following domain. Solve for the function at nodes 2, 3, 5 and 6 using FEM

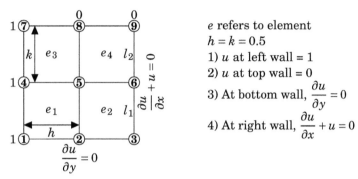

e refers to element

$h = k = 0.5$

1) u at left wall = 1

2) u at top wall = 0

3) At bottom wall, $\dfrac{\partial u}{\partial y} = 0$

4) At right wall, $\dfrac{\partial u}{\partial x} + u = 0$

Solution :

The domain has been discretized into four area elements and the number of unknowns are four.

Let us first consider a rectangular element in its local coordinate system $-1 \le \xi \le 1$ and $-1 \le \eta \le 1$.

The weight functions at the nodes are

$$N_a = \frac{(1-\xi)(1-\eta)}{4} \qquad N_b = \frac{(1+\xi)(1-\eta)}{4}$$

$$N_c = \frac{(1-\xi)(1+\eta)}{4} \qquad N_d = \frac{(1+\xi)(1+\eta)}{4}$$

For an internal node, we will consider integral I_2 and transform the same to local coordinates

$$I_2 = -\int_{A_e} \left(\frac{\partial N_i}{\partial x} \frac{\partial u}{\partial x} + \frac{\partial N_i}{\partial y} \frac{\partial u}{\partial y} \right) dA = -\frac{hk}{4} \int_{-1}^{1}\int_{-1}^{1} \left(\frac{\partial N_i}{\partial \xi} \frac{\partial u}{\partial \xi} \frac{4}{h^2} + \frac{\partial N_i}{\partial \eta} \frac{\partial u}{\partial \eta} \frac{4}{k^2} \right) d\xi d\eta$$

where

$$\frac{\partial u}{\partial \xi} = \frac{\partial N_a}{\partial \xi} u_a + \frac{\partial N_b}{\partial \xi} u_b + \frac{\partial N_c}{\partial \xi} u_c + \frac{\partial N_d}{\partial \xi} u_d$$

$$\frac{\partial u}{\partial \eta} = \frac{\partial N_a}{\partial \eta} u_a + \frac{\partial N_b}{\partial \eta} u_b + \frac{\partial N_c}{\partial \eta} u_c + \frac{\partial N_d}{\partial \eta} u_d$$

Then the residual integral for a node i in the element is

$$I_2 = \frac{hk}{4} \int_{-1}^{1}\int_{-1}^{1} \left(\frac{\partial N_i}{\partial \xi} \sum_j \frac{\partial N_j}{\partial \xi} u_j \frac{4}{h^2} + \frac{\partial N_i}{\partial \eta} \sum_j \frac{\partial N_j}{\partial \eta} u_j \frac{4}{k^2} \right) d\xi d\eta$$

We have to evaluate the integrals $\int_{-1}^{1}\int_{-1}^{1}\frac{\partial N_i}{\partial \xi}\frac{\partial N_j}{\partial \xi}d\xi d\eta$ and $\int_{-1}^{1}\int_{-1}^{1}\frac{\partial N_i}{\partial \eta}\frac{\partial N_j}{\partial \eta}d\xi d\eta$ The first partial derivatives are

$$\frac{\partial N_a}{\partial \xi}=-\frac{1-\eta}{4}; \quad \frac{\partial N_b}{\partial \xi}=\frac{1-\eta}{4}; \quad \frac{\partial N_c}{\partial \xi}=-\frac{1+\eta}{4}; \quad \frac{\partial N_d}{\partial \xi}=\frac{1+\eta}{4};$$

$$\frac{\partial N_a}{\partial \eta}=-\frac{1-\xi}{4}; \quad \frac{\partial N_b}{\partial \eta}=-\frac{1+\xi}{4}; \quad \frac{\partial N_c}{\partial \eta}=\frac{1-\xi}{4}; \quad \frac{\partial N_d}{\partial \eta}=\frac{1+\xi}{4};$$

On evaluating the integrals, we build the residual matrix for the element as

$$I_2=-\frac{k}{6h}\begin{pmatrix} 2 & -2 & 1 & -1 \\ -2 & 2 & -1 & 1 \\ 1 & -1 & 2 & -2 \\ -1 & 1 & -2 & 2 \end{pmatrix}\begin{pmatrix} u_a \\ u_b \\ u_c \\ u_d \end{pmatrix}-\frac{h}{6k}\begin{pmatrix} 2 & 1 & -2 & -1 \\ 1 & 2 & -1 & -2 \\ -2 & -1 & 2 & 1 \\ -1 & -2 & 1 & 2 \end{pmatrix}\begin{pmatrix} u_a \\ u_b \\ u_c \\ u_d \end{pmatrix}$$

Now we shall consider individual elements.

Element 1 consists of 4 nodes 1, 2, 4 and 5 which are a, b, c and d respectively in the local coordinates. Function values at 2 and 5 are unknown and the local elemental equation are

$$\text{node 2} \quad -\frac{k}{6h}(-2+2u_2-1+u_5)-\frac{h}{6k}(1+2u_2-1-2u_5)$$

$$-\frac{k}{6h}(-3+2u_2+u_5)-\frac{h}{6k}(2u_2-2u_5)$$

$$\text{node 5} \quad -\frac{k}{6h}(-2+u_2-1+2u_5)-\frac{h}{6k}(-1-2u_2+1+2u_5)$$

$$-\frac{k}{6h}(-3+u_2+2u_5)-\frac{h}{6k}(-2u_2+2u_5)$$

Element 2 is composed of 2, 3, 5 and 6 which are a, b, c and d respectively in the local coordinates. All the four are unknowns and the elemental equations are same as stated earlier which are

$$\text{node 2} \quad -\frac{k}{6h}(2u_2-2u_3+u_5-u_6)-\frac{h}{6k}(2u_2+u_3-2u_5-u_6)$$

$$\text{node 3} \quad -\frac{k}{6h}(-2u_2+2u_3-u_5+u_6)-\frac{h}{6k}(u_2+2u_3-u_5-2u_6)$$

$$\text{node 5} \quad -\frac{k}{6h}(u_2-u_3+2u_5-2u_6)-\frac{h}{6k}(-2u_2-u_3+2u_5+u_6)$$

$$\text{node 6} \quad -\frac{k}{6h}(-u_2+u_3-2u_5+2u_6)-\frac{h}{6k}(-u_2-2u_3+u_5+2u_6)$$

Element 3 is composed of nodes 4, 5, 7 and 8 which are a, b, c and d respectively in the local coordinates. Function values at 5 is unknown and the local elemental equation is

$$\text{node 5} \quad -\frac{k}{6h}(-3+2u_5)-\frac{h}{6k}(2u_5)$$

Element 4 is composed of nodes 5, 6, 8 and 9 which are a, b, c and d respectively in the local coordinates. Function values at 5 and 6 are unknown and the local elemental equation are

$$\text{node 5} \quad -\frac{k}{6h}(2u_5-2u_6)-\frac{h}{6k}(2u_5+u_6)$$

$$\text{node 6} \quad -\frac{k}{6h}(2u_6 - 2u_5) - \frac{h}{6k}(2u_6 + u_5)$$

Boundary conditions

Now we have to consider the boundary conditions. The bottom wall is subjected to Neumann boundary condition $\frac{\partial u}{\partial y} = 0$. This boundary condition has already been represented in the discretized equations. The right wall is subjected to Robin condition and integral I_1 has to be evaluated at the right boundary which is

$$I_1 = \int_s N_i . \nabla u \, ds = \int N_i \frac{\partial u}{\partial x} dy = -\int N_i u \, dy$$

As the integral is applied over a line, we need to consider the line element alone. We have two such linear elements on the boundary, l_1 - nodes 3 and 6 and l_2 nodes 6 and 9. The above integral has already been evaluated in Chapter 11.6.3. From element l_1 we have nodal equation for node 6

$$\text{node 6} \quad -\frac{k}{6} 2u_6$$

and from element l_2 we have

$$\text{node 3} \quad -\frac{k}{6}(u_6 + 2u_3)$$

$$\text{node 6} \quad -\frac{k}{6}(2u_6 + u_3)$$

Nodal equations

We shall assemble all the elemental equations to obtain the following set of linear equations

$$\text{node 2} \quad u_2 = \frac{1}{8}(2u_5 + u_3 + 2u_6 + 3)$$

$$\text{node 3} \quad u_3 = \frac{(u_2 + 2u_5 + u_6(1-k))}{4 + 2k}$$

$$\text{node 5} \quad u_5 = \frac{1}{8}(3 + u_2 + u_3 + u_6)$$

$$\text{node 6} \quad u_6 = \frac{(2u_2 + 2u_5 + u_3(1-k))}{8 + 4k}$$

On solving the above set of equations we get $u_2 = 0.6130$, $u_3 = 0.3580$, $u_5 = 0.5271$ and $u_6 = 0.2459$. We have to refine the grid for obtaining an acceptable solution. FEM can be programmed to take into account of such grid refinement.

FEM program would consist of three stages

1. Preprocessing - discretizing the domain
2. Processing - determine the elemental equations, assemble these equations and solve the system of equations
3. Post-processing - analyze the solution, plot the results etc.

For that matter, the above three stages form the core for any method including FDM and FVM. The examples that have been presented thus far, the discretization has been straightforward. Such grids are also called as structured grids. But for an arbitrary irregular domain structured grids may not be possible and we may have to use unstructured grids or meshes. The reader would have to refer to advanced books for programming aspects of FEM. Nevertheless we briefly introduce some aspects (the program to be discussed is only for illustration). We shall consider a two dimensional domain discretized using triangular elements.

Preprocessing: Constructing mesh, creating nodes, element and boundary database can be obtained using any preprocessing tool. MATLAB itself has an inbuilt preprocessing tool available. To demonstrate the application of FEM, we take up a simple strategy to construct grid and necessary databases.

We shall consider a rectangular domain. Firstly, nodes are distributed uniformly throughout the domain (similar to FDM) n nodes along x coordinate and m nodes along y coordinate. Nodes are numbered similar to Example 13.11. If i and j are indices along x and y coordinate respectively, the node can be numbered using the formula $(j-1)n-i$. In addition to this, we should know the element properties (type of element -linear triangular, quadrilateral etc) and its associated nodes. We shall consider the domain to be discretized into triangular elements. A MATLAB program has been written to produce the array containing coordinates of the nodes and element array.

Program 13.1: *Construct list of nodes over a rectangular domain*

```
 1  function [p,t] = constructnodes(n,m)
 2  %   Input    n,m : no of nodes along x and y
 3  %   Output   p   : coordinates of nodes
 4  %                  p(:,1)   x coordinate
 5  %                  p(:,2)   y coordinate
 6  %            t   : stores node numbers associated with element
 7  %                  t(:,1),t(:,2),t(:,3) are node numbers arranged
 8  %                  in counter clockwise direction
 9  x = 0:1/(n-1):1;    y = 0:1/(m-1):1;      % nodes
10  [Y,X] = meshgrid(y,x);  % mesh
11  nn = n*m; p = zeros(nn,2);       % initialize nodes
12  count = 1;               % initialize node count
13  for i=1:n
14      for j=1:m
15          p(count,1) = X(i,j);
16          p(count,2) = Y(i,j);
17          count = count + 1;
18      end
19  end
20  nqe = 2*(n-1)*(m-1);     % no of elements
21  t = zeros(nqe,3);        % initialize elements
22  % t stores the node numbers associated with the element
23  count = 1;               % initialize element count
24  for j=1:m-1
25      for i=1:n-1
```

```
26          t(count,1)  =  (j-1)*n+i;
27          t(count,2)  =  (j-1)*n+i+1;
28          t(count,3)  =  j*n+i+1;
29          t(count+1,1)  =  (j-1)*n+i;
30          t(count+1,2)  =  j*n+i+1;
31          t(count+1,3)  =  j*n+i;
32          count  =  count + 2;
33      end
34 end
```

A sample output is given below.

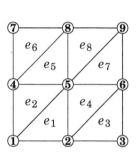

node number	x	y
1	0	0
2	0.5	0
3	1	0
4	0	0.5
5	0.5	0.5
6	1	0.5
7	0	1
8	0.5	1
9	1	1

Element numbers	Node numbers		
1	1	2	5
2	1	5	4
3	2	3	6
4	2	6	5
5	4	5	8
6	4	8	7
7	5	6	9
8	5	9	8

In addition to this we must also have a database of the boundary elements, its associated nodes and the respective condition applicable over the element. For a two dimensional domain, boundaries are represented by line elements. There can be two types of boundary conditions viz. Dirichlet or Neumann/ Robin boundary condition. We shall store these boundary conditions separately. The database associated with Dirichlet boundary condition would store the node number and the value fixed at the node. We shall follow the following convention

```
% Dirichlet
bd(:,1) stores node number
bd(:,2) stores value associated with the node
```

A Neumann condition is specified as

$$\frac{\partial u}{\partial n} + c_1 u = c_2$$

Hence, the database associated with Neumann boundary condition should store the nodes associated with the line element and values of c_1 and c_2. We shall follow the following convention for programming

```
% Neumann
% br(:,1) stores first node number
% br(:,2) stores second node number
% br(:,3) stores c1
% br(:,4) stores c2
```

We shall take up these two database when taking up specific examples.

Processing: We have to construct a system of linear equations and it is always better to consider element by element. Like wise, it would be better to treat each term of the equation independently and later assembled together. Determining all the local matrices involves evaluation of shape functions and quadrature rules. Hence, separate sub programs for quadrature of a term over an element can be written. Hence, the local coordinate system representation becomes useful.

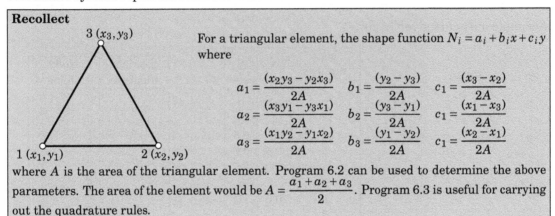

Recollect

For a triangular element, the shape function $N_i = a_i + b_i x + c_i y$ where

$$a_1 = \frac{(x_2 y_3 - y_2 x_3)}{2A} \quad b_1 = \frac{(y_2 - y_3)}{2A} \quad c_1 = \frac{(x_3 - x_2)}{2A}$$

$$a_2 = \frac{(x_3 y_1 - y_3 x_1)}{2A} \quad b_2 = \frac{(y_3 - y_1)}{2A} \quad c_1 = \frac{(x_1 - x_3)}{2A}$$

$$a_3 = \frac{(x_1 y_2 - y_1 x_2)}{2A} \quad b_3 = \frac{(y_1 - y_2)}{2A} \quad c_1 = \frac{(x_2 - x_1)}{2A}$$

where A is the area of the triangular element. Program 6.2 can be used to determine the above parameters. The area of the element would be $A = \dfrac{a_1 + a_2 + a_3}{2}$. Program 6.3 is useful for carrying out the quadrature rules.

First let us consider the Laplacian term, $\displaystyle\int_{A_e} N_i \nabla^2 u \, dA$. We have already shown that this term can be reduced to

$$-\int_{A_e} \frac{\partial N_i}{\partial x} \sum_{j=1}^{3} \frac{\partial N_j}{\partial x} u_j \, dA + \int_{A_e} \frac{\partial N_i}{\partial y} \sum_{j=1}^{3} \frac{\partial N_j}{\partial y} u_j \, dA$$

The first partial derivatives for a triangular element are constant and hence the above integral becomes

$$\mathbf{K} = -A \begin{pmatrix} b_1^2 & b_1 b_2 & b_1 b_3 \\ b_1 b_2 & b_2^2 & b_2 b_3 \\ b_1 b_3 & b_2 b_3 & b_3^2 \end{pmatrix} - A \begin{pmatrix} c_1^2 & c_1 c_2 & c_1 c_3 \\ c_1 c_2 & c_2^2 & c_2 c_3 \\ c_1 c_3 & c_2 b_3 & c_3^2 \end{pmatrix}$$

$$= -A \begin{pmatrix} b_1 \\ b_2 \\ b_3 \end{pmatrix} \begin{pmatrix} b_1 & b_2 & b_3 \end{pmatrix} - A \begin{pmatrix} c_1 \\ c_2 \\ c_3 \end{pmatrix} \begin{pmatrix} c_1 & c_2 & c_3 \end{pmatrix}$$

The integral related to source term $\displaystyle\int_{A_e} N_i q \, dA$ can be evaluated using quadrature rules. The simplest would be to use the corners of three nodes for evaluating the integral. Then the source term simply reduces to

$$\mathbf{Q}^T = \frac{A}{3} \begin{pmatrix} q_1 & q_2 & q_3 \end{pmatrix}$$

Now, we have to consider integrals related to Neumann and Robin condition. A general boundary condition involves

$$\frac{\partial u}{\partial n} + c_1 u = c_2 \quad \text{or} \quad \frac{\partial u}{\partial n} = -c_1 u + c_2$$

If $c_1 = 0$, we get Neumann condition. Now the line integral $I_1 = \int_S N_i . \nabla u \, ds$ is considered which becomes $\int_S N_i (-c_1 u + c_2) ds$. This integral is applied over the edges over which such boundary conditions are specified. Let us consider one element subjected to the above condition. Then the integral to be evaluated becomes

$$-c_1 \int_S \begin{pmatrix} N_1^2 & N_1 N_2 \\ N_1 N_2 & N_2^2 \end{pmatrix} \begin{pmatrix} u_1 \\ u_2 \end{pmatrix} ds + c_2 \int_S \begin{pmatrix} N_1 \\ N_2 \end{pmatrix} ds$$

N_1 and N_2 are the shape functions for a line element. Hence, the integral becomes

$$-\frac{c_1 L}{6} \begin{pmatrix} 2 & 1 \\ 1 & 2 \end{pmatrix} \begin{pmatrix} u_1 \\ u_2 \end{pmatrix} + \frac{c_2 L}{2} \begin{pmatrix} 1 \\ 1 \end{pmatrix}$$

where L is the length of the line element.

Assembly: All the local elemental terms have to be assembled together to form a global set of equations. Assembly remains the most crucial aspect of FEM. Unlike, one dimensional problem, the matrix is not tridiagonal. However matrix is sparse i.e. large number of zero elements in the matrix. Thus one needs to store only the non zero values of the matrix. MATLAB provides inbuilt functions to handle sparse matrices. However, we shall leave sparse matrices to the interest of the reader.

A simple way to assemble the equations is to consider the full matrix directly. We shall initialize the full matrix to zero. We start calculating the local matrices for every element and add the full matrix directly. After assembling all the equations, Dirichlet boundary conditions is also fed into the matrix. These equations are then solved using any of the methods that have been presented earlier. As we have used full matrix, the method would be computationally intensive. However, using sparse matrix algebra can tremendously reduce the computational cost.

A MATLAB function has been given below to solve two dimensional Poisson equation based on what has been discussed till now.

Program 13.2: *FEM to solve 2D Poisson equation*

```
function u = fempoisson(p,t,bd,br,q)
% Input    p : coordinates of nodes
%          t : triangular elements
%          bd: nodes with Dirichlet BC
%          br: nodes with Neumann/ Robin BC
%          q : source value at nodes
```

```
 7 % Output  u : solution
 8 nn = size(p,1);             % no. of nodes
 9 nte = size(t,1);            % no. pf elements
10 A = zeros(nn);              % initialize matrix A
11 B = zeros(nn,1);            % initialize vector B
12 nbd = size(bd,1);           % no. of dirichlet boundary nodes
13 nbr = size(br,1);           % no. of Neumann boundary elements
14 for i=1:nte                 % loop for laplace and source term
15 [a,b,c,Area] = trlocalcoeff(p(t(i,1),:),p(t(i,2),:) ...
16                            ,p(t(i,3),:));
17 K = -Area*(b'*b+c'*c);         %% elemental laplacian matrix
18 % updating global matrix
19 A(t(i,1),t(i,1)) = A(t(i,1),t(i,1))+K(1,1);
20 A(t(i,1),t(i,2)) = A(t(i,1),t(i,2))+K(1,2);
21 A(t(i,1),t(i,3)) = A(t(i,1),t(i,3))+K(1,3);
22 A(t(i,2),t(i,1)) = A(t(i,2),t(i,1))+K(2,1);
23 A(t(i,2),t(i,2)) = A(t(i,2),t(i,2))+K(2,2);
24 A(t(i,2),t(i,3)) = A(t(i,2),t(i,3))+K(2,3);
25 A(t(i,3),t(i,1)) = A(t(i,3),t(i,1))+K(3,1);
26 A(t(i,3),t(i,2)) = A(t(i,3),t(i,2))+K(3,2);
27 A(t(i,3),t(i,3)) = A(t(i,3),t(i,3))+K(3,3);
28 % % updating source term
29 B(t(i,1)) = B(t(i,1)) + q(t(i,1))*Area/3;
30 B(t(i,2)) = B(t(i,2)) + q(t(i,2))*Area/3;
31 B(t(i,3)) = B(t(i,3)) + q(t(i,3))*Area/3;
32 end
33 % Neumann and Robin condition
34 for i=1:nbr
35 length = sqrt((p(br(i,1),1)-p(br(i,2),1))^2 ...
36        + (p(br(i,1),2)-p(br(i,2),2))^2);
37   c1 = br(i,3); c2 = br(i,4);
38 A(br(i,1),br(i,1)) = A(br(i,1),br(i,1))-c1*length/3;
39 A(br(i,1),br(i,2)) = A(br(i,1),br(i,2))-c1*length/6;
40 A(br(i,2),br(i,2)) = A(br(i,2),br(i,2))-c1*length/3;
41 A(br(i,2),br(i,1)) = A(br(i,2),br(i,1))-c1*length/6;
42 B(br(i,1)) = B(br(i,1)) - c2*length/2;
43 B(br(i,2)) = B(br(i,2)) - c2*length/2;
44 end
45 % Dirichlet condition
46 for i=1:nbd
47    A(bd(i,1),:) = 0; A(bd(i,1),bd(i,1)) = 1;
48    B(bd(i,1)) = bd(i,2);
49 end
50 u = A\B;             % solve
```

The above program can be used to solve a two dimensional Poisson equation over an arbitrary domain, provided the required database is already available. Now we shall take up some cases considered earlier.

Poisson equation over square domain, Example 13.2

The boundaries are subjected to Dirichlet boundary condition $u = 0$ and the source term

$q = 100\sin(\pi x)\cos(\pi x)$. MATLAB program is provided below to solve this problem using FEM.

```
n = 41; m = 41; % no of nodes
% construct nodes and elements
[p,t] = constructnodes(n,m);
% construct boundary condition database
bdcount = 1; % Dirichlet condition database
j = 1;
for i=1:n    % bottom boundary
       bd(bdcount,1) = (j-1)*n+i; % node number
       bd(bdcount,2) = 0;        % value
       bdcount = bdcount + 1;
end
i= n;        % right boundary
for j=2:m
       bd(bdcount,1) = (j-1)*n+i; % node number
       bd(bdcount,2) = 0;        % value
       bdcount = bdcount + 1;
end
i= 1;  % left boundary
for j=2:m
       bd(bdcount,1) = (j-1)*n+i; % node number
       bd(bdcount,2) = 0;        % value
       bdcount = bdcount + 1;
end
j = m;
for i=2:n-1    % bottom boundary
       bd(bdcount,1) = (j-1)*n+i; % node number
       bd(bdcount,2) = 0;        % value
       bdcount = bdcount + 1;
end
% source term at the nodes
q = -100*sin(pi*p(:,1)).*sin(pi*p(:,2));
u = fempoisson(p,t,bd,[],q);  % solve using FEM
% [] indicates empty matrix
% post processing
X = 0:0.01:1; Y = 0:0.01:1;
[X,Y] = meshgrid(X,Y);
Z = griddata(p(:,1),p(:,2),u,X,Y); % data in grid form
contour(X,Y,Z);            % contour plot
```

The solution obtained is comparable to the FDM solution. For a grid of 41×41, the function value at $(0.5, 0.5)$, $u(0.5, 0.5) = 5.06866$

Laplace equation over irregular domain, Example 13.10 FEM is suitable for an irregular domain. For the present case, we have first created the nodes and elements in the transformed coordinates after which the nodes are transformed to Cartesian coordinates.

```
n = 5; m = 5;
[p1,t] = constructnodes(n,m);
p(:,1) = p1(:,1);
p(:,2) = p1(:,2).*(0.5-0.25*p1(:,1));
```

Now, we shall create the boundary condition database and solve the equations using FEM.

```
bdcount = 1; brcount = 1;
j = 1;
for i=1:n-1    % bottom  boundary
        br(brcount,1) = (j-1)*n+i;
        br(brcount,2) = (j-1)*n+i+1;
        br(brcount,3) = 0;
        br(brcount,4) = 0;
        brcount = brcount + 1;
end
% right  boundary
i= n;
for j=1:m-1
        br(brcount,1) = (j-1)*n+i;
        br(brcount,2) = (j)*n+i;
        br(brcount,3) = 0;
        br(brcount,4) = 0;
        brcount = brcount + 1;
end
i= 1;   % left boundary
for j=1:m
        bd(bdcount,1) = (j-1)*n+i;
        bd(bdcount,2) = 1;
        bdcount = bdcount + 1;
end
j = m;
for i=1:n-1    % top boundary
        br(brcount,1) = (j-1)*n+i;
        br(brcount,2) = (j-1)*n+i+1;
        br(brcount,3) = 1;
        br(brcount,4) = 0;
        brcount = brcount + 1;
end
q = 0*p;
u1 = fempoisson(p,t,bd,br,q);
% post-processing
Xi =0:0.01:1; Eta = 0:0.01:1;
[Xi,Eta] = meshgrid(Xi,Eta);
X = Xi; Y = 0.5*Eta-0.25*Eta.*Xi;
 Z = griddata(p(:,1),p(:,2),u1,X,Y);
figure;contour(X,Y,Z)
```

We have demonstrated how a same set of programs can be applied to wide range of problems. The programs presented in the section are not optimized for computation. They may be used by the reader to solve simple problems. We have not dealt with the pre-processing stage of FEM. However, this is the most important stage of FEM and the reader should refer to a specialized references on **FEM**. The quality of the mesh would affect the accuracy of the solution and often in practice, most of the effort goes into producing quality meshes.

FVM

Let us consider the Poisson equation over a square domain. The domain is to be divided into finite volumes. For FVM, the weight over a volume is equal to 1. The residual equation applied to one finite volume is given by

$$R = \int_{A_e} \left(\frac{\partial^2 u}{\partial x^2} + \frac{\partial^2 u}{\partial y^2} + q \right) dA = \int_s \frac{\partial u}{\partial n}.nds + \int_{A_e} qdA$$

The above integral is related to the first derivatives normal to the boundary of the volume similar to FEM. For a rectangular domain the following shows the details

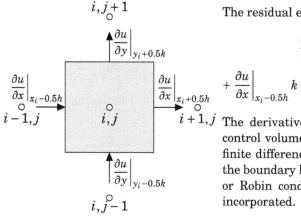

The residual equation for node i,j will be

$$\left.\frac{\partial u}{\partial y}\right|_{y_i+0.5k} h + \left.\frac{\partial u}{\partial y}\right|_{y_i-0.5k} h$$

$$+ \left.\frac{\partial u}{\partial x}\right|_{x_i-0.5h} k + \left.\frac{\partial u}{\partial x}\right|_{x_i+0.5h} k + q_{i,j}hk = 0$$

The derivatives at the boundaries of the control volume can be approximated using finite difference schemes. If derivatives at the boundary has been provided (Neumann or Robin condition), they can be directly incorporated.

We shall take up an example illustrating FVM.

Example 13.12

Solve Example 13.11 using FVM.

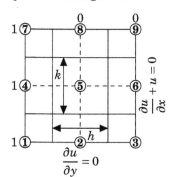

$h = k = 0.5$

1) u at left wall = 1

2) u at top wall = 0

3) At bottom wall, $\dfrac{\partial u}{\partial y} = 0$

4) At right wall, $\dfrac{\partial u}{\partial x} + u = 0$

Solution :

We have to determine the function value at four nodes. We shall write down the nodal equations. The derivatives at the boundary of the domain are to be determined using central differences.

Node 5 Internal node

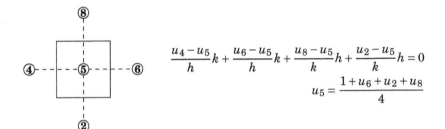

$$\frac{u_4 - u_5}{h}k + \frac{u_6 - u_5}{h}k + \frac{u_8 - u_5}{k}h + \frac{u_2 - u_5}{k}h = 0$$

$$u_5 = \frac{1 + u_6 + u_2 + u_8}{4}$$

Node 2 Bottom wall

$$\frac{u_1 - u_2}{h}\frac{k}{2} + \frac{u_3 - u_2}{h}\frac{k}{2} + \frac{u_5 - u_2}{k}h = 0$$

$$u_2 = \frac{1 + 2u_5 + u_3}{4}$$

Node 6 Right wall

$$\frac{u_5 - u_6}{h}k + \frac{u_9 - u_6}{k}\frac{h}{2} + \frac{u_3 - u_6}{k}\frac{h}{2} - u_6 k = 0$$

$$u_6 = \frac{2u_5 + u_3}{4 + 2k}$$

Node 3 Right bottom corner

$$\frac{u_2 - u_3}{h}\frac{k}{2} + \frac{u_6 - u_3}{k}\frac{h}{2} - u_3\frac{k}{2} = 0$$

$$u_3 = \frac{u_2 + u_6}{2 + k}$$

On solving the above set of equations we get $u_2 = 0.4286$, $u_3 = 0.4286$, $u_5 = 0.1429$ and $u_6 = 0.1429$. The solution obtained using FVM is different from FEM. However, on grid refinement both the methods would yield the same solution. We encourage the reader to write a program for FVM and compare the same with the FDM/FEM solutions obtained earlier.

Concluding remarks

In this chapter we have covered various techniques of numerically solving Poisson and Laplace equations. We have also attempted problems in cylindrical coordinates and also problems involving irregular boundaries to show that FDM and other techniques are quite capable of obtaining solutions in these cases.

Chapter *14*

Advection and diffusion equations

This chapter considers an important class of problems where advection and diffusion come together. Numerical schemes for advection and diffusion terms have different limitations. Hence it is a challenge to evolve schemes where both are present.

Stability and accuracy are dependent on the specific schemes used for spatial and temporal discretization. Hence the study of stability issues and error analysis using methods such as von Neumann stability analysis and Taylor series are important. These are discussed in sufficient detail so that the reader can look independently at such issues in new situations he/she may come across. Many worked examples bring out the features of different numerical schemes.

14.1   Introduction

Some examples of PDE involving time are

$$\frac{\partial u}{\partial t} = a\frac{\partial u}{\partial x} \qquad \text{advection equation}$$

$$\frac{\partial u}{\partial t} = \frac{\partial^2 u}{\partial x^2} \qquad \text{Diffusion equation}$$

$$\frac{\partial u}{\partial t} = a\frac{\partial u}{\partial x} + \frac{\partial^2 u}{\partial x^2} \qquad \text{Advection-diffusion equation}$$

$$\frac{\partial^2 u}{\partial t^2} = a^2\frac{\partial^2 u}{\partial x^2} \qquad \text{wave equation}$$

A PDE with time as one of its independent variable is an initial value problem. The right hand side of the above equations are function of space variables alone. The terms involving spatial derivatives can be discretized using FDM, collocation, FEM, FVM and other suitable methods. Let us consider the diffusion equation whose spatial derivative is discretized using second order central difference. Then we have a set of ODEs with time as the independent variable and the function values at the nodes as the dependent variable.

$$\frac{du_i}{dt} = \frac{u_{i+1} - 2u_i + u_{i-1}}{h^2}$$

As we are already familiar with a set of IVPs, the above set of ODEs can be integrated with respect to time using the methods discussed in Chapter 10. Such an approach to solve time dependent PDEs is also referred to as *Method of lines*. However, being an initial value problem, one has to consider both stability and accuracy of the numerical solution. It has to be noted that the spatial discretization affects the stability of the solution. The present chapter would introduce numerical methods applicable to advection equation and diffusion equation with attention directed to errors and stability.

14.2   The advection equation

It is instructive to consider the solution of a first order PDE of form given by Equation 12.1. Specifically we consider the special form of the equation, the advection equation, given by

$$\frac{\partial u}{\partial t} + a\frac{\partial u}{\partial x} = 0 \qquad\qquad (14.1)$$

where a is a constant representing the advection velocity. The characteristic direction is given by $\xi = x - at$. The solution is given by $u = f(\xi)$. If the initial solution is $u = f(x)$ the same will be translated to the right by a distance equal to at_0 at a later time t_0. Figure 14.1 shows this for an initial pulse located at $x_0 = b$. It is found, without change of shape, at $x = b + at_0$ at a later time t_0. Thus the pulse translates to the right, with velocity equal to the advection velocity.

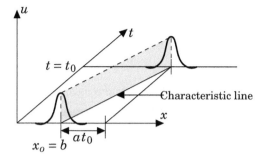

Figure 14.1: Solution of advection equation

Since the solution to the advection equation is very simple, it is possible to use this equation to study the usefulness of various finite difference schemes, in solving advection as well as other PDEs.

14.2.1 Finite difference schemes for advection equation

Even though the advection equation is very simple and has a straight forward analytical solution, the number of different numerical schemes that may be used to solve it are many. We will consider a few of them here. We shall discretize the space domain by taking uniformly spaced nodes with a spacing of Δx and choose a time step of Δt. The function value is represented as u_i^k where i represents node number along space and k represents node number along time. The computational domain for the problem is then visualized as in Figure 14.2.

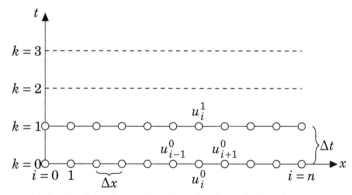

Figure 14.2: Nodal scheme for discretization of the advection equation

Different numerical schemes are compared based on their stability characteristics as well as on the nature and consequence of errors introduced by them.

1. Explicit FTCS (Forward in Time - Centered in Space) scheme:

Figure 14.3 shows the nodes involved in the computational scheme. We use forward difference for the time derivative while using the central difference for the space derivative. This means the finite difference approximation is first order accurate in time and second order accurate in space. An explicit scheme means the above integration process is the same as Euler method.

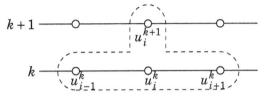

Figure 14.3: *Nodes involved in the FTCS scheme*

The scheme is written down easily as

$$\frac{u_i^{k+1}-u_i^k}{\Delta t}+a\frac{u_{i+1}^k-u_{i-1}^k}{2\Delta x}=0 \text{ or } u_i^{k+1}=u_i^k-\underbrace{\frac{C}{2}\left(u_{i+1}^k-u_{i-1}^k\right)}_{\text{Central difference in space}} \tag{14.2}$$

where $C=\dfrac{a\Delta t}{\Delta x}$ is known as the Courant number. Now we shall look at the stability of the numerical scheme. For this purpose we derive the analytical solution to the problem. Let the solution to the advection equation be represented by $u(x,t)=A(t)e^{j\omega x}$ where ω is the wave number. We then have

$$\frac{\partial u}{\partial x}=j\omega A(t)e^{j\omega x}$$

$$\frac{\partial u}{\partial t}=\frac{dA}{dt}e^{j\omega x}$$

Substitute these in Equation 14.1 to get the following equation.

$$\frac{dA}{dt}+ja\omega A=0 \tag{14.3}$$

This is an initial value problem having the solution

$$A=A_0e^{-ja\omega t} \tag{14.4}$$

where A_0 is the initial value of A at $t=0$. We see thus that the solution to the advection equation is given by

$$u(x,t)=A_0e^{j\omega(x-at)} \tag{14.5}$$

The above solution to the advection equation is consistent with that described in Section 10.6. Also observe that the term $x-at$ is related to the characteristic line. We shall now substitute the exact values of u in Equation 14.2. We have $u_i^k=A_0e^{j\omega(i\Delta x-ak\Delta t)}$, $u_{i+1}^k=A_0e^{j\omega((i+1)\Delta x-ak\Delta t)}=e^{j\omega(i\Delta x-ak\Delta t)}e^{j\omega\Delta x}=u_i^ke^{j\omega\Delta x}$ and $u_{i-1}^k=u_i^ke^{-j\omega\Delta x}$. On simplification Equation 14.2 becomes

$$\frac{u_i^{k+1}}{u_i^k}=1-C\frac{\left(e^{j\omega\Delta x}-e^{-j\omega\Delta x}\right)}{2}=1-jC\sin(\omega\Delta x) \tag{14.6}$$

This is nothing but the magnification factor arising out of the FTCS scheme. For stability magnification factor must be less than 1. The magnitude of the ratio is equal to $\sqrt{1 + C^2 \sin^2(\omega x)}$ which is equal to or greater than 1 for all C. Hence the FTCS scheme is *unconditionally unstable* for all values of Δx and is not useful. The stability analysis presented here is known as *von Neumann stability analysis*[1].

2. Explicit FTFS and FTBS schemes leading to the upwind scheme:

Now let us consider FTFS scheme (Forward in Time - Forward in Space). We approximate both the time and space derivatives by forward differences to get the following.

$$\frac{u_i^{k+1} - u_i^k}{\Delta t} + a \frac{u_{i+1}^k - u_i^k}{\Delta x} = 0 \text{ or } u_i^{k+1} = u_i^k - C\left(u_{i+1}^k - u_i^k\right) \tag{14.7}$$

Both derivatives are first order accurate and hence the truncation errors are proportional to the step size along the two directions. The nodes involved in the FTFS scheme are shown in Figure 14.4.

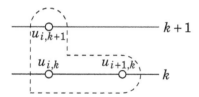

$k+1$

$u_{i,k+1}$

$u_{i,k}$ $\quad u_{i+1,k}$

k

Figure 14.4: *Nodes involved in the FTFS scheme*

Using the exact solution, we have $u_i^k = A_0 e^{j\omega(i\Delta x - ak\Delta t)}$. Also, we have $u_{i+1}^k = A_0 e^{j\omega((i+1)\Delta x - ak\Delta t)} = e^{j\omega(i\Delta x - ak\Delta t)} e^{j\omega\Delta x} = u_i^k e^{j\omega\Delta x}$. Introducing these in the FTFS scheme and after simplification we get

$$\frac{u_i^{k+1}}{u_i^k} = 1 + C\left(1 - e^{j\omega\Delta x}\right) \tag{14.8}$$

This is nothing but the magnification factor arising out of the FTFS scheme. The maximum magnitude of the ratio is equal to $(1+C)^2 + C^2$ and for stability should be less than 1. Hence stability requires that $C < 0$ and $|C| \leq 1$. This simply means that positive advection velocity $a > 0$ makes FTFS unconditionally unstable. In fact the requirement that $C < 0$ leads to the FTBS (Forward in Time Backward in Space) given by

$$u_i^{k+1} = u_i^k - C\left(u_i^k - u_{i-1}^k\right) \tag{14.9}$$

which is stable for $a > 0$ and $|C| \leq 1$. This is known as Courant-Friedrichs-Lewy[2] condition or CFL condition. The FTFS scheme is stable when $a < 0$. When $C = 1$, we get $u_i^{k+1} = u_{i-1}^k$,

---

[1]after John von Neumann, 1903-1957, a Hungarian-born American mathematician

[2]after Richard Courant, 1888-1972, Kurt Otto Friedrichs, 1901-1982, and Hans Lewy, 1904-1988, all German American mathematicians

which means the disturbance moves exactly by one grid spacing in one time step. The method becomes unstable if the disturbance moves more than one grid spacing!

The spatial differencing has to be done according to the direction of velocity a i.e. use FTFS for $a < 0$ and FTBS for $a > 0$. The scheme is referred to as upwind scheme. Mathematically upwind scheme can be written as

$$u_i^{k+1} = u_i^k - \left(\max[-C, 0] u_{i+1}^k - |C| u_i^k + \max[0, C] u_{i-1}^k \right)$$

The upwind scheme involves the nodes shown in Figure 14.5.

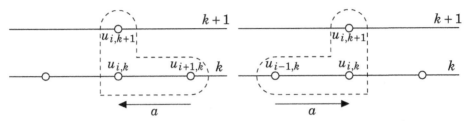

Figure 14.5: *Nodes involved in the Upwind scheme with $a < 0$ and $a > 0$*

Upwind scheme is *conditionally stable* with $C \leq 1$.

Example 14.1

Solve the advection equation using the upwind scheme with $a = 1$. Initial shape of the solution is specified to be a Gaussian defined by $u(x, 0) = e^{-100(x-0.5)^2}$ for $0 \leq x \leq 1$. Use a Courant number of 0.5.

Solution :

We choose $\Delta x = 0.02$ and get $\Delta t = 0.01$ using $C = 0.5$. We then make use of Equation 14.9 to get

$$u_i^{k+1} = u_i^k - \frac{\left(u_i^k - u_{i-1}^k \right)}{2} = \frac{u_{i-1}^k + u_i^k}{2}$$

where $0 < i < 50$ and $k \geq 0$. At the two end nodes we use periodic boundary condition such that

$$u_0^{k+1} = \frac{u_{49}^k + u_0^k}{2} \quad \text{and} \quad u_{50}^{k+1} = \frac{u_{49}^k + u_{50}^k}{2}$$

The periodic boundary condition ensures that the pulse will reappear from the left boundary when once it crosses the right boundary. A MATLAB program has been written to carry out the computation

```
dx = 0.02;        % step size
n = 1/dx+1;       % no of nodes
x = 0:dx:1;       % nodes
dt = 0.5*dx;      % time step
c = dt/dx;        % Courant number
u = exp(-100*(0.5-x').^2); un = u; % initial velocity
for t=1:10        % loop for time integration
       un(1) = u(1)-c*(u(n)-u(n-1));   % periodic
       un(n) = un(1);                   % boundary conditions
```

```
    for  i=2:n-1    % loop  for  internal  nodes
        un(i)  =  u(i)-c*(u(i)-u(i-1));
    end
    u  =  un;     % update  u
end
```

We have obtained the solution over several time steps to make a sketch of the solution for larger time also. Figure 14.6 shows the state of affairs. We notice that the position of the peak is close to that given by the exact solution i.e $\xi = 0.5$. However the amplitude of the numerical solution keeps decreasing with time.

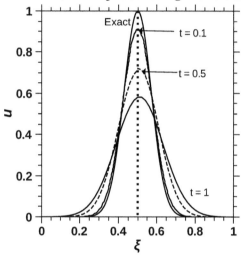

Figure 14.6: *Evolution of solution with time computed using the upwind scheme with $C = 0.5$*

Error analysis of Upwind scheme: Upwind scheme is only first order accurate in space. Let us consider advection with $a > 0$. We have to use backward difference for space derivative.

$$\frac{u_i^{k+1} - u_i^k}{\Delta t} + a\frac{u_i^k - u_{i-1}^k}{\Delta x} = 0 \tag{14.10}$$

Perform Taylor series expansion of u_{i-1}^k and u_i^{k+1} about u_i to get

$$u_{i-1}^k = u_i^k - \frac{\partial u}{\partial x}\Delta x + \frac{1}{2!}\frac{\partial^2 u}{\partial x^2}\Delta x^2 + O(\Delta x^3)$$

$$u_i^{k+1} = u_i^k + \frac{\partial u}{\partial t}\Delta t + \frac{1}{2!}\frac{\partial^2 u}{\partial t^2}\Delta t^2 + O(\Delta t^3)$$

Replacing u_{i-1}^k and u_i^{k+1} in Equation 14.10 with those from the Taylor series expansion we get

$$\underbrace{\frac{\partial u}{\partial t} + a\frac{\partial u}{\partial x}}_{\text{advection}} = \underbrace{-\frac{1}{2}\frac{\partial^2 u}{\partial t^2}\Delta t + \frac{1}{2}a\Delta x\frac{\partial^2 u}{\partial x^2}}_{\text{error}} + O(\Delta t^2, \Delta x^2) \tag{14.11}$$

Upwind scheme solves the above PDE rather than the advection equation. The above may
be simplified as follows. Differentiate Equation 14.11 with respect to t and get

$$\frac{\partial^2 u}{\partial t^2} + a\frac{\partial^2 u}{\partial x \partial t} = -\frac{\Delta t}{2}\frac{\partial^3 u}{\partial t^3} + a\frac{\Delta x}{2}\frac{\partial^3 u}{\partial x^2 \partial t} + O(\Delta t^2, \Delta x^2) \tag{14.12}$$

Differentiate Equation 14.11 with respect to x after multiplying it by a and get

$$a\frac{\partial^2 u}{\partial t \partial x} + a^2\frac{\partial^2 u}{\partial x^2} = -a\frac{\Delta t}{2}\frac{\partial^3 u}{\partial t^2 \partial x} + O(\Delta t^2, \Delta x^2) \tag{14.13}$$

Subtracting Equation 14.13 from 14.12 we get

$$\frac{\partial^2 u}{\partial t^2} = a^2\frac{\partial^2 u}{\partial x^2} + O(\Delta t, \Delta x) \tag{14.14}$$

Using this back in Equation 14.11 the error term simplifies to

$$\epsilon = \underbrace{\frac{a\Delta x}{2}(1-C)\frac{\partial^2 u}{\partial x^2}}_{\text{diffusive/dissipative}} + O(\Delta t^2, \Delta x^2) \tag{14.15}$$

The error term is in the form of diffusion or dissipation term. This term causes error in
amplitude of the function. The diffusion term is the reason for reduction in amplitude and
increase in the width of the pulse (spreading) seen in the previous example. The diffusive
nature of upwind scheme increases in the presence of sharp gradients of u. Also, the
diffusion error is related to C and step size Δx. Including terms of second order leads to
dispersion error given by[3]

$$\epsilon = \underbrace{-a\frac{\Delta x^2}{6}(2C^2 - 3C + 1)}_{\text{dispersion error}} \tag{14.16}$$

Figure 14.7 shows the effect of step size and Courant number on the solution. C should
be close to 1 for minimum error. This plot uses ξ along the abscissa so that the disturbance
is stationary. The amplitude and the width of the pulse are used to bring out the effects of
dispersion and dissipation. In fact when $C = 1$ the diffusion error and dispersive error are
reduced to zero i.e. the numerical solution is the same as the exact analytical solution.
This deduction is hardly surprising because the numerical computation is performed
along the characteristic line. It should also be remembered that the numerical solution
becomes unstable for $C > 1$. The error is significantly reduced by choosing smaller step
size. But one has to understand that smaller the step size more the computation time
(Δt is restricted by C). For upwind scheme, dispersion component of the error is small
(higher order) compared to the diffusion error. We shall discuss the relevance of dispersion
later when handling higher order methods. We have analyzed Euler scheme for time
integration, however one can improve the accuracy of time integration by using higher
order schemes such as RK4, ABM4 etc.

---

[3]For details see Analysis of numerical dissipation and dispersion at www.mathematik.uni-
dortmund.de/ kuzmin/cfdintro/lecture10.pdf

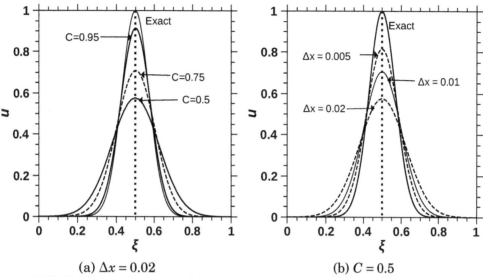

(a) $\Delta x = 0.02$ (b) $C = 0.5$

Figure 14.7: *Evolution of solution with time computed using the upwind scheme at $t = 1$*

3. Explicit Lax-Friedrichs scheme:

The FTCS scheme may be made stable by replacing the first term on the right hand side of Equation 14.2 by the mean value of the function at nodes $i - 1$ and $i + 1$. Thus we have

$$u_i^{k+1} = \frac{\left(u_{i-1}^k + u_{i+1}^k\right)}{2} - \frac{C}{2}\left(u_{i+1}^k - u_{i-1}^k\right) \tag{14.17}$$

The nodes involved in this scheme are shown in Figure 14.8.

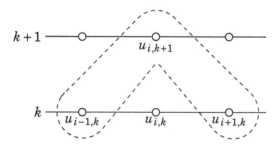

Figure 14.8: *Nodes involved in the Lax-Friedrichs scheme*

Even though the stability of the scheme is improved i.e. the scheme is stable as long as $C \le 1$ (the reader may show this using von Neumann stability analysis), it is seen that the above may be rewritten as

$$u_i^{k+1} = u_i^k + \frac{\left(u_{i-1}^k - 2u_i^k + u_{i+1}^k\right)}{2} - \frac{C}{2}\left(u_{i+1}^k - u_{i-1}^k\right)$$

This may further be recast as

$$\frac{u_i^{k+1} - u_i^k}{\Delta t} = \left(\frac{\Delta x^2}{2\Delta t}\right)\frac{\left(u_{i-1}^k - 2u_i^k + u_{i+1}^k\right)}{\Delta x^2} - \left(\frac{a}{2\Delta x}\right)\left(u_{i+1}^k - u_{i-1}^k\right)$$

which is equivalent to the PDE given by

$$\underbrace{\frac{\partial u}{\partial t} + a\frac{\partial u}{\partial x}}_{\text{advection}} = \underbrace{\left(\frac{\Delta x^2}{2\Delta t}\right)\frac{\partial^2 u}{\partial x^2} + O(\Delta x^3)}_{\text{error}} \tag{14.18}$$

Thus the Lax-Friedrichs scheme represents the solution to a slightly different PDE which has introduced a second derivative term with a small coefficient $\dfrac{\Delta x^2}{2\Delta t}$. This artificial diffusion term introduces a small amount of 'spreading' of the function as it propagates in time. A detailed error analysis of the Lax-Friedrichs scheme leads to the following equation

$$\epsilon = \underbrace{\frac{\Delta x^2}{2\Delta t}\left(1-C^2\right)\frac{\partial^2 u}{\partial x^2}}_{\text{diffusive/dissipative}} - \underbrace{\frac{a\Delta x^2}{2}(1-C^2)\frac{\partial^3 u}{\partial x^3}}_{\text{dispersive}} \tag{14.19}$$

This means Lax[4] Friedrichs is also first order accurate in space with leading error term being dissipative. Like upwind scheme, both dispersive and dissipative errors are zero when $C = 1$.

Example 14.2

Solve Example 14.1 by the use of Lax-Friedrichs scheme. Use Courant number of $C = 0.5$ and a spatial step of $\Delta x = 0.02$ with an advection speed of $a = 1$.

Solution :

With $C = 0.5$, $\Delta x = 0.02$ and $a = 1$, we have $\Delta t = \dfrac{C\Delta x}{a} = \dfrac{0.5 \times 0.02}{1} = 0.01$. The algorithm takes the form

$$u_i^{k+1} = 0.5\left(u_{i-1}^k + u_{i+1}^k\right) - 0.25\left(u_{i+1}^k - u_{i-1}^k\right)$$

The initial solution is given by the Gaussian profile $u_i^0 = e^{-100(x_i - 0.5)^2}$ where $0 \le i \le 50$. Periodic boundary condition is applied by the use of the following:

$$u_0^{k+1} = u_{50}^{k+1} = 0.5\left(u_{49}^k + u_1^k\right) - 0.25\left(u_1^k - u_{49}^k\right)$$

Consider the initial peak of the profile located at $i = 25$. Let us see what happens to the function value at the peak. A short extract of a table is given below to indicate the state of affairs after one time step, i.e. at $k = 1$.

x	$u(x,0)$	$u(x,0.0025)$
0.485	0.97775	0.96810
0.490	0.99005	0.98269
0.495	0.99750	0.99254
0.500	1	0.99750
0.505	0.99750	0.99751
0.510	0.99005	0.99257
0.515	0.97775	0.98273

---

[4]after Peter David Lax, born 1926, an Hungarian born American mathematician

Figure 14.9 shows the solution at various time intervals.

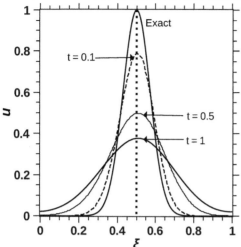

t = 0.1

t = 0.5

t = 1

Exact

Figure 14.9: *Numerical solution using the Lax-Friedrichs scheme in Example 14.2*

The Lax-Friedrichs scheme is more diffusive compared to the upwind scheme. The artificial diffusion term which is proportional to $\dfrac{\Delta x^2}{2\Delta t} = \dfrac{0.02^2}{2 \times 0.01} = 0.02$. If the step size were to be reduced to $\Delta x = 0.005$, the artificial diffusion term would be 0.005 and hence one would expect smaller error.

4. Explicit Lax Wendroff scheme:

The Lax Wendroff[5] scheme, an improved scheme that achieves second order accuracy along both space and time directions, is based on Taylor expansion. Consider x fixed and t alone is varying. We then have

$$u(x, t + \Delta t) = u(x, t) + \frac{\partial u}{\partial t}\Delta t + \frac{1}{2}\frac{\partial^2 u}{\partial t^2}\Delta t^2 + \dots \tag{14.20}$$

Making use of the advection equation we have $\dfrac{\partial u}{\partial t} = -a\dfrac{\partial u}{\partial x}$ and $\dfrac{\partial^2 u}{\partial t^2} = \dfrac{\partial}{\partial t}\left(\dfrac{\partial u}{\partial t}\right) = \dfrac{\partial}{\partial t}\left(-a\dfrac{\partial u}{\partial x}\right) = -a\dfrac{\partial}{\partial x}\left(\dfrac{\partial u}{\partial t}\right) = a^2\dfrac{\partial}{\partial x}\left(\dfrac{\partial u}{\partial x}\right) = a^2\dfrac{\partial^2 u}{\partial x^2}$. Hence Equation 14.20 may be rearranged as

$$\frac{\partial u}{\partial t} = \frac{u(x, t + \Delta t) - u(x, t)}{\Delta t} - \frac{a^2}{2}\frac{\partial^2 u}{\partial x^2}\Delta t \tag{14.21}$$

Substituting this into the advection equation, using FD representation for the derivatives, we get the Lax Wendroff scheme given below.

$$u_i^{k+1} = u_i^k - \frac{C}{2}\left(u_{i+1}^k - u_{i-1}^k\right) + \frac{C^2}{2}\left(u_{i-1}^k - 2u_i^k + u_{i+1}^k\right) \tag{14.22}$$

---

[5]after Burton Wendroff, born 1930, an American applied mathematician

We may perform von Neumann stability analysis for this scheme to get

$$G = \frac{u_i^{k+1}}{u_i^k} = (1 - C^2) + C^2 \cos(jw\Delta x) + jC\sin(jw\Delta x) \qquad (14.23)$$

The gain factor may then be written as

$$|G| = 1 - C^2(1 - C^2)[1 - \cos(jw\Delta x)]^2 \qquad (14.24)$$

Stability is conditional and guaranteed for $C \le 1$. If we choose $C = 1$, we at once see that Equation 14.22 becomes

$$u_i^{k+1} = u_{i-1}^k \qquad (14.25)$$

which is the same as the exact solution!

Error analysis: Lax Wendroff scheme is second order in space and time. Now we shall look at the leading order truncation error terms. Expanding the functions using Taylor series we get

$$u_{i+1} = u_i + \frac{\partial u}{\partial x}\Delta x + \frac{1}{2!}\frac{\partial^2 u}{\partial x^2}\Delta x^2 + \frac{1}{3!}\frac{\partial^3 u}{\partial x^3}\Delta x^3 + O(\Delta x^4)$$

$$u_{i-1} = u_i - \frac{\partial u}{\partial x}\Delta x + \frac{1}{2!}\frac{\partial^2 u}{\partial x^2}\Delta x^2 - \frac{1}{3!}\frac{\partial^3 u}{\partial x^3}\Delta x^3 + O(\Delta x^4)$$

Substituting the above expressions into Equation 14.22 we get

$$\frac{\partial u}{\partial t} + a\frac{\partial u}{\partial x} = -\frac{1}{3!}\frac{\partial^3 u}{\partial t^3}\Delta t^2 + \frac{1}{3!}a\frac{\partial^3 u}{\partial x^3}\Delta x^2 + O(\Delta x^3)$$

Noting that $\dfrac{\partial^3 u}{\partial t^3}$ is equal to $a^3\dfrac{\partial^3 u}{\partial x^3}$, we get

$$\underbrace{\frac{\partial u}{\partial t} + a\frac{\partial u}{\partial x}}_{\text{advection}} = \underbrace{-\frac{1}{6}a(1 - C^2)\frac{\partial^3 u}{\partial x^3} + O(\Delta x^3)}_{\text{error}}$$

The leading order error term is proportional to $\dfrac{\partial^3 u}{\partial x^3}$ which is dispersive in nature and the dissipation error is smaller. Figure 14.10 shows the solution obtained using Lax Wendroff scheme. A much welcome improvement in the solution is that the amplitude is very close to the exact solution unlike previously discussed methods. However, the solution contains unphysical features or *wiggles*. Also position of the peak is shifted with respect to the exact solution. These unsavory features are caused due to dispersive errors.

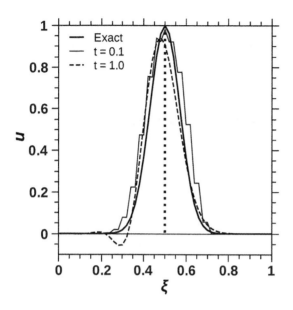

Figure 14.10: Numerical solution using the Lax-Wendroff scheme to Example 14.2, $C = 0.5$, $\Delta x = 0.02$

5. Crank Nicolson scheme:

The last scheme we consider here is the unconditionally stable Crank Nicolson[6] scheme. This scheme is implicit unlike others that have been discussed until now. This scheme uses central difference along both time and space directions by satisfying the advection equation at $t + \dfrac{\Delta t}{2}$. Crank Nicolson scheme is the same as trapezoidal scheme applied to stiff initial value problems. The nodes used in this scheme are shown in Figure 14.11.

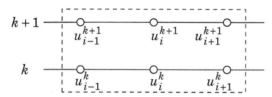

Figure 14.11: Nodes involved in the CN method

The time derivative is written as $\dfrac{\partial u}{\partial t} = \dfrac{u_i^{k+1} - u_i^k}{\Delta t}$ which is central difference approximation for the derivative at $t = t + \dfrac{\Delta t}{2}$ or $k + \dfrac{1}{2}$. The space derivative is evaluated using central differences and by averaging the derivatives at k and $k+1$ thus obtaining the space derivative essentially at $k + \dfrac{1}{2}$. Thus we have $\dfrac{\partial u}{\partial x} = \dfrac{u_{i+1}^k - u_{i-1}^k}{4\Delta x} + \dfrac{u_{i+1}^{k+1} - u_{i-1}^{k+1}}{4\Delta x}$. The advection equation may then be written as

$$-\frac{C}{4}u_{i-1}^{k+1} + u_i^{k+1} + \frac{C}{4}u_{i+1}^{k+1} = u_i^k - \frac{C}{4}\left(u_{i+1}^k - u_{i-1}^k\right) \tag{14.26}$$

---

[6]after John Crank, 1916-2006, British mathematical physicist and Phyllis Nicolson, 1917-1968, British mathematician

Note that the nodal equation is in implicit form since all three function values on the left hand side are unknown. We may perform stability analysis using the exact solution derived earlier. We see that the following hold:

$$u_{i+1}^k - u_{i-1}^k = u_i^k \left(e^{j\omega\Delta x} - e^{-j\omega\Delta x} \right)$$

$$u_{i+1}^{k+1} - u_{i-1}^{k+1} = u_i^{k+1} \left(e^{j\omega\Delta x} - e^{-j\omega\Delta x} \right)$$

These in Equation 14.26 yield us the following.

$$G = \frac{u_i^{k+1}}{u_i^k} = \frac{1 - \frac{C}{2}\sin(\omega\Delta x)}{1 + \frac{C}{2}\sin(\omega\Delta x)} \tag{14.27}$$

We see that $|G| = 1$ for all ω and C. Hence the Crank Nicolson scheme is unconditionally stable. Typically we choose $C = 2$ to get the following nodal equation.

$$-u_{i-1}^{k+1} + 2u_i^{k+1} + u_{i+1}^{k+1} = u_{i-1}^k + 2u_i^k - u_{i+1}^k \tag{14.28}$$

Error analysis On performing Taylors expansion of all the terms about $u_i^{k+1/2}$ we get

$$\frac{\partial u}{\partial t} + a\frac{\partial u}{\partial x} = \underbrace{-\frac{a\Delta x^2}{24}(4 + C^2)\frac{\partial^3 u}{\partial x^3}}_{error}$$

$$\underbrace{\phantom{\frac{\partial u}{\partial t} + a\frac{\partial u}{\partial x}}}_{advection}$$

The leading error term being proportional to $\partial^3 u/\partial x^3$, Crank Nicolson also suffers from dispersion errors.

Example 14.3

Solve Example 14.1 by the use of Crank Nicolson scheme. Use Courant number of $C = 2$ and a spatial step of $\Delta x = 0.02$ with an advection speed of $a = 1$.

Solution :

With $C = 2$, $\Delta x = 0.02$ and $a = 1$, we have $\Delta t = \dfrac{C\Delta x}{a} = \dfrac{2 \times 0.02}{1} = 0.04$. Periodic boundary conditions are made use of. We start with the Gaussian specified in Example 14.1 and perform the calculations for two time steps. The results are shown in tabular form for $0.5 \leq x \leq 0.7$ where the function variation is significant.

x	$u(x,0)$	$u(x,0.04)$	$u(x,0.08)$	$u_E(x,0.08)$
0.50	1.0000	0.8752	0.5737	0.5273
0.52	0.9608	0.9734	0.7507	0.6977
0.54	0.8521	0.9979	0.8964	0.8521
0.56	0.6977	0.9450	0.9821	0.9608
0.58	0.5273	0.8282	0.9918	1.0000
0.60	0.3679	0.6730	0.9268	0.9608
0.62	0.2369	0.5082	0.8042	0.8521
0.64	0.1409	0.3574	0.6504	0.6977
0.66	0.0773	0.2347	0.4919	0.5273
0.68	0.0392	0.1443	0.3492	0.3679
0.70	0.0183	0.0834	0.2335	0.2369

This data is also shown as a plot in Figure 14.12.

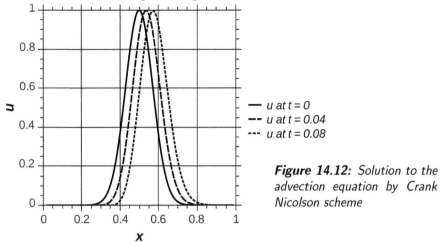

Figure 14.12: Solution to the advection equation by Crank Nicolson scheme

We see that the Crank Nicolson scheme performs well in solving the advection equation.

Dispersion and dissipation errors

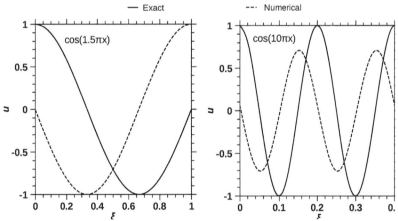

Figure 14.13: Numerical solution using the Lax-Wendroff scheme to Example 14.2, $C = 0.5$, $\Delta x = 0.02$

We shall discuss about dissipation and dispersive errors. Figure 14.13 indicate the exact and numerical solution of sinusoidal functions using Lax Wendroff scheme. Observe that the numerical solution of both sinusoidal curves are out of phase with respect to the exact solution. The phase difference between the solution and exact curve is a function of the wavelength of the sinusoidal curve. Also, dissipation is a function of wavelength. Sinusoidal functions of different wavelengths travel at different speeds. We know that any periodic function can be represented as a Fourier series and different terms in the series travel at different rates and also undergo some amount of dissipation. This causes distortion of the original function.

Dispersion error becomes substantial in the presence of sharp gradients in the function. Let us consider an example with an initial shape given by the "top hat" function as

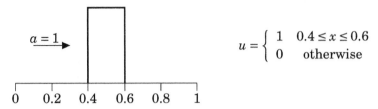

$$u = \begin{cases} 1 & 0.4 \leq x \leq 0.6 \\ 0 & \text{otherwise} \end{cases}$$

Figure 14.14 compares different methods applied to the top hat function for identical parameters. The results are self explanatory. Upwind and Lax Friedrichs scheme suffer from strong diffusive errors where as Lax Wendroff and Crank Nicolson methods have dominant dispersive errors. The problem with dispersive errors is the presence of large undesirable unphysical oscillations.

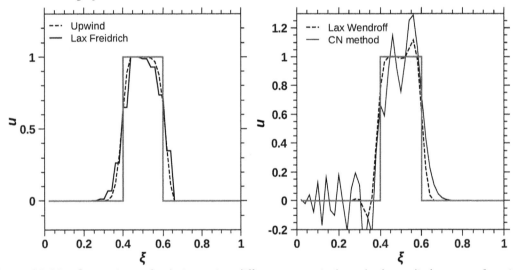

Figure 14.14: *Comparison of solution using different numerical methods applied to step function with Courant number 0.8 and $\Delta x = 0.1$*

All the methods that have been discussed for solving advection equation suffer from either dissipative or dispersion errors. Though implicit schemes are unconditionally stable, the accuracy of implicit scheme is no better than the explicit schemes. In some cases having artificial dissipation in the solution procedure stabilizes the numerical solution especially when treating sharp function variations such as shocks. A solution procedure has to be chosen such that the solution is devoid of wiggles. Godunov's theorem states that only first order numerical schemes can produce such solutions. All higher order methods suffer from dispersive errors and would produce non-physical solutions. However, use of flux limiter with higher order schemes can produce solutions without unphysical wiggles. Interested reader should refer to advanced texts.

14.2.2 Advection equation with varying a:

The solution of advection equation when the advection velocity varies with space and time may be handled by the numerical schemes that have been considered above. Consider the advection equation given by

$$\frac{\partial u}{\partial t} + a(x,t)\frac{\partial u}{\partial x} = 0 \tag{14.29}$$

We know that u = constant requires the following to hold.

$$\frac{\partial u}{\partial t}dt + \frac{\partial u}{\partial x}dx = 0$$

In other words we have, on using Equation 14.29

$$\frac{dx}{dt} = -\frac{\frac{\partial u}{\partial t}}{\frac{\partial u}{\partial x}} = a(x,t) \tag{14.30}$$

Thus the characteristic is such that its slope is nothing but the advection velocity. If it varies with space and time the characteristic is a curve instead of a straight line. If it varies only with space it is a straight line but with different slopes at different locations.

Consider Example 14.1 again but with the advection velocity $a(x,t) = 1 + 0.4x$ in the range $0 \le x \le 1$. We run through the upwind scheme again but with the nodal equations given by

$$u_i^{k+1} = u_i^k - (1 + 0.4x_i)\frac{\left(u_i^k - u_{i-1}^k\right)}{2}$$

where we have taken $C = 0.5$. We use $\Delta x = 0.02$ so that $\Delta t = 0.01$. Results are calculated for several time steps and the results are shown plotted in Figure 14.15. We observe

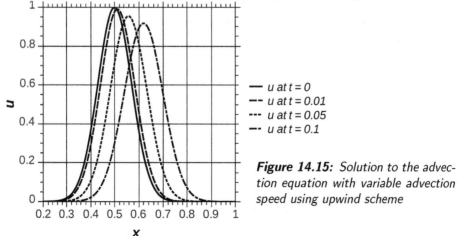

— u at $t = 0$
-- u at $t = 0.01$
··· u at $t = 0.05$
—· u at $t = 0.1$

Figure 14.15: Solution to the advection equation with variable advection speed using upwind scheme

that the symmetry of the initial shape is lost as time proceeds. The upwind scheme also exhibits a reduction in the peak amplitude with time.

14.2.3   Nonlinear advection equation

We come across nonlinear advection equation while treating fluid flow problems.

$$\frac{\partial u}{\partial t} + u\frac{\partial u}{\partial x} = 0$$

Here u stands for the fluid velocity, in this case, a function of x, t. The above equation is also a form of inviscid Burger equation. One can directly use explicit schemes. In such a case, velocity at previous time step can be used as wave velocity. The upwind scheme discussed in the earlier section can be employed.

> The solution of advection equation has introduced many different schemes that may be used in the numerical solution of such an equation. Advection terms may occur in heat diffusion problem when there is a physical movement of the medium. This will be considered at the end of next section dealing with parabolic equations. Advection equation is also a precursor to the wave equation that will be considered later under hyperbolic equations.

14.3   Parabolic PDE: Transient diffusion equation

The simplest case of a parabolic equation is the one dimensional diffusion equation. In Cartesian coordinates, we have the equation

$$\frac{\partial^2 u}{\partial x^2} = \frac{1}{\alpha}\frac{\partial u}{\partial t} \tag{14.31}$$

In this equation u may stand for temperature, α for thermal diffusivity, a property of the material. This equation also represents the case of mass transport. In that case u may stand for concentration, α for mass diffusivity. We have to specify initial and boundary conditions to obtain the solution. Typically the solution is required for $t > 0$ and $0 \le x \le L$ (say).

Non-dimensionalization: We may easily standardize the diffusion equation by introducing $\xi = \dfrac{x}{L}$ and $\tau = \dfrac{\alpha t}{L^2}$ to get the diffusion equation in the standard form.

$$\frac{\partial^2 u}{\partial \xi^2} = \frac{\partial u}{\partial \tau} \tag{14.32}$$

The above equation is valid for $\tau > 0$ and $0 \le \xi \le 1$. The initial variation of u is specified as $u(\xi, 0) = f(\xi)$. Different types of boundary conditions may be specified at the two ends as summarized below.

$$u(0,\tau) = 0 \quad \text{and} \quad u(1,\tau) = c_1 \cdots\cdots\cdots (Case1)$$

$$u(0,\tau) = 0 \quad \text{and} \quad \left.\frac{\partial u}{\partial \xi}\right|_{\tau=1} = 0 \cdots\cdots\cdots (Case2) \tag{14.33}$$

$$u(0,\tau) = 0 \quad \text{and} \quad \frac{\partial u}{\partial \xi} + c_2 u(1,\tau) = c_3 \cdots\cdots (Case3)$$

where $c_1 - c_3$ are specified constants.

14.3.1 Explicit formulation

Simplest formulation uses an explicit scheme (counterpart of Euler scheme used in the case of ODE). The time derivative is written as a forward difference formula while the second derivative with respect to space is written using central differences, based on the values at the beginning of the time step. Thus we have, for an internal node, the following equation.

$$\frac{u_i^{k+1} - u_i^k}{\Delta \tau} = \frac{u_{i-1}^k - 2u_i^k + u_{i+1}^k}{\Delta \xi^2}$$

which may be rearranged to obtain the following explicit expression for u_i^{k+1}.

$$u_i^{k+1} = u_i^k + Fo_e \left(u_{i-1}^k - 2u_i^k + u_{i+1}^k \right) \tag{14.34}$$

where $Fo_e = \left(\dfrac{\Delta \tau}{\Delta \xi^2} \right)$ is known as the elemental Fourier number[7]. The node connectivities involved in the explicit scheme are shown in Figure 14.16. The solution may be updated, starting with the initial solution, by the use of the above explicit expression, by simple arithmetic operations. However the method is subject to stability condition as we shall see a little later.

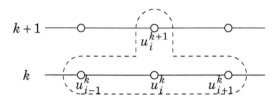

Figure 14.16: Nodes involved in the explicit formulation for FD solution of diffusion equation

Boundary conditions

Dirichlet boundary conditions are easy to apply. Neumann or the Robin condition requires some effort. Consider Case 3 in Equation 14.33. We may either use a second order accurate three point backward difference formula or use the concept of a ghost point that uses central differences to approximate the derivative. In the latter case, we have

$$u_{N+1}^k = u_{N-1}^k - 2c_1 \Delta \xi u_N^k + 2c_2 \Delta \xi \tag{14.35}$$

Treating the boundary node as an internal node, we may then get the following nodal equation.

$$u_N^{k+1} = u_N^k + Fo_e \left(2u_{N-1}^k - 2u_N^k - 2c_1 \Delta \xi u_N^k + 2c_2 \Delta \xi \right) \tag{14.36}$$

---

[7]in terms of x and t elemental Fourier number is $Fo_e = \dfrac{\alpha \Delta t}{\Delta x^2}$

Stability of explicit method

Assume that the discretization has been done as explained above. Since the solution is a function of time and space, we shall assume that the solution is subject to a perturbation given by $u_i^k = A(\tau)\cos(\omega\xi_i)$ where where $A(\tau)$ is the time varying amplitude and ω is the spatial frequency of the disturbance.[8] We then have the following:

$$u_{i-1}^k = A(\tau)\cos[\omega(\xi_i - \Delta\xi)] \;=\; A(\tau)[\cos\omega\xi_i \cos\omega\Delta\xi - \sin\omega\xi_i \sin\omega\Delta\xi]$$
$$u_{i+1}^k = A(\tau)\cos[\omega(\xi_i + \Delta\xi)] \;=\; A(\tau)[\cos\omega\xi_i \cos\omega\Delta\xi + \sin\omega\xi_i \sin\omega\Delta\xi]$$
$$u_i^{k+1} \;=\; A(\tau + \Delta\tau)\cos(\omega\xi_i)$$

We substitute these in Equation 14.34, and after simplification we get

$$\frac{A(\tau + \Delta\tau)}{A(\tau)} = 1 - 2Fo_e(1 - \cos\omega\Delta\xi) \tag{14.37}$$

Note that the maximum value of $1 - \cos\omega\Delta\xi$ is 2, the amplitude ratio $|G| = \left|\dfrac{A(\tau + \Delta\tau)}{A(\tau)}\right|$ is less than unity (this is the necessary condition for stability) only if $Fo_e < 0.5$. Thus the explicit method is conditionally stable.

Error analysis

We expand u_i^{k+1}, u_{i+1}^k and u_{i-1}^k about u_i^k and replace the same in Equation 14.34, to get

$$\underbrace{\frac{\partial u}{\partial t} - \alpha\frac{\partial^2 u}{\partial x^2}}_{\text{diffusion}} = \underbrace{-\frac{1}{2}\frac{\partial^2 u}{\partial t^2}\Delta t + \frac{\alpha\Delta x^2}{12}\frac{\partial^4 u}{\partial x^4}}_{\text{error}}$$

Instead of solving the diffusion equation, explicit scheme solves the above PDE. All the error terms are of even order and hence they are also diffusive in nature. As $\Delta x \to 0$, all the truncation errors tend to zero and we approach the exact solution.

Example 14.4

1D heat equation is satisfied in the region $0 \le \xi \le 0.5$ with the initial condition given by a linear variation of u from $u = 0$ at $\xi = 0$ to $u = 1$ at $\xi = 0.5$. The left boundary is subject to the boundary condition $u = 0$ for $\tau > 0$ while the right boundary is subject to the condition $\dfrac{\partial u}{\partial \xi} = 0$ for $\tau > 0$. Obtain the solution by the use of three different values of $Fo_e = 0.25, 0.50$ and 0.75. Calculate the solution up to $\tau = 0.03$ in each case and comment on the results. Use $\Delta\xi = 0.1$.

Solution :

We choose $\Delta\xi = 0.1$ to get six equi-spaced nodes covering the domain from end to end. Explicit formulation presented immediately above is made use of. In case we choose

---

[8]The cosine form chosen here for the spatial behavior is based on the fact that any function may be written as a weighted sum of its Fourier components.

$Fo_e = 0.25$ the step size along the time direction is obtained as $\Delta\tau = 0.25 \times 0.1^2 = 0.0025$. The number of explicit steps to reach $\tau = 0.03$ is 12. We start with the initial set $u_0^0 = 0, u_1^0 = 0.2, u_2^0 = 0.4, u_3^0 = 0.6, u_4^0 = 0.8, u_5^0 = 1$ and obtain the nodal values after one time step as shown below, using Equations 14.34 and 14.36.

$$u_1^1 = 0.2 + 0.25(0 - 2 \times 0.2 + 0.4) = 0.2$$
$$u_2^1 = 0.4 + 0.25(0.2 - 2 \times 0.4 + 0.6) = 0.4$$
$$u_3^1 = 0.6 + 0.25(0.4 - 2 \times 0.6 + 0.8) = 0.6$$
$$u_4^1 = 0.8 + 0.25(0.6 - 2 \times 0.8 + 1) = 0.8$$
$$u_5^1 = 1 + 0.25(2 \times 0.8 - 2 \times 1) = 0.9$$

The calculations may proceed along the same lines for subsequent time steps. Table below shows the results. First four columns show the calculations with $\Delta\tau = 0.0025$. Subsequently the results at the end of every fourth explicit step is shown.

ξ	$\tau = 0$	$\tau = 0.0025$	$\tau = 0.005$	$\tau = 0.0075$	$\tau = 0.01$	$\tau = 0.02$	$\tau = 0.03$
0	0	0.0000	0.0000	0.0000	0.0000	0.0000	0.0000
0.1	0.2	0.2000	0.2000	0.2000	0.2000	0.1949	0.1827
0.2	0.4	0.4000	0.4000	0.4000	0.3984	0.3794	0.3510
0.3	0.6	0.6000	0.6000	0.5938	0.5844	0.5374	0.4892
0.4	0.8	0.8000	0.7750	0.7500	0.7266	0.6465	0.5809
0.5	1	0.9000	0.8500	0.8125	0.7813	0.6858	0.6132

We next choose a bigger step size along the time direction ($\delta\tau = 0.005$ - twice that in the above case) corresponding to $Fo_e = 0.5$. Calculations proceed as before and are tabulated below.

ξ	$\tau = 0$	$\tau = 0.005$	$\tau = 0.01$	$\tau = 0.015$	$\tau = 0.02$	$\tau = 0.025$	$\tau = 0.03$
0	0	0.0000	0.0000	0.0000	0.0000	0.0000	0.0000
0.1	0.2	0.2000	0.2000	0.2000	0.2000	0.1875	0.1875
0.2	0.4	0.4000	0.4000	0.4000	0.3750	0.3750	0.3438
0.3	0.6	0.6000	0.6000	0.5500	0.5500	0.5000	0.5000
0.4	0.8	0.8000	0.7000	0.7000	0.6250	0.6250	0.5625
0.5	1	0.8000	0.8000	0.7000	0.7000	0.6250	0.6250

Finally we choose a time step size of ($\delta\tau = 0.0075$ - thrice that in the first case) corresponding to $Fo_e = 0.75$. The calculations proceed similarly as in the previous two cases. Results are tabulated as below.

ξ	$\tau = 0$	$\tau = 0.0075$	$\tau = 0.03$	$\tau = 0.06$
0	0	0.0000	0.0000	0.0000
0.1	0.2	0.2000	0.2000	1.0691
0.2	0.4	0.4000	0.2734	-1.7081
0.3	0.6	0.6000	0.6844	3.4038
0.4	0.8	0.8000	0.2516	-3.4752
0.5	1	0.7000	0.9813	4.6702

We make a plot of the results at $\tau = 0.03$ obtained using the three values of the elemental Fourier number.

We infer from the figure that the last case corresponding to $\Delta\tau = 0.0075$ leads to an unstable solution. Case with $\Delta\tau = 0.005$ is stable but shows minor oscillations. The most reliable solution is the case with $\Delta\tau = 0.0025$. These observations are in agreement with the stability condition presented earlier.

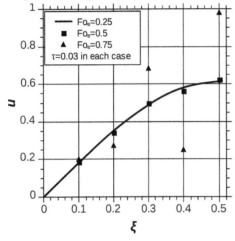

Figure 14.17: *Solution at $\tau = 0.03$ obtained by the explicit method using different time steps*

14.3.2   Implicit formulation

In order to overcome the stability constraint on the time step we explore the possibility of a method that has better stability. Implicit method uses backward difference for the time derivative and uses central difference for the space derivative, using the as yet unknown function values at $k+1$. The formulation becomes fully implicit. Nodal equation takes the form

$$Fo_e u_{i-1}^{k+1} - (1 + 2Fo_e) u_i^{k+1} + Fo_e u_{i+1}^{k+1} = -u_i^k \tag{14.38}$$

When we assemble all the nodal equations we get a tri-diagonal system that may be solved by TDMA.

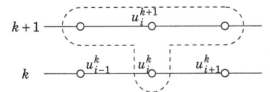

Figure 14.18: *Nodes involved in the implicit formulation for FD solution of diffusion equation*

On performing stability analysis we get the amplification factor

$$|G| = \left| \frac{A(\tau + \Delta\tau)}{A(\tau)} \right| = \left| \frac{1}{1 + 2Fo_e(1 - \cos\omega\Delta\xi)} \right| \tag{14.39}$$

which is less than 1 for all values of Fo_e. Hence, implicit method is unconditionally stable. Implicit method is first order accurate in time and second order accurate in space. The implicit method is same as Backward Euler method discussed earlier in the chapter on IVPs.

14.3.3 Crank Nicolson or semi-implicit scheme

Semi-implicit or CN method uses central difference in both space and time. The second derivative with respect to space is taken as the mean of the second derivatives at k and $k + 1$. The time derivative, in effect takes on the central difference form.

$$\frac{u_{i-1}^{k+1} - 2u_i^{k+1} + u_{i+1}^{k+1}}{2\Delta\xi^2} + \frac{u_{i-1}^k - 2u_i^k + u_{i+1}^k}{2\Delta\xi^2} = \frac{u_i^{k+1} - u_i^k}{\Delta\tau} \tag{14.40}$$

This may be recast as

$$Fo_e u_{i-1}^{k+1} - 2(1 + Fo_e)u_i^{k+1} + Fo_e u_{i+1}^{k+1} = 2(1 - Fo_e)u_i^k - Fo_e(u_{i-1}^k + u_{i+1}^k) \tag{14.41}$$

Stability of the CN method

A stability analysis may be made as we did in the case of the explicit method. Introducing a disturbance of form $u = A(\tau)\cos(\omega\xi)$, substituting in the nodal equation above, we can show that

$$|G| = \left|\frac{A(\tau + \Delta\tau)}{A(\tau)}\right| = \left|\frac{1 + \frac{Fo_e}{2}(2\cos\omega\Delta\xi - 2)}{1 - \frac{Fo_e}{2}(2\cos\omega\Delta\xi - 2)}\right| \tag{14.42}$$

which is less than unity for all Fo_e. Hence the CN method is unconditionally stable. If Fo_e is chosen as 1, the nodal equations become

$$u_{i-1}^{k+1} - 4u_i^{k+1} + u_{i+1}^{k+1} = -(u_{i-1}^k + u_{i+1}^k) \tag{14.43}$$

Figure 14.19 shows the nodes involved in the CN method ($Fo_e = 1$). When the nodal equations are all assembled end to end we have a tri-diagonal system solvable by TDMA.

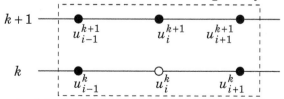

Figure 14.19: *Nodes involved (indicated by dotted line) in the CN method for the finite difference solution of the 1D diffusion equation. In the special case $Fo_e = 1$ node i at k is not involved as indicated by open circle.*

Boundary conditions

Application of boundary conditions are simple for Dirichlet type. Robin and Neumann boundary conditions can be treated in the same way as discussed earlier in BVPs. Robin and Neumann conditions can be applied in two ways - ghost cell approach or the three point difference formula. Consider a ghost node to the right, as usual, assuming that the right boundary has the Robin boundary condition specified as $\dfrac{\partial u}{\partial \xi} + c_1 u = c_2$. We then have

$$u_{N+1}^k = u_{N-1}^k - 2c_1\Delta\xi u_N^k + 2c_2\Delta\xi \qquad (14.44)$$

$$u_{N+1}^{k+1} = u_{N-1}^{k+1} - 2c_1\Delta\xi u_N^{k+1} + 2c_2\Delta\xi \qquad (14.45)$$

These are substituted into Equation 14.43 after setting $i = N$ and simplified to get

$$u_{N-1}^{k+1} - (2 + c_1\Delta\xi)u_N^{k+1} = -(u_{N-1}^k - c_1\Delta\xi u_N^k - 2c_2\Delta\xi) \qquad (14.46)$$

The nodal equation is in a form convenient for the application of TDMA. The difference between the above equation and BVPs is that an additional transient term is present in the equation.

Example 14.5

A semi-infinite wall is initially maintained at uniform temperature $u = 0$ (non-dimensional temperature). The surface of the wall is instantaneously exposed to a step change in temperature $u = 1$. The governing equation is the standard diffusion equation. Obtain the temperature distribution in the wall by using explicit, implicit and CN scheme up to $\tau = 0.05$ and comment on the results. Use $\Delta\xi = 0.1$ and $\alpha = 1$. Compare the results with the exact solution.

Background :

The exact solution for the present problem can be determined using method of similarity. Method of similarity converts a PDE to a ODE by using a similarity variable. The similarity variable for the present problem is $\eta = \dfrac{x}{2\sqrt{\alpha t}}$ and the transformed ODE is

$$\frac{d^2u}{d\eta^2} + 2\eta\frac{du}{d\eta} = 0$$

The exact solution for the above problem is

$$u = 1 - \frac{2}{\sqrt{\pi}}\int_0^\eta e^{-\eta^2}\,d\eta = \text{erfc}(\eta)$$

where erfc is the complementary error function. Numerically, the above ODE can be solved using the method discussed in Section 11.2.3. However, we shall solve the PDE using FDM.

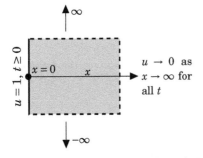

A semi-infinite solid is bounded at one end and extends to infinity in other direction. It may not be possible to simulate the entire domain in the current form. The boundary condition at $x \to \infty$ can be modeled as a finite domain with the boundary condition at the other end as Neumann condition $\dfrac{\partial u}{\partial x} = 0$. The size of the domain should be such that u is negligible at this end.

Solution :

The size of the domain for the present problem is taken as $x = 2$ and the step size would be 0.1. Let us take $Fo_e = 0.5$, then $dt = 0.005$.

Discretization for interior nodes:

$$\beta Fo_e \left(u_{i-1}^{k+1} - 2u_i^{k+1} + u_{i+1}^{k+1} \right) + (1-\beta)Fo_e \left(u_{i-1}^k - 2u_i^k + u_{i+1}^k \right) = u_i^{k+1} - u_i^k$$

where β is a parameter used to decide the time integration scheme. $\beta = 1$ is fully implicit, $\beta = 0$ is fully explicit and $\beta = 0.5$ is Crank Nicolson.

Discretization at last node:

$$2\beta Fo_e \left(u_{n-1}^{k+1} - u_n^{k+1} \right) + 2(1-\beta)Fo_e \left(u_{n-1}^k - u_n^k \right) = u_n^{k+1} - u_n^k$$

MATLAB program has been written to solve the problem

```
n = 21;                    % no of nodes
x = 0:2/(n-1):2;           % nodes
dx = 2/(n-1);              % step size
u  = zeros(n,1); u(1) = 1;        % initial condition
a = zeros(n,1);            % TDMA coefficient a
a1 = zeros(n,1);           % TDMA coefficient a'
a2 = zeros(n,1);           % TDMA coefficient a''
b = zeros(n,1);            % TDMA constant b
fou = 0.5;                 % Fourier number
dt = fou*dx^2;             % time step
beta = 0.5;                % CN scheme
% beta = 0;                % explicit
% beta = 1;                % implicit
t = 5;                     % no. of time steps
a(1) = 1;   b(1) = 1;      % u(x=0)
for i=2:n-1                % internal nodes
    a(i) = 2*beta + 1/fou;
    a1(i) = beta;
    a2(i) = beta;
end
a(n) = 2*beta + 1/fou;  % last node
a2(n) = 2*beta;
unew = u;
for j=1:t                  % loop for time integration
    for i=2:n-1            % interior nodes
        b1(i) = (1-beta)*(u(i-1)+u(i+1)-2*u(i)) + u(i)/fou;
    end
    b(n) = (1-beta)*(2*u(n-1)-2*u(n)) + u(n)/fou;
    unew(:,j+1) = tdma(a2,a,a1,b);        % solve for u
    u = unew(:,j+1);       % update time
end
uexact = erfc(x/(2*sqrt(t*dt)));          % exact solution
```

The following figure shows the temperature distribution at $t = 0.05$ obtained using the three time integration schemes.

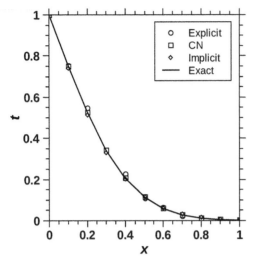

Figure 14.20: *Temperature distribution in the wall at $t = 0.05$ ($Fo_e = 0.5$ and $\Delta x = 0.1$).*

All the three integration schemes capture the trend satisfactorily. The following table shows the maximum difference between numerical simulation and exact solution for various step sizes and Fourier numbers at $t = 0.05$.

Δx	Fo_e	Explicit	CN	Implicit
0.1	0.5	0.02174	0.00252	0.01452
0.1	0.25	0.00611	0.00266	0.00776
0.05	0.5	0.00611	0.00067	0.00363
0.05	0.25	0.00153	0.00067	0.00194
0.05	0.125	0.00078	0.00068	0.00110
0.025	0.25	0.00039	0.00017	0.00049
0.025	0.125	0.00020	0.00017	0.00027

Both time step and space step are important for accuracy of numerical simulation. CN being second order accurate, converges to the true solution much faster than explicit and implicit schemes which are only first order accurate. Notice, that for a given step size, reduction in time step does not improve the solution. At this stage one has to refine the spatial grid.

One has to keep in mind that the temperature at last node should be negligible. If that is not the case, the domain will have to be extended.

Pseudo transient method: Methods for solving transient boundary value problems can be used to solve steady state boundary value problems also. At very long periods of time, the change in the function value would be negligible and would reach steady state. This approach of solving boundary value problems is known as "*pseudo transient method*". The method is equivalent to solving the boundary value problem using Gauss Seidel iteration scheme. The time parameter Fo_e plays the role of the relaxation parameter.

Example 14.6

Initial nodal values follow a triangular distribution with $u = 0.5$ at the two ends and $u = 1$ at the middle. For $\tau > 0$ the two ends are subject to Neumann condition in the form $\dfrac{\partial u}{\partial \xi} = 0$.

Obtain the transient field in the physical domain using CN method if the equation governing u is the standard diffusion equation with $0 \leq \xi \leq 1$. Use a spatial step of $\Delta \xi = 0.1$.

Solution :

Choosing $Fo_e = 1$ the corresponding time step is given by $\Delta \tau = 0.1^2 = 0.01$. We note that there is symmetry with respect to $\xi = 0.5$ and hence we need to consider only the region $0 \leq \xi \leq 0.5$ which will contain 6 nodes. With the initial condition specified in the problem, we have the following table.

i	ξ	$u(\tau = 0)$	a	a'	a''	b
0	0	0.5	4	2	0	1.2000
1	0.1	0.6	4	1	1	1.2000
2	0.2	0.7	4	1	1	1.4000
3	0.3	0.8	4	1	1	1.6000
4	0.4	0.9	4	1	1	1.8000
5	0.5	1	4	0	2	1.8000

With these starting values the coefficients appearing in the tri-diagonal matrix are obtained and shown in the table. The boundary conditions at $\xi = 0$ and $\xi = 0.5$ are satisfied by choosing c_1 and c_2 as zero in Equation 14.46. We use TDMA, take one time step and obtain the following at $\tau = 0.01$.

ξ	P	Q	$u(\tau = 0.01)$
0	0.5	0.3000	0.6152
0.1	0.2857	0.4286	0.6303
0.2	0.2692	0.4923	0.7061
0.3	0.2680	0.5608	0.7939
0.4	0.2680	0.6326	0.8697
0.5	0.0000	0.8848	0.8848

This procedure may be continued to obtain the solution till a desired value of τ is reached.

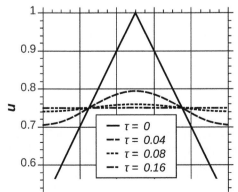

Figure 14.21: Transient solution in Example 14.6

The results are plotted as shown in Figure 14.21. The results are plotted for the whole domain from end to end. The initial triangular variation will eventually become a uniform

value or reach *steady state* equal to the mean of $u = 0.75$. At $\tau = 0.16$ it appears to be close
to this value.

Now we shall consider an example with Robin boundary condition.

Example 14.7

Diffusion equation is satisfied in the range $0 \leq \xi \leq 1$ with the initial value of $u = 1$ throughout.
For $\tau > 0$ u is maintained constant at zero at the left boundary while Robin condition is
specified at the right boundary with $c_1 = 1$ and $c_2 = 0.5$. Obtain numerically the solution to
the problem using Crank Nicolson method with $\Delta\xi = 0.1$.

Solution :

With $\Delta\xi = 0.1$ the number of nodes from end to end is 11. Node 0 corresponds to the
left boundary while node 10 corresponds to the right boundary. At the left boundary the
boundary condition is easily taken as $u_0 = 0$. At the right boundary Robin condition needs
to be applied. Based on Equation 14.46, by taking $c_1 = 1$ and $c_2 = 0.5$, we have the right
boundary nodal equation given by

$$u_9^{k+1} - 2.2u_{10}^{k+1} = -(u_9^k - 0.1u_{10}^k - 0.1)$$

The interior nodal equations from $i = 1$ to $i = 9$ may be written down based on Equation
14.43. Solution to the equations are obtained by the use of TDMA as explained in Example
14.6. With $Fo_e = 1$ and $\Delta\xi = 0.1$ the time step is $\Delta\tau = 0.01$. Table below shows the results
of TDMA for different τ's, from the beginning. As $\tau \to \infty$ the solution tends to a linear
variation of u with $u_0 = 0$ and $u_{10} = 0.25$. This represents the steady state for the system.

ξ	$\tau = 0$	$\tau = 0.02$	$\tau = 0.05$	$\tau = 0.1$	$\tau = 0.2$	$\tau \to \infty$
0	1	0.0000	0.0000	0.0000	0.0000	0.0000
0.1	1	0.4226	0.2614	0.1794	0.1172	0.0250
0.2	1	0.7624	0.4947	0.3488	0.2303	0.0500
0.3	1	0.9141	0.6814	0.4996	0.3354	0.0750
0.4	1	0.9709	0.8134	0.6255	0.4292	0.1000
0.5	1	0.9901	0.8958	0.7233	0.5088	0.1250
0.6	1	0.9954	0.9387	0.7923	0.5723	0.1500
0.7	1	0.9941	0.9531	0.8339	0.6182	0.1750
0.8	1	0.9857	0.9474	0.8507	0.6462	0.2000
0.9	1	0.9646	0.9265	0.8459	0.6562	0.2250
1	1	0.9311	0.8928	0.8222	0.6492	0.2500

The results are also shown graphically in Figure 14.22

CN scheme is second order accurate in time and space where as explicit and implicit scheme are
first order accurate in time and second order accurate in space. Hence in practice, CN scheme is
preferred for the solution of equations with diffusion terms.

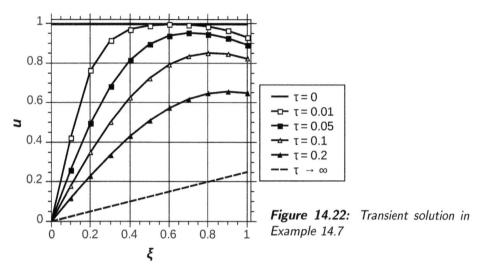

Figure 14.22: *Transient solution in Example 14.7*

14.4 Advection with diffusion

We have considered advection equation and diffusion equation, independently, in the previous sections. There are many applications in which both advection and diffusion take place simultaneously. Such applications are to be found in transport of a pollutant in a river or atmosphere, transient heat transfer in a moving medium such as in welding, mechanics of fluids etc. In the present section we will consider the simplest of such cases, viz. advection and diffusion in one dimension.

Consider the advection diffusion equation given by

$$\frac{\partial u}{\partial t} + a\frac{\partial u}{\partial x} = \alpha\frac{\partial^2 u}{\partial x^2} \tag{14.47}$$

where a is the constant advection velocity and α is the diffusivity. If the problem is one of heat transfer α will be the thermal diffusivity and u the temperature. If the problem refers to pollutant transport α will be the mass diffusivity and u the concentration of the pollutant.

Non dimensionalization: The above equations can be non dimensionalized as

$$\frac{\partial u}{\partial \tau} + Pe\frac{\partial u}{\partial \xi} = \frac{\partial^2 u}{\partial \xi^2}$$

where $\xi = \dfrac{x}{L}$ and $\tau = \dfrac{\alpha t}{L^2}$. The parameter $Pe = \dfrac{aL}{\alpha}$ is known as the Peclet number. Pe can be rewritten as ratio of Courant number and Fourier number i.e. $Pe = \dfrac{C}{Fo_e}$, which signifies the relative importance of advection and diffusion. As $Pe \to 0$, we move towards pure diffusion where as $Pe \to \infty$ the solution tends to pure advection. Numerical solution of the advection diffusion equation may easily be accomplished by the numerical schemes that have already been discussed in great detail. Unfortunately, a scheme that is good for diffusion may not be good for advection and vice versa.

14.4.1   FTCS

Explicit FTCS scheme is suitable for diffusion where as it is not suitable for advection. Nevertheless, we shall briefly consider discretization of advection diffusion equation using FTCS.

$$\frac{u_i^{k+1} - u_i^k}{\Delta t} + a\frac{u_{i+1}^k - u_{i-1}^k}{2\Delta x} = \alpha\frac{u_{i+1}^k - 2u_i^k + u_{i-1}^k}{\Delta x^2}$$

$$\text{or}\quad u_i^{k+1} = (1 - 2Fo_e)u_i^k + (Fo_e - C/2)u_{i+1}^k + (Fo_e + C/2)u_{i-1}^k$$

The above equation is similar to Lax Friedrichs scheme where an artificial diffusion term was added. The physical diffusion in the present case is equivalent to the artificial diffusion term. When $Fo_e = \dfrac{1}{2}$, the above discretization scheme becomes the same as Lax Friedrichs scheme. For numerical stability, $C^2 \leq 2Fo_e \leq 1$ should be met. This is consistent with our deductions for pure advection and pure diffusion cases. For a advection dominated problem ($Pe \gg 1$), the domain has to be discretized into very small steps to maintain stability. Hence explicit FTCS is of academic interest and is not used in practice.

14.4.2   Upwind for advection and central difference for diffusion

We consider first an explicit method - the upwinding scheme that was presented earlier while solving the advection equation. Upwinding is used for the advection term and central difference is used for the diffusion term. The idea of treating each term differently is known as *operator splitting*. Time derivative is written using forward difference. Finite difference form of Equation 14.47 may then be shown to lead to the following nodal equation.

$$u_i^{k+1} = (C + Fo_e)u_{i-1}^k + (1 - C - 2Fo_e)u_i^k + Fo_e u_{i+1}^k \tag{14.48}$$

Stability of upwinding scheme is limited by $C + 2Fo_e \leq 1$ i.e. coefficient of $u_i^k \geq 0$.

An example is considered below to demonstrate the use of upwind scheme.

Example 14.8

Initially the function is distributed linearly with $u = 0$ at $x = 0$ and $u = 1$ at $x = 1$. For $t > 0$ the boundaries continue to remain at the values given above. The governing equation is $u_t + au_x = \alpha u_{xx}$ where $a = \alpha = 1$. Obtain the solution at various times as also the steady state solution. Use the upwind scheme.

Solution :

We choose $Fo_e = 0.25$ and make use of Equation 14.48 to obtain all the interior nodal equations. At the boundary nodes we apply the Dirichlet conditions specified in the problem. We choose $\Delta x = 0.1$, and with $Fo_e = 0.25$ we obtain $\Delta t = 0.25\Delta x^2 = 0.25 \times 0.1^2 = 0.0025$. The Courant number is then given by $C = \dfrac{a\Delta x}{\Delta t} = \dfrac{0.0025}{0.1} = 0.025$. With this value of C Equation 14.48 takes the form

$$u_i^{k+1} = 0.275u_{i-1}^k + 0.475u_i^k + 0.25u_{i+1}^k$$

Starting with the initial solution $u(x, t = 0) = x$ we see that $u_i^0 = x_i$ and hence

$$
\begin{aligned}
u_i^1 &= 0.275x_{i-1} + 0.475x_i + 0.25x_{i+1} \\
&= 0.275(i-1)\Delta x + 0.475i\Delta x + 0.25(i+1)\Delta x \\
&= 0.0275(i-1) + 0.0475i + 0.025(i+1)
\end{aligned}
$$

where $1 \le i \le 9$ and the boundary conditions take care of the end nodes such that $u_0^1 = 0$ and $u_{10}^1 = 1$. Thus the solution at $t = \Delta t = 0.0025$ is obtained. Solution is extended to larger t by repeated application of the upwind algorithm. Eventually the solution tends to the steady state solution which is the solution to the equation with the time derivative term set to zero i.e. of the equation $au_x = \alpha u_{xx}$ subject to the stated boundary conditions. This equation has an exact solution obtained easily by noting that the equation is in variable separable form. The reader is encouraged to show that the solution is

$$
u(x) = \frac{(e^{xPe} - 1)}{(e^{Pe} - 1)}
$$

The following table shows the function history for a few representative times.

x	u	u	u	u	u	u	Steady
x	$t = 0$	$t = 0.0025$	$t = 0.01$	$t = 0.05$	$t = 0.1$	$t = 0.5$	State
0.00	0.0000	0.0000	0.0000	0.0000	0.0000	0.0000	0.0000
0.10	0.1000	0.0975	0.0929	0.0811	0.0734	0.0629	0.0612
0.20	0.2000	0.1975	0.1906	0.1683	0.1530	0.1321	0.1289
0.30	0.3000	0.2975	0.2901	0.2603	0.2384	0.2081	0.2036
0.40	0.4000	0.3975	0.3900	0.3558	0.3290	0.2917	0.2862
0.50	0.5000	0.4975	0.4900	0.4540	0.4247	0.3837	0.3775
0.60	0.6000	0.5975	0.5900	0.5546	0.5257	0.4847	0.4785
0.70	0.7000	0.6975	0.6900	0.6580	0.6323	0.5958	0.5900
0.80	0.8000	0.7975	0.7905	0.7652	0.7458	0.7179	0.7132
0.90	0.9000	0.8975	0.8927	0.8783	0.8676	0.8523	0.8495
1.00	1.0000	1.0000	1.0000	1.0000	1.0000	1.0000	1.0000

Solution at $t = 0.5$ is already close to the steady state solution. The data presented in the table is also shown plotted in Figure 14.23.

14.4.3   Crank Nicolson scheme

Crank Nicolson scheme - a semi-implicit method which is unconditionally stable as shown earlier - may be applied to rewrite Equation 14.47 as

$$
\begin{aligned}
-\left(\frac{C}{4} + \frac{Fo_e}{2}\right)u_{i-1}^{k+1} + (1 + Fo_e)u_i^{k+1} + \left(\frac{C}{4} - \frac{Fo_e}{2}\right)u_{i+1}^{k+1} \\
= \left(\frac{C}{4} + \frac{Fo_e}{2}\right)u_{i-1}^k + (1 - Fo_e)u_i^k - \left(\frac{C}{4} - \frac{Fo_e}{2}\right)u_{i+1}^k
\end{aligned}
\tag{14.49}
$$

If we choose $Fo_e = 1$ the above further simplifies to

$$
-\left(1 + \frac{C}{2}\right)u_{i-1}^{k+1} + 4u_i^{k+1} + \left(1 - \frac{C}{2}\right)u_{i+1}^{k+1} = \left(1 + \frac{C}{2}\right)u_{i-1}^k + \left(1 - \frac{C}{2}\right)u_{i+1}^k
\tag{14.50}
$$

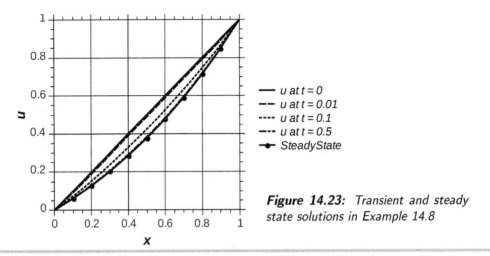

Figure 14.23: Transient and steady state solutions in Example 14.8

Example 14.9

Advection diffusion equation $u_t + u_x = u_{xx}$ is satisfied in the region $0 \le x \le 1$. The initial distribution of the function is given for $0 < x < 1$ as $u(x,0) = 0$ throughout. Boundary conditions are specified as $u(0,t) = 0$ and $u(1,t) = 1$ for $t>0$. Obtain the transient solution using the Crank Nicolson method.

Solution :

We choose $Fo_e = 1$ and make use of Equation 14.50 to obtain all the interior nodal equations. At the boundary nodes we apply the Dirichlet conditions specified in the problem. We choose $\Delta x = 0.1$, and with $Fo_e = 1$ we obtain $\Delta t = \Delta x^2 = 0.1^2 = 0.01$. We then have the Courant number given by $C = \dfrac{\Delta x}{\Delta t} = \dfrac{0.1}{0.01} = 0.1$. With this value of C Equation 14.50 takes the form

$$-1.05u_{i-1}^{k+1} + 4u_i^{k+1} + 0.95u_{i+1}^{k+1} = 1.05u_{i-1}^k + 0.95u_{i+1}^k$$

The initial function is defined by $u_i = 0$ where $x_i = i\Delta x = 0.1i$ for $0 \le i \le 10$. The resulting nodal equations are in tri-diagonal form. The coefficient matrix is tabulated below. These incorporate $u_0 = 0$ and $u_{10} = 1$ for all t.

i	a_i	a_i'	a_i''	d_i
1	4	0.95	0	0
2	4	0.95	1.05	0
3	4	0.95	1.05	0
4	4	0.95	1.05	0
5	4	0.95	1.05	0
6	4	0.95	1.05	0
7	4	0.95	1.05	0
8	4	0.95	1.05	0
9	4	0	1.05	1.9

The last column shows the non-homogeneous terms arising out of the right hand side terms, for the starting time step. This column alone will change from time step to time step. The solution obtained by TDMA after the first time step is shown below.

x	$u(t=0)$	a	a'	a''	d	P	Q	$u(\Delta t)$
0	0	1			Boundary condition→			0
0.1	0.0000	4	0.95	0	0	0.2375	0	0.0000
0.2	0.0000	4	0.95	1.05	0	0.2533	0	0.0000
0.3	0.0000	4	0.95	1.05	0	0.2544	0	0.0001
0.4	0.0000	4	0.95	1.05	0	0.2545	0	0.0005
0.5	0.0000	4	0.95	1.05	0	0.2545	0	0.0021
0.6	0.0000	4	0.95	1.05	0	0.2545	0	0.0084
0.7	0.0000	4	0.95	1.05	0	0.2545	0	0.0330
0.8	0.0000	4	0.95	1.05	0	0.2545	0	0.1295
0.9	0.0000	4	0	1.05	1.9	0	0.5090	0.5090
1		1	1		Boundary condition→			1

We may extend the solution to larger times by the repeated application of TDMA, changing b values after each time step. Eventually the solution tends to the steady state solution which has already been obtained in the previous example.[9] Alternately it may be obtained numerically by the use of FDM. The nodal equations are obtained as

$$-\left(1+\frac{\Delta x}{2}\right)u_{i-1}+2u_i-\left(1-\frac{\Delta x}{2}\right)u_{i+1}=0$$

or

$$-1.1u_{i-1}+2u_i-0.9u_{i+1}=0$$

Again we may use TDMA for the solution. The exact, FDM and the solution obtained by solving for the nodal values using the transient as $t\to\infty$ all are equally acceptable as indicated by the following table.

	Steady state solution		
	Transient	Steady State	
x	$t\to\infty$	FDM	Exact
0	0.000	0.000	0.000
0.1	0.061	0.061	0.061
0.2	0.129	0.129	0.129
0.3	0.204	0.204	0.204
0.4	0.287	0.286	0.286
0.5	0.378	0.377	0.378
0.6	0.479	0.478	0.478
0.7	0.590	0.590	0.590
0.8	0.714	0.713	0.713
0.9	0.850	0.849	0.849
1	1.000	1.000	1.000

A plot of the transient and the steady state solutions are given in Figure 14.24.

---

[9]The steady state solution in Examples 14.8 and 14.9 is the same even though the initial conditions are different

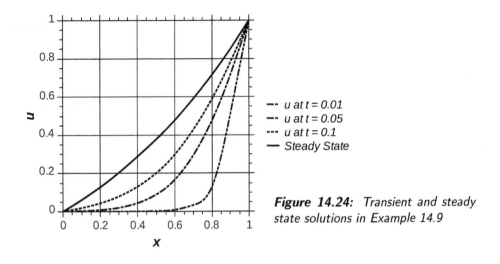

Figure 14.24: *Transient and steady state solutions in Example 14.9*

We have seen that CN scheme applied to pure advection equation produces unphysical oscillations. Consider the initial distribution given by step function

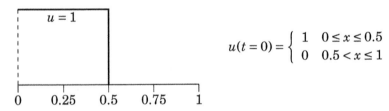

$$u(t=0) = \begin{cases} 1 & 0 \le x \le 0.5 \\ 0 & 0.5 < x \le 1 \end{cases}$$

The following figure shows the function simulated using Crank Nicolson scheme for advection dominated flow.

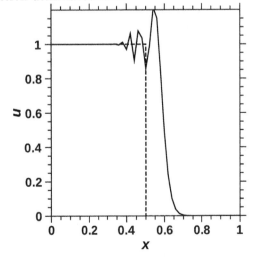

Figure 14.25: *Function simulated using Crank Nicolson scheme for $Pe = 1000$, $C = 0.5$*

It is evident that CN scheme applied for both advection and diffusion does not perform well for high Peclet numbers. CN scheme being a higher order method causes unphysical oscillations for advection dominated problems. Rather, we can use upwind scheme for advection and Crank Nicolson scheme for diffusion. This is also known as *operator splitting* where each term in PDE is treated separately. The advection term can be either treated explicitly or implicitly. The following equation represents the explicit upwind formulation for advection and CN formulation for diffusion.

$$-\frac{Fo_e}{2}u_{i-1}^{k+1} + (1+Fo_e)u_i^{k+1} - \frac{Fo_e}{2}u_{i+1}^{k+1}$$
$$= \left(\frac{C}{2} + \frac{Fo_e}{2}\right)u_{i-1}^k + \left(1 - Fo_e - \frac{C}{2}\right)u_i^k + \frac{Fo_e}{2}u_{i+1}^k \quad (14.51)$$

Diffusion term is easier to treat than advection term. The main limitation of upwind scheme as stated earlier is its strong dissipative nature. On the other hand higher order advection schemes suffer from dispersive errors. However, one can use Total Variation Diminishing (TVD) schemes for advection which are higher order in nature but also suppress unphysical oscillations. These schemes can be incorporated by using operator splitting.

14.5 Advection equation in multi dimensions

Advection equation in two dimensions is given by

$$\frac{\partial u}{\partial t} + a_x \frac{\partial u}{\partial x} + a_y \frac{\partial u}{\partial y} = 0 \quad (14.52)$$

where a_x and a_y are the velocity of the wave along x and y direction. Let us look at some numerical schemes applicable to two dimensions.

14.5.1 Upwind scheme

First order upwind scheme is the simplest numerical scheme. Let us consider a case where both a_x and a_y are positive. Then the advection equation has to be discretized using backward differences.

$$\frac{u_{i,j}^{k+1} - u_{i,j}^k}{\Delta t} + a_x \frac{u_{i,j}^k - u_{i-1,j}^k}{\Delta x} + a_y \frac{u_{i,j}^k - u_{i,j-1}^k}{\Delta y} = 0$$
$$\text{or} \quad u_{i,j}^{k+1} = C_x u_{i-1,j}^k + C_y u_{i,j-1}^k + (1 - C_x - C_y)u_{i,j}^k$$

where $C_x = \dfrac{a_x \Delta t}{\Delta x}$ and $C_y = \dfrac{a_y \Delta t}{\Delta y}$ are the Courant numbers along x and y direction. For stability we require $C_x \geq 0$ and $C_y \geq 0$. Also we require $C_x + C_y \leq 1$ for numerical stability. In general, upwind equation in two dimensions can be summarized as

$$u_{i,j}^{k+1} = -max[0, -C_x]u_{i+1,j}^k - max[0, C_x]u_{i-1,j}^k - max[0, -C_y]u_{i,j+1}^k$$
$$-max[0, C_y]u_{i,j-1}^k + (1 - |C_x| - |C_y|)u_{i,j}^k$$

We shall take up an example to check the performance of upwinding scheme.

Example 14.10

Solve 2D advection equation using upwind scheme with $a_x = a_y = 1$. The initial shape of the solution is a Gaussian profile given by $u(x,y,0) = e^{-100[(x-0.5)^2+(y-0.5)^2]}$

Solution :

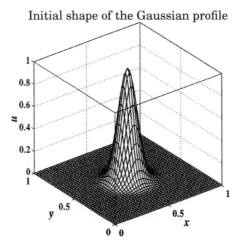

Initial shape of the Gaussian profile

The magnitude of advection velocity is equal in both x and y directions and therefore the advection velocity would be equal to $\sqrt{2}$ along a direction that makes an angle of $45°$ with respect to the horizontal. We shall consider the domain as $0 \leq x, y \leq 1$ with periodic boundary conditions applied at the boundaries. The step size of the discretization is 0.02. From stability considerations, we have $C_x = C_y \leq 0.5$. MATLAB program has been written to perform the calculations.

```
dx = 0.02;                % step size
n = 1/dx+1;               % no. of nodes
x = 0:dx:1; y = x;        % nodes
[X,Y] = meshgrid(x,y);    % mesh
u = exp(-100*((0.5-X).^2+(0.5-Y).^2));  % initial  profile
dt = 0.5*dx;              % time step
c = dt/dx;                % Courant number
un = u;
for t=1:10                % time step
    % corner nodes
    un(n,n) = u(n,n)-c*(2*u(n,n)-u(n-1,n)-u(n,n-1));
    un(1,n) = u(1,n)-c*(2*u(1,n)-u(n-1,n)-u(n,n-1));
    un(1,1) = u(1,1)-c*(2*u(1,1)-u(n-1,1)-u(n-1,1));
    un(n,1) = u(n,1)-c*(2*u(n,1)-u(n,n-1)-u(n-1,1));
    for i=2:n-1
    % boundary nodes
    un(1,i) = u(1,i)-c*(2*u(1,i)-u(n-1,i)-u(n,i-1));
    un(n,i) = un(1,i);
    un(i,1) = u(i,1)-c*(2*u(i,1)-u(i,n-1)-u(i-1,n));
    un(i,n) = un(i,1);
        for j=2:n-1   % internal nodes
        un(i,j) = u(i,j)-c*(2*u(i,j)-u(i-1,j)-u(i,j-1));
```

```
            end
        end
        u = un;        % update function
    end
```

The following figure indicates the solution at two different time steps.

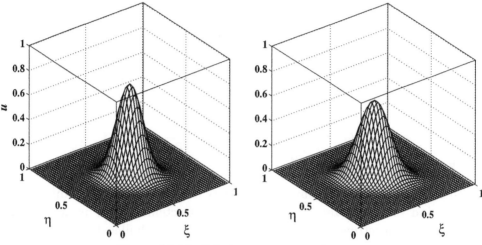

Figure 14.26: *Solution at $t = 0.1$ and $t = 0.2$*

One can clearly see that the peak of the solution has dissipated to a large extent. This is a known artifact of upwind scheme. Following figure shows the spread of the solution with time.

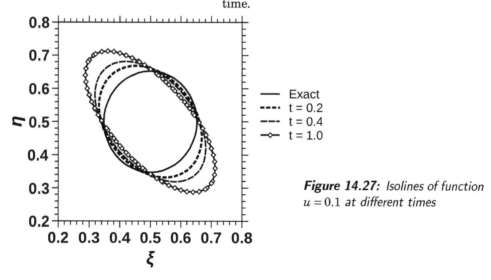

Figure 14.27: *Isolines of function $u = 0.1$ at different times*

It is interesting to note that the rate of diffusion varies with direction. The diffusion is higher along a direction that is perpendicular to the advection velocity. We discuss the source of such errors in what follows.

Error analysis of upwind scheme in two dimensions We perform Taylor series expansion of terms $u_{i,j-1}^k$, $u_{i-1,j-1}^k$ and $u_{i,j}^{k+1}$ and replace the same in the advection equation to get

$$\frac{\partial u}{\partial t} + a_x \frac{\partial u}{\partial x} + a_y \frac{\partial u}{\partial y} = \underbrace{-\frac{1}{2}\frac{\partial^2 u}{\partial t^2}\Delta t + \frac{a_x \Delta x}{2}\frac{\partial^2 u}{\partial x^2} + \frac{a_y \Delta y}{2}\frac{\partial^2 u}{\partial y^2}}_{error}$$

The second derivative of u with respect to time $\dfrac{\partial^2 u}{\partial t^2}$ may be obtained by differentiating the advection equation with respect to time. Then we replace this term in the above expression to get

$$\epsilon = \frac{a_x \Delta x}{2}(1 - C_x)\frac{\partial^2 u}{\partial x^2} + \frac{a_y \Delta y}{2}(1 - C_y)\frac{\partial^2 u}{\partial y^2} - a_x a_y \Delta t \frac{\partial^2 u}{\partial x \partial y}$$

The last term in the above expression is an anisotropic term (a term which makes diffusion vary with direction). When the advection velocity is along x and y, the anisotropic error term of upwind scheme is zero. But when the advection velocity is at an angle, the diffusion becomes anisotropic. In order to overcome this, we shall look at operator splitting algorithms.

14.5.2   Operator splitting and upwind scheme

Operator splitting is a simple technique which overcomes the effects of anisotropic diffusion. In this method we consider advection along x and y directions independently. We first consider advection along x (or y) alone and then correct the solution by considering advection along other direction. Operator splitting applied to upwind scheme (a_x and a_y are positive) is written down as

$$u_{i,j}^* = u_{i,j}^k + C_x\left(u_{i,j}^k - u_{i-1,j}^k\right) \qquad \text{along } x \text{ direction}$$

$$u_{i,j}^{k+1} = u_{i,j}^* + C_y\left(u_{i,j}^* - u_{i,j-1}^*\right) \qquad \text{along } y \text{ direction}$$

In the above equation, the first step acts like a predictor and the second step as a corrector. Effectively, the operator splitting can be written as

$$u_{i,j}^{k+1} = (1 - C_x)(1 - C_y)u_{i,j}^k + C_x(1 - C_y)u_{i-1,j}^k + C_y(1 - C_x)u_{i,j-1}^k + C_x C_y u_{i-1,j-1}^k \qquad (14.53)$$

The last term in the above equation corresponds to $u_{i-1,j-1}$ and this makes a huge difference between the simple upwind scheme and the operator splitting based upwind scheme. For stability requirements, $C_x \leq 1$ and $C_y \leq 1$. Figure 14.28 shows the spread of the solution (curve is isoline of $u = 0.1$) applied to Example 14.10 using Operator splitting method. Unlike the normal upwind method, the operator splitting shows diffusion uniformly in all the directions.

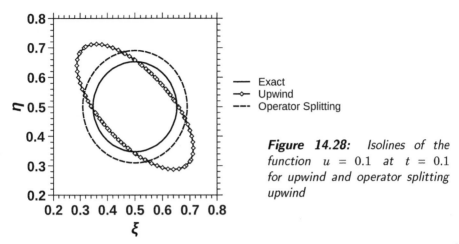

Figure 14.28: *Isolines of the function $u = 0.1$ at $t = 0.1$ for upwind and operator splitting upwind*

Error analysis: Let us perform Taylor series expansion of all the terms about $u_{i,j}^k$. Then the actual PDE we are solving would be

$$\underbrace{\frac{\partial u}{\partial t} + a_x \frac{\partial u}{\partial x} + a_y \frac{\partial u}{\partial y}}_{\text{advection}} = \underbrace{-\frac{1}{2}\frac{\partial^2 u}{\partial t^2}\Delta t + \frac{a_x \Delta x}{2}\frac{\partial^2 u}{\partial x^2} + \frac{a_y \Delta y}{2}\frac{\partial^2 u}{\partial y^2} + a_x a_y \Delta t \frac{\partial^2 u}{\partial x \partial y}}_{\text{error}}$$

Rewriting $\dfrac{\partial^2 u}{dt^2}$ as before, the error becomes

$$\epsilon = \frac{a_x \Delta x}{2}(1 - C_x)\frac{\partial^2 u}{\partial x^2} + \frac{a_y \Delta y}{2}\left(1 - C_y\right)\frac{\partial^2 u}{\partial y^2}$$

The anisotropic diffusion term is absent from the solution. Let us try and understand the physical implications of advection in two dimensions.

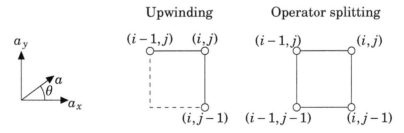

The above figure shows the nodes involved using upwind scheme and the operator splitting scheme. For a scheme to yield exact solution, the first derivative in the advection term should also be oriented along the advection velocity. However, in upwind scheme the diagonal nodes are completely disregarded and the derivative is not oriented along the direction of advection. By splitting the operator, the diagonal nodes have been implicitly included in the solution procedure. One must realize that the discretized equation of the operator splitting algorithm is same as bilinear interpolation across the rectangular region.

The above method can easily be extended to three dimensions. The advantage of this method is that one can apply an one dimensional advection algorithm over multiple dimensions.

14.6 Diffusion equation in multi dimensions

Heat equation in two dimensions is given by

$$\frac{\partial^2 u}{\partial x^2} + \frac{\partial^2 u}{\partial y^2} = \frac{1}{\alpha}\frac{\partial u}{\partial t} \tag{14.54}$$

Assume that the physical domain is a rectangle of width L_x and height L_y. Discretization of the equation is easily accomplished by finite differences.

14.6.1 Explicit formulation

Explicit formulation is only of academic interest and is presented here because of its simplicity. Let i,j represent a node where the indices indicate x and y positions of a node. Step sizes along the two directions are taken as Δx and Δy such that the x and y coordinates of the node are given by $x_i = i\Delta x$ and $y_j = j\Delta y$. Step size along the time direction is taken as Δt. Time is indicated by superscript k such that $t_k = k\Delta t$. Using forward difference for time derivative and central differences for the space derivatives, the explicit formulation yields the following for an interior node:

$$\frac{u_{i-1,j}^k - 2u_{i,j}^k + u_{i+1,j}^k}{\Delta x^2} + \frac{u_{i,j-1}^k - 2u_{i,j}^k + u_{i,j+1}^k}{\Delta y^2} = \frac{1}{\alpha}\frac{u_{i,j}^{k+1} - u_{i,j}^k}{\Delta t} \tag{14.55}$$

This equation leads to an explicit relation for the nodal function value at $t + \Delta t$ or $k+1$ in the form

$$u_{i,j}^{k+1} = u_{i,j}^k + \frac{\alpha\Delta t}{\Delta x^2}\left(u_{i-1,j}^k - 2u_{i,j}^k + u_{i+1,j}^k\right) + \frac{\alpha\Delta t}{\Delta y^2}\left(u_{i1,j-1}^k - 2u_{i,j}^k + u_{i,j+1}^k\right)$$

which may be recast as

$$
\begin{aligned}
u_{i,j}^{k+1} &= u_{i,j}^k + Fo_{e,x}\left(u_{i-1,j}^k - 2u_{i,j}^k + u_{i+1,j}^k\right) + Fo_{e,y}\left(u_{i,j-1}^k - 2u_{i,j}^k + u_{i,j+1}^k\right) \\
\text{or } u_{i,j}^{k+1} &= \left(1 - 2Fo_{e,x} - 2Fo_{e,y}\right)u_{i,j}^k + Fo_{e,x}\left(u_{i-1,j}^k + u_{i+1,j}^k\right) \\
&+ Fo_{e,y}\left(u_{i,j-1}^k + u_{i,j+1}^k\right)
\end{aligned}
\tag{14.56}
$$

where $Fo_{e,x} = \dfrac{\alpha\Delta t}{\Delta x^2}$ and $Fo_{e,y} = \dfrac{\alpha\Delta t}{\Delta y^2}$ are the elemental Fourier numbers. The stability of explicit scheme would require $Fo_{e,x} + Fo_{e,y} \leq \dfrac{1}{2}$ i.e. the coefficient of $u_{i,j}^k \geq 0$.

14.6.2 Implicit and Crank Nicolson schemes

Similar to one dimensional diffusion cases, the discretization can be carried out using implicit schemes. The discretized equation for implicit scheme would be

$$\left(1 + 2Fo_{e,x} + 2Fo_{e,y}\right)u_{i,j}^{k+1} - Fo_{e,x}\left(u_{i-1,j}^{k+1} + u_{i+1,j}^{k+1}\right) - Fo_{e,y}\left(u_{i,j-1}^{k+1} + u_{i,j+1}^{k+1}\right) = u_{i,j}^{k}$$

Each discretized equation contains 5 unknowns. The above equations can be solved using any method discussed earlier. Crank Nicolson scheme would also lead to similar set of equations. However, solving the above equations would become laborious. On the contrary, one can use operator splitting algorithms such as ADI which can simplify the cost of computation to a great extent.

14.6.3 ADI method

We have already presented the ADI method as applicable to the elliptic case. ADI method may be extended to the parabolic case without much difficulty.[10] It requires time (operator) splitting in which the time step is taken as a sequence of two half time steps (first ADI step from t to $t + \dfrac{\Delta t}{2}$ or k to $k + \dfrac{1}{2}$; second ADI step from $t + \dfrac{\Delta t}{2}$ to $t + \Delta t$ or $k + \dfrac{1}{2}$ to $k + 1$). During the first half time step x direction is treated implicitly while the y direction is treated explicitly. In the next half time step the y direction is treated implicitly while the x direction is treated explicitly. In either step we reduce the formulation to *one dimension in space* and hence it is possible to use line by line calculation using TDMA.

Nodal equations

As far as internal nodes are concerned the nodal equations in the first ADI step are written down as under.

$$\frac{u_{i-1,j}^{k+\frac{1}{2}} - 2u_{i,j}^{k+\frac{1}{2}} + u_{i+1,j}^{k+\frac{1}{2}}}{\Delta x^2} + \frac{u_{i1,j-1}^{k} - 2u_{i,j}^{k} + u_{i,j+1}^{k}}{\Delta y^2} = \frac{1}{\alpha}\frac{u_{i,j}^{k+\frac{1}{2}} - u_{i,j}^{k}}{\frac{\Delta t}{2}}$$

which may be rearranged as

$$Fo_{e,x}u_{i-1,j}^{k+\frac{1}{2}} - 2(1+Fo_{e,x})u_{i,j}^{k+\frac{1}{2}} + Fo_{e,x}u_{i+1,j}^{k+\frac{1}{2}} =$$
$$-Fo_{e,y}u_{i,j-1}^{k} - 2(1-Fo_{e,y})u_{i,j}^{k} - Fo_{e,y}u_{i,j+1}^{k} \qquad (14.57)$$

The boundary conditions also need to be treated in a similar fashion. Analogously the nodal equations in the second ADI step are written down as under.

$$Fo_{e,y}u_{i,j-1}^{k+1} - 2(1+Fo_{e,y})u_{i,j}^{k+1} + Fo_{e,x}u_{i,j+1}^{k+1} =$$

[10]D.W. Peaceman and H.H. Rachford, "The numerical solution of parabolic and elliptic differential equations", SIAM, Vol. 3, pp. 28-41

$$-Fo_{e,x}u_{i-1,j}^{k+\frac{1}{2}} - 2(1-Fo_{e,x})u_{i,j}^{k+\frac{1}{2}} \quad - \quad Fo_{e,x}u_{i+1,j}^{k+\frac{1}{2}} \qquad (14.58)$$

The ADI method is unconditionally stable. In case the step sizes are the same along x and y we have $Fo_{e,x} = Fo_{e,y} = Fo_e$. We may choose $Fo_e = 1$ and simplify the above to get the following nodal equations.

$$\textbf{ADI Step 1: } u_{i-1,j}^{k+\frac{1}{2}} - 4u_{i,j}^{k+\frac{1}{2}} + u_{i+1,j}^{k+\frac{1}{2}} = -\left(u_{i,j-1}^{k} - u_{i,j+1}^{k}\right) \qquad (14.59)$$

$$\textbf{ADI Step 2: } u_{i,j-1}^{k+1} - 4u_{i,j}^{k+1} + u_{i,j+1}^{k+1} = -\left(u_{i-1,j}^{k+\frac{1}{2}} - u_{i+1,j}^{k+\frac{1}{2}}\right) \qquad (14.60)$$

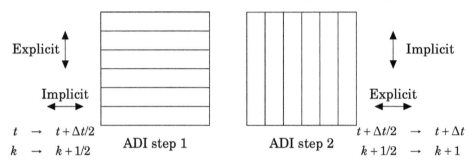

Both these form a set of linear equations in tri diagonal form and hence are solved using TDMA. In the example that follows we shall demonstrate the treatment of boundary conditions.

Example 14.11

The heat equation in 2 dimensions is valid in a square region which is defined by the corners at $\xi = 0, \eta = 0$; $\xi = 0, \eta = 1$; $\xi = 1, \eta = 0$ and $\xi = 1, \eta = 1$. The heat equation is in the standard form $u_{\xi\xi} + u_{\eta\eta} = u_\tau$. Initially $u = 1$ throughout the physical domain. For $\tau > 0$ the left, bottom and right sides are maintained at $u = 0$ while Robin condition is specified along the top as $u_\eta + c_1 u = c_2$ with $c_1 = 1$ and $c_2 = 0.5$. Obtain the transient temperature field within the domain by ADI method.

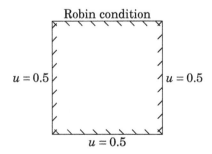

Solution :

We choose $\Delta\xi = \Delta\eta = 0.1$ and $Fo_e = 1$. The step size along the time direction is then given by $\Delta\tau = \Delta\xi^2 = 0.1^2 = 0.01$. In the ADI method we take two half steps of 0.005 each to cover one time step. We note that the nodal system will be such that $0 \le i \le 10$ and $0 \le j \le 10$. However there is symmetry with respect to $i = 5$ and hence we need to consider the nodes in the range $0 \le i \le 5$ and $0 \le j \le 10$ only.

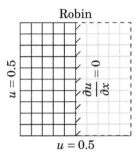

```
n = 6; m = 2*n-1;              % no of nodes along x and y
x = 0:0.5/(n-1):0.5; y = 0:1/(n-1):1;    % nodes
h = 0.5/(n-1);  Foe = 1; dt = Foe*h^2;  % step size
u = ones(n,m); uo = u;  % initialize function
c1 = 1; c2 = 0.5;              % coefficients for Robin condition
ax = zeros(n,1);  % coefficient of i,j along x
ay = zeros(m,1);  % coefficient of i,j along y
bx = zeros(n,1);  % coefficient of i+1,j (along x)
cx = zeros(n,1);  % coefficient of i-1,j (along x)
by = zeros(m,1);  % coefficient of i,j+1 (along y)
cy = zeros(m,1);  % coefficient of i,j-1 (along y)
dx = zeros(n,1);  % constant term (along x)
dy = zeros(m,1);  % constant term (along y)
```

Interior nodal equations follow those given in Equations 14.59 and 14.60. We now look at the boundary nodes.

Node on top boundary

Top boundary corresponds to $j = 10$. We have the Robin condition specified there. In the first ADI step the boundary condition is applied in an explicit fashion. Thus we have, using the concept of ghost node

$$\frac{u_{i,11}^k - u_{i,9}^k}{2\Delta\eta} + c_1 u_{i,10}^k = c_2 \quad \text{or} \quad u_{i,11}^k = u_{i,9}^k - 2c_1\Delta\eta u_{i,10}^k + 2c_2\Delta\eta$$

We substitute this in the nodal equation of ADI Step 1, treating the boundary node as an internal node to get

$$u_{i-1,10}^{k+\frac{1}{2}} - 4u_{i,10}^{k+\frac{1}{2}} + u_{i+1,10}^{k+\frac{1}{2}} = -\left(2u_{i,9}^k - 2c_1\Delta\eta u_{i,10}^k + 2c_2\Delta\eta\right)$$

In the second ADI step the boundary condition is applied in an implicit fashion. Thus we have, using the concept of ghost node

$$\frac{u_{i,11}^{k+1} - u_{i,9}^{k+1}}{2\Delta\eta} + c_1 u_{i,10}^{k+1} = c_2 \quad \text{or} \quad u_{i,11}^{k+1} = u_{i,9}^{k+1} - 2c_1\Delta\eta u_{i,10}^{k+1} + 2c_2\Delta\eta$$

We substitute this in the nodal equation of ADI Step 2, treating the boundary node as an internal node to get

$$2u_{i,9}^{k+1} - (4+2c_1\Delta\eta)u_{i,10}^{k+1} = -\left(u_{i-1,10}^{k+\frac{1}{2}} + u_{i+1,10}^{k+\frac{1}{2}} + 2c_2\Delta\eta\right)$$

Node on right boundary

Right boundary corresponds to $i = 5$. We have the Neumann condition specified there. In the first ADI step the boundary condition is applied in an implicit fashion. Thus we have, using the concept of ghost node, $u_{6,j}^{k+\frac{1}{2}} = u_{4,j}^{k+\frac{1}{2}}$. Treating the boundary node as an internal node, we then get

$$2u_{4,j}^{k+\frac{1}{2}} - 4u_{5,j}^{k+\frac{1}{2}} = -\left(u_{5,j-1}^{k} - u_{5,j+1}^{k}\right)$$

In the second ADI step the boundary condition is applied in an explicit fashion. Thus we have, using the concept of ghost node, $u_{6,j}^{k+\frac{1}{2}} = u_{4,j}^{k+\frac{1}{2}}$. Treating the boundary node as an internal node, we then get

$$u_{5,j-1}^{k+1} - 4u_{5,j}^{k+1} - u_{5,j+1}^{k+1} = -2u_{4,j}^{k+\frac{1}{2}}$$

Node at right corner $i = 5, j = 10$

In this case two ghost nodes are introduced as shown in Figure 14.29. In the first ADI step we treat the ghost node $6, 10$ implicitly and ghost node $5, 11$ explicitly.

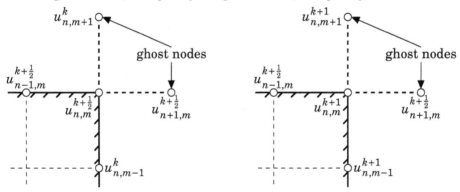

Note: in the example $n = 5$ and $m = 10$

Figure 14.29: *Corner node in Example 14.11*

We thus have the following two relations:

$$u_{6,10}^{k+\frac{1}{2}} = u_{4,10}^{k+\frac{1}{2}}$$
$$u_{5,11}^{k} = u_{5,9}^{k} - 2c_1\Delta\eta u_{5,10}^{k} + 2c_2\Delta\eta$$

Now treat the corner node as an interior node, introduce the above to get the following equation in the first ADI step.

$$2u_{4,10}^{k+\frac{1}{2}} - 4u_{5,10}^{k+\frac{1}{2}} = -2\left(u_{5,9}^{k} - c_1\Delta\eta u_{5,10}^{k} + c_2\Delta\eta\right)$$

In the second ADI step we treat the ghost node $6, 10$ explicitly and ghost node $5, 11$ implicitly. We thus have the following two relations:

$$u_{6,10}^{k+\frac{1}{2}} = u_{4,10}^{k+\frac{1}{2}}$$
$$u_{5,11}^{k+1} = u_{5,9}^{k+1} - 2c_1\Delta\eta u_{5,10}^{k+1} + 2c_2\Delta\eta$$

Now treat the corner node as an interior node, introduce the above to get the following equation in the first ADI step.

$$2u_{5,9}^{k+1} - (4 + 2c_1\Delta\eta)u_{5,10}^{k+1} = -2\left(u_{4,9}^{k+\frac{1}{2}} + c_2\Delta\eta\right)$$

```
% TDMA coefficients for ADI step 1
ax(1) = 1; dx(1) = 0.5;
ax(2:n-1) = 2+2*Foe;
bx(2:n-1) = Foe;
cx(2:n-1) = Foe;
ax(n) = 2+2*Foe;
cx(n) = 2*Foe;
% TDMA coefficients for ADI step 2
ay(1) = 1; dy(1) = 0.5;
ay(2:m-1) = 2+2*Foe;
by(2:m-1) = Foe;
cy(2:m-1) = Foe;
ay(m) = 2+2*Foe+2*Foe*c1*h;
cy(m) = 2*Foe;
% time integration
for t=1:5
    % ADI STEP 1
    for j=2:m
        if(j==m)            % top boundary
            dx(2:n-1) = 2*Foe*uo(2:n-1,m-1) ...
                + 2*(1-Foe-Foe*c1*h)*uo(2:n-1,m) + 2*Foe*c2*h;
            dx(n) = 2*Foe*uo(n,m-1) ...
                + 2*(1-Foe-Foe*c1*h)*uo(n,m) + 2*Foe*c2*h;
        else                % interior line
            dx(2:n-1) = Foe*(uo(2:n-1,j+1)+uo(2:n-1,j-1)) ...
                + 2*(1-Foe)*uo(2:n-1,j);
            dx(n) = Foe*(uo(n,j+1)+uo(n,j-1)) ...
                + 2*(1-Foe)*uo(n,j);
        end
        u(:,j) = tdma(cx,ax,bx,dx);
    end
    %ADI step 2
    for i=2:n
        if(i==n)            % symmetric boundary
            dy(2:m-1) = 2*Foe*uo(n-1,2:m-1) ...
                + 2*(1-Foe)*uo(n,2:m-1);
            dy(m) = 2*Foe*uo(n-1,m) ...
                + 2*(1-Foe)*uo(n,m) + 2*c2*Foe*h;
        else                % interior line
            dy(2:m-1) = Foe*(uo(i+1,2:m-1)+uo(i-1,2:m-1)) ...
```

```
                    +  2*(1-Foe)*uo(i,2:m-1);
        dy(m)  =  Foe*(uo(i+1,m)+uo(i-1,m))  ...
                    +  2*(1-Foe)*uo(i,m)  +  2*c2*Foe*h;
        u(i,:)  =  tdma(cy,ay,by,dy);
      end
    end
    uo = u;        % update function
end
```

The function values after one time step have been tabulated below

$i \setminus j$	0	1	2	3	4	5
10	0.5	0.73635	0.97271	0.97271	0.97271	0.94876
9	0.5	0.74634	0.99268	0.99268	0.99268	0.99862
8	0.5	0.74901	0.99803	0.99803	0.99803	0.99862
7	0.5	0.74971	0.99943	0.99943	0.99943	0.99862
6	0.5	0.74984	0.99967	0.99967	0.99967	0.99862
5	0.5	0.74964	0.99927	0.99927	0.99927	0.99862
4	0.5	0.74871	0.99741	0.99741	0.99741	0.99862
3	0.5	0.74519	0.99038	0.99038	0.99038	0.99862
2	0.5	0.73205	0.96410	0.96410	0.96410	0.99862
1	0.5	0.68301	0.86603	0.86603	0.86603	0.74931
0	0.5	0.5	0.5	0.5	0.5	0.5

The temperature field at $t = 1$ have been tabulated below

$i \setminus j$	0	1	2	3	4	5
10	0.5	0.50019	0.50036	0.50049	0.50058	0.50061
9	0.5	0.50020	0.50039	0.50053	0.50063	0.50066
8	0.5	0.50021	0.50040	0.50055	0.50065	0.50068
7	0.5	0.50021	0.50040	0.50054	0.50064	0.50068
6	0.5	0.50020	0.50038	0.50052	0.50061	0.50064
5	0.5	0.50018	0.50034	0.50047	0.50055	0.50058
4	0.5	0.50015	0.50029	0.50040	0.50047	0.50050
3	0.5	0.50012	0.50023	0.50032	0.50037	0.50039
2	0.5	0.50008	0.50016	0.50022	0.50026	0.50027
1	0.5	0.50004	0.50008	0.50011	0.50013	0.50014
0	0.5	0.5	0.5	0.5	0.5	0.5

The solution is close to the steady state solution. The transient method discussed above can be applied to solve a steady state solution (*pseudo-transient method*). In fact, the method discussed above is the same as the one considered to solve elliptic problems.

It is important to perform space and time step sensitivity study to ascertain convergence of the solution. We leave it to the interest of the reader.

All the methods that have been discussed so far can be applied in other coordinate systems as well. In such a case the spatial discretization remains similar to that considered in elliptic problems with a suitable time integration scheme. The spatial discretization can be performed using FDM, FEM or FVM. As such, methods that have been discussed to solve a system of initial value ODEs such as Runge Kutta, ABM, implicit methods can be used to perform time integration.

Concluding remarks

In this chapter we have presented simple explicit schemes, implicit and semi-implicit schemes and hybrid schemes for the solution of advection equation, diffusion equation and advection diffusion equation. No numerical scheme is of either universal applicability or capable of giving satisfactory solution in all these cases. Hence a number of specific schemes are discussed applicable to advection, diffusion and advection-diffusion equations.

We have also presented detailed von Neumann stability analysis of various numerical schemes presented here. Accuracy characteristics of the schemes have been brought out by use of Taylor series based method.

14.A Suggested reading

1. **S. Patankar** *Numerical heat transfer and fluid flow* Taylor and Francis, 1980
2. **H.K. Versteeg** and **W. Malalasekera** *An introduction to computational fluid dynamics: the finite volume method* Prentice Hall, 2007
3. **C. Hirsch** *Numerical computation of internal and external flows: the fundamentals of Computational Fluid Dynamics* Butterworth -Heinemann, 2007

Chapter 15

Wave equation

Waves occur in various forms such as longitudinal acoustic waves in a medium such as air, transverse waves in a string, waves in a membrane and as electromagnetic radiation. All these are governed by the wave equation with the dependent variable being pressure or displacement or electric field etc. Wave equation is hyperbolic in nature and exhibits real and distinct characteristic directions in the case of waves in one dimension. Hence it is possible to solve analytically the wave equation in one dimension. Wave equation is also amenable to analytical solutions using the method of separation of variables. However emphasis here is on numerical solution of the wave equation in different circumstances. We also consider waves in two dimensions, such as those that are possible in a vibrating membrane or a diaphragm. Speakers and musical instruments use vibrating diaphragms to produce sound.

15.1 Introduction

Consider the wave equation in one dimension i.e.

$$a^2\frac{\partial^2 u}{\partial x^2} = \frac{\partial^2 u}{\partial t^2} \tag{15.1}$$

where a is the wave speed. This equation may be factored as under:

$$\left(a\frac{\partial}{\partial x} - \frac{\partial}{\partial t}\right)\left(a\frac{\partial}{\partial x} + \frac{\partial}{\partial t}\right)u = 0 \tag{15.2}$$

Here the operator $\left(a^2\frac{\partial^2}{\partial x^2} - \frac{\partial^2}{\partial t^2}\right)$ has been written as the product of two operators acting on u. Hence the general solution should be the sum of two solutions of the advection equation with speeds a and $-a$. Thus we should have

$$u(x,t) = f_1(x - at) + f_2(x + at) \tag{15.3}$$

The first part remains constant along the characteristic $x - at = \xi =$ constant and the second part remains constant along the characteristic $x + at = \eta =$ constant. Along the first characteristic we have $\frac{dx}{dt} = a$ and along the second characteristic we have $\frac{dx}{dt} = -a$. Consider a point $x = x_0, t = 0$. Then we see that $\xi = x_0$ along the first characteristics and $\eta = x_0$ along the second characteristics. Set of characteristics for the wave equation are shown in Figure 15.1. The full lines represent $\xi =$ constant while the dashed lines represent $\eta =$ constant.

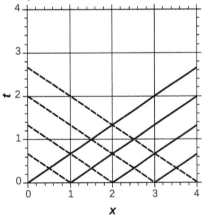

Figure 15.1: *Set of characteristics for the wave equation*

The wave equation can support two initial conditions and two boundary conditions because of second derivatives with respect to space and time that occur in it. In case the domain stretches from $x = -\infty$ to $x = \infty$, we may specify both u as well as $\frac{\partial u}{\partial t}$ i.e. the initial displacement and initial velocity for all x. In case the domain is finite, $0 \le x \le L$ we have to, in addition, specify two boundary conditions.

Example 15.1

Consider an infinite string of lineal density $m = 0.1\ kg/m$ under a tension of $T = 2.5\ N$. Determine wave speed when a small transverse displacement is set up in the string.

Solution :

When the transverse displacement is small we may assume the tension in the string to remain constant. Consider a small length dx of string as shown in Figure 15.2.

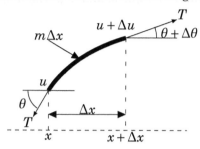

Figure 15.2: *Transverse displacement of a string under constant tension*

Resolve forces along the u direction to obtain the net force acting on the element of string in the transverse direction as

$$
\begin{aligned}
F &= T[\sin(\theta + \Delta\theta) - \sin\theta] \\
&= T[\sin\theta\cos\Delta\theta + \cos\theta\sin\Delta\theta - \sin\theta] \approx \Delta\theta
\end{aligned}
$$

where we have set $\cos\theta \approx 1$ and $\sin\theta \approx \theta$ in the limit of small displacement. We note also that $\tan\theta \approx \sin\theta \approx \theta = \dfrac{\partial u}{\partial x}$. Hence we write the net transverse force as

$$
F = T\Delta\theta \approx T\frac{\partial^2 u}{\partial x^2}\Delta x
$$

Applying Newton's second law we than have

$$
m\Delta x\frac{\partial^2 u}{\partial t^2} = T\frac{\partial^2 u}{\partial x^2}\Delta x
$$

or

$$
\frac{\partial^2 u}{\partial t^2} = \frac{T}{m}\frac{\partial^2 u}{\partial x^2} = a^2\frac{\partial^2 u}{\partial x^2}
$$

where the wave speed is identified as $a = \sqrt{\dfrac{T}{m}}$. In the present case the wave speed is calculated as

$$
a = \sqrt{\frac{2.5}{0.1}} = 5\ m/s
$$

15.2   General solution of the wave equation

Consider the case where the domain stretches from $x = -\infty$ to $x = \infty$. At $t = 0$ the initial shape is specified with zero velocity throughout. Consider the solution as given previously by $u = f_1(x - at) + f_2(x + at) = f_1(\xi) + f_2(\eta)$. We then have

$$
\frac{\partial u}{\partial x} = \frac{df_1}{d\xi} + \frac{df_2}{d\eta}; \qquad \frac{\partial u}{\partial t} = -a\frac{df_1}{d\xi} + a\frac{df_2}{d\eta};
$$

$$\frac{\partial^2 u}{\partial x^2} = \frac{d^2 f_1}{d\xi^2} + \frac{d^2 f_2}{d\eta^2}; \qquad \frac{\partial^2 u}{\partial t^2} = a^2\frac{d^2 f_1}{d\xi^2} + a^2\frac{d^2 f_2}{d\eta^2} \qquad (15.4)$$

The wave equation is identically satisfied. Hence $u = f_1(x-at) + f_2(x+at) = f_1(\xi) + f_2(\eta)$ is a general solution to the wave equation. The solution may be interpreted as being made up of two waves, one propagating towards the right $[f_1(x-at)]$ and the other propagating towards the left $[f_2(x+at)]$. Consider the case where the initial shape of the disturbance is given by $u(x,0) = f(x)$ for $-\infty \le x \le \infty$. Solution at later time is given by

$$u(x,t) = \frac{f(x-at) + f(x+at)}{2} = \frac{f(\xi) + f(\eta)}{2} \qquad (15.5)$$

The initial disturbance divides into two equal parts, one moving towards the right and the other moving towards the left as shown in Figure 15.3.

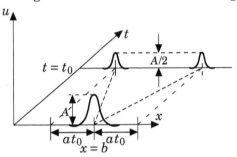

Figure 15.3: *Right and left going waves in an infinite domain*

15.2.1 d'Alembert's solution to the one dimensional wave equation

Now we consider the general case where the initial shape as well as initial velocity is prescribed. We start with the general solution given by Equation 15.3. From this solution we get the velocity as

$$\left.\frac{\partial u}{\partial t}\right|_{x,t} = -a\frac{d f_1}{d\xi} + a\frac{d f_2}{d\eta} \qquad (15.6)$$

If we let $t = 0$ in the above (then $\xi = \eta = x$), we get the initial velocity given by

$$\left.\frac{\partial u}{\partial t}\right|_{x,0} = -a\frac{d f_1}{dx} + a\frac{d f_2}{dx} = g(x) \qquad (15.7)$$

We may integrate this equation with respect to x to get

$$-af_1(x) + af_2(x) = \int g(x)dx + h(t=0) \quad \text{or} \quad -f_1(x) + f_2(x) = \frac{1}{a}\int_{x_0}^{x} g(s)ds \qquad (15.8)$$

where x_0 is an arbitrary point within the domain. Also we have, from the initial condition defining the shape as

$$f_1(x) + f_2(x) = f(x) \qquad (15.9)$$

Adding Equations 15.8 and 15.9 and rearranging we get

$$f_2(x) = \frac{f(x)}{2} + \frac{1}{2a}\int_{x_0}^{x} g(s)ds \qquad (15.10)$$

Subtracting Equation 15.8 from Equation 15.9 and rearranging we get

$$f_1(x) = \frac{f(x)}{2} - \frac{1}{2a} \int_{x_0}^{x} g(s)ds \qquad (15.11)$$

It is clear, in general, that we can replace x by $x-at$ in Equation 15.11 to get the first part of the general solution and that we can replace x by $x+at$ in Equation 15.10 to get the second part of the general solution. We then combine these two to get the general solution as

$$u(x,t) = \frac{f(x-at)+f(x+at)}{2} + \frac{1}{2a} \int_{x-at}^{x+at} g(s)ds \qquad (15.12)$$

This last result is known as the d'Alembert's[1] solution to the one dimensional wave equation. d'Alembert's solution shows that the solution is influenced by the zone of influence associated with the point, as indicated in Figure 15.4.

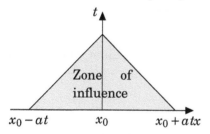

Figure 15.4: *Zone of influence arising out of d'Alembert's solution*

15.2.2 Boundary conditions

If the two ends are rigid no displacement is possible at the ends and hence Dirichlet conditions $u(0,t) = u(L,t) = 0$ may be specified. This condition can be visualized by a string tied to a pole. A wave traveling towards a fixed node, gets reflected. The reflected wave moves with the same speed and amplitude but is out of phase with respect to the original wave as illustrated in the figure below.

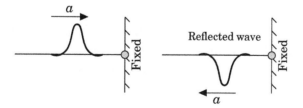

Free nodes are free to move about the equilibrium position. A wave reflected from the free node has the same speed, amplitude and phase as the original wave as illustrated below.

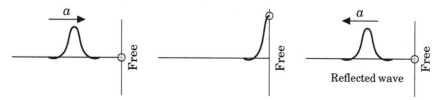

---

[1]after Jean-Baptiste le Rond d'Alembert, 1717-1783, a French mathematician

This boundary condition can be realized by tying the string to a ring where the ring can slide freely on the pole. Mathematically, a free boundary means the first derivative of displacement $\frac{\partial u}{\partial x} = 0$ at this node which is nothing but Neumann boundary condition.

Another possible boundary condition is a combination of fixed and free conditions where there is a small resistance to the displacement at this boundary. This will occur at the interface of two media characterized by different wave speeds.

Then part of the initial wave is reflected back into the first medium where as the other part of the disturbance is transmitted into the second medium.

Non-dimensionalization

The wave equation may be written in terms of non-dimensional variables $X = \frac{x}{L_{ch}}$ and $\tau = \frac{at}{L_{ch}}$ where L_{ch} is a characteristic length. In case the string is of finite length L stretched between two supports, we may set $L_{ch} = L$ and obtain the wave equation in the standard form

$$\frac{\partial u}{\partial X^2} = \frac{\partial u}{\partial \tau^2} \tag{15.13}$$

15.2.3   A useful theorem

A very useful theorem is derived based on the general solution of the one dimensional wave equation. Consider four points P, Q, R, S as shown in Figure 15.5. The lines joining the points are characteristics of the wave equation. Two of the characteristics are right going while the other two are left going.

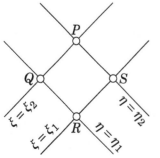

Figure 15.5: *Four points in the t, x plane lying on characteristics*

From the general solution to the one-dimensional wave equation we have the following identities.

$$u_P = f_1(x_P - at_P) + f_2(x_P + at_P); \quad u_Q = f_1(x_Q - at_Q) + f_2(x_Q + at_Q);$$
$$u_R = f_1(x_R - at_R) + f_2(x_R + at_R); \quad u_S = f_1(x_S - at_S) + f_2(x_S + at_S) \quad (15.14)$$

Since the points lie on the characteristics, we also have the following identities.

$$x_P - at_P = x_Q - at_Q; \quad x_P + at_P = x_S + at_S;$$
$$x_R - at_R = x_S - at_S; \quad x_R + at_R = x_Q + at_Q \quad (15.15)$$

With these Equations 15.14 may be recast as follows.

$$u_P = f_1(x_Q - at_Q) + f_2(x_S + at_S); \quad u_Q = f_1(x_Q - at_Q) + f_2(x_Q + at_Q)$$
$$u_R = f_1(x_S - at_S) + f_2(x_Q + at_Q); \quad u_S = f_1(x_S - at_S) + f_2(x_S + at_S) \quad (15.16)$$

Thus we have

$$u_P + u_R = u_Q + u_S \quad \text{or} \quad u_P = u_Q + u_S - u_R \quad (15.17)$$

An example is considered now to demonstrate the use of the above theorem.

Example 15.2

Initial shape of a string stretched between two supports is the sinusoidal distribution given by $u = 0.05 \sin\left(\dfrac{\pi x}{L}\right)$. Initial velocity is zero for all x. The tension and mass per unit length of string are given as $T = 2.5\,N$ and $m = 0.1\,kg/m$. The length of the string between supports is $L = 0.1\,m$. Determine the displacement of the string at $x = \dfrac{L}{4}$ at $t = 0.0075\,s$.

Solution :

The wave speed in this case is $a = \sqrt{\dfrac{2.5}{0.1}} = 5\,m/s$. The characteristics passing through point P with $x_P = \dfrac{L}{4} = \dfrac{0.1}{4} = 0.025\,m$ and $t_P = 0.0075\,s$ are given by

$$\xi_P = x_P - at_P = 0.025 - 5 \times 0.0075 = -0.0125$$
$$\eta_P = x_P + at_P = 0.25 + 5 \times 0.0075 = 0.0625$$

The construction shown in Figure 15.6 gives the characteristics required in solving the problem. The characteristic $\xi = -0.0125$ is outside the domain i.e. $0 \le \xi \le 0.1$ and hence the general solution cannot be used directly to obtain the displacement at this point. The characteristic $\xi = -0.0125$ hits the left boundary at the point Q. With $x_Q = 0$ we get $t_Q = -\dfrac{\xi_Q}{a} = -\dfrac{-0.0125}{5} = 0.0025$. Consider now the characteristic line passing through $x = 0$ which corresponds to $\xi = 0$. Points R and S are chosen to lie on this characteristic. We note that $\eta_R = \eta_Q = 5 \times 0.0025 = 0.0125$. Hence we have $x_R + at_R = 0.0125$. Since $\xi_R = 0$ we also have $x_R - at_R = 0$ or $at_R = x_R$. Thus we have $x_R = \dfrac{0.0125}{2} = 0.00625$ and $t_R = \dfrac{0.00625}{5} = 0.00125$. Similarly we can show that $x_S = 0.03125$ and $t_S = 0.00625$. Hence the solution required at P is obtained by the use of the theorem presented earlier, using the four points P,Q,R,S shown in Figure 15.6.

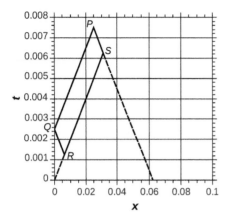

Figure 15.6: *Construction of characteristics for Example 15.2*

Since point Q lies at the left boundary $u_Q = 0$ for all t. The displacement at point R is obtained using the general solution as

$$u_R = \frac{f(\xi_R) + f(\eta_R)}{2} = \frac{0 + 0.05\sin\left(\pi \cdot \frac{0.00625}{0.1}\right)}{2} = 0.009755$$

Displacement at point S may again be obtained from the general solution as

$$u_S = \frac{f(\xi_S) + f(\eta_S)}{2} = \frac{0 + 0.05\sin\left(\pi \cdot \frac{0.0625}{0.1}\right)}{2} = 0.023097$$

Using the theorem we than have

$$u_P = u_Q + u_S - u_R = 0 + 0.023097 - 0.009755 = 0.013342$$

We consider another example to demonstrate the use of the d'Alembert's solution. Initially the string is in its equilibrium position with the velocity specified at each point on the string.

Example 15.3

A string $L = 4\,m$ long is stretched between two rigid supports. The tension in the spring is $T = 100\,N$ and the lineal density is $m = 0.64\,kg/m$. The string is initially in its equilibrium position and the transverse velocity is given by the formula $g(x,0) = 0.1x(L-x)\,m/s$. Calculate the displacement at the midpoint of the string (i.e. at $x = 2\,m$) as a function of time.

Solution :

With the data provided in the problem the wave speed is calculated as

$$a = \sqrt{\frac{100}{0.64}} = 12.5\,m/s$$

The integral involved in the d'Alembert's solution is obtained as

$$\int_{\xi}^{\eta} g(s)ds = \int_{\xi}^{\eta} 0.1s(L-s)ds = 0.1\left(\frac{L}{2}(\eta^2 - \xi^2) + \frac{\eta^3 - \xi^3}{3}\right)$$

Since the string is initially at the equilibrium position, $f(x) = 0$ and hence the solution is given by

$$u(x,t) = \frac{1}{2a}\int_{\xi}^{\eta} g(s)ds = 0.004\left(\frac{L}{2}(\eta^2 - \xi^2) + \frac{\eta^3 - \xi^3}{3}\right)$$

where u is in m. At any time, we have,

$$\xi = x - 12.5t; \quad \text{and} \quad \eta = x + 12.5t$$

Using these the solution may be simplified as

$$u(x,t) = 0.1\left[tx(L-x) - \frac{a^2 t^3}{3}\right]$$

Also we may then obtain the velocity by differentiating the above with respect to t as

$$\left.\frac{\partial u}{\partial t}\right|_{x,t} = 0.1\left[x(L-x) - a^2 t^2\right]$$

The displacement at mid span may be obtained by substituting $x = 2\,m$ in the expression for u to get

$$u(2,t) = 0.1\left[4t - \frac{a^2 t^3}{3}\right]$$

Mid span displacement has been calculated from $t = 0$, the equilibrium position to $t = 0.16\,s$ when the string reaches the maximum displacement position and is shown in the following table. We may verify , using the expression given above for the velocity of the string, that the velocity is zero at $t = 0.16\,s$. For $t > 0.16\,s$ the displacement decreases and is again at the equilibrium position at $t = 0.32\,s$. The motion of the string then continues in the negative direction and returns to the equilibrium position at $t = 0.64\,s$ i.e. after one period of oscillation. The motion is periodic and repeats itself with this period.

x	t	ξ	η	u	t	ξ	η	u
2	0	2	2	0	0.09	0.875	3.125	0.0322
2	0.01	1.875	2.125	0.004	0.1	0.75	3.25	0.0348
2	0.02	1.75	2.25	0.008	0.11	0.625	3.375	0.0371
2	0.03	1.625	2.375	0.0119	0.12	0.5	3.5	0.039
2	0.04	1.5	2.5	0.0157	0.13	0.375	3.625	0.0406
2	0.05	1.375	2.625	0.0193	0.14	0.25	3.75	0.0417
2	0.06	1.25	2.75	0.0229	0.15	0.125	3.875	0.0424
2	0.07	1.125	2.875	0.0262	0.16	0	4	0.0427
2	0.08	1	3	0.0293				

The data is also shown plotted in Figure 15.7.

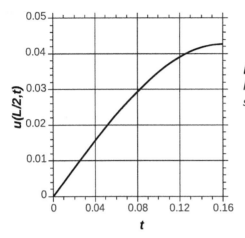

Figure 15.7: *Displacement history at mid span of the string in Example 15.3*

15.3   Numerical solution of the one dimensional wave equation

We now direct our attention to the numerical solution of the one dimensional wave equation using the finite difference method. We present first an explicit scheme followed by a discussion of an implicit method.

15.3.1   Explicit scheme

Since the wave equation involves second derivatives with respect to both space and time, the least we can do is to apply central differences along both the space and time directions. Letting subscript i represent node number along x and superscript k represent nodes along t, central difference in time and central difference in space (CTCS) may be used to recast the governing PDE in terms of nodal equation. With Δx and Δt as the steps along x and t respectively, we have

$$\frac{u_{i-1}^k - 2u_i^k + u_{i+1}^k}{\Delta x^2} = \frac{1}{a^2}\frac{u_i^{k-1} - 2u_i^k + u_i^{k+1}}{\Delta t^2} \tag{15.18}$$

This may be recast as an explicit relation between u_i^{k+1} and other nodal values as

$$u_i^{k+1} = 2(1 - C^2)u_i^k - u_i^{k-1} + C^2(u_{i-1}^k + u_{i+1}^k) \tag{15.19}$$

where $C = \dfrac{a\Delta t}{\Delta x}$ is the Courant number. We see that displacements at $k-1$ and k are required for determining the displacement at $k+1$.

First time step: Let us assume that we are starting at $k = 0$ when both the displacement and velocity are given. We may then approximate the nodal velocity by central difference in time, using a ghost node at $k - 1 = 0 - 1 = -1$. We then have

$$\frac{u_i^1 - u_i^{-1}}{2\Delta t} = g_i \tag{15.20}$$

where g_i represents the nodal velocity. We can solve this for u_i^{-1} as

$$u_i^{-1} = u_i^1 - 2g_i \Delta t \tag{15.21}$$

We substitute this in Equation 15.19 with $k = 1$ to get

$$u_i^1 = (1 - C^2)u_i^0 + g_i \Delta t + \frac{C^2}{2}(u_{i-1}^0 + u_{i+1}^0) \tag{15.22}$$

Thus we are able to calculate explicitly the nodal displacements after one time step. The computations can then progress for higher times simply be the use of Equation 15.19. Note that only nodal displacements and not velocities are involved in calculations beyond the first time step.

15.3.2 Stability analysis of the explicit scheme

Stability of the explicit scheme is studied by the use of familiar von Neumann stability analysis. We assume that the solution is given by $u(x,t) = A(t)e^{j\omega x}$ where the amplitude is a function of time. We assume that the time variation is such that the magnification is given by $\frac{A(t+\Delta t)}{A(t)} = \lambda$ or $\frac{A(t)}{A(t-\Delta t)} = \lambda$. Substituting the above in Equation 15.19 we get

$$\lambda A(t)e^{j\omega x} = 2(1 - C^2)e^{j\omega x} - \frac{A(t)}{\lambda}e^{j\omega x} + C^2 A(t)e^{j\omega x}(e^{j\omega \Delta x} + e^{-j\omega \Delta x})$$

This may be simplified, after canceling the common $e^{j\omega x}$ term, to read

$$\lambda^2 - 2\left(1 - 2C^2 \sin^2\left(\frac{\omega \Delta x}{2}\right)\right)\lambda + 1 = 0$$

$$\text{or} \qquad \lambda^2 - 2\alpha\lambda + 1 = 0 \tag{15.23}$$

where we have represented the circular function in terms of half angle and set $\alpha = \left(1 - 2C^2 \sin^2\left(\frac{\omega \Delta x}{2}\right)\right)$. We see at once that α is bound by 1 and $1 - 2C^2$. The former is of no consequence. However if $\alpha = 1 - 2C^2$, the solution to the above quadratic for λ is

$$\lambda = \alpha \pm \sqrt{\alpha^2 - 1} = 1 - 2C^2 \pm 2C\sqrt{(C^2 - 1)} \tag{15.24}$$

If $C > 1$ the amplification factor $\lambda > 1$ and the numerical scheme is unstable. For example, if $C = \sqrt{2}$, we get $\lambda = -3 \pm \sqrt{8}$ and one of the roots is greater than 1. If $C = 1$, λ is real (equal to -1) with the absolute value equal to 1 and hence the numerical scheme is stable. If $C < 1$, λ is complex with the absolute value of the amplification factor equal to 1. For example, if $C = 0.5$ we have $\lambda = -0.5 \pm j\sqrt{0.75}$. The magnitude of this complex number is seen to be 1. Thus the explicit scheme is conditionally stable for $C \leq 1$.

Example 15.4

Waves are set up in the region $0 \leq x \leq 1$ with an initial displacement given by $u(x,0) = f(x) = 0.01\{1 - \cos(2\pi x)\}$ and initial velocity $h(x) = 0.02x(1-x)$. Compute the shape of the wave at $t = 0.3$ using the explicit method with $C = 0.5$, $C = 1$ and $C = 1.5$. Comment on the results in comparison with the exact d'Alembert's solution. The wave equation is specified to be $u_{xx} = u_{tt}$.

Solution :

We shall first obtain the exact solution. Using Equation 15.12 and the initial displacement and velocity functions prescribed in the problem, we get

$$u(x,t) = 0.005\left(2 - \cos(2\pi\xi) - \cos(2\pi\eta)\right) + 0.01\left(\frac{\eta^2 - \xi^2}{2} + \frac{\eta^3 - \xi^3}{3}\right)$$

We choose $\Delta x = 0.1$ so that there are 11 nodes along the x direction. The exact solution is calculated at these points. At the two ends we impose zero displacement conditions. At $t = 0.3$ the ξ and η values are within range (i.e. $0 \leq \xi$ or $\eta \leq 1$) for all x except $x = 0.1$, $x = 0.2$, $x = 0.8$ and $x = 0.9$. Because of symmetry in the initial values, the displacement at $x = 0.1$ is the same as that at $x = 0.9$ and the displacement at $x = 0.2$ is the same as that at $x = 0.8$. Thus it is enough if we evaluate the displacements at $x = 0.1$, $t = 0.3$ and $x = 0.2$, $t = 0.3$ using the theorem introduced earlier. At all other points the d'Alembert's solution may be directly applied.

The four points required in the case of $x = 0.1$, $t = 0.3$ are shown below

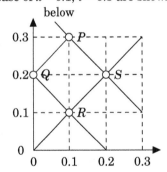

The four points required in the case of $x = 0.2$, $t = 0.3$ are shown below

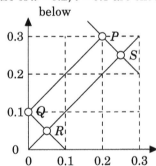

Calculations for these two points are tabulated below.

x_P	ξ_R	η_R	ξ_S	η_S	u_S	u_R	u_Q	u_P
0.1 (0.9)	0	0.2	0	0.4	0.00963	0.00363	0	0.00600
0.2 (0.8)	0	0.1	0	0.5	0.01083	0.00100	0	0.00983

We may calculate all the other nodal displacements (u_E indicates Exact) and tabulate the results as shown below.

x	ξ	η	u_E	
0			0	
0.1	-0.20000	0.40000	0.00600	} Calculated
0.2	-0.10000	0.50000	0.00983	above
0.3	0.00000	0.60000	0.01013	
0.4	0.10000	0.70000	0.00876	
0.5	0.20000	0.80000	0.00823	
0.6	0.30000	0.90000	0.00876	
0.7	0.40000	1.00000	0.01013	
0.8	0.50000	1.10000	0.00983	} Calculated
0.9	0.60000	1.20000	0.00600	above
1			0	

FDM solution Now consider the numerical solution based on the explicit formulation. The time step size is determined by the value of the chosen Courant number. In case $C = 0.5$ we have $\Delta t = 0.05$ and hence $t = 0.3$ is reached after 6 time steps. In case $C = 1$ we have $\Delta t = 0.1$ and hence $t = 0.3$ is reached after 3 time steps. In case $C = 1.5$ we have $\Delta t = 0.15$ and hence $t = 0.3$ is reached after 2 time steps. First time step uses Equation 15.22 while subsequent time steps use Equation 15.19. The state of affairs at $t = 0.3$, obtained with the three values of C are shown in the table below.

	$u(x,0.3)$		
x	$C = 0.5$	$C = 1$	$C = 1.5$
0	0	0	0
0.1	0.00612	0.00601	0.00356
0.2	0.00978	0.00985	0.01191
0.3	0.00993	0.01015	0.01004
0.4	0.00893	0.00878	0.00848
0.5	0.00846	0.00825	0.00787
0.6	0.00893	0.00878	0.00848
0.7	0.00993	0.01015	0.01004
0.8	0.00978	0.00985	0.01191
0.9	0.00612	0.00601	0.00356
1	0	0	0

It is seen from the table that the solution is erratic for $C = 1.5$. This is expected since the CFL condition is violated and the numerical scheme is unstable. Both the results for $C = 0,5$ and $C = 1$ appear to be close to the exact solution. The above procedure has been programmed in Matlab.

```
n = 11;                    % no. of nodes
x = [0:1/(n-1):1]';        % nodes
c = 1; dx = 1/(n-1); dt = c*dx;  % c = Courant number
u0 = zeros(n,1);           % displacement at k-1
u1 =  0.01*(1-cos(2*pi*x)); % displacement at k
u2 = zeros(n,1);           % displacement at k+1
```

```
v   = 0.02*x.*(1-x);          % initial velocity
% first time step
for i=2:n-1                   % calculate new displacement
    u2(i) = c^2*(u1(i-1)+u1(i+1))/2 + (1-c^2)*u1(i) + v(i)*dt;
end
u0 = u1; u1 = u2;             % update displacements
for k=2:3                     % loop for higher time steps
    for i=2:n-1               % calculate new displacement
        u2(i) = c^2*(u1(i-1)+u1(i+1)-2*u1(i)) + 2*u1(i)-u0(i);
    end
    u0 = u1;  u1 = u2;  % update displacements
end
```

The results are also shown plotted in Figure 15.8. Figure 15.9 shows the displacement history of midpoint $x = 0.5$. One can clearly see the divergence of displacement for $C = 1.5$.

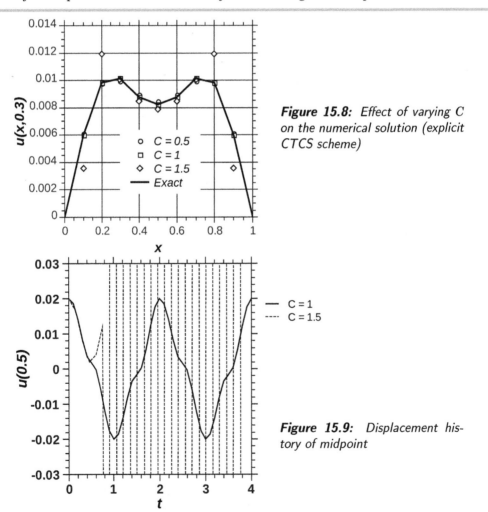

Figure 15.8: *Effect of varying C on the numerical solution (explicit CTCS scheme)*

Figure 15.9: *Displacement history of midpoint*

Modes of vibration

In section 3.1.3, we have seen that the solution of the wave equation is an eigenvalue problem. The wave equation can be solved analytically by separation of variables where the displacement is written as a product of two functions $u(x,t) = v(x)w(t)$ to reduce the partial differential equation to a set of ODE. Thus

$$\frac{d^2v}{dx^2} = -\frac{\lambda^2}{a^2}v$$

$$\frac{d^2w}{dt^2} = -\lambda^2 w$$

The disturbance due to the wave can be represented as a weighted sum of modes of vibration given by

$$u(x) = \sum_{n=1}^{\infty} A_n \sin\frac{\lambda_n x}{a} \sin\lambda_n t \qquad (15.25)$$

where $\lambda_n = \frac{n a \pi}{L}$ is n^{th} mode of vibration. The modes of vibration can be determined by solving the following eigenvalue problem

$$\frac{d^2v}{dx^2} + kv = 0$$

where $k = \frac{\lambda^2}{a^2}$ or $\lambda = a\sqrt{k}$.

Example 15.5

Determine the fundamental mode of vibration of a string of length 1, fixed at $x = 0$ and free at $x = 1$. The wave velocity in the string is 1.

Solution :

We have to solve the following eigenvalue problem

$$\frac{d^2v}{dx^2} + kv = 0$$

We shall discretize the derivative term using FDM. The domain is discretized using $n = 21$ equal elements of step size h i.e. $n + 1$ nodes.
For interior nodes, the discretization becomes

$$\frac{v_{i+1} - 2v_i + v_{i-1}}{h^2} + kv_i = 0$$

The first node is fixed and hence $v_0 = 0$. Hence, the discretization equation at node $i = 1$ becomes

$$\frac{v_2 - 2v_1}{h^2} + kv_1 = 0$$

The last node is free and hence Neumann boundary condition is applicable at this node. Using the ghost node approach, we obtain the following equation

$$\frac{2v_{n-1} - 2v_n}{h^2} + kv_n = 0$$

Writing the system of linear equations in matrix form

$$
\frac{1}{h^2}
\begin{pmatrix}
-2 & 1 & 0 & \cdots & 0 & 0 \\
1 & -2 & 1 & \cdots & 0 & 0 \\
\cdots & \cdots & \cdots & \cdots & \cdots & \cdots \\
0 & 0 & 0 & \cdots & 2 & -2
\end{pmatrix}
\underbrace{}_{\mathbf{A}}
\begin{Bmatrix}
u_1 \\ u_2 \\ u_3 \\ \vdots \\ u_{n-1} \\ u_n
\end{Bmatrix}
+ k
\begin{Bmatrix}
u_1 \\ u_2 \\ u_3 \\ \vdots \\ u_{n-1} \\ u_n
\end{Bmatrix}
$$

Now the fundamental mode corresponds to the smallest eigenvalue of the matrix $\mathbf{A}$. Note the eigenvalue of matrix $\mathbf{A}$ will be equal to $-k$. The smallest eigenvalue can be determined using inverse power shift method. We shall use MATLAB to determine the fundamental mode of the above system.

```
n = 21;                    % no. of nodes
A = zeros(n-1);            % initialize matrix
h = 1/(n-1);               % step size
A(1,1) = -2/h^2; A(1,2) = 1/h^2;          % node at i=1
for i=2:n-2                % interior nodes
    A(i,i-1) = 1/h^2; A(i,i) = -2/h^2; A(i,i+1) = 1/h^2;
end
A(n-1,n-1) = -2/h^2; A(n-1,n-2) = 2/h^2; % last node
A = inv(A);                % inverse of A
uguess = ones(n-1,1);      % guess values of eigenvector
[e1,u1] = powermethod(A ,eguess ,1e-6);   % Program 3.1
k1 = -1/e1;                % 1/e
```

On solving, we get $k_1 = 2.4661$ or $\lambda_1 = 0.4999\pi$ which is in good agreement with the exact solution ($\lambda_1 = 0.5\pi$). The eigenvector corresponds to the shape of the string for fundamental frequency. We can now determine the next mode using inverse power method combined with deflation.

```
A = A-e1*u1*u1';           % power method with deflation
[e2,u2] = powermethod(A,uguess,1e-6);
k2 = -1/e2;                % 1/e
```

We obtain $k_2 = 22.104$ or $\lambda_2 = 1.4965\pi$ (exact value $\lambda_2 = 1.5\pi$). Figure 15.10 shows the shape of the string for the fundamental and first overtones. Similarly higher modes of vibration can also be determined. However, the accuracy of higher modes of vibration would be large compared to the fundamental mode. Note: one has to perform grid sensitivity study to ascertain convergence of the solution.

The number of modes captured in the numerical solution depends on the spatial discretization. Spatial discretization should be such that all desired wavelengths are captured by the numerical solution. Otherwise the shape of the wave will not be preserved.

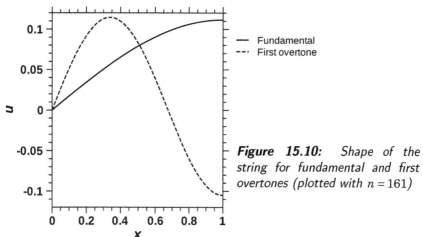

Figure 15.10: *Shape of the string for fundamental and first overtones (plotted with $n = 161$)*

15.3.3 Implicit scheme

Now we turn our attention to derive an implicit scheme for the solution of the wave equation in one dimension. Consider time levels $t + \Delta t$ and t i,e. $k + 1$ and k. The second derivative with respect to space at $t + \dfrac{\Delta t}{2}$ or at $k + \dfrac{1}{2}$ is given by

$$\left.\frac{\partial^2 u}{\partial x^2}\right|_{i,k+\frac{1}{2}} = \frac{u_{i-1}^{k+1} - 2u_i^{k+1} + u_{i+1}^{k+1}}{2\Delta x^2} + \frac{u_{i-1}^{k} - 2u_i^{k} + u_{i+1}^{k}}{2\Delta x^2} \tag{15.26}$$

Consider time levels t and $t - \Delta t$ i,e. k and $k - 1$. The second derivative with respect to space at $t - \dfrac{\Delta t}{2}$ or at $k - \dfrac{1}{2}$ is given by

$$\left.\frac{\partial^2 u}{\partial x^2}\right|_{i,k-\frac{1}{2}} = \frac{u_{i-1}^{k} - 2u_i^{k} + u_{i+1}^{k}}{2\Delta x^2} + \frac{u_{i-1}^{k-1} - 2u_i^{k-1} + u_{i+1}^{k-1}}{2\Delta x^2} \tag{15.27}$$

The implicit scheme is based on the mean of the second derivative expressions given above i.e.the second derivative is approximated as

$$\left.\frac{\partial^2 u}{\partial x^2}\right|_{i,k} = \frac{\left.\frac{\partial^2 u}{\partial x^2}\right|_{i,k+\frac{1}{2}} + \left.\frac{\partial^2 u}{\partial x^2}\right|_{i,k-\frac{1}{2}}}{2}$$

Thus we have

$$\left.\frac{\partial^2 u}{\partial x^2}\right|_{i,k} = \frac{(u_{i-1}^{k+1} - 2u_i^{k+1} + u_{i+1}^{k+1}) + 2(u_{i-1}^{k} - 2u_i^{k} + u_{i+1}^{k}) + (u_{i-1}^{k-1} - 2u_i^{k-1} + u_{i+1}^{k-1})}{4\Delta x^2} \tag{15.28}$$

The second derivative with respect to time is evaluated as in the explicit scheme. Hence the implicit formulation is written down as

$$-u_{i-1}^{k+1} + 2\left(1 + \frac{2}{C^2}\right)u_i^{k+1} - u_{i+1}^{k+1} = 2\left[u_{i-1}^{k} - 2\left(1 - \frac{2}{C^2}\right)u_i^{k} + u_{i+1}^{k}\right]$$

$$+u_{i-1}^{k-1} - 2\left(1 + \frac{2}{C^2}\right)u_i^{k-1} + u_{i+1}^{k-1} \tag{15.29}$$

which is in tri-diagonal form. For the very first time step we use the central difference formula for the velocity as done earlier to write the nodal equations as

$$-u_{i-1}^1 + 2\left(1 + \frac{2}{C^2}\right)u_i^1 - u_{i+1}^1 = \left[f_{i-1} - 2\left(1 - \frac{2}{C^2}\right)f_i + f_{i+1}\right]$$
$$-\Delta t \left[g_{i-1} - 2\left(1 + \frac{2}{C^2}\right)g_i + g_{i+1}\right] \tag{15.30}$$

15.3.4 Stability of implicit scheme

von Neumann stability analysis may be performed in a straight forward manner. Introducing the gain factor λ as the ratio $\dfrac{A(t+\Delta t)}{A(t)}$ and assuming the space dependence to be $e^{j\omega x}$, Equation 15.30 may be shown to lead to

$$\lambda^2 + 2\lambda \left[\frac{1 - \frac{2}{C^2} - \cos\omega\Delta x}{1 + \frac{2}{C^2} - \cos\omega\Delta x}\right] + 1 = 0 \tag{15.31}$$

Expressing $\cos\omega\Delta x$ in terms of half angle, we then get

$$\lambda^2 + 2\lambda \left[\frac{\sin^2\frac{\omega\Delta x}{2} - \frac{1}{C^2}}{\sin^2\frac{\omega\Delta x}{2} + \frac{1}{C^2}}\right] + 1 = 0 \tag{15.32}$$

Solution to this quadratic is written down, after simplification as

$$\lambda = \frac{\left[\sin^2\frac{\omega\Delta x}{2} - \frac{1}{C^2}\right] \pm j\frac{2}{C^2}\sin\frac{\omega\Delta x}{2}}{\left[\sin^2\frac{\omega\Delta x}{2} + \frac{1}{C^2}\right]} \tag{15.33}$$

The magnitude of λ (λ is complex and hence the magnitude is the square root of sum of squares of real and imaginary parts) is seen to be unity for all C. Hence the implicit scheme is unconditionally stable. In what follows we solve an example using the implicit scheme with $C = 1$. The nodal equations are then given by the following.

$$-u_{i-1}^{k+1} + 6u_i^{k+1} - u_{i+1}^{k+1} = 2\left[u_{i-1}^k + 2u_i^k + u_{i+1}^k\right] + u_{i-1}^{k-1} - 6u_i^{k-1} + u_{i+1}^{k-1} \tag{15.34}$$

and

$$-u_{i-1}^1 + 6u_i^1 - u_{i+1}^1 = [f_{i-1} + 2f_i + f_{i+1}] - \Delta t[g_{i-1} - 6g_i + g_{i+1}] \tag{15.35}$$

Example 15.6
Redo Example 15.4 using the implicit scheme with $C = 1$. Compare the solution at $t = 0.3$ with that obtained by the explicit scheme and the exact solution.

Solution :
We choose $\Delta x = 0.05$ and hence $\Delta t = 0.05$ with $C = 1$. Zero boundary conditions are imposed at $x = 0$ and $x = 1$. For the first time step we use the $f(x)$ and $g(x)$ specified in Example 15.4. TDMA is made use of to solve for the nodal values of the displacement at $t = 0.05$ and $0 < i < 10$ using nodal equations based on Equation 15.35. Table below gives the results at the end of first time step.

x_i	$f(x)$	$g(x)$	a_i	a_i'	a_i''	b_i	P_i	Q_i	u_i^1
0	0.00000	0.00000	...	...	...	...	...	...	0.00000
0.05	0.00049	0.00095	6	1	0	0.00308	0.16667	0.00051	0.00091
0.1	0.00191	0.00180	6	1	1	0.00880	0.17143	0.00160	0.00237
0.15	0.00412	0.00255	6	1	1	0.01758	0.17157	0.00329	0.00453
0.2	0.00691	0.00320	6	1	1	0.02859	0.17157	0.00547	0.00722
0.25	0.01000	0.00375	6	1	1	0.04076	0.17157	0.00793	0.01019
0.3	0.01309	0.00420	6	1	1	0.05290	0.17157	0.01044	0.01315
0.35	0.01588	0.00455	6	1	1	0.06385	0.17157	0.01275	0.01582
0.4	0.01809	0.00480	6	1	1	0.07253	0.17157	0.01463	0.01794
0.45	0.01951	0.00495	6	1	1	0.07811	0.17157	0.01591	0.01930
0.5	0.02000	0.00500	6	1	1	0.08003	0.17157	0.01646	0.01977
0.55	0.01951	0.00495	6	1	1	0.07811	0.17157	0.01623	0.01930
0.6	0.01809	0.00480	6	1	1	0.07253	0.17157	0.01523	0.01794
0.65	0.01588	0.00455	6	1	1	0.06385	0.17157	0.01357	0.01582
0.7	0.01309	0.00420	6	1	1	0.05290	0.17157	0.01140	0.01315
0.75	0.01000	0.00375	6	1	1	0.04076	0.17157	0.00895	0.01019
0.8	0.00691	0.00320	6	1	1	0.02859	0.17157	0.00644	0.00722
0.85	0.00412	0.00255	6	1	1	0.01758	0.17157	0.00412	0.00453
0.9	0.00191	0.00180	6	1	1	0.00880	0.17157	0.00222	0.00237
0.95	0.00049	0.00095	6	0	1	0.00308	0.00000	0.00091	0.00091
1	0.00000	0.00000	...	...	...	...	...	...	0.00000

Extension to time $t = 0.1$ requires the values of u at $t = 0.05$ as well as $t = 0$ as shown in the table below.

x_i	u_i^0	u_i^1	a_i	a_i'	a_i''	b_i	P_i	Q_i	u_i^2
0	0.0000	0.0000	...	...	...	...	...	...	...
0.05	0.00312	0.00326	6	1	0	0.01285	0.16667	0.00214	0.00315
0.1	0.00589	0.00632	6	1	1	0.02469	0.17143	0.00460	0.00605
0.15	0.00789	0.00861	6	1	1	0.03487	0.17157	0.00677	0.00846
0.2	0.00938	0.00993	6	1	1	0.04039	0.17157	0.00809	0.00984
0.25	0.01064	0.01059	6	1	1	0.04140	0.17157	0.00849	0.01019
0.3	0.01177	0.01093	6	1	1	0.03992	0.17157	0.00831	0.00992
0.35	0.01276	0.01113	6	1	1	0.03762	0.17157	0.00788	0.00941
0.4	0.01353	0.01125	6	1	1	0.03550	0.17157	0.00744	0.00891
0.45	0.01402	0.01132	6	1	1	0.03406	0.17157	0.00712	0.00857
0.5	0.01418	0.01134	6	1	1	0.03355	0.17157	0.00698	0.00845
0.55	0.01402	0.01132	6	1	1	0.03406	0.17157	0.00704	0.00857
0.6	0.01353	0.01125	6	1	1	0.03550	0.17157	0.00730	0.00891
0.65	0.01276	0.01113	6	1	1	0.03762	0.17157	0.00771	0.00941
0.7	0.01177	0.01093	6	1	1	0.03992	0.17157	0.00817	0.00992
0.75	0.01064	0.01059	6	1	1	0.04140	0.17157	0.00851	0.01019
0.8	0.00938	0.00993	6	1	1	0.04039	0.17157	0.00839	0.00984
0.85	0.00789	0.00861	6	1	1	0.03487	0.17157	0.00742	0.00846
0.9	0.00589	0.00632	6	1	1	0.02469	0.17157	0.00551	0.00605
0.95	0.00312	0.00326	6	0	1	0.01285	0.00000	0.00315	0.00315
1	0.0000	0.0000	...	...	...	...	...	...	...

The state of affairs at $t = 0.3$ obtained by the implicit scheme with $C = 1$, explicit scheme with $C = 1$ and the exact are shown in the following table.

x_i	u_i at $t=0.3$		
	Implicit	Explicit	Exact
0	0.00000	0.00000	0.00000
0.1	0.00605	0.00601	0.00600
0.2	0.00984	0.00985	0.00983
0.3	0.00992	0.01015	0.01013
0.4	0.00891	0.00878	0.00876
0.5	0.00845	0.00825	0.00823
0.6	0.00891	0.00878	0.00876
0.7	0.00992	0.01015	0.01013
0.8	0.00984	0.00985	0.00983
0.9	0.00605	0.00601	0.00600
1	0.00000	0.00000	0.00000

It is seen that the three solutions agree closely with each other.

15.4 Waves in a diaphragm

Transverse waves may be set up in a diaphragm which is tightly stretched and clamped at the periphery, as in a drum. Diaphragm may be rectangular, circular or of any other shape. We consider first a diaphragm in the form of a rectangle of length L_x and width L_y. Boundary conditions are then specified as $u(\pm L_x/2, y, t) = 0$ and $u(x, \pm L_y/2, t) = 0$. Let the stretching force per unit length in the diaphragm be T while the mass of diaphragm per unit area is m. Initially the displacement of the diaphragm is specified as $u(x, y, 0) = f(x, y)$. It is required to study the variation of displacement as a function of time.

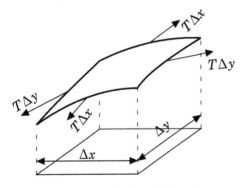

Figure 15.11: *Waves in a diaphragm under constant tension*

Consider an element $dxdy$ of the diaphragm that has undergone a *small* displacement away from the equilibrium position, as shown in Figure 15.11. The net force acting on the element in the transverse direction is given by

$$\text{Net transverse force on the element} = T dy \frac{\partial^2 u}{\partial x^2} dx + T dx \frac{\partial^2 u}{\partial y^2} dy \qquad (15.36)$$

This must equal the product of mass of the element and the acceleration i.e.

$$\text{Mass of element} \times \text{acceleration} = m dx dy \frac{\partial^2 u}{\partial t^2} \qquad (15.37)$$

Equating these two and removing the common factor $dxdy$ results in the two dimensional wave equation in cartesian frame of reference.

$$\frac{\partial^2 u}{\partial x^2} + \frac{\partial^2 u}{\partial y^2} = \nabla^2 u = \frac{1}{a^2}\frac{\partial^2 u}{\partial t^2} \tag{15.38}$$

In the above expression the wave speed a is given by $a = \sqrt{\dfrac{T}{m}}$ and ∇^2 is the Laplace operator. In case the diaphragm is circular it is convenient to represent the wave equation in cylindrical coordinates. All one has to do is to recast the Laplacian in cylindrical coordinates. Thus, for a circular diaphragm, we have

$$\frac{1}{r}\frac{\partial}{\partial r}\left(r\frac{\partial u}{\partial r}\right) + \frac{1}{r^2}\frac{\partial^2 u}{\partial \theta^2} = \frac{1}{a^2}\frac{\partial^2 u}{\partial t^2} \tag{15.39}$$

where a point P is represented by r, θ in cylindrical coordinates. An one dimensional form for the wave equation is possible in cylindrical coordinates, if the displacement is a function of r alone. Then the wave equation becomes

$$\frac{1}{r}\frac{\partial}{\partial r}\left(r\frac{\partial u}{\partial r}\right) = \frac{1}{a^2}\frac{\partial^2 u}{\partial t^2} \tag{15.40}$$

This may be recast in the following alternate form.

$$\frac{\partial^2 u}{\partial r^2} + \frac{1}{r}\frac{\partial u}{\partial r} = \frac{1}{a^2}\frac{\partial^2 u}{\partial t^2} \tag{15.41}$$

15.4.1 Explicit scheme for one dimensional waves in a circular diaphragm

The explicit scheme is the familiar CTCS scheme used earlier in the case of one dimensional wave equation in cartesian frame of reference. The spatial derivatives appearing in Equation 15.41 are approximated as follows using the already known function values.

$$\frac{\partial^2 u}{\partial r^2} = \frac{u_{i-1}^k - 2u_i^k + u_{i+1}^k}{\Delta r^2}$$
$$\frac{1}{r}\frac{\partial u}{\partial r} = \frac{1}{r_i}\frac{u_{i+1}^k - u_{i-1}^k}{2\Delta r} \tag{15.42}$$

Note that both the first and second derivatives use central differences and hence are second order accurate. The time derivative is approximated again by central differences as

$$\frac{1}{a^2}\frac{\partial^2 u}{\partial t^2} = \frac{u_i^{k-1} - 2u_i^k + u_i^{k+1}}{a^2\Delta t^2} \tag{15.43}$$

Using these expressions, solving for the displacement at $k+1$, after some simplification, we get the required nodal equation.

$$u_i^{k+1} = C^2\left(1 - \frac{\Delta r}{2r_i}\right)u_{i-1}^k + 2(1-C^2)u_i^k + C^2\left(1 + \frac{\Delta r}{2r_i}\right)u_{i+1}^k - u_i^{k-1} \qquad (15.44)$$

where C is the familiar Courant number given by $C = \dfrac{a\Delta t}{\Delta r}$. Above equation is valid for nodes $1 \le i \le (n-1)$. Node $i = n$ corresponds to $r = R$, the radius of the circular diaphragm where $u = 0$. The node at center i.e. $i = 0$ requires special attention since $r_0 = 0$ there. The first derivative term should vanish at $r = 0$ and hence the term $\dfrac{1}{r}\dfrac{\partial u}{\partial r}$ is in the form $\dfrac{0}{0}$. Using l'Hospital's rule we get this ratio as $\dfrac{\partial^2 u}{\partial r^2}$ and hence the wave equation at the origin becomes

$$2\frac{\partial^2 u}{\partial r^2} = \frac{1}{a^2}\frac{\partial^2 u}{\partial t^2}$$

In order to obtain finite difference analog of the second derivative, we use a ghost node at $i = -1$ where the function value u_{-1}^k is obtained using symmetry condition as u_1^k. The second derivative is then approximated as

$$\frac{\partial^2 u}{\partial r^2} = \frac{u_{-1}^k - 2u_0^k + u_1^k}{\Delta r^2} = \frac{2u_1^k - 2u_0^k}{\Delta r^2}$$

The nodal equation may then be approximated as

$$u_0^{k+1} = 4C^2 u_1^k + 2(1 - 2C^2)u_0^k - u_0^{k-1} \qquad (15.45)$$

The very first time step requires the use of initial displacement and velocity to eliminate u_i^{-1}. The reader is encouraged to work this out.

Example 15.7

A circular diaphragm of unit radius is stretched such that the wave speed is unity. Initially the diaphragm is displaced with $f(x) = 0.01(1 - r^2)$ and let go with zero velocity. Obtain the shape of the diaphragm as a function of time using explicit CTCS scheme.

Solution :

We discretize the domain with $\Delta r = 0.1$ such that there are 11 nodes. Node 10 corresponds to the fixed periphery of the diaphragm. We choose $C = 0.5$ such that $\Delta t = \dfrac{\Delta r}{2} = 0.05$. The initial condition imposes a displacement but no velocity. Hence $u_i^{-1} = u_i^1$ and the nodal equation for the first step becomes

$$u_i^1 = \frac{C^2}{2}\left(1 - \frac{\Delta r}{2r_i}\right)u_{i-1}^0 + (1-C^2)u_i^0 + \frac{C^2}{2}\left(1 + \frac{\Delta r}{2r_i}\right)u_{i+1}^0$$

where $1 \le i \le 9$. For the node at the center the nodal equation for the first time step is given by

$$u_0^1 = 2C^2 u_1^0 + (1 - 2C^2)u_0^0$$

With $f(r = 0) = 0.01$ and $f(r = 0.1) = 0.01(1 - 0.1^2) = 0.0099$, we have $u_0^1 = 2 \times 0.5^2 \times 0.0099 +$
$(1 - 2 \times 0.5^2) \times 0.01 = 0.0098$. For the node at $i = 1$ we have $u_1^1 = \dfrac{1}{2} \times 0.5^2 \left(1 - \dfrac{0.1}{2 \times 0.1}\right) 0.0099 +$
$(1 - 0.5^2)0.01 + \dfrac{1}{2}\left(1 + \dfrac{0.1}{2 \times 0.1}\right) 0.0099 = 0.0097$. Similarly we may evaluate all the nodal
displacements. For calculating the nodal displacements further we use Equations 15.44
and 15.45. The calculations are tabulated as shown below.

r	u_i^0	u_i^1	u_i^2
0	0.01000	0.00995	0.00980
0.1	0.00990	0.00985	0.00970
0.2	0.00960	0.00955	0.00940
0.3	0.00910	0.00905	0.00890
0.4	0.00840	0.00835	0.00820
0.5	0.00750	0.00745	0.00730
0.6	0.00640	0.00635	0.00620
0.7	0.00510	0.00505	0.00490
0.8	0.00360	0.00355	0.00340
0.9	0.00190	0.00185	0.00171
1	0.00000	0.00000	0.00000

The data is shown in the form of a plot in Figure 15.12 where the shape of the diaphragm
is shown for t up to 0.5. The above procedure has been programmed in Matlab.

```
n = 11;                        % no of nodes
r = [0:1/(n-1):1]';            % radial nodes
c = 0.5;    dr = 1/(n-1); dt = c*dr; % c = Courant number
u0 = zeros(n,1); u2 = u0;      % displacement at k-1 and k+1
u1 = 0.01*(1-r^2);             % displacement at k
%v = zeros(n,1);                        % initial velocity
% calculate displacement for first time step
for i=2:n-1    % internal nodes
    u2(i) =0.5*(c^2*(1-dr/(2r(i))*u1(i-1)+2*(1-c^2)*u1(i) ...
          +c^2*(1+dr/(2r(i))*u1(i+1));
end
u2(1) =  2*c^2*u1(2)+(1-2*c^2)*u0(1);  % u(r=0)
u0 = u1; u1 = u2;              % update displacements
for k=2:100                    % loop for higher time steps
    for i=2:n-1                % internal nodes
        u2(i) = c^2*(1-dr/(2r(i))*u1(i-1)+2*(1-c^2)*u1(i) ...
              +c^2*(1+dr/(2r(i))*u1(i+1)-u0(i)
    end
    u2(1) =  4*c^2*u1(2)+2*(1-2*c^2)*u1(1)-u0(1);
    u0 = u1; u1 = u2;          % update displacements
end
```

Example 15.7 has considered the solution of the wave equation in a diaphragm. The initial
displacement variation was axi-symmetric and led to an one-dimensional wave propagation

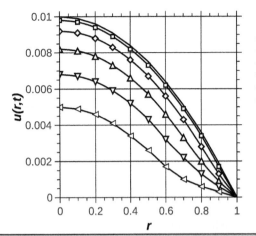

Figure 15.12: *Displacement history for a circular diaphragm*

problem in r, t coordinates. The initial shape of the disturbed diaphragm was basically a parabola with $u(0,0) = 0.01$ and $u(1,0) = 0$. This shape may be split into a number of Fourier components in terms of Bessel functions with respect to r and circular functions with respect to t. Each component is referred to as a mode. Various modes are defined by a set of orthogonal functions. These solutions are obtained easily by the method of separation of variables.

The shape of the surface will vary with some periodicity determined by the combination of modes that go into the solution. In order to clarify these ideas we consider next an example where the initial shape is specified by the first mode given by $f(r) = J_0(\lambda_0 r)$ where $\lambda_0 = 2.4048$ is the parameter that characterizes the first mode. Here J_0 is the Bessel function of first kind and order 0, familiar to us as the solution of Laplace equation in cylindrical coordinates. Shape of the displaced surface has a temporal variation given by $\cos \lambda_0 t$ so that the period of the wave is $\tau = \dfrac{2\pi}{\lambda_0} = \dfrac{2\pi}{2.4048} = 2.6127$. Hence the displacement which is at its maximum at $t = 0$ will reach zero value at $t = \dfrac{\tau}{4} = 0.6532$ (after a quarter period).

Example 15.8

Initial displacement distribution of a circular diaphragm is given by its fundamental mode given by $u(r,0) = f(r) = 0.01 J_0(\lambda_0 r)$ and the velocity is zero everywhere. Solve the wave equation from $t = 0$ to $t = \dfrac{\tau}{4} = 0.6532$ by an explicit method.

Solution :

The problem has been solved using a spreadsheet program that has Bessel function as a built in function that may be called using the function call BESSEL(x, v) with $x = 2.4048r$ and $v = 0$ in the present case. We choose a spatial step size of $\Delta r = 0.1$ and a temporal step size of $\Delta t = \dfrac{\tau}{40} = \dfrac{2.4048}{40} = 0.06532$ so that a quarter period is covered by taking 10 time steps from the beginning. The initial displacement distribution is evaluated using the spreadsheet Table 15.1.

In MATLAB, inbuilt subroutines to evaluate Bessel functions are available. The initial condition can be specified as

```
u1 = 0.01*besselj(0,2.4048*r)
```

The rest of the code from the earlier example can be used to solve this problem. Figure 15.13 indicates the initial displacement of the circular membrane. With the step sizes

Table 15.1

r	$f(r) = 0.01 \times$ $BESSEL(2.4048x, 0)$	r	$f(r) = 0.01 \times$ $BESSEL(2.4048x, 0)$
0	0.010000	0.6	0.005435
0.1	0.009856	0.7	0.004076
0.2	0.009430	0.8	0.002680
0.3	0.008741	0.9	0.001303
0.4	0.007817	1	0.000000
0.5	0.006699		

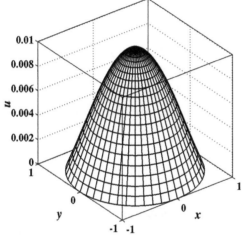

Figure 15.13: Initial displacement in circular membrane corresponding to fundamental mode

chosen above, the value of Courant number is given by $C = \dfrac{0.06532}{0.1} = 0.6532$. The nodal equations given by Equations 15.44 and 15.45 are made use of. For the very first time step nodal equations are modified as explained in Example 15.7. The shape of the diaphragm as a function of time is evaluated using the explicit scheme and the results are shown plotted in Figure 15.14. It is seen that the diaphragm has arrived at the equilibrium position with zero displacement everywhere at $t = \dfrac{\tau}{4}$. However the velocity is non-zero all along the diaphragm except at the periphery where it is zero.

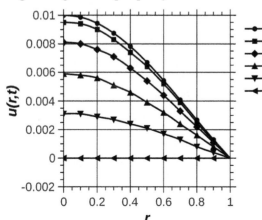

Figure 15.14: Displacement history for a circular diaphragm from $t = 0$ to $t = \dfrac{\tau}{4}$

15.4.2   Waves in two dimensions - waves in a rectangular diaphragm or plate

Waves set up in a rectangular diaphragm (length of rectangle L_x and width of rectangle L_y or plate satisfy the wave equation in two dimensions (Equation 15.38). We choose spatial steps of Δx and Δy along the two directions and a time step of Δt. We make use of explicit CTCS scheme. We then have

$$\frac{u_{i-1,j}^k - 2u_{i,j}^k + u_{i+1,j}^k}{\Delta x^2} + \frac{u_{i,j-1}^k - 2u_{i,j}^k + u_{i,j+1}^k}{\Delta y^2} = \frac{1}{a^2}\frac{u_{i,j}^{k-1} - 2u_{i,j}^k + u_{i,j}^{k-1}}{\Delta t^2}$$

which may be simplified to obtain the following nodal equation.

$$u_{i,j}^{k+1} = C_x^2(u_{i-1,j}^k + u_{i+1,j}^k) + C_y^2(u_{i,j-1}^k + u_{i,j+1}^k) - u_{i,j}^{k-1} + 2(1 - C_x^2 - C_y^2)u_{i,j}^k \qquad (15.46)$$

In the above subscript i identifies the position along x, subscript j identifies the position along y, $C_x = \dfrac{a\Delta t}{\Delta x}$ and $C_y = \dfrac{a\Delta t}{\Delta y}$. In case the step sizes are identical along x and y we have $C_x^2 = C_y^2 = C^2$ and the above simplifies to

$$u_{i,j}^{k+1} = C^2(u_{i-1,j}^k + u_{i+1,j}^k + u_{i,j-1}^k + u_{i,j+1}^k) - u_{i,j}^{k-1} + 2(1 - 2C^2)u_{i,j}^k \qquad (15.47)$$

Stability requires that $1 - 2C^2 \geq 0$ (coefficient of $u_{i,j}^k$ in Equation 15.47) or $C \leq \dfrac{1}{\sqrt{2}}$.

The diaphragm is clamped all along its four edges where zero displacement boundary conditions are imposed. In fact, the boundary conditions allow vibration of the diaphragm for selected modes with unique mode shapes and the solution to the wave equation is dependent on the strength of each of these modes. Each mode is characterized by an ordered pair of mode numbers n_x and n_y (both are integers) such that $\left(\dfrac{n_x}{L_x}\right)^2 + \left(\dfrac{n_y}{L_y}\right)^2 = \dfrac{4}{\lambda^2}$ where λ is the wavelength of the wave. For example, if the diaphragm is a square of side L each, the simplest wave that can occur is the one that corresponds to $n_x = n_y = 1$ such that $\lambda = \sqrt{2}L$.

Example 15.9

A square diaphragm of size 1×1 is fixed along all its edges. Initial displacement of the diaphragm is given by $u(x,y,0) = f(x,y) = 0.01\sin(\pi x)\sin(\pi y)$ with zero velocity everywhere. The stretching force in the diaphragm is such that the wave speed is unity. Obtain the shape of the diaphragm as a function of time for two time steps (using explicit scheme) of suitable size.

Solution :

In this case $L_x = L_y = L = 1$ and $a = 1$. We choose equal step sizes such that $\Delta x = \Delta y = 0.1$. We make use of the maximum possible value of $C = \dfrac{1}{\sqrt{2}} = 0.707107$ and obtain time step size of $\Delta t = 0.070711$.

```
nx = 11; ny = 11;          % no. of nodes along x and y
x = 0:1/(nx-1):1; y = 0:1/(ny-1):1;   % x and y coordinates
dx = 1/(nx-1); dy = 1/(ny-1);         % step size
```

```
[Y,X] = meshgrid(y,x);              % nodes
u0 = zeros(nx,ny);                  % displacement at k-1
u1 = 0.01*sin(pi*X).*sin(pi*Y);     % displacement at k
u2 = u0;                            % displacement at k+1
cx = 1/sqrt(2); dt = cx*dx; cy = dt/dy; % Courant number
```

For the first time step the nodal equations are given by

$$u^1_{i,j} = \frac{C^2}{2}(u^0_{i-1,j}+u^0_{i+1,j}+u^0_{i,j-1}+u^0_{i,j+1})+(1-2C^2)u^0_{i,j}$$

With $C^2 = \frac{1}{2}$ this expression simplifies to

$$u^1_{i,j} = \frac{(u^0_{i-1,j}+u^0_{i+1,j}+u^0_{i,j-1}+u^0_{i,j+1})}{4}$$

Consider node $1,1$ corresponding to $x = 0.1, y = 0.1$. Initial displacements at the nearest neighbours are given by

$j\downarrow\ i\rightarrow$	0	1	2
0		0.000000	
1	0.000000	0.000955	0.001816
2		0.001816	

After the first time step we have $t = 0+\Delta t = 0.070711$ and the displacement at $1,1$ is given by

$$u^1_{1,1} = \frac{0+0+0.001816+0.001816}{4} = 0.000908$$

Similarly we may obtain all other nodal displacements.

```
% first time step
for i=2:nx-1
    for j=2:ny-1
        u2(i,j) = (1-cx^2-cy^2)*u1(i,j)    ...
            + 0.5*cy^2*(u1(i+1,j) +u1(i-1,j)) ...
            + 0.5*cx^2*(u1(i,j+1)+u1(i,j-1));
    end
end
u0 = u1; u1 = u2;        % update displacements
```

For the next time step we make use of the following nodal equations.

$$u^2_{i,j} = \frac{(u^1_{i-1,j}+u^1_{i+1,j}+u^1_{i,j-1}+u^1_{i,j+1})}{2} - u^0_{i,j}$$

Again, for the node $1,1$ we have

$j\downarrow\ i\rightarrow$	0	1	2
0		0.000000	
1	0.000000	0.000908	0.001727
2		0.001727	

and $u_{1,1}^0 = 0.000955$. Using these we get

$$u_{1,1}^1 = \frac{0+0+0.001727+0.001727}{2} - 0.000955 = 0.000773$$

Similar calculations may be made for all other interior nodes to get the shape of the diaphragm at $t = 2 \times 0.070711 = 0.141422$.

```
% higher time steps
for t=1:1000
    for i=2:nx-1
        for j=2:ny-1
            u2(i,j) = 2*(1-cx^2-cy^2)*u1(i,j) ...
                    + cy^2*(u1(i+1,j) + u1(i-1,j)) ...
                    + cx^2*(u1(i,j+1)+u1(i,j-1)) - u0(i,j);
        end
    end
    u0 = u1; u1 = u2;   % update displacements
end
```

The results have been plotted for displacement at $y = 0.5$ or $x = 0.5$ and for two time steps as shown in Figure 15.15. Figure 15.16 shows the shape of the diaphragm initially and after half time period.

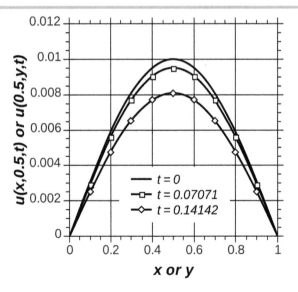

Figure 15.15: Displacement history for a square diaphragm at midplane, $y = 0.5$ or $x = 0.5$

15.4.3   Waves in two dimensions - waves in a circular diaphragm or plate

Displacement in this case can vary with both r and θ and the governing equation is given by Equation 15.39. We discretize the physical domain by using step size of Δr along the radial direction and $\Delta\theta$ along the θ direction. Note that $\Delta\theta$ has to be chosen such that $n\Delta\theta = 2\pi$ where n is an integer. Let i and j identify the node position along r and θ. The

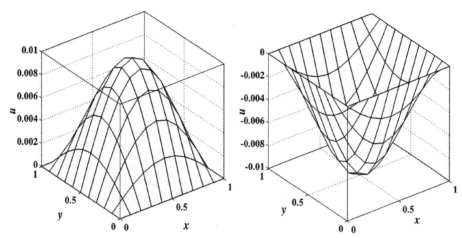

Figure 15.16: *Initial deflection in the diaphragm and deflection after one half period*

spatial derivatives appearing in Equation 15.39 are approximated as follows using the already known function values.

$$
\begin{aligned}
\frac{\partial^2 u}{\partial r^2} &= \frac{u_{i-1,j}^k - 2u_{i,j}^k + u_{i+1,j}^k}{\Delta r^2} \\
\frac{1}{r}\frac{\partial u}{\partial r} &= \frac{1}{r_i}\frac{u_{i+1,j}^k - u_{i-1,j}^k}{2\Delta r} \\
\frac{1}{r^2}\frac{\partial^2 u}{\partial \theta^2} &= \frac{1}{r_i^2}\frac{u_{i,j-1}^k - 2u_{i,j}^k + u_{i,j+1}^k}{\Delta \theta^2}
\end{aligned}
\tag{15.48}
$$

Finite difference form of the time derivative is obtained as in the case of a rectangular diaphragm. We then have the following for the nodal equation.

$$
\begin{aligned}
u_{i,j}^{k+1} &= C_r^2\left(1 - \frac{\Delta r}{r_i}\right)u_{i-1,j}^k + 2\left\{1 - C_r^2\left(1 + \left[\frac{\Delta r}{r_i \Delta \theta}\right]^2\right)\right\}u_{i,j}^k + C_r^2\left(1 + \frac{\Delta r}{r_i}\right)u_{i+1,j}^k \\
&\quad + \left(\frac{C_r \Delta r}{r_i \Delta \theta}\right)^2 (u_{i,j-1}^k + u_{i,j+1}^k) - u_{i,j}^{k-1}
\end{aligned}
\tag{15.49}
$$

Stability of the scheme requires that $1 - C_r^2\left(1 + \left[\frac{\Delta r}{r_{min}\Delta \theta}\right]^2\right) \geq 1$ (coefficient of u_i^k should be positive). This may be recast in the form $C_r^2 \leq \dfrac{1}{1 + \dfrac{\Delta r^2}{r_{min}^2 \Delta \theta^2}}$. This may further be simplified to read

$$
\Delta t \leq \frac{1}{\sqrt{\frac{1}{\Delta r^2} + \frac{1}{r_{min}^2 \Delta \theta^2}}}
\tag{15.50}
$$

The node at the center of the diaphragm requires special treatment. Locally one may use the wave equation in Cartesian frame of reference. We consider the node at the center

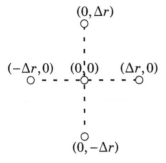

$(0, \Delta r)$

$(-\Delta r, 0)$ $(0,0)$ $(\Delta r, 0)$

$(0, -\Delta r)$

Figure 15.17: *Node at the center of a circular diaphragm*

and four nearest neighbors as shown in Figure 15.17. Note that all nodal spacings are the same and equal to Δr. The spatial derivatives required are calculated as

$$
\frac{\partial^2 u}{\partial x^2} = \frac{u^k(-\Delta r, 0) - 2u^k(0,0) + u^k(\Delta r, 0)}{\Delta r^2}
$$
$$
\frac{\partial^2 u}{\partial y^2} = \frac{u^k(0, -\Delta r) - 2u^k(0,0) + u^k(0, \Delta r)}{\Delta r^2} \qquad (15.51)
$$
$$
\frac{\partial^2 u}{\partial t^2} = \frac{u^{k-1}(0,0) - 2u^k(0,0) + u^{k+1}(0,0)}{\Delta t^2}
$$

Using these the nodal equation for the center node turns out to be

$$
u^{k+1}(0,0) = C^2[u^k(-\Delta r, 0) + u^k(\Delta r, 0) + u^k(0, -\Delta r) + u^k(0, \Delta r)]
$$
$$
-2(1 - C^2)u^k(0,0) - u^{k-1}(0,0) \qquad (15.52)
$$

Note that the nodal values have been indicated with the x, y coordinates instead of r, θ coordinates. However, it is easy to see that this equation may be recast in r, θ coordinates as given below.

$$
u^{k+1}(0,0) = C^2\left[u^k(\Delta r, \pi) + u^k(\Delta r, 0) + u^k\left(\Delta r, \frac{3\pi}{2}\right) + u^k\left(\Delta r, \frac{\pi}{2}\right) \right]
$$
$$
-2(1 - C^2)u^k(0,0) - u^{k-1}(0,0) \qquad (15.53)
$$

Example 15.10

A circular diaphragm of radius $r = 1$ is given an initial displacement of $u = 0.01 \sin(r\pi) \cos\theta$ with zero velocity everywhere. The wave velocity in the membrane is unity. Obtain the displacement history of the membrane.

Solution :

We shall discretize the radius into 10 equal segments. We shall discretize each quarter of the membrane into 3 equal segments such that we have 13 nodes along θ direction as shown in the following figure .

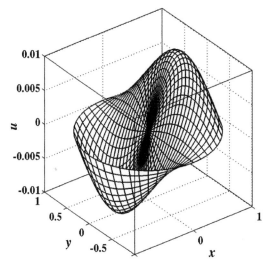

Initial displacement of circular diaphragm.

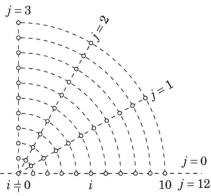

$j = 3$

$j = 2$

$j = 1$

$j = 0$

$i = 0$ i $10\ j = 12$

Note: Only a quadrant is depicted in
the figure

$\Delta r = 0.1$, $\Delta\theta = \dfrac{2\pi}{12}$

As solution is periodic in θ, $u_{i,0} = u_{i,12}$

At $r = 0$, we use the Cartesian representation as discussed earlier.

The time step has been chosen according to the stability requirements.

$$C_r = \sqrt{\dfrac{1}{1 + \dfrac{1}{\Delta\theta^2}}}$$

```
nt=3; nr = 11; nthe = 4*nt+1;   % no of nodes along r and
                      %theta
r = 0:1/(nr-1):1; the = 0:2*pi/(nthe-1):2*pi; % nodes
dr = 1/(nr-1); dthe = 2*pi/(nthe-1);   % step size
u0 = zeros(nr,nthe);            % displacement at k-1
u1 = 0.01*sin(R*pi).*cos(The);       % displacement at k
u2 = u0;                 % displacement at k+1
cr = 1/sqrt(1+1/dthe^2); dt = cr*dr;  % Courant number
```

At first time step, we have zero initial velocity, therefore we have $u_{i,j}^{+1} = u_{i,j}^{-1}$. Accordingly,
the equations are modified as below

```
% r=0, center node
u2(1,1)=u1(1,1)*(1-2*cr^2) + cr^2*0.5*(u1(2,1) + u1(2,1+nt)...
    + u1(2,2*nt+1) +  u1(2,1+nt*3)));
u2(1,2:nthe) = u2(1,1);
for i=2:nr-1   % loop for interior nodes
    j = 1;     % θ=0 or θ=2π
    u2(i,j) = (1-cr^2-cr^2*dr^2/(r(i)^2*dthe^2))*u1(i,j)+ ...
        0.5*cr^2*(u1(i+1,j)+u1(i-1,j)) + ...
        0.5*cr^2*dr*(u1(i+1,j)-u1(i-1,j))/(2*r(i))   ...
```

```
         +0.5*cr^2*dr^2*(u1(i,j+1)+u1(i,nthe-1))/(r(i)^2*dthe^2)
                                    %;
      u2(i,nthe) = u2(i,1);
      for  j=2:nthe-1        %  0<θ<2π
         u2(i,j)=(1-cr^2-cr^2*dr^2/(r(i)^2*dthe^2))*u1(i,j)+ ...
         0.5*cr^2*(u1(i+1,j)+u1(i-1,j)) + ...
         0.5*cr^2*dr*(u1(i+1,j)-u1(i-1,j))/(2*r(i))        ...
         +0.5*cr^2*dr^2*(u1(i,j+1)+u1(i,j-1))/(r(i)^2*dthe^2);
      end
   end
   u0 = u1; u1 = u2;         % update displacements
```

For subsequent time steps, we use the discretized equations as discussed earlier.

```
for  t=1:50        %  loop  for  time
%  r=0,  center  node
u2(1,1)  =  2*u1(1,1)*(1-2*cr^2)+cr^2*(u1(2,1)+u1(2,1+nt)...
    +  u1(2,2*nt+1)  +   u1(2,1+nt*3))  -  u0(1,1);
u2(1,2:nthe) = u2(1,1);
for  i=2:nr-1    %  loop  for  internal  nodes
    j  =  1;        %  θ=0  or  θ=2π
    u2(i,j)=2*(1-cr^2-cr^2*dr^2/(r(i)^2*dthe^2))*u1(i,j)+...
    cr^2*(u1(i+1,j)+u1(i-1,j))  -  u0(i,j)  +...
    cr^2*dr*(u1(i+1,j)-u1(i-1,j))/(2*r(i))        ...
    +cr^2*dr^2*(u1(i,j+1)+u1(i,nthe-1))/(r(i)^2*dthe^2);
    u2(i,nthe)  =  u2(i,1);
    for  j=2:nthe-1        %  0<θ<2π
       u2(i,j)=2*(1-cr^2-cr^2*dr^2/(r(i)^2*dthe^2))*u1(i,j)...
           +  cr^2*(u1(i+1,j)+u1(i-1,j))  -  u0(i,j)  +  ...
           cr^2*dr*(u1(i+1,j)-u1(i-1,j))/(2*r(i))        ...
           +  cr^2*dr^2*(u1(i,j+1)+u1(i,j-1))/(r(i)^2*dthe^2)  ;
    end
end
u0 = u1; u1 = u2;         % update  displacements
end
```

Figure 15.18 shows the displacement history of the circular membrane along $\theta = 0$. Figure 15.19 shows the shape of the diaphragm at $t = 0.5102$

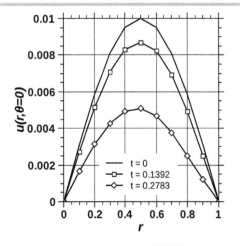

Figure 15.18: *Displacement history of circular membrane along $\theta = 0$*

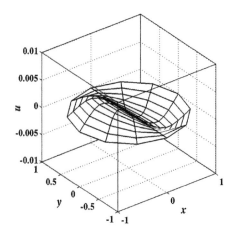

Figure 15.19: *Circular diaphragm at $t = 0.5102$*

Concluding remarks

This chapter has been a short introduction to solution of the wave equation. We have considered only one dimensional and two dimensional problems representing transverse waves on a string or the motion of a diaphragm. We have presented numerical schemes that may be used for the solution of the wave equation in these cases. We discuss but without proof about the stability aspects of the numerical schemes. Wherever possible we have drawn attention to underlying physical aspects of the problem under consideration.

Waves of high intensity such as shock waves that occur in supersonic flow are considered in specialized books and the interested reader may consult them.

Exercise IV

IV.1 Types of partial differential equations

Ex IV.1: Consider the first order PDE $(2x + y)u_x + (x + 2y)u_y = 0$. Sketch a characteristic line that passes through the point $x = 0.5$, $y = 0.5$.
(Hint: Characteristic line is obtained as the solution of an ODE. The ODE may be solved by a suitable numerical method.)

Ex IV.2: Discuss the nature of the PDE $u_{xx} + 6u_{xy} + 7u_{yy} + u_x + 3u_y + 4u = y$? Reduce it to the standard[2] or canonical form by suitable change of variables.

Ex IV.3: Reduce the PDE $9u_{xx} + 6u_{xy} + 4u_{yy} + u_x + 3x = 0$ to the standard form. What is the nature of this PDE?

Ex IV.4: Discuss the nature of the PDE $4u_{xx} + 5u_{xy} + 9u_{yy} + u_x + 3u_y + 4u = y$? Reduce it to the standard form.

IV.2 Laplace and Poisson equations

Ex IV.5: Figure IV.1 shows some nodal temperatures that have been obtained (function values are shown for these) and a few that need to be obtained (node numbers are shown for these). Obtain these by writing appropriate nodal equations if Laplace equation is satisfied within the domain. The nodal spacing is the same and equal to 0.05 along both the x and y directions.

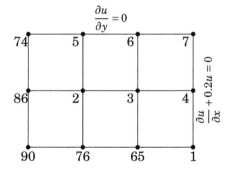

$h = k = 0.05$

Figure IV.1: Figure showing data for Exercise IV.5

Reader may try FDM, FEM and FVM formulations for solving the problem.

---

[2]See, for example, Alan Jefferey, "Advanced Engineering Mathematics", Academic Press, 2002.

Ex IV.6: Consider a rectangular region of width $w = 1$ and height $h = 2$ in which a certain function satisfies the Poisson equation $u_{xx} + u_{yy} = f(x,y)$. The source term (the right hand term) is given by $f(x,y) = \cos\left(\dfrac{\pi x}{w}\right)\cos\left(\dfrac{\pi y}{h}\right)$ where the origin is placed at the center of the domain with the four corners at $x = \dfrac{w}{2}, y = \dfrac{h}{2}, x = -\dfrac{w}{2}, y = \dfrac{h}{2}, x = \dfrac{w}{2}, y = -\dfrac{h}{2}$ and $x = -\dfrac{w}{2}, y = -\dfrac{h}{2}$. The boundary conditions are specified as given below:

$$u_x = 0.5; x = -\frac{w}{2}, -\frac{h}{2} \le y \le \frac{h}{2} : u_x = -0.5; x = \frac{w}{2}, -\frac{h}{2} \le y \le \frac{h}{2}$$

$$u_y = 1; y = -\frac{h}{2}, -\frac{w}{2} \le x \le \frac{w}{2} : u_y = -1; y = \frac{h}{2}, -\frac{w}{2} \le x \le \frac{w}{2}$$

Obtain numerically the solution by FDM using ADI method.

Ex IV.7: A tube (made of a dielectric) of square section $0.1 \times 0.1\ m$ has walls that are $0.02\ m$ thick. The inner surface of the tube is at a potential of $100\ V$ while the outer surface is connected to the ground. The electric potential is known to satisfy the Laplace equation within the tube material. Obtain the solution by discretization the domain suitably. Use an iterative scheme of your choice to obtain the nodal potentials. Make a contour plot showing equipotential lines.

Ex IV.8: Solve Poisson equation $u_{xx} + u_{yy} = 0.5 + 1.5x$ in two dimensions in a rectangular domain of length 10 units and width 1.5 units. The origin is placed at a corner and the long side is oriented along x. Use 5 nodes along the short side and a suitable number of nodes along the longer side. One of the short sides is specified with Dirichlet condition $u = 1$ and the other sides are specified Robin condition in the form $u_n + 0.2u + 0.1 = 0$ where u_n represents the normal derivative. (a) Obtain nodal values of u. Make use of SOR method with a suitable relaxation parameter. (b) Redo the problem if the Robin condition is recast as $u_n + 0.2u^{1.25} + 0.1 = 0$.

Ex IV.9: Solve for nodal values of a certain function $u(x,y)$ that satisfies the Laplace equation in a rectangular domain of sides $l = 2$ and $b = 1$. One of the longer edges is placed along the x axis and one of the shorter edges is placed along the y axis. The boundary conditions are specified as under:

$$\frac{\partial u}{\partial x} = 1 \text{ for } x = 0, 0 \le y \le 1 : \frac{\partial u}{\partial x} = 0 \text{ for } x = 2, 0 \le y \le 1$$

$$u = 1 \text{ for } y = 0, 0 \le x \le 2 : \frac{\partial u}{\partial y} = 0.5u \text{ for } y = 1, 0 \le x \le 2$$

Use suitable step sizes so as to get results accurate to four digits after decimals.

Ex IV.10: A channel of unit width has a $90°$ bend in it. The channel extends 2 channel widths as shown in the Figure IV.2 and a certain function $u(x,y)$ satisfies the Laplace equation within the channel.[3] u is zero along the shorter boundary while it is equal to 1 along the longer boundary. Also the function varies linearly between 0 and 1 along the other two boundaries. Obtain the solution to Laplace equation by finite differences. Use suitable step sizes along the two directions. Locate the origin according to convenience and use symmetry to your

---

[3]If u is identified as the stream function, the problem represents ideal fluid flow in a channel with $90°$ bend. See a book on fluid flow such as Frank M White, "Fluid Mechanics", Tata McGraw-Hill Education, 1994

advantage. Solution is required good to 5 digits after decimals. This may be ascertained by a grid sensitivity study. Make a contour plot.

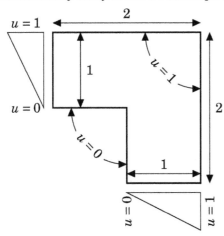

Figure IV.2: Stream function data for Exercise IV.10

Ex IV.11: Poisson equation in the form $\dfrac{1}{r}\dfrac{\partial}{\partial r}\left(r\dfrac{\partial u}{\partial r}\right) + \dfrac{1}{r^2}\dfrac{\partial^2 u}{\partial \theta^2} = 0.5$ is satisfied by a function $u(r,\theta)$ in a cylinder defined by the boundary at $r = 1$. Robin condition $\dfrac{\partial u}{\partial r} = 0.1(u - 0.4)$ is specified along one half of the boundary between $\theta = 0$ and $\theta = \pi$. The other half of the boundary is imposed Dirichlet condition in the form $u = 0.4$. Obtain numerical solution using finite differences. Would you be able to use the ADI method in this case?

Ex IV.12: Poisson equation in the form $\dfrac{1}{r}\dfrac{\partial}{\partial r}\left(r\dfrac{\partial u}{\partial r}\right) + \dfrac{1}{r^2}\dfrac{\partial^2 u}{\partial \theta^2} = \cos\left(\dfrac{\pi(r-1)}{2}\right)$ is satisfied by a function $u(r,\theta)$ in an annulus defined by inner boundary $r = 1$ and outer boundary $r = 2$. Neumann condition is specified at the inner boundary as $\dfrac{\partial u}{\partial r} = 0$ and Robin condition $\dfrac{\partial u}{\partial r} = 0.1(u^4 - 0.4)$ is specified along one half of the outer boundary between $\theta = 0$ and $\theta = \pi$. The other half of the outer boundary is imposed Dirichlet condition in the form $u = 0.4$. Obtain numerical solution using finite differences.

Ex IV.13: Poisson equation in the form $u_{xx} + u_{yy} + 0.04(x^2 + y^2) = 0$ is satisfied in a right angled triangular region as shown in the Figure IV.3. The long side is aligned with the x axis and the short side with the y axis. Right angled vertex is positioned at the origin. The following boundary conditions have been specified:

$$u \;=\; 1;\; y = 0,\; 0 \le x \le 4:\; u = 0.5;\; x = 0,\; 0 \le y \le 3$$

$$\frac{du}{dn} \;=\; 0;\; \text{Normal derivative on the diagonal}$$

Obtain the solution numerically by FDM. Also solve the problem using FEM.

Ex IV.14: $\nabla^2 u - 0.2u = 0$ governs the variation of a function u over a thin semi-circular disk of radius 0.5. Dirichlet condition $u = 1$ is specified along the flat edge of the plate that is aligned with the y axis and the center of the semi-circular disk is located at the origin. Over the semi-circular edge Robin condition is specified as $u_r + 0.1u = 0$. Obtain numerical solution by FDM. Use such step lengths that the solution is good to at least three digits after the decimal point.

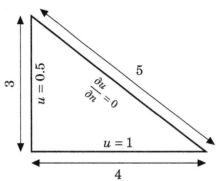

Figure IV.3: Data for Exercise IV.13

IV.3   Advection and diffusion equations

Ex IV.15: A function that satisfies advection equation in one dimension $u_t - 0.2u_x = 0$, is initially specified as $u(x,0) = 50(x-0.4)(0.6-x)$ and zero outside the interval $0.4 \leq x \leq 0.6$. Obtain the solution for a few suitably chosen space and time steps using different schemes introduced in the text. Make comments based on the results. Compare each numerically obtained solution with the exact solution.

Ex IV.16: Consider advection in one dimension of a triangular pulse with height 1 and width 0.05. The advection speed is equal to unity and in the positive direction. Compare the performance of different schemes in the quality of the solution, in comparison with the exact solution. Make plots such that the pulse is stationary. Use periodic condition as explained in the text.

Ex IV.17: A wave travels to the right on a string of infinite length. The velocity of the wave is discontinuous and is equal to 1 *m/s* to the left of the origin and 2 *m/s* to the right of the origin. At $t = 0$ a triangular pulse of height 0.1 *m* and width 0.02 *m* is located at $x = -0.2\,m$. Describe the state of affairs for $t > 0$. Use a numerical scheme of your choice to obtain the solution. Specifically find out what happens when the pulse reaches the origin.

Ex IV.18: Transient temperature field in a plate follows the equation $\dfrac{\partial u}{\partial t} = \dfrac{\partial^2 u}{\partial x^2} + \cos\left(\dfrac{\pi x}{2}\right)$ where $0 \leq x \leq 1$ and $t > 0$. The boundary conditions have been specified as $u(0,t) = 0$ and $\dfrac{\partial u}{\partial x} = 0$ at $x = 1$ for $t > 0$. At $t = 0$ the entire plate is at zero temperature. Obtain numerically the solution for $0 < t \leq 1$ by (a) explicit scheme and (b) Crank Nicolson scheme. What step sizes are required to get the solution, good to three digits after decimals, in each of these methods? Obtain also the steady solution to the problem. How close is the transient at $\tau = 1$ to the steady solution?

Ex IV.19: Transient heat transfer in a plate is governed by the equation $\dfrac{\partial u}{\partial t} = \dfrac{\partial^2 u}{\partial x^2} - u$ where $0 \leq \xi \leq 1$ and $t > 0$. The boundary conditions have been specified as $u(0,t) = 1$ and $\dfrac{\partial u}{\partial x} = 0$ at $x = 1$ for all t. Initially $u(x,0) = 1$ for $0 \leq x \leq 1$. Obtain numerically the solution for t up to 1 by (a) explicit scheme and (b) Crank Nicolson scheme. What step sizes are required to get the solution, good to three digits after decimals, in each of these methods? Obtain also the steady solution to the problem. How close is the transient at $t = 1$ to the steady solution?

Ex IV.20: Transient temperature field in a circular disk of unit radius is governed by the equation $\dfrac{\partial u}{\partial t} = \dfrac{1}{r}\dfrac{\partial}{\partial r}\left(r\dfrac{\partial u}{\partial r}\right) + (1 - r^2)$. Initially $u = 1$ throughout the disk. For $t > 0$ the periphery of the disk is maintained at a uniform u of 0.2. Obtain the transient solution till the steady state by FDM. Results are desired convergent to 4 digits after decimals.

Ex IV.21: Perform von Karman stability analysis of an explicit method applied to the case of transient heat transfer in two dimensions. Obtain therefrom the appropriate stability conditions.

Ex IV.22: Discuss the stability of ADI scheme when applied to heat equation in two and three dimensions.

Ex IV.23: Advection with diffusion is governed by the equation $u_t - 0.5u_x = u_{xx} + x(1-x)$. Boundary conditions are specified as $u(0,0) = 1$ and $u_x = 0$ at $x = 1$ and for $t > 0$. Initial condition is specified by the function $u(x,0) = 0.5(1 + e^{-x})$. Obtain the solution to steady state using an appropriate numerical scheme.

Ex IV.24: Transient temperature field in a plate follows the equation $\dfrac{\partial u}{\partial t} = \dfrac{\partial^2 u}{\partial x^2} + \cos\left(\dfrac{\pi x}{2}\right)$ where $0 \le x \le 1$ and $t > 0$. The boundary conditions have been specified as $u(0,t) = 0$ and $\dfrac{\partial u}{\partial x} + 0.5u^4 = 0$ at $x = 1$ for $t > 0$. At $t = 0$ the entire plate is at zero temperature. Obtain numerically the solution for $0 < t \le 1$ by a numerical scheme of your choice. Indicate how you would take care of the nonlinear boundary condition.

Ex IV.25: A sphere of unit radius is initially at a uniform temperature of $u = 1$ throughout. It starts cooling by losing heat to a cold ambient by radiation. The appropriate PDE has been derived and is given by $\dfrac{\partial u}{\partial t} = \dfrac{1}{r^2}\dfrac{\partial}{\partial r}\left(r^2\dfrac{\partial u}{\partial r}\right)$. Radiation boundary condition at $r = 1$ is specified as $\dfrac{\partial u}{\partial r} + 0.2u^4 = 0$. Obtain the transient temperature field numerically using a numerical scheme of your choice. Indicate how you would deal with the node at the center of the sphere.

Ex IV.26: $u_t = \nabla^2 u - 0.2u$ governs the transient variation of a function u over a thin semi-circular disk of radius 0.5. Dirichlet condition $u = 1$ is specified along the flat edge of the plate that is aligned with the y axis and the center of the semi-circular disk is located at the origin. Over the semi-circular edge Robin condition is specified as $u_r + 0.1u = 0$. Initially $u = 1$ throughout the domain. Obtain numerical solution by FDM till $t = 1$.

Ex IV.27: A cylindrical annulus has an inner radius of 0.5 and an outer radius of 1. Heat is being uniformly generated within the annulus. The inner boundary is insulated so that Neumann condition is applicable there i.e. $u_r(0.5,t) = 0$. At the second boundary Dirichlet condition is specified as $u(1,t) = 0.5$. Initially the annulus is at zero temperature throughout. The governing equation is given as $\dfrac{\partial u}{\partial t} = \dfrac{1}{r}\dfrac{\partial}{\partial r}\left(r\dfrac{\partial u}{\partial r}\right) + \left(1 - r^2\right)$. Obtain the solution till $t = 5$ by FDM.

Ex IV.28: A long rectangular duct has outer dimensions of 4×2 and a uniform thickness of 0.5. Initially the duct cross section is at a uniform temperature of zero. For $t > 0$ heat is generated in the duct such that the governing equation is of form $u_t = u_{xx} + u_{yy} + 0.5(2 - x)(1 - y)$. Inner boundary is specified with Neumann condition i.e. normal derivative $u_n = 0$ while the outer boundary is maintained at zero temperature. Obtain the solution by finite differences. Take advantage of any symmetry present in the problem.

Ex IV.29: One dimensional heat equation appropriate to a material of temperature dependent thermal conductivity is given by $\dfrac{\partial u}{\partial t} = \dfrac{\partial}{\partial x}\left((1+u)\dfrac{\partial u}{\partial x}\right) - 2u$. In a certain application x is limited to the region $0 \le x \le 0.5$. Initially the entire domain has a uniform temperature of $u = 1$. For $t > 0$ temperature at $x = 0$ remains unaltered while the temperature at $x = 0.5$ takes on a value of $u = 0.5$ and remains fixed at this value. Obtain the transient temperature field by FDM using the Crank Nicolson scheme.

IV.4   Wave equation

Ex IV.30: Consider waves in an infinitely long uniform cross sectioned string under tension. The wave speed has been specified as unity. At $t = 0$ the string is displaced from its equilibrium position such that the displacement profile is $y(x) = 0.1\cos(5\pi x)$ for $-0.1 \le x \le 0.1$ and zero for all x outside this range. The string is let go with zero velocity. Obtain solution numerically and compare it with the exact solution.

Ex IV.31: A string of uniform cross section and $1\,m$ long has a mass of $250\,g$. It is stretched between rigid supports with a tension of $9\,N$. What is the wave speed? The string has an initial displacement given by $u(x) = 0.02\sin(\pi x)$ and is released from rest. Obtain the shape of the string after 0.125 times the period of oscillation. (a) Use d'Alembert's solution for this purpose. (b) Use finite difference solution and obtain the answer. Will the nodal spacing have a role to play in the solution? Explain giving reasons.

Ex IV.32: Consider the wave equation with damping given by $u_{xx} = u_{tt} + \gamma u_t$ in the region $0 \le x \le 1$. The wave speed is unity and γ is the damping parameter. The system is allowed to perform motion form a position of rest but with an initial displacement given by $u(x,0) = 0.02x(1-x)$. Allow the damping parameter to vary between 0 and 0.5. Use FDM to obtain the solution. Study the effect of γ on the variation of the displacement at $x = 0.5$ with time. You may assume that the displacement remains zero at $x = 0$ and $x = 1$ for all t.

Ex IV.33: A rectangular diaphragm of mass per unit area $m = 0.1\,kg/m^2$ is stretched with a uniform force per unit length of $T = 20\,N/m$. The length and width of the diaphragm are $L_x = 0.25\,m$ and $L_y = 0.375\,m$. What is the wave speed? What is the frequency of the fundamental mode of vibration of the diaphragm?

Initial displacement of the diaphragm is $u(x,y,0) = xy(x - 0.25)(y - 0.375)$ with the origin located at left bottom corner. Obtain displacement throughout the diaphragm as a function of time, by a suitable numerical scheme, if the diaphragm is allowed to vibrate starting from rest. Comment on the nodal spacing vis a vis the accuracy of the displacement calculated by the numerical scheme.

Ex IV.34: A string of nonuniform density is stretched between two rigid supports L apart. The density (mass per unit length) varies according to the relation $m(x) = m_0\left(1 + a\dfrac{x}{L}\right)$ where m_0 and a are specified constants. Formulate the governing PDE for this case assuming that the tension T in the string remains constant as the string is set into vibration.

Consider the specific case $m_0 = 0.1\,kg/m$, $a = 1$, $T = 10\,N$ and $L = 0.5\,m$. The initial displacement of the string is given by the function $u(x,0) = 0.04x(0.5-x)$. Obtain the solution using a numerical scheme of your choice. How does the solution differ from that for a string of uniform density? Assume that the string is initially at rest.

Ex IV.35: A circular membrane of variable thickness of radius R has the mass per unit area varying with radius as $m(r) = m_0\left(1 - a\dfrac{r}{R}\right)$. The membrane is firmly stretched such that the periphery is attached to a rigid ring and the stretching force per unit length is equal to T. Formulate the applicable PDE that describes axi-symmetric vibrations in the membrane.

In a certain case $m_0 = 0.1 \, kg/m^2$, $T = 100 \, N/m$, $R = 0.2 \, m$ and $a = 0.25$. The membrane is initially disturbed from the equilibrium position by the displacement function given by $u(r,0) = 0.1\left(1 - \dfrac{r}{R}\right)$ and allowed to execute transverse vibrations starting from rest. Obtain the solution numerically using a scheme of your choice. How does the solution differ from that for a membrane of uniform density?

Ex IV.36: Solve $u_{tt} + 0.4u_t = u_{xx}$ if $u(x,0) = 0$ and

$$u_t(x,0) = \begin{cases} 0 & 0 \le x < 0.4 \\ \dfrac{(x-0.4)(0.6-x)}{5} & 0.4 \le x \le 0.6 \\ 0 & 0.6 < x \le 1 \end{cases}$$

Chapter 16

Beyond the book - the way ahead

The book has presented essential ideas behind computational methods, useful especially for engineering applications and hence is, at best, an introduction to computational methods in engineering. Many worked examples have been presented to bring home the applications of the many methods that have been dealt in the book. Application examples have been broadly from the area of mechanics - structural mechanics, vibrations, fluid mechanics and heat transfer. Some examples have treated analysis of electrical circuits also.

The book should benefit two types of readers.

- Those who want to learn the techniques and use these in their own research activity involving specific problems that require development of optimized code.
- Those who want to learn just enough to understand how commercial codes work. They will benefit by the basic knowledge gained from the book so that they can use the commercial codes efficiently. They will also avoid making costly mistakes in using commercial codes.

The "way forward" will address the needs of both these types of readers.

16.1 Researchers developing their own code

This class of readers should refer to books that deal with the theory of ordinary and partial differential equations to benefit fully from the present book. A thorough understanding of analytical methods will provide the *insight* they need for independent research. They should also refer to more advanced books on numerical methods. Domain specific books that deal with numerical and computational methods should be consulted for assessing the *state of the art* in their chosen field.

Several topics such as integral transforms, Fourier analysis, integral equations, grid generation etc. have not been covered in the present book. Specialized books exist on these topics and the reader is advised to consult them.

This class of reader would benefit by the excellent support available in computational resources such as MATLAB. He/she may develop fairly complex computational tools using MATLAB. Capability to use spreadsheet programs will help in exploratory analysis.

In addition, this class of readers will benefit, by looking at applications of computational methods, presented in domain specific peer reviewed journals.

16.2   Users of commercial codes

A good grounding in computational methods is essential for those who use many commercial software available as general purpose and domain specific computational tools. A good understanding of the physical problem being modeled is essential in obtaining proper solutions using the relevant commercial code. Grasp of the underlying physics is essential in getting physically meaningful numerical solutions. Most mistakes occur due to use of wrong initial and boundary conditions. Before using any software they should consult the concerned manual to know about the capabilities as well as limitations of the concerned software. It is good practice to use the software to analyze simple cases which are easy to interpret and move on to more complex cases. Validation exercise should be taken up before using any computational tool for solving new problems. Most commercial codes are supplied with test cases that have been validated independently and hence may be used by the beginner to test his own skills in using the software.

16.3   Where does one look for help?

There are excellent books and web resources for the interested reader. One very useful book that may be referred to is the text book by S. Chapra and R. Canale, *"Numerical Methods for Engineers"*, McGraw-Hill, 6th edition, 2009. Both authors are Professors of Civil Engineering and bring in an engineering point of view to the material being presented. As far as specialized books are concerned, the book *"Fundamentals of Computational Fluid Dynamics"*, Hermosa Publishers,1998 by P.J. Roache is one that takes the reader beyond what is in the present book in solving PDEs of interest in Fluid Dynamics. The book *"An introduction to computational fluid dynamics"* 2nd Edition by H. K. Versteeg and W. Malalasekera, Pearson/Prentice Hall, 2007 is a detailed exposition on the finite volume method. Those interested in the finite element method may look at *"An Introduction to the Finite Element Method"* 3rd edition, McGraw-Hill, 2005 by J. N. Reddy as the starting point.

Resources on the net are plenty and a few useful sites are given below.

16.3.1 Free resources

- **Numerical recipes:** at http://www.nr.com/ provide algorithms and programs for several commonly used functions.
- **Netlib Repository:** http://www.netlib.org/ is a repository of library of subroutines maintained at University of Tennessee, Knoxville and Oakridge National Laboratory managed by University of Tennessee Battelle.
- **GNU operating system:** http://www.gnu.org/software/gsl/ provides free software useful for numerical computations.
- **Fortran Library:** http://www.fortranlib.com/ is a directory of Fortran resources including books, tutorials, compilers, and free software.
- **MATLAB:** http://www.mathworks.fr/matlabcentral/fileexchange/ provides a platform for the users to share their MATLAB programs and also find useful programs developed by others.

16.3.2 Licensed resources

- **IBM - ESSL:** Look up www.ibm.com/systems/software/essl/ and access Engineering and Scientific Subroutine Library (ESSL) and Parallel ESSL
- **Numerical Analysis Group:** The NAG Fortran Library provides a large number of subroutines that may be used for computing.

Index

Printed and bound by CPI Group (UK) Ltd, Croydon, CR0 4YY

03/10/2024

01040326-0008